W9-CYV-056

INTRODUCTION TO CHEMICAL ENGINEERING THERMODYNAMICS

McGraw-Hill Chemical Engineering Series

BUILDING THE LITERATURE OF A PROFESSION

Fifteen prominent chemical engineers first met in New York more than 60 years ago to plan a continuing literature for their rapidly growing profession. From industry came such pioneer practitioners as Leo H. Baekeland, Arthur D. Little, Charles L. Reese, John V. N. Dorr, M. C. Whitaker, and R. S. McBride. From the universities came such eminent educators as William H. Walker, Alfred H. White, D. D. Jackson, J. H. James, Warren K. Lewis, and Harry A. Curtis. H. C. Parmelee, then editor of *Chemical and Metallurgical Engineering*, served as chairman and was joined subsequently by S. D. Kirkpatrick as consulting editor.

After several meetings, this committee submitted its report to the McGraw-Hill Book Company in September 1925. In the report were detailed specifications for a correlated series of more than a dozen texts and reference books which have since become the McGraw-Hill Series in Chemical Engineering and which became the cornerstone of the chemical engineering curriculum.

From this beginning there has evolved a series of texts surpassing by far the scope and longevity envisioned by the founding Editorial Board. The McGraw-Hill Series in Chemical Engineering stands as a unique historical record of the development of chemical engineering education and practice. In the series one finds the milestones of the subject's evolution: industrial chemistry, stoichiometry, unit operations and processes, thermodynamics, kinetics, and transfer operations.

Chemical engineering is a dynamic profession, and its literature continues to evolve. McGraw-Hill and its consulting editors remain committed to a publishing policy that will serve, and indeed lead, the needs of the chemical engineering profession during the years to come.

THE SERIES

Bailey and Ollis: *Biochemical Engineering Fundamentals*
Bennett and Myers: *Momentum, Heat, and Mass Transfer*
Beveridge and Schechter: *Optimization: Theory and Practice*
Carberry: *Chemical and Catalytic Reaction Engineering*
Churchill: *The Interpretation and Use of Rate Data—The Rate Concept*
Clarke and Davidson: *Manual for Process Engineering Calculations*
Coughanowr and Koppel: *Process Systems Analysis and Control*
Daubert: *Chemical Engineering Thermodynamics*
Fahien: *Fundamentals of Transport Phenomena*
Finlayson: *Nonlinear Analysis in Chemical Engineering*
Gates, Katzer, and Schuit: *Chemistry of Catalytic Processes*
Holland: *Fundamentals of Multicomponent Distillation*
Holland and Liapis: *Computer Methods for Solving Dynamic Separation Problems*
Johnson: *Automatic Process Control*
Johnstone and Thring: *Pilot Plants, Models, and Scale-Up Methods in Chemical Engineering*
Katz, Cornell, Kobayashi, Poettmann, Vary, Elenbaas, and Weinaug: *Handbook of Natural Gas Engineering*
King: *Separation Processes*
Klinzing: *Gas-Solid Transport*
Knudsen and Katz: *Fluid Dynamics and Heat Transfer*
Luyben: *Process Modeling, Simulation, and Control for Chemical Engineers*
McCabe, Smith, J. C., and Harriott: *Unit Operations of Chemical Engineering*
Mickley, Sherwood, and Reed: *Applied Mathematics in Chemical Engineering*
Nelson: *Petroleum Refinery Engineering*
Perry and Green (Editors): *Chemical Engineers' Handbook*
Peters: *Elementary Chemical Engineering*
Peters and Timmerhaus: *Plant Design and Economics for Chemical Engineers*
Probstein and Hicks: *Synthetic Fuels*
Ray: *Advanced Process Control*
Reid, Prausnitz, and Sherwood: *The Properties of Gases and Liquids*
Resnick: *Process Analysis and Design for Chemical Engineers*
Satterfield: *Heterogeneous Catalysis in Practice*
Sherwood, Pigford, and Wilke: *Mass Transfer*
Smith, B. D.: *Design of Equilibrium Stage Processes*
Smith, J. M.: *Chemical Engineering Kinetics*
Smith, J. M., and Van Ness: *Introduction to Chemical Engineering Thermodynamics*
Thompson and Ceckler: *Introduction to Chemical Engineering*
Treybal: *Mass Transfer Operations*
Valle-Riestra: *Project Evolution in the Chemical Process Industries*
Van Ness and Abbott: *Classical Thermodynamics of Nonelectrolyte Solutions: With Applications to Phase Equilibria*
Van Winkle: *Distillation*
Volk: *Applied Statistics for Engineers*
Walas: *Reaction Kinetics for Chemical Engineers*
Wei, Russell, and Swartzlander: *The Structure of the Chemical Processing Industries*
Whitwell and Toner: *Conservation of Mass and Energy*

INTRODUCTION TO CHEMICAL ENGINEERING THERMODYNAMICS

Fourth Edition

J. M. Smith

Professor of Chemical Engineering
University of California, Davis

H. C. Van Ness

Institute Professor of Chemical Engineering
Rensselaer Polytechnic Institute

McGraw-Hill Book Company

New York St. Louis San Francisco Auckland Bogotá Hamburg
London Madrid Mexico Milan Montreal New Delhi Panama Paris São Paulo
Singapore Sydney Tokyo Toronto

INTRODUCTION TO CHEMICAL ENGINEERING THERMODYNAMICS
INTERNATIONAL EDITION

1st Printing 1987

This book was set in Times Roman.
The editors were Sanjeev Rao and John Morriss. The production supervisor was Marietta Breitwieser. Project supervision was done by Albert Harrison, Harley Editorial Services.

Library of Congress Cataloging in Publication Data

Smith, J. M. (Joseph Mauck), 1916–
Introduction to chemical engineering thermodynamics.

(McGraw-Hill series in chemical engineering)
Includes bibliographical references and index.
1. Thermodynamics. 2. Chemical engineering.
I. Van Ness, H. C. (Hendrick C.) II. Title III. Series.
TP149.S582 1987 660.2′969 86–7184
ISBN 0–07–058703–5

When ordering this title use ISBN 0–07–100303–7

Printed and Bound in Singapore by Fong & Sons Printers Pte Ltd

CONTENTS

PREFACE

The purpose of this text is to provide an introductory treatment of thermodynamics from a chemical-engineering viewpoint. We have sought to present material so that it may be readily understood by the average undergraduate, while at the same time maintaining the standard of rigor demanded by sound thermodynamic analysis.

The justification for a separate text for chemical engineers is no different now than it has been for the past thirty-seven years during which the first three editions have been in print. The same thermodynamic principles apply regardless of discipline. However, these abstract principles are more effectively taught when advantage is taken of student commitment to a chosen branch of engineering. Thus, applications indicating the usefulness of thermodynamics in chemical engineering not only stimulate student interest, but also provide a better understanding of the fundamentals themselves.

The first two chapters of the book present basic definitions and a development of the first law as it applies to nonflow and simple steady-flow processes. Chapters 3 and 4 treat the pressure-volume-temperature behavior of fluids and certain heat effects, allowing early application of the first law to important engineering problems. The second law and some of its applications are considered in Chap. 5. A treatment of the thermodynamic properties of pure fluids in Chap. 6 allows application in Chap. 7 of the first and second laws to flow processes in general and in Chaps. 8 and 9 to power production and refrigeration processes. Chapters 10 through 15, dealing with fluid mixtures, treat topics in the special domain of chemical engineering thermodynamics. In Chap. 10 we present the simplest possible descriptions of mixture behavior, with application to vapor/liquid equilibrium. This is expanded in Chaps. 11 and 12 to a general treatment of vapor/liquid equilibrium for systems at modest pressures. Chapter 13 is devoted to solution thermodynamics, providing a comprehensive exposition of the thermodynamic properties of fluid mixtures. The application of equations of state in thermodynamic calculations, particularly in vapor/liquid equilibrium, is discussed in

Chap. 14. Chemical-reaction equilibrium is covered at length in Chap. 15. Finally, Chap. 16 deals with the thermodynamic analysis of real processes. This material affords a review of much of the practical subject matter of thermodynamics.

Although the text contains much introductory material, and is intended for undergraduate students, it is reasonably comprehensive, and should also serve as a useful reference source for practicing chemical engineers.

We gratefully acknowledge the contributions of Professor Charles Muckenfuss, of Debra L. Saucke, and of Eugene N. Dorsi, whose efforts produced computer programs for calculation of the thermodynamic properties of steam and ultimately the Steam Tables of App. C. We would also like to thank the reviewers of this edition: Stanley M. Walas, University of Kansas; Robert G. Squires, Purdue University; Professor Donald Sundstrom, University of Connecticut; and Professor Michael Mohr, Massachusetts Institute of Technology. Most especially, we acknowledge the contributions of Professor M. M. Abbott, whose creative ideas are reflected in the structure and character of this fourth edition, and who reviewed the entire manuscript.

J. M. Smith
H. C. Van Ness

INTRODUCTION TO CHEMICAL ENGINEERING THERMODYNAMICS

CHAPTER

ONE

INTRODUCTION

1.1 THE SCOPE OF THERMODYNAMICS

The word *thermodynamics* means heat power, or power developed from heat, reflecting its origin in the analysis of steam engines. As a fully developed modern science, thermodynamics deals with transformations of energy of all kinds from one form to another. The general restrictions within which all such transformations are observed to occur are known as the first and second laws of thermodynamics. These laws cannot be proved in the mathematical sense. Rather, their validity rests upon experience.

Given mathematical expression, these laws lead to a network of equations from which a wide range of practical results and conclusions can be deduced. The universal applicability of this science is shown by the fact that it is employed alike by physicists, chemists, and engineers. The basic principles are always the same, but the applications differ. The chemical engineer must be able to cope with a wide variety of problems. Among the most important are the determination of heat and work requirements for physical and chemical processes, and the determination of equilibrium conditions for chemical reactions and for the transfer of chemical species between phases.

Thermodynamic considerations by themselves are not sufficient to allow calculation of the *rates* of chemical or physical processes. Rates depend on both driving force and resistance. Although driving forces are thermodynamic variables, resistances are not. Neither can thermodynamics, a macroscopic-property formulation, reveal the microscopic (molecular) mechanisms of physical or chemical processes. On the other hand, knowledge of the microscopic behavior of matter can be useful in the calculation of thermodynamic properties. Such property values are essential to the practical application of thermodynamics; numerical results of thermodynamic analysis are accurate only to the extent that the required data are accurate. The chemical engineer must deal with many chemical species and their mixtures, and experimental data are often unavailable. Thus one must make effective use of correlations developed from a limited data base, but generalized to provide estimates in the absence of data.

The application of thermodynamics to any real problem starts with the identification of a particular body of matter as the focus of attention. This quantity of matter is called the *system*, and its thermodynamic state is defined by a few measurable macroscopic properties. These depend on the fundamental dimensions of science, of which length, time, mass, temperature, and amount of substance are of interest here.

1.2 DIMENSIONS AND UNITS

The *fundamental* dimensions are *primitives*, recognized through our sensory perceptions and not definable in terms of anything simpler. Their use, however, requires the definition of arbitrary scales of measure, divided into specific *units* of size. Primary units have been set by international agreement, and are codified as the International System of Units (abbreviated SI, for Système International).

The *second*, symbol s, is the SI unit of time, defined as the duration of 9,192,631,770 cycles of radiation associated with a specified transition of the cesium atom. The *meter*, symbol m, is the fundamental unit of length, defined as the distance light travels in a vacuum during 1/299,792,458 of a second. The *kilogram*, symbol kg, is the mass of a platinum/iridium cylinder kept at the International Bureau of Weights and Measures at Sèvres, France. The unit of temperature is the *kelvin*, symbol K, equal to 1/273.16 of the thermodynamic temperature of the triple point of water. A more detailed discussion of temperature, the characteristic dimension of thermodynamics, is given in Sec. 1.4. The measure of the amount of substance is the *mole*, symbol mol, defined as the amount of substance represented by as many elementary entities (e.g., molecules)

Table 1.1 Prefixes for SI units

Fraction or multiple	Prefix	Symbol
10^{-9}	nano	n
10^{-6}	micro	μ
10^{-3}	milli	m
10^{-2}	centi	c
10^{3}	kilo	k
10^{6}	mega	M
10^{9}	giga	G

as there are atoms in 0.012 kg of carbon-12. This is equivalent to the "gram mole" commonly used by chemists.

Decimal multiples and fractions of SI units are designated by prefixes. Those in common use are listed in Table 1.1. Thus we have, for example, that 1 cm = 10^{-2} m and 1 kg = 10^3 g.

Other systems of units, such as the English engineering system, use units that are related to SI units by fixed conversion factors. Thus, the foot (ft) is defined as 0.3048 m, the pound *mass* (lb_m) as 0.45359237 kg, and the pound mole (lb mol) as 453.59237 mol.

1.3 FORCE

The SI unit of force is the *newton*, symbol N, derived from Newton's second law, which expresses force F as the product of mass m and acceleration a:

$$F = ma$$

The newton is defined as the force which when applied to a mass of 1 kg produces an acceleration of 1 m s^{-2}; thus the newton is a *derived* unit representing 1 kg m s^{-2}.

In the English engineering system of units, force is treated as an additional independent dimension along with length, time, and mass. The pound *force* (lb_f) is defined as that force which accelerates 1 pound *mass* 32.1740 feet per second per second. Newton's law must here include a dimensional proportionality constant if it is to be reconciled with this definition. Thus, we write

$$F = \frac{1}{g_c} ma$$

whence†

$$1(\text{lb}_f) = \frac{1}{g_c} \times 1(\text{lb}_m) \times 32.1740(\text{ft})(\text{s})^{-2}$$

and

$$g_c = 32.1740(\text{lb}_m)(\text{ft})(\text{lb}_f)^{-1}(\text{s})^{-2}$$

The pound *force* is equivalent to 4.4482216 N.

Since force and mass are different concepts, a pound *force* and a pound *mass* are different quantities, and their units cannot be cancelled against one another. When an equation contains both units, (lb_f) and (lb_m), the dimensional constant g_c must also appear in the equation to make it dimensionally correct.

Weight properly refers to the force of gravity on a body, and is therefore correctly expressed in newtons or in pounds *force*. Unfortunately, standards of

† Where English units are employed, parentheses enclose the abbreviations of all units.

mass are often called "weights," and the use of a balance to compare masses is called "weighing." Thus, one must discern from the context whether force or mass is meant when the word "weight" is used in a casual or informal way.

Example 1.1 An astronaut weighs 730 N in Houston, Texas, where the local acceleration of gravity is $g = 9.792 \text{ m s}^{-2}$. What is the mass of the astronaut, and what does he weigh on the moon, where $g = 1.67 \text{ m s}^{-2}$?

SOLUTION Letting $a = g$, we write Newton's law as

$$F = mg$$

whence

$$m = \frac{F}{g} = \frac{730 \text{ N}}{9.792 \text{ m s}^{-2}} = 74.55 \text{ N m}^{-1} \text{ s}^2$$

Since the newton N has the units kg m s^{-2}, this result simplifies to

$$m = 74.55 \text{ kg}$$

This *mass* of the astronaut is independent of location, but his *weight* depends on the local acceleration of gravity. Thus on the moon his weight is

$$F_{\text{moon}} = mg_{\text{moon}} = 74.55 \text{ kg} \times 1.67 \text{ m s}^{-2}$$

or

$$F_{\text{moon}} = 124.5 \text{ kg m s}^{-2} = 124.5 \text{ N}$$

To work this problem in the English engineering system of units, we convert the astronaut's weight to (lb_f) and the values of g to $(\text{ft})(\text{s})^{-2}$. Since 1 N is equivalent to $0.224809(\text{lb}_f)$ and 1 m to 3.28084(ft), we have:

$$\text{Weight of astronaut in Houston} = 164.1(\text{lb}_f)$$

$$g_{\text{Houston}} = 32.13 \qquad \text{and} \qquad g_{\text{moon}} = 5.48(\text{ft})(\text{s})^{-2}$$

Newton's law here gives

$$m = \frac{Fg_c}{g} = \frac{164.1(\text{lb}_f) \times 32.1740(\text{lb}_m)(\text{ft})(\text{lb}_f)^{-1}(\text{s})^{-2}}{32.13(\text{ft})(\text{s})^{-2}}$$

or

$$m = 164.3(\text{lb}_m)$$

Thus the astronaut's mass in (lb_m) and weight in (lb_f) in Houston are *numerically* almost the same, but on the moon this is not the case:

$$F_{\text{moon}} = \frac{mg_{\text{moon}}}{g_c} = \frac{(164.3)(5.48)}{32.1740} = 28.0(\text{lb}_f)$$

1.4 TEMPERATURE

The most common method of temperature measurement is with a liquid-in-glass thermometer. This method depends on the expansion of fluids when they are heated. Thus a uniform tube, partially filled with mercury, alcohol, or some other fluid, can indicate degree of "hotness" simply by the length of the fluid column. However, numerical values are assigned to the various degrees of hotness by arbitrary definition.

For the Celsius scale, the ice point (freezing point of water saturated with air at standard atmospheric pressure) is zero, and the steam point (boiling point of pure water at standard atmospheric pressure) is 100. We may give a thermometer a numerical scale by immersing it in an ice bath and making a mark for zero at the fluid level, and then immersing it in boiling water and making a mark for 100 at this greater fluid level. The distance between the two marks is divided into 100 equal spaces called *degrees*. Other spaces of equal size may be marked off below zero and above 100 to extend the range of the thermometer.

All thermometers, regardless of fluid, read the same at zero and 100 if they are calibrated by the method described, but at other points the readings do not usually correspond, because fluids vary in their expansion characteristics. An arbitrary choice could be made, and for many purposes this would be entirely satisfactory. However, as will be shown, the temperature scale of the SI system, with its kelvin unit, symbol K, is based on the ideal gas as thermometric fluid. Since the definition of this scale depends on the properties of gases, detailed discussion of it is delayed until Chap. 3. We note, however, that this is an absolute scale, and depends on the concept of a lower limit of temperature.

Kelvin temperatures are given the symbol T; Celsius temperatures, given the symbol t, are defined in relation to Kelvin temperatures by

$$t\text{°C} = T\,\text{K} - 273.15$$

The unit of Celsius temperature is the degree Celsius, °C, equal to the kelvin. However, temperatures on the Celsius scale are 273.15 degrees lower than on the Kelvin scale. This means that the lower limit of temperature, called absolute zero on the Kelvin scale, occurs at −273.15°C.

In practice it is the *International Practical Temperature Scale of 1968* (IPTS-68) which is used for calibration of scientific and industrial instruments.† This scale has been so chosen that temperatures measured on it closely approximate ideal-gas temperatures; the differences are within the limits of present accuracy of measurement. The IPTS-68 is based on assigned values of temperature for a number of reproducible equilibrium states (defining fixed points) and on *standard instruments* calibrated at these temperatures. Interpolation between the fixed-point temperatures is provided by formulas that establish the relation between readings of the standard instruments and values of the international practical temperature. The defining fixed points are specified phase-equilibrium states of pure substances,‡ as given in Table 1.2.

† The English-language text of the definition of IPTS-68, as agreed upon by the International Committee of Weights and Measures, is published in *Metrologia*, **5**:35–44, 1969; see also ibid., **12**:7–17, 1976.

‡ See Secs. 2.7 and 2.8.

Table 1.2 Assigned values for fixed points of the IPTS-68

Equilibrium state†	T_{68}/K	t_{68}/°C
Equilibrium between the solid, liquid, and vapor phases of equilibrium hydrogen (triple point of equilibrium hydrogen)	13.81	−259.34
Equilibrium between the liquid and vapor phases of equilibrium hydrogen at 33,330.6 Pa	17.042	−256.108
Equilibrium between the liquid and vapor phases of equilibrium hydrogen (boiling point of equilibrium hydrogen)	20.28	−252.87
Equilibrium between the liquid and vapor phases of neon (boiling point of neon)	27.102	−246.048
Equilibrium between the solid, liquid, and vapor phases of oxygen (triple point of oxygen)	54.361	−218.789
Equilibrium between the liquid and vapor phases of oxygen (boiling point of oxygen)	90.188	−182.962
Equilibrium between the solid, liquid, and vapor phases of water (triple point of water)	273.16	0.01
Equilibrium between the liquid and vapor phases of water (boiling point of water)	373.15	100.00
Equilibrium between the solid and liquid phases of zinc (freezing point of zinc)	692.73	419.58
Equilibrium between the solid and liquid phases of silver (freezing point of silver)	1,235.08	961.93
Equilibrium between the solid and liquid phases of gold (freezing point of gold)	1,337.58	1,064.43

† Except for the triple points and one equilibrium point (17.042 K), temperatures are for equilibrium states at 1(atm).

The standard instrument used from −259.34 to 630.74°C is the platinum-resistance thermometer, and from 630.74 to 1064.43°C the platinum–10 percent rhodium/platinum thermocouple is used. Above 1064.43°C the temperature is defined by Planck's radiation law.

In addition to the Kelvin and Celsius scales two others are in use by engineers in the United States: the Rankine scale and the Fahrenheit scale. The Rankine scale is directly related to the Kelvin scale by

$$T(\text{R}) = 1.8T\ \text{K}$$

and is an absolute scale.

The Fahrenheit scale is related to the Rankine scale by an equation analogous to the relation between the Celsius and Kelvin scales.

$$t(°\text{F}) = T(\text{R}) - 459.67$$

Thus the lower limit of temperature on the Fahrenheit scale is −459.67(°F). The relation between the Fahrenheit and Celsius scales is given by

$$t(°\text{F}) = 1.8t°\text{C} + 32$$

This gives the ice point as 32(°F) and the normal boiling point of water as 212(°F).

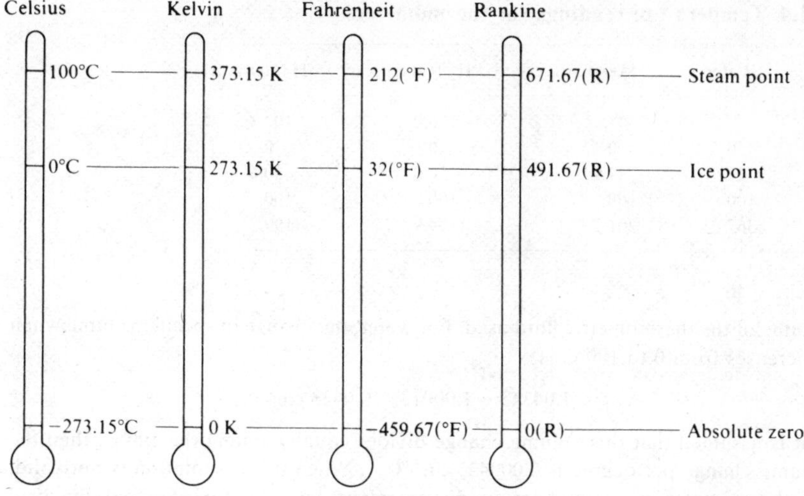

Figure 1.1 Relations among temperature scales.

The Celsius degree and the kelvin represent the same temperature *interval*, as do the Fahrenheit degree and the rankine. However, 1°C (or 1 K) is equivalent to 1.8(°F) [or 1.8(R)]. The relationships among the four temperature scales are shown in Fig. 1.1. In thermodynamics, when temperature is referred to without qualification, absolute temperature is implied.

Example 1.2 Table 1.3 lists the specific volumes of water, mercury, hydrogen at 1(atm), and hydrogen at 100(atm) for a number of temperatures on the International Practical Temperature Scale. Assume that each substance is the fluid in a thermometer, calibrated at the ice and steam points as suggested at the beginning of this section. To determine how good these thermometers are, calculate what each reads at the true temperatures for which data are given.

SOLUTION In calibrating a thermometer as specified, one assumes that each degree is represented by a fixed scale length. This is equivalent to the assumption that each degree of temperature change is accompanied by a fixed change in volume or specific

Table 1.3 Specific volumes in $cm^3\,g^{-1}$

t/°C	Water	Mercury	H_2 1(atm)	H_2 100(atm)
−100			7,053	76.03
0	1.00013	0.073554	11,125	118.36
50	1.01207	0.074223	13,161	139.18
100	1.04343	0.074894	15,197	159.71
200	1.1590	0.076250	19,266	200.72

Table 1.4 Temperature readings for thermometers

t/°C	Water	Mercury	H_2 1(atm)	H_2 100(atm)
−100			−100.0	−102.3
0	0	0	0	0
50	27.6	49.9	50.0	50.4
100	100	100	100	100
200	367	201.2	199.9	199.2

volume of the thermometric fluid used. For water, the change in specific volume when t increases from 0 to 100°C is

$$1.04343 - 1.00013 = 0.0433 \text{ cm}^3$$

If it is assumed that this volume change divides equally among the 100°C, then the volume change per degree is 0.000433 $\text{cm}^3\ {}^\circ\text{C}^{-1}$. When this assumption is not valid, the thermometer gives readings in disagreement with the International Practical Temperature Scale.

The change in specific volume of water between 0 and 50°C is

$$1.01207 - 1.00013 = 0.01194 \text{ cm}^3$$

If each degree on the water thermometer represents 0.000433 cm^3, the number of these degrees represented by a volume change of 0.01194 cm^3 is 0.01194/0.000433, or 27.6(degrees). Thus the water thermometer reads 27.6(degrees) when the actual temperature is 50°C.

At 200°C, the specific volume of water is 1.1590 cm^3, and the change between 0 and 200°C is $1.1590 - 1.00013 = 0.1589 \text{ cm}^3$. Thus the water thermometer reads 0.1589/0.000433, or 367(degrees), when the true temperature is 200°C. Table 1.4 gives all the results obtained by similar calculations.

Each thermometer reads the true Celsius temperature at 0 and 100 because each was calibrated at these points. At other points, however, the readings may differ from the true values of the temperature. Water is seen to be a singularly poor thermometric fluid. Mercury, on the other hand, is good, which accounts for its widespread use in thermometers. Hydrogen at 1(atm) makes a very good thermometric fluid, but is not practical for general use. Hydrogen at 100(atm) is no more practical and is less satisfactory.

1.5 DEFINED QUANTITIES; VOLUME

We have seen that in the international system of units force is defined through Newton's law. Convenience dictates the introduction of a number of other defined quantities. Some, like volume, are so common as to require almost no discussion. Others, requiring detailed explanation, are treated in the following sections.

Volume V is a quantity representing the product of three lengths. The volume of a substance, like its mass, depends on the amount of material considered. Specific or molar volume, on the other hand, is defined as volume per unit mass or per mole, and is therefore independent of the total amount of material considered. Density ρ is the reciprocal of specific or molar volume.

1.6 PRESSURE

The pressure P of a fluid on a surface is defined as the normal force exerted by the fluid per unit area of the surface. If force is measured in N and area in m^2, the unit is the newton per square meter or $N\ m^{-2}$, called the pascal, symbol Pa, the basic SI unit of pressure. In the English engineering system the most common unit is the pound *force* per square inch (psi).

The primary standard for the measurement of pressure derives from its definition. A known force is balanced by a fluid pressure acting on a known area; whence $P = F/A$. The apparatus providing this direct pressure measurement is the dead-weight gauge. A simple design is shown in Fig. 1.2. The piston is carefully fitted to the cylinder so that the clearance is small. Weights are placed on the pan until the pressure of the oil, which tends to make the piston rise, is just balanced by the force of gravity on the piston and all that it supports. With the force of gravity given by Newton's law, the pressure of the oil is

$$P = \frac{F}{A} = \frac{mg}{A}$$

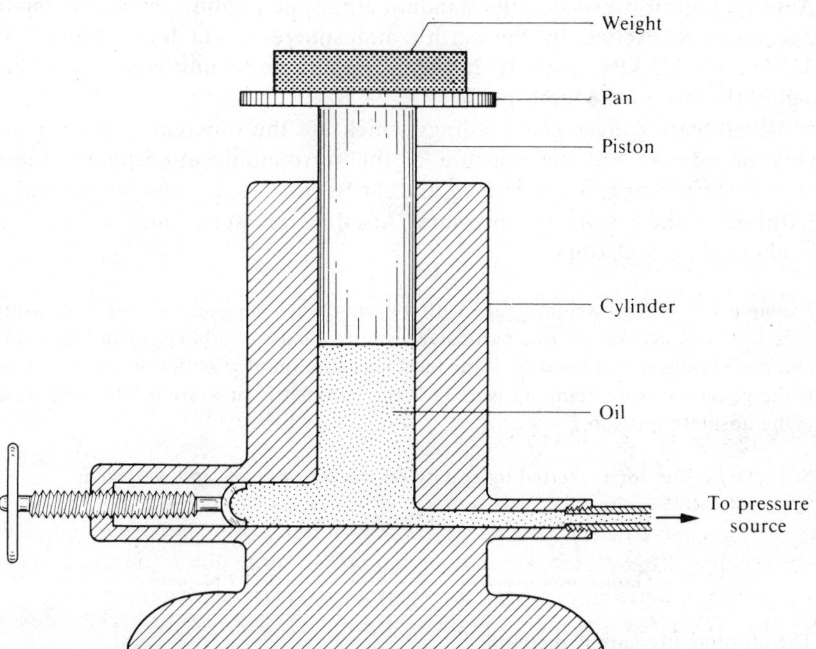

Figure 1.2 Dead-weight gauge.

where m is the mass of the piston, pan, and weights, g is the local acceleration of gravity, and A is the cross-sectional area of the piston. Gauges in common use, such as Bourdon gauges, are calibrated by comparison with dead-weight gauges.

Since a vertical column of a given fluid under the influence of gravity exerts a pressure at its base in direct proportion to its height, pressure is also expressed as the equivalent height of a fluid column. This is the basis for the use of manometers for pressure measurement. Conversion of height to force per unit area follows from Newton's law applied to the force of gravity acting on the mass of fluid in the column. The mass m is given by

$$m = Ah\rho$$

where A is the cross-sectional area of the column, h is its height, and ρ is the fluid density. Therefore

$$P = \frac{F}{A} = \frac{mg}{A} = \frac{Ah\rho g}{A} = h\rho g$$

The pressure to which a fluid height corresponds depends on the density of the fluid, which depends on its identity and temperature, and on the local acceleration of gravity. Thus the *torr* is the pressure equivalent of 1 millimeter of mercury at 0°C in a standard gravitational field and is equal to 133.322 Pa.

Another unit of pressure is the standard atmosphere (atm), the approximate average pressure exerted by the earth's atmosphere at sea level, defined as 101,325 Pa, 101.325 kPa, or 0.101325 MPa. The *bar*, an SI unit equal to 10^5 Pa, is roughly the size of the atmosphere.

Most pressure gauges give readings which are the difference between the pressure of interest and the pressure of the surrounding atmosphere. These readings are known as *gauge* pressures, and can be converted to *absolute* pressures by addition of the barometric pressure. Absolute pressures must be used in thermodynamic calculations.

Example 1.3 A dead-weight gauge with a 1-cm-diameter piston is used to measure pressures very accurately. In a particular instance a mass of 6.14 kg (including piston and pan) brings it into balance. If the local acceleration of gravity is 9.82 m s^{-2}, what is the gauge pressure being measured? If the barometric pressure is 748(torr), what is the absolute pressure?

SOLUTION The force exerted by gravity on the piston, pan, and weights is

$$F = mg = (6.14)(9.82) = 60.295 \text{ N}$$

$$\text{Gauge pressure} = \frac{F}{A} = \frac{60.295}{(1/4)(\pi)(1)^2} = 76.77 \text{ N cm}^{-2}$$

The absolute pressure is therefore

$$P = 76.77 + (748)(0.013332) = 86.74 \text{ N cm}^{-2}$$

or

$$P = 867.4 \text{ kPa}$$

Example 1.4 At 27°C a manometer filled with mercury reads 60.5 cm. The local acceleration of gravity is 9.784 m s^{-2}. To what pressure does this height of mercury correspond?

SOLUTION From the equation in the preceding text,

$$P = h\rho g$$

At 27°C the density of mercury is 13.53 g cm^{-3}. Then

$$P = 60.5 \text{ cm} \times 13.53 \text{ g cm}^{-3} \times 9.784 \text{ m s}^{-2}$$
$$= 8{,}009 \text{ g m s}^{-2} \text{ cm}^{-2}$$

or

$$P = 8.009 \text{ kg m s}^{-2} \text{ cm}^{-2} = 8.009 \text{ N cm}^{-2}$$

or

$$P = 80.09 \text{ kPa} = 0.8009 \text{ bar}$$

1.7 WORK

Work W is done whenever a force acts through a distance. The quantity of work done is defined by the equation

$$dW = F\,dl \tag{1.1}$$

where F is the component of the force acting in the direction of the displacement dl. This equation must be integrated if the work for a finite process is required.

In engineering thermodynamics an important type of work is that which accompanies a change in volume of a fluid. Consider the compression or expansion of a fluid in a cylinder caused by the movement of a piston. The force exerted by the piston on the fluid is equal to the product of the piston area and the pressure of the fluid. The displacement of the piston is equal to the volume change of the fluid divided by the area of the piston. Equation (1.1) therefore becomes

$$dW = PA\,d\frac{V}{A}$$

or, since A is constant,

$$dW = P\,dV \tag{1.2}$$

Integrating,

$$W = \int_{V_1}^{V_2} P\,dV \tag{1.3}$$

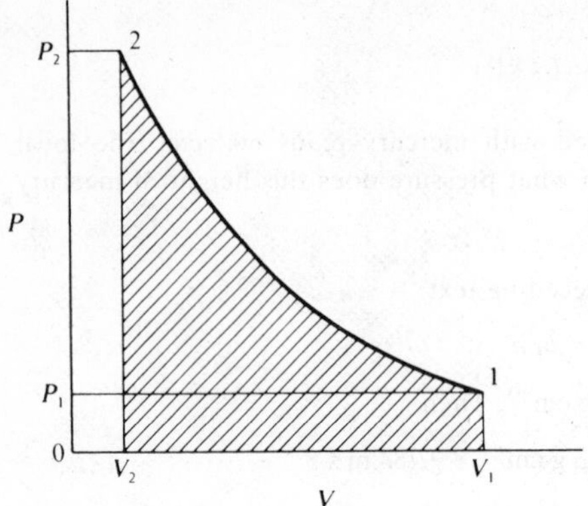

Figure 1.3 *PV* diagram.

Equation (1.3) is an expression for the work done as a result of a finite compression or expansion process.† This kind of work can be represented as an area on a pressure-vs.-volume (*PV*) diagram, such as is shown in Fig. 1.3. In this case a gas having an initial volume V_1 at pressure P_1 is compressed to volume V_2 at pressure P_2 along the path shown from 1 to 2. This path relates the pressure at any point during the process to the volume. The work required for the process is given by Eq. (1.3) and is represented on Fig. 1.3 by the area under the curve. The SI unit of work is the newton-meter or joule, symbol J. In the English engineering system the unit often used is the foot-pound *force* (ft lb_f).

1.8 ENERGY

The general principle of conservation of energy was established about 1850. The germ of this principle as it applies to mechanics was implicit in the work of Galileo (1564–1642) and Isaac Newton (1642–1726). Indeed, it follows almost automatically from Newton's second law of motion once work is defined as the product of force and displacement. No such concept existed until 1826, when it was introduced by the French mathematician J. V. Poncelet at the suggestion of G. G. Coriolis, a French engineer. The word *force* (or the Latin *vis*) was used not only in the sense described by Newton in his laws of motion, but also was applied to the quantities we now define as work and potential and kinetic energy. This ambiguity precluded for some time the development of any general principle of mechanics beyond Newton's laws of motion.

Several useful relationships follow from the definition of work as a quantitative and unambiguous physical entity. If a body of mass m is acted upon by the force F during a differential interval of time dt, the displacement of the body is dl. The work done by the force F is given by Eq. (1.1), which when combined

† However, see Sec. 2.9 for limitations on its application.

with Newton's second law becomes

$$dW = ma\,dl$$

By definition the acceleration is $a = du/dt$, where u is the velocity of the body. Thus

$$dW = m\frac{du}{dt}\,dl$$

which may be written

$$dW = m\frac{dl}{dt}\,du$$

Since the definition of velocity is $u = dl/dt$, the expression for work becomes

$$dW = mu\,du$$

This equation may now be integrated for a finite change in velocity from u_1 to u_2:

$$W = m\int_{u_1}^{u_2} u\,du = m\left(\frac{u_2^2}{2} - \frac{u_1^2}{2}\right)$$

or

$$W = \frac{mu_2^2}{2} - \frac{mu_1^2}{2} = \Delta\left(\frac{mu^2}{2}\right) \tag{1.4}$$

Each of the quantities $\frac{1}{2}mu^2$ in Eq. (1.4) is a *kinetic energy*, a term introduced by Lord Kelvin† in 1856. Thus, by definition,

$$E_K = \tfrac{1}{2}mu^2 \tag{1.5}$$

Equation (1.4) shows that the work done *on* a body in accelerating it from an initial velocity u_1 to a final velocity u_2 is equal to the change in kinetic energy of the body. Conversely, if a moving body is decelerated by the action of a resisting force, the work done *by* the body is equal to its change in kinetic energy. In the SI system of units with mass in kg and velocity in m s^{-1}, kinetic energy E_K has the units of kg m^2 s^{-2}. Since the newton is the composite unit kg m s^{-2}, E_K is measured in newton-meters or joules. In accord with Eq. (1.4), this is the unit of work.

In the English engineering system, kinetic energy is expressed as $\frac{1}{2}mu^2/g_c$, where g_c has the value 32.1740 and the units $(\mathrm{lb_m})(\mathrm{ft})(\mathrm{lb_f})^{-1}(\mathrm{s})^{-2}$. Thus the unit of kinetic energy in this system is

$$E_K = \frac{mu^2}{2g_c} = \frac{(\mathrm{lb_m})(\mathrm{ft})^2(\mathrm{s})^{-2}}{(\mathrm{lb_m})(\mathrm{ft})(\mathrm{lb_f})^{-1}(\mathrm{s})^{-2}} = (\mathrm{ft\ lb_f})$$

Dimensional consistency here requires the inclusion of g_c.

† Lord Kelvin, or William Thomson (1824–1907), was an English physicist who, along with the German physicist Rudolf Clausius (1822–1888), laid the foundations for the modern science of thermodynamics.

If a body of mass m is raised from an initial elevation z_1 to a final elevation z_2, an upward force at least equal to the weight of the body must be exerted on it, and this force must move through the distance $z_2 - z_1$. Since the weight of the body is the force of gravity on it, the minimum force required is given by Newton's law as

$$F = ma = mg$$

where g is the local acceleration of gravity. The minimum work required to raise the body is the product of this force and the change in elevation:

$$W = F(z_2 - z_1) = mg(z_2 - z_1)$$

or

$$W = mz_2g - mz_1g = \Delta(mzg) \tag{1.6}$$

We see from Eq. (1.6) that the work done *on* the body in raising it is equal to the change in the quantity *mzg.* Conversely, if the body is lowered against a resisting force equal to its weight, the work done *by* the body is equal to the change in the quantity *mzg.* Equation (1.6) is similar in form to Eq. (1.4), and both show that the work done is equal to the change in a quantity which describes the condition of the body in relation to its surroundings. In each case the work performed can be recovered by carrying out the reverse process and returning the body to its initial condition. This observation leads naturally to the thought that, if the work done on a body in accelerating it or in elevating it can be subsequently recovered, then the body by virtue of its velocity or elevation must contain the ability or capacity to do this work. This concept proved so useful in rigid-body mechanics that the capacity of a body for doing work was given the name *energy,* a word derived from the Greek and meaning "in work." Hence the work of accelerating a body is said to produce a change in its *kinetic energy,* or

$$W = \Delta E_K = \Delta\left(\frac{mu^2}{2}\right)$$

and the work done on a body in elevating it is said to produce a change in its *potential energy,* or

$$W = \Delta E_P = \Delta(mzg)$$

Thus potential energy is defined as

$$E_P = mzg \tag{1.7}$$

This term was first proposed in 1853 by the Scottish engineer William Rankine (1820–1872). In the SI system of units with mass in kg, elevation in m, and the acceleration of gravity in m s^{-2}, potential energy has the units of kg m^2 s^{-2}. This is the newton-meter or joule, the unit of work, in agreement with Eq. (1.6).

In the English engineering system, potential energy is expressed as mzg/g_c. Thus the unit of potential energy in this system is

$$E_P = \frac{mzg}{g_c} = \frac{(\mathrm{lb_m})(\mathrm{ft})(\mathrm{ft})(\mathrm{s})^{-2}}{(\mathrm{lb_m})(\mathrm{ft})(\mathrm{lb_f})^{-1}(\mathrm{s})^{-2}} = (\mathrm{ft\ lb_f})$$

Again, g_c must be included for dimensional consistency.

In any examination of physical processes, an attempt is made to find or to define quantities which remain constant regardless of the changes which occur. One such quantity, early recognized in the development of mechanics, is mass. The great utility of the law of conservation of mass as a general principle in science suggests that further principles of conservation should be of comparable value. Thus the development of the concept of energy logically led to the principle of its conservation in mechanical processes. If a body is given energy when it is elevated, then the body should conserve or retain this energy until it performs the work of which it is capable. An elevated body, allowed to fall freely, should gain in kinetic energy what it loses in potential energy so that its capacity for doing work remains unchanged. For a freely falling body, we should be able to write:

$$\Delta E_K + \Delta E_P = 0$$

or

$$\frac{mu_2^2}{2} - \frac{mu_1^2}{2} + mz_2 g - mz_1 g = 0$$

The validity of this equation has been confirmed by countless experiments. Success in application to freely falling bodies led to the generalization of the principle of energy conservation to apply to all *purely mechanical processes*. Ample experimental evidence to justify this generalization was readily obtained.

Other forms of mechanical energy besides kinetic and gravitational potential energy are possible. The most obvious is potential energy of configuration. When a spring is compressed, work is done by an external force. Since the spring can later perform this work against a resisting force, the spring possesses capacity for doing work. This is potential energy of configuration. Energy of the same form exists in a stretched rubber band or in a bar of metal deformed in the elastic region.

To increase the generality of the principle of conservation of energy in mechanics, we look upon work itself as a form of energy. This is clearly permissible, because both kinetic- and potential-energy changes are equal to the work done in producing them [Eqs. (1.4) and (1.6)]. However, work is energy in transit and is never regarded as residing in a body. When work is done and does not appear simultaneously as work elsewhere, it is converted into another form of energy.

The body or assemblage on which attention is focused is called the *system*. All else is called the *surroundings*. When work is done, it is done by the surroundings on the system, or vice versa, and energy is transferred from the surroundings to the system, or the reverse. It is only during this transfer that the form of energy known as work exists. In contrast, kinetic and potential energy reside with the system. Their values, however, are measured with reference to the surroundings, i.e., kinetic energy depends on velocity with respect to the surroundings, and potential energy depends on elevation with respect to a datum level. *Changes* in kinetic and potential energy do not depend on these reference conditions, provided they are fixed.

Example 1.5 An elevator with a mass of 2,500 kg rests at a level of 10 m above the base of an elevator shaft. It is raised to 100 m above the base of the shaft, where the cable holding it breaks. The elevator falls freely to the base of the shaft and strikes a strong spring. The spring is designed to bring the elevator to rest and, by means of a catch arrangement, to hold the elevator at the position of maximum spring compression. Assuming the entire process to be frictionless, and taking $g = 9.8 \text{ m s}^{-2}$, calculate:

(*a*) The potential energy of the elevator in its initial position relative to the base of the shaft.

(*b*) The work done in raising the elevator.

(*c*) The potential energy of the elevator in its highest position relative to the base of the shaft.

(*d*) The velocity and kinetic energy of the elevator just before it strikes the spring.

(*e*) The potential energy of the compressed spring.

(*f*) The energy of the system consisting of the elevator and spring (1) at the start of the process, (2) when the elevator reaches its maximum height, (3) just before the elevator strikes the spring, and (4) after the elevator has come to rest.

SOLUTION Let subscript 1 designate the initial conditions; subscript 2, conditions when the elevator is at its highest position; and subscript 3, conditions just before the elevator strikes the spring.

(*a*) By Eq. (1.7),

$$E_{P_1} = mz_1 g = (2{,}500)(10)(9.8) = 245{,}000 \text{ J}$$

(*b*) $W = \int_{z_1}^{z_2} F\,dl = \int_{z_1}^{z_2} mg\,dl = mg(z_2 - z_1)$

whence

$$W = (2{,}500)(9.8)(100 - 10) = 2{,}205{,}000 \text{ J}$$

(*c*) $E_{P_2} = mz_2 g = (2{,}500)(100)(9.8) = 2{,}450{,}000 \text{ J}$

Note that $W = E_{P_2} - E_{P_1}$.

(*d*) From the principle of conservation of mechanical energy, one may write that the sum of the kinetic- and potential-energy changes during the process from conditions 2 to 3 is zero; that is,

$$\Delta E_{K_{2\to 3}} + \Delta E_{P_{2\to 3}} = 0$$

or

$$E_{K_3} - E_{K_2} + E_{P_3} - E_{P_2} = 0$$

However, E_{K_2} and E_{P_3} are zero. Therefore

$$E_{K_3} = E_{P_2} = 2{,}450{,}000 \text{ J}$$

Since $E_{K_3} = \frac{1}{2}mu_3^2$,

$$u_3^2 = \frac{2E_{K_3}}{m} = \frac{(2)(2{,}450{,}000)}{2{,}500}$$

whence

$$u_3 = 44.27 \text{ m s}^{-1}$$

(*e*)
$$\Delta E_{P_{\text{spring}}} + \Delta E_{K_{\text{elevator}}} = 0$$

Since the initial potential energy of the spring and the final kinetic energy of the elevator are zero, the final potential energy of the spring must equal the kinetic energy of the elevator just before it strikes the spring. Thus the final potential energy of the spring is 2,450,000 J.

(*f*) If the elevator and the spring together are taken as the system, the initial energy of the system is the potential energy of the elevator, or 245,000 J. The total energy of the system can change only if work is transferred between it and the surroundings. As the elevator is raised, work is done on the system by the surroundings in the amount of 2,205,000 J. Thus the energy of the system when the elevator reaches its maximum height is 245,000 + 2,205,000 = 2,450,000 J. Subsequent changes occur entirely within the system, with no work transfer between the system and surroundings. Hence the total energy of the system remains constant at 2,450,000 J. It merely changes from potential energy of position (elevation) of the elevator to kinetic energy of the elevator to potential energy of configuration of the spring.

This example serves to illustrate the application of the law of conservation of mechanical energy. However, the entire process is assumed to occur without friction; the results obtained are exact only for such an idealized process.

During the period of development of the law of conservation of mechanical energy, heat was not generally recognized as a form of energy, but was considered an indestructible fluid called *caloric*. This concept was so firmly entrenched that no connection was made between heat resulting from friction and the established forms of energy, and the law of conservation of energy was limited in application to frictionless mechanical processes. Such a limitation is no longer appropriate; the concept that heat like work is energy in transit gained acceptance during the years following 1850, largely on account of the classic experiments of J. P. Joule (1818–1889), a brewer of Manchester, England. These experiments are considered in detail in Chap. 2, but first we examine some of the characteristics of heat.

1.9 HEAT

We know from experience that a hot object brought in contact with a cold object becomes cooler, whereas the cold object becomes warmer. A reasonable view is that something is transferred from the hot object to the cold one, and we call that something heat Q. Two theories of heat developed by the Greek philosophers have been in contention until modern times. The one most generally accepted until the middle of the nineteenth century was that heat is a weightless and indestructible substance called caloric. The other represented heat as connected in some way with motion, either of the ultimate particles of a body or of some medium permeating all matter. This latter view was held by Francis Bacon, Newton, Robert Boyle, and others during the seventeenth century. Without the concept of energy this view could not be exploited, and by the middle of the eighteenth century the caloric theory of heat gained ascendancy. However, a few men of science did retain the other view, notably Benjamin Thompson† (1753–1814) and Sir Humphrey Davy (1778–1829). Both submitted experimental evidence contrary to the caloric theory of heat, but their work went unheeded. Moreover, the steam engine, a working example of the conversion of heat into work, had been perfected by James Watt (1736–1819) and was in common use at the time.

One notable advance in the theory of heat was made by Joseph Black (1728–1799), a Scottish chemist and a collaborator of James Watt. Prior to Black's time no distinction was made between heat and temperature, just as no distinction was made between force and work. Temperature was regarded as the measure of the quantity of heat or caloric in a body, and a thermometer reading was referred to as a "number of degrees of heat." In fact, the word *temperature* still had its archaic meaning of *mixture* or *blend*. Thus a given temperature indicated a given mixture or blend of caloric with matter. Black correctly recognized temperature as a property which must be carefully distinguished from quantity of heat. In addition, he showed experimentally that different substances of the same mass vary in their *capacity* to absorb heat when they are warmed through the same temperature range. Moreover, he was the discoverer of *latent* heat. In spite of the difficulty of explaining these phenomena by the caloric theory, Black supported this theory throughout his life. Here the matter rested until near the middle of the nineteenth century.

Among the early champions of the energy concept of heat were Mohr, Mayer, and Helmholtz in Germany; Colding, a Dane; and especially James P. Joule in England. Joule presented the experimental evidence which conclusively demonstrated the energy theory, and thus made possible the generalization of the law of conservation of energy to include heat. The concept of heat as a form of energy is now universally accepted and is implicit in the modern science of thermodynamics.

One of the most important observations about heat is that it always flows from a higher temperature to a lower one. This leads to the concept of temperature as the driving force for the transfer of energy as heat. More precisely, the rate of heat transfer from one body to another is proportional to the temperature difference between the two bodies; when there is no temperature difference, there is no net transfer of heat. In the thermodynamic sense, heat is never regarded

† Better known as Count Rumford. Born in Woburn, Mass., unsympathetic to the American cause during the Revolution, he spent most of his extraordinary life in Europe.

as being stored within a body. Like work, it exists only as energy in transit from one body to another, or between a system and its surroundings. When energy in the form of heat is added to a body, it is stored not as heat but as kinetic and potential energy of the atoms and molecules making up the body. Not surprisingly, the energy theory of heat did not prevail until the atomic theory of matter was well established.

In spite of the transient nature of heat, it is often thought of in terms of its effects on the body from which or to which it is transferred. As a matter of fact, until about 1930 definitions of the quantitative units of heat were based on the temperature changes of a unit mass of water. Thus the *calorie* was long defined as that quantity of heat which must be transferred to one gram of water to raise its temperature one degree Celsius. Likewise, the *British thermal unit*, or (Btu), was defined as that quantity of heat which must be transferred to one pound *mass* of water to raise its temperature one degree Fahrenheit. Although these definitions provide a "feel" for the size of heat units, they depend on the accuracy of experiments made with water and are thus subject to change with each increasingly accurate measurement. The calorie and (Btu) are now recognized as units of energy, and are defined in relation to the joule, the only SI unit of energy. It is defined as 1 N m, and is therefore equal to the mechanical work done when a force of one newton acts through a distance of one meter. All other energy units are defined as multiples of the joule. The foot-pound *force*, for example, is equivalent to 1.3558179 J, and the calorie to 4.1840 J. The SI unit of power is the watt, symbol W, defined as an energy rate of one joule per second.

Appendix A gives an extensive table of conversion factors for energy as well as for other units.

PROBLEMS

1.1 Using data given in Table 1.3, confirm one of the results given in the last three columns of Table 1.4.

1.2 Pressures up to 3,000 bar are measured with a dead-weight gauge. The piston diameter is 0.35 cm. What is the approximate mass in kg of the weights required?

1.3 Pressures up to 3,000(atm) are measured with a dead-weight gauge. The piston diameter is 0.14(in). What is the approximate mass in $(\mathrm{lb_m})$ of the weights required?

1.4 A mercury manometer at 20°C and open at one end to the atmosphere reads 38.72 cm. The local acceleration of gravity is 9.790 m s^{-2}. Atmospheric pressure is 99.24 kPa. What is the absolute pressure in kPa being measured?

1.5 A mercury manometer at 75(°F) and open at one end to the atmosphere reads 16.81(in). The local acceleration of gravity is 32.143(ft)$(\mathrm{s})^{-2}$. Atmospheric pressure is 29.48(in Hg). What is the absolute pressure in (psia) being measured?

1.6 An instrument to measure the acceleration of gravity on Mars is constructed of a spring from which is suspended a mass of 0.24 kg. At a place on earth where the local acceleration of gravity is 9.80 m s^{-2}, the spring extends 0.61 cm. When the instrument package is landed on Mars, it radios the information that the spring is extended 0.20 cm. What is the Martian acceleration of gravity?

1.7 A group of engineers has landed on the moon, and would like to determine the mass of several unusual rocks. They have a spring scale calibrated to read pounds *mass* at a location where the acceleration of gravity is $32.20(\text{ft})(\text{s})^{-2}$. One of the moon rocks gives a reading of 25 on the scale. What is its mass? What is its weight on the moon? Take $g_{\text{moon}} = 5.47(\text{ft})(\text{s})^{-2}$.

1.8 A gas is confined by a piston, 5(in) in diameter, on which rests a weight. The mass of the piston and weight together is $60(\text{lb}_m)$. The local acceleration of gravity is $32.13(\text{ft})(\text{s})^{-2}$, and atmospheric pressure is 30.16(in Hg).

(*a*) What is the force in (lb_f) exerted on the gas by the atmosphere, the piston, and the weight, assuming no friction between the piston and cylinder?

(*b*) What is the pressure of the gas in (psia)?

(*c*) If the gas in the cylinder is heated, it expands, pushing the piston and weight upward. If the piston and weight are raised 15(in), what is the work done by the gas in (ft lb_f)? What is the change in potential energy of the piston and weight?

1.9 A gas is confined by a piston, 10 cm in diameter, on which rests a weight. The mass of the piston and weight together is 30 kg. The local acceleration of gravity is 9.805 m s^{-2}, and atmospheric pressure is 101.22 kPa.

(*a*) What is the force in newtons exerted on the gas by the atmosphere, the piston, and the weight, assuming no friction between the piston and cylinder?

(*b*) What is the pressure of the gas in kPa?

(*c*) If the gas in the cylinder is heated, it expands, pushing the piston and weight upward. If the piston and weight are raised 40 cm, what is the work done by the gas in kJ? What is the change in potential energy of the piston and weight?

1.10 Verify that the SI unit of kinetic and potential energy is the joule.

1.11 An automobile having a mass of 1,500 kg is traveling at 25 m s^{-1}. What is its kinetic energy in kJ? How much work must be done to bring it to a stop?

1.12 Liquid water at 0°C and atmospheric pressure has a density of 1.000 g cm^{-3}. At the same conditions, ice has a density of 0.917 g cm^{-3}. How much work is done at these conditions by 1 kg of ice as it melts to liquid water?

CHAPTER

TWO

THE FIRST LAW AND OTHER BASIC CONCEPTS

2.1 JOULE'S EXPERIMENTS

During the years 1840–1878, J. P. Joule† carried out careful experiments on the nature of heat and work. These experiments are fundamental to an understanding of the first law of thermodynamics and of the underlying concept of energy.

In their essential elements Joule's experiments were simple enough, but he took elaborate precautions to ensure accuracy. In his most famous series of experiments, he placed measured amounts of water in an insulated container and agitated the water with a rotating stirrer. The amounts of work done on the water by the stirrer were accurately measured, and the temperature changes of the water were carefully noted. He found that a fixed amount of work was required per unit mass of water for every degree of temperature rise caused by the stirring. The original temperature of the water could then be restored by the transfer of heat through simple contact with a cooler object. Thus Joule was able to show conclusively that a quantitative relationship exists between work and heat and, therefore, that heat is a form of energy.

† For a fascinating account of Joule's celebrated experiments, see T. W. Chalmers, *Historic Researches*, chap. II, Scribner, New York, 1952.

2.2 INTERNAL ENERGY

In experiments such as those conducted by Joule, energy is added to the water as work, but is extracted from the water as heat. The question arises as to what happens to this energy between the time it is added to the water as work and the time it is extracted as heat. Logic suggests that this energy is contained in the water in another form, a form which we define as *internal energy U*.

The internal energy of a substance does not include any energy that it may possess as a result of its macroscopic position or movement. Rather it refers to the energy of the molecules making up the substance, which are in ceaseless motion and possess kinetic energy of translation; except for monatomic molecules, they also possess kinetic energy of rotation and of internal vibration. The addition of heat to a substance increases this molecular activity, and thus causes an increase in its internal energy. Work done on the substance can have the same effect, as was shown by Joule.

In addition to kinetic energy, the molecules of any substance possess potential energy because of interactions among their force fields. On a submolecular scale there is energy associated with the electrons and nuclei of atoms, and bond energy resulting from the forces holding atoms together as molecules. Although absolute values of internal energy are unknown, this is not a disadvantage in thermodynamic analysis, because only *changes* in internal energy are required.

The designation of this form of energy as *internal* distinguishes it from kinetic and potential energy which the substance may possess as a result of its macroscopic position or motion, and which can be thought of as *external* forms of energy.

2.3 FORMULATION OF THE FIRST LAW OF THERMODYNAMICS

The recognition of heat and internal energy as forms of energy suggests a generalization of the law of conservation of mechanical energy (Sec. 1.8) to apply to heat and internal energy as well as to work and external potential and kinetic energy. Indeed, the generalization can be extended to still other forms, such as surface energy, electrical energy, and magnetic energy. This generalization was at first no more than a postulate, but without exception all observations of ordinary processes support it.† Hence it has achieved the stature of a law of nature, and is known as the first law of thermodynamics. One formal statement is as follows: *Although energy assumes many forms, the total quantity of energy is constant, and when energy disappears in one form it appears simultaneously in other forms.*

In application of the first law to a given process, the sphere of influence of the process is divided into two parts, the *system* and its *surroundings*. The part

† For nuclear-reaction processes, the Einstein equation applies, $E = mc^2$, where c is the velocity of light. Here, mass is transformed into energy, and the laws of conservation of mass and energy combine to state that mass and energy together are conserved.

in which the process occurs is taken as the system. Everything not included in the system constitutes the surroundings. The system may be of any size depending on the particular conditions, and its boundaries may be real or imaginary, rigid or flexible. Frequently a system is made up of a single substance; in other cases it may be very complicated. In any event, the equations of thermodynamics are written with reference to some well-defined system. This allows one to focus attention on the particular process of interest and on the equipment and material directly involved in the process.

However, the first law applies to the system *and* surroundings, and not to the system alone. In its most basic form, the first law may be written:

$$\Delta(\text{energy of the system}) + \Delta(\text{energy of surroundings}) = 0 \qquad (2.1)$$

Changes may occur in internal energy of the system, in potential and kinetic energy of the system as a whole, or in potential and kinetic energy of finite parts of the system. Likewise, the energy change of the surroundings may consist of increases or decreases in energy of various forms.

In the thermodynamic sense, heat and work refer to energy *in transit* across the boundary between the system and its surroundings. These forms of energy can never be stored. To speak of heat or work as being contained in a body or system is wrong; energy is stored in its potential, kinetic, and internal forms. These forms reside with material objects and exist because of the position, configuration, and motion of matter. The transformations of energy from one form to another and the transfer of energy from place to place often occur through the mechanisms of heat and work.

If the boundary of a system does not permit the transfer of mass between the system and its surroundings, the system is said to be *closed*, and its mass is necessarily constant. For such systems all energy passing across the boundary between system and surroundings is transferred as heat and work. Thus the total energy change of the surroundings equals the net energy transferred to or from it as heat and work, and the second term of Eq. (2.1) may be replaced by

$$\Delta(\text{energy of surroundings}) = \pm Q \pm W$$

The choice of signs used with Q and W depends on which direction of transfer is regarded as positive.

The first term of Eq. (2.1) may be expanded to show energy changes in various forms. If the mass of the system is constant and if only internal-, kinetic-, and potential-energy changes are involved,

$$\Delta(\text{energy of the system}) = \Delta U + \Delta E_K + \Delta E_P$$

With these substitutions, Eq. (2.1) becomes

$$\Delta U + \Delta E_K + \Delta E_P = \pm Q \pm W \qquad (2.2)$$

The traditional choice of signs on the right-hand side of Eq. (2.2) makes the numerical value of heat positive when it is transferred to the system from the surroundings, and the numerical value of work positive for the *opposite* direction of transfer. With this understanding, Eq. (2.2) becomes†

$$\Delta U + \Delta E_K + \Delta E_P = Q - W \tag{2.3}$$

In words, Eq. (2.3) states that the total energy change of the system is equal to the heat added to the system minus the work done by the system. This equation applies to the changes which occur in a constant-mass system over a period of time.

Closed systems often undergo processes that cause no changes in external potential or kinetic energy, but only changes in internal energy. For such processes, Eq. (2.3) reduces to

$$\boxed{\Delta U = Q - W} \tag{2.4}$$

Equation (2.4) applies to processes involving *finite* changes in the system. For *differential* changes this equation is written:

$$\boxed{dU = dQ - dW} \tag{2.5}$$

Equation (2.5) is useful when U, Q, and W are expressed as functions of process variables, and like Eq. (2.4) applies to closed systems which undergo changes in *internal* energy only. The system must of course be clearly defined, as illustrated in the examples of this and later chapters.

The units used in Eqs. (2.3) through (2.5) must be the same for all terms. In the SI system the energy unit is the joule. Other energy units still in use are the calorie, the foot-pound *force*, and the (Btu).

2.4 THE THERMODYNAMIC STATE AND STATE FUNCTIONS

In thermodynamics we distinguish between two types of quantities: those which depend on path and those which do not. Actually, both types are in everyday use. Consider for example an automobile trip from New York to San Francisco. The straight-line distance between these two cities is fixed; it does not depend on the path or route taken to get from one to the other. On the other hand, such measurements as miles traveled and fuel consumed definitely depend on the path. So it is in thermodynamics; both types of quantities are used.

There are many examples of quantities which do not depend on path; among them are temperature, pressure, and specific volume. We know from experience that fixing two of these quantities automatically fixes all other such properties of a homogeneous pure substance and, therefore, determines the condition or

† Those who prefer consistency over tradition make both heat and work positive for transfer to the system from the surroundings. Eq. (2.2), then becomes

$$\Delta U + \Delta E_K + \Delta E_P = Q + W$$

state of the substance. For example, nitrogen gas at a temperature of 300 K and a pressure of 10^5 kPa (1 bar) has a definite specific volume or density, a definite viscosity, a definite thermal conductivity; in short it has a definite set of properties. If this gas is heated or cooled, compressed or expanded, and then returned to its initial conditions, it is found to have exactly the same set of properties as before. These properties do not depend on the past history of the substance nor on the path it followed in reaching a given state. They depend only on present conditions, however reached. Such quantities are known as *state functions.* When two of them are fixed or held at definite values for a homogeneous pure substance, the *thermodynamic state* of the substance is fixed.

For systems more complicated than a simple homogeneous pure substance, the number of properties or state functions that must be arbitrarily specified in order to define the state of the system may be different from two. The method of determining this number is the subject of Sec. 2.8.

Internal energy and a number of other thermodynamic variables (defined later) are state functions and are, therefore, properties of the system. Since state functions can be expressed mathematically as functions of thermodynamic coordinates such as temperature and pressure, their values can always be identified with points on a graph. The differential of a state function is spoken of as an infinitesimal *change* in the property. The integration of such a differential results in a finite difference between two values of the property. For example,

$$\int_{P_1}^{P_2} dP = P_2 - P_1 = \Delta P \qquad \text{and} \qquad \int_{U_1}^{U_2} dU = U_2 - U_1 = \Delta U$$

Work and heat, on the other hand, are not state functions. Since they depend on path, they cannot be identified with points on a graph, but rather are represented by areas, as shown in Fig. 1.3. The differentials of heat and work are not referred to as changes, but are regarded as infinitesimal *quantities* of heat and work. When integrated, these differentials give not a finite change but a finite quantity. Thus

$$\int dQ = Q \qquad \text{and} \qquad \int dW = W$$

Experiment shows that processes which accomplish the same change in state by different paths in a closed system require, in general, different amounts of heat and work, but that *the difference $Q - W$ is the same for all such processes.* This gives experimental justification to the statement that internal energy is a state function. Equation (2.4) yields the same value of ΔU regardless of the path followed, provided only that the change in the system is always from the same initial to the same final state.

Another difference between state functions and heat or work is that a state function represents a property of a system and always has a value. Work and heat appear only when changes are caused in a system by a process, which requires time. Although the time required for a process cannot be predicted by thermodynamics alone, nevertheless the passage of time is inevitable whenever heat is transferred or work is accomplished.

The internal energy of a system, like its volume, depends on the quantity of material involved; such properties are said to be *extensive.* In contrast, temperature and pressure, the principal thermodynamic coordinates for homogeneous fluids, are independent of the quantity of material making up the system, and are known as *intensive* properties.

The first-law equations may be written for systems containing any quantity of material; the values of Q, W, and the energy terms then refer to the entire system. More often, however, we write the equations of thermodynamics for a representative unit amount of material, either a unit mass or a mole. We can then deal with properties such as volume and internal energy on a unit basis, in which case they become intensive properties, independent of the quantity of material actually present. Thus, although the total volume and total internal energy of an arbitrary quantity of material are extensive properties, specific and molar volume (or density) and specific and molar internal energy are intensive. Writing Eqs. (2.4) and (2.5) for a representative unit amount of the system puts all of the terms on a unit basis, but this does not make Q and W into thermodynamic properties or state functions. Multiplication of a quantity on a unit basis by the total mass (or total moles) of the system gives the total quantity.

Internal energy (through the enthalpy, defined in Sec. 2.5) is useful for the calculation of heat and work quantities for such equipment as heat exchangers, evaporators, distillation columns, pumps, compressors, turbines, engines, etc., because it is a state function. The tabulation of all possible Q's and W's for all possible processes is impossible. But the intensive state functions, such as specific volume and specific internal energy, are properties of matter, and they can be measured and their values tabulated as functions of temperature and pressure for a particular substance for future use in the calculation of Q or W for any process involving that substance. The measurement, correlation, and use of these state functions is treated in detail in later chapters.

Example 2.1 Water flows over a waterfall 100 m in height. Consider 1 kg of the water, and assume that no energy is exchanged between the 1 kg and its surroundings.

(*a*) What is the potential energy of the water at the top of the falls with respect to the base of the falls?

(*b*) What is the kinetic energy of the water just before it strikes bottom?

(*c*) After the 1 kg of water enters the river below the falls, what change has occurred in its state?

SOLUTION Taking the 1 kg of water as the system, and noting that it exchanges no energy with its surroundings, we may set Q and W equal to zero and write Eq. (2.3) as

$$\Delta U + \Delta E_K + \Delta E_P = 0$$

This equation applies to each part of the process.

(*a*) From Eq. (1.7),

$$E_P = mzg = 1 \text{ kg} \times 100 \text{ m} \times 9.8066 \text{ m s}^{-2}$$

where g has been taken as the standard value. This gives

$$E_P = 980.66 \text{ N m} \qquad \text{or} \qquad 980.66 \text{ J}$$

(b) During the free fall of the water no mechanism exists for the conversion of potential or kinetic energy into internal energy. Thus ΔU must be zero, and

$$\Delta E_K + \Delta E_P = E_{K_2} - E_{K_1} + E_{P_2} - E_{P_1} = 0$$

For practical purposes we may take $E_{K_1} = E_{P_2} = 0$. Then

$$E_{K_2} = E_{P_1} = 980.66 \text{ J}$$

(c) As the 1 kg of water strikes bottom and mixes with other falling water to form a river, there is much turbulence, which has the effect of converting kinetic energy into internal energy. During this process, ΔE_P is essentially zero, and Eq. (2.3) becomes

$$\Delta U + \Delta E_K = 0 \qquad \text{or} \qquad \Delta U = E_{K_2} - E_{K_3}$$

However, the river velocity is assumed small, and therefore E_{K_3} is negligible. Thus

$$\Delta U = E_{K_2} = 980.66 \text{ J}$$

The overall result of the process is the conversion of potential energy of the water into internal energy of the water. This change in internal energy is manifested by a temperature rise of the water. Since energy in the amount of 4,184 J kg^{-1} is required for a temperature rise of 1°C in water, the temperature increase is 980.66/4,184 = 0.234°C, if there is no heat transfer with the surroundings.

Example 2.2 A gas is confined in a cylinder by a piston. The initial pressure of the gas is 7 bar, and the volume is 0.10 m^3. The piston is held in place by latches in the cylinder wall. The whole apparatus is placed in a total vacuum. What is the energy change of the apparatus if the retaining latches are removed so that the gas suddenly expands to double its initial volume? The piston is again held by latches at the end of the process.

SOLUTION Since the question concerns the entire apparatus, the system is taken as the gas, piston, and cylinder. No work is done during the process, because no force external to the system moves, and no heat is transferred through the vacuum surrounding the apparatus. Hence Q and W are zero, and the total energy of the system remains unchanged. Without further information we can say nothing about the distribution of energy among the parts of the system. This may well be different than the initial distribution.

Example 2.3 If the process described in Example 2.2 is repeated, not in a vacuum but in air at standard atmospheric pressure of 101.3 kPa, what is the energy change of the apparatus? Assume the rate of heat exchange between the apparatus and the surrounding air slow compared with the rate at which the process occurs.

SOLUTION The system is chosen exactly as before, but in this case work is done by the system in pushing back the atmosphere. This work is given by the product of the force exerted by the atmospheric pressure on the piston and the displacement of the piston. If the area of the piston is A, the force is $F = P_{\text{atm}} A$. The displacement of the piston is equal to the volume change of the gas divided by the area of the piston, or $\Delta l = \Delta V/A$. The work done by the system on the surroundings, according to Eq. (1.1), is then

$$W = F\,\Delta l = P_{\text{atm}}\,\Delta V$$

$$W = (101.3)(0.2 - 0.1) = 10.13 \text{ kPa m}^3$$

or

$$W = 10.13 \text{ kN m} = 10.13 \text{ kJ}$$

Heat transfer between the system and surroundings is also possible in this case, but the problem is worked for the instant after the process has occurred and before appreciable heat transfer has had time to take place. Thus Q is assumed to be zero in Eq. (2.3), giving

$$\Delta(\text{energy of the system}) = Q - W = 0 - 10.13 = -10.13 \text{ kJ}$$

The total energy of the system has *decreased* by an amount equal to the work done on the surroundings.

Example 2.4 When a system is taken from state a to state b in Fig. 2.1 along path *acb*, 100 J of heat flows into the system and the system does 40 J of work. How much heat flows into the system along path *aeb* if the work done by the system is 20 J? The system returns from b to a along the path *bda*. If the work done on the system is 30 J, does the system absorb or liberate heat? How much?

Solution We presume that the system changes only in its internal energy and that Eq. (2.4) is applicable. For path *acb*,

$$\Delta U_{ab} = Q_{acb} - W_{acb} = 100 - 40 = 60 \text{ J}$$

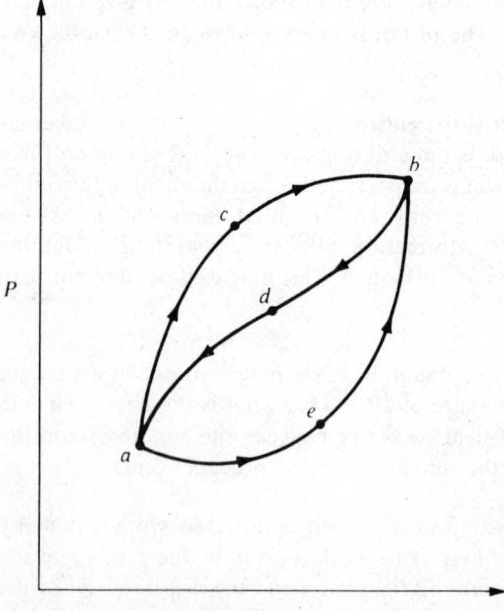

Figure 2.1 Diagram for Example 2.4.

This is the internal energy change for the state change from *a* to *b* by *any* path. Thus for path *aeb*,

$$\Delta U_{ab} = 60 = Q_{aeb} - W_{aeb} = Q_{aeb} - 20$$

whence

$$Q_{aeb} = 80 \text{ J}$$

For path *bda*,

$$\Delta U_{ba} = -\Delta U_{ab} = -60 = Q_{bda} - W_{bda} = Q_{bda} - (-30)$$

thus

$$Q_{bda} = -60 - 30 = -90 \text{ J}$$

Heat is therefore liberated from the system.

2.5 ENTHALPY

In addition to internal energy a number of other thermodynamic functions are in common use because of their practical importance. Enthalpy (en-thal′-py) is introduced in this section, and others are treated later. Enthalpy is explicitly defined for any system by the mathematical expression

$$\boxed{H \equiv U + PV} \tag{2.6}$$

where U = internal energy
P = absolute pressure
V = volume

The units of all terms of this equation must be the same. The product PV has the units of energy, as does U; therefore H also has units of energy. In the SI system the basic unit of pressure is the pascal or N m^{-2} and, for volume, the m^3. Thus the PV product has the unit N m or joule. In the English engineering system a common unit for the PV product is the (ft lb$_f$), which arises when pressure is in (lb$_f$)(ft)$^{-2}$ with volume in (ft)^3. This result is usually converted to (Btu) through division by 778.16 for use in Eq. (2.6), because the common English engineering unit for U and H is the (Btu).

Since U, P, and V are all state functions, H as defined by Eq. (2.6) must also be a state function. In differential form Eq. (2.6) may be written

$$dH = dU + d(PV) \tag{2.7}$$

This equation applies whenever a differential change occurs in the system. Integration of Eq. (2.7) gives

$$\Delta H = \Delta U + \Delta(PV) \tag{2.8}$$

an equation applicable whenever a finite change occurs in the system. Equations (2.6) through (2.8) may be written for any amount of material, though they are often applied to a unit mass or to a mole. Like volume and internal energy, enthalpy is an extensive property; specific or molar enthalpy is of course intensive.

Enthalpy is useful as a thermodynamic property because the $U + PV$ group appears frequently, particularly in problems involving flow processes, as illustrated in Sec. 2.6. The calculation of a numerical value for ΔH is carried out in the following example.

Example 2.5 Calculate ΔU and ΔH for 1 kg of water when it is vaporized at the constant temperature of 100°C and the constant pressure of 101.33 kPa. The specific volumes of liquid and vapor water at these conditions are 0.00104 and 1.673 $m^3\ kg^{-1}$. For this change, heat in the amount of 2,256.9 kJ is added to the water.

SOLUTION The kilogram of water is taken as the system, because it alone is of interest. We imagine the fluid contained in a cylinder by a frictionless piston which exerts a constant pressure of 101.33 kPa. As heat is added, the water expands from its initial to its final volume, doing work on the piston. By Eq. (1.3),

$$W = P\,\Delta V = 101.33\ \text{kPa} \times (1.673 - 0.001)\ \text{m}^3$$

whence

$$W = 169.4\ \text{kPa m}^3 = 169.4\ \text{kN m}^{-2}\ \text{m}^3 = 169.4\ \text{kJ}$$

Since $Q = 2{,}256.9$ kJ, Eq. (2.4) gives

$$\Delta U = Q - W = 2{,}256.9 - 169.4 = 2{,}087.5\ \text{kJ}$$

With P constant, Eq. (2.8) becomes

$$\Delta H = \Delta U + P\,\Delta V$$

But $P\,\Delta V = W$. Therefore

$$\Delta H = \Delta U + W = Q = 2{,}256.9\ \text{kJ}$$

2.6 THE STEADY-STATE FLOW PROCESS

The application of Eqs. (2.4) and (2.5) is restricted to nonflow (constant mass) processes in which only internal-energy changes occur. Far more important industrially are processes which involve the steady-state flow of a fluid through equipment. For such processes the more general first-law expression [Eq. (2.3)] must be used. However, it may be put in more convenient form. The term *steady state* implies that conditions at all points in the apparatus are constant with time. For this to be the case, all rates must be constant, and there must be no accumulation of material or energy within the apparatus over the period of time considered. Moreover, the total mass flow rate must be the same at all points along the path of flow of the fluid.

Consider the general case of a steady-state-flow process as represented in Fig. 2.2. A fluid, either liquid or gas, flows through the apparatus from section 1 to section 2. At section 1, the entrance to the apparatus, conditions in the fluid are denoted by subscript 1. At this point the fluid has an elevation above an

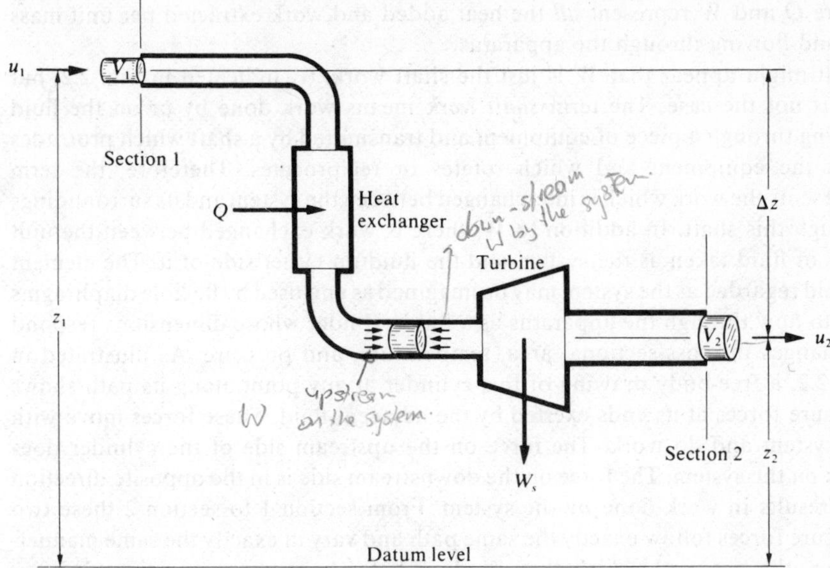

Figure 2.2 Steady-state flow process.

arbitrary datum level of z_1, an average velocity u_1, a specific volume V_1, a pressure P_1, an internal energy U_1, etc. Similarly, the conditions in the fluid at section 2, the exit of the apparatus, are denoted by subscript 2.

The system is taken as a unit mass of the fluid, and we consider the overall changes which occur in this unit mass of fluid as it flows through the apparatus from section 1 to section 2. The energy of the unit mass may change in all three of the forms taken into account by Eq. (2.3), that is, potential, kinetic, and internal. The kinetic-energy change of a unit mass of fluid between sections 1 and 2 follows from Eq. (1.5):

$$\Delta E_K = \tfrac{1}{2}u_2^2 - \tfrac{1}{2}u_1^2 = \tfrac{1}{2}\Delta u^2$$

In this equation u represents the average velocity of the flowing fluid, defined as the volumetric flow rate divided by the cross-sectional area.† As a result of Eq. (1.7) we have for the potential-energy change of a unit mass of fluid between sections 1 and 2

$$\Delta E_P = z_2 g - z_1 g = g\,\Delta z$$

Equation (2.3) now becomes

$$\Delta U + \frac{\Delta u^2}{2} + g\,\Delta z = Q - W \tag{2.9}$$

† The development of the expression $\frac{1}{2}u^2$ for kinetic energy in terms of the *average* fluid velocity is considered in detail in Chap. 7.

where Q and W represent *all* the heat added and work extracted per unit mass of fluid flowing through the apparatus.

It might appear that W is just the shaft work W_s indicated in Fig. 2.2, but this is not the case. The term *shaft work* means work done by or on the fluid flowing through a piece of equipment and transmitted by a shaft which protrudes from the equipment and which rotates or reciprocates. Therefore, the term represents the work which is interchanged between the system and its surroundings through this shaft. In addition to W_s there is work exchanged between the unit mass of fluid taken as the system and the fluid on either side of it. The element of fluid regarded as the system may be imagined as enclosed by flexible diaphragms and to flow through the apparatus as a fluid cylinder whose dimensions respond to changes in cross-sectional area, temperature, and pressure. As illustrated in Fig. 2.2, a free-body drawing of this cylinder at any point along its path shows pressure forces at its ends exerted by the adjacent fluid. These forces move with the system and do work. The force on the upstream side of the cylinder does work *on* the system. The force on the downstream side is in the opposite direction and results in work done *by* the system. From section 1 to section 2 these two pressure forces follow exactly the same path and vary in exactly the same manner. Hence, the net work which they produce between these two sections is zero. However, the terms representing work done by these pressure forces as the fluid enters and leaves the apparatus do not, in general, cancel. In Fig. 2.2 the unit mass of fluid is shown just before it enters the apparatus. This cylinder of fluid has a volume V_1 equal to its specific volume at the conditions existing at section 1. If its cross-sectiional area is A_1, its length is V_1/A_1. The force exerted on its upstream face is P_1A_1, and the work done by this force in pushing the cylinder into the apparatus is

$$W_1 = P_1A_1\frac{V_1}{A_1} = P_1V_1$$

This represents work done *on* the system by the surroundings. At section 2 work is done *by* the system on the surroundings as the fluid cylinder emerges from the apparatus. This work is given by

$$W_2 = P_2A_2\frac{V_2}{A_2} = P_2V_2$$

Since W in Eq. (2.9) represents *all* the work done *by* the unit mass of fluid, it is equal to the algebraic sum of the shaft work and the entrance and exit work quantities; that is,

$$W = W_s + P_2V_2 - P_1V_1$$

In combination with this result, Eq. (2.9) becomes

$$\Delta U + \frac{\Delta u^2}{2} + g\,\Delta z = Q - W_s - P_2V_2 + P_1V_1$$

or

$$\Delta U + \Delta(PV) + \frac{\Delta u^2}{2} + g\,\Delta z = Q - W_s$$

But by Eq. (2.8),

$$\Delta U + \Delta(PV) = \Delta H$$

Therefore,

$$\boxed{\Delta H + \frac{\Delta u^2}{2} + g\,\Delta z = Q - W_s} \tag{2.10a}$$

This equation is the mathematical expression of the first law for a steady-state-flow process. All the terms are expressions for energy per unit mass of fluid; in the SI system of units, energy is expressed in joules or in some multiple of the joule. For the English engineering system of units, this equation must be reexpressed to include the dimensional constant g_c in the kinetic- and potential-energy terms:

$$\boxed{\Delta H + \frac{\Delta u^2}{2g_c} + \frac{g}{g_c}\,\Delta z = Q - W_s} \tag{2.10b}$$

Here, the usual unit for ΔH and Q is the (Btu), whereas kinetic energy, potential energy, and work are usually expressed as (ft lb_f). Therefore the factor 778.16(ft lb_f)$(\text{Btu})^{-1}$ must be used with the appropriate terms to put them all in consistent units of either (ft lb_f) or (Btu).

For many of the applications considered in thermodynamics, the kinetic- and potential-energy terms are very small compared with the others and may be neglected. In such a case Eq. (2.10) reduces to

$$\Delta H = Q - W_s \tag{2.11}$$

This expression of the first law for a steady-flow process is analogous to Eq. (2.4) for a nonflow process. Here, however, the enthalpy rather than the internal energy is the thermodynamic property of importance.

Equations (2.10) and (2.11) are universally used for the solution of problems involving the steady-state flow of fluids through equipment. For most such applications values of the enthalpy must be available. Since H is a state function and a property of matter, its values depend only on point conditions; once determined, they may be tabulated for subsequent use whenever the same sets of conditions are encountered again. Thus Eq. (2.10) may be applied to laboratory processes designed specifically for the determination of enthalpy data.

One such process employs a flow calorimeter. A simple example of this device is illustrated schematically in Fig. 2.3. Its essential feature is an electric heater immersed in a flowing fluid. The apparatus is designed so that the kinetic- and potential-energy changes of the fluid from section 1 to section 2 (Fig. 2.3) are

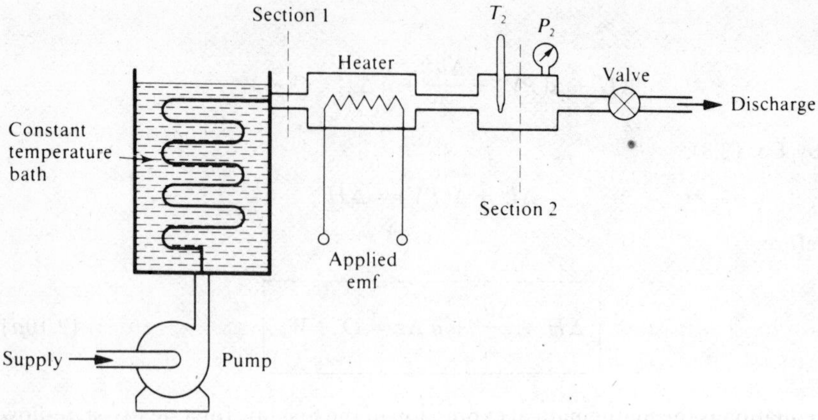

Figure 2.3 Flow calorimeter.

negligible. This requires merely that the two sections be at the same elevation and that the velocities be small. Furthermore, no shaft work is accomplished between sections 1 and 2. Hence Eq. (2.10) reduces to

$$\Delta H = H_2 - H_1 = Q$$

Heat is added to the fluid from the electric resistance heater; the rate of energy input is determined from the resistance of the heater and the current passing through it. The entire apparatus is well insulated. In practice there are a number of details which need attention, but in principle the operation of the flow calorimeter is simple. Measurements of the rate of heat input and the rate of flow of the fluid allow calculation of values of ΔH between sections 1 and 2.

As an example, consider the measurement of enthalpies of H_2O, both as liquid and as vapor. Liquid water is supplied to the apparatus by the pump. The constant-temperature bath might be filled with a mixture of crushed ice and water to maintain a temperature of 0°C. The coil which carries the test fluid, in this case, water, through the constant-temperature bath is made long enough so that the fluid emerges essentially at the bath temperature of 0°C. Thus the fluid at section 1 is always liquid water at 0°C. The temperature and pressure at section 2 are measured by suitable instruments. Values of the enthalpy of H_2O for various conditions at section 2 may be calculated by the equation

$$H_2 = H_1 + Q$$

where Q is the heat added by the resistance heater per unit mass of water flowing. Clearly, H_2 depends not only on Q but also on H_1. The conditions at section 1 are always the same, i.e., liquid water at 0°C, except that the pressure varies from run to run. However, pressure has a negligible effect on the properties of liquids unless very high pressures are reached, and for practical purposes H_1 may be considered a constant. Absolute values of enthalpy, like absolute values of internal energy, are unknown. An arbitrary value may therefore be assigned to H_1 as the *basis* for all other enthalpy values. If we set $H_1 = 0$ for liquid water at 0°C, then the values of H_2 are given by

$$H_2 = H_1 + Q = 0 + Q = Q$$

These results may be tabulated along with the corresponding conditions of T and P existing at section 2 for a large number of runs. In addition, specific-volume measurements may be made for these same conditions, and these may be tabulated. Corresponding values of the internal energy of water may be calculated by Eq. (2.6), $U = H - PV$, and these numbers too may be tabulated. In this way tables of thermodynamic properties may be compiled over the entire useful range of conditions. The most widely used such tabulation is for H_2O and is known as the *steam tables.*†

The enthalpy may be taken as zero for some other state than liquid at 0°C. The choice is arbitrary. The equations of thermodynamics, such as Eq. (2.10), apply to *changes* of state, for which the enthalpy *differences* are independent of where the origin of values is placed. However, once an arbitrary zero point is selected for the enthalpy, an arbitrary choice cannot be made for the internal energy, for values of internal energy are then calculable from the enthalpy by Eq. (2.6).

Example 2.6 For the flow calorimeter just discussed, the following data are taken with water as the test fluid:

$$\text{Flow rate} = 4.15 \text{ g s}^{-1}$$

$$t_1 = 0°\text{C} \qquad t_2 = 300°\text{C} \qquad P_2 = 3 \text{ bar}$$

$$\text{Rate of heat addition from resistance heater} = 12{,}740 \text{ W}$$

It is observed that the water is completely vaporized in the process. Calculate the enthalpy of steam at 300°C and 3 bar based on $H = 0$ for liquid water at 0°C.

SOLUTION If Δz and Δu^2 are negligible and if W_s and H_1 are zero, then $H_2 = Q$, and

$$H_2 = \frac{12{,}740 \text{ J s}^{-1}}{4.15 \text{ g s}^{-1}} = 3{,}070 \text{ J g}^{-1}$$

Example 2.7 Air at 1 bar and 25°C enters a compressor at low velocity, discharges at 3 bar, and enters a nozzle in which it expands to a final velocity of 600 m s^{-1} at the initial conditions of pressure and temperature. If the work of compression is 240 kJ per kilogram of air, how much heat must be removed during compression?

SOLUTION Since the air returns to its initial conditions of T and P, the overall process produces no change in enthalpy of the air. Moreover, the potential-energy change of the air is presumed negligible. Neglecting also the initial kinetic energy of

† Steam tables are given in App. C. Tables for various other substances are found in the literature. A discussion of compilations of thermodynamic properties appears in Chap. 6.

the air, we write Eq. (2.10*a*) as

$$Q = \frac{u_2^2}{2} + W_s$$

The kinetic-energy term is evaluated as follows:

$$\tfrac{1}{2}u_2^2 = \tfrac{1}{2}(600)^2 = 180{,}000 \text{ m}^2 \text{ s}^{-2}$$

or

$$\tfrac{1}{2}u_2^2 = 180{,}000 \text{ N m kg}^{-1} = 180 \text{ kJ kg}^{-1}$$

Then

$$Q = 180 - 240 = -60 \text{ kJ kg}^{-1}$$

Thus, heat must be removed in the amount of 60 kJ for each kilogram of air compressed.

Example 2.8 Water at 200(°F) is pumped from a storage tank at the rate of 50(gal)(min)$^{-1}$. The motor for the pump supplies work at the rate of 2(hp). The water passes through a heat exchanger, where it gives up heat at the rate of 40,000(Btu)(min)$^{-1}$, and is delivered to a second storage tank at an elevation 50(ft) above the first tank. What is the temperature of the water delivered to the second tank?

SOLUTION This is a steady-flow process for which Eq. (2.10*b*) applies. The initial and final velocities of water in the storage tanks are negligible, and the term $\Delta u^2/2g_c$ may be omitted. The remaining terms are expressed in units of (Btu)(lb$_m$)$^{-1}$ through use of appropriate conversion factors. At 200(°F) the density of water is 60.1(lb$_m$)(ft)$^{-3}$, and 1(ft)^3 is equivalent to 7.48(gal); thus the mass flow rate is

$$(50)(60.1/7.48) = 402(\text{lb}_m)(\text{min})^{-1}$$

from which we obtain

$$Q = -40{,}000/402 = -99.50(\text{Btu})(\text{lb}_m)^{-1}$$

Since 1(hp) is equivalent to 42.41(Btu)(min)$^{-1}$, the shaft work is

$$W_s = -(2)(42.41)/(402) = -0.21(\text{Btu})(\text{lb}_m)^{-1}$$

If the local acceleration of gravity is taken as the standard value of 32.174(ft)(s)$^{-2}$, the potential-energy term becomes

$$\frac{g}{g_c}\Delta z = \left(\frac{32.174}{32.174}\right)\frac{(50)}{(778.16)} = 0.06(\text{Btu})(\text{lb}_m)^{-1}$$

Equation (2.10*b*) now yields ΔH:

$$\Delta H = Q - W_s - \frac{g}{g_c}\Delta z = -99.50 - (-0.21) - 0.06$$

$$\Delta H = -99.35(\text{Btu})(\text{lb}_m)^{-1}$$

The enthalpy of water at 200(°F) is given in the steam tables as 168.09(Btu)(lb$_m$)$^{-1}$. Thus

$$\Delta H = H_2 - H_1 = H_2 - 168.09 = -99.35$$

and

$$H_2 = 168.09 - 99.35 = 68.74(\text{Btu})(\text{lb}_m)^{-1}$$

The temperature of water having this enthalpy is found from the steam tables to be

$$t_2 = 100.74(°\text{F})$$

In this example W_s and $(g/g_c)\,\Delta z$ are small compared with Q, and for practical purposes they could be neglected.

2.7 EQUILIBRIUM

Equilibrium is a word denoting a static condition, the absence of change. In thermodynamics it is taken to mean not only the absence of change but the absence of any *tendency* toward change on a macroscopic scale. Thus a system at equilibrium is one which exists under such conditions that there is no tendency for a change in state to occur. Since any tendency toward change is caused by a driving force of one kind or another, the absence of such a tendency indicates also the absence of any driving force. Hence a system at equilibrium may be described as one in which all forces are in exact balance. Whether a change actually occurs in a system not at equilibrium depends on resistance as well as on driving force. Many systems undergo no measurable change even under the influence of large driving forces, because the resistance is very large.

Different kinds of driving forces tend to bring about different kinds of change. Mechanical forces such as pressure on a piston tend to cause energy transfer as work; temperature differences tend to cause the flow of heat; chemical potentials tend to cause substances to react chemically or to be transferred from one phase to another. At equilibrium all such forces are in balance. Often we are content to deal with systems at partial equilibrium. In many applications of thermodynamics, chemical reactions are of no concern. For example, a mixture of hydrogen and oxygen at ordinary conditions is not in chemical equilibrium, because of the large driving force for the formation of water. In the absence of chemical reaction, this system may well be in thermal and mechanical equilibrium, and purely physical processes may be analyzed without regard to the possible chemical reaction.

2.8 THE PHASE RULE

As mentioned earlier, the state of a pure homogeneous fluid is fixed whenever two intensive thermodynamic properties are set at definite values. However, for more complex systems this number is not necessarily two. For example, a mixture of steam and liquid water in equilibrium at 101.33 kPa can exist only at 100°C. It is impossible to change the temperature without also changing the pressure if vapor and liquid are to continue to exist in equilibrium; one cannot exercise independent control over these two variables for this system. The number of independent variables that must be arbitrarily fixed to establish the *intensive* state of a system, i.e., the *degrees of freedom F* of the system, is given by the celebrated phase rule of J. Willard Gibbs,† who deduced it by theoretical reasoning in 1875. It is presented here without proof in the form applicable to nonreacting systems:‡

$$\boxed{F = 2 - \pi + N} \tag{2.12}$$

where π = number of phases, and N = number of chemical species.

The intensive state of a system at equilibrium is established when its temperature, pressure, and the compositions of all phases are fixed. These are therefore phase-rule variables, but they are not all independent. The phase rule gives the number of variables from this set which must be arbitrarily specified to fix all remaining phase-rule variables.

A *phase* is a homogeneous region of matter. A gas or a mixture of gases, a liquid or a liquid solution, and a solid crystal are examples of phases. A phase need not be continuous; examples of discontinuous phases are a gas dispersed as bubbles in a liquid, a liquid dispersed as droplets in another liquid with which it is immiscible, and a crystalline solid dispersed in either a gas or liquid. In each case a dispersed phase is distributed throughout a continuous phase. An abrupt change in properties always occurs at the boundary between phases. Various phases can coexist, but they *must be in equilibrium* for the phase rule to apply. An example of a system at equilibrium which is made up of three phases is a boiling saturated solution of a salt in water with excess salt crystals present. The three phases are crystalline salt, the saturated aqueous solution, and the vapor generated by boiling.

The phase-rule variables are *intensive* properties, which are independent of the extent of the system and of the individual phases. Thus the phase rule gives the same information for a large system as for a small one and for different relative amounts of the phases present. Moreover, the only compositions that are phase-rule variables are those of the individual phases. Overall or total compositions are not phase-rule variables when more than one phase is present.

The minimum number of degrees of freedom for any system is zero. When $F = 0$, the system is invariant, and Eq. (2.12) becomes $\pi = 2 + N$. This value of π is the maximum number of phases which can coexist at equilibrium for a system containing N chemical species. When $N = 1$, this number is 3, and we have a triple point. For example, the triple point of water, where liquid, vapor, and the common form of ice exist together in equilibrium, occurs at 0.01°C and 0.00610 bar. Any change from these conditions causes at least one phase to disappear.

† Josiah Willard Gibbs (1839–1903), American mathematical physicist.

‡ The justification of the phase rule for nonreacting systems is given in Sec. 12.2, and the phase rule for reacting systems is considered in Sec. 15.8.

Example 2.9 How many degrees of freedom has each of the following systems?

(*a*) Liquid water in equilibrium with its vapor.

(*b*) Liquid water in equilibrium with a mixture of water vapor and nitrogen.

(*c*) A liquid solution of alcohol in water in equilibrium with its vapor.

SOLUTION

(*a*) The system contains a single chemical species. There are two phases (one liquid and one vapor). Thus

$$F = 2 - \pi + N = 2 - 2 + 1 = 1$$

This result is in agreement with the well-known fact that at a given pressure water has but one boiling point. Temperature or pressure, but not both, may be specified for a system consisting of water in equilibrium with its vapor.

(*b*) In this case two chemical species are present. Again there are two phases. Thus

$$F = 2 - \pi + N = 2 - 2 + 2 = 2$$

We see from this example that the addition of an inert gas to a system of water in equilibrium with its vapor changes the characteristics of the system. Now temperature and pressure may be independently varied, but once they are fixed the system described can exist in equilibrium only at a particular composition of the vapor phase. (If nitrogen is taken to be negligibly soluble in water, we need not consider the composition of the liquid phase.)

(*c*) Here $N = 2$, and $\pi = 2$. Thus

$$F = 2 - \pi + N = 2 - 2 + 2 = 2$$

The phase-rule variables are temperature, pressure, and the phase compositions. The composition variables are either the weight or mole fractions of the species in a phase, and they must sum to unity for each phase. Thus fixing the mole fraction of the water in the liquid phase automatically fixes the mole fraction of the alcohol. These two compositions cannot both be arbitrarily specified.

2.9 THE REVERSIBLE PROCESS

The development of thermodynamics is facilitated by the introduction of a special kind of nonflow process characterized as *reversible.* A process is reversible *when its direction can be reversed at any point by an infinitesimal change in external conditions.*

To indicate the nature of reversible processes, we examine the simple expansion of a gas in a piston/cylinder arrangement. The apparatus is shown in Fig. 2.4, and is imagined to exist in an evacuated space. The gas trapped inside the cylinder is chosen as the system; all else is the surroundings. Expansion processes result when mass is removed from the piston. To make the process as simple as possible, we assume that the piston slides within the cylinder without friction and that the piston and cylinder neither absorb nor transmit heat. Moreover, because the density of the gas in the cylinder is low and because the mass of gas is small, we ignore the effects of gravity on the contents of the

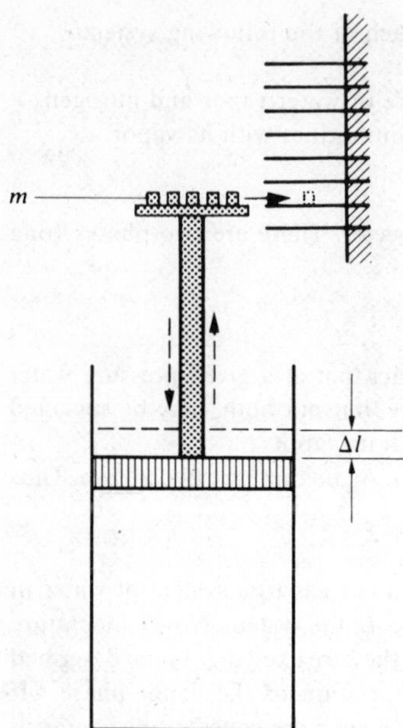

Figure 2.4 Expansion of a gas.

cylinder. This means that gravity-induced pressure gradients in the gas are considered very small relative to its pressure and that changes in potential energy of the gas are taken as negligible in comparison with the potential-energy changes of the piston assembly.

The piston in Fig. 2.4 confines the gas at a pressure just sufficient to balance the weight of the piston and all that it supports. This is a condition of equilibrium, for the system has no tendency to change. Mass must be removed from the piston if it is to rise. We imagine first that a mass m is suddenly slid from the piston to a shelf (at the same level). The piston assembly accelerates upward, reaching its maximum velocity at the point where the upward force on the piston just balances its weight. Its momentum then carries it to a higher level, where it reverses direction. If the piston were held in this position of maximum elevation, its potential-energy increase would very nearly equal the work done by the gas during the initial stroke. However, when unconstrained, the piston assembly oscillates, with decreasing amplitude, ultimately coming to rest at a new equilibrium position at a level Δl above its initial position.

The oscillations of the piston assembly are damped out because the viscous nature of the gas gradually converts gross directed motion of the molecules into chaotic molecular motion. This *dissipative* process transforms some of the work initially done by the gas in accelerating the piston back into internal energy of the gas. Once the process is initiated, no *infinitesimal* change in external conditions can reverse its direction; the process is *irreversible.*

All processes carried out in finite time with real substances are accompanied in some degree by dissipative effects of one kind or another, and all are therefore irreversible. However, we can *imagine* processes that are free of dissipative effects. For the expansion process of Fig. 2.4, they have their origin in the sudden removal of a finite mass from the piston. The resulting imbalance of forces acting on the piston causes its acceleration, and leads to its subsequent oscillation. The sudden removal of smaller mass increments reduces but does not eliminate this dissipative effect. Even the removal of an infinitesimal mass leads to piston oscillations of infinitesimal amplitude and a consequent dissipative effect. However, one may imagine a process wherein small mass increments are removed one after another at a rate such that the piston's rise is continuous, with oscillation only at the end of the process.

The limiting case of removal of a succession of infinitesimal masses from the piston is approximated when the mass m in Fig. 2.4 is replaced by a pile of powder, blown in a very fine stream from the piston. During this process, the piston rises at a uniform but very slow rate, and the powder collects in storage at ever higher levels. The system is never more than differentially displaced either from internal equilibrium or from equilibrium with its surroundings. If the removal of powder from the piston is stopped and the direction of transfer of powder is reversed, the process reverses direction and proceeds backward along its original path. Both the system and its surroundings are ultimately restored to their initial conditions. The original process is *reversible.*

Without the assumption of a frictionless piston, we cannot imagine a reversible process. If the piston sticks because of friction, a finite mass must be removed before the piston breaks free. Thus the equilibrium condition necessary to reversibility is not maintained. Moreover, friction between two sliding parts is a mechanism for the dissipation of mechanical energy into internal energy.

Our discussion has centered on a single nonflow process, the expansion of a gas in a cylinder. The opposite process, compression of a gas in a cylinder, is described in exactly the same way. There are, however, many processes which are driven by other-than-mechanical forces. For example, heat flow occurs when a temperature difference exists, electricity flows under the influence of an electromotive force, and chemical reactions occur because a chemical potential exists. In general, a process is reversible when the net force driving it is only differential in size. Thus heat is transferred reversibly when it flows from a finite object at temperature T to another such object at temperature $T - dT$.

The concept of a reversible chemical reaction may be illustrated by the decomposition of calcium carbonate, which when heated forms calcium oxide and carbon dioxide gas. At equilibrium, this system exerts a definite decomposition pressure of CO_2 for a given temperature. When the pressure falls below this value, $CaCO_3$ decomposes. Assume now that a cylinder is fitted with a frictionless piston and contains $CaCO_3$, CaO, and CO_2 in equilibrium. It is immersed in a constant-temperature bath, as shown in Fig. 2.5, with the temperature adjusted to a value such that the decomposition pressure is just sufficient to balance the weight on the piston. The system is in mechanical equilibrium, the temperature of the system is equal to that of the bath, and the chemical reaction is held in balance by the pressure of the CO_2. Any change of conditions, however slight,

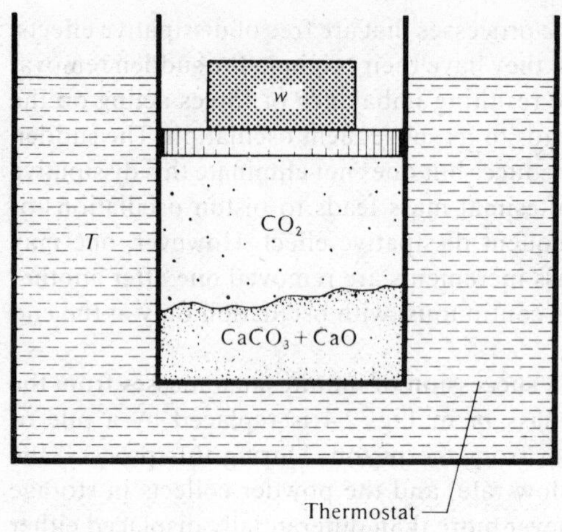

Figure 2.5 Reversibility of a chemical reaction.

upsets the equilibrium and causes the reacton to proceed in one direction or the other. If the weight is differentially increased, the CO_2 pressure rises differentially, and CO_2 combines with CaO to form $CaCO_3$, allowing the weight to fall slowly. The heat given off by this reaction raises the temperature in the cylinder, and heat flows to the bath. Decreasing the weight differentially sets off the opposite chain of events. The same results are obtained if the temperature of the bath is raised or lowered. If the temperature of the bath is raised differentially, heat flows into the cylinder and calcium carbonate decomposes. The CO_2 generated causes the pressure to rise differentially, which in turn raises the piston and weight. This continues until the $CaCO_3$ is completely decomposed. The process is reversible, for the system is never more than differentially displaced from equilibrium, and only a differential lowering of the temperature of the bath causes the system to return to its initial state.

Chemical reactions can sometimes be carried out in an electrolytic cell, and in this case they can be held in balance by an applied potential difference. If such a cell consists of two electrodes, one of zinc and the other of platinum, immersed in an aqueous solution of hydrochloric acid, the reaction that occurs is

$$Zn + 2HCl \rightleftharpoons H_2 + ZnCl_2$$

The cell is held under fixed conditions of temperature and pressure, and the electrodes are connected externally to a potentiometer. If the electromotive force produced by the cell is exactly balanced by the potential difference of the potentiometer, the reaction is held in equilibrium. The reaction may be made to proceed in the forward direction by a slight decrease in the opposing potential difference, and it may be reversed by a corresponding increase in the potential difference above the emf of the cell.

In summary, a reversible process is frictionless; it is never more than differentially removed from equilibrium, and therefore traverses a succession of equilibrium states; the driving forces are differential in magnitude; its direction can be reversed at any point by a differential change in external conditions, causing the process to retrace its path, leading to restoration of the initial state of the system and its surroundings.

In Sec. 1.6 we derived an equation for the work of compression or expansion of a gas caused by the differential displacement of a piston in a cylinder:

$$dW = P\,dV \tag{1.2}$$

The work appearing in the *surroundings* is given by this equation only when certain characteristics of the reversible process are realized. The first requirement is that the system be no more than infinitesimally displaced from a state of *internal* equilibrium characterized by uniformity of temperature and pressure. The system then always has an identifiable set of properties, including pressure P. The second requirement is that the system be no more than infinitesimally displaced from mechanical equilibrium with its surroundings. In this event, the internal pressure P is never more than minutely out of balance with the external force, and we may make the substitution $F = PA$ that transforms Eq. (1.1) into Eq. (1.2). Processes for which these requirements are met are said to be *mechanically reversible.* For such processes, Eq. (1.3) correctly yields the work appearing in the surroundings:

$$W = \int_{V_1}^{V_2} P\,dV \tag{1.3}$$

The reversible process is ideal in that it can never be fully realized; it represents a limit to the performance of actual processes. In thermodynamics, the calculation of work is usually made for reversible processes, because of their tractability to mathematical analysis. The choice is between these calculations and no calculations at all. Results for reversible processes in combination with appropriate efficiencies yield reasonable approximations of the work for actual processes.

Example 2.10 A horizontal piston-and-cylinder arrangement is placed in a constant-temperature bath. The piston slides in the cylinder with negligible friction, and an external force holds it in place against an initial gas pressure of 14 bar. The initial gas volume is 0.03 m^3. The external force on the piston is reduced gradually, allowing the gas to expand until its volume doubles. Experiment shows that under these conditions the volume of the gas is related to its pressure in such a way that the product PV is constant. Calculate the work done in moving the external force.

How much work would be done if the external force were suddenly reduced to half its initial value instead of being gradually reduced?

SOLUTION The process, carried out as first described, is mechanically reversible, and Eq. (1.3) is applicable. If $PV = k$, then $P = k/V$, and

$$W = k\int_{V_1}^{V_2} \frac{dV}{V} = k \ln \frac{V_2}{V_1}$$

But

$$V_1 = 0.03\ \text{m}^3 \qquad V_2 = 0.06\ \text{m}^3$$

and

$$k = PV = P_1V_1 = (14 \times 10^5)(0.03) = 42{,}000 \text{ J}$$

Therefore

$$W = 42{,}000 \ln 2 = 29{,}112 \text{ J}$$

The final pressure is

$$P_2 = \frac{k}{V_2} = \frac{42{,}000}{0.06} = 700{,}000 \text{ Pa} \quad \text{or} \quad 7 \text{ bar}$$

In the second case, after half the initial force has been removed, the gas undergoes a sudden expansion against a constant force equivalent to a pressure of 7 bar. Eventually the system returns to an equilibrium condition identical with the final state attained in the reversible process. Thus ΔV is the same as before, and the net work accomplished equals the equivalent external pressure times the volume change, or

$$W = (7 \times 10^5)(0.06 - 0.03) = 21{,}000 \text{ J.}$$

This process is clearly irreversible, and compared with the reversible process is said to have an efficiency of

$$\frac{21{,}000}{29{,}112} = 0.721 \quad \text{or} \quad 72.1\%$$

Example 2.11 The piston-and-cylinder arrangement shown in Fig. 2.6 contains nitrogen gas trapped below the piston at a pressure of 7 bar. The piston is held in

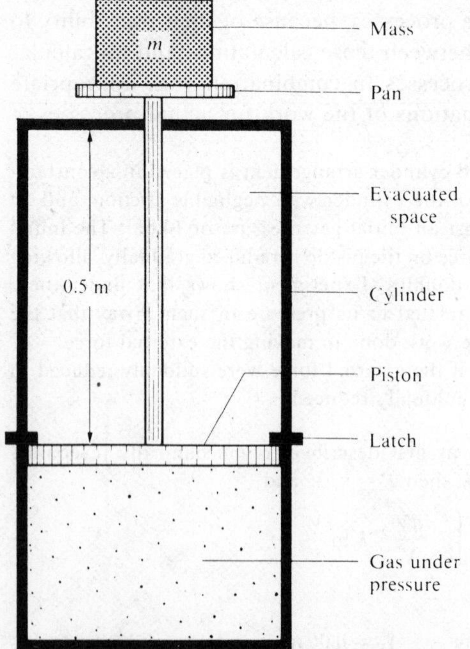

Figure 2.6 Diagram for Example 2.11.

place by latches. The space behind the piston is evacuated. A pan is attached to the piston rod and a mass m of 45 kg is fastened to the pan. The piston, piston rod, and pan together have a mass of 23 kg. The latches holding the piston are released, allowing the piston to rise rapidly until it strikes the top of the cylinder. The distance moved by the piston is 0.5 m. The local acceleration of gravity is 9.8 m s^{-2}. Discuss the energy changes that occur because of this process.

SOLUTION This example serves to illustrate some of the difficulties encountered when irreversible nonflow processes are analyzed. We take the gas alone as the system. According to the basic definition, the work done by the gas on the surroundings is equal to $\int P'\,dV$, where P' is the pressure exerted on the face of the piston by the gas. Because the expansion is very rapid, pressure gradients exist in the gas, and neither P' nor the integral can be evaluated. However, we can avoid the calculation of W by returning to Eq. (2.1). The total energy change of the system (the gas) is its internal-energy change. For $Q = 0$, the energy changes of the surroundings consist of potential-energy changes of the piston, rod, pan, and mass m and of internal-energy changes of the piston, rod, and cylinder. Therefore, Eq. (2.1) may be written

$$\Delta U_{\text{sys}} + (\Delta U_{\text{surr}} + \Delta E_{P_{\text{surr}}}) = 0$$

The potential-energy term is

$$\Delta E_{P_{\text{surr}}} = (45 + 23)(9.8)(0.5) = 333.2 \text{ N m}$$

Therefore

$$\Delta U_{\text{sys}} + \Delta U_{\text{surr}} = -333.2 \text{ N m} = -333.2 \text{ J}$$

and one cannot determine the individual internal-energy changes which occur in the piston-and-cylinder assembly.

2.10 NOTATION; CONSTANT-VOLUME AND CONSTANT-PRESSURE PROCESSES

To this point, extensive properties have been represented by plain uppercase letters, such as U and V, without specification of the amount of material to which they apply. Henceforth we denote by these symbols only *specific* or *molar* properties. For a system of mass m or of n moles, we write mU or nU, mV or nV, etc., indicating explicitly the amount of material in the system. Thus, for a closed system of n moles, Eq. (2.5) is replaced by

$$d(nU) = dQ - dW \tag{2.13}$$

where Q and W always represent *total* heat and work, whatever the value of n.

The work of a mechanically reversible, nonflow process is given by

$$dW = P\,d(nV) \tag{2.14}$$

whence Eq. (2.13) becomes

$$d(nU) = dQ - P\,d(nV) \tag{2.15}$$

This is the general first-law equation for a mechanically reversible, nonflow process. If in addition the process occurs at constant volume, then

$$dQ = d(nU) \qquad \text{(const } V\text{)} \tag{2.16}$$

Integration yields

$$Q = n\,\Delta U \qquad \text{(const } V\text{)} \tag{2.17}$$

Thus for a mechanically reversible, constant-volume, nonflow process, the heat transferred is equal to the internal-energy change of the system.

Equation (2.6), which defines the enthalpy, may be written

$$nH = nU + P(nV)$$

For an infinitesimal, constant-pressure change of state,

$$d(nH) = d(nU) + P\,d(nV)$$

Combining this with Eq. (2.15) gives

$$dQ = d(nH) \qquad \text{(const } P\text{)} \tag{2.18}$$

Integration yields

$$Q = n\,\Delta H \qquad \text{(const } P\text{)} \tag{2.19}$$

Thus for a mechanically reversible, constant-pressure, nonflow process, the heat transferred equals the enthalpy change of the system. Comparison of the last two equations with Eqs. (2.16) and (2.17) shows that the enthalpy plays a role in constant-pressure processes analogous to the internal energy in constant-volume processes.

2.11 HEAT CAPACITY

We remarked earlier that heat is often thought of in relation to its effect on the object to which or from which it is transferred. This is the origin of the idea that a body has a capacity for heat. The smaller the temperature change in a body caused by the transfer of a given quantity of heat, the greater its capacity. Indeed, a *heat capacity* might be defined as

$$C = \frac{dQ}{dT}$$

The difficulty with this is that it makes C, like Q, a path-dependent quantity rather than a state function. However, it does suggest the possibility that more than one heat capacity might be usefully defined.

There are in fact two heat capacities in common use for homogeneous fluids; although their names belie the fact, both are state functions, defined unambiguously in relation to other state functions:

Heat capacity at constant volume

$$C_V \equiv \left(\frac{\partial U}{\partial T}\right)_V \tag{2.20}$$

Heat capacity at constant pressure

$$C_P \equiv \left(\frac{\partial H}{\partial T}\right)_P \tag{2.21}$$

These definitions accommodate both molar heat capacities and specific heat capacities (usually called specific heats), depending on whether U and H are molar or specific properties.

Although the definitions of C_V and C_P make no reference to any process, each allows an especially simple description of a particular process. Thus, if we have a constant-volume process, Eq. (2.20) may be written

$$dU = C_V\,dT \qquad \text{(const } V\text{)} \tag{2.22}$$

Integration yields

$$\Delta U = \int_{T_1}^{T_2} C_V\,dT \qquad \text{(const } V\text{)} \tag{2.23}$$

For a mechanically reversible, constant-volume process, this result may be combined with Eq. (2.17) to give

$$Q = n\,\Delta U = n\int_{T_1}^{T_2} C_V\,dT \qquad \text{(const } V\text{)} \tag{2.24}$$

Consider now the case in which the volume varies during the process, but is the same at the end as at the beginning. Such a process cannot rightly be called one of constant volume, even though $V_2 = V_1$ and $\Delta V = 0$. However, changes in state functions or properties are independent of path and are, therefore, the same for all processes which lead from the same initial to the same final conditions. Hence, property changes for this case may be calculated from the equations for a truly constant-volume process leading from the same initial to the same final conditions. For such processes Eq. (2.23) gives $\Delta U = \int C_V\,dT$, because U, C_V, and T are all state functions or properties. On the other hand, Q does depend on path, and Eq. (2.24) is a valid expression for Q only for a *constant-volume* process. For the same reason, W is in general zero only for a constant-volume process. This discussion illustrates the reason for the careful distinction made between state functions and heat and work. The principle that state functions are independent of path is an important and useful concept. Thus for the calculation of property changes an actual process may be replaced by any other process which accomplishes the same change in state. Such an alternative process may be selected, for example, because of its simplicity.

For a constant-pressure process, Eq. (2.21) may be written

$$dH = C_P\,dT \qquad \text{(const } P\text{)} \tag{2.25}$$

whence

$$\Delta H = \int_{T_1}^{T_2} C_P \, dT \qquad \text{(const } P\text{)} \tag{2.26}$$

Combination with Eq. (2.19) for a mechanically reversible, constant-pressure process gives

$$Q = n\,\Delta H = n \int_{T_1}^{T_2} C_P \, dT \qquad \text{(const } P\text{)} \tag{2.27}$$

Since H, C_P, and T are all state functions, Eq. (2.26) applies to any process for which $P_2 = P_1$ whether or not it is actually carried out at constant pressure. However, it is only for the mechanically reversible, constant-pressure path that heat and work can be calculated by the equations $Q = n\,\Delta H$, $Q = n \int C_P \, dT$, and $W = Pn\,\Delta V$.

Example 2.12 An ideal gas is one for which PV/T is a constant regardless of the changes it undergoes. Such a gas has a volume of 0.02271 $m^3\,mol^{-1}$ at 0°C and 1 bar. In the following problem, air may be considered an ideal gas with the constant heat capacities

$$C_V = (5/2)R \qquad \text{and} \qquad C_P = (7/2)R$$

where $R = 8.314\ J\,mol^{-1}\,K^{-1}$. Thus

$$C_V = 20.785 \qquad \text{and} \qquad C_P = 29.099\ J\,mol^{-1}\,K^{-1}$$

The initial conditions of the air are 1 bar and 25°C. It is compressed to 5 bar and 25°C by two different mechanically reversible processes. Calculate the heat and work requirements and ΔU and ΔH of the air for each path:

(*a*) Cooling at constant pressure followed by heating at constant volume.
(*b*) Heating at constant volume followed by cooling at constant pressure.

SOLUTION In each case we take the system as 1 mol of air contained in an imaginary piston-and-cylinder arrangement. Since the processes considered are mechanically reversible, the piston is imagined to move in the cylinder without friction. The initial volume of air is

$$V_1 = (0.02271)\left(\frac{298.15}{273.15}\right) = 0.02479\ m^3$$

The final volume is

$$V_2 = V_1 \frac{P_1}{P_2} = (0.02479)\left(\frac{1}{5}\right) = 0.004958\ m^3$$

(*a*) In this case during the first step the air is cooled at the constant pressure of 1 bar until the final volume of 0.004958 m^3 is reached. During the second step the volume is held constant at this value while the air is heated to its final state. The temperature of the air at the end of the cooling step is

$$T = (298.15)\left(\frac{0.004958}{0.02479}\right) = 59.63\ K$$

For this step the pressure is constant. By Eq. (2.27),

$$Q = \Delta H = C_P \, \Delta T = (29.099)(59.63 - 298.15) = -6{,}941 \text{ J}$$

Since $\Delta U = \Delta H - \Delta(PV) = \Delta H - P\,\Delta V$, then

$$\Delta U = -6{,}941 - (1 \times 10^5)(0.004958 - 0.02479) = -4{,}958 \text{ J}$$

In the second step the air is heated at constant volume. By Eq. (2.24),

$$\Delta U = Q = C_V \, \Delta T = (20.785)(298.15 - 59.63) = 4{,}958 \text{ J}$$

The complete process represents the sum of its steps. Hence

$$Q = -6{,}941 + 4{,}958 = -1{,}983 \text{ J}$$

and

$$\Delta U = -4{,}958 + 4{,}958 = 0$$

Since the first law applies to the entire process, $\Delta U = Q - W$, and therefore

$$0 = -1{,}983 - W$$

Whence

$$W = -1{,}983 \text{ J}$$

Equation (2.8), $\Delta H = \Delta U + \Delta(PV)$, also applies to the entire process. But $T_1 = T_2$, and therefore $P_1 V_1 = P_2 V_2$. Hence $\Delta(PV) = 0$, and

$$\Delta H = \Delta U = 0$$

(*b*) Two different steps are used in this case to reach the same final state of the air. In the first step the air is heated at a constant volume equal to its initial value until the final pressure of 5 bar is reached. During the second step the air is cooled at the constant pressure of 5 bar to its final state. The air temperature at the end of the first step is

$$T = (298.15)(5/1) = 1{,}490.75 \text{ K}$$

For this step the volume is constant, and

$$Q = \Delta U = C_V \, \Delta T = (20.785)(1{,}490.75 - 298.15) = 24{,}788 \text{ J}$$

For the second step pressure is constant, and

$$Q = \Delta H = C_P \, \Delta T = (29.099)(298.15 - 1{,}490.75) = -34{,}703 \text{ J}$$

Also

$$\Delta U = \Delta H - \Delta(PV) = \Delta H - P\,\Delta V$$

$$\Delta U = -34{,}703 - (5 \times 10^5)(0.004958 - 0.02479) = -24{,}788 \text{ J}$$

For the two steps combined,

$$Q = 24{,}788 - 34{,}703 = -9{,}915 \text{ J}$$

$$\Delta U = 24{,}788 - 24{,}788 = 0$$

$$W = Q - \Delta U = -9{,}915 - 0 = -9{,}915 \text{ J}$$

and as before

$$\Delta H = \Delta U = 0$$

The property changes ΔU and ΔH calculated for the given change in state are the same for both paths. On the other hand the answers to parts (*a*) and (*b*) show that Q and W depend on the path.

Example 2.13 Calculate the internal-energy and enthalpy changes that occur when air is changed from an initial state of 40(°F) and 10(atm), where its molar volume is $36.49(\text{ft})^3(\text{lb mol})^{-1}$ to a final state of 140(°F) and 1(atm). Assume for air that PV/T is constant and that $C_V = 5$ and $C_P = 7(\text{Btu})(\text{lb mol})^{-1}(°\text{F})^{-1}$.

SOLUTION Since property changes are independent of the process that brings them about, we can base calculations on a simple two-step, mechanically reversible process in which 1(lb mol) of air is

(*a*) cooled at constant volume to the final pressure, and
(*b*) heated at constant pressure to the final temperature.

The absolute temperatures here are on the Rankine scale:

$$T_1 = 40 + 459.67 = 499.67(\text{R})$$

$$T_2 = 140 + 459.67 = 599.67(\text{R})$$

Since $PV = kT$, the ratio T/P is constant for step (*a*). The intermediate temperature between the two steps is therefore

$$T' = (499.67)(1/10) = 49.97(\text{R})$$

and the temperature changes for the two steps are

$$\Delta T_a = 49.97 - 499.67 = -449.70(\text{R})$$

and

$$\Delta T_b = 599.67 - 49.97 = 549.70(\text{R})$$

For step (*a*), Eq. (2.23) becomes

$$\Delta U_a = C_V \, \Delta T_a$$

whence

$$\Delta U_a = (5)(-449.70) = -2{,}248.5(\text{Btu})(\text{lb mol})^{-1}$$

For step (*b*), Eq. (2.26) becomes

$$\Delta H_b = C_P \, \Delta T_b$$

whence

$$\Delta H_b = (7)(549.70) = 3{,}847.9(\text{Btu})(\text{lb mol})^{-1}$$

For step (*a*), Eq. (2.8) becomes

$$\Delta H_a = \Delta U_a + V \, \Delta P_a$$

Whence

$$\Delta H_a = -2{,}248.5 + 36.49(1 - 10)(2.7195) = -3{,}141.6(\text{Btu})$$

The factor 2.7195 converts the PV product from (atm)(ft)^3, which is an energy unit, into (Btu). For step (b), Eq. (2.8) becomes

$$\Delta U_b = \Delta H_b - P\,\Delta V_b$$

The final volume of the air is given by

$$V_2 = V_1 \frac{P_1 T_2}{P_2 T_1}$$

from which we find that $V_2 = 437.93(\text{ft})^3$. Therefore

$$\Delta U_b = 3{,}847.9 - (1)(437.93 - 36.49)(2.7195) = 2{,}756.2(\text{Btu})$$

For the two steps together,

$$\Delta U = -2{,}248.5 + 2{,}756.2 = 507.7(\text{Btu})$$

and

$$\Delta H = -3{,}141.6 + 3{,}847.9 = 706.3(\text{Btu})$$

PROBLEMS

2.1 An insulated and nonconducting container filled with 10 kg of water at 20°C is fitted with a stirrer. The stirrer is made to turn by gravity acting on a weight of mass 25 kg. The weight falls slowly through a distance of 10 m in driving the stirrer. Assuming that all work done on the weight is transferred to the water and that the local acceleration of gravity is 9.8 m s^{-2}, determine:
(a) The amount of work done on the water.
(b) The internal-energy change of the water.
(c) The final temperature of the water.
(d) The amount of heat that must be removed from the water to return it to its initial temperature.
(e) The total energy change of the universe because of (1) the process of lowering the weight, (2) the process of cooling the water back to its initial temperature, and (3) both processes together.

2.2 Rework Prob. 2.1 taking into account that the container changes in temperature along with the water and has a heat capacity equivalent to 3 kg of water. Work the problem in two ways: (a) taking the water and container as the system, and (b) taking the water alone as the system.

2.3 Comment on the feasibility of cooling your kitchen in the summer by opening the door to the electrically powered refrigerator.

2.4 Liquid water at 100°C and 1 bar has an internal energy (on an arbitrary scale) of 419.0 kJ kg^{-1} and a specific volume of 1.044 cm^3 g^{-1}.

(a) What is its enthalpy?

(b) The water is brought to the vapor state at 200°C and 800 kPa, where its enthalpy i 2,838.6 kJ kg^{-1} and its specific volume is 260.79 cm^3 g^{-1}. Calculate ΔU and ΔH for the process.

2.5 With respect to 1 kg of a substance,

(a) How much change in elevation must it undergo to change its potential energy by 1 kJ?

(b) Starting from rest, to what velocity must it accelerate so that its kinetic energy is 1 kJ?

(c) What conclusions are indicated by these results?

2.6 Heat in the amount of 5 kJ is added to a system while its internal energy decreases by 10 kJ. How much energy is transferred as work? For a process causing the same change of state but for which the work is zero, how much heat is transferred?

2.7 A block of copper weighing 0.2 kg has an initial temperature of 400 K; 4 kg of water initially at 300 K is contained in a perfectly insulated tank, also made of copper and weighing 0.5 kg. The copper block is immersed in the water and allowed to come to equilibrium. What is the change in internal energy of the copper block and of the water? What is the change in energy of the entire system, including the tank? Ignore effects of expansion and contraction, and assume that the specific heats are constant at 4.184 J g^{-1} K^{-1} for water and 0.380 J g^{-1} K^{-1} for copper.

2.8 In the preceding problem, suppose that the copper block is dropped into the water from a height of 50 m. Assuming no loss of water from the tank, what is the change in internal energy of the water?

2.9 Nitrogen flows at steady state through a horizontal, insulated pipe with inside diameter of 2(in) [5.08 cm]. A pressure drop results from flow through a partially opened valve. Just upstream from the valve the pressure is 80(psia) [551.6 kPa], the temperature is 100(°F) [37.8°C], and the average velocity is $15(\mathrm{ft})(\mathrm{s})^{-1}$ [4.57 m s^{-1}]. If the pressure just downstream from the valve is 20(psia) [137.9 kPa], what is the temperature? Assume for nitrogen that $PV/T = \text{const}$, $C_V = (5/2)R$, and $C_P = (7/2)R$. (Find R values in App. A.)

2.10 Liquid water at 70(°F) [294.26 K] flows in a straight horizontal pipe in which there is no exchange of either heat or work with the surroundings. Its velocity is $30(\mathrm{ft})(\mathrm{s})^{-1}$ [9.144 m s^{-1}] in a pipe with an internal diameter of 1(in) [2.54 cm] until it flows into a section where the pipe diameter abruptly increases. What is the enthalpy change of the water if the downstream diameter is 1.5(in) [3.81 cm]? If it is 3(in) [7.62 cm]? What is the maximum change in enthalpy for an enlargement in the pipe?

2.11 Water flows through a horizontal coil heated from the outside by high-temperature flue gases. As it passes through the coil the water changes state from 2(atm) [202.66 kPa] and 180(°F) [82.2°C] to 1(atm) [101.33 kPa] and 250(°F) [121.1°C]. Its entering velocity is $10(\mathrm{ft})(\mathrm{s})^{-1}$ [3.05 m s^{-1}] and its exit velocity is $600(\mathrm{ft})(\mathrm{s})^{-1}$ [182.9 m s^{-1}]. Determine the heat transferred through the coil per unit mass of water. Enthalpies of the inlet and outlet water streams are:

Inlet: $148.0(\mathrm{Btu})(\mathrm{lb_m})^{-1}$ [344.2 kJ kg^{-1}]
Outlet: $1{,}168.8(\mathrm{Btu})(\mathrm{lb_m})^{-1}$ [2,718.5 kJ kg^{-1}]

2.12 Steam flows at steady state through a converging, insulated nozzle, 10(in) [25.4 cm] long and with an inlet diameter of 2(in) [5.08 cm]. At the nozzle entrance (state 1), the temperature and pressure are 600(°F) [312.56°C] and 100(psia) [689.5 kPa] and the velocity is $100(\mathrm{ft})(\mathrm{s})^{-1}$ [30.5 m s^{-1}]. At the nozzle exit (state 2), the steam temperature and pressure are 450(°F) [232.22°C] and 50(psia) [344.75 kPa]. The enthalpy values are:

$$H_1 = 1{,}329.6(\mathrm{Btu})(\mathrm{lb_m})^{-1}\ [3{,}092.5\ \mathrm{kJ\ kg^{-1}}]$$

$$H_2 = 1{,}259.6(\mathrm{Btu})(\mathrm{lb_m})^{-1}\ [2{,}929.7\ \mathrm{kJ\ kg^{-1}}]$$

What is the velocity of the steam at the nozzle exit, and what is the exit diameter?

2.13 A system consisting of *n*-butane and propane exists as two phases in vapor/liquid equilibrium at 10 bar and 323 K. The mole fraction of propane is about 0.67 in the vapor phase and about 0.40 in the liquid phase. Additional pure propane is added to the system, which is brought again to equilibrium at the same temperature and pressure, with both liquid and vapor phases still present. What is the effect of the addition of propane on the mole fractions of propane in the vapor and liquid phases?

2.14 In a natural gasoline fractionation system there are usually six chemical species present in appreciable quantities: methane, ethane, propane, isobutane, *n*-butane, and *n*-pentane. A mixture of these species is placed in a closed vessel from which all air has been removed. If the temperature and pressure are fixed so that both liquid and vapor phases exist at equilibrium, how many additional phase-rule variables must be chosen to fix the compositions of both phases?

If the temperature and pressure are to remain the same, is there any way that the composition of the total contents of the vessel can be changed (by adding or removing material) without affecting the compositions of the liquid and vapor phases?

2.15 In the following take $C_V = 20.8$ and $C_P = 29.1$ J mol^{-1} $°C^{-1}$ for nitrogen gas:

(*a*) Five moles of nitrogen at 80°C is contained in a rigid vessel. How much heat must be added to the system to raise its temperature to 300°C if the vessel has a negligible heat capacity? If the mass of the vessel is 100 kg and if its heat capacity is 0.5 J g^{-1} $°C^{-1}$, how much heat is required?

(*b*) Three moles of nitrogen at 230°C is contained in a piston/cylinder arrangement. How much heat must be extracted from this system, which is kept at constant pressure, to cool it to 80°C if the heat capacity of the piston and cylinder is neglected?

2.16 In the following take $C_V = 5$ and $C_P = 7(\text{Btu})(\text{lb mol})^{-1}(°\text{F})^{-1}$ for nitrogen gas:

(*a*) Five pound moles of nitrogen at 100(°F) is contained in a rigid vessel. How much heat must be added to the system to raise its temperature to 400(°F) if the vessel has a negligible heat capacity? If the vessel weighs $250(\text{lb}_m)$ and has a heat capacity of $0.12(\text{Btu})(\text{lb}_m)^{-1}(°\text{F})^{-1}$, how much heat is required?

(*b*) Three pound moles of nitrogen at 450(°F) is contained in a piston/cylinder arrangement. How much heat must be extracted from this system, which is kept at constant pressure, to cool it to 100(°F) if the heat capacity of the piston and cylinder is neglected?

2.17 The internal energy U^t of an amount of gas is given by the equation,

$$U^t = 1.5\ PV^t$$

where P is in (psia) and V^t is in $(\text{ft})^3$. The gas undergoes a mechanically reversible process from an initial state at 1,500(psia) and 500(R). During the process V^t is constant and equal to $10(\text{ft})^3$ and P increases by 50 percent. Determine values for Q and ΔH^t in (Btu) for the process.

2.18 The internal energy U^t of an amount of gas is given by the equation,

$$U^t = 0.01\ PV^t$$

where P is in kPa, V^t is in m^3. The gas undergoes a mechanically reversible process from an initial state at 10,000 kPa and 280 K. During the process V^t is constant and equal to 0.3 m^3 and P increases by 50 percent. Determine values for Q and ΔH^t in kJ for the process.

2.19 The path followed by a gas during a particular mechanically reversible process is described by the equation

$$P + aV^t = c$$

where a and c are constants. In the initial state, $P_1 = 60$ bar and $V_1^t = 0.002\ \text{m}^3$; in the final state, $P_2 = 20$ bar and $V_2^t = 0.004\ \text{m}^3$. During the process, heat in the amount of 5,000 J is transferred to the gas. Determine W and ΔU^t for the process. Suppose the gas followed a different path connecting the *same* initial and final states. Which of the quantities Q, W, and ΔU^t must be unchanged? Why?

2.20 A particular substance undergoes a mechanically reversible process, expanding from an initial state of 20 bar to a final state of 8 bar. The path for the process is described by the equation

$$P = \frac{0.036}{V^t} - 4$$

where P is in bar and V^t is in m^3. If ΔU^t for the change of state is $-1{,}400$ J, determine W, Q, and ΔH^t.

2.21 One kilogram of air is heated reversibly at constant pressure from an initial state of 300 K and 1 bar until its volume triples. Calculate W, Q, ΔU, and ΔH for the process. Assume that air obeys the relation $PV/T = 83.14\ \text{bar cm}^3\ \text{mol}^{-1}\ \text{K}^{-1}$ and that $C_P = 29\ \text{J mol}^{-1}\ \text{K}^{-1}$.

CHAPTER

THREE

VOLUMETRIC PROPERTIES OF PURE FLUIDS

3.1 THE PVT BEHAVIOR OF PURE SUBSTANCES

Thermodynamic properties, such as internal energy and enthalpy, from which one calculates the heat and work requirements of industrial processes, are not directly measurable. They can, however, be calculated from volumetric data. To provide part of the background for such calculations, we describe in this chapter the pressure-volume-temperature (*PVT*) behavior of pure fluids. Moreover, these *PVT* relations are important in themselves for such purposes as the metering of fluids and the sizing of vessels and pipelines.

Homogeneous fluids are normally divided into two classes, liquids and gases. However, the distinction cannot always be sharply drawn, because the two phases become indistinguishable at what is called the *critical point.* Measurements of the vapor pressure of a pure solid at temperatures up to its triple point and measurements of the vapor pressure of the pure liquid at temperatures above the triple point lead to a pressure-vs.-temperature curve such as the one made up of lines 1-2 and 2-C in Fig. 3.1. The third line (2-3) shown on this graph gives the solid/liquid equilibrium relationship. These three curves represent the conditions of P and T required for the coexistence of two phases and thus are boundaries for the single-phase regions. Line 1-2, the sublimation curve, separates the solid and gas regions; line 2-3, the fusion curve, separates the solid and liquid regions; line 2-C, the vaporization curve, separates the liquid and gas regions. The three curves meet at the triple point, where all three phases coexist in equilibrium. According to the phase rule [Eq. (2.12)], the triple point is invariant. If the system exists along any of the two-phase lines of Fig. 3.1, it is univariant, whereas in

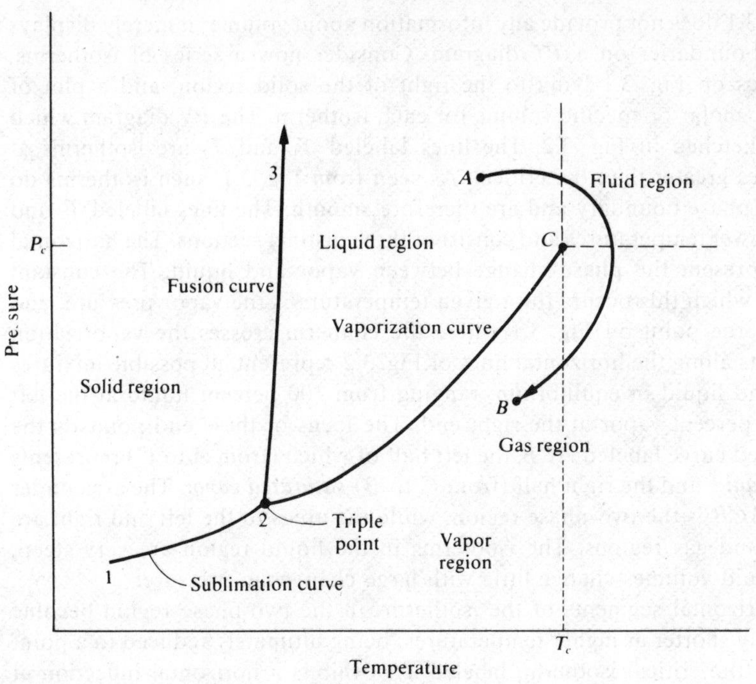

Figure 3.1 *PT* diagram for a pure substance.

the single-phase regions it is divariant. Although the fusion curve 2-3 continues upward indefinitely, the vaporization curve 2-*C* terminates at point *C*, the critical point. The coordinates of this point are the critical pressure P_c and the critical temperature T_c, the highest temperature and pressure at which a pure material can exist in vapor/liquid equilibrium. The fluid region, existing at higher temperatures and pressures, is marked off by dashed lines, which do not represent phase transitions, but rather are limits fixed by the meanings accorded the *words* liquid and gas. A phase is generally considered a liquid if it can be vaporized by reduction in pressure at constant temperature. A phase is considered a gas if it can be condensed by reduction of temperature at constant pressure. Since the fluid region fits neither of these definitions, it is neither a gas nor a liquid. The gas region is sometimes divided into two parts, as shown by the dotted line of Fig. 3.1. A gas to the left of this line, which can be condensed either by compression at constant temperature or by cooling at constant pressure, is called a vapor.

Because of the existence of the critical point, a path can be drawn from the liquid region to the gas region that does not cross a phase boundary; e.g., the path from *A* to *B* in Fig. 3.1. This path represents a gradual transition from the liquid to the gas region. On the other hand, a path crossing phase boundary 2-*C* includes a vaporization step, where an abrupt change of properties occurs.

Figure 3.1 does not provide any information about volume; it merely displays the phase boundaries on a *PT* diagram. Consider now a series of isotherms, vertical lines on Fig. 3.1 lying to the right of the solid region, and a plot of pressure vs. molar or specific volume for each isotherm. The *PV* diagram which results is sketched in Fig. 3.2. The lines labeled T_1 and T_2 are isotherms at temperatures greater than the critical. As seen from Fig. 3.1, such isotherms do not cross a phase boundary and are therefore smooth. The lines labeled T_3 and T_4 are for lower temperatures and consist of three distinct sections. The horizontal sections represent the phase change between vapor and liquid. The constant pressure at which this occurs for a given temperature is the vapor pressure, and is given by the point on Fig. 3.1 where the isotherm crosses the vaporization curve. Points along the horizontal lines of Fig. 3.2 represent all possible mixtures of vapor and liquid in equilibrium, ranging from 100 percent liquid at the left end to 100 percent vapor at the right end. The locus of these end points is the dome-shaped curve labeled *ACB*, the left half of which (from *A* to *C*) represents *saturated liquid*, and the right half (from *C* to *B*) *saturated vapor*. The area under the dome *ACB* is the two-phase region, while the areas to the left and right are the liquid and gas regions. The isotherms in the liquid region are very steep, because liquid volumes change little with large changes in pressure.

The horizontal segments of the isotherms in the two-phase region become progressively shorter at higher temperatures, being ultimately reduced to a point at *C*. Thus, the critical isotherm, labeled T_c, exhibits a horizontal inflection at the critical point *C* at the top of the dome. Here the liquid and vapor phases cannot be distinguished from one another, because their properties are the same.

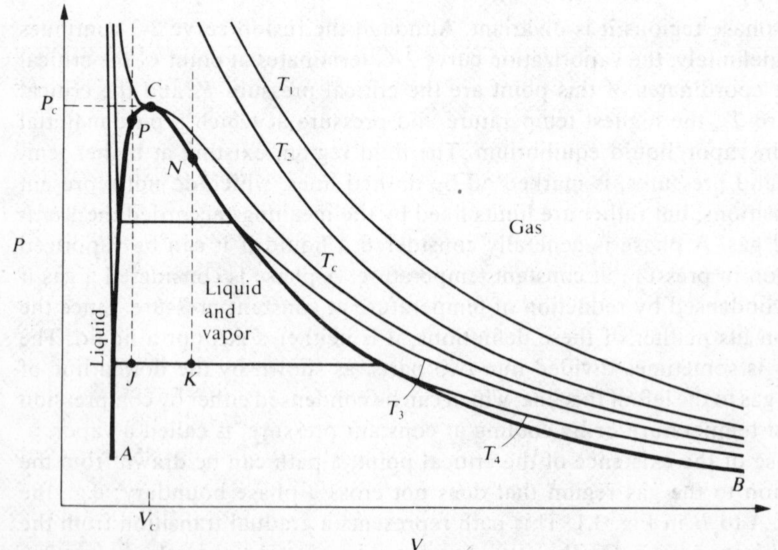

Figure 3.2 *PV* diagram for a pure fluid.

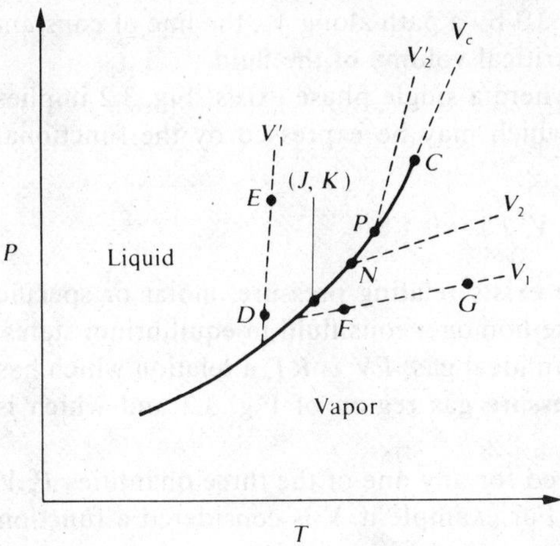

Figure 3.3 *PT* diagram showing the vapor-pressure curve for a pure substance and constant-volume lines in the single-phase regions.

The physical significance of the critical point becomes evident from the changes that occur when a pure substance is heated in a sealed upright tube of constant volume. Such changes follow vertical lines in Fig. 3.2. They are also shown on the *PT* diagram of Fig. 3.3, where the vaporization curve of Fig. 3.1 appears as a solid line. The dashed lines are constant-volume paths in the single-phase regions only. If the tube is filled with either liquid or gas, the heating process produces changes described by these lines, for example by the change from *D* to *E* (liquid region) and by the change from *F* to *G* (vapor region). The corresponding vertical lines on Fig. 3.2 lie to the left and to the right of *ACB*.

If the tube is only partially filled with liquid (the remainder being vapor in equilibrium with the liquid), heating at first causes changes described by the vapor-pressure curve (solid line) of Fig. 3.3. If the meniscus separating the two phases is initially near the bottom of the tube, liquid vaporizes, and the meniscus recedes to the bottom of the tube and disappears as the last drop of liquid vaporizes. For example in Fig. 3.3, one such path is from (J, K) to N; it then follows the line of constant molar volume V_2 upon further heating. If the meniscus is originally near the top of the tube, the liquid expands upon heating until it completely fills the tube. One such process is represented by the path from (J, K) to P; it then follows the line of constant molar volume V_2' with continued heating. The two paths are also shown by the dashed lines of Fig. 3.2, the first passing through points *K* and *N*, and the second through *J* and *P*.

A unique filling of the tube, with a particular intermediate meniscus level, causes the path of the heating process to coincide with the vapor-pressure curve of Fig. 3.3 all the way to its end at the critical point *C*. On Fig. 3.2 the path is a vertical line passing through the critical point. Physically, heating does not produce much change in the level of the meniscus. As the critical point is approached, the meniscus becomes indistinct, then hazy, and finally disappears as the system changes from two phases (as represented by the vapor-pressure curve) to a single phase (as represented by the region above *C*). Further heating produces changes represented in Fig. 3.3 by a path along V_c, the line of constant molar volume corresponding to the critical volume of the fluid.

For the regions of the diagram where a single phase exists, Fig. 3.2 implies a relation connecting P, V, and T which may be expressed by the functional equation:

$$f(P, V, T) = 0$$

This means that an *equation of state* exists relating pressure, molar or specific volume, and temperature for any pure homogeneous fluid in equilibrium states. The simplest equation of state is for an ideal gas, $PV = RT$, a relation which has approximate validity for the low-pressure gas region of Fig. 3.2 and which is discussed in detail in Sec. 3.3.

An equation of state may be solved for any one of the three quantities P, V, or T as a function of the other two. For example if V is considered a function of T and P, then $V = V(T, P)$, and

$$dV = \left(\frac{\partial V}{\partial T}\right)_P dT + \left(\frac{\partial V}{\partial P}\right)_T dP \tag{3.1}$$

The partial derivatives in this equation have definite physical meanings and are measurable quantities. For liquids they are related to two commonly tabulated properties:

1. The volume expansivity

$$\beta \equiv \frac{1}{V}\left(\frac{\partial V}{\partial T}\right)_P \tag{3.2}$$

2. The isothermal compressibility

$$\kappa \equiv -\frac{1}{V}\left(\frac{\partial V}{\partial P}\right)_T \tag{3.3}$$

Combination of Eqs. (3.1) through (3.3) provides the general equation

$$\frac{dV}{V} = \beta\, dT - \kappa\, dP \tag{3.4}$$

The isotherms for the liquid phase on the left side of Fig. 3.2 are very steep and closely spaced. Thus both $(\partial V/\partial P)_T$ and $(\partial V/\partial T)_P$, and hence both β and κ, are small. This characteristic behavior of liquids (outside the region of the critical point) suggests an idealization, commonly employed in fluid mechanics and known as the *incompressible fluid*, for which β and κ are both zero. No real fluid is in fact incompressible, but the idealization is nevertheless useful, because it often provides a sufficiently realistic model of liquid behavior for practical purposes. The incompressible fluid cannot be described by an equation of state relating V to T and P, because V is constant.

For real liquids β and κ are weak functions of temperature and pressure. Thus for small changes in T and P little error is introduced if we regard them as constant. Then Eq. (3.4) may be integrated to give

$$\ln \frac{V_2}{V_1} = \beta(T_2 - T_1) - \kappa(P_2 - P_1) \tag{3.5}$$

This is a different order of approximation than the assumption of an incompressible fluid.

Example 3.1 For acetone at 20°C and 1 bar,

$$\beta = 1.487 \times 10^{-3}\ °\mathrm{C}^{-1}$$

$$\kappa = 62 \times 10^{-6}\ \mathrm{bar}^{-1}$$

and

$$V = 1.287\ \mathrm{cm^3\ g^{-1}}$$

Find:

(*a*) The value of $(\partial P/\partial T)_V$.

(*b*) The pressure generated when acetone is heated at constant volume from 20°C and 1 bar to 30°C.

(*c*) The volume change when acetone is changed from 20°C and 1 bar to 0°C and 10 bar.

SOLUTION

(*a*) The derivative $(\partial P/\partial T)_V$ is determined by application of Eq. (3.4) to the case for which V = const and $dV = 0$:

$$\beta\, dT - \kappa\, dP = 0 \qquad (\text{const } V)$$

or

$$\left(\frac{\partial P}{\partial T}\right)_V = \frac{\beta}{\kappa} = \frac{1.487 \times 10^{-3}}{62 \times 10^{-6}} = 24\ \mathrm{bar}\ °\mathrm{C}^{-1}$$

(*b*) If β and κ are assumed constant in the 10°C temperature interval, then the equation derived in (*a*) may be written (V = const):

$$\Delta P = \frac{\beta}{\kappa}\Delta T = (24)(10) = 240\ \mathrm{bar}$$

and

$$P_2 = P_1 + \Delta P = 1 + 240 = 241\ \mathrm{bar}$$

(*c*) Direct substitution into Eq. (3.5) gives

$$\ln \frac{V_2}{V_1} = (1.487 \times 10^{-3})(-20) - (62 \times 10^{-6})(9) = -0.0303$$

Whence

$$\frac{V_2}{V_1} = 0.9702$$

and

$$V_2 = (0.9702)(1.287) = 1.249 \text{ cm}^3 \text{ g}^{-1}$$

which gives

$$\Delta V = V_2 - V_1 = 1.249 - 1.287 = -0.038 \text{ cm}^3 \text{ g}^{-1}$$

3.2 THE VIRIAL EQUATION

Figure 3.2 indicates the complexity of the *PVT* behavior of a pure substance and suggests the difficulty of its description by an equation. However, for the gas region alone relatively simple equations often suffice. For an isotherm such as T_1 we note from Fig. 3.2 that as P increases V decreases. Thus the PV product for a gas or vapor should be much more nearly constant than either of its members. This suggests the representation of PV along an isotherm by a power series expansion in P:

$$PV = a + bP + cP^2 + \cdots$$

If we let $b = aB'$, $c = aC'$, etc., this equation becomes

$$PV = a(1 + B'P + C'P^2 + \cdots) \tag{3.6}$$

where a, B', C', etc., are constants for a given temperature and a given chemical species.

In principle, the right-hand side of Eq. (3.6) is an infinite series. However, in practice a finite number of terms is used. In fact, *PVT* data show that at low pressures truncation after two terms provides satisfactory results. In general, the greater the pressure range, the larger the number of terms required.

Parameters B', C', etc., are functions of temperature and the identity of the chemical species; parameter a, however, is the same function of temperature for all species. Data taken for various gases at a specific constant temperature (fixed by use of a reproducible state such as the triple point of water or the normal boiling point of water) show that plots of PV vs. P have the same limiting value of PV as $P \to 0$ for all gases. For $P \to 0$, Eq. (3.6) becomes

$$\lim_{P \to 0} (PV) \equiv (PV)^* = a$$

Thus, a is the same for all gases and depends on temperature only. Whence

$$(PV)^* = a = f(T)$$

It is this remarkable property of gases that makes them valuable in thermometry, for the limiting values of $(PV)^*$ are used to establish a temperature scale which is independent of the identity of the gas used as thermometric fluid. One need only fix the form of the functional relationship to T and define a quantitative scale; both steps are completely arbitrary. The simplest procedure, and the one adopted internationally, is:

1. Fix the functional relationship so that $(PV)^*$ is directly proportional to T

$$(PV)^* = a = RT \tag{3.7}$$

   where R is the proportionality constant.
2. Assign a value of 273.16 K to the temperature of the triple point of water

$$(PV)_t^* = R \times 273.16 \text{ K} \tag{3.8}$$

   where the subscript t denotes the value at the triple point of water.

Division of Eq. (3.7) by Eq. (3.8) gives

$$\frac{(PV)^*}{(PV)_t^*} = \frac{T \text{ K}}{273.16 \text{ K}}$$

or

$$\boxed{T \text{ K} = 273.16 \frac{(PV)^*}{(PV)_t^*}} \tag{3.9}$$

Equation (3.9) establishes the Kelvin temperature scale throughout the temperature range for which limiting values of PV as $P \to 0$ [values of $(PV)^*$] are experimentally accessible.

The state of a gas at the limiting condition where $P \to 0$ deserves some discussion. As the pressure on a gas is decreased, the individual molecules become more and more widely separated. The volume of the molecules themselves becomes a smaller and smaller fraction of the total volume occupied by the gas. Furthermore, the forces of attraction between molecules become ever smaller because of the increasing distances between them. In the limit, as the pressure approaches zero, the molecules are separated by infinite distances. Their volumes become negligible compared with the total volume of the gas, and the intermolecular forces approach zero. A gas which meets these conditions is said to be ideal, and the temperature scale established by Eq. (3.9) is known as the ideal-gas temperature scale.

The proportionality constant R in Eq. (3.7) is called the *universal gas constant.* Its numerical value is determined by means of Eq. (3.8) from experimental PVT data for gases:

$$R = \frac{(PV)_t^*}{273.16 \text{ K}}$$

Since PVT data cannot in fact be taken at a pressure approaching zero, data taken at finite pressures are extrapolated to the zero-pressure state. The currently accepted value of $(PV)_t^*$ is 22,711.6 $\text{cm}^3 \text{ bar mol}^{-1}$. Figure 3.4 shows how this

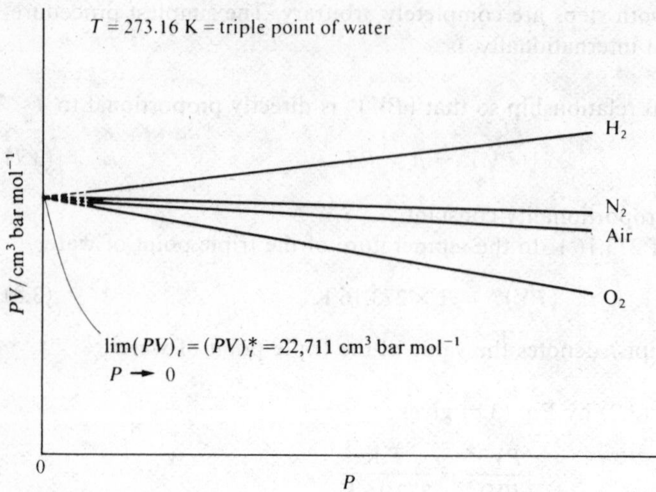

Figure 3.4 The limit of PV as $P \to 0$ is independent of the gas.

determination is made. Its leads to the following value of R:

$$R = \frac{22{,}711.6 \text{ cm}^3 \text{ bar mol}^{-1}}{273.16 \text{ K}} = 83.144 \text{ cm}^3 \text{ bar mol}^{-1} \text{ K}^{-1}$$

Through the use of conversion factors, R may be expressed in various units. Commonly used values are given in App. A.

With the establishment of the ideal-gas temperature scale, the constant a in Eq. (3.6) may be replaced by RT, in accord with Eq. (3.7). Thus Eq. (3.6) becomes

$$Z \equiv \frac{PV}{RT} = 1 + B'P + C'P^2 + D'P^3 + \cdots \tag{3.10}$$

where the ratio PV/RT is called the *compressibility factor* and is given the symbol Z. An alternative and equivalent expression for Z, which is also in common use, is

$$Z = 1 + \frac{B}{V} + \frac{C}{V^2} + \frac{D}{V^3} + \cdots \tag{3.11}$$

Both of these equations are known as *virial expansions*, and the parameters B', C', D', etc., and B, C, D, etc., are called *virial coefficients.* Parameters B' and B are second virial coefficients; C' and C are third virial coefficients; etc. For a given gas the virial coefficients are functions of temperature only.

Many other equations of state have been proposed for gases, but the virial equations are the only ones having a firm basis in theory. The methods of statistical mechanics allow derivation of the virial equations and provide physical significance to the virial coefficients. Thus, for the expansion in $1/V$, the term B/V arises on account of interactions between pairs of molecules; the C/V^2 term, on account of three-body interactions; etc. Since two-body interactions are many times more common than three-body interactions, and three-body interactions are many times more numerous than four-body interactions, etc., the contributions to Z of the successively higher-ordered terms fall off rapidly.

The two sets of coefficients in Eqs. (3.10) and (3.11) are related as follows:

$$B' = \frac{B}{RT}$$

$$C' = \frac{C - B^2}{(RT)^2}$$

$$D' = \frac{D - 3BC + 2B^3}{(RT)^3}$$

etc.

The first step in the derivation of these relations is elimination of P on the right-hand side of Eq. (3.10) through use of Eq. (3.11). The resulting equation is a power series in $1/V$ which is compared term by term with Eq. (3.11). This comparison provides the equations relating the two sets of virial coefficients. They hold exactly only for the two virial expansions as infinite series. For the truncated forms of the virial equations treated in Sec. 3.4, these relations are only approximate.

3.3 THE IDEAL GAS

Since the terms B/V, C/V^2, etc., of the virial expansion [Eq. (3.11)] arise on account of molecular interactions, the virial coefficients B, C, etc., would be zero if no such interactions existed. The virial expansion would then reduce to

$$Z = 1 \qquad \text{or} \qquad PV = RT$$

For a real gas, molecular interactions *do* exist, and exert an influence on the observed behavior of the gas. As the pressure of a real gas is reduced at constant temperature, V increases and the contributions of the terms B/V, C/V^2, etc., decrease. For a pressure approaching zero, Z approaches unity, not because of any change in the virial coefficients, but because V becomes infinite. Thus in the limit as the pressure approaches zero, the equation of state assumes the same simple form as for the hypothetical case of $B = C = \cdots = 0$; that is

$$Z = 1 \qquad \text{or} \qquad PV = RT$$

We know from the phase rule that the internal energy of a real gas is a function of pressure as well as of temperature. This pressure dependency arises as a result of forces between the molecules. If such forces did not exist, no energy would be required to alter the average intermolecular distance, and therefore no energy would be required to bring about volume and pressure changes in a gas at constant temperature. We conclude that, in the absence of molecular interactions, the internal energy of a gas depends on temperature only. These considerations of the behavior of a hypothetical gas in which no molecular forces exist and of a real gas in the limit as pressure approaches zero lead to the definition of an *ideal gas* as one whose macroscopic behavior is characterized by:

1. The equation of state

$$\boxed{PV = RT} \tag{3.12}$$

2. An internal energy that is a function of temperature only, and as a result of Eq. (2.20) a heat capacity C_V which is also a function of temperature only.

The ideal gas is a model fluid that is useful because it is described by simple equations frequently applicable as good approximations for actual gases. In engineering calculations, gases at pressures up to a few bars may often be considered ideal. The remainder of this section is therefore devoted to the development of thermodynamic relationships for ideal gases.

The Constant-Volume Process

The equations which apply to a mechanically reversible constant-volume process were developed in Sec. 2.10. No simplification results for an ideal gas. Thus for one mole:

$$dU = dQ = C_V\,dT \tag{3.13}$$

For a finite change,

$$\Delta U = Q = \int C_V\,dT \tag{3.14}$$

Since both the internal energy and C_V of an ideal gas are functions of temperature only, ΔU for an ideal gas may *always* be calculated by $\int C_V\,dT$, regardless of the kind of process causing the change. This is demonstrated in Fig. 3.5, which shows a graph of internal energy as a function of molar volume with temperature as a parameter. Since U is independent of V at constant temperature, a plot of U vs. V at constant temperature is a horizontal line. For different temperatures, U has different values, and there is a separate line for each temperature. Two such lines are shown in Fig. 3.5, one for temperature T_1 and one for temperature T_2. The dashed line connecting points a and b represents a constant-volume process for which the temperature increases from T_1 to T_2 and the internal energy changes by $\Delta U = U_2 - U_1$. This change in internal energy is given by Eq. (3.14) as $\Delta U = \int C_V\,dT$. The dashed lines connecting points a and c and points a and d

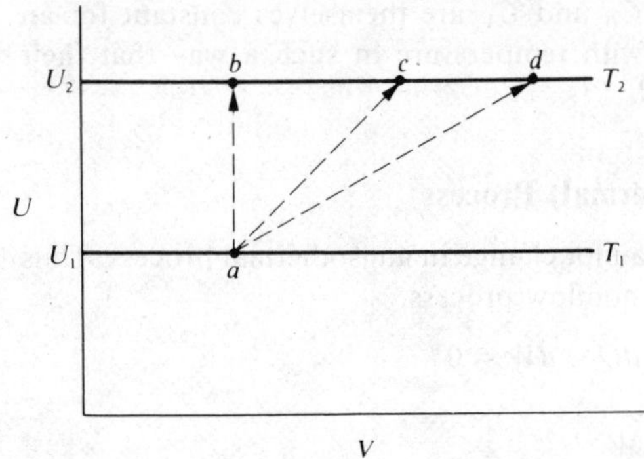

Figure 3.5 Internal energy changes for an ideal gas.

represent other processes not occurring at constant volume but which also lead from an initial temperature T_1 to a final temperature T_2. The graph clearly shows that the change in U for these processes is the same as for the constant-volume process, and it is therefore given by the same equation, namely, $\Delta U = \int C_V\,dT$. However, ΔU is *not* equal to Q for these processes, because Q depends not only on T_1 and T_2 but also on the path followed.

The Constant-Pressure (Isobaric) Process

The equations which apply to a mechanically reversible, constant-pressure nonflow process were developed in Sec. 2.10. For one mole,

$$dH = dQ = C_P\,dT \tag{3.15}$$

and

$$\Delta H = Q = \int C_P\,dT \tag{3.16}$$

Because the internal energy of an ideal gas is a function of temperature only, both enthalpy and C_P also depend on temperature alone. This is evident from the definition $H = U + PV$, or $H = U + RT$ for an ideal gas, and from Eq. (2.21). Therefore, just as $\Delta U = \int C_V\,dT$ for any process involving an ideal gas, so $\Delta H = \int C_P\,dT$ not only for constant-pressure processes but for *all* finite processes.

These expressions for ΔU and ΔH and the definition of enthalpy imply a simple relationship between C_P and C_V for an ideal gas; since

$$dH = dU + R\,dT$$

then from Eqs. (3.13) and (3.15)

$$C_P\,dT = C_V\,dT + R\,dT$$

and

$$\boxed{C_P = C_V + R} \tag{3.17}$$

This equation does *not* imply that C_P and C_V are themselves constant for an ideal gas, but only that they vary with temperature in such a way that their difference is equal to the constant R.

The Constant-Temperature (Isothermal) Process

The internal energy of an ideal gas cannot change in an isothermal process. Thus for one mole of an ideal gas in any nonflow process,

$$dU = dQ - dW = 0$$

and

$$Q = W$$

For a mechanically reversible nonflow process and with $P = RT/V$, we have immediately that

$$Q = W = \int P\,dV = \int RT\,\frac{dV}{V}$$

Integration at constant temperature from the initial volume V_1 to the final volume V_2 gives

$$Q = W = RT \ln \frac{V_2}{V_1} \tag{3.18}$$

Since $P_1/P_2 = V_2/V_1$ for the isothermal process, Eq. (3.18) may also be written:

$$Q = W = RT \ln \frac{P_1}{P_2} \tag{3.19}$$

The Adiabatic Process

An adiabatic process is one for which there is no heat transfer between the system and its surroundings; that is, $dQ = 0$. Therefore, application of the first law to one mole of an ideal gas in mechanically reversible nonflow processes gives

$$dU = -dW = -P\,dV$$

Since the change in internal energy for any process involving an ideal gas is given by Eq. (3.13), this becomes

$$C_V\,dT = -P\,dV$$

Substituting RT/V for P and rearranging, we get

$$\frac{dT}{T} = -\frac{R}{C_V}\frac{dV}{V} \tag{3.20}$$

If the ratio of heat capacities C_P/C_V is designated by γ, then in view of Eq. (3.17),

$$\gamma = \frac{C_V + R}{C_V} = 1 + \frac{R}{C_V}$$

or

$$\frac{R}{C_V} = \gamma - 1 \tag{3.21}$$

Substitution in Eq. (3.20) gives

$$\frac{dT}{T} = -(\gamma - 1)\frac{dV}{V}$$

If γ is constant,† integration yields

$$\ln\frac{T_2}{T_1} = -(\gamma - 1)\ln\frac{V_2}{V_1}$$

or

$$\boxed{\frac{T_2}{T_1} = \left(\frac{V_1}{V_2}\right)^{\gamma-1}} \tag{3.22}$$

This equation relates temperature and volume for a mechanically reversible adiabatic process involving an ideal gas with constant heat capacities. The analogous relationships between temperature and pressure and between pressure and volume can be obtained from Eq. (3.22) and the ideal-gas equation. Since $P_1V_1/T_1 = P_2V_2/T_2$, we may eliminate V_1/V_2 from Eq. (3.22), obtaining:

$$\boxed{\frac{T_2}{T_1} = \left(\frac{P_2}{P_1}\right)^{(\gamma-1)/\gamma}} \tag{3.23}$$

A comparison of Eqs. (3.22) and (3.23) shows that

$$\left(\frac{V_1}{V_2}\right)^{\gamma-1} = \left(\frac{P_2}{P_1}\right)^{(\gamma-1)/\gamma}$$

or

$$\boxed{P_1V_1^{\gamma} = P_2V_2^{\gamma} = PV^{\gamma} = \text{const}} \tag{3.24}$$

The work of an adiabatic process may be obtained from the relation

$$-dW = dU = C_V\,dT \tag{3.25}$$

If C_V is constant, integration gives

$$W = -\Delta U = -C_V\,\Delta T \tag{3.26}$$

† The assumption that γ is constant for an ideal gas is equivalent to the assumption that the heat capacities themselves are constant. This is the only way that the ratio $C_P/C_V = \gamma$ and the difference $C_P - C_V = R$ can *both* be constant. However, since both C_P and C_V increase with temperature, their ratio γ is less sensitive to temperature than the heat capacities themselves.

Alternative forms of Eq. (3.26) are obtained if C_V is eliminated by Eq. (3.21):

$$W = -C_V\,\Delta T = \frac{-R\,\Delta T}{\gamma - 1} = \frac{RT_1 - RT_2}{\gamma - 1}$$

Since $RT_1 = P_1V_1$ and $RT_2 = P_2V_2$, this expression may also be written

$$W = \frac{P_1V_1 - P_2V_2}{\gamma - 1} \tag{3.27}$$

If V_2 is not known, as is usually the case, it can be eliminated from Eq. (3.27) by Eq. (3.24). This leads to the expression

$$W = \frac{P_1V_1}{\gamma - 1}\left[1 - \left(\frac{P_2}{P_1}\right)^{(\gamma-1)/\gamma}\right] = \frac{RT_1}{\gamma - 1}\left[1 - \left(\frac{P_2}{P_1}\right)^{(\gamma-1)/\gamma}\right] \tag{3.28}$$

The same result is obtained when the relation between P and V given by Eq. (3.24) is used for integration of the expression $W = \int P\,dV$.

Equations (3.22) through (3.28) are for ideal gases with constant heat capacities. They also require the process to be mechanically reversible as well as adiabatic. Processes which are adiabatic but not mechanically reversible are *not* described by these equations.

As applied to real gases, Eqs. (3.22) through (3.28) often yield satisfactory approximations, provided the deviations from ideality are not too great. For monatomic gases, $\gamma = 1.67$; approximate values of γ are 1.4 for diatomic gases and 1.3 for simple polyatomic gases such as CO_2, SO_2, NH_3, and CH_4.

The Polytropic Process

This is the general case for which no specific conditions other than mechanical reversibility are imposed. Thus only the *general* equations applying to an ideal gas in a nonflow process apply. For one mole, these are:

$$dU = dQ - dW \qquad \Delta U = Q - W \qquad \text{(first law)}$$

$$dW = P\,dV \qquad W = \int P\,dV$$

$$dU = C_V\,dT \qquad \Delta U = \int C_V\,dT$$

$$dH = C_P\,dT \qquad \Delta H = \int C_P\,dT$$

Values for Q cannot be determined directly, but must be obtained through the first law. Substitution for dU and dW gives

$$dQ = C_V\,dT + P\,dV \tag{3.29}$$

and

$$Q = \int C_V\,dT + \int P\,dV \tag{3.30}$$

Since the first law has been used for the calculation of Q, the work must be calculated directly from the integral $\int P\,dV$.

The equations developed in this section have been *derived* for mechanically reversible nonflow processes involving ideal gases. However, those equations which relate state functions only are valid for ideal gases regardless of the process and apply equally to reversible and irreversible flow and nonflow processes, because changes in state functions depend only on the initial and final states of the system. On the other hand, an equation for Q or W is specific to the case considered in its derivation.

The work of an *irreversible* process is calculated by a two-step procedure. First, W is determined for a mechanically reversible process that accomplishes the same change of state. Second, this result is multiplied or divided by an efficiency to give the actual work. If the process produces work, the reversible value is too large and must be multiplied by an efficiency. If the process requires work, the reversible value is too small and must be divided by an efficiency.

Applications of the concepts and equations developed in this section are illustrated in the examples that follow. In particular, the work of irreversible processes is treated in Example 3.3.

Example 3.2 Air is compressed from an initial condition of 1 bar and 25°C to a final state of 5 bar and 25°C by three different mechanically reversible processes:

(*a*) Heating at constant volume followed by cooling at constant pressure.

(*b*) Isothermal compression.

(*c*) Adiabatic compression followed by cooling at constant volume.

At these conditions, air may be considered an ideal gas with the constant heat capacities, $C_V = (5/2)R$ and $C_P = (7/2)R$.

Calculate the work required, heat transferred, and the changes in internal energy and enthalpy of the air for each process.

SOLUTION In each case the system is taken as 1 mol of air, contained in an imaginary frictionless piston-and-cylinder arrangement. For $R = 8.314\ \text{J mol}^{-1}\ \text{K}^{-1}$

$$C_V = 20.785 \qquad \text{and} \qquad C_P = 29.099\ \text{J mol}^{-1}\ \text{K}^{-1}$$

The initial and final conditions of the air are identical with those of Example 2.12. It was shown there that

$$V_1 = 0.02479 \qquad \text{and} \qquad V_2 = 0.004958\ \text{m}^3$$

(*a*) This part of the problem is identical with part (*b*) of Example 2.12. However, it may now be solved in a simpler manner. The temperature at the end of the constant-volume heating step was calculated in Example 2.12 as 1,490.75 K. Also for this step $W = 0$ and therefore

$$Q = \Delta U = C_V\,\Delta T = 24{,}788\ \text{J}$$

Moreover,

$$\Delta H = C_P\,\Delta T = (29.099)(1{,}490.75 - 298.15) = 34{,}703 \text{ J}$$

For the second step at constant pressure, Eq. (3.16) yields

$$Q = \Delta H = C_P\,\Delta T = (29.099)(298.15 - 1{,}490.75) = -34{,}703 \text{ J}$$

$$\Delta U = C_V\,\Delta T = (20.785)(298.15 - 1{,}490.75) = -24{,}788 \text{ J}$$

and

$$W = Q - \Delta U = -34{,}703 - (-24{,}788) = -9{,}915 \text{ J}$$

For the entire process,

$$\Delta U = 24{,}788 - 24{,}788 = 0$$

$$\Delta H = 34{,}703 - 34{,}703 = 0$$

$$Q = 24{,}788 - 34{,}703 = -9{,}915 \text{ J}$$

and

$$W = -9{,}915 - 0 = -9{,}915 \text{ J}$$

(*b*) For the isothermal compression of an ideal gas,

$$\Delta U = \Delta H = 0$$

Equation (3.19) gives

$$Q = W = RT \ln \frac{P_1}{P_2} = (8.314)(298.15) \ln \frac{1}{5} = -3{,}990 \text{ J}$$

(*c*) The initial adiabatic compression of the air takes it to its final volume of 0.004958 m^3. The temperature and pressure at this point are given by Eqs. (3.22) and (3.24):

$$T_2 = T_1 \left(\frac{V_1}{V_2}\right)^{\gamma-1} = (298.15)\left(\frac{0.02479}{0.004958}\right)^{0.4} = 567.57 \text{ K}$$

and

$$P_2 = P_1 \left(\frac{V_1}{V_2}\right)^{\gamma} = (1)\left(\frac{0.02479}{0.004958}\right)^{1.4} = 9.52 \text{ bar}$$

For this step $Q = 0$. Hence

$$\Delta U = -W = C_V\,\Delta T = (20.785)(567.57 - 298.15) = 5{,}600 \text{ J}$$

and

$$\Delta H = C_P\,\Delta T = (29.099)(567.57 - 298.15) = 7{,}840 \text{ J}$$

For the second step $\Delta V = 0$ and $W = 0$; therefore

$$Q = \Delta U = C_V\,\Delta T = (20.785)(298.15 - 567.57) = -5{,}600 \text{ J}$$

and

$$\Delta H = C_P\,\Delta T = (29.099)(298.15 - 567.57) = -7{,}840 \text{ J}$$

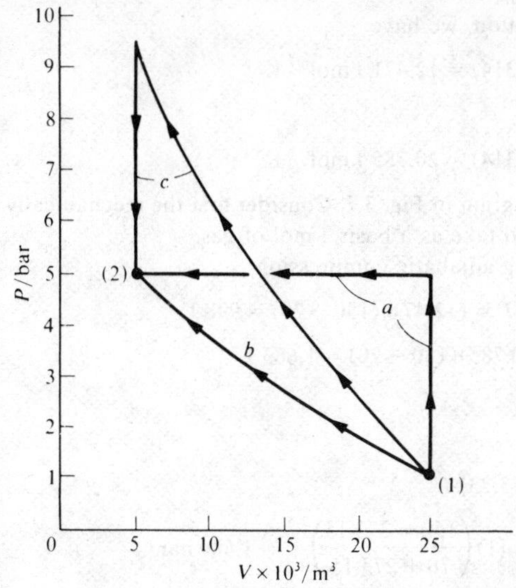

Figure 3.6 Diagram for Example 3.2.

For the entire process,

$$\Delta U = 5{,}600 - 5{,}600 = 0$$

$$\Delta H = 7{,}840 - 7{,}840 = 0$$

$$Q = 0 - 5{,}600 = -5{,}600 \text{ J}$$

and

$$W = -5{,}600 + 0 = -5{,}600 \text{ J}$$

Figure 3.6 shows these processes sketched on a *PV* diagram.

A comparison of the answers to the three parts of this problem shows that the property changes ΔU and ΔH are the same regardless of the path for which they are calculated. On the other hand, Q and W depend on path.

The work for each of these mechanically reversible processes can also be calculated by $W = \int P\,dV$. The value of this integral is proportional to the area below the curve on the *PV* diagram representing the path of the process. The relative sizes of these areas correspond to the numerical values of W.

Example 3.3 An ideal gas undergoes the following sequence of mechanically reversible processes:

(a) From an initial state of 70°C and 1 bar, it is compressed adiabatically to 150°C.

(b) It is then cooled from 150 to 70°C at constant pressure.

(c) Finally, it is expanded isothermally to its original state.

Calculate W, Q, ΔU, and ΔH for each of the three processes and for the entire cycle. Take $C_V = (3/2)R$ and $C_P = (5/2)R$.

If these processes are carried out *irreversibly* but so as to accomplish exactly the same *changes of state* (i.e., the same changes in P, T, U, and H), then the values of Q and W are different. Calculate values of Q and W for an efficiency of 80 percent for each step.

SOLUTION From the given information, we have

$$C_V = (3/2)(8.314) = 12.471 \text{ J mol}^{-1} \text{ K}^{-1}$$

and

$$C_P = (5/2)(8.314) = 20.785 \text{ J mol}^{-1} \text{ K}^{-1}$$

The cycle is represented on a PV diagram in Fig. 3.7. Consider first the mechanically reversible operation of the cycle, and take as a basis 1 mol of gas.

(*a*) For an ideal gas undergoing adiabatic compression,

$$\Delta U = -W = C_V \,\Delta T = (12.471)(150 - 70) = 998 \text{ J}$$

$$\Delta H = C_P \,\Delta T = (20.785)(150 - 70) = 1{,}663 \text{ J}$$

and

$$Q = 0$$

Pressure P_2 can be found from Eq. (3.23)

$$P_2 = P_1 \left(\frac{T_2}{T_1}\right)^{\gamma/(\gamma-1)} = (1)\left(\frac{150 + 273.15}{70 + 273.15}\right)^{2.5} = 1.689 \text{ bar}$$

(*b*) Equation (3.16) is applicable to the constant-pressure process:

$$\Delta H = Q = C_P \,\Delta T = (20.785)(70 - 150) = -1{,}663 \text{ J}$$

Also

$$\Delta U = C_V \,\Delta T = (12.471)(70 - 150) = -998 \text{ J}$$

By the first law,

$$W = Q - \Delta U = -1{,}663 - (-998) = -665 \text{ J}$$

(*c*) For ideal gases ΔU and ΔH are zero for an isothermal process. Since $P_3 = P_2$, Eq. (3.19) gives

$$Q = W = RT \ln \frac{P_3}{P_1} = (8.314)(343.15) \ln \frac{1.689}{1} = 1{,}495 \text{ J}$$

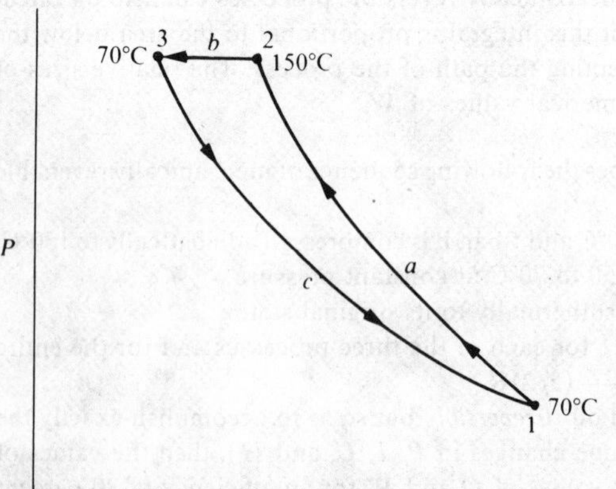

Figure 3.7 Diagram for Example 3.3.

For the entire process,

$$Q = 0 - 1{,}663 + 1{,}495 = -168 \text{ J}$$

$$W = -998 - 665 + 1{,}495 = -168 \text{ J}$$

$$\Delta U = 998 - 998 + 0 = 0$$

and

$$\Delta H = 1{,}663 - 1{,}663 + 0 = 0$$

The property changes ΔU and ΔH both are zero for the entire cycle, because the initial and final states are identical. Note also that $Q = W$ for the cycle. This follows from the first law with $\Delta U = 0$.

If the same changes of state are carried out by irreversible processes, the property changes for the steps are identical with those already calculated. However, the values of Q and W are different.

(*a*) This step can no longer be adiabatic. For mechanically reversible, adiabatic compression, W was -998 J. If the process is 80 percent efficient compared with this, then

$$W = \frac{-998}{0.80} = -1{,}248 \text{ J}$$

Since ΔU is still 998 J, by the first law,

$$Q = \Delta U + W = 998 - 1{,}248 = -250 \text{ J}$$

(*b*) The work for the mechanically reversible cooling process was -665 J. For the irreversible process,

$$W = \frac{-665}{0.80} = -831 \text{ J}$$

and

$$Q = \Delta U + W = -998 - 831 = -1{,}829 \text{ J}$$

(*c*) As work is done *by* the system in this step, the irreversible work is less than the reversible work:

$$W = (0.80)(1{,}495) = 1{,}196 \text{ J}$$

and

$$Q = \Delta U + W = 0 + 1{,}196 = 1{,}196 \text{ J}$$

For the entire cycle, ΔU and ΔH are again zero, but

$$Q = -250 - 1{,}829 + 1{,}196 = -883 \text{ J}$$

and

$$W = -1{,}248 - 831 + 1{,}196 = -883 \text{ J}$$

A summary of these results is given in the following table. All values are in joules.

	Mechanically reversible				Irreversible			
	ΔU	ΔH	Q	W	ΔU	ΔH	Q	W
Step *a*	998	1,663	0	−998	998	1,663	−250	−1,248
Step *b*	−998	−1,663	−1,663	−665	−998	−1,663	−1,829	−831
Step *c*	0	0	1,495	1,495	0	0	1,196	1,196
Cycle	0	0	−168	−168	0	0	−883	−883

The cycle is one which requires work and produces an equal amount of heat. The striking feature of the comparison shown in the table is that the total work required when the cycle consists of three irreversible steps is more than five times the total work required when the steps are mechanically reversible, even though each irreversible step is 80 percent efficient.

Example 3.4 A 0.4-kg mass of nitrogen at 27°C is held in a vertical cyclinder by a frictionless piston. The weight of the piston makes the pressure of the nitrogen 0.35 bar higher than that of the surrounding atmosphere, which is at 1 bar and 27°C. Thus the nitrogen is initially at a pressure of 1.35 bar, and is in mechanical and thermal equilibrium with its surroundings. Consider the following sequence of processes:

(*a*) The apparatus is immersed in an ice/water bath and is allowed to come to equilibrium.

(*b*) A variable force is slowly applied to the piston so that the nitrogen is compressed reversibly at the constant temperature of 0°C until the gas volume reaches one-half that at the end of step *a*. At this point the piston is held in place by latches.

(*c*) The apparatus is removed from the ice/water bath and comes to thermal equilibrium in the surrounding atmosphere at 27°C.

(*d*) The latches are removed, and the apparatus is allowed to return to complete equilibrium with its surroundings.

Sketch the entire cycle on a *PV* diagram, and calculate Q, W, ΔU, and ΔH for the nitrogen for each step of the cycle. Nitrogen may be considered an ideal gas for which $C_V = (5/2)R$ and $C_P = (7/2)R$.

SOLUTION At the end of the cycle the nitrogen returns to its initial conditions of 27°C and 1.35 bar. The steps making up the cycle are

(*a*) $$27°\text{C, 1.35 bar} \xrightarrow{\text{const } P} 0°\text{C, 1.35 bar}$$

(*b*) $$0°\text{C}, V_2 \xrightarrow{\text{const } T} 0°\text{C}, V_3 = \tfrac{1}{2}V_2$$

(*c*) $$0°\text{C}, V_3 \xrightarrow{\text{const } V} 27°\text{C}, V_4 = V_3$$

(*d*) $$27°\text{C}, V_4 \xrightarrow{T_4 = T_1} 27°\text{C, 1.35 bar}$$

(*a*) In this step, represented by the horizontal line marked *a* in Fig. 3.8, the nitrogen is cooled at constant pressure. The process is mechanically reversible, even though the heat transfer occurs irreversibly as the result of a finite temperature

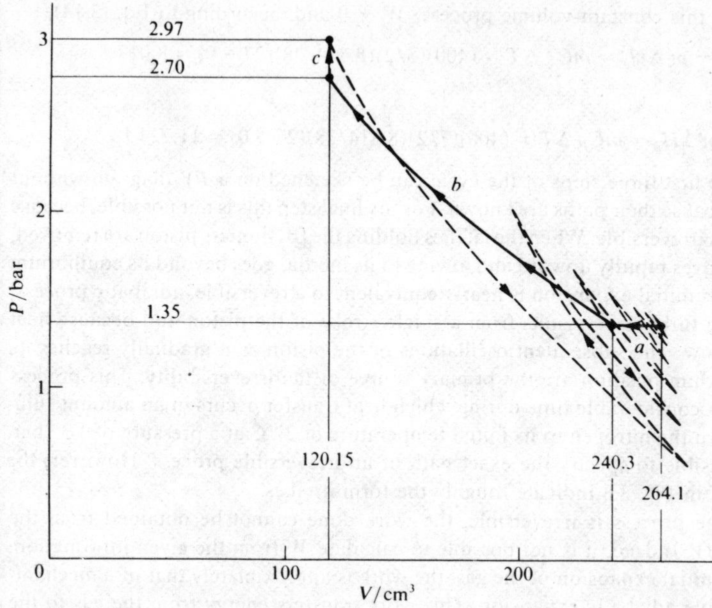

Figure 3.8 Diagram for Example 3.4.

difference. Thus for the mass m of nitrogen

$$W_a = m \int P\,dV = mP\,\Delta V = \frac{mR\,\Delta T}{M}$$

With $R = 8.314\ \text{J mol}^{-1}\ \text{K}^{-1}$, $m = 400$ g, and the molar mass (molecular weight) $M = 28$, we have

$$W_a = \frac{(400)(8.314)(0-27)}{28} = -3{,}207\ \text{J}$$

and

$$Q_a = m\,\Delta H_a = mC_P\,\Delta T = (400)(7/2)(8.314/28)(0-27) = -11{,}224\ \text{J}$$

From the first law,

$$m\,\Delta U_a = Q_a - W_a = -11{,}224 - (-3{,}207) = -8{,}017\ \text{J}$$

The internal-energy change may also be evaluated from Eq. (3.14):

$$m\,\Delta U_a = mC_V\,\Delta T = (400)(5/2)(8.314/28)(0-27) = -8{,}017\ \text{J}$$

(*b*) The process carried out here is an isothermal compression shown by curve *b* in Fig. 3.8. Since the internal energy cannot change at constant temperature,

$$\Delta U_b = \Delta H_b = 0$$

and under conditions of mechanical reversibility,

$$Q_b = W_b = \frac{mRT}{M}\ln\frac{V_3}{V_2} = \frac{(400)(8.314)(273.15)}{28}\ln\frac{1}{2} = -22{,}487\ \text{J}$$

(c) For this constant-volume process, $W_c = 0$ and, according to Eq. (3.14),

$$Q_c = m\,\Delta U_c = mC_V\,\Delta T = (400)(5/2)(8.314/28)(27-0) = 8{,}017 \text{ J}$$

In addition,

$$m\,\Delta H_c = mC_P\,\Delta T = (400)(7/2)(8.314/28)(27-0) = 11{,}224 \text{ J}$$

(d) The first three steps of the cycle can be sketched on a PV diagram without difficulty, because their paths are known. For the final step this is not possible, because the process is irreversible. When the latches holding the frictionless piston are removed, the piston moves rapidly upward and, owing to its inertia, goes beyond its equilibrium position. This initial expansion is nearly equivalent to a reversible, adiabatic process, because little turbulence results from a single stroke of the piston and because heat transfer is slow. The subsequent oscillations of the piston as it gradually reaches its final equilibrium position are the primary source of the irreversibility. This process goes on for a considerable time during which heat transfer occurs in an amount sufficient to return the nitrogen to its initial temperature of 27°C at a pressure of 1.35 bar. It is not possible to specify the exact path of an irreversible process. However, the dashed lines in Fig. 3.8 indicate roughly the form it takes.

Since the process is irreversible, the work done cannot be obtained from the integral $\int P\,dV$. Indeed, it is not possible to calculate W from the given information. During the initial expansion of the gas, the work is approximately that of a mechanically reversible adiabatic expansion. This work transfers energy from the gas to the surroundings, where it pushes back the atmosphere and increases the potential energy of the piston. If the piston were held at its position of maximum travel, the major part of the irreversibility would be avoided, and the work could be calculated to a good approximation by the equations for a reversible adiabatic expansion. However, as the process actually occurs, the oscillating piston causes turbulence or stirring in both the gas and the atmosphere, and there is no way to know the extent of either. This makes impossible the calculation of either Q or W.

Unlike work and heat, the property changes of the system for step d can be computed, since they depend solely on the initial and final states, and these are known. The internal energy and enthalpy of an ideal gas are functions of temperature only. Therefore, ΔU_d and ΔH_d are zero, because the initial and final temperatures are both 27°C. The first law applies to irreversible as well as to reversible processes, and for step d it becomes

$$\Delta U_d = Q_d - W_d = 0$$

or

$$Q_d = W_d$$

Although neither Q_d nor W_d can be calculated, they clearly are equal. Step d results in net energy changes consisting of elevation of the piston and atmosphere and a compensating decrease in the internal energy of the surrounding atmosphere.

Example 3.5 Air is flowing at a steady rate through a horizontal insulated pipe which contains a partly closed valve. The conditions of the air upstream from the valve are 20°C and 6 bar, and the downstream pressure is 3 bar. The line leaving the valve is enough larger than the entrance line that the kinetic-energy change of the air in flowing through the valve is negligible. If air is regarded as an ideal gas, what is the temperature of the air some distance downstream from the valve?

SOLUTION Flow through a partly closed valve is known as a *throttling process.* Since flow is at a steady rate, Eq. (2.10) applies. The line is insulated, making Q small; moreover, the potential-energy and kinetic-energy changes are negligible. Since no shaft work is accomplished, $W_s = 0$. Hence, Eq. (2.10) reduces to

$$\Delta H = 0$$

Thus, for an ideal gas,

$$\Delta H = \int_{T_1}^{T_2} C_P\, dT = 0$$

and

$$T_2 = T_1$$

The result that $\Delta H = 0$ is general for a throttling process, because the assumptions of negligible heat transfer and potential- and kinetic-energy changes are usually valid. If the fluid is an ideal gas, no temperature change occurs. The throttling process is inherently irreversible, but this is immaterial to the calculation of $\Delta H = \int C_P\, dT$, which is a property relation of general validity for an ideal gas.

3.4 APPLICATION OF THE VIRIAL EQUATION

The two forms of the virial expansion given by Eqs. (3.10) and (3.11) are infinite series. For engineering purposes their use is practical only where convergence is very rapid, that is, where no more than two or three terms are required to yield reasonably close approximations to the values of the series. This is realized for gases and vapors at low to moderate pressures.

Figure 3.9 shows a compressibility-factor graph for methane. Values of the compressibility factor Z (as calculated from PVT data for methane by the defining equation $Z = PV/RT$) are plotted against pressure for various constant temperatures. The resulting isotherms show graphically what the virial expansion in P is intended to represent analytically. All isotherms originate at the value $Z = 1$ for $P = 0$. In addition the isotherms are nearly straight lines at low pressures. Thus the tangent to an isotherm at $P = 0$ is a good approximation of the isotherm for a finite pressure range. Differentiation of Eq. (3.10) for a given temperature gives

$$\frac{dZ}{dP} = B' + 2C'P + 3D'P^2 + \cdots$$

from which

$$\left(\frac{dZ}{dP}\right)_{P=0} = B'$$

Thus the equation of the tangent line is

$$Z = 1 + B'P$$

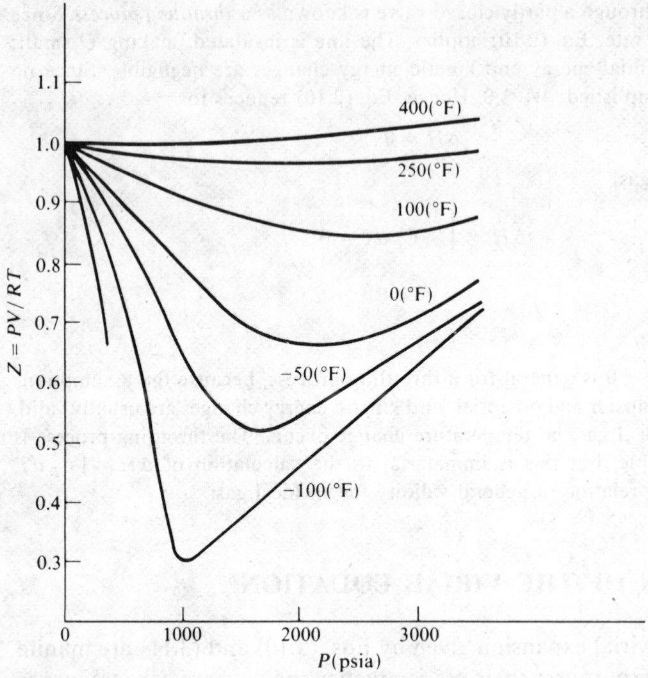

Figure 3.9 Compressibility-graph factor for methane.

a result also given by truncation of Eq. (3.10) to two terms. In addition we may use the approximate relation $B' = B/RT$ to express the equation for Z in terms of the coefficient B:

$$\boxed{Z = \frac{PV}{RT} = 1 + \frac{BP}{RT}} \tag{3.31}$$

Since Eq. (3.11) may also be truncated to two terms for application at low pressures,

$$Z = \frac{PV}{RT} = 1 + \frac{B}{V} \tag{3.32}$$

a question arises as to which equation provides the better representation of low-pressure *PVT* data. Experience shows that Eq. (3.31) is at least as accurate as Eq. (3.32). Moreover, it is much more convenient for use in most applications, because it may be solved explicitly for either pressure or volume. Thus when the virial equation is truncated to two terms, Eq. (3.31) is preferred. This equation satisfactorily represents the *PVT* behavior of most vapors at subcritical temperatures up to a pressure of about 15 bar. At higher temperatures it is appropriate for gases over an increasing pressure range as the temperature increases. Values of B, the second virial coefficient, depend on the nature of the gas and on temperature. Experimental values are available for a number of gases. Moreover, estimation of second virial coefficients is possible where no data are available, as discussed in Sec. 3.6.

For pressures above the range of applicability of Eq. (3.31) but below about 50 bar, the virial equation truncated to three terms usually provides excellent results. In this case Eq. (3.11), the expansion in $1/V$, is far superior to Eq. (3.10). Thus when the virial equation is truncated to three terms, the appropriate form is

$$\boxed{Z = \frac{PV}{RT} = 1 + \frac{B}{V} + \frac{C}{V^2}} \tag{3.33}$$

This equation is explicit in pressure, but cubic in volume. Solution for V is usually done by an iterative scheme with a calculator.

Values of C, like those of B, depend on the identity of the gas and on the temperature. However, much less is known about third virial coefficients than about second virial coefficients, though data for a number of gases can be found in the literature. Since virial coefficients beyond the third are rarely known and since the virial expansion with more than three terms becomes unwieldy, virial equations of more than three terms are rarely used. Alternative equations are described in Secs. 3.5 and 3.6, which follow.

Example 3.6 Reported values for the virial coefficients of isopropanol vapor at 200°C are:

$$B = -388 \text{ cm}^3 \text{ mol}^{-1}$$

$$C = -26{,}000 \text{ cm}^6 \text{ mol}^{-2}$$

Calculate V and Z for isopropanol vapor at 200°C and 10 bar by:

(*a*) The ideal-gas equation.
(*b*) Equation (3.31).
(*c*) Equation (3.33).

SOLUTION The absolute temperature is $T = 473.15$ K, and the appropriate value of the gas constant is $R = 83.14 \text{ cm}^3 \text{ bar mol}^{-1} \text{ K}^{-1}$.

(*a*) By the ideal-gas equation,

$$V = \frac{RT}{P} = \frac{(83.14)(473.15)}{10} = 3{,}934 \text{ cm}^3 \text{ mol}^{-1}$$

and of course $Z = 1$.

(*b*) Solving Eq. (3.31) for V, we find

$$V = \frac{RT}{P} + B = 3{,}934 - 388 = 3{,}546 \text{ cm}^3 \text{ mol}^{-1}$$

Whence

$$Z = \frac{PV}{RT} = \frac{V}{RT/P} = \frac{3{,}546}{3{,}934} = 0.9014$$

(*c*) To facilitate iteration, we write Eq. (3.33) as

$$V_{i+1} = \frac{RT}{P}\left(1 + \frac{B}{V_i} + \frac{C}{V_i^2}\right)$$

where subscript i denotes the iteration number. For the first iteration, $i = 0$, and

$$V_1 = \frac{RT}{P}\left(1 + \frac{B}{V_0} + \frac{C}{V_0^2}\right)$$

where V_0 is an initial estimate of the molar volume. For this we use the ideal-gas value, which gives

$$V_1 = 3{,}934\left(1 - \frac{388}{3{,}934} - \frac{26{,}000}{(3{,}934)^2}\right) = 3{,}539$$

The second iteration depends on this result:

$$V_2 = \frac{RT}{P}\left(1 + \frac{B}{V_1} + \frac{C}{V_1^2}\right)$$

whence

$$V_2 = 3{,}934\left(1 - \frac{388}{3{,}539} - \frac{26{,}000}{(3{,}539)^2}\right) = 3{,}495$$

Iteration continues until the difference $V_{i+1} - V_i$ is insignificant, and leads after five iterations to the final value,

$$V = 3{,}488 \text{ cm}^3 \text{ mol}^{-1}$$

from which $Z = 0.8866$. In comparison with this result, the ideal-gas value is 13 percent too high and Eq. (3.31) gives a value 1.7 percent too high.

3.5 CUBIC EQUATIONS OF STATE

For an accurate description of the *PVT* behavior of fluids over wide ranges of temperature and pressure, an equation of state more comprehensive than the virial equation is required. Such an equation must be sufficiently general to apply to liquids as well as to gases and vapors. Yet is must not be so complex as to present excessive numerical or analytical difficulties in application.

Polynomial equations that are cubic in molar volume offer a compromise between generality and simplicity that is suitable to many purposes. Cubic equations are in fact the simplest equations capable of representing both liquid and vapor behavior. The first general cubic equation of state was proposed by J. D. van der Waals† in 1873:

$$P = \frac{RT}{V - b} - \frac{a}{V^2} \tag{3.34}$$

† Johannes Diderik van der Waals (1837–1923), Dutch physicist who won the 1910 Nobel Prize for physics.

Here, a and b are positive constants; when they are zero, the ideal-gas equation is recovered.

Given values of a and b for a particular fluid, one can calculate P as a function of V for various values of T. Figure 3.10 is a schematic PV diagram showing three such isotherms. Superimposed is the curve representing states of saturated liquid and saturated vapor. For the isotherm $T_1 > T_c$, pressure is a monotonically decreasing function with increasing molar volume. The critical isotherm (labeled T_c) contains the horizontal inflection at C characteristic of the critical point. For the isotherm $T_2 < T_c$, the pressure decreases rapidly in the liquid region with increasing V; after crossing the saturated-liquid line, it goes through a minimum, rises to a maximum, and then decreases, crossing the saturated-vapor line and continuing into the vapor region. Experimental isotherms do not exhibit this smooth transition from the liquid to the vapor region; rather, they contain a horizontal segment within the two-phase region where saturated liquid and saturated vapor coexist in varying proportions at the saturation or

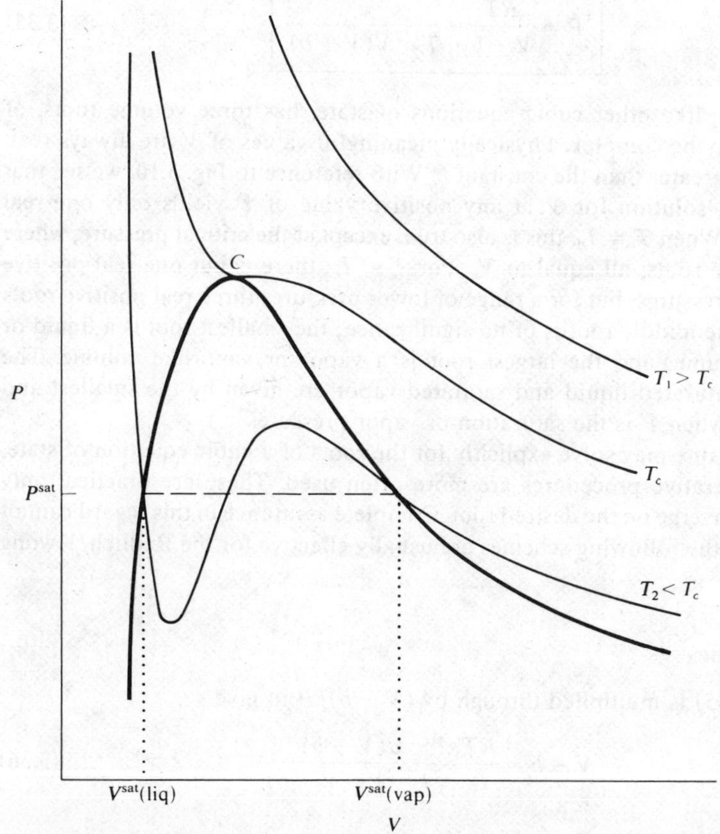

Figure 3.10 Isotherms as given by a cubic equation of state.

vapor pressure. This behavior, shown by the dashed line of Fig. 3.10, cannot be represented analytically, and we accept as inevitable the unrealistic behavior of equations of state in the two-phase region.

Actually, the *PV* behavior predicted in this region by proper cubic equations of state is not wholly fictitious. When the pressure is decreased on saturated liquid devoid of vapor-nucleation sites in a carefully controlled experiment, vaporization does not occur, and the liquid phase persists alone to pressures well below its vapor pressure. Similarly, raising the pressure on a saturated vapor in a suitable experiment does not cause condensation, and the vapor persists alone to pressures well above the vapor pressure. These nonequilibrium or metastable states of superheated liquid and subcooled vapor are approximated by those portions of the *PV* isotherm which lie in the two-phase region adjacent to the saturated-liquid and saturated-vapor states.

The modern development of cubic equations of state started in 1949 with publication of the Redlich/Kwong equation:†

$$\boxed{P = \frac{RT}{V-b} - \frac{a}{T^{1/2}V(V+b)}} \tag{3.35}$$

This equation, like other cubic equations of state, has three volume roots, of which two may be complex. Physically meaningful values of V are always real, positive, and greater than the constant b. With reference to Fig. 3.10, we see that when $T > T_c$, solution for V at any positive value of P yields only one real positive root. When $T = T_c$, this is also true, except at the critical pressure, where there are three roots, all equal to V_c. For $T < T_c$, there is but one real positive root at high pressures, but for a range of lower pressures three real positive roots exist. Here, the middle root is of no significance; the smallest root is a liquid or liquidlike volume, and the largest root is a vapor or vaporlike volume. The volumes of saturated liquid and saturated vapor are given by the smallest and largest roots when P is the saturation or vapor pressure.

Although one may solve explicitly for the roots of a cubic equation of state, in practice iterative procedures are more often used. These are practical only when they converge on the desired root. Complete assurance in this regard cannot be given, but the following schemes are usually effective for the Redlich/Kwong equation.

Vapor Volumes

Equation (3.35) is multiplied through by $(V - b)/P$ to give

$$V - b = \frac{RT}{P} - \frac{a(V-b)}{T^{1/2}PV(V+b)} \tag{3.36}$$

† Otto Redlich and J. N. S. Kwong, *Chem. Rev.*, **44**: 233, 1949.

For iteration, we write

$$V_{i+1} = \frac{RT}{P} + b - \frac{a(V_i - b)}{T^{1/2}PV_i(V_i + b)} \tag{3.37}$$

The ideal-gas equation provides a suitable initial value, $V_0 = RT/P$.

Liquid Volumes

Equation (3.35) is put into standard polynomial form:

$$V^3 - \frac{RT}{P}V^2 - \left(b^2 + \frac{bRT}{P} - \frac{a}{PT^{1/2}}\right)V - \frac{ab}{PT^{1/2}} = 0$$

An iteration scheme results when this is written

$$V_{i+1} = \frac{1}{c}\left(V_i^3 - \frac{RT}{P}V_i^2 - \frac{ab}{PT^{1/2}}\right) \tag{3.38}$$

where

$$c = b^2 + \frac{bRT}{P} - \frac{a}{PT^{1/2}} \tag{3.39}$$

For an initial value, take $V_0 = b$.

The constants in an equation of state may of course be evaluated by a fit to available *PVT* data. For simple cubic equations of state, however, suitable estimates come from the critical constants T_c and P_c. Since the critical isotherm exhibits a horizontal inflection at the critical point, we may impose the mathematical conditions:

$$\left(\frac{\partial P}{\partial V}\right)_{T;\text{cr}} = \left(\frac{\partial^2 P}{\partial V^2}\right)_{T;\text{cr}} = 0$$

where the subscript cr indicates application at the critical point. Differentiation of Eq. (3.34) or Eq. (3.35) yields expressions for both derivatives, which may be equated to zero for $P = P_c$, $T = T_c$, and $V = V_c$. The equation of state may itself be written for the critical conditions, providing three equations in the five constants P_c, V_c, T_c, a, and b. Of the several ways to treat these equations, experience shows the most suitable to be elimination of V_c to yield expressions relating a and b to P_c and T_c:

The van der Waals equation

$$a = \frac{27R^2T_c^2}{64P_c} \qquad b = \frac{RT_c}{8P_c}$$

The Redlich/Kwong equation

$$a = \frac{0.42748R^2T_c^{2.5}}{P_c} \tag{3.40}$$

$$b = \frac{0.08664RT_c}{P_c} \tag{3.41}$$

Although these equations may not yield the best possible values, they give values that are reasonable and which can almost always be determined, because critical temperatures and pressures (in contrast to extensive *PVT* data) are usually known. A list of values of T_c and P_c is provided in App. B.

The inherent limitations of cubic equations of state are discussed by Abbott.† Equations of greater overall accuracy are necessarily more complex, as is illustrated by the Benedict/Webb/Rubin equation:

$$P = \frac{RT}{V} + \frac{B_0RT - A_0 - C_0/T^2}{V^2} + \frac{bRT - a}{V^3}$$

$$+ \frac{a\alpha}{V^6} + \frac{c}{V^3T^2}\left(1 + \frac{\gamma}{V^2}\right)\exp\frac{-\gamma}{V^2} \tag{3.42}$$

where A_0, B_0, C_0, a, b, c, α, and γ are all constants for a given fluid. This equation and its modifications, despite their complexity, are widely used in the petroleum and natural-gas industries for light hydrocarbons and a few other commonly encountered gases.

Example 3.7 Given that the vapor pressure of methyl chloride at 60°C is 13.76 bar, use the Redlich/Kwong equation to calculate the molar volumes of saturated vapor and saturated liquid at these conditions.

SOLUTION We evaluate the constants a and b by Eqs. (3.40) and (3.41) with values of T_c and P_c taken from App. B:

$$a = \frac{(0.42748)(83.14)^2(416.3)^{2.5}}{66.8}$$

$$= 1.56414 \times 10^8 \text{ cm}^6 \text{ bar mol}^{-2} \text{ K}^{1/2}$$

and

$$b = \frac{(0.08664)(83.14)(416.3)}{66.8} = 44.891 \text{ cm}^3 \text{ mol}^{-1}$$

For evaluation of the molar volume of saturated vapor, we substitute known values into Eq. (3.37); this gives

$$V_{i+1} = 2{,}057.83 - \frac{622{,}784}{V_i}\left(\frac{V_i - 44.891}{V_i + 44.891}\right)$$

Iteration starts with $V_i = V_0 = RT/P = 2{,}012.94 \text{ cm}^3 \text{ mol}^{-1}$, and continues to convergence on the value

$$V = 1{,}712 \text{ cm}^3 \text{ mol}^{-1}$$

The experimental result is $1{,}635.6 \text{ cm}^3 \text{ mol}^{-1}$.

† M. M. Abbott, *AIChE J.*, **19**: 596, 1973; Adv. in Chem. Series 183, pp. 47–70, Am. Chem. Soc., Washington, D.C., 1979.

For evaluation of the molar volume of saturated liquid, we substitute known values into Eqs. (3.38) and (3.39); the resulting equation is

$$V_{i+1} = \frac{V_i^3 - 2{,}012.94\,V_i^2 - 2.79573 \times 10^7}{-530{,}405}$$

Iteration starts with $V_i = V_0 = b = 44.891\ \text{cm}^3\ \text{mol}^{-1}$, and continues to convergence on the value

$$V = 71.34\ \text{cm}^3\ \text{mol}^{-1}$$

The experimental result is $60.37\ \text{cm}^3\ \text{mol}^{-1}$.

3.6 GENERALIZED CORRELATIONS FOR GASES

An alternative form of the Redlich/Kwong equation is obtained by multiplication of Eq. (3.35) by V/RT:

$$Z = \frac{1}{1-h} - \frac{a}{bRT^{1.5}}\left(\frac{h}{1+h}\right)$$

where

$$h \equiv \frac{b}{V} = \frac{b}{ZRT/P} = \frac{bP}{ZRT}$$

Elimination of a and b in these equations by Eqs. (3.40) and (3.41) gives

$$Z = \frac{1}{1-h} - \frac{4.9340}{T_r^{1.5}}\left(\frac{h}{1+h}\right) \tag{3.43a}$$

$$h = \frac{0.08664 P_r}{Z T_r} \tag{3.43b}$$

where $T_r \equiv T/T_c$ and $P_r \equiv P/P_c$ are called *reduced temperature* and *reduced pressure*.

This pair of equations is arranged for convenient iterative solution for the compressibility factor Z for any gas at any conditions T_r and P_r. For an initial value of $Z = 1$, h is calculated by Eq. (3.43b). With this value of h, Eq. (3.43a) yields a new value of Z for substitution into Eq. (3.43b). This procedure is continued until a new iteration produces a change in Z less than some small preset tolerance. The process does not converge for liquids.

Equations of state which express Z as a function of T_r and P_r are said to be *generalized*, because of their general applicability to all gases. An alternative to the use of an equation is a graph of Z vs. P_r which shows isotherms for various values of T_r. Such a *generalized chart* can be prepared from a generalized equation; alternatively, the isotherms may be drawn to provide the best fit of experimental *PVT* data for various gases. The advantage of a generalized correlation is that it allows the prediction of property values for gases from very limited information.

For use of the generalized Redlich/Kwong equation one needs only the critical temperature and critical pressure of the gas. This is the basis for the two-parameter *theorem of corresponding states*: All gases, when compared at the same reduced temperature and reduced pressure, have approximately the same compressibility factor, and all deviate from ideal-gas behavior to about the same degree.

Although use of an equation based on the two-parameter theorem of corresponding states provides far better results in general than the ideal-gas equation, significant deviations from experiment still exist for all but the *simple fluids* argon, krypton, and xenon. Appreciable improvement results from the introduction of a third corresponding-states parameter, characteristic of molecular structure; the most popular such parameter is the *acentric factor* ω, introduced by K. S. Pitzer and coworkers.†

The acentric factor for a pure chemical species is defined with reference to its vapor pressure. Since the logarithm of the vapor pressure of a pure fluid is approximately linear in the reciprocal of absolute temperature, we may write

$$\frac{d \log P_r^{\text{sat}}}{d(1/T_r)} = a$$

where P_r^{sat} is the reduced vapor pressure, T_r is the reduced temperature, and a is the slope of a plot of $\log P_r^{\text{sat}}$ vs. $1/T_r$. If the two-parameter theorem of corresponding states were generally valid, the slope a would be the same for all pure fluids. This is observed not to be true; each fluid has its own characteristic value of a, which could in principle serve as a third corresponding-states parameter. However, Pitzer noted that all vapor-pressure data for the simple fluids (Ar, Kr, Xe) lie on the same line when plotted as $\log P_r^{\text{sat}}$ vs. $1/T_r$ and that the line passes through $\log P_r^{\text{sat}} = -1.0$ at $T_r = 0.7$. This is illustrated in Fig. 3.11. Data for other fluids define other lines whose locations can be fixed in relation to the line for the simple fluids (SF) by the difference:

$$\log P_r^{\text{sat}}(SF) - \log P_r^{\text{sat}}$$

The acentric factor is defined as this difference evaluated at $T_r = 0.7$:

$$\omega \equiv -1.0 - \log (P_r^{\text{sat}})_{T_r=0.7} \tag{3.44}$$

Therefore ω can be determined for any fluid from T_c, P_c, and a single vapor-pressure measurement made at $T_r = 0.7$. Values of ω and the critical constants T_c, P_c, and V_c for a number of fluids are listed in App. B.

The definition of ω makes its value zero for argon, krypton, and xenon, and experimental data yield compressibility factors for all three fluids that are correlated by the same curves when Z is represented as a function of T_r and P_r. Thus the basic premise of the three-parameter theorem of corresponding states is that all fluids having the same value of ω have the same value of Z when compared at the same T_r and P_r.

† The work of Pitzer et al. is fully described in G. N. Lewis and M. Randall, *Thermodynamics*, 2d ed., revised by K. S. Pitzer and L. Brewer, App. 1, McGraw-Hill, New York, 1961.

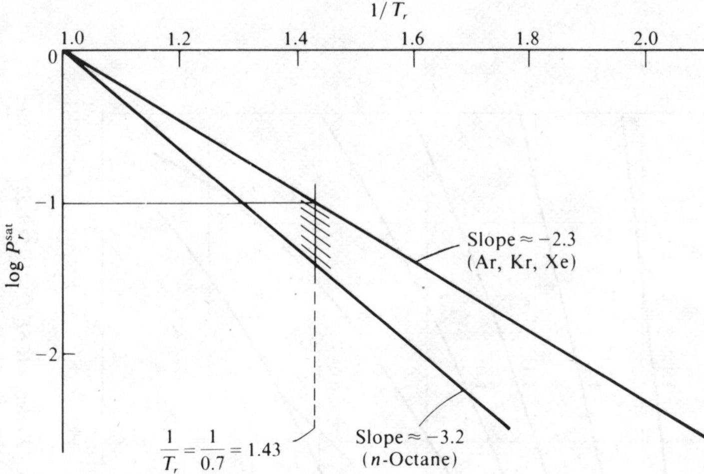

Figure 3.11 Approximate temperature dependence of reduced vapor pressure.

The correlation for Z developed by Pitzer and coworkers takes the form

$$Z = Z^0 + \omega Z^1 \tag{3.45}$$

where Z^0 and Z^1 are complex functions of both T_r and P_r. When $\omega = 0$, as is the case for the simple fluids, the second term disappears, and Z^0 becomes identical with Z. Thus a generalized correlation for Z as a function of T_r and P_r based on data for just argon, krypton, and xenon provides the relationship $Z^0 = F^0(T_r, P_r)$. This function is plotted in Figs. 3.12 and 3.13.

Equation (3.45) is a simple linear relation between Z and ω for given values of T_r and P_r. Experimental data for Z for the nonsimple fluids plotted vs. ω at constant T_r and P_r do indeed yield straight lines, and their slopes provide values for Z^1 from which the generalized function $Z^1 = f^1(T_r, P_r)$ can be constructed. The result is provided by Figs. 3.14 and 3.15.

Figures 3.12 and 3.13 for Z^0, based on data for the simple fluids, provide a complete *two*-parameter corresponding-states correlation for Z. Since the second term of Eq. (3.45) is a relatively small correction to this two-parameter correlation, its omission does not introduce large errors. Thus Figs. 3.12 and 3.13 may be used alone for quick but less precise estimates of Z than are obtained from the complete three-parameter correlation.

The Pitzer correlation provides reliable results for gases which are nonpolar or only slightly polar; for these, errors of no more than 2 or 3 percent are indicated. When applied to highly polar gases or to gases that associate, larger errors can be expected.

A disadvantage of the generalized compressibility-factor correlation is its graphical nature, but the complexity of the functions Z^0 and Z^1 precludes their general representation by simple equations. However, we can give approximate

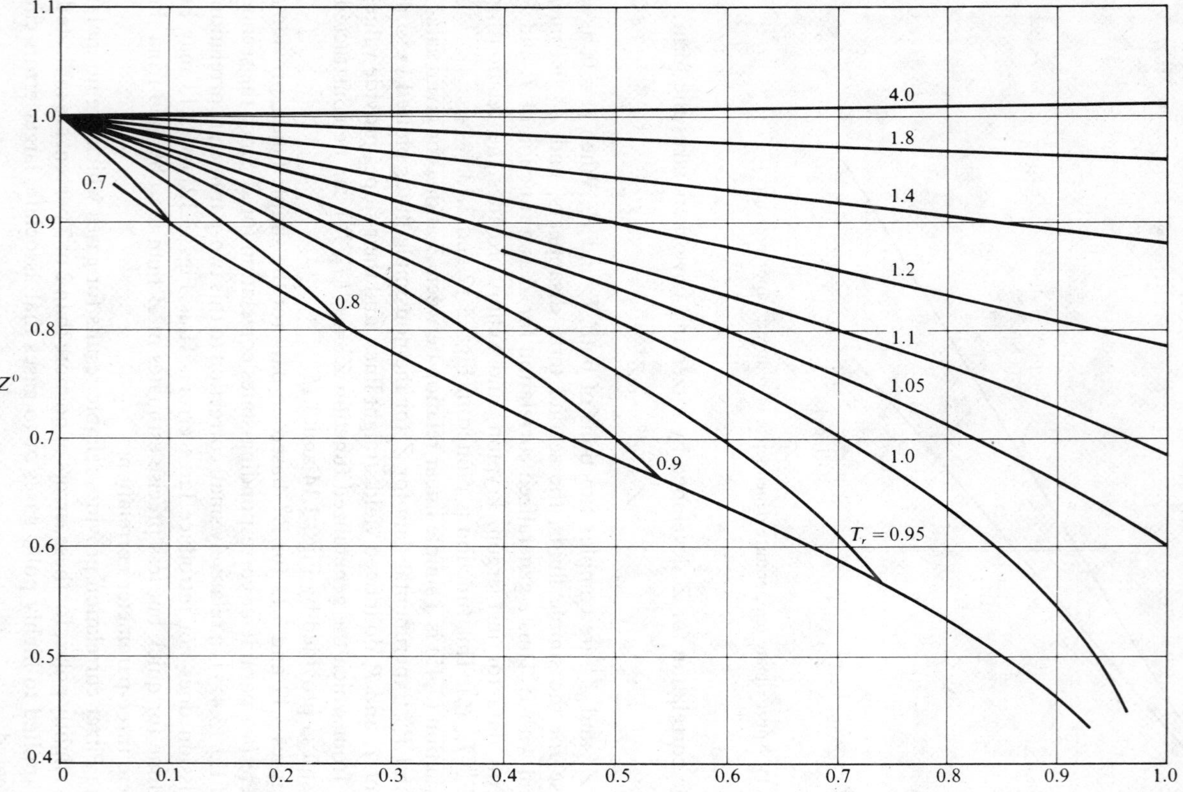

Figure 3.12 Generalized correlation for Z^0, $P_r < 1.0$. *(Based on data of B. I. Lee and M. G. Kesler, AIChE J.,* **21**: *510–527, 1975.)*

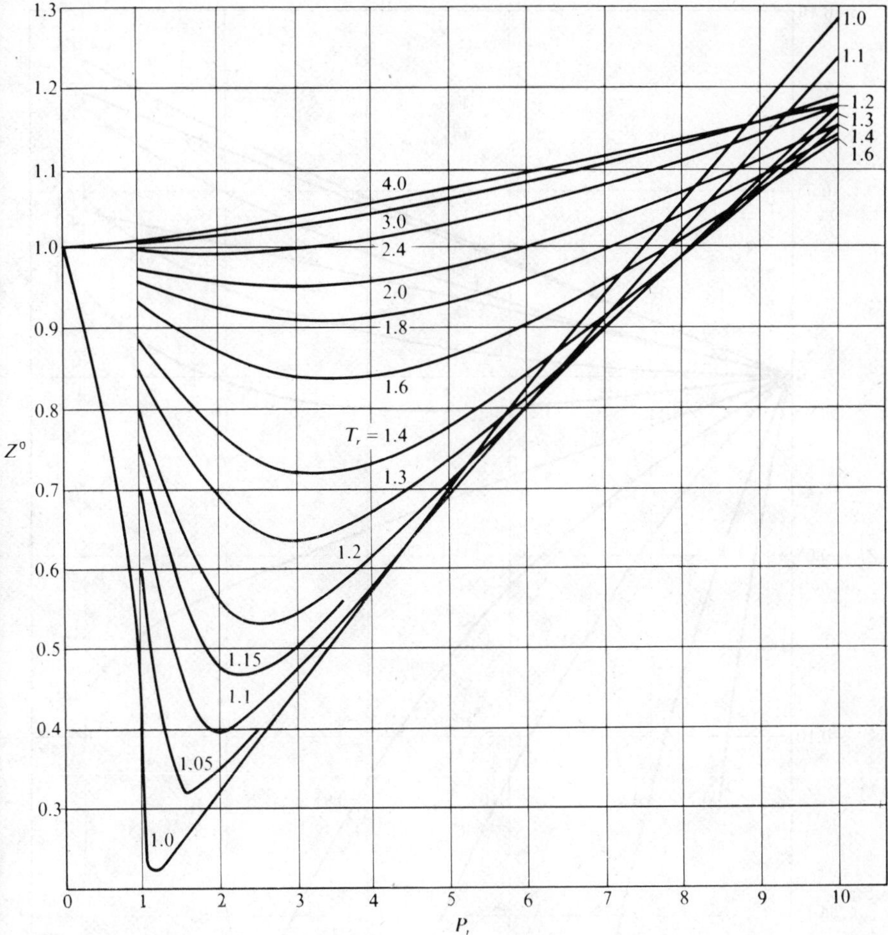

Figure 3.13 Generalized correlation for Z^0, $P_r > 1.0$. *(Based on data of B. I. Lee and M. G. Kesler,* ibid.*)*

analytical expression to these functions for a limited range of pressures. The basis for this is Eq. (3.31), the simplest form of the virial equation, which may be written

$$Z = 1 + \frac{BP}{RT} = 1 + \left(\frac{BP_c}{RT_c}\right)\frac{P_r}{T_r} \tag{3.46}$$

Thus, Pitzer and coworkers proposed a second correlation, which expresses the quantity BP_c/RT_c as

$$\frac{BP_c}{RT_c} = B^0 + \omega B^1 \tag{3.47}$$

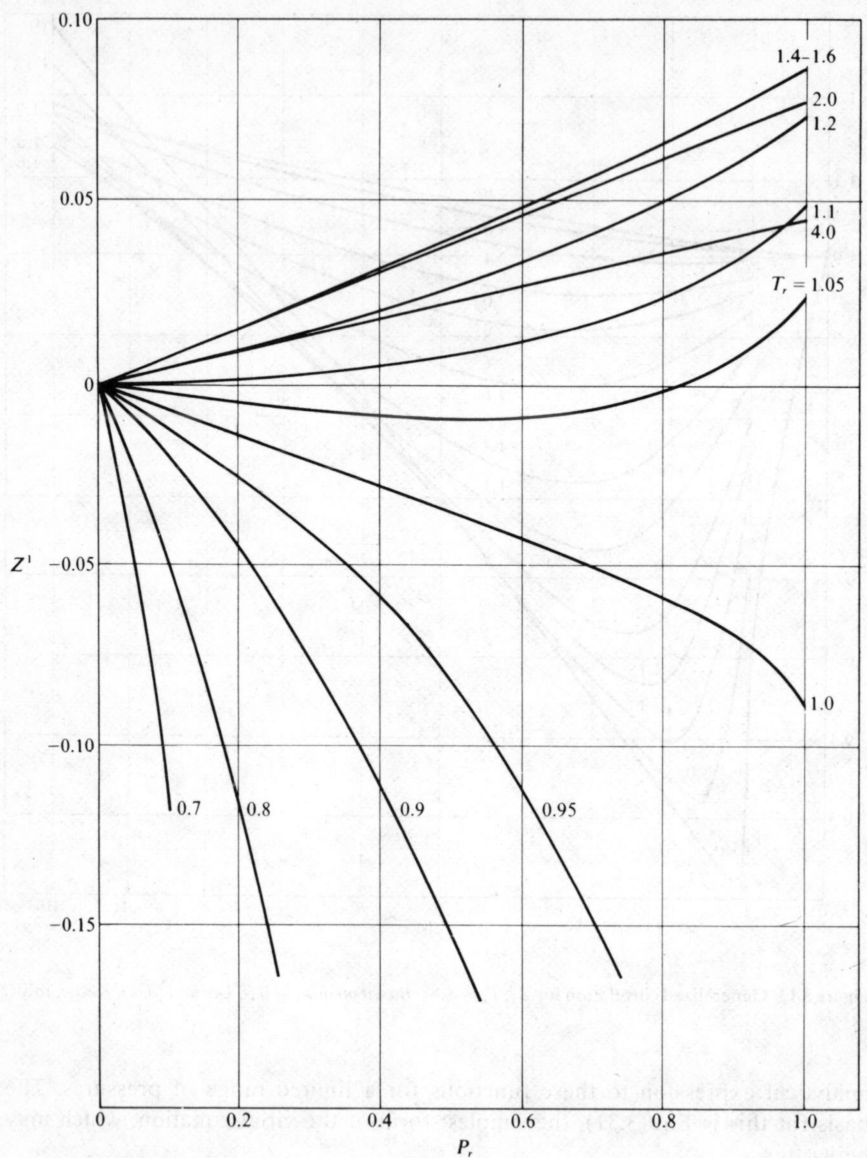

Figure 3.14 Generalized correlation for Z^1, $P_r < 1.0$. *(Based on data of B. I. Lee and M. G. Kesler,* ibid.*)*

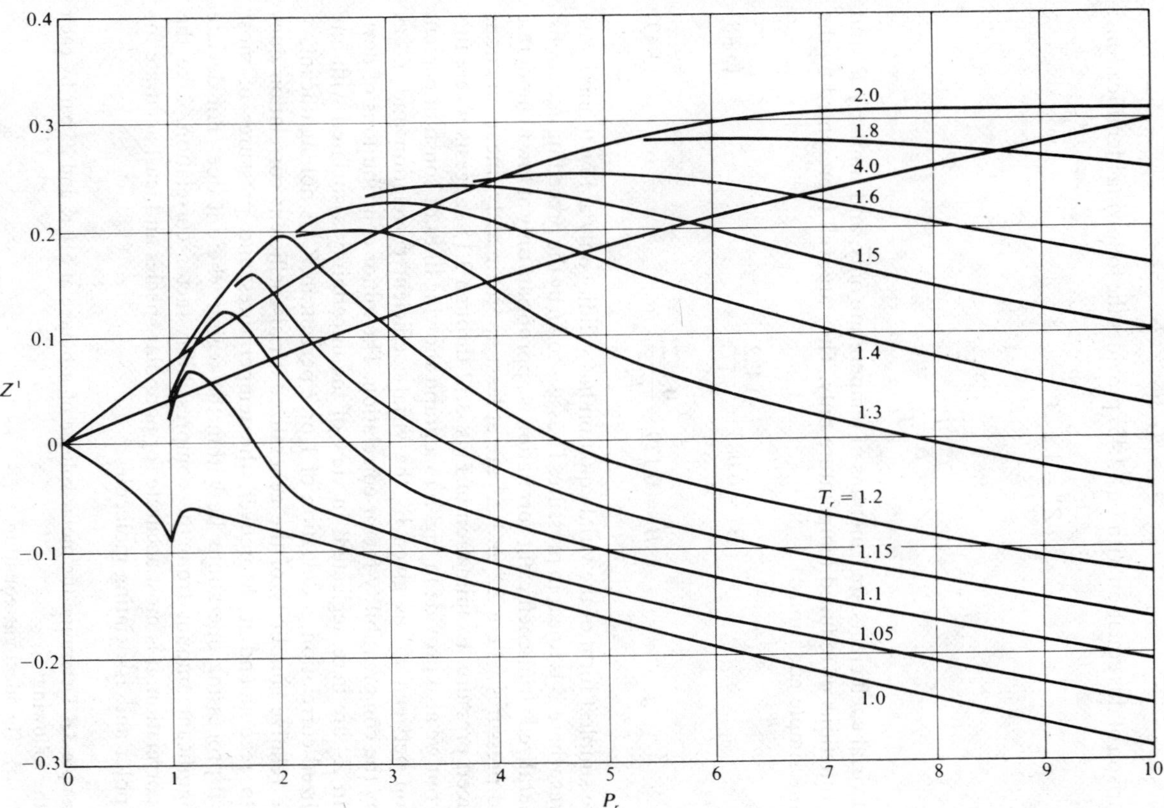

Figure 3.15 Generalized correlation for Z^1, $P_r > 1.0$. *(Based on data of B. I. Lee and M. G. Kesler,* ibid.*)*

Combination of Eqs. (3.46) and (3.47) gives

$$Z = 1 + B^0 \frac{P_r}{T_r} + \omega B^1 \frac{P_r}{T_r}$$

Comparison of this equation with Eq. (3.45) provides the following identifications:

$$Z^0 = 1 + B^0 \frac{P_r}{T_r}$$

and

$$Z^1 = B^1 \frac{P_r}{T_r}$$

Second virial coefficients are functions of temperature only, and similarly B^0 and B^1 are functions of reduced temperature only. They are well represented by the following simple equations:†

$$B^0 = 0.083 - \frac{0.422}{T_r^{1.6}} \tag{3.48}$$

$$B^1 = 0.139 - \frac{0.172}{T_r^{4.2}} \tag{3.49}$$

The simplest form of the virial equation has validity only at low to moderate pressures where Z is linear in pressure. Thus as shown by the preceding equations, the generalized virial-coefficient correlation is appropriate only at low to moderate reduced pressures where Z^0 and Z^1 are at least approximately linear functions of reduced pressure. Examination of Figs. 3.12 through 3.15 suggests where this is true, but we also provide in Fig. 3.16 a graph showing lines of constant percent deviation between Z^0 as given by the virial-coefficient correlation and Z^0 as given by the compressibility-factor correlation. The minor contributions of deviations in Z^1 are here neglected. In view of the uncertainty associated with any generalized correlation, a deviation of 1 or 2 percent in Z^0 is not significant.

The relative simplicity of the generalized virial-coefficient correlation does much to recommend it. Moreover, the temperatures and pressures of most chemical-processing operations lie within the region where it does not deviate by a significant amount from the compressibility-factor correlation. Like the parent correlation, it is most accurate for nonpolar species and least accurate for highly polar and associating molecules.

Example 3.8 Determine the molar volume of *n*-butane at 510 K and 25 bar by each of the following:

(*a*) The ideal-gas equation.

(*b*) The generalized compressibility-factor correlation.

(*c*) The generalized virial-coefficient correlation.

† M. M. Abbott, personal communication.

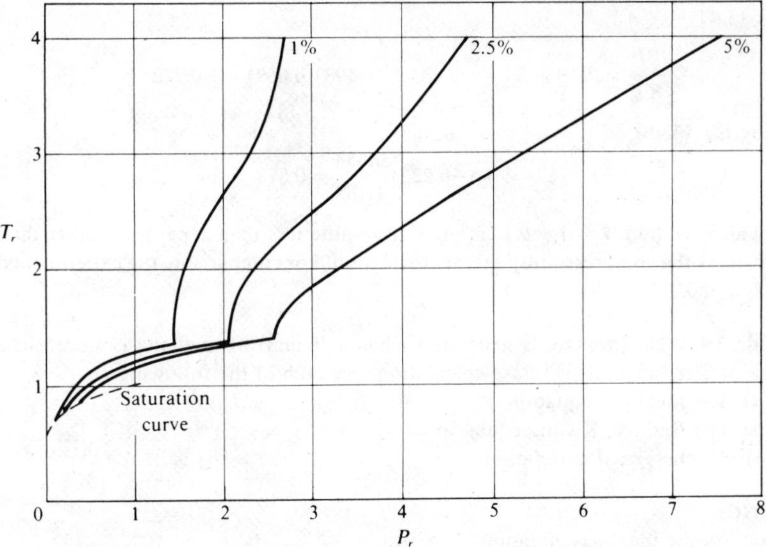

Figure 3.16 Plot of T_r vs. P_r showing lines of constant percent deviation between values of Z^0 as calculated by the virial-coefficient correlation and by the compressibility-factor correlation.

SOLUTION

(*a*) By the ideal-gas equation,

$$V = \frac{RT}{P} = \frac{(83.14)(510)}{25} = 1{,}696.1 \text{ cm}^3 \text{ mol}^{-1}$$

(*b*) Taking values of T_c and P_c from App. B, we find:

$$T_r = \frac{510}{425.2} = 1.198 \qquad P_r = \frac{25}{38.0} = 0.658$$

Figures 3.12 and 3.14 then provide:

$$Z^0 = 0.865 \qquad Z^1 = 0.038$$

Thus, by Eq. (3.45) with $\omega = 0.193$,

$$Z = Z^0 + \omega Z^1 = 0.865 + (0.193)(0.038) = 0.872$$

and

$$V = \frac{ZRT}{P} = \frac{(0.872)(83.14)(510)}{25} = 1{,}479.0 \text{ cm}^3 \text{ mol}^{-1}$$

If we take $Z = Z^0 = 0.865$, in accord with the two-parameter corresponding states correlation, then $V = 1{,}467.1 \text{ cm}^3 \text{ mol}^{-1}$, which is less than 1 percent lower than the value given by the three-parameter correlation.

(*c*) Values of B^0 and B^1 are given by Eqs. (3.48) and (3.49):

$$B^0 = -0.233 \qquad B^1 = 0.059$$

By Eq. (3.47)

$$\frac{BP_c}{RT_c} = B^0 + \omega B^1 = -0.233 + (0.193)(0.059) = -0.222$$

Then by Eq. (3.46),

$$Z = 1 + (-0.222)\frac{0.658}{1.198} = 0.878$$

from which we find $V = 1{,}489.1\ \text{cm}^3\ \text{mol}^{-1}$, a value less than 1 percent higher than that given by the compressibility-factor correlation. For comparison, the experimental value is 1,480.7.

Example 3.9 What pressure is generated when 1(lb mol) of methane is stored in a volume of $2(\text{ft})^3$ at 122(°F)? Base calculations on each of the following:

(*a*) The ideal-gas equation.

(*b*) The Redlich/Kwong equation.

(*c*) A generalized correlation.

SOLUTION

(*a*) By the ideal-gas equation,

$$P = \frac{RT}{V} = \frac{(0.7302)(122 + 459.67)}{2} = 212.4(\text{atm})$$

(*b*) For the Redlich/Kwong equation, we calculate values of a and b by Eqs. (3.40) and (3.41):

$$a = \frac{(0.42748)(0.7302)^2(343.1)^{2.5}}{45.4} = 10{,}945.4(\text{atm})(\text{ft})^6(\text{R})^{1/2}$$

and

$$b = \frac{(0.08664)(0.7302)(343.1)}{45.4} = 0.4781(\text{ft})^3$$

where values of T_c and P_c from App. B have been converted to (R) and (atm). Substitution of known values into Eq. (3.35) now gives:

$$P = \frac{(0.7302)(581.67)}{2 - 0.4781} - \frac{10{,}945.4}{(581.67)^{1/2}(2)(2 + 0.4781)} = 187.5(\text{atm})$$

(*c*) Since the pressure here is high, the generalized compressibility-factor correlation is the proper choice. In the absence of a known value for P_r, an iterative procedure is based on the following equation:

$$P = \frac{ZRT}{V} = \frac{Z(0.7302)(581.67)}{2} = 212.4Z$$

Since $P = P_c P_r = 45.4 P_r$, this equation becomes

$$Z = \frac{45.5 P_r}{212.4} = 0.2138 P_r$$

or

$$P_r = \frac{Z}{0.2138}$$

One now assumes a starting value for Z, say $Z = 1$. This gives $P_r = 4.68$, and allows a new value of Z to be calculated by Eq. (3.45) from values read from Figs. 3.13 and 3.15 at the reduced temperature of $T_r = 581.67/343.1 = 1.695$. With this new value of Z, a new value of P_r is calculated, and the procedure continues until no significant change occurs from one step to the next. The final value of Z so found is 0.885 at $P_r = 4.14$. This may be confirmed by substitution of values for Z^0 and Z^1 from Figs. 3.13 and 3.15 read at $P_r = 4.14$ and $T_r = 1.695$ into Eq. (3.45). Since $\omega = 0.007$, we have

$$Z = Z^0 + \omega Z^1 = 0.884 + (0.007)(0.25) = 0.885$$

and

$$P = \frac{ZRT}{V} = \frac{(0.885)(0.7302)(581.67)}{2} = 188.9(\text{atm})$$

Since the acentric factor is here so small, the two- and three-parameter compressibility-factor correlations are little different. Both the Redlich/Kwong equation and the generalized compressibility-factor correlation give answers very close to the experimental value of 185(atm). The ideal-gas equation yields a result that is high by 14.6 percent.

Example 3.10 A mass of 500 g of gaseous ammonia is contained in a 30,000-cm^3 vessel immersed in a constant-temperature bath at 65°C. Calculate the pressure of the gas by each of the following:

(*a*) The ideal-gas equation.
(*b*) The Redlich/Kwong equation.
(*c*) A generalized correlation.

SOLUTION The molar volume of ammonia in the vessel is given by

$$V = \frac{V^t}{n} = \frac{V^t}{m/M}$$

where n is the total number of moles and m is the mass of ammonia in the vessel of total volume V^t and M is the molar mass of ammonia. Thus

$$V = \frac{30{,}000}{500/17.02} = 1{,}021.2 \text{ cm}^3 \text{ mol}^{-1}$$

(*a*) By the ideal-gas equation,

$$P = \frac{RT}{V} = \frac{(83.14)(65 + 273.15)}{1{,}021.2} = 27.53 \text{ bar}$$

(*b*) For application of the Redlich/Kwong equation, we first evaluate a and b by Eqs. (3.40) and (3.41):

$$a = \frac{(0.42748)(83.14)^2(405.6)^{2.5}}{112.8} = 8.679 \times 10^7 \text{ bar cm}^6 \text{ K}^{1/2}$$

and

$$b = \frac{(0.08664)(83.14)(405.6)}{112.8} = 25.90 \text{ cm}^3$$

where values of T_c and P_c are from App. B. Substitution of known values into Eq. (3.35) now gives:

$$P = \frac{(83.14)(338.15)}{1{,}021.2 - 25.9} - \frac{8.679 \times 10^7}{(338.15)^{1/2}(1{,}021.2)(1{,}021.2 + 25.9)}$$
$$= 23.83 \text{ bar}$$

(*c*) Since the reduced pressure here is low ($\simeq 0.2$), we use the generalized virial-coefficient correlation. For a reduced temperature of $T_r = 338.15/405.6 = 0.834$, values of B^0 and B^1 as given by Eqs. (3.48) and (3.49) are

$$B^0 = -0.482 \qquad B^1 = -0.232$$

Substitution into Eq. (3.47) with $\omega = 0.250$ yields

$$\frac{BP_c}{RT_c} = -0.482 + (0.250)(-0.232) = -0.540$$

and

$$B = \frac{-0.540RT_c}{P_c} = \frac{-(0.540)(83.14)(405.6)}{112.8} = -161.4 \text{ cm}^3 \text{ mol}^{-1}$$

Solving Eq. (3.31) for *P*, we obtain

$$P = \frac{RT}{V - B} = \frac{(83.14)(338.15)}{1{,}021.2 + 161.4} = 23.77 \text{ bar}$$

An iterative solution here is not necessary, because *B* is independent of *P*.

We can use this result to check our initial assumption as to the adequacy of the generalized virial-coefficient correlation. At the calculated reduced pressure of $P_r = 23.77/112.8 = 0.211$, values for Z^0 and Z^1 from Figs. 3.12 and 3.14 are

$$Z^0 = 0.867 \qquad Z^1 = -0.092$$

Whence

$$Z = 0.867 - (0.250)(0.092) = 0.844$$

from which we find $P = 23.24$ bar. An additional iteration produces no further refinement of this result, which is just over 2 percent lower than the value calculated by the virial-coefficient correlation.

Experimental data indicate that the pressure is 23.82 bar at the given conditions. Thus the ideal-gas equation yields an answer that is high by about 15 percent, whereas the other two methods give answers in substantial agreement with experiment, even though ammonia is a polar molecule.

3.7 GENERALIZED CORRELATIONS FOR LIQUIDS

Although the molar volumes of liquids can be calculated by means of generalized cubic equations of state, the results are not of high accuracy. However, generalized equations are available for the calculation of molar volumes of *saturated* liquids.

The following equation, proposed by Rackett,† is an example:

$$V^{sat} = V_c Z_c^{(1-T_r)^{0.2857}} \tag{3.50}$$

The only data required are the critical constants, given in App. B. Results are usually accurate to 1 or 2 percent.

Lydersen, Greenkorn, and Hougen‡ developed a general method for estimation of liquid volumes, based on the principle of corresponding states. It applies to liquids just as the two-parameter compressibility-factor correlation applies to gases, but is based on a correlation of reduced density as a function of reduced temperature and pressure. Reduced density is defined as

$$\rho_r = \frac{\rho}{\rho_c} = \frac{V_c}{V} \tag{3.51}$$

where ρ_c is the density at the critical point. The generalized correlation is shown in Fig. 3.17. This figure may be used directly with Eq. (3.51) for determination of liquid volumes if the value of the critical volume is known. A better procedure is to make use of a single known liquid volume (state 1) by the identity,

$$V_2 = V_1 \frac{\rho_{r_1}}{\rho_{r_2}} \tag{3.52}$$

where V_2 = required volume

V_1 = known volume

ρ_{r_1}, ρ_{r_2} = reduced densities read from Fig. 3.17

This method gives good results and requires only experimental data that are usually available. Figure 3.17 makes clear the increasing effects of both temperature and pressure on liquid density as the critical point is approached.

Example 3.11 (*a*) Estimate the density of saturated liquid ammonia at 310 K.

(*b*) Estimate the density of liquid ammonia at 310 K and 100 bar.

SOLUTION

(*a*) We apply the Rackett equation at the reduced temperature,

$$T_r = \frac{310}{405.6} = 0.7643$$

With $V_c = 72.5$ and $Z_c = 0.242$ (from App. B), we get

$$V^{sat} = V_c Z_c^{(1-T_r)^{0.2857}} = (72.5)(0.242)^{(0.2357)^{0.2857}}$$

$$V^{sat} = 28.35 \text{ cm}^3 \text{ mol}^{-1}$$

This compares with the experimental value of 29.14 $\text{cm}^3 \text{ mol}^{-1}$, and is in error by 2.7 percent.

† H. G. Rackett, *J. Chem. Eng. Data*, **15**: 514, 1970; see also C. F. Spencer and S. B. Adler, ibid., **23**: 82, 1978 for a review of available equations.

‡ A. L. Lydersen, R. A. Greenkorn, and O. A. Hougen, "Generalized Thermodynamic Properties of Pure Fluids," *Univ. Wisconsin, Eng. Expt. Sta. Rept. 4, 1955.*

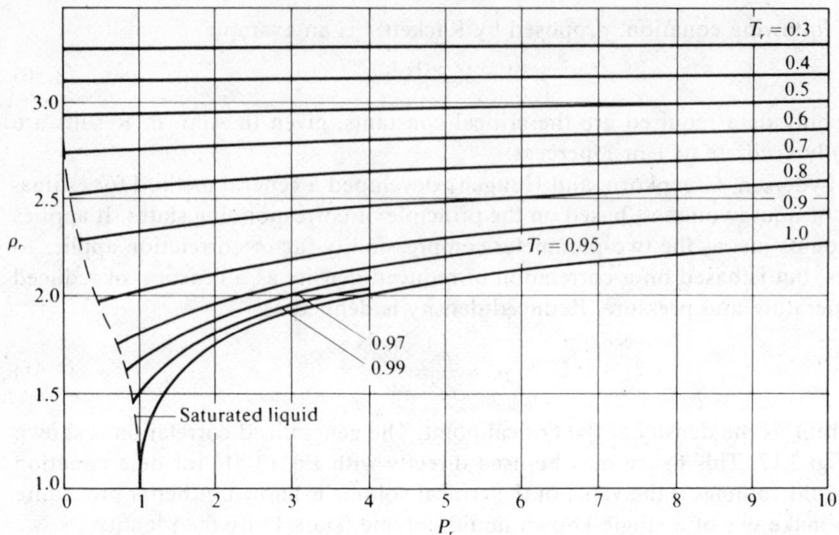

Figure 3.17 Generalized density correlation for liquids. (*Based on A. L. Lydersen, R. A. Greenkorn, and O. A. Hougen, Generalized Thermodynamic Properties of Pure Fluids, Univ. Wisconsin, Eng. Expt. Sta. Rept. 4, 1955.)*

(*b*) The reduced conditions are

$$T_r = 0.764 \qquad P_r = \frac{100}{112.8} = 0.887$$

From Fig. 3.17, we have $\rho_r = 2.38$. Substituting this value along with V_c into Eq. (3.51) gives

$$V = \frac{V_c}{\rho_r} = \frac{72.5}{2.38} = 30.5 \text{ cm}^3 \text{ mol}^{-1}$$

In comparison with the experimental value of 28.6 cm^3 mol^{-1}, this result is in error by 6.6 percent.

If we start with the experimental value of 29.14 cm^3 mol^{-1} for saturated liquid at 310 K, Eq. (3.52) may be used. For the saturated liquid at $T_r = 0.764$, we find from Fig. 3.17 that $\rho_{r_1} = 2.34$. Substitution of known values into Eq. (3.52) gives

$$V_2 = V_1 \frac{\rho_{r_1}}{\rho_{r_2}} = (29.14)\left(\frac{2.34}{2.38}\right) = 28.65 \text{ cm}^3 \text{ mol}^{-1}$$

This result is in essential agreement with the experimental value.

PROBLEMS

3.1 An incompressible fluid is contained in an insulated cylinder fitted with a frictionless piston. Can energy as work be transferred to the fluid? What is the change in internal energy of the fluid when the pressure is increased from P_1 to P_2?

3.2 Express the volume expansivity and the isothermal compressibility as functions of density ρ and its partial derivatives. For water at 50°C and 1 bar, $\kappa = 44.18 \times 10^{-6}$ bar^{-1}. To what pressure must water be compressed at 50°C to change its density by 1 percent? Assume that κ is independent of P.

3.3 Five kilograms of liquid carbon tetrachloride undergo a mechanically reversible, isobaric change of state at 1 bar during which the temperature changes from 0 to 20°C. Determine ΔV^t, W, Q, ΔH^t, and ΔU^t. The following properties for liquid carbon tetrachloride at 1 bar and 0°C may be assumed independent of temperature: $\beta = 1.2 \times 10^{-3}$ K^{-1}, $C_P = 0.84$ kJ kg^{-1} K^{-1}. The density at 0°C and 1 bar is 1,590 kg m^{-3}.

3.4 One mole of an ideal gas, $C_P = (7/2)R$ and $C_V = (5/2)R$, expands from $P_1 = 10$ bar and $V_1 = 0.005$ m^3 to $P_2 = 1$ bar by each of the following paths:

(*a*) Constant volume.

(*b*) Constant temperature.

(*c*) Adiabatically.

Assuming mechanical reversibility, calculate W, Q, ΔU, and ΔH for each process. Sketch each path on a single PV diagram.

3.5 An ideal gas, $C_P = (5/2)R$ and $C_V = (3/2)R$, is changed from $P_1 = 1$ bar and $V_1^t = 10$ m^3 to $P_2 = 10$ bar and $V_2^t = 1$ m^3 by the following mechanically reversible processes:

(*a*) Isothermal compression.

(*b*) Adiabatic compression followed by cooling at constant pressure.

(*c*) Adiabatic compression followed by cooling at constant volume.

(*d*) Heating at constant volume followed by cooling at constant pressure.

(*e*) Cooling at constant pressure followed by heating at constant volume.

Calculate Q, W, ΔU^t, and ΔH^t for each of these processes, and sketch the paths of all processes on a single PV diagram.

3.6 A rigid, nonconducting tank with a volume of 4 m^3 is divided into two equal parts by a thin membrane. On one side of the membrane the tank contains nitrogen gas at 5 bar and 80°C, and the other side is a perfect vacuum. The membrane ruptures and the gas fills the tank. What is the final temperature of the gas? How much work is done? Is the process reversible? Describe a reversible process by which the gas can be returned to its initial state. How much work is done? Assume nitrogen an ideal gas for which $C_P = (7/2)R$ and $C_V = (5/2)R$.

3.7 An ideal gas, $C_P = (7/2)R$ and $C_V = (5/2)R$, undergoes the following mechanically reversible changes in a series of nonflow processes:

(*a*) From an initial state of 104(°F) [40°C] and 21.75(psia) [150 kPa], it is compressed adiabatically to 87(psia) [600 kPa].

(*b*) It is then cooled to 104(°F) [40°C] at a constant pressure of 87(psia) [600 kPa].

(*c*) Finally, the gas is expanded isothermally to its original state.

Calculate Q, W, ΔU, and ΔH for each of the three processes and for the cycle.

Repeat these calculations for exactly the same changes of state accomplished irreversibly with an efficiency for each process of 80 percent compared with the corresponding mechanically reversible process.

3.8 One cubic meter of an ideal gas at 500 K and 2,000 kPa expands to ten times its initial volume as follows:

(*a*) By a mechanically reversible, isothermal process.

(*b*) By a mechanically reversible, adiabatic process.

(*c*) By an adiabatic, irreversible process in which expansion is against a restraining pressure of 100 kPa.

For each case calculate the final temperature, pressure, and the work done by the gas. $C_p =$ 21 J mol^{-1} K^{-1}.

3.9 A perfectly insulated, rigid cylinder of 0.5-m^3 volume is divided in half by a weightless, frictionless piston of high thermal conductivity that is initially held in place by latches. An ideal gas at 100 kPa and 300 K is on one side of the piston and the same ideal gas at 900 kPa and 300 K is on the other.

(*a*) What are the final equilibrium temperature and pressure after release of the piston?

(*b*) Suppose a rod attached to the piston extends through an end of the cylinder and acts against a constant resisting force equivalent to 100 kPa. What are the final equilibrium T and P upon release of the piston if $C_V = (5/2)R$?

3.10 One pound mole of air, initially at 248(°F) [120°C] and 8(atm) [8.11 bar], undergoes the following mechanically reversible changes. It expands isothermally to a pressure such that when it is cooled at constant volume to 68(°F) [20°C] its final pressure is 3(atm) [3.04 bar]. If air is assumed an ideal gas for which $C_P = (7/2)R$ and $C_V = (5/2)R$, calculate W, Q, ΔU, and ΔH.

3.11 An ideal gas is flowing in steady state through a horizontal tube. No heat is added and no shaft work is done. The cross-sectional area of the tube changes with length, and this causes the velocity to change. Derive an equation relating the temperature to the velocity of the gas. If nitrogen at 140°C flows past one section of the tube at a velocity of 2 m s^{-1}, what is its temperature at another section where its velocity is 40 m s^{-1}? $C_P = (7/2)R$.

3.12 One mole of an ideal gas, initially at 40°C and 1 bar, is changed to 120°C and 15 bar by three different mechanically reversible processes:

(*a*) The gas is first heated at constant volume until its temperature is 120°C; then it is compressed isothermally until its pressure is 15 bar.

(*b*) The gas is first heated at constant pressure until its temperature is 120°C; then it is compressed isothermally to 15 bar.

(*c*) The gas is first compressed isothermally to 15 bar; then it is heated at constant pressure to 120°C.

Calculate Q, W, ΔU, and ΔH in each case. Take $C_P = (7/2)R$ and $C_V = (5/2)R$. Repeat, with $C_P = (5/2)R$ and $C_V = (3/2)R$.

3.13 One mole of an ideal gas, initially at 20°C and 1 bar, undergoes the following mechanically reversible changes. It is compressed isothermally to a point such that when it is heated at constant volume to 100°C its final pressure is 10 bar. Calculate Q, W, ΔU, and ΔH for the process. Take $C_P = (7/2)R$ and $C_V = (5/2)R$.

3.14 Figure P3.14 depicts two mechanically reversible processes undergone by 1 mol of an ideal gas. Curves T_a and T_b are isotherms, paths 2-3 and 5-6 are isobars, and paths 3-1 and 6-4 are at constant volume. Show that W and Q are the same for processes 1-2-3-1 and 4-5-6-4.

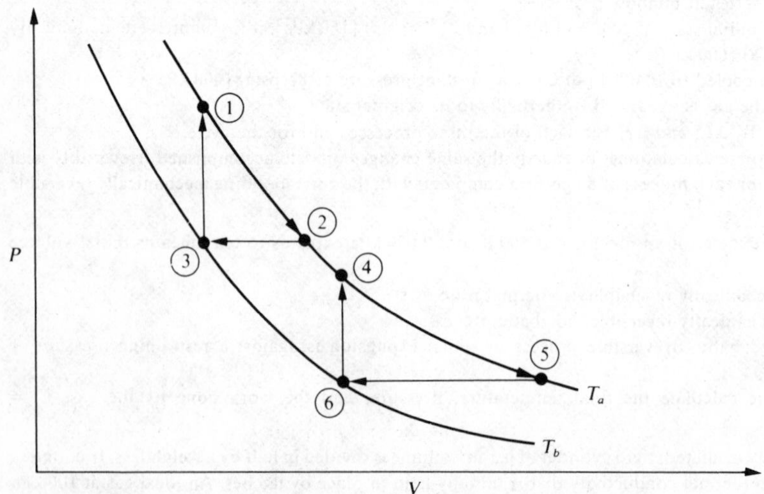

Figure P3.14

3.15 A particular quantity of an ideal gas $[C_V = (5/2)R]$ undergoes the following mechanically reversible steps that together form a cycle. The gas, initially at 1 bar and 300 K, is compressed isothermally to 3 bar. It is then heated at constant P to a temperature of 900 K. Finally, it is cooled at constant volume to its initial state with the extraction of 1,300 J as heat. Determine Q and W for each step of the cycle and for the complete cycle.

3.16 An existing process consists of two steps:
(a) One mole of air at $T_1 = 900$ K and $P_1 = 3$ bar is cooled at constant volume to $T_2 = 300$ K.
(b) The air is then heated at constant pressure until its temperature reaches 900 K.
It is proposed to replace this two-step process by a single isothermal expansion of the air from 900 K and 3 bar to some final pressure P. What is the value of P that makes the work of the proposed process equal to that of the existing process? Assume mechanical reversibility and treat air as an ideal gas with $C_P = (7/2)R$ and $C_V = (5/2)R$.

3.17 Derive an equation for the work of mechanically reversible, isothermal compression of 1 mol of a gas from an initial volume V_1 to a final volume V_2 when the equation of state is

$$P(V - b) = RT$$

where b is a positive constant.

Derive an equation for the work of mechanically reversible, isothermal compression of 1 mol of a gas from an initial pressure P_1 to a final pressure P_2 when the equation of state is the virial expansion [Eq. (3.10)] truncated to

$$Z = 1 + B'P$$

How do these two results compare with the corresponding equations for an ideal gas?

3.18 A substance for which κ is a constant undergoes an isothermal, mechanically reversible process from initial state (P_1, V_1) to final state (P_2, V_2), where V is molar volume.

(a) Starting with the definition of κ, show that the path of the process is described by

$$V = A \exp(-\kappa P)$$

where A depends on T only.

(b) Determine an exact expression which gives the isothermal work done on 1 mol of this constant-κ substance when κ and the initial and final pressures and molar volumes are known.

3.19 An empirical equation, $PV^\delta = \text{const}$, where δ is a constant, is sometimes used to relate P and V for *any* mechanically reversible process. Assuming the validity of this equation for an ideal gas, show that

$$W = \frac{RT_1}{\delta - 1}\left[1 - \left(\frac{P_2}{P_1}\right)^{(\delta-1)/\delta}\right]$$

If the process is isothermal, $\delta = 1$. Show that this equation reduces in this case to the isothermal-work equation,

$$W = RT \ln \frac{P_1}{P_2}$$

3.20 For methyl chloride at 125°C the virial coefficients are

$$B = -207.5 \text{ cm}^3 \text{ mol}^{-1}$$

$$C = 18{,}200 \text{ cm}^6 \text{ mol}^{-2}$$

Calculate the work of mechanically reversible, isothermal compression of 1 mol of methyl chloride from 1 bar to 60 bar at 125°C. Base calculations on the following forms of the virial equation:

(a) $$Z = 1 + \frac{B}{V} + \frac{C}{V^2}$$

(b) $$Z = 1 + B'P + C'P^2$$

where

$$B' = \frac{B}{RT} \quad \text{and} \quad C' = \frac{C - B^2}{(RT)^2}$$

Why don't both equations give exactly the same result?

3.21 Calculate Z and V for methanol vapor at 200°C and 10 bar by the following equations:
(*a*) The truncated virial equation (3.33) with the following experimental values of virial coefficients:

$$B = -219 \text{ cm}^3 \text{ mol}^{-1} \qquad C = -17{,}300 \text{ cm}^6 \text{ mol}^{-2}$$

(*b*) The truncated virial equation (3.31), with a value of B from the generalized Pitzer correlation.
(*c*) The Redlich/Kwong equation, with estimates of a and b from Eqs. (3.40) and (3.41).

3.22 Calculate Z and V for ethane at 50°C and 12 bar by the following equations:
(*a*) The truncated virial equation (3.33) with the following experimental values of virial coefficients:

$$B = -156.7 \text{ cm}^3 \text{ mol}^{-1} \qquad C = 9{,}650 \text{ cm}^6 \text{ mol}^{-2}$$

(*b*) The truncated virial equation (3.31), with a value of B from the generalized Pitzer correlation.
(*c*) The Redlich/Kwong equation, with estimates of a and b from Eqs. (3.40) and (3.41).

3.23 Calculate Z and V for sulfur hexafluoride at 100°C and 15 bar by the following equations:
(*a*) The truncated virial equation (3.33) with the following experimental values of virial coefficients:

$$B = -163.4 \text{ cm}^3 \text{ mol}^{-1} \qquad C = 12{,}120 \text{ cm}^6 \text{ mol}^{-2}$$

(*b*) The truncated virial equation (3.31), with a value of B from the generalized Pitzer correlation.
(*c*) The Redlich/Kwong equation, with estimates of a and b from Eqs. (3.40) and (3.41).
For sulfur hexafluoride, $T_c = 318$ K, $P_c = 37.6$ bar, $V_c = 198$ cm^3 mol^{-1}, and $\omega = 0.286$.

3.24 Determine Z and V for steam at 250°C and 2,000 kPa by the following:
(*a*) The truncated virial equation (3.33) with the following experimental values of virial coefficients:

$$B = -152.5 \text{ cm}^3 \text{ mol}^{-1} \qquad C = -5{,}800 \text{ cm}^6 \text{ mol}^{-2}$$

(*b*) The truncated virial equation (3.31), with a value of B from the generalized Pitzer correlation.
(*c*) The steam tables.

3.25 Calculate the molar volume of saturated liquid and the molar volume of saturated vapor by the Redlich/Kwong equation for one of the following and compare results with values found by suitable generalized correlations.

(*a*) Propane at 40°C where $P^{sat} = 13.71$ bar.
(*b*) Propane at 50°C where $P^{sat} = 17.16$ bar.
(*c*) Propane at 60°C where $P^{sat} = 21.22$ bar.
(*d*) Propane at 70°C where $P^{sat} = 25.94$ bar.
(*e*) *n*-Butane at 100°C where $P^{sat} = 15.41$ bar.
(*f*) *n*-Butane at 110°C where $P^{sat} = 18.66$ bar.
(*g*) *n*-Butane at 120°C where $P^{sat} = 22.38$ bar.
(*h*) *n*-Butane at 130°C where $P^{sat} = 26.59$ bar.
(*i*) Isobutane at 90°C where $P^{sat} = 16.54$ bar.
(*j*) Isobutane at 100°C where $P^{sat} = 20.03$ bar.
(*k*) Isobutane at 110°C where $P^{sat} = 24.01$ bar.
(*l*) Isobutane at 120°C where $P^{sat} = 28.53$ bar.
(*m*) Chlorine at 60°C where $P^{sat} = 18.21$ bar.
(*n*) Chlorine at 70°C where $P^{sat} = 22.49$ bar.
(*o*) Chlorine at 80°C where $P^{sat} = 27.43$ bar.
(*p*) Chlorine at 90°C where $P^{sat} = 33.08$ bar.
(*q*) Sulfur dioxide at 80°C where $P^{sat} = 18.66$ bar.
(*r*) Sulfur dioxide at 90°C where $P^{sat} = 23.31$ bar.
(*s*) Sulfur dioxide at 100°C where $P^{sat} = 28.74$ bar.
(*t*) Sulfur dioxide at 110°C where $P^{sat} = 35.01$ bar.

3.26 Calculate the following:
(*a*) The volume occupied by 20 kg of ethane at 50°C and 30 bar.
(*b*) The mass of ethane contained in a 0.3-m^3 cylinder at 60°C and 130 bar.

3.27 To a good approximation, what is the molar volume of ethanol vapor at 900(°F) [482.22°C] and 900(psia) [6,206 kPa]? How does this result compare with the ideal-gas value?

3.28 A 0.4-m^3 vessel is used to store liquid propane at its vapor pressure. Safety considerations dictate that at a temperature of 320 K the liquid must occupy no more than 75 percent of the total volume of the vessel. For these conditions, determine the mass of vapor and the mass of liquid in the vessel. At 320 K the vapor pressure of propane is 16.0 bar.

3.29 A 1,000-$(ft)^3$ [28.32-m^3] tank contains 500$(ft)^3$ [14.16 m^3] of liquid *n*-butane in equilibrium with its vapor at 77(°F) [25°C]. Determine a good estimate of the mass of *n*-butane vapor in the tank. The vapor pressure of *n*-butane at the given temperature is 2.40(atm) [2.43 bar].

3.30 Calculate the mass of ethane contained in a 0.5-$(ft)^3$ [0.0142-m^3] vessel at 140(°F) [60°C] and 2,000(psia) [13,790 kPa].

If 10(lb_m) [4.54 kg] of ethane is contained in a 0.5-$(ft)^3$ [0.0142-m^3] vessel, at what temperature does it exert a pressure of 3,000(psia) [20,480 kPa]?

3.31 To what pressure does one fill a 0.1-m^3 vessel at 25°C in order to store 25 kg of ethylene in it?

3.32 If 1 kg of water in a 0.03-m^3 container is heated to 450°C what pressure is developed?

3.33 A 0.3-m^3 vessel holds ethane vapor at 18°C and 2,500 kPa. If it is heated to 200°C, what pressure is developed?

3.34 What is the pressure in a 0.45-m^3 vessel when it is charged with 8 kg of carbon dioxide at 40°C?

3.35 A rigid vessel, filled to one-half its volume with liquid nitrogen at its normal boiling point (−195.8°C), is allowed to warm to 25°C. What pressure is developed? The molar volume of liquid nitrogen at its normal boiling point is 34.7 $cm^3\ mol^{-1}$.

3.36 The specific volume of isobutane liquid at 300 K and 4 bar is 1.824 $cm^3\ g^{-1}$. Estimate the specific volume at 400 K and 60 bar.

3.37 The density of liquid *n*-pentane is 0.630 g cm^{-3} at 18°C and 1 bar. Estimate its density at 150°C and 100 bar.

3.38 Estimate the density of liquid ethanol at 190°C and 190 bar.

3.39 Estimate the volume change of vaporization for ammonia at 20°C. At this temperature the vapor pressure of ammonia is 857 kPa.

3.40 *PVT* data may be taken by the following procedure. A mass m of a substance of molar mass M is introduced into a thermostated vessel of known total volume V^t. The system is allowed to equilibrate, and the temperature T and pressure P are measured.

(*a*) Approximately what percentage errors are allowable in the measured variables (m, M, V^t, T, and P) if the maximum allowable error in the calculated compressibility factor Z is ±1 percent?

(*b*) Approximately what percentage errors are allowable in the measured variables if the maximum allowable error in calculated values of the second virial coefficient B is ±1 percent? Assume that $Z \simeq 0.9$ and that values of B are calculated by Eq. (3.32).

3.41 For a gas described by the Redlich/Kwong equation [Eq. (3.35)] and for a temperature greater than T_c, develop expressions for the two limiting slopes,

$$\lim_{P\to 0}\left(\frac{\partial Z}{\partial P}\right)_T \qquad \lim_{P\to\infty}\left(\frac{\partial Z}{\partial P}\right)_T$$

The expressions should contain the temperature T and the Redlich/Kwong parameters a and/or b. Note that in the limit as $P \to 0$, $V = \infty$, and that in the limit as $P \to \infty$, $V = b$.

3.42 One mole of an ideal gas with constant heat capacities undergoes an arbitrary mechanically reversible process. Show that

$$\Delta U = \frac{1}{\gamma - 1}\,\Delta(PV)$$

3.43 The *PVT* behavior of a certain gas is described by the equation of state

$$P(V-b) = RT$$

where b is a constant. If in addition C_V is constant, show that

(*a*) U is a function of T only.
(*b*) $\gamma = \text{const.}$
(*c*) For a mechanically reversible adiabatic process, $P(V-b)^{\gamma} = \text{const.}$

3.44 A certain gas is described by the equation of state

$$PV = RT + \left(b - \frac{\theta}{RT}\right)P$$

Here, b is a constant and θ is a function of T only. For this gas, determine expressions for the isothermal compressibility κ and the thermal pressure coefficient $(\partial P/\partial T)_V$. These expressions should contain only T, P, θ, $d\theta/dT$, and constants.

3.45 Methane gas is stored in a 0.1-m^3 tank at 1,500 kPa and 25°C. Gas is allowed to flow from the tank through a partially opened valve into a gas holder where the pressure is constant at 115 kPa. When the pressure in the tank has dropped to 750 kPa, calculate:

(*a*) The mass of methane in the gas holder if the process takes place slowly enough that the temperature is constant.
(*b*) The mass of methane in the gas holder and its temperature if the process occurs so rapidly that heat transfer is negligible, i.e., there is no heat transfer either between parts of the system or between the system and the surroundings.
(*c*) Would the answers to part (*b*) be different if the pressure in the gas holder were 300 kPa? Assume that methane is an ideal gas for which $\gamma = 1.31$.

CHAPTER

FOUR

HEAT EFFECTS

Heat transfer is one of the fundamental operations of the chemical industry. Consider, for example, the manufacture of ethylene glycol (an antifreeze agent) by the oxidation of ethylene to ethylene oxide and its subsequent hydration to glycol. The catalytic oxidation process is most effective when carried out at temperatures in the neighborhood of 250°C. Therefore the reactants, ethylene and air, are heated to this temperature before they enter the reactor, and design of the preheater requires calculation of the heat required. The reactions of ethylene with oxygen in the catalyst bed are combustion processes that tend to raise the temperature. However, heat is removed from the reactor, and the temperature does not rise much above 250°C. Higher temperatures promote the production of CO_2, an undesired product. Design of the reactor requires knowledge of the amount of heat that must be transferred, and this is determined by the heat effects associated with the chemical reactions. The ethylene oxide formed is hydrated to glycol by absorption in water. This is accompanied by evolution of heat as a result of the phase change, the formation of a solution, and the hydration reaction between the dissolved ethylene oxide and water. Finally, the glycol is recovered from the water by distillation, a process requiring vaporization of a liquid and resulting in the separation of a solution into its components.

All of the important heat effects are illustrated by this relatively simple chemical manufacturing process. In contrast to *sensible* heat effects, which are characterized by temperature changes, the heat effects of chemical reaction, phase transition, and the formation and separation of solutions are determined from experimental measurements made at constant temperature. In this chapter we apply thermodynamics to the evaluation of most of the heat effects that accompany physical and chemical operations. However, the heat effects of mixing processes, which depend on the thermodynamic properties of mixtures, are treated in Chap. 13.

4.1 SENSIBLE HEAT EFFECTS

Heat transfer to a system in which there are no phase transitions, no chemical reactions, and no changes in composition causes the temperature of the system to change. Our object here is to develop relations between the quantity of heat transferred and the resulting temperature change.

When the system is a homogeneous substance of constant composition, the phase rule indicates that fixing the values of two intensive properties establishes its state. The molar or specific enthalpy of a substance may therefore be expressed as a *function* of two other state variables. Arbitrarily selecting these as temperature and pressure, we write

$$H = H(T, P)$$

whence

$$dH = \left(\frac{\partial H}{\partial T}\right)_P dT + \left(\frac{\partial H}{\partial P}\right)_T dP$$

As a result of Eq. (2.21) this becomes

$$dH = C_P\, dT + \left(\frac{\partial H}{\partial P}\right)_T dP$$

The final term may be set equal to zero in two circumstances:

1. For any constant-pressure process, regardless of the substance.
2. Whenever the enthalpy of the substance is independent of pressure, regardless of the process. This is exactly true for ideal gases and approximately true for low-pressure gases, for solids, and for liquids outside the critical region.

In either case,

$$dH = C_P\, dT$$

and

$$\Delta H = \int_{T_1}^{T_2} C_P\, dT \tag{4.1}$$

Moreover, $Q = \Delta H$ both for mechanically reversible, constant-pressure, nonflow processes [Eq. (2.19)] and for the transfer of heat in steady-flow exchangers where ΔE_P and ΔE_K are negligible and $W_s = 0$.

Similarly, we may express the molar or specific internal energy as a function of temperature and molar or specific volume:

$$U = U(T, V)$$

whence

$$dU = \left(\frac{\partial U}{\partial T}\right)_V dT + \left(\frac{\partial U}{\partial V}\right)_T dV$$

As a result of Eq. (2.20) this becomes

$$dU = C_V\, dT + \left(\frac{\partial U}{\partial V}\right)_T dV$$

Again, there are two instances for which the final term may be set equal to zero:

1. For any constant-volume process, regardless of substance.
2. Whenever the internal energy is independent of volume, regardless of the process. This is exactly true for ideal gases and incompressible fluids.

In either case,

$$dU = C_V\, dT$$

and

$$\Delta U = \int_{T_1}^{T_2} C_V\, dT \tag{4.2}$$

The integrals of Eqs. (4.1) and (4.2), particularly the former, must frequently be evaluated. The most common direct engineering application is to steady-flow heat transfer where the equation

$$Q = \Delta H = \int_{T_1}^{T_2} C_P\, dT \tag{4.3}$$

often applies. In general, integration requires knowledge of the temperature dependence of the heat capacity.

As shown in Chap. 6, *ideal-gas heat capacities*, rather than the actual heat capacities of gases, are used in the evaluation of thermodynamic properties such as internal energy and enthalpy. The reason is that thermodynamic-property evaluation is conveniently accomplished in two steps: first, calculation of ideal-gas values from ideal-gas heat capacities; second, calculation from PVT data of the differences between real-gas and ideal-gas values. A real gas becomes ideal in the limit as $P \to 0$; if it were to remain ideal when compressed to a finite pressure, its state would remain that of an ideal-gas. Gases in these hypothetical *ideal-gas states* have properties that reflect their individuality just as do real gases. Ideal-gas heat capacities (designated by C_P^{ig} and C_V^{ig}) are therefore different for different gases; although functions of temperature, they are independent of pressure.

The temperature dependence may be shown graphically, as illustrated in Fig. 4.1, where C_P^{ig}/R is plotted vs. temperature for argon, nitrogen, water, and carbon dioxide. More commonly, however, temperature dependence is given by an empirical equation; the two simplest expressions of practical value are

$$\frac{C_P^{ig}}{R} = \alpha + \beta T + \gamma T^2$$

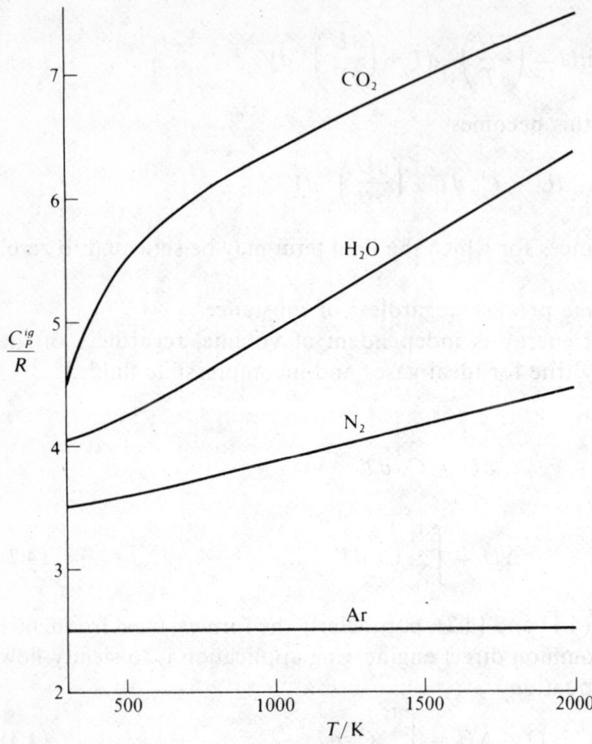

Figure 4.1 Ideal-gas heat capacities of argon, nitrogen, water, and carbon dioxide as functions of temperature.

and

$$\frac{C_P^{ig}}{R} = a + bT + cT^{-2}$$

where α, β, and γ and a, b, and c are constants characteristic of the particular gas. With the exception of the last term, these equations are of the same form. We therefore combine them to provide a single expression:

$$\frac{C_P^{ig}}{R} = A + BT + CT^2 + DT^{-2} \tag{4.4}$$

where either C or D is zero, depending on the gas considered. Since the ratio C_P^{ig}/R is dimensionless, the units of C_P^{ig} are governed by the choice of R. Values of the constants are given in Table 4.1 for a number of common organic and inorganic gases. More accurate but more complex equations are found in the literature.†

† See C. A. Passut and R. P. Danner, *Ind. Eng. Chem. Proc. Des. Dev.*, **11**: 543, 1972; P. K. Huang and T. E. Daubert, Ibid., **13**: 193, 1974.

Table 4.1 Heat capacities of gases in the ideal-gas state†

Constants for the equation $C_P^{ig}/R = A + BT + CT^2 + DT^{-2}$ $\quad T$ (kelvins) from 298 K to T_{max}

Chemical species		T_{max}	A	10^3B	10^6C	$10^{-5}D$
Paraffins:						
Methane	CH_4	1,500	1.702	9.081	−2.164	
Ethane	C_2H_6	1,500	1.131	19.225	−5.561	
Propane	C_3H_8	1,500	1.213	28.785	−8.824	
n-Butane	C_4H_{10}	1,500	1.935	36.915	−11.402	
iso-Butane	C_4H_{10}	1,500	1.677	37.853	−11.945	
n-Pentane	C_5H_{12}	1,500	2.464	45.351	−14.111	
n-Hexane	C_6H_{14}	1,500	3.025	53.722	−16.791	
n-Heptane	C_7H_{16}	1,500	3.570	62.127	−19.486	
n-Octane	C_8H_{18}	1,500	8.163	70.567	−22.208	
1-Alkenes:						
Ethylene	C_2H_4	1,500	1.424	14.394	−4.392	
Propylene	C_3H_6	1,500	1.637	22.706	−6.915	
1-Butene	C_4H_8	1,500	1.967	31.630	−9.873	
1-Pentene	C_5H_{10}	1,500	2.691	39.753	−12.447	
1-Hexene	C_6H_{12}	1,500	3.220	48.189	−15.157	
1-Heptene	C_7H_{14}	1,500	3.768	56.588	−17.847	
1-Octene	C_8H_{16}	1,500	4.324	64.960	−20.521	
Miscellaneous organics:						
Acetaldehyde	C_2H_4O	1,000	1.693	17.978	−6.158	
Acetylene	C_2H_2	1,500	6.132	1.952	......	−1.299
Benzene	C_6H_6	1,500	−0.206	39.064	−13.301	
1,3-Butadiene	C_4H_6	1,500	2.734	26.786	−8.882	
Cyclohexane	C_6H_{12}	1,500	−3.876	63.249	−20.928	
Ethanol	C_2H_6O	1,500	3.518	20.001	−6.002	
Ethylbenzene	C_8H_{10}	1,500	1.124	55.380	−18.476	
Ethylene oxide	C_2H_4O	1,000	−0.385	23.463	−9.296	
Formaldehyde	CH_2O	1,500	2.264	7.022	−1.877	
Methanol	CH_4O	1,500	2.211	12.216	−3.450	
Toluene	C_7H_8	1,500	0.290	47.052	−15.716	
Styrene	C_8H_8	1,500	2.050	50.192	−16.662	
Miscellaneous inorganics						
Air		2,000	3.355	0.575	......	−0.016
Ammonia	NH_3	1,800	3.578	3.020	......	−0.186
Bromine	Br_2	3,000	4.493	0.056	......	−0.154
Carbon monoxide	CO	2,500	3.376	0.557	......	−0.031
Carbon dioxide	CO_2	2,000	5.457	1.045	......	−1.157
Carbon disulfide	CS_2	1,800	6.311	0.805	......	−0.906
Chlorine	Cl_2	3,000	4.442	0.089	......	−0.344
Hydrogen	H_2	3,000	3.249	0.422	......	0.083
Hydrogen sulfide	H_2S	2,300	3.931	1.490	......	−0.232
Hydrogen chloride	HCl	2,000	3.156	0.623	......	0.151
Hydrogen cyanide	HCN	2,500	4.736	1.359	......	−0.725
Nitrogen	N_2	2,000	3.280	0.593	......	0.040
Dinitrogen oxide	N_2O	2,000	5.328	1.214	......	−0.928
Nitric oxide	NO	2,000	3.387	0.629	......	0.014
Nitrogen dioxide	NO_2	2,000	4.982	1.195	......	−0.792
Dinitrogen tetroxide	N_2O_4	2,000	11.660	2.257	......	−2.787
Oxygen	O_2	2,000	3.639	0.506	......	−0.227
Sulfur dioxide	SO_2	2,000	5.699	0.801	......	−1.015
Sulfur trioxide	SO_3	2,000	8.060	1.056	......	−2.028
Water	H_2O	2,000	3.470	1.450	......	0.121

† Selected from H. M. Spencer, *Ind. Eng. Chem.*, **40**: 2152, 1948; K. K. Kelley, *U.S. Bur. Mines Bull.*, 584, 1960; L. B. Pankratz, *U.S. Bur. Mines Bull.*, 672, 1982.

As a result of Eq. (3.17), the two ideal-gas heat capacities are related:

$$\frac{C_V^{ig}}{R} = \frac{C_P^{ig}}{R} - 1 \tag{4.5}$$

Thus the temperature dependence of C_V^{ig}/R is readily found from the equation for C_P^{ig}/R.

The effects of temperature on C_P^{ig} or C_V^{ig} are determined by experiment, most often from spectroscopic data and knowledge of molecular structure by the methods of statistical mechanics. Where experimental data are not available, methods of estimation are employed, as described by Reid, Prausnitz, and Sherwood.† Ideal-gas heat capacities increase smoothly with increasing temperature toward an upper limit, which is reached when all translational, rotational, and vibrational modes of molecular motion are fully excited.

Although ideal-gas heat capacities are exactly correct for real gases only at zero pressure, real gases rarely depart significantly from ideality up to several bars, and therefore C_P^{ig} and C_V^{ig} are usually good approximations for the heat capacities of real gases at low pressures.

Example 4.1 The constants in Table 4.1 require use of Kelvin temperatures in Eq. (4.4). Equations of the same form may also be developed for use with temperatures in °C, (R), and (°F), but the constants are different. The molar heat capacity of methane in the ideal-gas state is given in Table 4.1 as

$$\frac{C_P^{ig}}{R} = 1.702 + 9.081 \times 10^{-3}T - 2.164 \times 10^{-6}T^2$$

where T is in kelvins. Develop an equation for C_P^{ig}/R for temperatures in °C.

SOLUTION The relation between the two temperature scales is

$$T\,\text{K} = t\,°\text{C} + 273.15$$

Therefore

$$\frac{C_P^{ig}}{R} = 1.702 + 9.081 \times 10^{-3}(t + 273.15) - 2.164 \times 10^{-6}(t + 273.15)^2$$

or

$$\frac{C_P^{ig}}{R} = 4.021 + 7.899 \times 10^{-3}t - 2.164 \times 10^{-6}t^2$$

Example 4.2 Calculate the heat required to raise the temperature of 1 mol of methane from 260 to 600°C in a flow process at a pressure of approximately 1 bar.

SOLUTION For the application of Eq. (4.3) when the expression for C_P^{ig}/R is from Table 4.1, we need temperatures in kelvins:

$$T_1 = 533.15\ \text{K} \qquad T_2 = 873.15\ \text{K}$$

† R. C. Reid, J. M. Prausnitz, and T. K. Sherwood, *The Properties of Gases and Liquids*, 3d ed., chap. 7, McGraw-Hill, New York, 1977.

Then

$$Q = R\int_{533.15}^{873.15}(1.702 + 9.081\times10^{-3}T - 2.164\times10^{-6}T^2)\,dT$$

Integration gives

$$Q = 2{,}378.8\ \mathrm{R} = (2{,}378.8)(8.314) = 19{,}780\ \mathrm{J}$$

The same result is obtained when Eq. (4.3) is applied with the equation developed in Example 4.1 for C_P^{ig}/R with temperatures in °C.

As a matter of convenience, we define a *mean heat capacity*:

$$C_{P_{\mathrm{mh}}} \equiv \frac{\int_{T_1}^{T_2} C_P\,dT}{T_2 - T_1} \tag{4.6}$$

The subscript "mh" denotes a mean value specific to enthalpy calculations, and distinguishes this mean heat capacity from a similar quantity introduced in the next chapter.

When Eq. (4.4), written not just for an ideal gas but in general, is substituted for C_P in Eq. (4.6), integration gives:

$$C_{P_{\mathrm{mh}}}/R = A + BT_{\mathrm{am}} + \frac{C}{3}(4T_{\mathrm{am}}^2 - T_1T_2) + \frac{D}{T_1T_2} \tag{4.7}$$

where $T_{\mathrm{am}} \equiv (T_1 + T_2)/2$ is the arithmetic-mean temperature. Thus the integration required for evaluation of enthalpy changes has been accomplished, and ΔH is given by

$$\Delta H = C_{P_{\mathrm{mh}}}(T_2 - T_1) \tag{4.8}$$

a result that follows from Eqs. (4.1) and (4.6).

Example 4.3 Rework Example 4.2, applying Eq. (4.7).

SOLUTION With values of the constants taken from Table 4.1 and with

$$T_{\mathrm{am}} = (533.15 + 873.15)/2 = 703.15\ \mathrm{K}$$

Eq. (4.7) becomes

$$\frac{C_{P_{\mathrm{mh}}}^{ig}}{R} = 1.702 + (9.081\times10^{-3})(703.15)$$

$$- \frac{2.164\times10^{-6}}{3}[(4)(703.15)^2 - (873.15)(533.15)]$$

$$= 6.997$$

By Eqs. (4.3) and (4.8)

$$Q = \Delta H = (6.997)(8.314)(873.15 - 533.15) = 19{,}780\ \mathrm{J}$$

Given T_1 and T_2, the calculation of Q or ΔH is straightforward. Less direct is the calculation of T_2, given T_1 and Q or ΔH. Here, an iteration scheme is useful. Solving Eq. (4.8) for T_2 gives

$$T_2 = \frac{\Delta H}{C_{P_{\text{mh}}}} + T_1 \tag{4.9}$$

One assumes an initial value of T_2 for purposes of calculating $C_{P_{\text{mh}}}$ by Eq. (4.7). Substitution of the resulting value into Eq. (4.9) provides a new value of T_2 from which to reevaluate $C_{P_{\text{mh}}}$. Iteration continues in like fashion to convergence on a final value of T_2.

Example 4.4 What is the final temperature when 0.4×10^6(Btu) are added to 25(lb mol) of ammonia initially at 500(°F) in a steady-flow process at approximately 1(atm)?

SOLUTION If ΔH is the enthalpy change for 1(lb mol), $Q = n\,\Delta H$, and

$$\Delta H = \frac{Q}{n} = \frac{0.4 \times 10^6}{25} = 16{,}000(\text{Btu})(\text{lb mol})^{-1}$$

In Eq. (4.9), if ΔH is in (Btu)(lb mol)$^{-1}$ and $C_{P_{\text{mh}}}$ is in (Btu)(lb mol)$^{-1}$(R)$^{-1}$, the first term on the right has units of (R). On the other hand, T_1 and T_2 are most conveniently expressed in kelvins. We therefore write Eq. (4.9) as

$$T_2 = \frac{\Delta H}{1.8 C^{ig}_{P_{\text{mh}}}} + T_1 \tag{A}$$

where the divisor 1.8 changes the units of the term to kelvins.

Substituting the constants for ammonia from Table 4.1 into Eq. (4.7), we get

$$C^{ig}_{P_{\text{mh}}} = R\left(3.578 + 3.020 \times 10^{-3} T_{\text{am}} - \frac{0.186 \times 10^5}{T_1 T_2}\right) \tag{B}$$

With

$$R = 1.986(\text{Btu})(\text{lb mol})^{-1}(\text{R})^{-1}$$

$$T_1 = \frac{500 + 459.67}{1.8} = 533.15 \text{ K}$$

and

$$T_{\text{am}} = \left(\frac{533.15 + T_2}{2}\right) \quad \text{K}$$

we may calculate $C^{ig}_{P_{\text{mh}}}$ for any value of T_2. Iteration between Eqs. (A) and (B) starts with a value $T_2 \geq T_1$, and converges on the final value,

$$T_2 = 1{,}250.10 \text{ K} \qquad \text{or} \qquad 1{,}790.51(°\text{F})$$

Gas mixtures of constant composition may be treated in exactly the same way as pure gases. An ideal gas, by definition, is a gas whose molecules have no influence on one another. This means that each gas exists in a mixture independent of the others, and that its properties are unaffected by the presence of different

Table 4.2 Heat capacities of solids†

Constants for the equation $C_P/R = A + BT + DT^{-2}$
T (kelvins) from 298 K to T_{max}

Chemical species	T_{max}	A	$10^3 B$	$10^{-5} D$
CaO	2,000	6.104	0.443	−1.047
$CaCO_3$	1,200	12.572	2.637	−3.120
$Ca(OH)_2$	700	9.597	5.435	
CaC_2	720	8.254	1.429	−1.042
$CaCl_2$	1,055	8.646	1.530	−0.302
C (graphite)	2,000	1.771	0.771	−0.867
Cu	1,357	2.677	0.815	0.035
CuO	1,400	5.780	0.973	−0.874
Fe(α)	1.043	−0.111	6.111	1.150
Fe_2O_3	960	11.812	9.697	−1.976
Fe_3O_4	850	9.594	27.112	0.409
FeS	411	2.612	13.286	
I_2	386.8	6.481	1.502	
NH_4Cl	458	5.939	16.105	
Na	371	1.988	4.688	
NaCl	1,073	5.526	1.963	
NaOH	566	0.121	16.316	1.948
$NaHCO_3$	400	5.128	18.148	
S (rhombic)	368.3	4.114	−1.728	−0.783
SiO_2 (quartz)	847	4.871	5.365	−1.001

† Selected from K. K. Kelley, *U.S. Bur. Mines Bull.* 584, 1960; L. B. Pankratz, *U.S. Bur. Mines Bull.* 672, 1982.

molecules. Thus one calculates the ideal-gas heat capacity of a gas mixture by taking the molar average of the heat capacities of the individual species. Consider 1 mol of gas mixture consisting of species A, B, and C, and let y_A, y_B, and y_C represent the mole fractions of these species. The molar heat capacity of the mixture in the ideal-gas state is given by

$$C^{ig}_{P_{\text{mixture}}} = y_A C^{ig}_{P_A} + y_B C^{ig}_{P_B} + y_C C^{ig}_{P_C} \tag{4.10}$$

where $C^{ig}_{P_A}$, $C^{ig}_{P_B}$, and $C^{ig}_{P_C}$ are the molar heat capacities of pure A, B, and C in the ideal-gas state.

As with gases, data for the heat capacities of solids and liquids come from experiment. The temperature dependence of C_P for solids and liquids can also be expressed by equations of the form of Eq. (4.4). Data for a few solids are given in Table 4.2, and for a few liquids, in Table 4.3. Data for specific heats (C_P on a unit-mass basis) of many solids and liquids are given by Perry and Green.†

† R. H. Perry and D. Green, *Perry's Chemical Engineers' Handbook*, 6th ed., sec. 3, McGraw-Hill, New York, 1984.

Table 4.3 Heat capacities of liquids†

Constants for the equation $C_P/R = A + BT + CT^2$
T from 273.15 to 373.15 K

Chemical species	A	$10^3 B$	$10^6 C$
Ammonia	22.626	−100.75	192.71
Aniline	15.819	29.03	−15.80
Benzene	−0.747	67.96	−37.78
1,3-Butadiene	22.711	−87.96	205.79
Carbon tetrachloride	21.155	−48.28	101.14
Chlorobenzene	11.278	32.86	−31.90
Chloroform	19.215	−42.89	83.01
Cyclohexane	−9.048	141.38	−161.62
Ethanol	33.866	−172.60	349.17
Ethylene oxide	21.039	−86.41	172.28
Methanol	13.431	−51.28	131.13
n-Propanol	41.653	−210.32	427.20
Sulfur trioxide	−2.930	137.08	−84.73
Toluene	15.133	6.79	16.35
Water	8.712	1.25	−0.18

† Based on correlations presented by J. W. Miller, Jr., G. R. Schorr, and C. L. Yaws, *Chem. Eng.*, **83**(23): 129, 1976.

4.2 HEAT EFFECTS ACCOMPANYING PHASE CHANGES OF PURE SUBSTANCES

When a pure substance is liquefied from the solid state or vaporized from the liquid at constant pressure, there is no change in temperature but there is a definite transfer of heat from the surroundings to the substance. These heat effects are commonly called the latent heat of fusion and the latent heat of vaporization. Similarly, there are heats of transition accompanying the change of a substance from one solid state to another; for example, the heat absorbed when rhombic crystalline sulfur changes to the monoclinic structure at 95°C and 1 bar is 360 J for each gram-atom.

The characteristic feature of all these processes is the coexistence of two phases. According to the phase rule, a two-phase system consisting of a single species is univariant, and its intensive state is determined by the specification of just one intensive property. Thus the latent heat accompanying a phase change is a function of temperature only, and is related to other system properties by an exact thermodynamic equation:

$$\boxed{\Delta H = T\,\Delta V \frac{dP^{\text{sat}}}{dT}} \qquad (4.11)$$

where, for a pure species at temperature T,

ΔH = latent heat

ΔV = volume change accompanying the phase change

P^{sat} = vapor pressure

The derivation of this equation, known as the Clapeyron equation, is given in Chap. 6.

When Eq. (4.11) is applied to the vaporization of a pure liquid, dP^{sat}/dT is the slope of the vapor pressure-vs.-temperature curve at the temperature of interest, ΔV is the difference between molar volumes of saturated vapor and saturated liquid, and ΔH is the latent heat of vaporization. Thus values of ΔH may be calculated from vapor-pressure and volumetric data.

Latent heats may also be measured calorimetrically. Experimental values are available at selected temperatures for many substances. For example, extensive lists are given by Perry and Green.† However, such data are frequently unavailable at the temperature of interest, and in many cases the data necessary for application of Eq. (4.11) are also not known. In this event approximate methods are used for estimates of the heat effect accompanying a phase change. Since heats of vaporization are by far the most important from a practical point of view, they have received most attention. The methods developed are for two purposes:

1. Prediction of the heat of vaporization at the normal boiling point.‡
2. Estimation of the heat of vaporization at any temperature from the known value at a single temperature.

A useful method for prediction of the heat of vaporization at the normal boiling point is the equation proposed by Riedel:§

$$\frac{\Delta H_n/T_n}{R} = \frac{1.092(\ln P_c - 1.013)}{0.930 - T_{r_n}} \tag{4.12}$$

where T_n = normal boiling point

ΔH_n = molar latent heat of vaporization at T_n

P_c = critical pressure, bar

T_{r_n} = reduced temperature at T_n

Since $\Delta H_n/T_n$ has the dimensions of the gas constant R, the units of this ratio are governed by the choice of units for R.

Eq. (4.12) is surprisingly accurate for an empirical expression; errors rarely exceed 5 percent. Applied to water it gives

$$\Delta H_n/T_n = R\left[\frac{1.092(\ln 220.5 - 1.013)}{0.930 - 0.577}\right] = 13.52R$$

† R. H. Perry and D. Green, op. cit., sec. 3.

‡ The convention with respect to the normal boiling point is that it refers to 1 standard atmosphere, defined as 101,325 Pa.

§ L. Riedel, *Chem. Ing. Tech.*, **26**: 679, 1954.

Taking $R = 8.314$ J mol^{-1} K^{-1} and the normal boiling point of water as 100°C or 373.15 K, we get

$$\Delta H_n = (13.52)(8.314)(373.15) = 41{,}940 \text{ J mol}^{-1}$$

This corresponds to 2,328 J g^{-1}, whereas the experimental value is 2,257 J g^{-1}; the error is 3.2 percent.

Estimates of the latent heat of vaporization of a pure liquid at any temperature from the known value at a single temperature may be based on a known experimental value or on a value estimated by Eq. (4.12). The method proposed by Watson† has found wide acceptance:

$$\frac{\Delta H_2}{\Delta H_1} = \left(\frac{1 - T_{r_2}}{1 - T_{r_1}}\right)^{0.38} \tag{4.13}$$

This equation is both simple and reliable; its use is illustrated in the following example.

Example 4.5 Given that the latent heat of vaporization of water at 100°C is 2,257 J g^{-1}, estimate the latent heat at 300°C.

SOLUTION Let

$$\Delta H_1 = \text{latent heat at 100°C} = 2{,}257 \text{ J g}^{-1}$$

$$\Delta H_2 = \text{latent heat at 300°C}$$

$$T_{r_1} = 373.15/647.1 = 0.577$$

$$T_{r_2} = 573.15/647.1 = 0.886$$

Then by Eq. (4.13),

$$\Delta H_2 = (2{,}257)\left(\frac{1 - 0.886}{1 - 0.577}\right)^{0.38} = (2{,}257)(0.270)^{0.38}$$

$$= 1{,}371 \text{ J g}^{-1}$$

The value given in the steam tables is 1,406 J g^{-1}.

4.3 THE STANDARD HEAT OF REACTION

The heat effects so far discussed have been for physical processes. Chemical reactions also are accompanied by the transfer of heat, by temperature changes during the course of reaction, or by both. These effects are manifestations of the differences in molecular structure, and therefore in energy, of the products and reactants. For example, the reactants in a combustion reaction possess greater

† K. M. Watson, *Ind. Eng. Chem.*, **35**: 398, 1943.

energy on account of their structure than do the products, and this energy must either be transferred to the surroundings as heat or result in products at an elevated temperature.

There are many different ways to carry out each of the vast number of possible chemical reactions, and each reaction carried out in a particular way is accompanied by a particular heat effect. Tabulation of all possible heat effects for all possible reactions is quite impossible. We therefore *calculate* the heat effects for other reactions from data for a particular kind of reaction carried out in a standard way. This reduces the required data to a minimum.

The amount of heat required for a specific chemical reaction depends on the temperatures of both the reactants and products. A consistent basis for treatment of reaction heat effects results when the *heat of reaction* is defined as the heat effect that results when all products and reactants are at the *same* temperature.

Consider the flow-calorimeter method for measurement of heats of combustion of fuel gases. The fuel is mixed with air at room temperature and ignited. Combustion takes place in a chamber surrounded by a cooling jacket through which water flows. In addition there is a long water-jacketed section in which the products of combustion are cooled to the temperature of the reactants. Whatever the details of this steady-flow process, the overall energy balance [Eq. (2.10)] reduces to

$$Q = \Delta H$$

No shaft work is produced by the process, and the calorimeter is built so that changes in potential and kinetic energy are negligible. Thus the heat Q absorbed by the water is identical with the enthalpy change caused by the combustion reaction, and universal practice is to designate the enthalpy change of reaction ΔH as the *heat of reaction.*

For purposes of data tabulation, we define the *standard* heat of the reaction,

$$aA + bB \rightarrow lL + mM$$

as the enthalpy change when a moles of A and b moles of B in their *standard states at temperature* T react to form l moles of L and m moles of M in their *standard states also at temperature* T. A *standard state* is the particular state of a species *at temperature* T defined by generally accepted reference conditions of pressure, composition, and physical state.

With respect to composition, the standard states used in this chapter are states of the *pure* species. For gases, the physical state is the ideal-gas state and for liquids and solids, the real state at the reference pressure and system temperature.

Historically, the reference pressure for the standard state, i.e., the *standard-state pressure,* has been 1 standard atmosphere (101,325 Pa), and most data tabulations are for this pressure. A change in the standard to 1 bar (10^5 Pa) is in progress, but for the purposes of this chapter, the change is of negligible consequence.

In summary, the standard states used in this chapter are:

1. *Gases.* The pure substance in the ideal-gas state at 1 bar or 1(atm).
2. *Liquids and solids.* The actual pure liquid or solid at 1 bar or 1(atm).

Property values in the standard state are denoted by the degree symbol (°). For example, C_P° is the standard-state heat capacity. Since the standard state for gases is the ideal-gas state, C_P° for gases is identical with C_P^{ig}, and the data of Table 4.1 apply to the standard state for gases. All conditions for a standard state are fixed except temperature, which is always the temperature of the system. Standard-state properties are therefore functions of temperature only.

The standard state chosen for gases is a hypothetical one, for at 1 bar or 1(atm) actual gases are not ideal. However, they seldom deviate much from ideality, and in most instances the ideal-gas state at 1 bar or 1(atm) may be regarded for practical purposes as the actual state of the gas at atmospheric pressure.

When a heat of reaction is given for a particular reaction, it applies for the stoichiometric coefficients as written. If each stoichiometric coefficient is doubled, the heat of reaction is doubled. For example, the ammonia-synthesis reaction may be written

$$\tfrac{1}{2}N_2 + \tfrac{3}{2}H_2 \rightarrow NH_3 \qquad \Delta H_{298}^\circ = -46{,}110\ \text{J}$$

or

$$N_2 + 3H_2 \rightarrow 2NH_3 \qquad \Delta H_{298}^\circ = -92{,}220\ \text{J}$$

The symbol ΔH_{298}° indicates that the heat of reaction is the *standard* value for a temperature of 298.15 K (25°C).

4.4 THE STANDARD HEAT OF FORMATION

The tabulation of data for just the *standard* heats of reaction for all of the vast number of possible reactions is impractical. Fortunately, the standard heat of any reaction can be calculated if the *standard heats of formation* of the compounds taking part in the reaction are known. A *formation* reaction is defined as a reaction which forms a single compound *from its constituent elements.* For example, the reaction $C + \frac{1}{2}O_2 + 2H_2 \rightarrow CH_3OH$ is the formation reaction for methanol. The reaction $H_2O + SO_3 \rightarrow H_2SO_4$ is *not* a formation reaction, because it forms sulfuric acid not from the elements but from other compounds. Formation reactions are always understood to result in the formation of 1 mol of the compound, and *the heat of formation is therefore based on 1 mol of the compound formed.*

Since heats of reaction at any temperature can be calculated from heat-capacity data if the value for one temperature is known, the tabulation of data can be reduced to the compilation of *standard heats of formation at a single temperature.* The usual choice for this temperature is 298.15 K or 25°C. The standard heat of formation of a compound at this temperature is represented by the symbol $\Delta H^\circ_{f_{298}}$. The superscript ° indicates that it is the standard value, the subscript f shows that it is a heat of formation, and the 298 is the approximate absolute temperature in kelvins. Tables of these values for common substances may be found in standard handbooks, but the most extensive compilations available are in specialized reference works.† An abridged list of values is given in Table 4.4.

When chemical equations are combined by addition, the standard heats of reaction may also be added to give the standard heat of the resulting reaction. This is possible because enthalpy is a property, and changes in it are independent of path. In particular, formation equations and standard heats of formation may always be combined to produce any desired equation (not itself a formation equation) and its accompanying standard heat of reaction. Equations written for this purpose often include an indication of the physical state of each reactant and product, i.e., the letter g, l, or s is placed in parentheses after the chemical formula to show whether it is a gas, a liquid, or a solid. This might seem unnecessary since a pure chemical species at a particular temperature and 1 bar or 1(atm) can usually exist only in one physical state. However, fictitious states are often assumed as a matter of convenience.

Consider the reaction $CO_2(g) + H_2(g) \rightarrow CO(g) + H_2O(g)$ at 25°C. This is a reaction commonly encountered in the chemical industry (the water-gas-shift reaction), though it takes place only at temperatures well above 25°C. However, the data used are for 25°C, and the initial step in any calculation of thermal effects concerned with this reaction is to evaluate the standard heat of reaction at 25°C. Since the reaction is actually carried out entirely in the gas phase at high temperature, convenience dictates that the standard states of all products and reactants at 25°C be taken as the ideal-gas state at 1 bar or 1(atm), even though water cannot actually exist as a gas at these conditions. The pertinent formation reactions are

$$CO_2(g):\quad C(s) + O_2(g) \rightarrow CO_2(g) \qquad \Delta H^\circ_{f_{298}} = -393{,}509 \text{ J}$$

$$H_2(g):\quad \text{Since hydrogen is an element} \qquad \Delta H^\circ_{f_{298}} = 0$$

$$CO(g):\quad C(s) + \tfrac{1}{2}O_2(g) \rightarrow CO(g) \qquad \Delta H^\circ_{f_{298}} = -110{,}525 \text{ J}$$

$$H_2O(g):\quad H_2(g) + \tfrac{1}{2}O_2(g) \rightarrow H_2O(g) \qquad \Delta H^\circ_{f_{298}} = -241{,}818 \text{ J}$$

† For example, see "TRC Thermodynamic Tables—Hydrocarbons" and "TRC Thermodynamic Tables—Non-hydrocarbons," serial publications of the Thermodynamics Research Center, Texas A & M Univ. System, College Station, Texas; "The NBS Tables of Chemical Thermodynamic Properties," J. Physical and Chemical Reference Data, vol. 11, supp. 2, 1982.

Table 4.4 Standard heats of formation at 25°C†

Joules per mole of the substance formed

Chemical species		State	ΔH°_{298}
Paraffins:			
Methane	CH_4	*g*	−74,520
Ethane	C_2H_6	*g*	−83,820
Propane	C_3H_8	*g*	−104,680
n-Butane	C_4H_{10}	*g*	−125,790
n-Pentane	C_5H_{12}	*g*	−146,760
n-Hexane	C_6H_{14}	*g*	−166,920
n-Heptane	C_7H_{16}	*g*	−187,780
n-Octane	C_8H_{18}	*g*	−208,750
1-Alkenes:			
Ethylene	C_2H_4	*g*	52,510
Propylene	C_3H_6	*g*	19,710
1-Butene	C_4H_8	*g*	−540
1-Pentene	C_5H_{10}	*g*	−21,280
1-Hexene	C_6H_{12}	*g*	−41,950
1-Heptene	C_7H_{14}	*g*	−62,760
Miscellaneous organics:			
Acetaldehyde	C_2H_4O	*g*	−166,190
Acetic acid	$C_2H_4O_2$	*l*	−484,500
Acetylene	C_2H_2	*g*	227,480
Benzene	C_6H_6	*g*	82,930
Benzene	C_6H_6	*l*	49,080
1,3-Butadiene	C_4H_6	*g*	109,240
Cyclohexane	C_6H_{12}	*g*	−123,140
Cyclohexane	C_6H_{12}	*l*	−156,230
1,2-Ethanediol	$C_2H_6O_2$	*l*	−454,800
Ethanol	C_2H_6O	*g*	−235,100
Ethanol	C_2H_6O	*l*	−277,690
Ethylbenzene	C_8H_{10}	*g*	29,920
Ethylene oxide	C_2H_4O	*g*	−52,630
Formaldehyde	CH_2O	*g*	−108,570
Methanol	CH_4O	*g*	−200,660
Methanol	CH_4O	*l*	−238,660
Methylcyclohexane	C_7H_{14}	*g*	−154,770
Methylcyclohexane	C_7H_{14}	*l*	−190,160
Styrene	C_8H_8	*g*	147,360
Toluene	C_7H_8	*g*	50,170
Toluene	C_7H_8	*l*	12,180

Table 4.4 Standard heats of formation at 25° C (*continued*)

Chemical species		State	ΔH°_{298}
Miscellaneous inorganics:			
Ammonia	NH_3	*g*	−46,110
Calcium carbide	CaC_2	*s*	−59,800
Calcium carbonate	$CaCO_3$	*s*	−1,206,920
Calcium chloride	$CaCl_2$	*s*	−795,800
Calcium chloride	$CaCl_2 \cdot 6H_2O$	*s*	−2,607,900
Calcium hydroxide	$Ca(OH)_2$	*s*	−986,090
Calcium oxide	CaO	*s*	−635,090
Carbon dioxide	CO_2	*g*	−393,509
Carbon monoxide	CO	*g*	−110,525
Hydrochloric acid	HCl	*g*	−92,307
Hydrogen cyanide	HCN	*g*	135,100
Hydrogen sulfide	H_2S	*g*	−20,630
Iron oxide	FeO	*s*	−272,000
Iron oxide (hematite)	Fe_2O_3	*s*	−824,200
Iron oxide (magnetite)	Fe_3O_4	*s*	−1,118,400
Iron sulfide (pyrite)	FeS_2	*s*	−178,200
Lithium chloride	$LiCl$	*s*	−408,610
Lithium chloride	$LiCl \cdot H_2O$	*s*	−712,580
Lithium chloride	$LiCl \cdot 2H_2O$	*s*	−1,012,650
Lithium chloride	$LiCl \cdot 3H_2O$	*s*	−1,311,300
Nitric acid	HNO_3	*l*	−174,100
Nitrogen oxides	NO	*g*	90,250
	NO_2	*g*	33,180
	N_2O	*g*	82,050
	N_2O_4	*g*	9,160
Sodium carbonate	Na_2CO_3	*s*	−1,130,680
Sodium carbonate	$Na_2CO_3 \cdot 10H_2O$	*s*	−4,081,320
Sodium chloride	$NaCl$	*s*	−411,153
Sodium hydroxide	$NaOH$	*s*	−425,609
Sulfur dioxide	SO_2	*g*	−296,830
Sulfur trioxide	SO_3	*g*	−395,720
Sulfur trioxide	SO_3	*l*	−441,040
Sulfuric acid	H_2SO_4	*l*	−813,989
Water	H_2O	*g*	−241,818
Water	H_2O	*l*	−285,830

† Taken from "TRC Thermodynamic Tables—Hydrocarbons," Thermodynamics Research Center, Texas A & M Univ. System, College Station, Texas; "The NBS Tables of Chemical Thermodynamic Properties," J. Physical and Chemical Reference Data, vol. 11, supp. 2, 1982.

These equations can be written so that their sum gives the desired reaction:

$$CO_2(g) \rightarrow C(s) + O_2(g) \qquad \Delta H^\circ_{f_{298}} = 393{,}509 \text{ J}\dagger$$

$$C(s) + \tfrac{1}{2}O_2(g) \rightarrow CO(g) \qquad \Delta H^\circ_{f_{298}} = -110{,}525 \text{ J}$$

$$H_2(g) + \tfrac{1}{2}O_2(g) \rightarrow H_2O(g) \qquad \Delta H^\circ_{f_{298}} = -241{,}818 \text{ J}$$

$$CO_2(g) + H_2(g) \rightarrow CO(g) + H_2O(g) \qquad \Delta H^\circ_{298} = 41{,}166 \text{ J}$$

The meaning of this result is that the enthalpy of 1 mol of CO plus 1 mol of H_2O is greater than the enthalpy of 1 mol of CO_2 plus 1 mol of H_2 by 41,166 J when each product and reactant is taken as the pure gas at 25°C in the ideal-gas state at 1 bar or 1(atm).

In this example the standard heat of formation of H_2O is available for its hypothetical standard state as a gas at 25°C. One might expect the value of the heat of formation of water to be listed for its actual state as a liquid at 1 bar or 1(atm) and 25°C. As a matter of fact, values for both states are given because they are both frequently used. This is true for many compounds that normally exist as liquids at 25°C and the standard-state pressure. Cases do arise, however, in which a value is given only for the standard state as a liquid or as an ideal gas when what is needed is the other value. Suppose that this were the case for the preceding example and that only the standard heat of formation of liquid H_2O is known. We must now include an equation for the physical change that transforms water from its standard state as a liquid into its standard state as a gas. The enthalpy change for this physical process is the difference between the heats of formation of water in its two standard states:

$$-241{,}818 - (-285{,}830) = 44{,}012 \text{ J}$$

This is approximately the latent heat of vaporization of water at 25°C. The sequence of steps is now:

$$CO_2(g) \rightarrow C(s) + O_2(g) \qquad \Delta H^\circ_{f_{298}} = 393{,}509 \text{ J}$$

$$C(s) + \tfrac{1}{2}O_2(g) \rightarrow CO(g) \qquad \Delta H^\circ_{f_{298}} = -110{,}525 \text{ J}$$

$$H_2(g) + \tfrac{1}{2}O_2(g) \rightarrow H_2O(l) \qquad \Delta H^\circ_{f_{298}} = -285{,}830 \text{ J}$$

$$H_2O(l) \rightarrow H_2O(g) \qquad \Delta H^\circ_{298} = 44{,}012 \text{ J}$$

$$CO_2(g) + H_2(g) \rightarrow CO(g) + H_2O(g) \qquad \Delta H^\circ_{298} = 41{,}166 \text{ J}$$

This result is of course in agreement with the original answer.

Example 4.6 Calculate the standard heat at 25°C for the following reaction:

$$4HCl(g) + O_2(g) \rightarrow 2H_2O(g) + 2Cl_2(g)$$

† The reaction as written is the reverse of the formation reaction of carbon dioxide; the sign is therefore opposite that given in Table 4.4 for its standard heat of formation.

SOLUTION Standard heats of formation from Table 4.4 are

$$\text{HCl:} \quad -92{,}307 \text{ J}$$

$$\text{H}_2\text{O:} \quad -241{,}818 \text{ J}$$

The following combination gives the desired result:

$$4\text{HCl}(g) \rightarrow 2\text{H}_2(g) + 2\text{Cl}_2(g) \qquad \Delta H^\circ_{298} = (4)(92{,}307)$$

$$2\text{H}_2(g) + \text{O}_2(g) \rightarrow 2\text{H}_2\text{O}(g) \qquad \Delta H^\circ_{298} = (2)(-241{,}818)$$

$$4\text{HCl}(g) + \text{O}_2(g) \rightarrow 2\text{H}_2\text{O}(g) + 2\text{Cl}_2(g) \qquad \Delta H^\circ_{298} = -114{,}408 \text{ J}$$

4.5 THE STANDARD HEAT OF COMBUSTION

Only a few formation reactions can actually be carried out, and therefore data for these reactions must usually be determined indirectly. One kind of reaction that readily lends itself to experiment is the combustion reaction, and many standard heats of formation come from standard heats of combustion, measured calorimetrically. A combustion reaction is defined as a reaction between an element or compound and oxygen to form specified combustion products. For organic compounds made up of carbon, hydrogen, and oxygen only, the products are carbon dioxide and water, but the state of the water may be either vapor or liquid. Data are always based on 1 mol of the substance burned.

A reaction such as the formation of *n*-butane:

$$4\text{C}(s) + 5\text{H}_2(g) \rightarrow \text{C}_4\text{H}_{10}(g)$$

cannot be carried out in practice. However, this equation results from combination of the following combustion reactions:

$$4\text{C}(s) + 4\text{O}_2(g) \rightarrow 4\text{CO}_2(g) \qquad \Delta H^\circ_{298} = (4)(-393{,}509)$$

$$5\text{H}_2(g) + 2\tfrac{1}{2}\text{O}_2(g) \rightarrow 5\text{H}_2\text{O}(l) \qquad \Delta H^\circ_{298} = (5)(-285{,}830)$$

$$4\text{CO}_2(g) + 5\text{H}_2\text{O}(l) \rightarrow \text{C}_4\text{H}_{10}(g) + 6\tfrac{1}{2}\text{O}_2(g) \qquad \Delta H^\circ_{298} = 2{,}877{,}396$$

$$4\text{C}(s) + 5\text{H}_2(g) \rightarrow \text{C}_4\text{H}_{10}(g) \qquad \Delta H^\circ_{298} = -125{,}790 \text{ J}$$

This is the value of the standard heat of formation of *n*-butane listed in Table 4.4.

4.6 EFFECT OF TEMPERATURE ON THE STANDARD HEAT OF REACTION

In the foregoing sections, standard heats of reaction are discussed for the base temperature of 298.15 K only. In this section we treat the calculation of standard heats of reaction at other temperatures from knowledge of the value at 298.15 K.

The general chemical reaction may be written as

$$|\nu_1|A_1 + |\nu_2|A_2 + \cdots \rightarrow |\nu_3|A_3 + |\nu_4|A_4 + \cdots$$

where the $|\nu_i|$ are stoichiometric coefficients and the A_i stand for chemical formulas. The species on the left are reactants; those on the right, products. We adopt a sign convention for ν_i that makes it

positive (+) *for products*

negative (−) *for reactants*

The ν_i with their accompanying signs are called stoichiometric *numbers.* For example, when the ammonia-synthesis reaction is written

$$N_2 + 3H_2 \rightarrow 2NH_3$$

then

$$\nu_{N_2} = -1 \qquad \nu_{H_2} = -3 \qquad \nu_{NH_3} = 2$$

This sign convention allows the definition of a standard heat of reaction to be expressed mathematically by the equation:

$$\Delta H^\circ = \sum \nu_i H_i^\circ \tag{4.14}$$

where H_i° is the enthalpy of species i in its standard state and the summation is over all products and reactants. The standard-state enthalpy of a chemical compound is equal to its heat of formation plus the standard-state enthalpies of its constituent elements. If we arbitrarily set the standard-state enthalpies of all elements equal to zero as the basis of calculation, then the standard-state enthalpy of each compound is its heat of formation. In this event, $H_i^\circ = \Delta H_{f_i}^\circ$ and Eq. (4.14) becomes

$$\Delta H^\circ = \sum \nu_i \, \Delta H_{f_i}^\circ \tag{4.15}$$

where the summation is over all products and reactants. This formalizes the procedure described in the preceding section for calculation of standard heats of other reactions from standard heats of formation. Applied to the reaction,

$$4HCl(g) + O_2(g) \rightarrow 2H_2O(g) + 2Cl_2(g)$$

Eq. (4.15) is written:

$$\Delta H^\circ = 2\Delta H_{f_{H_2O}}^\circ - 4\Delta H_{f_{HCl}}^\circ$$

With data from Table 4.4 for 298.15 K, this becomes

$$\Delta H_{298}^\circ = (2)(-241{,}818) - (4)(-92{,}307) = -114{,}408 \text{ J}$$

in agreement with the result of Example 4.6.

For a standard reaction, products and reactants are always at the same standard-state pressure of 1 bar or 1(atm). Standard-state enthalpies are therefore functions of temperature only, and their change with T is given by Eq. (2.25),

$$dH_i^\circ = C_{P_i}^\circ \, dT$$

where subscript i identifies a particular product or reactant. Multiplying by ν_i

and summing over all products and reactants gives

$$\sum \nu_i \, dH_i^\circ = \sum \nu_i C_{P_i}^\circ \, dT$$

Since ν_i is a constant,

$$\sum d(\nu_i H_i^\circ) = d \sum \nu_i H_i^\circ = \sum \nu_i C_{P_i}^\circ \, dT$$

The term $\sum \nu_i H_i^\circ$ is the standard heat of reaction, defined by Eq. (4.14). Similarly, we define the standard heat-capacity change of reaction as

$$\Delta C_P^\circ = \sum \nu_i C_{P_i}^\circ \tag{4.16}$$

As a result of these definitions, the preceding equation becomes

$$\boxed{d\Delta H^\circ = \Delta C_P^\circ \, dT} \tag{4.17}$$

This is the fundamental equation relating heats of reaction to temperature. It may be integrated between the limits of 298.15 K and temperature T:

$$\int_{\Delta H_{298}^\circ}^{\Delta H_T^\circ} d\Delta H^\circ = \int_{298.15}^{T} \Delta C_P^\circ \, dT$$

or

$$\Delta H_T^\circ = \Delta H_{298}^\circ + \Delta C_{P_{\text{mh}}}^\circ (T - 298.15) \tag{4.18}$$

If the temperature dependence of the heat capacity of each product and reactant is given by Eq. (4.4), then $\Delta C_{P_{\text{mh}}}^\circ$ is given by the analog of Eq. (4.7):

$$\frac{\Delta C_{P_{\text{mh}}}^\circ}{R} = \Delta A + (\Delta B) T_{\text{am}} + \frac{\Delta C}{3}(4T_{\text{am}}^2 - T_1 T_2) + \frac{\Delta D}{T_1 T_2} \tag{4.19}$$

where

$$\Delta A \equiv \sum \nu_i A_i$$

with analogous definitions for ΔB, ΔC, and ΔD. For use with Eq. (4.18), we set $T_1 = 298.15$.

Example 4.7 Calculate the standard heat of the methanol-synthesis reaction at 800°C:

$$CO(g) + 2H_2(g) \rightarrow CH_3OH(g)$$

SOLUTION Application of Eq. (4.15) to this reaction at 25°C with heat-of-formation data from Table 4.4 gives

$$\Delta H_{298}^\circ = -200{,}660 - (-110{,}525) = -90{,}135 \text{ J}$$

The value of $\Delta C_{P_{\text{mh}}}^\circ$ required for application of Eq. (4.18) is found from Eq. (4.19). The following constants are taken from Table 4.1:

i	ν_i	A	$B \times 10^3$	$C \times 10^6$	$D \times 10^{-5}$
CH_3OH	1	2.211	12.216	−3.450	0.000
CO	−1	3.376	0.557	0.000	−0.031
H_2	−2	3.249	0.422	0.000	0.083

By definition,

$$\Delta A = (1)(2.211) + (-1)(3.376) + (-2)(3.249) = -7.663$$

Similarly,

$$\Delta B = 10.815 \times 10^{-3}$$
$$\Delta C = -3.450 \times 10^{-6}$$
$$\Delta D = -0.135 \times 10^{5}$$

Substitution of these values along with $T_1 = 298.15$ K, $T_2 = 1{,}073.15$ K, and $R = 8.314$ J mol^{-1} K^{-1} into Eq. (4.19) gives

$$\Delta C^\circ_{P_{mh}} = -17.330 \text{ J K}^{-1}$$

Whence by Eq. (4.18)

$$\Delta H^\circ_{1073} = -90{,}135 - (17.330)(1{,}073.15 - 298.15)$$
$$= -103{,}566 \text{ J}$$

Integration of Eq. (4.17) between limits produces an equation suitable for the calculation of ΔH°_T when one wants a single value. The alternative is general integration to give

$$\Delta H^\circ_T = J + \int \Delta C^\circ_P \, dT \tag{4.20}$$

where J is the constant of integration. Evaluation of the final term requires an expression for the temperature dependence of ΔC°_P. Substitution of Eq. (4.4) into Eq. (4.16) leads to:

$$\frac{\Delta C^\circ_P}{R} = \Delta A + (\Delta B)T + (\Delta C)T^2 + \frac{\Delta D}{T^2} \tag{4.21}$$

Eliminating ΔC°_P from Eq. (4.20) and integrating, we get

$$\Delta H^\circ_T = J + R\left[(\Delta A)T + \frac{\Delta B}{2}T^2 + \frac{\Delta C}{3}T^3 - \frac{\Delta D}{T}\right] \tag{4.22}$$

Equation (4.22) provides a general method for the calculation of the standard heat of a particular reaction as a function of temperature. The integration constant J is evaluated by application of the equation at a temperature, usually 298.15 K, where ΔH°_T is known.

Example 4.8 Develop an equation giving the temperature dependence of the heat of the methanol-synthesis reaction:

$$CO(g) + 2H_2(g) \rightarrow CH_3OH(g)$$

SOLUTION The values of ΔA, ΔB, ΔC, and ΔD calculated in Example 4.7 are substituted into Eq. (4.22) to give:

$$\Delta H^\circ_T = J + 8.314\left(-7.663T + 5.408 \times 10^{-3}T^2 - 1.150 \times 10^{-6}T^3 + \frac{0.135 \times 10^5}{T}\right)$$

Also in Example 4.7 we found that $\Delta H^\circ_{298} = -90{,}135$ J. Substituting this value and $T = 298.15$ K into the preceding equation, we solve for J, finding

$$J = -75{,}259 \text{ J}$$

The general equation for ΔH°_T for the methanol-synthesis reaction is therefore:

$$\Delta H^\circ_T = -75{,}259 - 63.710T + 44.962 \times 10^{-3}T^2 - 9.561 \times 10^{-6}T^3 + \frac{1.122 \times 10^5}{T}$$

For $T = 1{,}073.15$ K, this equation gives $\Delta H^\circ_{1073} = -103{,}566$ J, the same result obtained in Example 4.7.

4.7 HEAT EFFECTS OF INDUSTRIAL REACTIONS

The preceding sections have dealt with the *standard* heat of reaction. Industrial reactions are rarely carried out under standard-state conditions. Furthermore, in actual reactions the reactants may not be present in stoichiometric proportions, the reaction may not go to completion, and the final temperature may differ from the initial temperature. Moreover, inert species may be present, and several reactions may occur simultaneously. Nevertheless, calculations of the heat effects of actual reactions are based on the principles already considered and are best illustrated by example.

Example 4.9 What is the maximum temperature that can be reached by the combustion of methane with 20 percent excess air? Both the methane and the air enter the burner at 25°C.

SOLUTION The reaction is

$$CH_4 + 2O_2 \rightarrow CO_2 + 2H_2O(g)$$

for which

$$\Delta H^\circ_{298} = -393{,}509 + (2)(-241{,}818) - (-74{,}520) = -802{,}625 \text{ J}$$

Since the maximum attainable temperature is sought, we assume complete adiabatic ($Q = 0$) combustion. With the additional assumptions that the kinetic- and potential-energy changes are negligible and that there is no shaft work, the overall energy balance for the process reduces to $\Delta H = 0$. For purposes of calculation of the final temperature,† any convenient path between the initial and final states may be used. The path chosen is indicated in the diagram. With one mole of methane burned as the basis for all calculations,

Moles O_2 required = 2.0
Moles excess O_2 = (0.2)(2.0) = 0.4
Moles N_2 entering = (2.4)(79/21) = 9.03

† This temperature is often called the theoretical flame temperature, because it is the maximum temperature attainable in the flame produced when the gas burns with the stated amount of air.

The gases leaving the burner contain 1 mol CO_2, 2 mol $H_2O(g)$, 0.4 mol O_2, and 9.03 mol N_2.

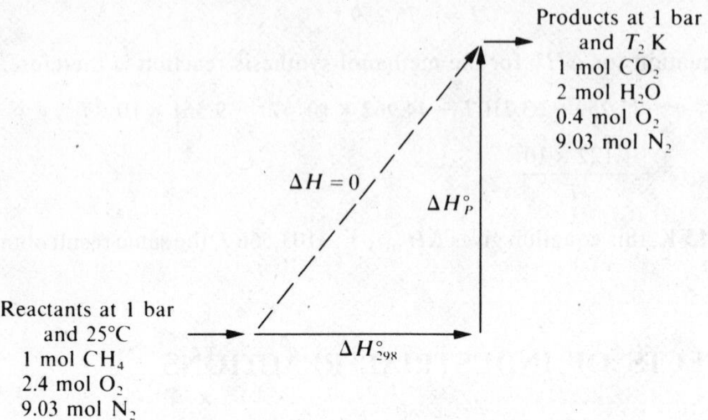

Since the enthalpy change must be the same regardless of path,

$$\Delta H_{298}^\circ + \Delta H_P^\circ = \Delta H = 0$$

The two terms on the left are

$$\Delta H_{298}^\circ = -802{,}625 \text{ J}$$

and

$$\Delta H_P^\circ = (\textstyle\sum n_i C_{P_{\mathrm{mh},i}}^\circ)(T_2 - 298.15)$$

where the summation runs over all product gases. Because the mean heat capacities depend on the final temperature, we set up an iteration scheme to solve for T_2. Combining the last three equations and solving for T_2 gives

$$T_2 = \frac{802{,}625}{\sum n_i C_{P_{\mathrm{mh},i}}^\circ} + 298.15 \qquad (A)$$

Since $C = 0$ in each heat-capacity equation for the product gases (Table 4.1), Eq. (4.7) yields

$$\sum n_i C_{P_{\mathrm{mh},i}}^\circ = R\left(\sum n_i A_i + (\textstyle\sum n_i B_i) T_{\mathrm{am}} + \frac{\sum n_i D_i}{T_1 T_2}\right)$$

With data from Table 4.1, we find:

$$\sum n_i A_i = (1)(5.457) + (2)(3.470) + (0.4)(3.639) + (9.03)(3.280)$$
$$= 43.489$$

Similarly,

$$\sum n_i B_i = 9.502 \times 10^{-3}$$
$$\sum n_i D_i = -0.645 \times 10^5$$

whence

$$\sum n_i C_{P_{\mathrm{mh},i}}^\circ = 8.314\left(43.489 + 9.502 \times 10^{-3} T_{\mathrm{am}} + \frac{-0.645 \times 10^5}{T_1 T_2}\right) \qquad (B)$$

with $T_1 = 298.15$ K. An initial value of $T_2 \geq 298.15$ K substituted into Eq. (B) yields a value for $\sum n_i C^\circ_{P_{mh,i}}$ which when substituted in Eq. (A) yields a new value for T_2. Continued iteration between Eqs. (A) and (B) yields a final value of

$$T_2 = 2{,}066 \text{ K} \qquad \text{or} \qquad 1{,}793^\circ\text{C}$$

Example 4.10 One method for the manufacture of "synthesis gas" (primarily a mixture of CO and H_2) is the catalytic reforming of CH_4 with steam at high temperature and atmospheric pressure:

$$CH_4(g) + H_2O(g) \rightarrow CO(g) + 3H_2(g)$$

The only other reaction which occurs to an appreciable extent is the water-gas-shift reaction:

$$CO(g) + H_2O(g) \rightarrow CO_2(g) + H_2(g)$$

If the reactants are supplied in the ratio, 2 mol steam to 1 mol CH_4, and if heat is supplied to the reactor so that the products reach a temperature of 1,300 K, the CH_4 is completely converted and the product stream contains 17.4 mole percent CO. Assuming the reactants to be preheated to 600 K, calculate the heat requirement for the reactor.

SOLUTION The standard heats of reaction at 25°C for the two reactions are calculated from the data of Table 4.4:

$$CH_4(g) + H_2O(g) \rightarrow CO(g) + 3H_2(g) \qquad \Delta H^\circ_{298} = 205{,}813 \text{ J}$$

$$CO(g) + H_2O(g) \rightarrow CO_2(g) + H_2(g) \qquad \Delta H^\circ_{298} = -41{,}166 \text{ J}$$

These two reactions may be added to give a third reaction:

$$CH_4(g) + 2H_2O(g) \rightarrow CO_2(g) + 4H_2(g) \qquad \Delta H^\circ_{298} = 164{,}647 \text{ J}$$

Any pair of these three reactions constitutes an independent set. The odd reaction is not independent, since it is obtained by combination of the other two. The reactions most convenient to work with here are:

$$CH_4(g) + H_2O(g) \rightarrow CO(g) + 3H_2(g) \qquad \Delta H^\circ_{298} = 205{,}813 \text{ J} \qquad (A)$$

$$CH_4(g) + 2H_2O(g) \rightarrow CO_2(g) + 4H_2(g) \qquad \Delta H^\circ_{298} = 164{,}647 \text{ J} \qquad (B)$$

We first determine the fraction of CH_4 converted by each of these reactions. As a basis for calculations, let 1 mol CH_4 and 2 mol steam be fed to the reactor. If x mol CH_4 reacts by Eq. (A), then $1 - x$ mol reacts by Eq. (B). On this basis the products of the reaction are:

CO:	x
H_2:	$3x + 4(1 - x) = 4 - x$
CO_2:	$1 - x$
H_2O:	$2 - x - 2(1 - x) = x$
Total:	5 mol products

The mole fraction of CO in the product stream is $x/5 = 0.174$; whence $x = 0.870$. Thus, on the basis chosen, 0.870 mol CH_4 reacts by Eq. (A) and 0.130 mol reacts by

Eq. (B). Furthermore, the amount of each species in the product stream is:

$$\text{Moles CO} = x = 0.87$$

$$\text{Moles } H_2 = 4 - x = 3.13$$

$$\text{Moles } CO_2 = 1 - x = 0.13$$

$$\text{Moles } H_2O = x = 0.87$$

We now devise a path, for purposes of calculation, to proceed from reactants at 600 K to products at 1,300 K. Since data are available for the standard heats of reaction at 25°C, the most convenient path is the one which includes the reactions at 25°C (298.15 K). This is shown schematically in the accompanying diagram.

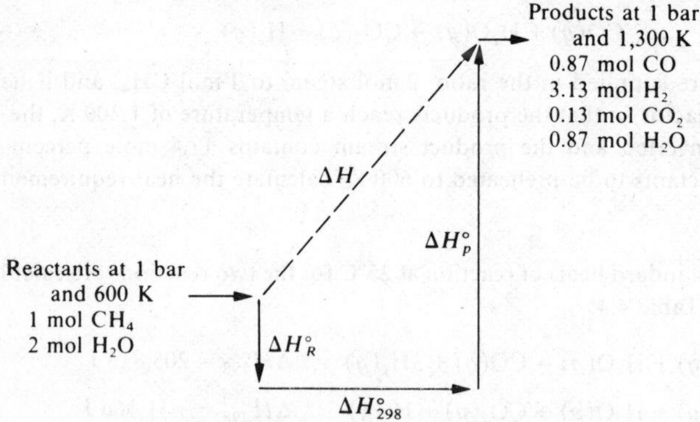

The dashed line represents the actual path for which the enthalpy change is ΔH. Since this enthalpy change is independent of path,

$$\Delta H = \Delta H_R^\circ + \Delta H_{298}^\circ + \Delta H_P^\circ$$

For the calculation of ΔH_{298}°, reactions (A) and (B) must both be taken into account. Since 0.87 mol CH_4 reacts by (A) and 0.13 mol reacts by (B),

$$\Delta H_{298}^\circ = (0.87)(205{,}813) + (0.13)(164{,}647) = 200{,}461 \text{ J}$$

The enthalpy change of the reactants as they are cooled from 600 to 298.15 K is given by:

$$\Delta H_R^\circ = \left(\sum n_i C_{P_{mh,i}}^\circ\right)(298.15 - 600)$$

where the mean heat capacities are calculated by Eq. (4.7):

$$\Delta H_R^\circ = \left[\underset{CH_4}{(1)(44.026)} + \underset{H_2O}{(2)(34.826)}\right](-301.85) = -34{,}314 \text{ J}$$

The enthalpy change of the products as they are heated from 298.15 to 1,300 K is calculated similarly:

$$\Delta H_P^\circ = \left(\sum n_i C_{P_{mh,i}}^\circ\right)(1{,}300 - 298.15)$$

$$\Delta H_P^\circ = \left[\underset{CO}{(0.87)(31.702)} + \underset{H_2}{(3.13)(29.994)} + \underset{CO_2}{(0.13)(49.830)} + \underset{H_2O}{(0.87)(38.742)}\right] \times (1{,}001.85) = 161{,}944 \text{ J}$$

Therefore,

$$\Delta H = -34{,}314 + 200{,}461 + 161{,}944 = 328{,}091 \text{ J}$$

The process is one of steady flow for which W_s, Δz, and $\Delta u^2/2$ are presumed negligible. Thus

$$Q = \Delta H = 328{,}091 \text{ J}$$

This result is on the basis of 1 mol CH_4 fed to the reactor. The factor for converting from J mol^{-1} to $(\text{Btu})(\text{lb mol})^{-1}$ is very nearly 0.43 (more exactly it is 0.429929). Therefore on the basis of 1(lb mol) CH_4 fed to the reactor, we have

$$Q = \Delta H = (328{,}091)(0.43) = 141{,}079(\text{Btu})$$

Example 4.11 A boiler is fired with a high-grade fuel oil (consisting only of hydrocarbons) having a standard heat of combustion of $-43{,}515 \text{ J g}^{-1}$ at 25°C with $CO_2(g)$ and $H_2O(l)$ as products. The temperature of the fuel and air entering the combustion chamber is 25°C. The air is assumed dry. The flue gases leave at 300°C, and their average analysis (on a dry basis) is 11.2 percent CO_2, 0.4 percent CO, 6.2 percent O_2, and 82.2 percent N_2. Calculate the fraction of the heating value of the oil that is transferred as heat to the boiler.

SOLUTION Take as a basis 100 mol dry flue gases, consisting of

CO_2	11.2 mol
CO	0.4 mol
O_2	6.2 mol
N_2	82.2 mol
Total	100.0 mol

This analysis, on a dry basis, does not take into account the H_2O vapor present in the flue gases. The amount of H_2O formed by the combustion reaction is found from an oxygen balance. The O_2 supplied in the air represents 21 mol percent of the air stream. The remaining 79 percent is N_2, which goes through the combustion process unchanged. Thus the 82.2 mol N_2 appearing in 100 mol dry flue gases is supplied with the air, and the O_2 accompanying this N_2 is:

$$\text{Moles } O_2 \text{ entering in air} = (82.2)(21/79) = 21.85$$

However,

$$\text{Moles } O_2 \text{ accounted for in the dry flue gases}$$
$$= 11.2 + 0.4/2 + 6.2 = 17.60$$

The difference between these figures is the moles of O_2 that react to form H_2O. Therefore on the basis of 100 mol dry flue gases,

$$\text{Moles } H_2O \text{ formed} = (21.85 - 17.60)(2) = 8.50$$

$$\text{Moles } H_2 \text{ in the fuel} = \text{moles of water formed} = 8.50$$

The amount of C in the fuel is given by a carbon balance:

$$\text{Moles C in flue gases} = \text{moles C in fuel}$$
$$= 11.2 + 0.4 = 11.60$$

These amounts of C and H_2 together give:

$$\text{Mass of fuel burned} = (8.50)(2) + (11.6)(12) = 156.2 \text{ g}$$

If this amount of fuel is burned completely to $CO_2(g)$ and $H_2O(l)$ at 25°C, the heat of combustion is

$$\Delta H^\circ_{298} = (-43{,}515)(156.2) = -6{,}797{,}040 \text{ J}$$

However, the reaction actually occurring does not represent complete combustion, and the H_2O is formed as vapor rather than as liquid. The 156.2 g of fuel is represented by the empirical formula $C_{11.6}H_{17}$, and the reaction is written:

$$C_{11.6}H_{17}(l) + 21.85O_2(g) + 82.2N_2(g) \rightarrow$$
$$11.2CO_2(g) + 0.4CO(g) + 8.5H_2O(g) + 6.2O_2(g) + 82.2N_2(g)$$

This equation is obtained by addition of the following reactions, for each of which the standard heat of reaction at 25°C is known:

$$C_{11.6}H_{17}(l) + 15.85O_2(g) \rightarrow 11.6CO_2(g) + 8.5H_2O(l)$$
$$\Delta H^\circ_{298} = -6{,}797{,}040 \text{ J}$$

$$8.5H_2O(l) \rightarrow 8.5H_2O(g) \qquad \Delta H^\circ_{298} = (44{,}012)(8.5) = 374{,}102 \text{ J}$$

$$0.4CO_2(g) \rightarrow 0.4CO(g) + 0.2O_2(g)$$
$$\Delta H^\circ_{298} = (282{,}984)(0.4) = 113{,}194 \text{ J}$$

$$6.2O_2(g) + 82.2N_2(g) \rightarrow 6.2O_2(g) + 82.2N_2(g)$$
$$\Delta H^\circ_{298} = 0$$

The sum of these reactions yields the actual reaction, and the sum of the ΔH°_{298} values gives the standard heat of the reaction occurring at 25°C:

$$\Delta H^\circ_{298} = -6{,}309{,}744 \text{ J}$$

This value is used as indicated in the accompanying diagram for calculation of the heat effect of the process considered.

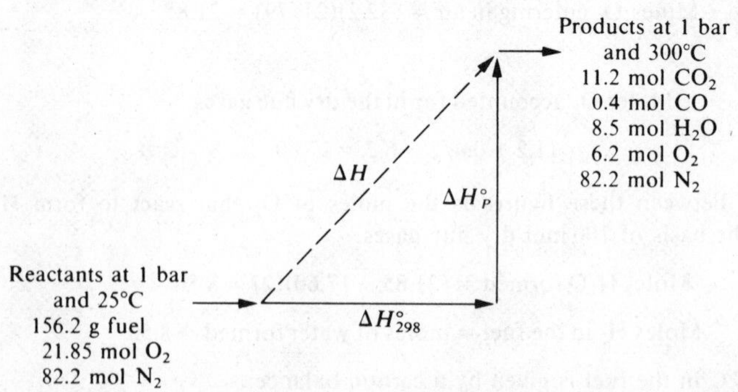

The actual process leading from reactants at 25°C to products at 300°C is represented by the dashed line in the diagram. For purposes of calculating ΔH for this process, we may use any convenient path. The one drawn with solid lines is a logical one, because the enthalpy changes for these steps are easily calculated, and ΔH°_{298} has already been evaluated. The enthalpy change caused by heating the products of reaction from 25 to 300°C is calculated with mean heat capacities by Eq. (4.7):

$$\Delta H^\circ_P = \left(\sum n_i C^\circ_{P_{\mathrm{mh},i}}\right)(573.15 - 298.15)$$

$$\Delta H^\circ_P = [(11.2)(43.675) + (0.4)(29.935) + (8.5)(34.690)$$
$$+ (6.2)(30.983) + (82.2)(29.612)](573.15 - 298.15)$$

$$\Delta H^\circ_P = 941{,}105 \text{ J}$$

Whence

$$\Delta H = \Delta H^\circ_{298} + \Delta H^\circ_P = -6{,}309{,}744 + 941{,}105 = -5{,}368{,}640 \text{ J}$$

Since the process is one of steady flow for which the shaft work and kinetic- and potential-energy terms in the energy balance [Eq. (2.10)] are zero or negligible, $\Delta H = Q$. Thus, $Q = -5{,}368.64$ kJ, and this amount of heat is transferred to the boiler for every 100 mol dry flue gases formed. This represents

$$\frac{5{,}368{,}640}{6{,}797{,}040}(100) = 79.0 \text{ percent}$$

of the higher heating value of the fuel.

In the foregoing examples of reactions that occur at approximately 1 bar, the reactants and products are for practical purposes in their standard states. For reactions at elevated pressures, this is not the case, and additional calculations are required to take into account the effect of pressure on the heat effects of reaction. The method of doing this is considered in Chap. 6. Suffice it to say at this point that the effect of pressure on the heat of reaction is usually small compared with the effect of temperature.

PROBLEMS

4.1 What is the heat required when 10 mol of ethylene is heated from 200 to 1,100°C at approximately atmospheric pressure in a steady-flow heat exchanger?

4.2 What is the heat required when 12 mol of 1-butene is heated from 250 to 1,200°C at approximately atmospheric pressure in a steady-flow heat exchanger?

4.3 What is the final temperature when heat in the amount of 1,100 kJ is added to 30 mol of SO_2 initially at 300°C in a steady-flow heat exchanger at approximately atmospheric pressure?

4.4 What is the final temperature when heat in the amount of 880 kJ is added to 25 mol of ammonia vapor initially at 260°C in a steady-flow heat exchanger at approximately atmospheric pressure?

4.5 What is the final temperature when heat in the amount of 10^6(Btu) [1.055×10^6 kJ] is added to 50(lb mol) [22.68 kg mol] of methane initially at 500(°F) [260°C] in a steady-flow heat exchanger at approximately atmospheric pressure?

4.6 If 350$(\mathrm{ft})^3(\mathrm{s})^{-1}$ [9.91 $\mathrm{m^3\,s^{-1}}$] of air at 77(°F) [25°C] and atmospheric pressure is preheated for a combustion process to 815(°F) [435°C], what rate of heat transfer is required?

4.7 How much heat is required when 10(tons) [9,070 kg] of $CaCO_3$ (calcite) is heated at atmospheric pressure from 95(°F) [35°C] to 1,580(°F) [860°C]?

4.8 If the heat capacity of a substance is correctly represented by an equation of the form

$$C_P = A + BT + CT^2$$

show that the error resulting when $C_{P_{mh}}$ is assumed equal to C_P evaluated at the arithmetic mean of the initial and final temperatures is $C(T_2 - T_1)^2/12$.

4.9 If the heat capacity of a substance is correctly represented by an equation of the form

$$C_P = A + BT + DT^{-2}$$

show that the error resulting when $C_{P_{mh}}$ is assumed equal to C_P evaluated at the arithmetic mean of the initial and final temperatures is $D(T_2 - T_1)^2/4T_1T_2T_{am}^2$.

4.10 A handbook value for the latent heat of vaporization of *n*-hexane at 25°C is 366.1 J g^{-1}. What approximately is the value at 150°C?

4.11 A handbook value for the latent heat of vaporization of benzene at 25°C is 433.3 J g^{-1}. What approximately is the value at 200°C?

4.12 A handbook value for the latent heat of vaporization of cyclohexane at 25°C is 392.5 J g^{-1}. What approximately is the value at 190°C?

4.13 A handbook value for the latent heat of vaporization of methyl ethyl ketone at 78.2°C is 443.2 J g^{-1}. What approximately is the value at 185°C?

4.14 A handbook value for the latent heat of vaporization of methanol at 64.7°C is 1,099.5 J g^{-1}. What approximately is the value at 175°C?

4.15 Calculate the latent heat of vaporization of ammonia at 320 K

(*a*) By Eqs. (4.12) and (4.13). T_n = 239.7 K.

(*b*) From the following handbook data for saturated ammonia:

T/K	P/bar	V^l/m^3 kg^{-1}	V^v/m^3 kg^{-1}
300	10.61	1.666×10^{-3}	0.121
310	14.24	1.710×10^{-3}	0.091
320	18.72	1.760×10^{-3}	0.069
330	24.20	1.815×10^{-3}	0.053
340	30.79	1.878×10^{-3}	0.041

The reported value is 1,065.7 kJ kg^{-1}.

4.16 Calculate the latent heat of vaporization of methanol at 300 K

(*a*) By Eqs. (4.12) and (4.13). T_n = 337.8 K.

(*b*) From the following handbook data for saturated methanol:

T/K	P/bar	V^l/m^3 kg^{-1}	V^v/m^3 kg^{-1}
280	0.0621	1.244×10^{-3}	11.62
290	0.1094	1.259×10^{-3}	6.778
300	0.1860	1.274×10^{-3}	4.095
310	0.3043	1.290×10^{-3}	2.566
320	0.4817	1.306×10^{-3}	1.661

The reported value is 1,167.3 kJ kg^{-1}.

4.17 Estimate the standard heat of formation of liquid ethyl benzene at 25°C. For ethyl benzene, T_n = 409.3 K, T_c = 617.1 K, and P_c = 36.1 bar.

4.18 A reversible compression of 1 mol of an ideal gas in a piston/cylinder device results in a pressure increase from 1 bar to P_2 and a temperature increase from 500 to 1,000 K. The path followed by the gas during compression is given by

$$PV^{1.5} = \text{const}$$

and the molar heat capacity of the gas is given by

$$C_P/R = 3.30 + 0.63 \times 10^{-3}T \qquad [T = \text{K}]$$

Determine the heat transferred during the process and the final pressure.

4.19 If the heat of combustion of urea, $(NH_2)_2CO(s)$, at 25°C is 631,660 J mol^{-1} when the products are $CO_2(g)$, $H_2O(l)$, and $N_2(g)$, what is the standard heat of formation of urea at 25°C?

4.20 Determine the standard heat of each of the following reactions at 25°C:

(a) $N_2(g) + 3H_2(g) \rightarrow 2NH_3(g)$
(b) $4NH_3(g) + 5O_2(g) \rightarrow 4NO(g) + 6H_2O(g)$
(c) $3NO_2(g) + H_2O(l) \rightarrow 2HNO_3(l) + NO(g)$
(d) $CaC_2(s) + H_2O(l) \rightarrow C_2H_2(g) + CaO(s)$
(e) $2Na(s) + 2H_2O(g) \rightarrow 2NaOH(s) + H_2(g)$
(f) $C_3H_8(g) \rightarrow C_2H_4(g) + CH_4(g)$
(g) $C_2H_4(g) + \frac{1}{2}O_2(g) \rightarrow \langle(CH_2)_2\rangle O(g)$
(h) $C_2H_2(g) + H_2O(g) \rightarrow \langle(CH_2)_2\rangle O(g)$
(i) $CH_4(g) + 2H_2O(g) \rightarrow CO_2(g) + 4H_2(g)$
(j) $CO_2(g) + 3H_2(g) \rightarrow CH_3OH(g) + H_2O(g)$
(k) $CH_3OH(g) + \frac{1}{2}O_2(g) \rightarrow HCHO(g) + H_2O(g)$
(l) $2H_2S(g) + 3O_2(g) \rightarrow 2H_2O(g) + 2SO_2(g)$
(m) $H_2S(g) + 2H_2O(g) \rightarrow 3H_2(g) + SO_2(g)$
(n) $N_2(g) + O_2(g) \rightarrow 2NO(g)$
(o) $CaCO_3(s) \rightarrow CaO(s) + CO_2(g)$
(p) $SO_3(g) + H_2O(l) \rightarrow H_2SO_4(l)$
(q) $C_2H_4(g) + H_2O(l) \rightarrow C_2H_5OH(l)$
(r) $CH_3CHO(g) + H_2(g) \rightarrow C_2H_5OH(g)$
(s) $C_2H_5OH(l) + O_2(g) \rightarrow CH_3COOH(l) + H_2O(l)$
(t) $C_2H_5CH{:}CH_2(g) \rightarrow CH_2{:}CHCH{:}CH_2(g) + H_2(g)$
(u) $C_4H_{10}(g) \rightarrow CH_2{:}CHCH{:}CH_2(g) + 2H_2(g)$
(v) $C_2H_5CH{:}CH_2(g) + \frac{1}{2}O_2(g) \rightarrow CH_2{:}CHCH{:}CH_2(g) + H_2O(g)$
(w) $2NH_3(g) + 3NO(g) \rightarrow 3H_2O(g) + \frac{5}{2}N_2(g)$
(x) $N_2(g) + C_2H_2(g) \rightarrow 2HCN(g)$
(y) $C_6H_5.C_2H_5(g) \rightarrow C_6H_5CH{:}CH_2(g) + H_2(g)$
(z) $C(s) + H_2O(l) \rightarrow H_2(g) + CO(g)$

4.21 What is the standard heat for the reaction of Prob. 4.20(a) at 550°C?

4.22 What is the standard heat for the reaction of Prob. 4.20(b) at 450°C?

4.23 What is the standard heat for the reaction of Prob. 4.20(j) at 500(°F) [260°C]?

4.24 What is the standard heat for the reaction of Prob. 4.20(l) at 800(°F) [426.7°C]

4.25 What is the standard heat for the reaction of Prob. 4.20(m) at 900 K?

4.26 What is the standard heat for the reaction of Prob. 4.20(n) at 1,500 K?

4.27 What is the standard heat for the reaction of Prob. 4.20(o) at 880°C?

4.28 What is the standard heat for the reaction of Prob. 4.20(r) at 400°C?

4.29 What is the standard heat for the reaction of Prob. 4.20(t) at 770(°F) [410°C]

4.30 What is the standard heat for the reaction of Prob. 4.20(u) at 700 K?

4.31 What is the standard heat for the reaction of Prob. 4.20(v) at 800 K?

4.32 What is the standard heat for the reaction of Prob. 4.20(w) at 400°C?

4.33 What is the standard heat for the reaction of Prob. 4.20(x) at 300°C?

4.34 What is the standard heat for the reaction of Prob. 4.20(y) at 1,535(°F) [835°C]?

4.35 Develop a general equation for the standard heat of reaction as a function of temperature for one of the reactions given in parts (*a*), (*b*), (*e*), (*f*), (*g*), (*h*), (*j*), (*k*), (*l*), (*m*), (*n*), (*o*), (*r*), (*t*), (*u*), (*v*), (*w*), (*x*), (*y*), and (*z*) of Prob. 4.20.

4.36 Hydrocarbon fuels can be produced from methanol by reactions such as the following, which yields 1-hexene:

$$6CH_3OH(g) \rightarrow C_6H_{12}(g) + 6H_2O(g)$$

Compare the standard heat of combustion at 25°C of $6CH_3OH(g)$ with the standard heat of combustion at 25°C of $C_6H_{12}(g)$, reaction products in both cases being $CO_2(g)$ and $H_2O(g)$.

4.37 Calculate the theoretical flame temperature when methane at 25°C is burned with

(*a*) The stoichiometric amount of air at 25°C.
(*b*) 25 percent excess air at 25°C.
(*c*) 50 percent excess air at 25°C.
(*d*) 100 percent excess air at 25°C.
(*e*) 50 percent excess air preheated to 500°C.

4.38 What is the standard heat of combustion of hexane gas at 25°C if the combustion products are $H_2O(l)$ and $CO_2(g)$?

4.39 A light fuel oil with an average chemical composition of C_9H_{15} is burned with oxygen in a bomb calorimeter. The heat evolved is measured as 47,730 J g^{-1} for the reaction at 25°C. Calculate the standard heat of combustion of the fuel oil at 25°C with $H_2O(g)$ and $CO_2(g)$ as products. Note that the reaction in the bomb occurs at constant volume, produces liquid water as a product, and goes to completion.

4.40 Methane gas is burned completely with 20 percent excess air at approximately atmospheric pressure. Both the methane and the air enter the furnace at 25°C saturated with water vapor, and the flue gases leave the furnace at 1,600°C. The flue gases then pass through a heat exchanger from which they emerge at 40°C. On the basis of 1 mol of methane, how much heat is lost from the furnace, and how much heat is transferred in the heat exchanger?

4.41 Ammonia gas enters the reactor of a nitric acid plant mixed with 25 percent more dry air than is required for the complete conversion of the ammonia to nitric oxide and water vapor. If the gases enter the reactor at 185(°F) [85°C], if conversion is 85 percent, if no side reactions occur, and if the reactor operates adiabatically, what is the temperature of the gases leaving the reactor? Assume ideal gases.

4.42 Sulfur dioxide gas is oxidized in 100 percent excess air with 80 percent conversion to sulfur trioxide. The gases enter the reactor at 770(°F) [410°C] and leave at 860(°F) [460°C]. How much heat must be transferred from the reactor on the basis of 1 (lb mol) [1 mol] of entering gas?

4.43 A fuel consisting of 75 mol percent ethane and 25 mol percent methane enters a furnace with 100 percent excess air at 25°C. If 10^6 kJ per kg mole of fuel is transferred as heat to boiler tubes, at what temperature do the flue gases leave the furnace? Assume complete combustion of the fuel.

4.44 The gas stream from a sulfur burner consists of 15 mole percent SO_2, 20 mole percent O_2, and 65 mole percent N_2. The gas stream at atmospheric pressure and 480°C enters a catalytic converter where 90 percent of the SO_2 is further oxidized to SO_3. On the basis of 1 mol of gas entering, how much heat must be removed from the converter so that the product gases leave at 480°C?

4.45 The gas-stream feed for the oxidation of ethylene to ethylene oxide is composed of 8 mole percent C_2H_4, 19 mole percent O_2, and 73 mole percent N_2. The feed stream at atmospheric pressure and 200°C enters a catalytic converter where 60 percent of the ethylene is converted to ethylene oxide and 30 percent is burned to carbon dioxide and water. On the basis of 1 mol of gas entering, how much heat must be removed from the converter so that the product gases leave at 260°C?

4.46 Hydrogen is produced by the reaction

$$CO(g) + H_2O(g) \rightarrow CO_2(g) + H_2(g)$$

The feed stream to the reactor is composed of 40 mole percent CO and 60 mole percent steam, and it enters the reactor at 150°C and atmospheric pressure. If 60 percent of the H_2O is converted to H_2

and if the product stream leaves the reactor at 450°C, how much heat must be transferred from the reactor?

4.47 A direct-fired drier burns a fuel oil with a net heating value of $19{,}000(\text{Btu})(\text{lb}_m)^{-1}$. (The net heating value is obtained when the products of combustion are $CO_2(g)$ and $H_2O(g)$.) The composition of the oil is 85 percent carbon, 12 percent hydrogen, 2 percent nitrogen, and 1 percent water by weight. The flue gases leave the drier at 400(°F), and a partial analysis shows that they contain 3 mole percent CO_2 and 11.8 mole percent CO on a dry basis. The fuel, air, and material being dried enter the drier at 77(°F). If the entering air is saturated with water and if 30 percent of the net heating value of the oil is allowed for heat losses (including the sensible heat carried out with the dried product), how much water is evaporated in the drier per (lb_m) of oil burned?

4.48 Propane is converted to ethylene and methane in a thermal cracking operation by the reaction

$$C_3H_8(g) \rightarrow C_2H_4(g) + CH_4(g)$$

Propane enters the cracker at 200°C at a rate of 1.25 kg s^{-1}, and heat transfer to the reactor is at the rate of 3,200 kJ s^{-1}. For 60 percent conversion of the propane, what is the temperature of the gas mixture leaving the cracker?

4.49 Chlorine is produced by the reaction

$$4HCl(g) + O_2(g) \rightarrow 2H_2O(g) + 2Cl_2(g)$$

The feed stream to the reactor consists of 67 mole percent HCl, 30 mole percent O_2, and 3 mole percent N_2, and it enters the reactor at 500°C. If the conversion of HCl is 75 percent and if the process is isothermal, how much heat must be transferred from the reactor per mole of the entering gas mixture?

4.50 A gas consisting of CO and N_2 is made by passing a mixture of flue gas and air through a bed of incandescent coke (assume pure carbon). The two reactions that occur both go to completion:

$$CO_2 + C \rightarrow 2CO$$

$$2C + O_2 \rightarrow 2CO$$

In a particular instance the flue gas that is mixed with air contains 13.7 mole percent CO_2, 3.4 mole percent CO, 5.1 mole percent O_2, and 77.8 mole percent N_2. The flue gas/air mixture is so proportioned that the heats of the two reactions cancel, and the temperature of the coke bed is therefore constant. If this temperature is 900°C, if the feed stream is preheated to 900°C, and if the process is adiabatic, what ratio of moles of flue gas to moles of air is required, and what is the composition of the gas produced?

4.51 A fuel gas consisting of 93 mole percent methane and 7 mole percent nitrogen is burned with 30 percent excess air in a continuous water heater. Both fuel gas and air enter dry at 25°C and atmospheric pressure. Water is heated at a rate of $75(\text{lb}_m)(\text{s})^{-1}$ [34.0 kg s^{-1}] from 59(°F) [15°C] to 185(°F) [85°C]. The flue gases leave the heater at 392(°F) [200°C]. Of the entering methane, two-thirds burns to carbon dioxide and one-third burns to carbon monoxide. What volumetric flow rate of fuel gas is required if there are no heat losses to the surroundings?

4.52 A process for the production of 1,3-butadiene results from the catalytic dehydrogenation of 1-butene according to the reaction

$$C_4H_8(g) \rightarrow C_4H_6(g) + H_2(g)$$

In order to suppress side reactions, the 1-butene feed stream is diluted with steam in the ratio of 12 moles of steam per mole of 1-butene. The reaction is carried out *isothermally* at 500°C, and at this temperature 30 percent of the 1-butene is converted to 1,3-butadiene. How much heat is transferred to the reactor per mole of entering 1-butene? Since the reaction is carried out at atmospheric pressure, the gases may be assumed ideal.

CHAPTER

FIVE

THE SECOND LAW OF THERMODYNAMICS

Thermodynamics is concerned with transformations of energy, and the laws of thermodynamics describe the bounds within which these transformations are observed to occur. The first law, stating that energy is conserved in any ordinary process, imposes no restriction on the process direction. Yet, all experience indicates the existence of a restriction. Its formulation completes the foundation for the science of thermodynamics and its concise statement constitutes the second law.

The differences between the two forms of energy, heat and work, provide some insight into the second law. In an energy balance, both work and heat are included as simple additive terms, implying that one unit of heat, a joule, is equivalent to the same unit of work. Although this is true with respect to an energy balance, experience teaches that there is a difference in quality between heat and work. This experience is summarized by the following facts.

Work is readily transformed into other forms of energy: for example, into potential energy by elevation of a weight, into kinetic energy by acceleration of a mass, into electrical energy by operation of a generator. These processes can be made to approach a conversion efficiency of 100 percent by elimination of friction, a dissipative process that transforms work into heat. Indeed, work is readily transformed completely into heat, as demonstrated by Joule's experiments.

On the other hand, all efforts to devise a process for the continuous conversion of heat completely into work or into mechanical or electrical energy have failed. Regardless of improvements to the devices employed, conversion efficiencies do not exceed about 40 percent. These low values lead to the conclusion that heat is a form of energy intrinsically less useful and hence less valuable than an equal quantity of work or mechanical or electrical energy.

Drawing further on our experience, we know that the flow of heat between two bodies always takes place from the hotter to the cooler body, and never in the reverse direction. This fact is of such significance that its restatement serves as an acceptable expression of the second law.

5.1 STATEMENTS OF THE SECOND LAW

The observations just described are results of the restriction imposed by the second law on the directions of actual processes. Many general statements may be made which describe this restriction and, hence, serve as statements of the second law. Two of the most common are:

1. No apparatus can operate in such a way that its *only* effect (in system and surroundings) is to convert heat absorbed by a system completely into work.
2. No process is possible which consists solely in the transfer of heat from one temperature level to a higher one.

Statement 1 does not imply that heat cannot be converted into work but that the process cannot leave both the system and its surroundings unchanged. Consider a system consisting of an ideal gas in a piston-and-cylinder assembly expanding reversibly at constant temperature. Work is produced equal to $\int P dV$, and for an ideal gas $\Delta U = 0$. Thus, according to the first law, the heat absorbed by the gas from the surroundings is equal to the work produced by the reversible expansion of the gas. At first this might seem a contradiction of statement 1, since in the surroundings the only result has been the complete conversion of heat into work. However, the second-law statement requires that there also be no change in the system, a requirement which has not been met.

This process is limited in another way, because the pressure of the gas soon reaches that of the surroundings, and expansion ceases. Therefore, the continuous production of work from heat by this method is impossible. If the original state of the system is restored in order to comply with the requirements of statement 1, energy from the surroundings in the form of work is needed to compress the gas back to its original pressure. At the same time energy as heat is transferred to the surroundings to maintain constant temperature: This reverse process requires at least the amount of work gained from the expansion; hence no net work is produced. Evidently, statement 1 may be expressed in an alternative way, viz.:

1*a*. It is impossible by a cyclic process to convert the heat absorbed by a system completely into work.

The word *cyclic* requires that the system be restored periodically to its original state. In the case of a gas in a piston-and-cylinder assembly the expansion and compression back to the original state constitute a complete cycle. If the process is repeated, it becomes a cyclic process. The restriction to a *cyclic* process in statement 1*a* amounts to the same limitation as that introduced by the words *only effect* in statement 1.

The second law does not prohibit the production of work from heat, but it does place a limit on the fraction of the heat that may be converted to work in any cyclic process. The partial conversion of heat into work is the basis for nearly all commercial production of power (water power is an exception). The development of a quantitative expression for the efficiency of this conversion is the next step in the treatment of the second law.

5.2 THE HEAT ENGINE

The classical approach to the second law is based on a *macroscopic* viewpoint of properties independent of any knowledge of the structure of matter or behavior of molecules. It arose from study of the *heat engine,* a device or machine that produces work from heat in a cyclic process. An example is a steam power plant in which the working fluid (steam) periodically returns to its original state. In such a power plant the cycle (in simple form) consists of the following steps:

1. Liquid water at approximately ambient temperature is pumped into a boiler.
2. Heat from a fuel (heat of combustion of a fossil fuel or heat from a nuclear reaction) is transferred in the boiler to the water, converting it to steam at high temperature and pressure.
3. Energy is transferred as shaft work from the steam to the surroundings by a device such as a turbine.
4. Exhaust steam from the turbine is condensed by the transfer of heat to cooling water, thus completing the cycle.

Essential to all heat-engine cycles are the absorption of heat at a high temperature, the rejection of heat at a lower temperature, and the production of work. In the theoretical treatment of heat engines, the two temperature levels which characterize their operation are maintained by *heat reservoirs,* bodies imagined capable of absorbing or rejecting an infinite quantity of heat without temperature change. In operation, the working fluid of a heat engine absorbs heat $|Q_H|$ from a hot reservoir, produces a net amount of work W, discards heat $|Q_C|$ to a cold reservoir, and returns to its initial state. The first law therefore reduces to

$$W = Q = |Q_H| - |Q_C| \tag{5.1}$$

Defining the *thermal efficiency* of the engine as

$$\eta \equiv \frac{\text{net work output}}{\text{heat input}}$$

we get

$$\eta \equiv \frac{W}{|Q_H|} = \frac{|Q_H| - |Q_C|}{|Q_H|}$$

or

$$\eta = 1 - \frac{|Q_C|}{|Q_H|} \tag{5.2}$$

Absolute-value signs are used with the heat quantities to make the equations independent of the sign convention for Q. We note that for η to be unity (100 percent thermal efficiency) $|Q_C|$ must be zero. No engine has ever been built for which this is true; some heat is always rejected to the cold reservoir. This result of engineering experience is the basis for statements 1 and 1*a* of the second law.

If a thermal efficiency of 100 percent is not possible for heat engines, what then determines the upper limit? One would certainly expect the thermal efficiency of a heat engine to depend on the degree of reversibility of its operation. Indeed, a heat engine operating in a completely reversible manner is very special, and is called a *Carnot engine.* The characteristics of such an ideal engine were first described by N. L. S. Carnot† in 1824. The four steps that make up a *Carnot cycle* are performed in the following order:

1. A system initially in thermal equilibrium with a cold reservoir at temperature T_C undergoes a *reversible* adiabatic process that causes its temperature to rise to that of a hot reservoir at T_H.
2. The system maintains contact with the hot reservoir at T_H, and undergoes a *reversible* isothermal process during which heat $|Q_H|$ is absorbed from the hot reservoir.
3. The system undergoes a *reversible* adiabatic process in the opposite direction of step 1 that brings its temperature back to that of the cold reservoir at T_C.
4. The system maintains contact with the reservoir at T_C, and undergoes a *reversible* isothermal process in the opposite direction of step 2 that returns it to its initial state with rejection of heat $|Q_C|$ to the cold reservoir.

A Carnot engine operates between two heat reservoirs in such a way that all heat absorbed is absorbed at the constant temperature of the hot reservoir and all heat rejected is rejected at the constant temperature of the cold reservoir. Any *reversible* engine operating between two heat reservoirs is a Carnot engine; an engine operating on a different cycle must necessarily transfer heat across finite temperature differences and therefore cannot be reversible.

Since a Carnot engine is reversible, it may be operated in reverse; the Carnot cycle is then traversed in the opposite direction, and it becomes a reversible

† Nicolas Léonard Sadi Carnot (1796–1832), a French engineer.

refrigeration cycle for which the quantities $|Q_H|$, $|Q_C|$, and $|W|$ are the same as for the engine cycle but are reversed in direction.

Carnot's theorem states that for two given heat reservoirs no engine can have a higher thermal efficiency than a Carnot engine. Consider a Carnot engine that absorbs heat $|Q_H|$ from a hot reservoir, produces work $|W|$, and discards heat $|Q_H| - |W|$ to a cold reservoir. Assume a second engine E *with a greater thermal efficiency* operating between the same heat reservoirs, absorbing heat $|Q'_H|$, producing the same work $|W|$, and discarding heat $|Q'_H| - |W|$. Then

$$\frac{|W|}{|Q'_H|} > \frac{|W|}{|Q_H|}$$

whence

$$|Q_H| > |Q'_H|$$

Let engine E drive the Carnot engine backward as a Carnot refrigerator, as shown schematically in Fig. 5.1. For the engine/refrigerator combination, the net heat extracted from the cold reservoir is

$$|Q_H| - |W| - (|Q'_H| - |W|) = |Q_H| - |Q'_H|$$

The net heat delivered to the hot reservoir is also $|Q_H| - |Q'_H|$. Thus, the sole result of the engine/refrigerator combination is the transfer of heat from tem-

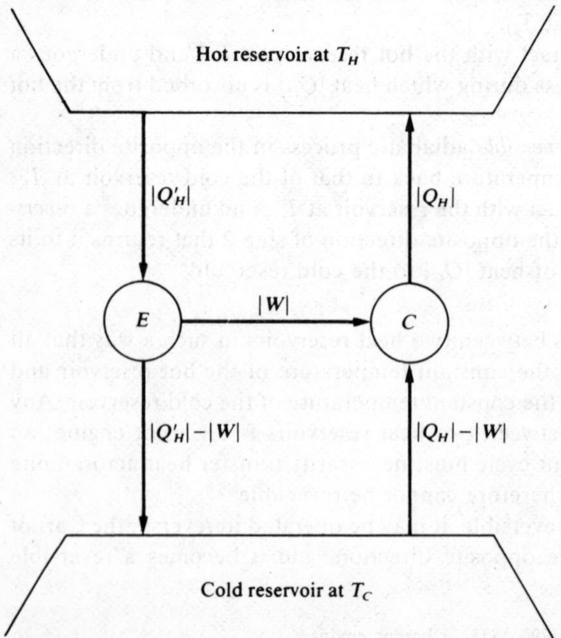

Figure 5.1 Engine E operating a Carnot refrigerator C.

perature T_C to the higher temperature T_H. Since this is in violation of statement 2 of the second law, the original premise that engine E has a greater thermal efficiency than the Carnot engine is false, and Carnot's theorem is proved. In similar fashion, one can prove a corollary to Carnot's theorem: All Carnot engines operating between heat reservoirs at the same two temperatures have the same thermal efficiency. These results show that the thermal efficiency of a Carnot engine depends only on the temperature levels T_H and T_C and not upon the working substance of the engine.

5.3 THERMODYNAMIC TEMPERATURE SCALES

In the preceding discussion we identified temperature levels by the Kelvin scale, established with ideal-gas thermometry. This does not preclude our taking advantage of the opportunity provided by the Carnot engine to establish a *thermodynamic* temperature scale that is truly independent of any material properties. Let θ represent temperature on some empirical scale that unequivocally identifies temperature levels. Consider now two Carnot engines, one operating between a hot reservoir at θ_H and a cold reservoir at temperature θ_C, and a second operating between the reservoir at θ_C and a still colder reservoir at θ_F, as shown in Fig. 5.2. The heat rejected by the first engine $|Q_C|$ is absorbed by the second; therefore the two engines working together constitute a third Carnot engine absorbing heat $|Q_H|$ from the reservoir at θ_H and rejecting heat $|Q_F|$ to the reservoir at θ_F. According to Carnot's theorem, the thermal efficiency of the first engine is a

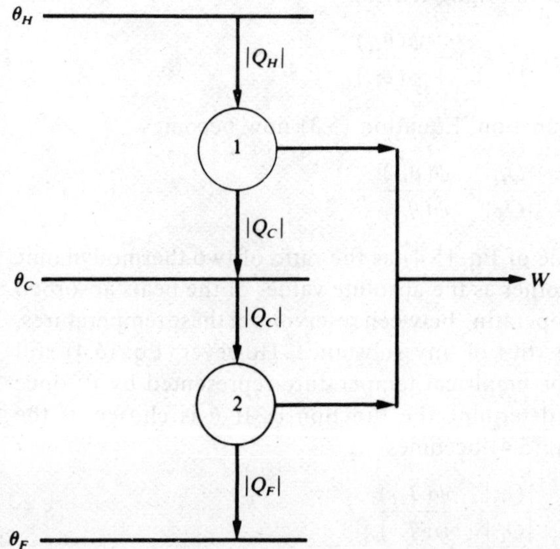

Figure 5.2 Carnot engines 1 and 2 together constitute a third Carnot engine.

function of θ_H and θ_C:

$$\eta = 1 - \frac{|Q_C|}{|Q_H|} = \phi(\theta_H, \theta_C)$$

Rearrangement gives

$$\frac{|Q_H|}{|Q_C|} = \frac{1}{1 - \phi(\theta_H, \theta_C)} = f(\theta_H, \theta_C) \tag{5.3}$$

where f is an unknown function.

For the second and third engines, equations of the same functional form apply:

$$\frac{|Q_C|}{|Q_F|} = f(\theta_C, \theta_F)$$

and

$$\frac{|Q_H|}{|Q_F|} = f(\theta_H, \theta_F)$$

As a result of the identity,

$$\frac{|Q_H|}{|Q_C|} \equiv \frac{|Q_H|/|Q_F|}{|Q_C|/|Q_F|}$$

we can also write,

$$f(\theta_H, \theta_C) = \frac{f(\theta_H, \theta_F)}{f(\theta_C, \theta_F)}$$

Since the arbitrary temperature θ_F does not appear on the left in this equation, it must cancel from the ratio on the right, leaving

$$f(\theta_H, \theta_C) = \frac{\psi(\theta_H)}{\psi(\theta_C)}$$

where ψ is another unknown function. Equation (5.3) now becomes

$$\frac{|Q_H|}{|Q_C|} = \frac{\psi(\theta_H)}{\psi(\theta_C)} \tag{5.4}$$

We may define the right side of Eq. (5.4) as the ratio of two thermodynamic temperatures; they are to each other as the absolute values of the heats absorbed and rejected by Carnot engines operating between reservoirs at these temperatures, quite independent of the properties of any substance. However, Eq. (5.4) still leaves us arbitrary choice of the empirical temperature represented by θ; once this choice is made, we must determine the function ψ. If θ is chosen as the Kelvin temperature T, then Eq. (5.4) becomes

$$\frac{|Q_H|}{|Q_C|} = \frac{\psi(T_H)}{\psi(T_C)} \tag{5.5}$$

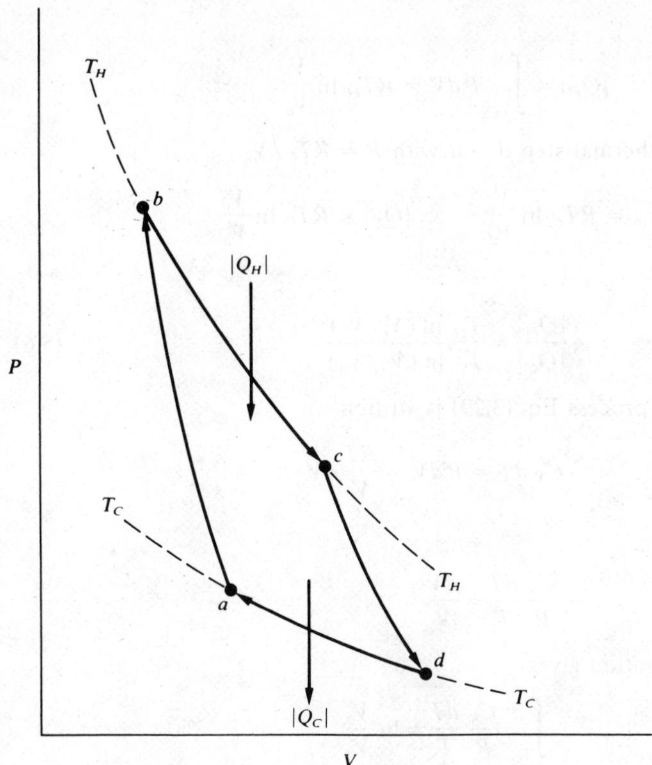

Figure 5.3 *PV* diagram showing Carnot cycle for an ideal gas.

5.4 CARNOT CYCLE FOR AN IDEAL GAS; THE KELVIN SCALE AS A THERMODYNAMIC TEMPERATURE SCALE

The cycle traversed by an ideal gas serving as the working fluid in a Carnot engine is shown by a *PV* diagram in Fig. 5.3. It consists of four reversible steps:

1. $a \rightarrow b$ Adiabatic compression until the temperature rises from T_C to T_H.
2. $b \rightarrow c$ Isothermal expansion to arbitrary point c with absorption of heat $|Q_H|$.
3. $c \rightarrow d$ Adiabatic expansion until the temperature decreases to T_C.
4. $d \rightarrow a$ Isothermal compression to the initial state with rejection of heat $|Q_C|$.

For any reversible process with an ideal gas as the system, the first law is given by Eq. (3.29):

$$dQ = C_V \, dT + P \, dV \tag{3.29}$$

For the isothermal step $b \rightarrow c$ with $P = RT_H/V$, Eq. (3.29) may be integrated to give:

$$|Q_H| = \int_{V_b}^{V_c} P\,dV = RT_H \ln \frac{V_c}{V_b}$$

Similarly, for the isothermal step $d \to a$ with $P = RT_C/V$,

$$Q_{ab} = RT_C \ln \frac{V_a}{V_d} \qquad |Q_C| = RT_C \ln \frac{V_d}{V_a}$$

Therefore

$$\frac{|Q_H|}{|Q_C|} = \frac{T_H}{T_C} \frac{\ln (V_c/V_b)}{\ln (V_d/V_a)} \tag{5.6}$$

For an adiabatic process Eq. (3.29) is written

$$-C_V\,dT = P\,dV = \frac{RT}{V}\,dV$$

or

$$-\frac{C_V}{R}\frac{dT}{T} = \frac{dV}{V}$$

For step $a \to b$, integration gives:

$$\int_{T_C}^{T_H} \frac{C_V}{R}\frac{dT}{T} = \ln \frac{V_a}{V_b}$$

Similarly, for step $c \to d$,

$$\int_{T_C}^{T_H} \frac{C_V}{R}\frac{dT}{T} = \ln \frac{V_d}{V_c}$$

Since the left-hand sides of these two equations are the same,

$$\ln \frac{V_a}{V_b} = \ln \frac{V_d}{V_c}$$

This may also be written

$$\ln \frac{V_c}{V_b} = \ln \frac{V_d}{V_a}$$

Equation (5.6) now becomes

$$\boxed{\frac{|Q_H|}{|Q_C|} = \frac{T_H}{T_C}} \tag{5.7}$$

Comparison of this result with Eq. (5.5) yields the simplest possible functional relation for ψ, namely, $\psi(T) = T$. We conclude that the Kelvin temperature scale, based on the properties of ideal gases, is in fact a thermodynamic scale, independent of the characteristics of any particular substance. Substitution of Eq. (5.7) into Eq. (5.2) gives

$$\boxed{\eta = 1 - \frac{T_C}{T_H}} \tag{5.8}$$

Equations (5.7) and (5.8) are known as *Carnot's equations.* In Eq. (5.7) the smallest possible value of $|Q_C|$ is zero; the corresponding value of T_C is the absolute zero of temperature on the Kelvin scale. As mentioned in Sec. 1.4, this occurs at −273.15°C. Equation (5.8) shows that the thermal efficiency of a Carnot engine can approach unity only when T_H approaches infinity or T_C approaches zero. On earth nature provides heat reservoirs at neither of these conditions; all heat engines therefore operate at thermal efficiencies less than unity. The cold reservoirs naturally available are the atmosphere, lakes and rivers, and the oceans, for which $T_C \simeq 300$ K. Practical hot reservoirs are objects such as furnaces maintained at high temperature by combustion of fossil fuels and nuclear reactors held at high temperature by fission of radioactive elements, for which $T_H \simeq 600$ K. With these values,

$$\eta = 1 - \frac{300}{600} = 0.5$$

This is a rough practical limit for the thermal efficiency of a Carnot engine; actual heat engines are irreversible, and their thermal efficiencies rarely exceed 0.35.

Example 5.1 A central power plant, rated at 800,000 kW, generates steam at 585 K and discards heat to a river at 295 K. If the thermal efficiency of the plant is 70 percent of the maximum possible value, how much heat is discarded to the river at rated power?

SOLUTION The maximum possible thermal efficiency is given by Eq. (5.8). Taking T_H as the steam-generation temperature and T_C as the river temperature, we get

$$\eta_{\text{max}} = 1 - \frac{295}{585} = 0.4957$$

The actual thermal efficiency is then

$$\eta = (0.7)(0.4957) = 0.3470$$

By definition

$$\eta = \frac{W}{|Q_H|}$$

Substituting for $|Q_H|$ by Eq. (5.1) gives

$$\eta = \frac{W}{W + |Q_C|}$$

which may be solved for $|Q_C|$:

$$|Q_C| = \left(\frac{1-\eta}{\eta}\right) W$$

Whence

$$|Q_C| = \left(\frac{1-0.347}{0.347}\right)(800{,}000) = 1{,}505{,}500 \text{ kW}$$

or

$$|Q_C| = 1{,}505{,}500 \text{ kJ s}^{-1}$$

This amount of heat would raise the temperature of a moderate-size river several degrees Celsius.

5.5 ENTROPY

Equation (5.7) for a Carnot engine may be written

$$\frac{|Q_H|}{T_H} = \frac{|Q_C|}{T_C}$$

If the heat quantities refer to the engine (rather than to the heat reservoirs), the numerical value of Q_H is positive and that of Q_C is negative. The equivalent equation written without absolute-value signs is therefore

$$\frac{Q_H}{T_H} = \frac{-Q_C}{T_C}$$

or

$$\frac{Q_H}{T_H} + \frac{Q_C}{T_C} = 0 \tag{5.9}$$

Thus for a complete cycle of a Carnot engine, the two quantities Q/T associated with the absorption and rejection of heat by the working fluid of the engine sum to zero. Since the working fluid of a Carnot engine periodically returns to its initial state, such properties as temperature, pressure, and internal energy return to their initial values even though they vary from one part of the cycle to another. The principal characteristic of a property is that the sum of its changes is zero for any complete cycle. Thus Eq. (5.9) suggests the existence of a property whose changes are here given by the quantities Q/T.

Further insight may be gained by study of an arbitrary reversible cyclic process, as represented schematically on a *PV* diagram in Fig. 5.4. We divide the entire closed area by a series of reversible adiabatic curves; since such curves cannot intersect (see Prob. 5.1), they may be drawn arbitrarily close to one another. A few of these curves are shown on the figure as long dashed lines. We connect adjacent adiabatic curves by two short reversible isotherms which approximate the curve of the general cycle as closely as possible. The approxima-

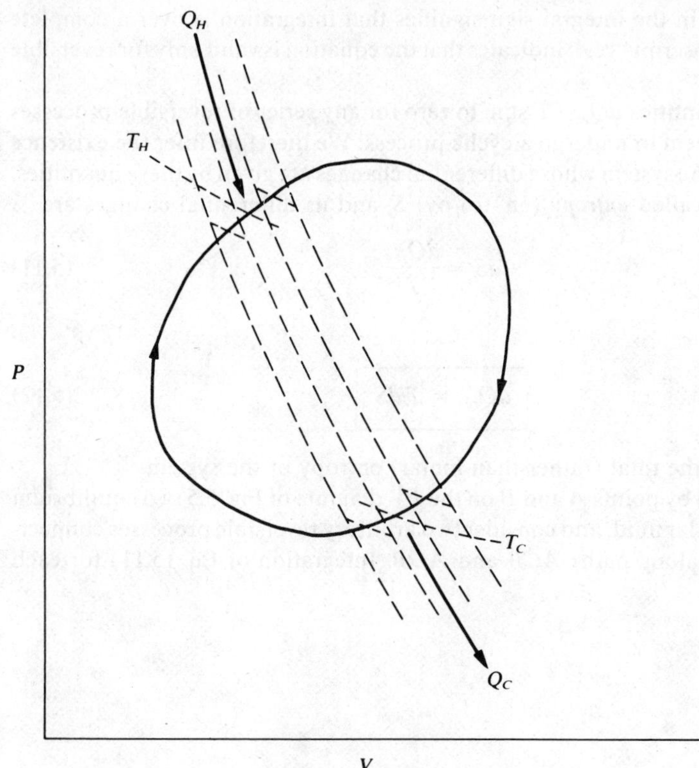

Figure 5.4 Schematic representation of an arbitrary cyclic process on a *PV* diagram.

tion clearly improves as the adiabatic curves are more closely spaced, and by making the separation arbitrarily small, we may approximate the original cycle as closely as we please. Each pair of adjacent adiabatic curves and their isothermal connecting curves represent a Carnot cycle for which Eq. (5.9) applies.

Each cycle has its own pair of isotherms T_H and T_C and associated heat quantities Q_H and Q_C. These are indicated on Fig. 5.4 for a representative cycle. When the adiabatic curves are so closely spaced that the isothermal steps are infinitesimal, the heat quantities become dQ_H and dQ_C, and Eq. (5.9) is written

$$\frac{dQ_H}{T_H} + \frac{dQ_C}{T_C} = 0$$

In this equation T_H and T_C are the absolute temperatures at which the quantities of heat dQ_H and dQ_C are transferred to the fluid of the cyclic process. Integration gives the sum of all quantities dQ/T for the entire cycle:

$$\oint \frac{dQ_{\text{rev}}}{T} = 0 \tag{5.10}$$

where the circle in the integral sign signifies that integration is over a complete cycle, and the subscript "rev" indicates that the equation is valid only for reversible cycles.

Thus the quantities dQ_{rev}/T sum to zero for any series of reversible processes that causes a system to undergo a cyclic process. We therefore infer the existence of a property of the system whose differential changes are given by these quantities. The property is called *entropy* (en′-tro-py) S, and its differential changes are

$$dS = \frac{dQ_{rev}}{T} \tag{5.11}$$

whence

$$\boxed{dQ_{rev} = T\,dS} \tag{5.12}$$

where S here is the total (rather than molar) entropy of the system.

We represent by points A and B on the PV diagram of Fig. 5.5 two equilibrium states of a particular fluid, and consider two arbitrary reversible processes connecting these points along paths ACB and ADB. Integration of Eq. (5.11) for each

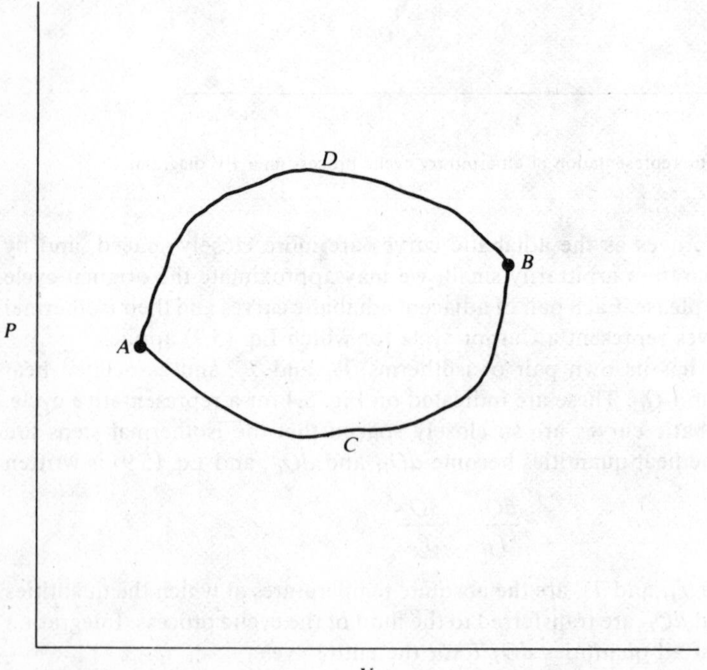

Figure 5.5 Two reversible paths joining equilibrium states A and B.

path gives

$$\Delta S = \int_{ACB} \frac{dQ_{\text{rev}}}{T}$$

and

$$\Delta S = \int_{ADB} \frac{dQ_{\text{rev}}}{T}$$

where in view of Eq. (5.10) the property change $\Delta S = S_B - S_A$ must be the same for the two paths. If the fluid is changed from state A to state B by an *irreversible* process, the entropy change must still be $\Delta S = S_A - S_B$, but experiment shows that this result is *not* given by $\int dQ/T$ evaluated for the irreversible process, because the calculation of entropy changes by this integral must in general be along reversible paths.

The entropy change of a *heat reservoir*, however, is always given by Q/T, where Q is the quantity of heat transferred to or from the reservoir at temperature T, whether the transfer is reversible or irreversible. The reason is that the effect of heat transfer on a heat reservoir is the same regardless of the temperature of the source or sink of the heat.

When a process is reversible and adiabatic, $dQ_{\text{rev}} = 0$; then by Eq. (5.11), $dS = 0$. Thus the entropy of a system is constant during a reversible adiabatic process, and the process is said to be *isentropic.*

This discussion of entropy can be summarized as follows:

1. The change in entropy of any system undergoing a *reversible* process is found by integration of Eq. (5.11):

$$\Delta S = \int \frac{dQ_{\text{rev}}}{T} \tag{A}$$

2. When a system undergoes an *irreversible* process from one equilibrium state to another, the entropy change of the system ΔS is still evaluated by Eq. (A). In this case Eq. (A) is applied to *an arbitrarily chosen reversible process* that accomplishes the same change of state. Integration is not carried out for the original irreversible path. Since entropy is a state function, the entropy changes of the irreversible and reversible processes are identical.
3. Entropy is useful precisely because it is a state function or *property.* It owes its existence to the second law, from which it arises in much the same way as internal energy does from the first law.

In the special case of a *mechanically reversible* process (Sec. 2.9), the entropy change of the system is correctly evaluated from $\int dQ/T$ applied to the actual process, even though the *heat transfer* between system and surroundings is irreversible. The reason is that it is immaterial, as far as the system is concerned, whether the temperature difference causing the heat transfer is differential (making the process reversible) or finite. The entropy change of a system *caused by the transfer of heat* can always be calculated by $\int dQ/T$, whether the heat transfer is accomplished reversibly or irreversibly. However, when a process is irreversible on account of finite differences in other driving forces, such as pressure, the entropy change is not caused solely by the heat transfer, and for its calculation one must devise a reversible means of accomplishing the same change of state.

This introduction to entropy through a consideration of heat engines is the classical approach, closely following its actual historical development. A complementary approach, based on molecular concepts and statistical mechanics, is considered briefly in Sec. 5.8.

5.6 ENTROPY CHANGES OF AN IDEAL GAS

By the first law written for one mole or a unit mass of fluid,

$$dU = dQ - dW$$

For a reversible process, this becomes

$$dU = dQ_{\text{rev}} - P\,dV$$

By the definition of enthalpy,

$$H = U + PV$$

whence

$$dH = dU + P\,dV + V\,dP$$

Substitution for dU gives

$$dH = dQ_{\text{rev}} - P\,dV + P\,dV + V\,dP$$

or

$$dQ_{\text{rev}} = dH - V\,dP \tag{5.13}$$

For an ideal gas, $dH = C_P^{ig}\,dT$ and $V = RT/P$; Eq. (5.13) then becomes

$$dQ_{\text{rev}} = C_P^{ig}\,dT - \frac{RT}{P}dP$$

or

$$\frac{dQ_{\text{rev}}}{T} = C_P^{ig}\frac{dT}{T} - R\frac{dP}{P}$$

As a result of Eq. (5.11), this may be written

$$dS = C_P^{ig}\frac{dT}{T} - R\frac{dP}{P} \tag{5.14}$$

Integration from an initial state at conditions T_1 and P_1 to a final state at conditions

T_2 and P_2 gives

$$\Delta S = \int_{T_1}^{T_2} C_P^{ig} \frac{dT}{T} - R \ln \frac{P_2}{P_1} \tag{5.15}$$

Although *derived* for a reversible process, this equation relates properties only, and is independent of the process causing the change of state. It is therefore a general equation for the calculation of entropy changes of an ideal gas.

Equation (4.4), giving the temperature dependence of the molar heat capacity C_P^{ig}, allows integration of the first term on the right of Eq. (5.15). For this purpose, we define a mean heat capacity for the integral by an equation analogous to Eq. (4.6):

$$C_{P_{ms}}^{ig} \equiv \frac{\int_{T_1}^{T_2} C_P^{ig}\, dT/T}{\ln (T_2/T_1)} \,. \tag{5.16}$$

Here, the subscript "ms" denotes a mean value specific to entropy calculations. When Eq. (4.4) is substituted for C_P^{ig} in Eq. (5.16), integration gives

$$\frac{C_{P_{ms}}^{ig}}{R} = A + BT_{lm} + T_{am} T_{lm} \left[C + \frac{D}{(T_1 T_2)^2} \right] \tag{5.17}$$

where T_{am} is the arithmetic-mean temperature, and T_{lm} is the logarithmic-mean temperature, defined as

$$T_{lm} \equiv \frac{T_2 - T_1}{\ln (T_2/T_1)}$$

Solving for the integral in Eq. (5.16), we get

$$\int_{T_1}^{T_2} C_P^{ig} \frac{dT}{T} = C_{P_{ms}}^{ig} \ln \frac{T_2}{T_1}$$

and Eq. (5.15) becomes

$$\boxed{\Delta S = C_{P_{ms}}^{ig} \ln \frac{T_2}{T_1} - R \ln \frac{P_2}{P_1}} \tag{5.18}$$

This equation for the entropy change of an ideal gas finds application in the next chapter.

Example 5.2 For an ideal gas with constant heat capacities undergoing a reversible adiabatic (and therefore isentropic) process, we found earlier that

$$\frac{T_2}{T_1} = \left(\frac{P_2}{P_1}\right)^{(\gamma-1)/\gamma} \tag{3.23}$$

Show that this same equation results from application of Eq. (5.18) with $\Delta S = 0$.

SOLUTION Since C_P^{ig} is constant, $C_{P_{ms}}^{ig} = C_P^{ig}$, and Eq. (5.18) can be written:

$$\ln \frac{T_2}{T_1} = \frac{R}{C_P^{ig}} \ln \frac{P_2}{P_1}$$

whence

$$\frac{T_2}{T_1} = \left(\frac{P_2}{P_1}\right)^{R/C_P^{ig}} \tag{A}$$

For an ideal gas Eq. (3.17) gives

$$C_P^{ig} = C_V^{ig} + R$$

Upon division by C_P^{ig} this becomes

$$1 = \frac{C_V^{ig}}{C_P^{ig}} + \frac{R}{C_P^{ig}} = \frac{1}{\gamma} + \frac{R}{C_P^{ig}}$$

where $\gamma = C_P^{ig}/C_V^{ig}$. Solving for R/C_P^{ig}, we get

$$\frac{R}{C_P^{ig}} = \frac{\gamma - 1}{\gamma}$$

This transforms Eq. (A) into Eq. (3.23), as required.

Example 5.3 Methane gas at 550 K and 5 bar undergoes a reversible adiabatic expansion to 1 bar. Assuming methane an ideal gas at these conditions, what is its final temperature?

SOLUTION For this process $\Delta S = 0$, and Eq. (5.18) becomes

$$\frac{C_{P_{ms}}^{ig}}{R} \ln \frac{T_2}{T_1} = \ln \frac{P_2}{P_1} = \ln \frac{1}{5} = -1.6094$$

Since $C_{P_{ms}}^{ig}$ depends on T_2, we rearrange this equation for iterative solution:

$$\ln \frac{T_2}{T_1} = \frac{-1.6094}{C_{P_{ms}}^{ig}/R}$$

whence

$$T_2 = T_1 \exp\left(\frac{-1.6094}{C_{P_{ms}}^{ig}/R}\right) \tag{A}$$

Here, $C_{P_{ms}}^{ig}/R$ is given by Eq. (5.17) with constants from Table 4.1:

$$\frac{C_{P_{ms}}^{ig}}{R} = 1.702 + 9.081 \times 10^{-3}\, T_{lm} - 2.164 \times 10^{-6} T_{am} T_{lm} \tag{B}$$

where

$$T_{am} = \frac{550 + T_2}{2}$$

and

$$T_{lm} = \frac{550 - T_2}{\ln(550/T_2)}$$

With an initial value of $T_2 < 550$, we find a value of $C^{ig}_{P_{ms}}/R$ from Eq. (*B*) for substitution into Eq. (*A*). This yields a new value of T_2 for Eq. (*B*), and the process continues to convergence on a final value of $T_2 = 411.34$ *K*.

5.7 PRINCIPLE OF THE INCREASE OF ENTROPY; MATHEMATICAL STATEMENT OF THE SECOND LAW

Consider two heat reservoirs, one at temperature T_H and a second at the lower temperature T_C. Let a quantity of heat $|Q|$ be transferred from the hotter to the cooler reservoir. The entropy decrease of the reservoir at T_H is

$$\Delta S_H = \frac{-|Q|}{T_H}$$

and the entropy increase of the reservoir at T_C is

$$\Delta S_C = \frac{|Q|}{T_C}$$

These two entropy changes are added to give

$$\Delta S_{\text{total}} = \Delta S_H + \Delta S_C = \frac{-|Q|}{T_H} + \frac{|Q|}{T_C}$$

or

$$\Delta S_{\text{total}} = |Q|\left(\frac{T_H - T_C}{T_H T_C}\right)$$

Since $T_H > T_C$, the *total* entropy change as a result of this irreversible process is positive. We note also that ΔS_{total} becomes smaller as the difference $T_H - T_C$ gets smaller. When T_H is only infinitesimally higher than T_C, the heat transfer is reversible, and ΔS_{total} approaches zero. Thus for the process of irreversible heat transfer, ΔS_{total} is always positive, approaching zero as the process becomes reversible.

Consider now adiabatic processes wherein no heat transfer occurs. We represent on the *PV* diagram of Fig. 5.6 an *irreversible*, adiabatic expansion of a fluid from an initial equilibrium state at point *A* to a final equilibrium state at point *B*. Now suppose the fluid is restored to its initial state by a *reversible* process. If the initial process results in an entropy change of the fluid, then there must be heat transfer during the reversible restoration process such that

$$\Delta S \equiv S_A - S_B = \int_B^A \frac{dQ_{\text{rev}}}{T}$$

The original irreversible process together with the reversible restoration process constitute a cycle for which $\Delta U = 0$ and for which the work is therefore

$$W = W_{\text{irr}} + W_{\text{rev}} = Q_{\text{rev}} = \int dQ_{\text{rev}}$$

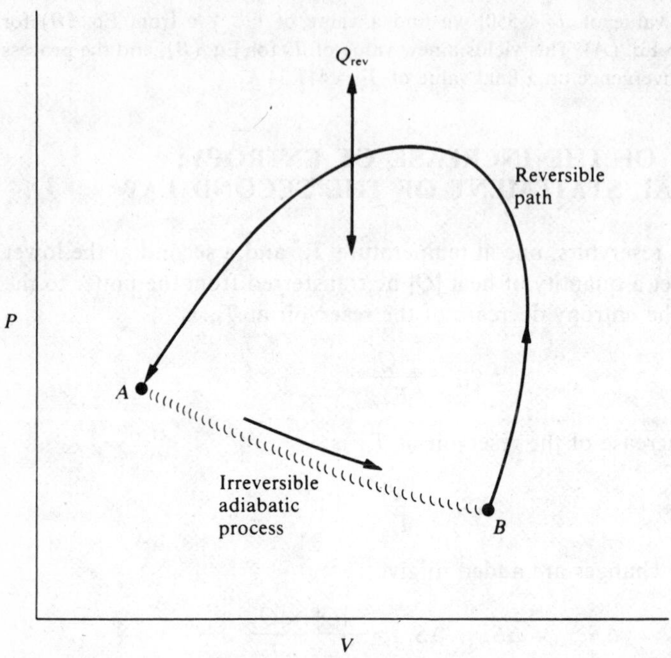

Figure 5.6 Cycle containing an irreversible adiabatic process A to B.

However, according to statement 1*a* of the second law, Q_{rev} cannot be directed *into* the system, for the cycle would then be a process for the complete conversion of heat into work. Thus, $\int dQ_{\text{rev}}$ is negative, and it follows that $S_A - S_B$ is also negative; whence $S_B > S_A$. Since the original irreversible process is adiabatic, the total entropy change as a result of this process is $\Delta S_{\text{total}} = S_B - S_A > 0$.

In arriving at this result, our presumption was that the original irreversible process results in an entropy change of the fluid. If we assume that the original process produces no entropy change of the fluid, then we can restore the system to its initial state by a simple reversible adiabatic process. This cycle is accomplished with no heat transfer and therefore with no net work. Thus the system is restored without leaving any change elsewhere, and this implies that the original process is reversible rather than irreversible.

We therefore have the same result for adiabatic processes that we found for heat transfer: ΔS_{total} is always positive, approaching zero as a limit when the process becomes reversible. This same conclusion can be demonstrated for any process whatever, and we therefore have the general equation:

$$\boxed{\Delta S_{\text{total}} \geq 0} \tag{5.19}$$

This is the mathematical statement of the second law. It affirms that every process proceeds in such a direction that the *total* entropy change associated with it is positive, the limiting value of zero being reached only by a reversible process. No process is possible for which the total entropy decreases.

Example 5.4 A steel casting [$C_P = 0.5$ kJ kg^{-1} K^{-1}] weighing 40 kg and at a temperature of 450°C is quenched in 150 kg of oil [$C_P = 2.5$ kJ kg^{-1} K^{-1}] at 25°C. If there are no heat losses, what is the change in entropy of (*a*) the casting, (*b*) the oil, and (*c*) both considered together?

SOLUTION The final temperature t of the oil and the steel casting is found by an energy balance. Since the change in energy of the oil and steel together must be zero,

$$(40)(0.5)(t-450)+(150)(2.5)(t-25)=0$$

Solution yields $t = 46.52$°C.

(*a*) Change in entropy of the casting:

$$\Delta S=\int\frac{dQ}{T}=\int\frac{C_P\,dT}{T}=C_P\ln\frac{T_2}{T_1}$$

$$\Delta S=(40)(0.5)\ln\frac{273.15+46.52}{273.15+450}=-16.33\text{ kJ K}^{-1}$$

(*b*) Change in entropy of the oil:

$$\Delta S=(150)(2.5)\ln\frac{273.15+46.52}{273.15+25}=26.13\text{ kJ K}^{-1}$$

(*c*) Total entropy change:

$$\Delta S_{\text{total}}=-16.33+26.13=9.80\text{ kJ K}^{-1}$$

We note that although the total entropy change is positive, the entropy of the casting has decreased.

Example 5.5 An inventor claims to have devised a process which takes in only saturated steam at 100°C and which by a complicated series of steps makes heat continuously available at a temperature level of 200°C. He claims further that, for every kilogram of steam taken into the process, 2,000 kJ of energy as heat is liberated at the higher temperature level of 200°C. Show whether or not this process is possible. In order to give the inventor the benefit of any doubt, assume cooling water available in unlimited quantity at a temperature of 0°C.

SOLUTION For any process to be theoretically possible, it must meet the requirements of the first and second laws of thermodynamics. The detailed mechanism need not be known in order to determine whether this is the case; only the overall result is required. If the results of the process satisfy the laws of thermodynamics, means for realizing them are theoretically possible. The determination of a mechanism is then a matter of ingenuity. Otherwise, the process is impossible, and no mechanism for carrying it out can be devised.

In the present instance, a continuous process takes saturated steam into some sort of apparatus, and heat is made continuously available at a temperature level of 200°C. Since cooling water is available at 0°C, maximum use can be made of the steam by cooling it to this temperature. We therefore assume that the steam is condensed and cooled to 0°C and is discharged from the process at this temperature and at atmospheric pressure. *All* the heat liberated in this operation cannot be made available at a temperature level of 200°C, because this would violate statement 2 of the second law. We must suppose that heat is also transferred to the cooling water at 0°C. Moreover, the process must satisfy the first law; thus by Eq. (2.11):

$$\Delta H = Q - W_s$$

where ΔH is the enthalpy change of the steam as it flows through the apparatus and Q is the total heat transfer between the apparatus and its surroundings. Since no shaft work is accomplished by the process, $W_s = 0$. The surroundings consist of the cooling water, which acts as a heat reservoir at the constant temperature of 0°C, and a heat reservoir at 200°C to which 2,000 kJ is transferred for each kilogram of steam entering the apparatus. The diagram of Fig. 5.7 pictures the overall results of the process.

The values of H and S for saturated steam at 100°C and for liquid water at 0°C are taken from the steam tables. The total heat transfer is

$$Q = -2{,}000 + Q_0$$

Thus on the basis of 1 kg of entering steam, the first law becomes

$$Q = -2{,}000 + Q_0 = \Delta H = 0.0 - 2{,}676.0 = -2{,}676.0 \text{ kJ}$$

whence

$$Q_0 = -676.0 \text{ kJ}$$

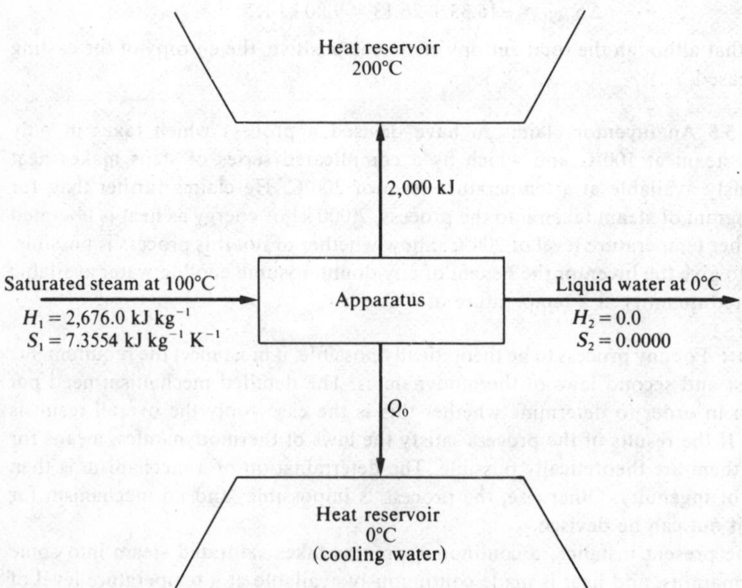

Figure 5.7 Process described in Example 5.5.

We now examine this result in the light of the second law to determine whether ΔS_{total} is greater than or less than zero for the process.

For 1 kg of steam,

$$\Delta S = 0.0000 - 7.3554 = -7.3554 \text{ kJ K}^{-1}$$

For the heat reservoir at 200°C,

$$\Delta S = \frac{2{,}000}{200 + 273.15} = 4.2270 \text{ kJ K}^{-1}$$

For the heat reservoir provided by the cooling water at 0°C,

$$\Delta S = \frac{676.0}{0 + 273.15} = 2.4748 \text{ kJ K}^{-1}$$

Thus

$$\Delta S_{\text{total}} = -7.3554 + 4.2270 + 2.4748 = -0.6536 \text{ kJ K}^{-1}$$

Since this result is negative, we conclude that the process as described is impossible; Eq. (5.19) requires that $\Delta S_{\text{total}} \geq 0$.

This does not mean that all processes of this general nature are impossible, but only that the inventor has claimed too much. Indeed, one can easily calculate the maximum amount of heat which can be transferred to the heat reservoir at 200°C, other conditions remaining the same. This calculation is left as an exercise.

5.8 ENTROPY FROM THE MICROSCOPIC VIEWPOINT (STATISTICAL THERMODYNAMICS)

Classical thermodynamics is based on a description of matter through such macroscopic properties as temperature and pressure. However, these properties are manifestations of the behavior of the countless *microscopic* particles, such as molecules, that make up a finite system. Evidently, one must seek an understanding of the fundamental nature of entropy in a microscopic description of matter. Because of the enormous number of particles contained in any system of interest, such a description must necessarily be statistical in nature. We present here a very brief indication of the statistical interpretation of entropy.†

Suppose an insulated container, partitioned into two equal volumes, contains Avogadro's number N_0 of molecules of an ideal gas in one section and no molecules in the other. When the partition is withdrawn, the molecules quickly distribute themselves uniformly throughout the total volume. The process is an adiabatic expansion that accomplishes no work. Therefore

$$\Delta U = C_V \, \Delta T = 0$$

and the temperature does not change. However, the pressure of the gas decreases

† An elementary account of statistical thermodynamics is given in H. C. Van Ness, *Understanding Thermodynamics*, chap. 7, Dover, New York, 1983.

by half, and the entropy change as given by Eq. (5.18) is

$$\Delta S = -R \ln \frac{P_2}{P_1} = R \ln 2$$

Since this is the total entropy change, the process is clearly irreversible.

Considering what happens at the molecular level, we note first that the process does not start until the partition is actually removed, and at that instant the molecules occupy only half the space available to them. In this momentary, initial state the molecules are not randomly distributed over the total volume to which they have access, but are crowded into just half the total volume. In this sense they are more ordered than they are in the final state of uniform distribution throughout the entire volume. Thus, the final state can be regarded as a more random, or less ordered, state than the initial state. From a microscopic point of view we therefore associate an entropy increase with an increase in randomness or a decrease in order at the molecular level.

These ideas were expressed mathematically by L. Boltzmann and J. W. Gibbs in terms of a quantity Ω, called the thermodynamic probability and defined as the *number of ways* that microscopic particles can be distributed among the "states" accessible to them. It is given by the general formula

$$\Omega = \frac{n!}{(n_1!)(n_2!)(n_3!)\cdots} \tag{5.20}$$

where n is the total number of particles, and n_1, n_2, n_3, etc., represent the numbers of particles in "states" 1, 2, 3, etc. The term "state" denotes the condition of the microscopic particles, and we use quotation marks to distinguish this idea of state from the usual thermodynamic meaning as applied to a macroscopic system. The thermodynamic probability is an extensive quantity, not to be identified with the mathematical probability, which is limited to values between 0 and 1. The mathematical probability is equal to Ω divided by the sum of all possible values of Ω.

With respect to our example there are but two "states," representing location in one half or the other of the container. The total number of particles is N_0 molecules, and initially they are all in a single "state." Thus

$$\Omega_1 = \frac{N_0!}{(N_0!)(0!)} = 1$$

This result confirms that initially there is just one way that the molecules can be distributed between the two accessible "states." They are all in a given "state," all in just one half of the container. For an assumed final condition of uniform distribution of the molecules between the two halves of the container, $n_1 = n_2 = N_0/2$, and

$$\Omega_2 = \frac{N_0!}{[(N_0/2)!]^2}$$

This expression gives a very large number for Ω_2, indicating that there are many ways for the molecules to be distributed equally between the two "states." Many other values of Ω_2 are possible, each one of which is associated with a particular *nonuniform* distribution of the molecules between the two halves of the container. The ratio of a particular Ω_2 to the sum of all possible values corresponds to the mathematical probability of that particular distribution.

The connection postulated by Boltzmann between entropy S and the thermodynamic probability Ω is given by the equation

$$S_2 - S_1 = k \ln \frac{\Omega_2}{\Omega_1} \tag{5.21}$$

where k is Boltzmann's constant, equal to R/N_0. Substitution of our values for Ω_1 and Ω_2 into this expression gives

$$S_2 - S_1 = k \ln \frac{N_0!}{[(N_0/2)!]^2} = k[\ln N_0! - 2\ln(N_0/2)!]$$

Since N_0 is very large, we take advantage of Stirling's formula for the logarithms of factorials of large numbers:

$$\ln X! = X \ln X - X$$

and as a result,

$$S_2 - S_1 = k\left[N_0 \ln N_0 - N_0 - 2\left(\frac{N_0}{2}\ln\frac{N_0}{2} - \frac{N_0}{2}\right)\right]$$

$$= kN_0 \ln \frac{N_0}{N_0/2} = kN_0 \ln 2 = R \ln 2$$

This is the same value for the entropy change obtained earlier from the classical thermodynamic formula for ideal gases.

In Eq. (5.21) S is the statistical average of values for many microscopic "states." If we were concerned with but a few particles distributed over a few "states," the statistical average would not be needed, because we could specify the possible distributions of the particles over the "states." However, for large collections of molecules and their many possible quantum states, the statistical approach is mandatory. Indeed, the concept we have used is not appropriate unless large numbers are involved. For example, if but two molecules (instead of N_0) were distributed between the sections, we could not assume with any confidence an equal number of molecules in each section. For a significant fraction of the time there would be two molecules in one section and none in the other.

Equation (5.20) is the basis for calculation of *absolute* entropies. In the case of an ideal gas, for example, it gives the probability Ω for the equilibrium distribution of molecules among the various quantum states determined by the translational, rotational, and vibrational energy levels of the molecules. When energy levels are assigned in accord with quantum mechanics, this procedure leads to a value for the energy as well as for the entropy. From these two quantities all other thermodynamic properties can be evaluated from definitions (of H, G, etc.). The data required are the bond distances and bond angles in the molecules, and the vibration frequencies associated with the various bonds. The procedure has been very successful in the evaluation of ideal-gas thermodynamic properties for molecules whose atomic structures are known. For nonideal gases and for liquids the molecules do not behave as independent particles, and account must be taken of the interactions between molecules. The difficulty lies in identification of the "states," particularly for liquids, and as a result the usefulness of the method is limited.

Equations (5.11) and (5.21) give *changes* in entropy; yet the previous paragraph discusses calculation of absolute entropies. These equations can be put on an absolute basis by application of the third law, discussed in the following section.

5.9 THE THIRD LAW OF THERMODYNAMICS

Measurements of heat capacities at very low temperatures provide data for the calculation from Eq. (5.11) of entropy changes down to 0 K. When these calculations are made for different crystalline forms of the same chemical species, the entropy at 0 K appears to be the same for all forms. When the form is noncrystalline, e.g., amorphous or glassy, calculations show that the entropy of the more random form is greater than that of the crystalline form. Such calculations, which are summarized elsewhere,† lead to the postulate that *the absolute entropy is zero for all perfect crystalline substances at absolute zero temperature.* While the essential ideas were advanced by Nernst and Planck at the beginning of the twentieth century, more recent studies at very low temperatures have increased our confidence in this postulate, which is now accepted as the third law.

If the entropy is zero at $T = 0$ K, then Eq. (5.11) lends itself to the calculation of absolute entropies. With $T = 0$ as the lower limit of integration, the absolute entropy of a gas at temperature T based on calorimetric data follows from Eq. (5.11) integrated to give:

$$S = \int_0^{T_f} \frac{(C_P)_s}{T}\,dT + \frac{\Delta H_f}{T_f} + \int_{T_f}^{T_v} \frac{(C_P)_l}{T}\,dT + \frac{\Delta H_v}{T_v} + \int_{T_v}^{T} \frac{(C_P)_g}{T}\,dT \qquad (5.22)$$

With respect to this equation,‡ we have supposed that there is no solid-state transition and thus no heat of transition. The only constant-temperature heat effects are those of fusion at T_f and vaporization at T_v. When a solid-phase transition occurs, a term $\Delta H_t / T_t$ is added.

If a substance is a perfect crystal at absolute zero temperature, each particle of the crystal is in its lowest quantum state, and there is but one way the particles can be arranged; the thermodynamic probability Ω is unity. If state 1 is chosen

† G. N. Lewis and M. Randall, *Thermodynamics*, 2d ed., chap. 12, McGraw-Hill, New York, 1961.

‡ Evaluation of the first term on the right is not a problem for crystalline substances because C_P / T remains finite as $T \to 0$.

to be absolute zero, Eq. (5.21) then becomes

$$S = k \ln \Omega \tag{5.23}$$

where S and Ω represent values at any finite temperature.

Both the classical and statistical equations [Eqs. (5.22) and (5.23)] yield absolute values of entropy. Equation (5.23) is known as the Boltzmann equation and, with Eq. (5.20) and quantum statistics, has been used for calculation of entropies in the ideal-gas state for many chemical species. Good agreement between these calculations and those based on calorimetric data provides some of the most impressive evidence for the validity of statistical mechanics and quantum theory. In some instances results based on Eq. (5.23) are considered more reliable because of uncertainties in heat-capacity data or about the crystallinity of the substance near absolute zero. Absolute entropies provide much of the data base for calculation of the equilibrium conversions of chemical reactions, as discussed in Chap. 15.

PROBLEMS

5.1 Prove that it is impossible for two lines representing reversible, adiabatic processes to intersect. (*Hint:* Assume that they do intersect, and complete the cycle with a line representing a reversible, isothermal process. Show that peformance of this cycle violates the second law.)

5.2 A Carnot engine receives 150 kJ s^{-1} of heat from a heat-source reservoir at 425°C and rejects heat to a heat-sink reservoir at 30°C. What are the power developed and the heat rejected?

5.3 The following heat engines produce power of 80,000 kW. Determine in each case the rates at which heat is absorbed from the hot reservoir and discarded to the cold reservoir.

(*a*) A Carnot engine operates between heat reservoirs at 600 and 300 K.

(*b*) A practical engine operates between the same heat reservoirs but with a thermal efficiency $\eta = 0.3$.

5.4 A particular power plant operates with a heat-source reservoir at 300°C and a heat-sink reservoir at 25°C. It has a thermal efficiency equal to 60 percent of the Carnot-engine thermal efficiency for the same temperatures.

(*a*) What is the thermal efficiency of the plant?

(*b*) To what temperature must the heat-source reservoir be raised to increase the thermal efficiency of the plant to 40 percent? Again η is 60 percent of the Carnot-engine value.

5.5 Large quantities of liquefied natural gas (LNG) are shipped by ocean tanker. At the unloading port provision is made for vaporization of the LNG so that it may be delivered to pipelines as gas. The LNG arrives in the tanker at atmospheric pressure and 113.7 K, and represents a possible heat sink for use as the cold reservoir of a heat engine. Assuming unloading of LNG as a vapor at the rate of 8,000 m^3 s^{-1}, as measured at 25°C and 1.0133 bar, and assuming the availability of an adequate heat source at 35°C, what is the maximum possible power obtainable and what is the rate of heat transfer from the heat source?

Assume that LNG at 25°C and 1.0133 bar is an ideal gas with a molar mass of 17. Also assume that the LNG vaporizes only, absorbing its latent heat of 512 kJ kg^{-1} at 113.7 K.

5.6 A quantity of an ideal gas, $C_P = (7/2)R$, at 20°C and 1 bar and having a volume of 70 m^3, is heated at constant pressure to 25°C by the transfer of heat from a heat reservoir at 40°C. Calculate the heat transfer to the gas, the entropy change of the heat reservoir, the entropy change of the gas, and ΔS_{total}. What is the irreversible feature of the process?

5.7 A rigid vessel of 0.05 m^3 volume contains an ideal gas, $C_V = (5/2)R$, at 500 K and 1 bar.

(a) If heat in the amount of 12,000 J is transferred to the gas, determine its entropy change.

(b) If the vessel is fitted with a stirrer that is rotated by a shaft so that work in the amount of 12,000 J is done on the gas, what is the entropy change of the gas if the process is adiabatic? What is ΔS_{total}? What is the irreversible feature of the process?

5.8 An ideal gas, $C_P = (7/2)R$, is heated in a steady-flow heat exchanger from 68(°F) [20°C] to 212(°F) [100°C] by another stream of the same ideal gas which enters at 356(°F) [180°C]. The flow rates of the two streams are the same, and heat losses from the exchanger are negligible.

(a) Calculate the molar entropy changes of the two gas streams for both parallel and countercurrent flow in the exchanger.

(b) What is ΔS_{total} in each case?

(c) Repeat parts (a) and (b) for countercurrent flow if the stream that is cooled enters an infinite heat exchanger at 212(°F) [100°C].

5.9 For an ideal gas with constant heat capacities, show that

(a) For a temperature increase from T_1 to T_2, ΔS of the gas is greater when the change occurs at constant pressure than when it occurs at constant volume.

(b) For a pressure change from P_1 to P_2, the sign of ΔS for an isothermal change is opposite that for a constant-volume change.

5.10 Imagine that a stream of fluid in steady-state flow serves as a heat source for an infinite set of Carnot engines, each of which absorbs a differential amount of heat from the fluid, causing its temperature to decrease by a differential amount, and each of which rejects a differential amount of heat to a heat reservoir at temperature T_0. As a result of the operation of the Carnot engines, the temperature T of the fluid decreases from T_1 to T_2. Equation (5.8) applies here in differential form, wherein η is defined as

$$\eta \equiv -dW/dQ$$

The minus sign is included because Q is heat transfer with respect to the flowing fluid. Show that the total work of the Carnot engines is given by

$$W = T_0 \Delta S - Q$$

where ΔS and Q both refer to the fluid.

In a particular instance the fluid is an ideal gas, $C_P = (7/2)R$, for which $T_1 = 500$ K and $T_2 = 350$ K. If $T_0 = 300$ K, what is the value of W in J mol^{-1}? How much heat is discarded to the heat reservoir at T_0? What is the entropy change of the heat reservoir? What is ΔS_{total}?

5.11 A piston/cylinder device contains 5 mol of an ideal gas, $C_P = (5/2)R$ and $C_V = (3/2)R$, at 20°C and 1 bar. The gas is compressed reversibly and adiabatically to 10 bar, where the piston is locked in position. The cylinder is then brought into thermal contact with a heat reservoir at 20°C, and heat transfer continues until the gas also reaches this temperature. Determine the entropy changes of the gas, the reservoir, and ΔS_{total}.

5.12 An ideal gas, $C_P = (7/2)R$ and $C_V = (5/2)R$, undergoes a cycle consisting of the following mechanically reversible steps:

(a) An adiabatic compression from P_1, V_1, T_1 to P_2, V_2, T_2.

(b) An isobaric expansion from P_2, V_2, T_2 to $P_3 = P_2$, V_3, T_3.

(c) An adiabatic expansion from P_3, V_3, T_3 to P_4, V_4, T_4.

(d) A constant-volume process from P_4, V_4, T_4 to P_1, $V_1 = V_4$, T_1.

Sketch this cycle on a PV diagram and determine its thermal efficiency if $T_1 = 500$ K, $T_2 = 800$ K, $T_3 = 2{,}000$ K, and $T_4 = 1{,}000$ K.

5.13 A reversible cycle executed by 1 mol of an ideal gas for which $C_P = (5/2)R$ and $C_V = (3/2)R$ consists of the following steps:

(a) Starting at 600 K and 2 bar, the gas is cooled at constant pressure to 300 K.

(b) From 300 K and 2 bar, the gas is compressed isothermally to 4 bar.

(c) The gas returns to its initial state along a path for which the product PT is constant.

What is the thermal efficiency of the cycle?

5.14 One mole of an ideal gas, $C_P = (7/2)R$ and $C_V = (5/2)R$, is compressed adiabatically in a piston/cylinder device from 1 bar and 40°C to 4 bar. The process is irreversible and requires 30 percent more work than a reversible, adiabatic compression from the same initial state to the same final pressure. What is the entropy change of the gas?

5.15 One mole of an ideal gas is compressed isothermally but irreversibly at 400 K from 3 bar to 7 bar in a piston/cylinder device. The work required is 35 percent greater than the work of reversible, isothermal compression. The heat transferred from the gas during compression flows to a heat reservoir at 300 K. Calculate the entropy changes of the gas, the heat reservoir, and ΔS_{total}.

5.16 If 10 mol of ethylene is heated from 200 to 1,000°C in a steady-flow process at approximately atmospheric pressure, what is its entropy change?

5.17 If 12 mol of 1-butene is heated from 250 to 1,200°C in a steady-flow process at approximately atmospheric pressure, what is its entropy change?

5.18 If heat in the amount of 1,300 kJ is added to 40 mol of SO_2 initially at 400°C in a steady-flow process at approximately atmospheric pressure, what is its entropy change?

5.19 If heat in the amount of 1,000 kJ is added to 30 mol of ammonia vapor initially at 250°C in a steady-flow process at approximately atmospheric pressure, what is its entropy change?

5.20 If heat in the amount of 5×10^5(Btu) [5.275×10^5 kJ] is added to 30(lb mol) [13.61 kg mol] of methane initially at 410(°F) [210°C] in a steady-flow process at approximately atmospheric pressure, what is its entropy change?

5.21 A device with no moving parts is claimed to provide a steady stream of chilled air at −20°C and 1 bar. The feed to the device is compressed air at 25°C and 4 bar. In addition to the stream of chilled air, a second stream of air flows at an equal mass rate from the device at 70°C and 1 bar. Are these claims in violation of the second law? Assume that air is an ideal gas for which $C_P = (7/2)R$.

5.22 An inventor has devised a complicated nonflow process in which 1 mol of air is the working fluid. The net effects of the process are claimed to be:

(*a*) A change in state of the air from 500 K and 2 bar to 350 K and 1 bar.

(*b*) The production of 2,000 J of work.

(*c*) The transfer of an undisclosed amount of heat to a heat reservoir at 300 K.

Determine whether the claimed performance of the process is consistent with the second law. Assume that air is an ideal gas for which $C_P = (7/2)R$.

5.23 Consider the heating of a house by a furnace, which serves as a heat-source reservoir at a high temperature T_F. The house acts as a heat-sink reservoir at temperature T, and heat $|Q|$ must be added to the house during a particular time interval to maintain this temperature. Heat $|Q|$ can of course be transferred directly from the furnace to the house, as is the usual practice. However, a third heat reservoir is readily available, namely, the surroundings at temperature T_0, which can serve as another heat source, thus reducing the amount of heat required from the furnace. Given that $T_F = 810$ K, $T = 295$ K, $T_0 = 265$ K, and $Q = 1{,}000$ kJ, determine the minimum amount of heat $|Q_F|$ which must be extracted from the heat-source reservoir (furnace) at T_F. No other sources of energy are available.

5.24 Consider the air conditioning of a house through use of solar energy. At a particular location experiment has shown that solar radiation allows a large tank of water to be maintained at 205°C. During a particular time interval, heat in the amount of 1,000 kJ must be extracted from the house to maintain its temperature at 20°C when the surroundings temperature is 32°C. Treating the tank of water, the house, and the surroundings as heat reservoirs, determine the minimum amount of heat that must be extracted from the tank of water by any device built to accomplish the required cooling of the house. No other sources of energy are available.

CHAPTER

SIX

THERMODYNAMIC PROPERTIES OF FLUIDS

The phase rule (Sec. 2.8) tells us that specification of a certain number of intensive properties of a system also establishes all other intensive properties at fixed values. However, the phase rule provides no information about how values for these other properties may be calculated.

The availability of numerical values for the thermodynamic properties is essential to the calculation of heat and work quantities for industrial processes. For example, the work requirement for a compressor designed to operate adiabatically and to raise the pressure of a gas from P_1 to P_2 is given by Eq. (2.10), which here becomes

$$-W_s = \Delta H = H_2 - H_1$$

when the small kinetic- and potential-energy changes of the gas are neglected. Thus, the shaft work is simply ΔH and depends only on the initial and final values of the enthalpy.

Our initial purpose in this chapter is to develop from the first and second laws the fundamental property relations which underlie the mathematical structure of thermodynamics. From these, we derive equations which allow calculation of enthalpy and entropy values from *PVT* and heat-capacity data. We then discuss the diagrams and tables by which both measured and calculated property values are presented for convenient use. Finally, we develop generalized correlations which allow estimates of property values to be made in the absence of complete experimental information.

6.1 RELATIONSHIPS AMONG THERMODYNAMIC PROPERTIES FOR A HOMOGENEOUS PHASE OF CONSTANT COMPOSITION

The first law for a closed system of *n* moles is given by Eq. (2.13):

$$d(nU) = dQ - dW \tag{2.13}$$

For the special case of a reversible process,

$$d(nU) = dQ_{\text{rev}} - dW_{\text{rev}}$$

and by Eqs. (2.14) and (5.12),

$$dW_{\text{rev}} = P\,d(nV)$$

and

$$dQ_{\text{rev}} = T\,d(nS)$$

These three equations combine to give

$$d(nU) = T\,d(nS) - P\,d(nV) \tag{6.1}$$

where *U*, *S*, and *V* are molar values of the internal energy, entropy, and volume.

This equation, combining the first and second laws, is *derived* for the special case of a reversible process. However, it contains only *properties* of the system. Properties depend on state alone, and not on the kind of process that produces the state. Therefore, Eq. (6.1) is not restricted in *application* to reversible processes. However, the restrictions placed on the *nature of the system* cannot be relaxed. Thus Eq. (6.1) applies to *any* process in a system of *constant mass* that results in a differential change from one equilibrium state to another. The system may consist of a single phase (a homogeneous system), or it may be made up of several phases (a heterogeneous system); it may be chemically inert, or it may undergo chemical reaction. The only requirements are that the system be closed and that the change occur between equilibrium states.

All of the *primary* thermodynamic properties—*P*, *V*, *T*, *U*, and *S*—are included in Eq. (6.1). Additional thermodynamic properties arise only by *definition* in relation to these primary properties. In Chap. 2 the enthalpy was defined as a matter of convenience by the equation:

$$\boxed{H \equiv U + PV} \tag{2.6}$$

Two additional properties, also defined for convenience, are the *Helmholtz energy*,

$$\boxed{A \equiv U - TS} \tag{6.2}$$

and the *Gibbs energy*,

$$\boxed{G \equiv H - TS} \tag{6.3}$$

Each of these defined properties leads directly to an equation like Eq. (6.1).

Upon multiplication by n, Eq. (2.6) becomes

$$nH = nU + P(nV)$$

Differentiation gives

$$d(nH) = d(nU) + P\,d(nV) + (nV)\,dP$$

When $d(nU)$ is replaced by Eq. (6.1), this reduces to

$$\boxed{d(nH) = T\,d(nS) + (nV)\,dP} \tag{6.4}$$

Similarly, we find from Eq. (6.2) that

$$d(nA) = d(nU) - T\,d(nS) - (nS)\,dT$$

Eliminating $d(nU)$ by Eq. (6.1), we get

$$\boxed{d(nA) = -P\,d(nV) - (nS)\,dT} \tag{6.5}$$

In analogous fashion, Eq. (6.3) together with Eq. (6.4) gives

$$\boxed{d(nG) = (nV)\,dP - (nS)\,dT} \tag{6.6}$$

Equations (6.4) through (6.6) have the same range of applicability as Eq. (6.1). All are written for the entire mass of any closed system.

Our immediate application of these equations is to one mole (or to a unit mass) of a homogeneous fluid of constant composition. For this case, they simplify to

$$dU = T\,dS - P\,dV \tag{6.7}$$

$$dH = T\,dS + V\,dP \tag{6.8}$$

$$dA = -P\,dV - S\,dT \tag{6.9}$$

$$dG = V\,dP - S\,dT \tag{6.10}$$

These *fundamental property relations* are general equations for a homogeneous fluid of constant composition.

Another set of equations follows from them by application of the criterion of exactness for a differential expression. If $F = F(x, y)$, then the total differential of F is defined as

$$dF = \left(\frac{\partial F}{\partial x}\right)_y dx + \left(\frac{\partial F}{\partial y}\right)_x dy$$

or

$$dF = M\,dx + N\,dy \tag{6.11}$$

where

$$M = \left(\frac{\partial F}{\partial x}\right)_y \qquad \text{and} \qquad N = \left(\frac{\partial F}{\partial y}\right)_x$$

By further differentiation we obtain

$$\left(\frac{\partial M}{\partial y}\right)_x = \frac{\partial^2 F}{\partial y\,\partial x} \qquad \text{and} \qquad \left(\frac{\partial N}{\partial x}\right)_y = \frac{\partial^2 F}{\partial x\,\partial y}$$

Since the order of differentiation in mixed second derivatives is immaterial, these equations give

$$\left(\frac{\partial M}{\partial y}\right)_x = \left(\frac{\partial N}{\partial x}\right)_y \tag{6.12}$$

When F is a function of x and y, the right-hand side of Eq. (6.11) is an *exact differential expression*; since Eq. (6.12) must then be satisfied, it serves as a criterion of exactness.

The thermodynamic properties U, H, A, and G are *known* to be functions of the variables on the right-hand sides of Eqs. (6.7) through (6.10); we may therefore write the relationship expressed by Eq. (6.12) for each of these equations:

$$\left(\frac{\partial T}{\partial V}\right)_S = -\left(\frac{\partial P}{\partial S}\right)_V \tag{6.13}$$

$$\left(\frac{\partial T}{\partial P}\right)_S = \left(\frac{\partial V}{\partial S}\right)_P \tag{6.14}$$

$$\left(\frac{\partial P}{\partial T}\right)_V = \left(\frac{\partial S}{\partial V}\right)_T \tag{6.15}$$

$$\left(\frac{\partial V}{\partial T}\right)_P = -\left(\frac{\partial S}{\partial P}\right)_T \tag{6.16}$$

These are known as *Maxwell's equations*.†

Equations (6.7) through (6.10) are the basis not only for derivation of the Maxwell equations but also of a large number of other equations relating thermodynamic properties. We develop here only a few expressions useful for evaluations of thermodynamic properties from experimental data. Their derivation requires application of Eqs. (6.8) and (6.16).

The most useful property relations for the enthalpy and entropy of a homogeneous phase result when these properties are expressed as functions of T and P. What we need to know is how H and S vary with temperature and pressure. This information is contained in the derivatives $(\partial H/\partial T)_P$, $(\partial S/\partial T)_P$, $(\partial H/\partial P)_T$, and $(\partial S/\partial P)_T$.

† After James Clark Maxwell (1831–1979), Scottish physicist.

Consider first the temperature derivatives. As a result of Eq. (2.21), which defines the heat capacity at constant pressure, we have

$$\left(\frac{\partial H}{\partial T}\right)_P = C_P \tag{2.21}$$

Another expression for this quantity is obtained by division of Eq. (6.8) by dT and restriction of the result to constant P:

$$\left(\frac{\partial H}{\partial T}\right)_P = T\left(\frac{\partial S}{\partial T}\right)_P$$

Combination of this equation with Eq. (2.21) gives

$$\left(\frac{\partial S}{\partial T}\right)_P = \frac{C_P}{T} \tag{6.17}$$

The pressure derivative of the entropy results directly from Eq. (6.16):

$$\left(\frac{\partial S}{\partial P}\right)_T = -\left(\frac{\partial V}{\partial T}\right)_P \tag{6.18}$$

The corresponding derivative for the enthalpy is found by division of Eq. (6.8) by dP and restriction to constant T:

$$\left(\frac{\partial H}{\partial P}\right)_T = T\left(\frac{\partial S}{\partial P}\right)_T + V$$

As a result of Eq. (6.18) this becomes

$$\left(\frac{\partial H}{\partial P}\right)_T = V - T\left(\frac{\partial V}{\partial T}\right)_P \tag{6.19}$$

Since the functional relations chosen here for H and S are

$$H = H(T, P) \qquad \text{and} \qquad S = S(T, P)$$

it follows that

$$dH = \left(\frac{\partial H}{\partial T}\right)_P dT + \left(\frac{\partial H}{\partial P}\right)_T dP$$

and

$$dS = \left(\frac{\partial S}{\partial T}\right)_P dT + \left(\frac{\partial S}{\partial P}\right)_T dP$$

Substituting for the partial derivatives in these two equations by Eqs. (2.21) and (6.17) through (6.19), we get

$$\boxed{dH = C_P\, dT + \left[V - T\left(\frac{\partial V}{\partial T}\right)_P\right] dP} \tag{6.20}$$

and

$$dS = C_P \frac{dT}{T} - \left(\frac{\partial V}{\partial T}\right)_P dP \tag{6.21}$$

These are general equations relating the enthalpy and entropy of homogeneous fluids of constant composition to temperature and pressure.

The coefficients of dT and dP in Eqs. (6.20) and (6.21) are evaluated from heat-capacity and PVT data. As an example of the application of these equations, we note that the PVT behavior of a fluid in the ideal-gas state is expressed by the equations:

$$PV^{ig} = RT$$

and

$$\left(\frac{\partial V^{ig}}{\partial T}\right)_P = \frac{R}{P}$$

where V^{ig} is the molar volume of an ideal gas at temperature T and pressure P. Substituting these equations into Eqs. (6.20) and (6.21) reduces them to

$$dH^{ig} = C_P^{ig}\, dT \tag{6.22}$$

and

$$dS^{ig} = C_P^{ig} \frac{dT}{T} - \frac{R}{P} dP \tag{6.23}$$

where the superscript "ig" denotes an ideal-gas value. These equations merely restate results derived for ideal gases in Chaps. 3 and 5.

Equations (6.18) and (6.19) are expressed in an alternative form by elimination of $(\partial V/\partial T)_P$ in favor of the volume expansivity β by Eq. (3.2):

$$\left(\frac{\partial S}{\partial P}\right)_T = -\beta V \tag{6.24}$$

and

$$\left(\frac{\partial H}{\partial P}\right)_T = (1 - \beta T)V \tag{6.25}$$

The pressure dependence of the internal energy is obtained by differentiation of the equation $U = H - PV$:

$$\left(\frac{\partial U}{\partial P}\right)_T = \left(\frac{\partial H}{\partial P}\right)_T - P\left(\frac{\partial V}{\partial P}\right)_T - V$$

Whence by Eqs. (6.25) and (3.3),

$$\left(\frac{\partial U}{\partial P}\right)_T = (\kappa P - \beta T)V \tag{6.26}$$

where κ is the isothermal compressibility. Equations (6.24) through (6.26), which require values of β and κ, are usually applied only to liquids.

For liquids not near the critical point, the volume itself is small, as are both β and κ. Thus at most conditions pressure has little effect on the entropy, enthalpy, and internal energy of liquids. For an *incompressible fluid* (Sec. 3.1), an idealization useful in fluid mechanics, both β and κ are zero. In this case both $(\partial S/\partial P)_T$ and $(\partial U/\partial P)_T$ are zero, and the entropy and internal energy are independent of P. However, the enthalpy of an incompressible fluid *is* a function of P, as is evident from Eq. (6.25).

When $(\partial V/\partial T)_P$ is replaced in Eqs. (6.20) and (6.21) in favor of the volume expansivity, they become

$$dH = C_P\, dT + V(1 - \beta T)\, dP \tag{6.27}$$

and

$$dS = C_P \frac{dT}{T} - \beta V\, dP \tag{6.28}$$

Since β and V are weak functions of pressure for liquids, they are usually assumed constant at appropriate average values for integration of the final terms of Eqs. (6.27) and (6.28).

Example 6.1 Determine the enthalpy and entropy changes for liquid water for a change of state from 1 bar and 25°C to 1,000 bar and 50°C. The following data for water are available.

t/°C	P/bar	C_P/J mol^{-1} K^{-1}	V/cm^3 mol^{-1}	β/K^{-1}
25	1	75.305	18.075	256×10^{-6}
25	1,000	....	17.358	366×10^{-6}
50	1	75.314	18.240	458×10^{-6}
50	1,000	....	17.535	568×10^{-6}

SOLUTION For application to the change of state described, Eqs. (6.27) and (6.28) require integration. Since enthalpy and entropy are state functions, the path of integration is arbitrary; the path most suited to the given data is shown in Fig. 6.1. Since the data indicate that C_P is a weak function of T and that both V and β are weak functions of P, integration with arithmetic averages is satisfactory. The integrated forms of Eqs. (6.27) and (6.28) that result are:

$$\Delta H = C_P^{\text{ave}}(T_2 - T_1) - V^{\text{ave}}(1 - \beta^{\text{ave}} T_2)(P_2 - P_1)$$

and

$$\Delta S = C_P^{\text{ave}} \ln \frac{T_2}{T_1} - \beta^{\text{ave}} V^{\text{ave}}(P_2 - P_1)$$

where for $P = 1$ bar

$$C_P^{\text{ave}} = \frac{75.305 + 75.314}{2} = 75.310 \text{ J mol}^{-1} \text{ K}^{-1}$$

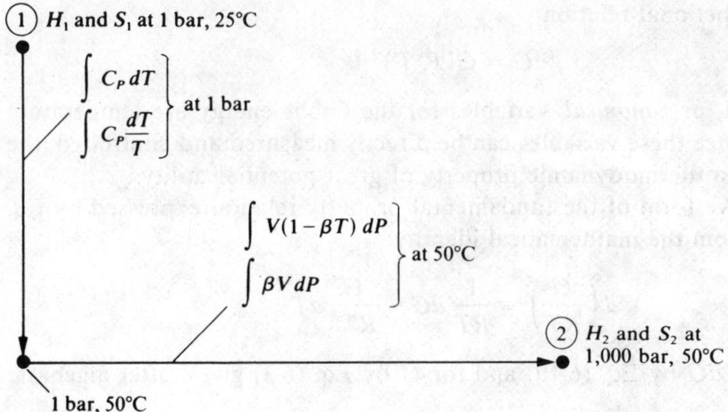

Figure 6.1

and for $t = 50°C$

$$V^{\text{ave}} = \frac{18.240 + 17.535}{2} = 17.888 \text{ cm}^3 \text{ mol}^{-1}$$

$$\beta^{\text{ave}} = \frac{458 + 568}{2} \times 10^{-6} \text{ K}^{-1}$$

Substitution of numerical values into the equation for ΔH gives

$$\Delta H = 75.310(323.15 - 298.15) + \frac{(17.888)[1 - (513 \times 10^{-6})(323.15)](1{,}000 - 1)}{10 \text{ cm}^3 \text{ bar J}^{-1}}$$

$$\Delta H = 1{,}883 + 1{,}491 = 3{,}374 \text{ J mol}^{-1}$$

Similarly for ΔS,

$$\Delta S = 75.310 \ln \frac{323.15}{298.15} - \frac{(513 \times 10^{-6})(17.888)(1{,}000 - 1)}{10 \text{ cm}^3 \text{ bar J}^{-1}}$$

$$\Delta S = 6.06 - 0.92 = 5.14 \text{ J mol}^{-1} \text{ K}^{-1}$$

Thus the effect of a pressure change of almost 1,000 bar on the enthalpy and entropy of liquid water is less than that of a temperature change of only 25°C.

6.2 RESIDUAL PROPERTIES

The fundamental property relations for homogeneous fluids of constant composition given by Eqs. (6.7) through (6.10) show that each of the thermodynamic properties U, H, A, and G is functionally related to a special pair of variables. In particular, Eq. (6.10),

$$dG = V \, dP - S \, dT \tag{6.10}$$

expresses the functional relation:

$$G = G(P, T)$$

Thus the special, or *canonical*, variables for the Gibbs energy are temperature and pressure. Since these variables can be directly measured and controlled, the Gibbs energy is a thermodynamic property of great potential utility.

An alternative form of the fundamental property relation expressed by Eq. (6.10) follows from the mathematical identity:

$$d\left(\frac{G}{RT}\right) \equiv \frac{1}{RT}\,dG - \frac{G}{RT^2}\,dT$$

Substitution for dG by Eq. (6.10) and for G by Eq. (6.3) gives, after algebraic reduction,

$$\boxed{d\left(\frac{G}{RT}\right) = \frac{V}{RT}\,dP - \frac{H}{RT^2}\,dT} \tag{6.29}$$

The advantage of this equation is that all terms are dimensionless; moreover, in contrast to Eq. (6.10), the enthalpy rather than the entropy appears on the right-hand side.

Equations such as Eq. (6.10) and (6.29) are too general for direct practical application, but they are readily applied in restricted form. Thus, from Eq. (6.29) we have immediately that:

$$\frac{V}{RT} = \left[\frac{\partial(G/RT)}{\partial P}\right]_T \tag{6.30}$$

and

$$\frac{H}{RT} = -T\left[\frac{\partial(G/RT)}{\partial T}\right]_P \tag{6.31}$$

When G/RT is known as a function of T and P, V/RT and H/RT follow by simple differentiation. The remaining properties are given by defining equations. In particular,

$$\frac{S}{R} = \frac{H}{RT} - \frac{G}{RT}$$

and

$$\frac{U}{RT} = \frac{H}{RT} - \frac{PV}{RT}$$

Thus, when we know how G/RT (or G) is related to its canonical variables, T and P, that is, when we are given $G/RT = G(T, P)$, we can evaluate all other thermodynamic properties by simple mathematical operations. The Gibbs energy therefore serves as a *generating function* for the other thermodynamic properties, and implicitly represents *complete* property information.

Unfortunately, we have no convenient experimental method for determining numerical values of G or G/RT, and the equations which follow directly from the Gibbs energy are of little practical use. However, the concept of the Gibbs energy as a generating function for other thermodynamic properties carries over to a closely related property for which numerical values *are* readily obtained. Thus we define the *residual* Gibbs energy as

$$G^R \equiv G - G^{ig} \tag{6.32}$$

where G and G^{ig} are the actual and the ideal-gas values of the Gibbs energy at the same temperature and pressure. We can define other residual properties in an analogous way. The residual volume, for example, is

$$V^R \equiv V - V^{ig} \tag{6.33}$$

whence

$$V^R = V - \frac{RT}{P}$$

Since $V = ZRT/P$, the residual volume and the compressibility factor are related:

$$V^R = \frac{RT}{P}(Z-1) \tag{6.34}$$

We can, in fact, write a general definition for residual properties:

$$\boxed{M^R \equiv M - M^{ig}} \tag{6.35}$$

where M is the molar value of any extensive thermodynamic property, for example, V, U, H, S, or G.

Equation (6.29), written for the special case of an ideal gas, becomes:

$$d\left(\frac{G^{ig}}{RT}\right) = \frac{V^{ig}}{RT}\,dP - \frac{H^{ig}}{RT^2}\,dT$$

Subtracting this equation from Eq. (6.29) gives:

$$\boxed{d\left(\frac{G^R}{RT}\right) = \frac{V^R}{RT}\,dP - \frac{H^R}{RT^2}\,dT} \tag{6.36}$$

This is a *fundamental property relation* for residual properties applicable to constant-composition fluids. From it we get immediately that:

$$\frac{V^R}{RT} = \left[\frac{\partial(G^R/RT)}{\partial P}\right]_T \tag{6.37}$$

and

$$\frac{H^R}{RT} = -T\left[\frac{\partial(G^R/RT)}{\partial T}\right]_P \tag{6.38}$$

In addition, the defining equation for the Gibbs energy, $G \equiv H - TS$, written for the special case of an ideal gas is $G^{ig} = H^{ig} - TS^{ig}$; by difference,

$$G^R = H^R - TS^R$$

from which we get the residual entropy:

$$\frac{S^R}{R} = \frac{H^R}{RT} - \frac{G^R}{RT} \tag{6.39}$$

Thus the residual Gibbs energy serves as a generating function for the other residual properties, and here we do have a direct link with experiment. It is provided by Eq. (6.37), written

$$d\left(\frac{G^R}{RT}\right) = \frac{V^R}{RT}\,dP \qquad \text{(const } T\text{)}$$

Integration from zero pressure to arbitrary pressure P gives

$$\frac{G^R}{RT} = \int_0^P \frac{V^R}{RT}\,dP \qquad \text{(const } T\text{)}$$

where at the lower limit we have set G^R/RT equal to zero on the basis that the zero-pressure state is an ideal-gas state. In view of Eq. (6.34), this result becomes,

$$\frac{G^R}{RT} = \int_0^P (Z-1)\frac{dP}{P} \qquad \text{(const } T\text{)} \tag{6.40}$$

When Eq. (6.40) is differentiated with respect to temperature in accord with Eq. (6.38), we get

$$\boxed{\frac{H^R}{RT} = -T\int_0^P \left(\frac{\partial Z}{\partial T}\right)_P \frac{dP}{P}} \qquad \text{(const } T\text{)} \tag{6.41}$$

Combining Eqs. (6.40) and (6.41) with Eq. (6.39) gives

$$\boxed{\frac{S^R}{R} = -T\int_0^P \left(\frac{\partial Z}{\partial T}\right)_P \frac{dP}{P} - \int_0^P (Z-1)\frac{dP}{P}} \qquad \text{(const } T\text{)} \tag{6.42}$$

The compressibility factor is by definition $Z = PV/RT$; values of Z and of $(\partial Z/\partial T)_P$ are calculated directly from experimental PVT data, and the two integrals in Eqs. (6.40) through (6.42) are evaluated by numerical or graphical methods. Alternatively, the two integrals are evaluated analytically when Z is expressed by an equation of state. Thus, given PVT data or an appropriate equation of state, we can evaluate H^R and S^R and hence all other residual properties. It is this direct connection with experiment that makes residual properties essential to the practical application of thermodynamics.

Applied to the enthalpy and entropy, Eq. (6.35) is written:

$$H = H^{ig} + H^R \tag{6.43}$$

and

$$S = S^{ig} + S^R \tag{6.44}$$

Thus, H and S are found from the corresponding ideal-gas and residual properties by simple addition. General expressions for H^{ig} and S^{ig} are obtained by integration of Eqs. (6.22) and (6.23) from an ideal-gas state at reference conditions T_0 and P_0 to the ideal-gas state at T and P:

$$H^{ig} = H_0^{ig} + \int_{T_0}^{T} C_P^{ig}\, dT$$

and

$$S^{ig} = S_0^{ig} + \int_{T_0}^{T} C_P^{ig} \frac{dT}{T} - R \ln \frac{P}{P_0}$$

Substitution into Eqs. (6.43) and (6.44) gives

$$H = H_0^{ig} + \int_{T_0}^{T} C_P^{ig}\, dT + H^R$$

and

$$S = S_0^{ig} + \int_{T_0}^{T} C_P^{ig} \frac{dT}{T} - R \ln \frac{P}{P_0} + S^R$$

In view of Eqs. (4.6) and (5.16), these are more simply expressed as

$$\boxed{H = H_0^{ig} + C_{P_{\mathrm{mh}}}^{ig}(T - T_0) + H^R} \tag{6.45}$$

and

$$\boxed{S = S_0^{ig} + C_{P_{\mathrm{ms}}}^{ig} \ln \frac{T}{T_0} - R \ln \frac{P}{P_0} + S^R} \tag{6.46}$$

where H^R and S^R are given by Eqs. (6.41) and (6.42).

Since the equations of thermodynamics which derive from the first and second laws do not permit calculation of absolute values for enthalpy and entropy, and since all we need in practice are relative values, the reference-state conditions T_0 and P_0 are selected for convenience, and values are assigned to H_0^{ig} and S_0^{ig} arbitrarily. The only data needed for application of Eqs. (6.45) and (6.46) are ideal-gas heat capacities and PVT data. Once V, H, and S are known at given conditions of T and P, the other thermodynamic properties follow from defining equations.

The true worth of the equations for ideal gases is now evident. They are important because they provide a convenient base for the calculation of real-gas properties. Although Eqs. (6.41) and (6.42) as written apply only to gases, residual properties have validity for liquids as well. However, the advantage of Eqs. (6.43) and (6.44) in application to gases is that H^R and S^R, the terms which contain all the complex calculations, are *residuals* that generally are quite small. They have the nature of corrections to the major terms, H^{ig} and S^{ig}. For liquids, this advantage is largely lost, because H^R and S^R must include the large enthalpy and entropy changes of vaporization. Property changes of liquids are usually calculated by integrated forms of Eqs. (6.27) and (6.28), as illustrated in Example 6.1.

Example 6.2 Calculate the enthalpy and entropy of saturated isobutane vapor at 360 K from the following information:

1. The vapor pressure of isobutane at 360 K is 15.41 bar.
2. Set $H_0^{ig} = 18{,}115.0 \text{ J mol}^{-1}$ and $S_0^{ig} = 295.976 \text{ J mol}^{-1} \text{ K}^{-1}$ for the ideal-gas reference state at 300 K and 1 bar.†
3. The ideal-gas heat capacity of isobutane vapor in the temperature range of interest is given by

$$C_P^{ig}/R = 1.7765 + 33.037 \times 10^{-3} T \qquad (T/\text{K})$$

4. Compressibility-factor data (values of Z) for isobutane vapor are as follows:†

P/bar	340 K	350 K	360 K	370 K	380 K
0.1	0.99700	0.99719	0.99737	0.99753	0.99767
0.5	0.98745	0.98830	0.98907	0.98977	0.99040
2	0.95895	0.96206	0.96483	0.96730	0.96953
4	0.92422	0.93069	0.93635	0.94132	0.94574
6	0.88742	0.89816	0.90734	0.91529	0.92223
8	0.84575	0.86218	0.87586	0.88745	0.89743
10	0.79659	0.82117	0.84077	0.85695	0.87061
12	......	0.77310	0.80103	0.82315	0.84134
14	......	......	0.75506	0.78531	0.80923
15.41	......	......	0.71727		

SOLUTION Calculation of H^R and S^R at 360 K and 15.41 bar by application of Eqs. (6.41) and (6.42) requires the evaluation of two integrals:

$$\int_0^P \left(\frac{\partial Z}{\partial T}\right)_P \frac{dP}{P} \qquad \text{and} \qquad \int_0^P (Z-1)\frac{dP}{P}$$

Graphical integration requires simple plots of both $(\partial Z/\partial T)_P/P$ and $(Z-1)/P$ vs. P. Values of $(Z-1)/P$ are calculated directly from the given compressibility-factor data at 360 K: The quantity $(\partial Z/\partial T)_P/P$ requires evaluation of the partial derivative

† R. D. Goodwin and W. M. Haynes, Nat. Bur. Stand. (U.S.), Tech. Note 1051, 1982.

$(\partial Z/\partial T)_P$, given by the slope of a plot of Z vs. T at constant pressure. For this purpose, separate plots are made of Z vs. T for each pressure at which compressibility-factor data are given, and a slope is determined at 360 K for each curve (for example, by construction of a tangent line at 360 K). The data for construction of the required plots are shown in the following table (values in parentheses are by extrapolation):

P bar	$-(Z-1)/P \times 10^2$ bar^{-1}	$(\partial Z/\partial T)_P/P \times 10^4$ K^{-1}bar^{-1}
0	(2.590)	(1.780)
0.1	2.470	1.700
0.5	2.186	1.514
2	1.759	1.293
4	1.591	1.290
6	1.544	1.395
8	1.552	1.560
10	1.592	1.777
12	1.658	2.073
14	1.750	2.432
15.41	(1.835)	(2.720)

The values of the two integrals are found to be

$$\int_0^P \left(\frac{\partial Z}{\partial T}\right)_P \frac{dP}{P} = 26.37 \times 10^{-4}\ \text{K}^{-1}$$

and

$$\int_0^P (Z-1)\frac{dP}{P} = -0.2596$$

Thus by Eq. (6.41)

$$\frac{H^R}{RT} = -(360)(26.37 \times 10^{-4}) = -0.9493$$

and by Eq. (6.42)

$$\frac{S^R}{R} = -0.9493 - (-0.2596) = -0.6897$$

For $R = 8.314$ J mol^{-1} K^{-1},

$$H^R = (-0.9493)(8.314)(360) = -2{,}841.3\ \text{J mol}^{-1}$$

and

$$S^R = (-0.6897)(8.314) = -5.734\ \text{J mol}^{-1}\ \text{K}^{-1}$$

Equations (4.7) and (5.17) for the mean heat capacities here become

$$C^{ig}_{P_{mh}}/R = A + BT_{am} = 1.7765 + 33.037 \times 10^{-3}T_{am}$$

and

$$C^{ig}_{P_{ms}}/R = A + BT_{lm} = 1.7765 + 33.037 \times 10^{-3}T_{lm}$$

With

$$T_{\text{am}} = (300 + 360)/2 = 330 \text{ K}$$

and

$$T_{\text{lm}} = \frac{360 - 300}{\ln (360/300)} = 329.09 \text{ K}$$

we get

$$C_{P_{\text{mh}}}^{ig}/R = 12.679 \qquad \text{and} \qquad C_{P_{\text{ms}}}^{ig}/R = 12.649$$

Finally, Eqs. (6.45) and (6.46) yield the required results:

$$H = H_0^{ig} + C_{P_{\text{mh}}}^{ig}(T - T_0) + H^R$$

$$= 18{,}115.0 + (12.679)(8.314)(360 - 300) - 2{,}841.3$$

$$= 21{,}598.5 \text{ J mol}^{-1}$$

and

$$S = S_0^{ig} + C_{P_{\text{ms}}}^{ig} \ln \frac{T}{T_0} - R \ln \frac{P}{P_0} + S^R$$

$$= 295.976 + (12.649)(8.314) \ln \frac{360}{300} - 8.314 \ln 15.41 - 5.734$$

$$= 286.676 \text{ J mol}^{-1} \text{ K}^{-1}$$

Although calculations have been carried out for just one state, enthalpies and entropies can be evaluated for any number of states, given adequate data. After having completed a set of calculations, one is not irrevocably committed to the particular values of H_0^{ig} and S_0^{ig} initially assigned. The scale of values for either the enthalpy or the entropy can be shifted by addition of a constant to all values. In this way one can give arbitrary values to H and S for some particular state so as to make the scales convenient for one purpose or another. A shift of scale does not affect differences in property values.

The accurate calculation of thermodynamic properties for construction of a table or diagram is an exacting task, seldom required of an engineer. However, engineers do make practical use of thermodynamic properties, and an understanding of the methods used for their calculation leads to an appreciation that some uncertainty is associated with every property value. There are two major reasons for inaccuracy. First, the experimental data are difficult to measure and are subject to error. Moreover, data are frequently incomplete, and are extended by interpolation and extrapolation. Second, even when reliable PVT data are available, a loss of accuracy occurs in the differentiation process required in the calculation of derived properties. This accounts for the fact that data of a high order of accuracy are required to produce enthalpy and entropy values suitable for engineering calculations.

6.3 TWO-PHASE SYSTEMS

The PT diagram of Fig. 3.1 shows curves representing phase boundaries for a pure substance. A phase transition at constant temperature and pressure occurs whenever one of these curves is crossed, and as a result the molar or specific values of the extensive thermodynamic properties change abruptly. Thus the molar or specific volume of a saturated liquid is very different from that for saturated vapor at the same T and P. This is true as well for internal energy, enthalpy, and entropy. The exception is the molar or specific Gibbs energy, which for a pure species does not change during a phase transition such as melting, vaporization, or sublimation. Consider a pure liquid in equilibrium with its vapor in a piston-and-cylinder arrangement at temperature T and the corresponding vapor pressure P^{sat}. If a differential amount of liquid is caused to evaporate at constant temperature and pressure, Eq. (6.6) reduces to $d(nG) = 0$ for the process. Since the number of moles n is constant, $dG = 0$, and this requires the molar (or specific) Gibbs energy of the vapor to be identical with that of the liquid. More generally, for two phases α and β of a pure species coexisting at equilibrium,

$$G^\alpha = G^\beta \tag{6.47}$$

where G^α and G^β are the molar Gibbs energies of the individual phases.

The Clapeyron equation, first introduced in Sec. 4.3, follows from this equality. If the temperature of a two-phase system is changed, then the pressure must also change in accord with the relation between vapor pressure and temperature if the two phases continue to coexist. Since Eq. (6.47) holds throughout this change, we have

$$dG^\alpha = dG^\beta$$

Substituting the expressions for dG^α and dG^β given by Eq. (6.10) yields

$$V^\alpha\, dP^{\text{sat}} - S^\alpha\, dT = V^\beta\, dP^{\text{sat}} - S^\beta\, dT$$

which upon rearrangement becomes

$$\frac{dP^{\text{sat}}}{dT} = \frac{S^\beta - S^\alpha}{V^\beta - V^\alpha} = \frac{\Delta S^{\alpha\beta}}{\Delta V^{\alpha\beta}}$$

The entropy change $\Delta S^{\alpha\beta}$ and the volume change $\Delta V^{\alpha\beta}$ are the changes which occur when a unit amount of a pure chemical species is transferred from phase α to phase β at constant temperature and pressure. Integration of Eq. (6.8) for this change yields the latent heat of phase transition:

$$\Delta H^{\alpha\beta} = T\,\Delta S^{\alpha\beta}$$

Thus, $\Delta S^{\alpha\beta} = \Delta H^{\alpha\beta}/T$, and substitution in the preceding equation gives

$$\frac{dP^{\text{sat}}}{dT} = \frac{\Delta H^{\alpha\beta}}{T\,\Delta V^{\alpha\beta}} \tag{6.48}$$

which is the Clapeyron equation. For the particularly important case of phase transition from liquid l to vapor v, it is written

$$\frac{dP^{\text{sat}}}{dT} = \frac{\Delta H^{lv}}{T\,\Delta V^{lv}} \tag{6.49}$$

Example 6.3 For vaporization at low pressures, one may introduce reasonable approximations into Eq. (6.49) by assuming that the vapor phase is an ideal gas and that the molar volume of the liquid is negligible compared with the molar volume of the vapor. How do these assumptions alter the Clapeyron equation?

SOLUTION The assumptions made are expressed by

$$\Delta V^{lv} = V^{v} = \frac{RT}{P^{\text{sat}}}$$

Equation (6.49) then becomes

$$\frac{dP^{\text{sat}}}{dT} = \frac{\Delta H^{lv}}{RT^2/P^{\text{sat}}}$$

or

$$\frac{dP^{\text{sat}}/P^{\text{sat}}}{dT/T^2} = \frac{\Delta H^{lv}}{R}$$

or

$$\Delta H^{lv} = -R\frac{d\ln P^{\text{sat}}}{d(1/T)}$$

This approximate equation, known as the Clausius/Clapeyron equation, relates the latent heat of vaporization directly to the vapor pressure curve. Specifically, it shows that ΔH^{lv} is proportional to the slope of a plot of $\ln P^{\text{sat}}$ vs. $1/T$. Experimental data for many substances show that such plots produce lines that are nearly straight. According to the Clausius/Clapeyron equation, this implies that ΔH^{lv} is almost constant, virtually independent of T. This is not true; ΔH^{lv} decreases monotonically with increasing temperature from the triple point to the critical point, where it becomes zero. The assumptions on which the Clausius/Clapeyron equation are based have approximate validity only at low pressures.

The Clapeyron equation is an exact thermodynamic relation, providing a vital connection between the properties of different phases. When applied to the calculation of latent heats of vaporization, its use presupposes knowledge of the vapor pressure-vs.-temperature relation. Since thermodynamics imposes no model of material behavior, either in general or for particular species, such relations are empirical. As noted in Example 6.3, a plot of $\ln P^{\text{sat}}$ vs. $1/T$ generally yields a line that is nearly straight, i.e.,

$$\ln P^{\text{sat}} = A - \frac{B}{T} \tag{6.50}$$

where A and B are constants for a given species. This equation gives a rough approximation of the vapor-pressure relation for the entire temperature range from the triple point to the critical point. Moreover, it is an excellent interpolation formula between values that are reasonably spaced.

The Antoine equation, which is more satisfactory for general use, has the form

$$\ln P^{\text{sat}} = A - \frac{B}{T+C} \tag{6.51}$$

A principal advantage of this equation is that values of the constants *A*, *B*, and *C* are readily available for many species.†

The accurate representation of vapor-pressure data over a wide temperature range requires an equation of greater complexity; an example is the Riedel equation:

$$\ln P^{\text{sat}} = A - \frac{B}{T} + D \ln T + FT^6 \tag{6.52}$$

where *A*, *B*, *D*, and *F* are constants.

When a system consists of saturated-liquid and saturated-vapor phases coexisting in equilibrium, the total value of any extensive property of the two-phase system is the sum of the total properties of the phases. Written for the volume, this relation is

$$nV = n^l V^l + n^v V^v$$

where V is the system volume on a molar basis and the total number of moles is $n = n^l + n^v$. Division by n gives

$$V = x^l V^l + x^v V^v$$

where x^l and x^v represent the fractions of the total system that are liquid and vapor. Since $x^l = 1 - x^v$,

$$V = (1 - x^v) V^l + x^v V^v$$

In this equation the properties V, V^l, and V^v may be either molar or unit-mass values. The mass or molar fraction of the system that is vapor x^v is called the *quality*. Analogous equations can be written for the other extensive thermodynamic properties. All of these relations may be summarized by the equation

$$M = (1 - x^v) M^l + x^v M^v \tag{6.53}$$

where M represents V, U, H, S, etc.

6.4 THERMODYNAMIC DIAGRAMS

A thermodynamic diagram represents the temperature, pressure, volume, enthalpy, and entropy of a substance on a single plot. (Sometimes data for all these variables are not included, but the term still applies.) The most common diagrams are: temperature/entropy, pressure/enthalpy (usually $\ln P$ vs. H), and enthalpy/entropy (called a *Mollier* diagram). The designations refer to the variables chosen for the coordinates. Other diagrams are possible, but are seldom used.

Figures 6.2 through 6.4 show the general features of the three common diagrams. These figures are based on data for water, but their general character

† S. Ohe, *Computer Aided Data Book of Vapor Pressure*, Data Book Publishing Co., Tokyo, 1976; T. Boublík, V. Fried, and E. Hála, *The Vapor Pressures of Pure Substances*, Elsevier, Amsterdam, 1984.

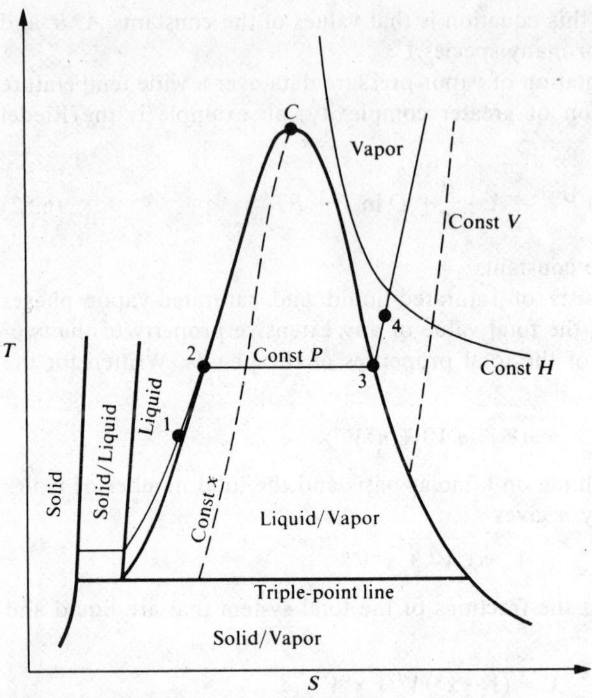

Figure 6.2 *TS* diagram.

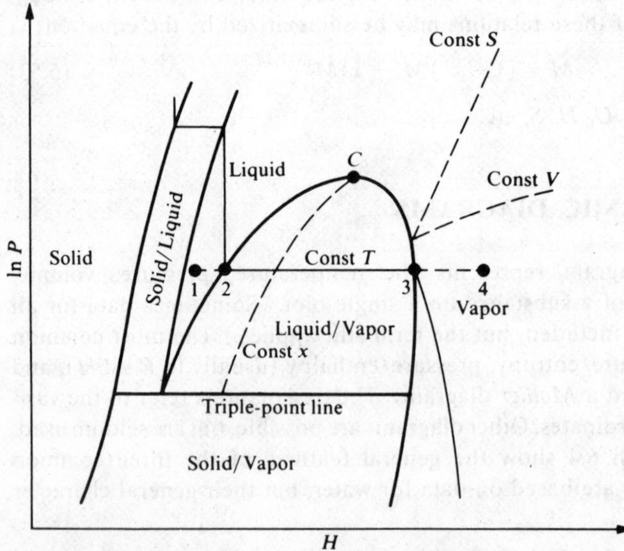

Figure 6.3 *PH* diagram.

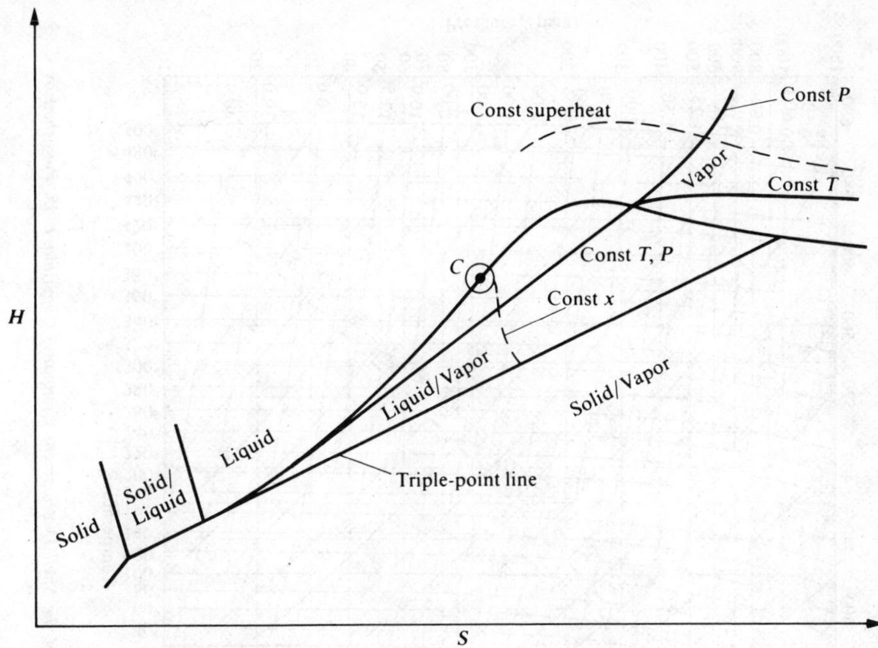

Figure 6.4 Mollier diagram.

is the same for all substances. The two-phase states, which fall on lines in the *PT* diagram of Fig. 3.1, lie over areas in these diagrams, and the triple point of Fig. 3.1 becomes a line. When lines of constant quality are shown in the liquid/vapor region, property values for two-phase mixtures are read directly from the diagram. The critical point is identified by the letter *C*, and the solid curve passing through this point represents the states of saturated liquid (to the left of *C*) and of saturated vapor (to the right of *C*). The Mollier diagram (Fig. 6.4) does not usually include volume data. In the vapor or gas region, lines for constant temperature and constant *superheat* appear. Superheat is a term used to designate the difference between the actual temperature and the saturation temperature at the same pressure.

Examples of specific thermodynamic diagrams are given for methane by the *PH* diagram of Fig. 6.5, for steam by the Mollier diagram on the inside of the back cover, for Freon-12 and ammonia by the *PH* diagrams of Figs. 9.3 and 9.4, and for air by the *TS* diagram of Fig. 9.8.

Paths of various processes are conveniently traced on a thermodynamic diagram. For example, consider the operation of the boiler in a steam power plant. The initial state is liquid water at a temperature below its boiling point; the final state is steam in the superheat region. As the water goes into the boiler and is heated, its temperature rises at constant pressure (line 1-2 in Figs. 6.2 and 6.3) until saturation is reached. From point 2 to point 3 the water vaporizes, the

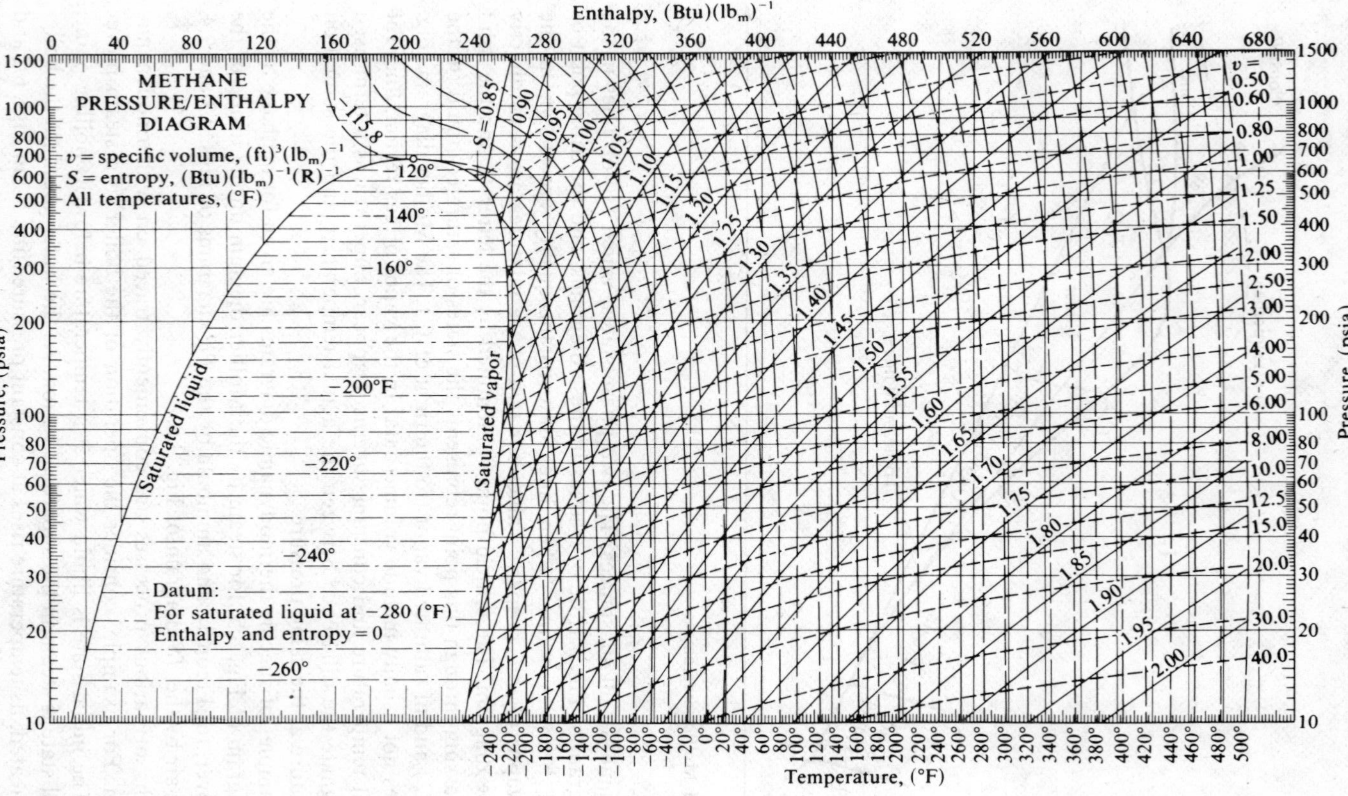

Figure 6.5 Pressure/enthalpy diagram for methane. *(Reproduced by permission of the Shell Development Company, Copyright 1945. Published by C. S. Matthews and C. O. Hurd, Trans. AIChE,* **42**: *55 (1946).)*

temperature remaining constant during the process. As more heat is added, the steam becomes superheated along line 3-4. On a pressure/enthalpy diagram (Fig. 6.3) the whole process is represented by a horizontal line corresponding to the boiler pressure. Since the compressibility of a liquid is small for temperatures well below T_c, the properties of liquids change very slowly with pressure. Thus on a *TS* diagram (Fig. 6.2), the constant-pressure lines in the liquid region lie very close together, and line 1-2 nearly coincides with the saturated-liquid curve. A reversible adiabatic process is isentropic and is therefore represented on a *TS* diagram by a vertical line. Hence the path followed by the fluid in reversible adiabatic turbines and compressors is simply a vertical line from the initial pressure to the final pressure. This is also true on the *HS* or Mollier diagram.

6.5 TABLES OF THERMODYNAMIC PROPERTIES

In many instances thermodynamic properties are reported in tables. The advantage is that, in general, data can be presented more accurately than in diagrams, but the need for interpolation is introduced.

The complete thermodynamic tables for saturated and superheated steam, both in SI and in English units, appear in App. C. Values are given at intervals close enough so that linear interpolation is satisfactory. The first table for each system of units presents the equilibrium properties of saturated liquid and vapor phases from the triple point to the critical point. The enthalpy and entropy are arbitrarily assigned values of zero for the saturated liquid state at the triple point. The second table is for the gas region, and gives properties of superheated steam at temperatures higher than the saturation temperature for a given pressure. Volume, internal energy, enthalpy, and entropy are tabulated as functions of pressure at various temperatures. The steam tables are the most thorough compilation of thermodynamic properties for any single material. However, extensive tables are available for certain other substances.†

Example 6.4 Superheated steam originally at P_1 and T_1 expands through a nozzle to an exhaust pressure P_2. Assuming the process is reversible and adiabatic and that equilibrium is attained, determine the state of the steam at the exit of the nozzle for the following conditions:

(*a*) $P_1 = 1{,}000$ kPa, $t_1 = 260$°C, and $P_2 = 200$ kPa.
(*b*) $P_1 = 150$(psia), $t_1 = 500$(°F), and $P_2 = 30$(psia).

† Comprehensive tables for a number of pure species, each in a separate volume, appear under the title *International Thermodynamic Tables of the Fluid State*, Pergamon Press, Oxford, starting 1972. Included are tables for argon, carbon dioxide, helium, methane, nitrogen, and propylene. Extensive data for ammonia appear in *J. Phys. Chem. Ref. Data*, 7: 635, 1978. Compilations done by the U.S. Bureau of Standards for ethane, ethylene, isobutane, *n*-butane, and propane are published as Technical Notes 684 (1976), 960 (1981), and 1051 (1982), and as Monographs 169 (1982), and 170 (1082).

SOLUTION Since the process is both reversible and adiabatic, the change in entropy of the steam is zero.

(*a*) The initial state of the steam is as follows (data from the SI steam tables):

$$t_1 = 260°\text{C}$$
$$P_1 = 1{,}000 \text{ kPa}$$
$$H_1 = 2{,}965.2 \text{ kJ kg}^{-1}$$
$$S_1 = 6.9860 \text{ kJ kg}^{-1} \text{ K}^{-1}$$

For the final state,

$$P_2 = 200 \text{ kPa}$$
$$S_2 = 6.9680 \text{ kJ kg}^{-1} \text{ K}^{-1}$$

Since the entropy of saturated vapor at 200 kPa is greater than S_2, the final state is in the two-phase region. Equation (6.53) applied to the entropy here becomes

$$S = (1 - x^v)S^l + x^v S^v$$

Whence

$$6.9680 = 1.5301(1 - x^v) + 7.1268x^v$$

where 1.5301 and 7.1268 are the entropies of saturated liquid and saturated vapor at 200 kPa. Solving, we get

$$x^v = 0.9716$$

On a mass basis, the mixture is 97.16 percent vapor and 2.84 percent liquid. Its enthalpy is obtained by further application of Eq. (6.53):

$$H = (0.0284)(504.7) + (0.9716)(2{,}706.7) = 2{,}644.2 \text{ kJ kg}^{-1}$$

(*b*) The initial state of the steam is as follows (data from the steam tables in English units):

$$t_1 = 500(°\text{F})$$
$$P_1 = 150(\text{psia})$$
$$H_1 = 1{,}274.3(\text{Btu})(\text{lb}_m)^{-1}$$
$$S_1 = 1.6602(\text{Btu})(\text{lb}_m)^{-1}(\text{R})^{-1}$$

In the final state,

$$P_2 = 30(\text{psia})$$
$$S_2 = 1.6602(\text{Btu})(\text{lb}_m)^{-1}(\text{R})^{-1}$$

Since the entropy of saturated vapor at 30(psia) is greater than S_2, the final state is in the two-phase region. Equation (6.53) applied to the entropy is written

$$S = (1 - x^v)S^l + x^v S^v$$

Whence

$$1.6602 = 0.3682(1 - x^v) + 1.6995x^v$$

where 0.3682 and 1.6995 are the entropies of saturated liquid and saturated vapor at

30(psia). Solving, we get

$$x^v = 0.9705$$

On a mass basis, the mixture is 97.05 percent vapor and 2.95 percent liquid. Its enthalpy follows from another application of Eq. (6.53):

$$H = (0.0295)(218.9) + (0.9705)(1{,}164.1) = 1{,}136.2(\text{Btu})(\text{lb}_\text{m})^{-1}$$

6.6 GENERALIZED CORRELATIONS OF THERMODYNAMIC PROPERTIES FOR GASES

Of the two kinds of data needed for evaluation of thermodynamic properties, heat capacities and *PVT* data, the latter are most frequently missing. Fortunately, the generalized methods developed in Sec. 3.6 for the compressibility factor are also applicable to residual properties.

Equations (6.41) and (6.42) are put into generalized form by substitution of the relationships,

$$P = P_c P_r \qquad T = T_c T_r$$

$$dP = P_c\, dP_r \qquad dT = T_c\, dT_r$$

The resulting equations are:

$$\frac{H^R}{RT_c} = -T_r^2 \int_0^{P_r} \left(\frac{\partial Z}{\partial T_r}\right)_{P_r} \frac{dP_r}{P_r} \tag{6.54}$$

and

$$\frac{S^R}{R} = -T_r \int_0^{P_r} \left(\frac{\partial Z}{\partial T_r}\right)_{P_r} \frac{dP_r}{P_r} - \int_0^{P_r} (Z-1)\frac{dP_r}{P_r} \tag{6.55}$$

The terms on the right-hand sides of these equations depend only on the upper limit P_r of the integrals and on the reduced temperature at which the integrations are carried out. Thus, H^R/RT_c and S^R/R may be evaluated once and for all at any reduced temperature and pressure from generalized compressibility-factor data.

The correlation for Z is based on Eq. (3.45),

$$Z = Z^0 + \omega Z^1$$

Differentiation yields

$$\left(\frac{\partial Z}{\partial T_r}\right)_{P_r} = \left(\frac{\partial Z^0}{\partial T_r}\right)_{P_r} + \omega\left(\frac{\partial Z^1}{\partial T_r}\right)_{P_r}$$

Substitution for Z and $(\partial Z/\partial T_r)_{P_r}$ in Eqs. (6.54) and (6.55) gives:

$$\frac{H^R}{RT_c} = -T_r^2 \int_0^{P_r} \left(\frac{\partial Z^0}{\partial T_r}\right)_{P_r} \frac{dP_r}{P_r} - \omega T_r^2 \int_0^{P_r} \left(\frac{\partial Z^1}{\partial T_r}\right)_{P_r} \frac{dP_r}{P_r}$$

and

$$\frac{S^R}{R} = -\int_0^{P_r}\left[T_r\left(\frac{\partial Z^0}{\partial T_r}\right)_{P_r} + Z^0 - 1\right]\frac{dP_r}{P_r} - \omega\int_0^{P_r}\left[T_r\left(\frac{\partial Z^1}{\partial T_r}\right)_{P_r} + Z^1\right]\frac{dP_r}{P_r}$$

The first integrals on the right-hand sides of these two equations are evaluated numerically or graphically for various values of T_r and P_r from the data of Figs. 3.12 and 3.13, and the integrals which follow ω in each equation are similarly evaluated from the data of Figs. 3.14 and 3.15. If the first terms on the right-hand sides of the preceding equations are represented by $(H^R)^0/RT_c$ and $(S^R)^0/R$ and if the terms which follow ω are represented by $(H^R)^1/RT_c$ and $(S^R)^1/R$, then we can write

$$\frac{H^R}{RT_c} = \frac{(H^R)^0}{RT_c} + \omega\frac{(H^R)^1}{RT_c} \tag{6.56}$$

and

$$\frac{S^R}{R} = \frac{(S^R)^0}{R} + \omega\frac{(S^R)^1}{R} \tag{6.57}$$

Calculated values of the quantities $(H^R)^0/RT_c$, $(H^R)^1/RT_c$, $(S^R)^0/R$, and $(S^R)^1/R$ are shown by plots of these quantities vs. P_r for various values of T_r in Figs. 6.6 through 6.13. These plots, together with Eqs. (6.56) and (6.57), allow estimation of the residual enthalpy and entropy on the basis of the three-parameter corresponding-states principle as developed by Pitzer (Sec. 3.6).

Figures 6.6, 6.7, 6.10, and 6.11 for $(H^R)^0/RT_c$ and $(S^R)^0/R$, used alone, provide two-parameter corresponding-states correlations that quickly yield coarse estimates of the residual properties.

As with the generalized compressibility-factor correlation, the complexity of the functions $(H^R)^0/RT_c$, $(H^R)^1/RT_c$, $(S^R)^0/R$, and $(S^R)^1/R$ preclude their general representation by simple equations. However, the correlation for Z based on generalized virial coefficients and valid at low pressures can be extended to the residual properties. The equation relating Z to the functions B^0 and B^1 is derived in Sec. 3.6 from Eqs. (3.46) and (3.47):

$$Z = 1 + B^0\frac{P_r}{T_r} + \omega B^1\frac{P_r}{T_r}$$

From this we find

$$\left(\frac{\partial Z}{\partial T_r}\right)_{P_r} = P_r\left(\frac{dB^0/dT_r}{T_r} - \frac{B^0}{T_r^2}\right) + \omega P_r\left(\frac{dB^1/dT_r}{T_r} - \frac{B^1}{T_r^2}\right)$$

Substituting these equations into Eqs. (6.54) and (6.55) gives

$$\frac{H^R}{RT_c} = -T_r\int_0^{P_r}\left[\left(\frac{dB^0}{dT_r} - \frac{B^0}{T_r}\right) + \omega\left(\frac{dB^1}{dT_r} - \frac{B^1}{T_r}\right)\right]dP_r$$

and

$$\frac{S^R}{R} = -\int_0^{P_r}\left(\frac{dB^0}{dT_r} - \omega\frac{dB^1}{dT_r}\right)dP_r$$

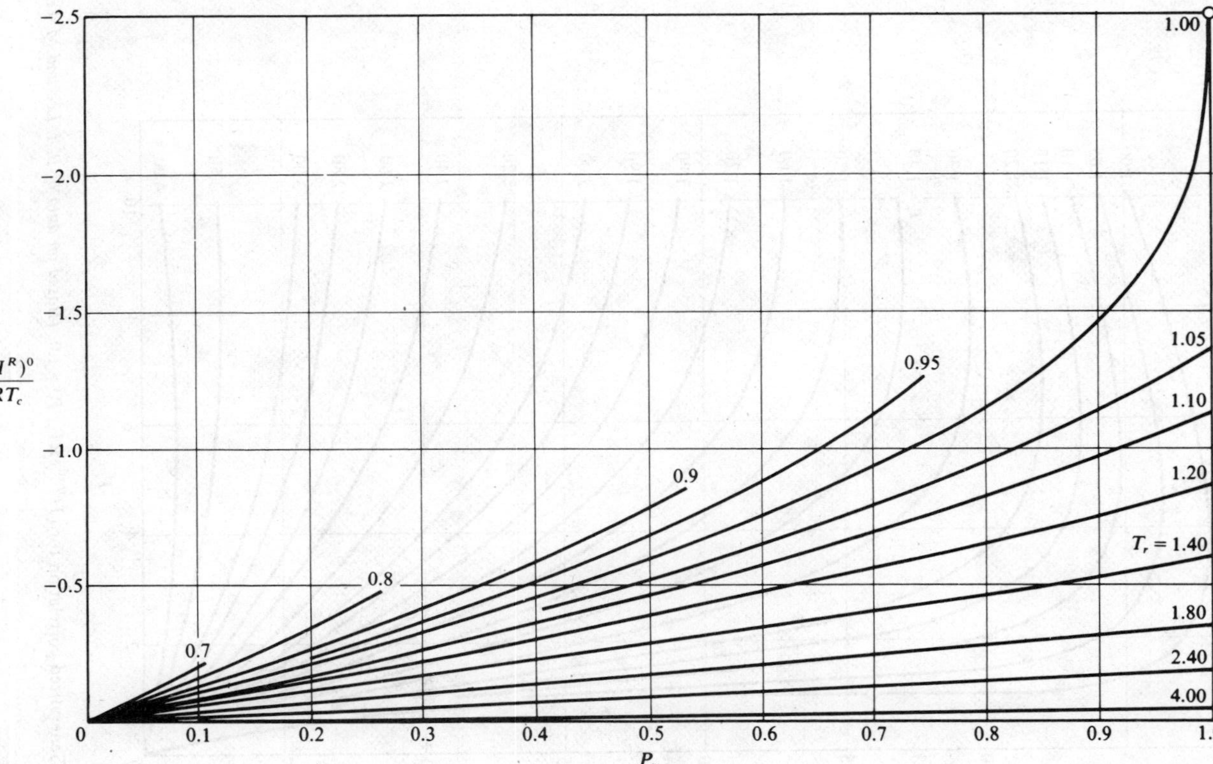

Figure 6.6 Generalized correlation for $(H^R)^0/RT_c$, $P_r < 1.0$. *(Based on data of B. I. Lee and M. G. Kesler, AIChE J.,* **21**: *510–527, 1975.)*

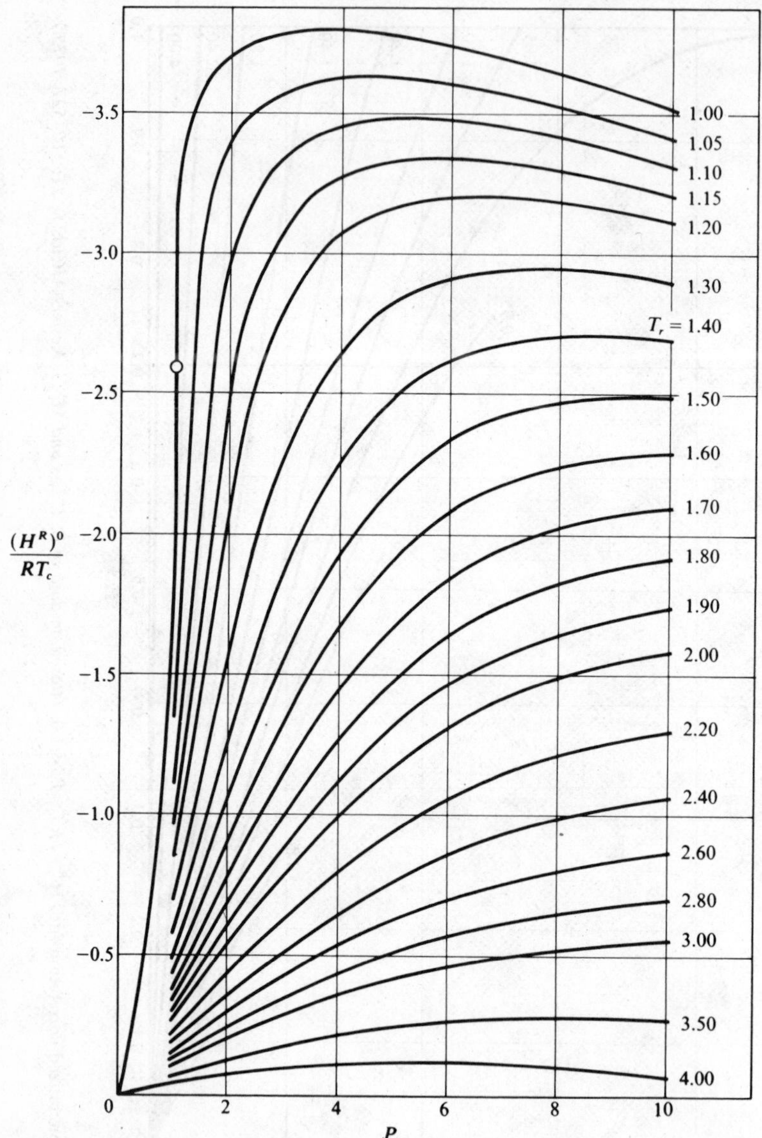

Figure 6.7 Generalized correlation for $(H^R)^0/RT_c$, $P_r > 1.0$. *(Based on data of B. I. Lee and M. G. Kesler,* ibid.*)*

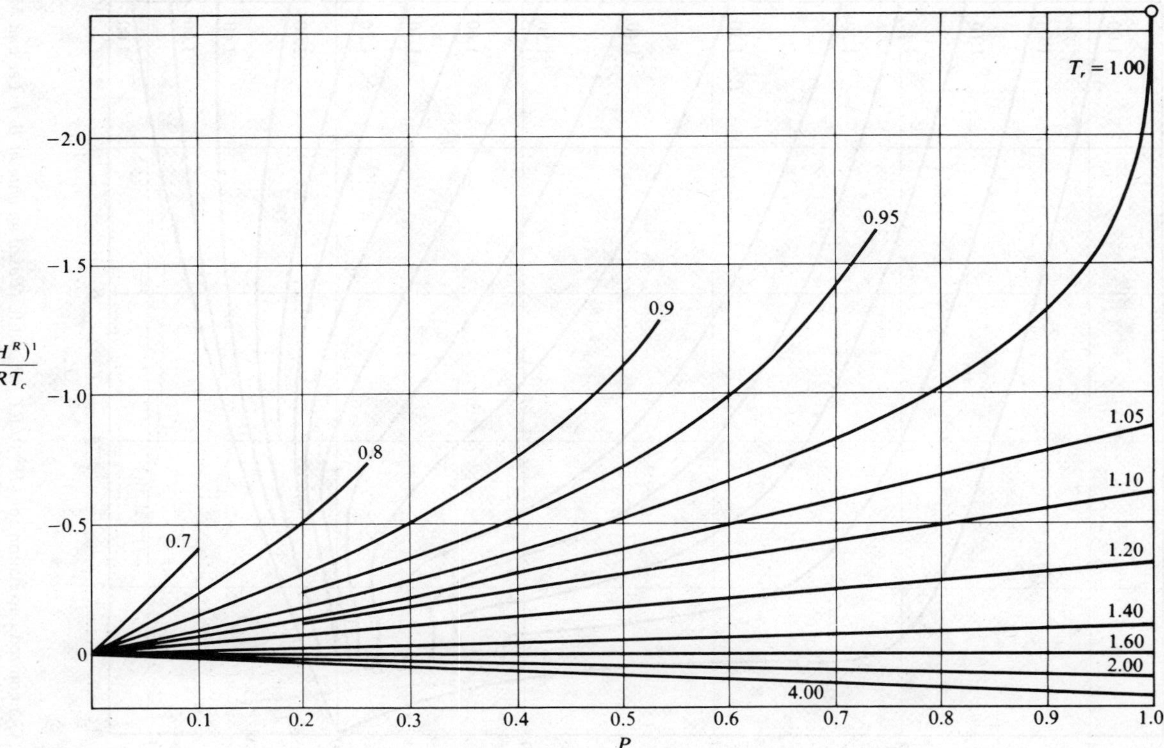

Figure 6.8 Generalized correlation for $(H^R)^1/RT_c$, $P_r < 1$. *(Based on data of B. I. Lee and M. G. Kesler,* ibid.*)*

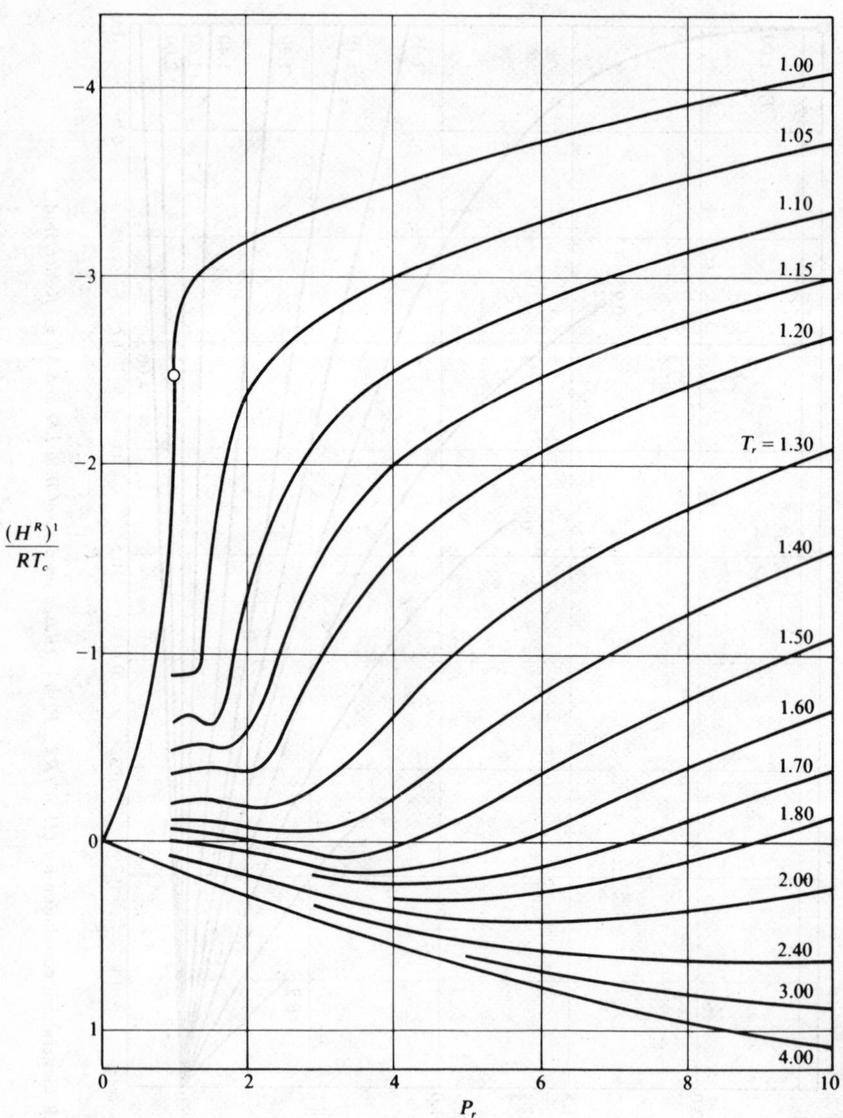

Figure 6.9 Generalized correlation for $(H^R)^1/RT_c$, $P_r > 1.0$. *(Based on data of B. I. Lee and M. G. Kesler,* ibid.*)*

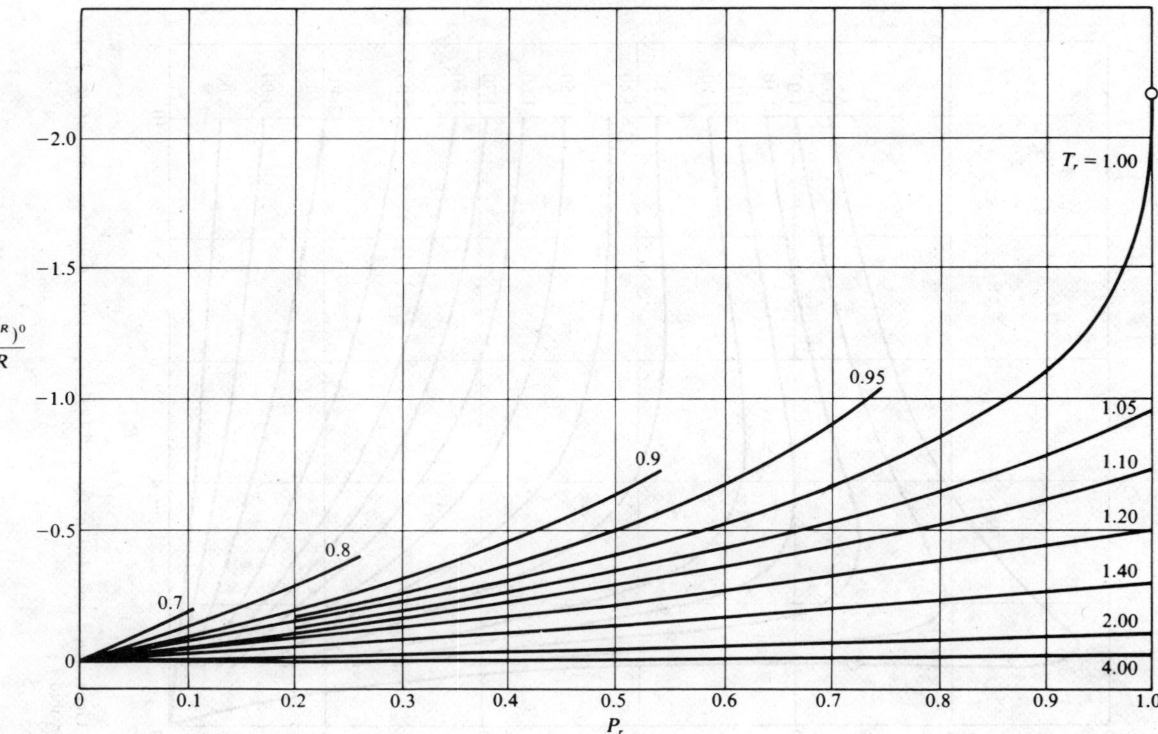

Figure 6.10 Generalized correlation for $(S^R)^0/R$, $P_r < 1.0$. *(Based on data of B. I. Lee and M. G. Kesler,* ibid.*)*

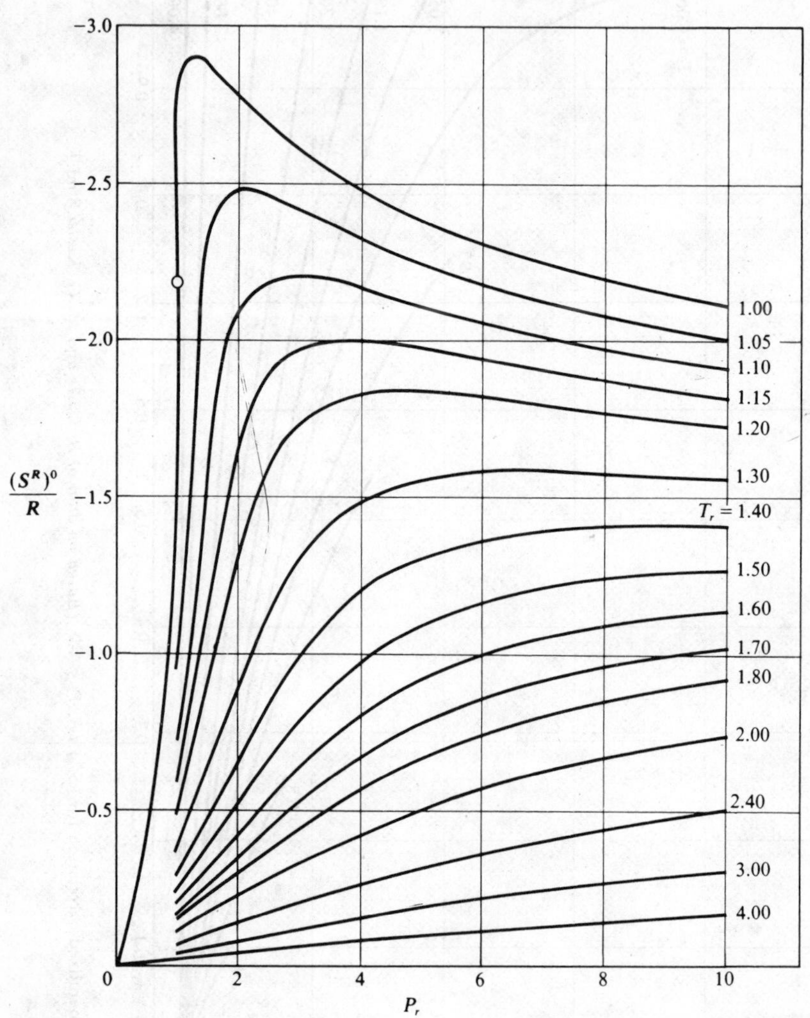

Figure 6.11 Generalized correlation for $(S^R)^0/R$, $P_r > 1.0$. *(Based on data of B. I. Lee and M. G. Kesler,* ibid.*)*

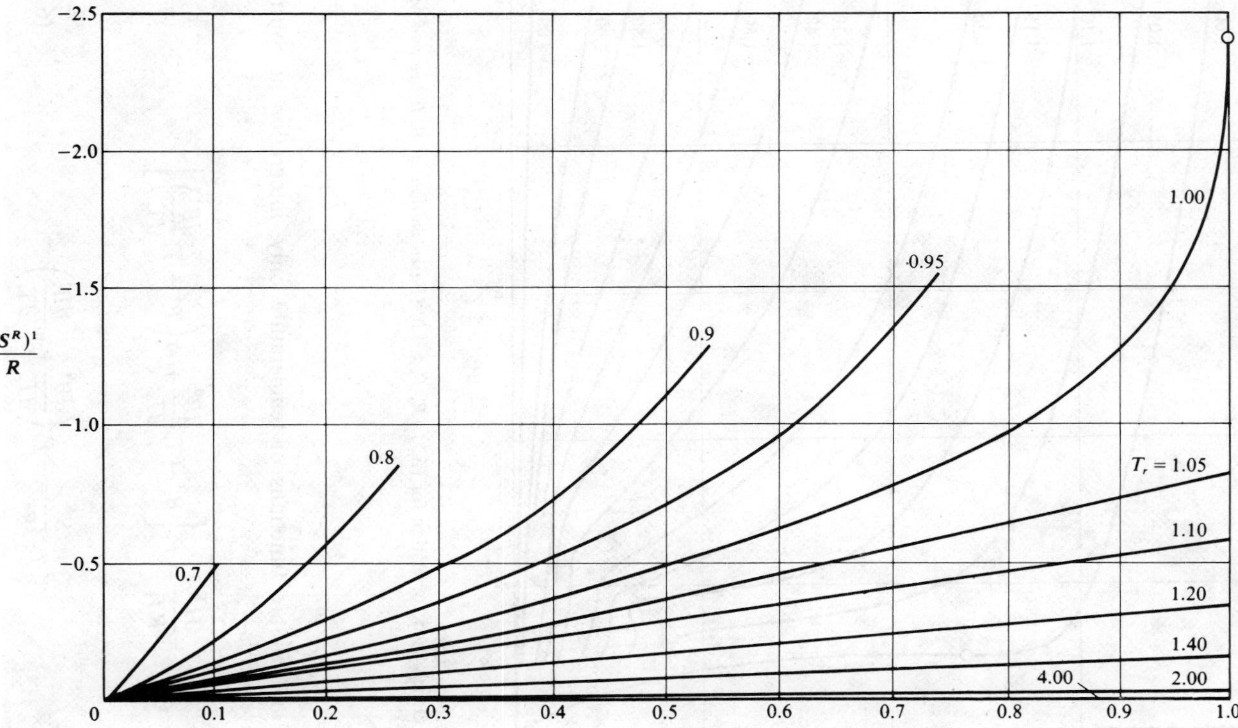

Figure 6.12 Generalized correlation for $(S^R)^1/R$, $P_r < 1.0$. *(Based on data of B. I. Lee and M. G. Kesler,* ibid.*)*

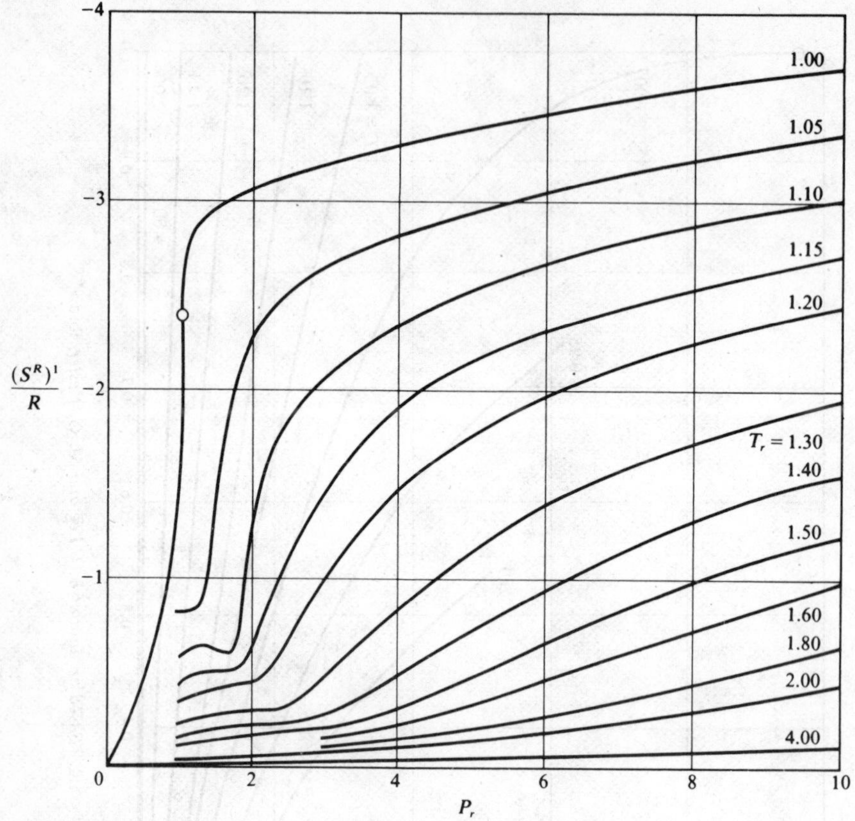

Figure 6.13 Generalized correlation for $(S^R)^1/R$, $P_r > 1.0$. *(Based on data of B. I. Lee and M. G. Kesler,* ibid.*)*

Since B^0 and B^1 are functions of temperature only, integration at constant temperature yields

$$\frac{H^R}{RT_c} = P_r\left[B^0 - T_r\frac{dB^0}{dT_r} + \omega\left(B^1 - T_r\frac{dB^1}{dT_r}\right)\right] \tag{6.58}$$

and

$$\frac{S^R}{R} = -P_r\left(\frac{dB^0}{dT_r} + \omega\frac{dB^1}{dT_r}\right) \tag{6.59}$$

The dependence of B^0 and B^1 on reduced temperature is provided by Eqs. (3.48) and (3.49). Differentiation of these equations gives expressions for dB^0/dT_r and dB^1/dT_r. Thus the four equations required for application of Eqs. (6.58) and (6.59) are:

$$B^0 = 0.083 - \frac{0.422}{T_r^{1.6}} \tag{3.48}$$

$$\frac{dB^0}{dT_r} = \frac{0.675}{T_r^{2.6}} \tag{6.60}$$

$$B^1 = 0.139 - \frac{0.172}{T_r^{4.2}} \tag{3.49}$$

$$\frac{dB^1}{dT_r} = \frac{0.722}{T_r^{5.2}} \tag{6.61}$$

Figure 3.16, drawn specifically for the compressibility-factor correlation, is also used as a guide to the reliability of the correlations of residual properties based on generalized second virial coefficients. However, all residual-property correlations are less precise than the compressibility-factor correlations on which they are based and are, of course, least reliable for strongly polar and associating molecules.

The generalized correlations for H^R and S^R, together with ideal-gas heat capacities, allow calculation of enthalpy and entropy values of gases at any temperature and pressure by Eqs. (6.45) and (6.46). For a change from state 1 to state 2, we write Eq. (6.45) for both states:

$$H_2 = H_0^{ig} + C_{P_{mh}}^{ig}(T_2 - T_0) + H_2^R$$

$$H_1 = H_0^{ig} + C_{P_{mh}}^{ig}(T_1 - T_0) + H_1^R$$

The enthalpy change for the process, $\Delta H = H_2 - H_1$, is given by the difference between these two equations:

$$\Delta H = C_{P_{mh}}^{ig}(T_2 - T_1) + H_2^R - H_1^R \tag{6.62}$$

Similarly, by Eq. (6.46) for the entropy, we get

$$\Delta S = C_{P_{ms}}^{ig} \ln \frac{T_2}{T_1} - R \ln \frac{P_2}{P_1} + S_2^R - S_1^R \tag{6.63}$$

The terms on the right-hand sides of Eqs. (6.62) and (6.63) are readily associated with steps in a *calculational path* leading from an initial to a final state of a system. Thus, in Fig. 6.14, the actual path from state 1 to state 2 (dashed line) is replaced by a three-step calculational path. Step $1 \rightarrow 1^{id}$ represents a hypothetical process that transforms a real gas into an ideal gas at T_1 and P_1. The enthalpy and entropy changes for this process are

$$H_1^{ig} - H_1 = -H_1^R$$

and

$$S_1^{ig} - S_1 = -S_1^R$$

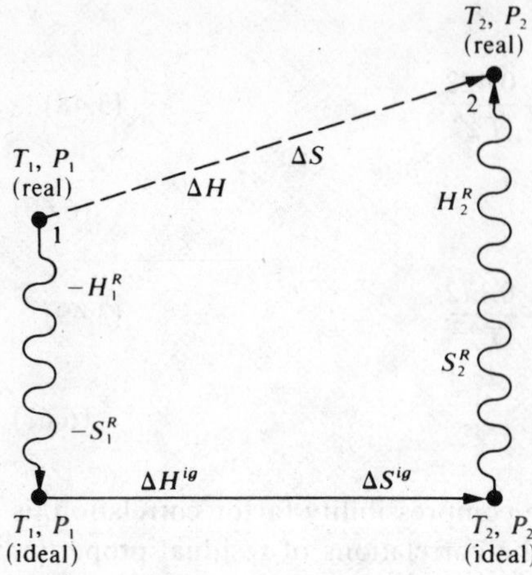

Figure 6.14 Calculational path for property changes ΔH and ΔS.

In step $1^{id} \to 2^{id}$ changes occur in the ideal-gas state from (T_1, P_1) to (T_2, P_2). For this process,

$$\Delta H^{ig} = H_2^{ig} - H_1^{ig} = C_{P_{mh}}^{ig}(T_2 - T_1) \tag{6.64}$$

and

$$\Delta S^{ig} = S_2^{ig} - S_1^{ig} = C_{P_{ms}}^{ig} \ln \frac{T_2}{T_1} - R \ln \frac{P_2}{P_1} \tag{6.65}$$

Finally, step $2^{id} \to 2$ is another hypothetical process that transforms the ideal gas back into a real gas at T_2 and P_2. Here,

$$H_2 - H_2^{ig} = H_2^R$$

and

$$S_2 - S_2^{ig} = S_2^R$$

Addition of the enthalpy and entropy changes for the three steps generates Eqs. (6.62) and (6.63).

Example 6.5 Estimate V, U, H, and S for 1-butene vapor at 200°C and 70 bar if H and S are set equal to zero for saturated liquid at 0°C. Assume that the only data available are:

$$T_c = 419.6 \text{ K} \qquad P_c = 40.2 \text{ bar} \qquad \omega = 0.187$$

$$T_n = 267 \text{ K} \qquad \text{(normal boiling point)}$$

$$C_P^{ig}/R = 1.967 + 31.630 \times 10^{-3}T - 9.837 \times 10^{-6}T^2 \qquad (T/\text{K})$$

SOLUTION The volume of 1-butene vapor at 200°C and 70 bar is calculated directly from the equation $V = ZRT/P$, where Z is given by Eq. (3.45) with values of Z^0

and Z^1 taken from Figs. 3.13 and 3.15. For the reduced conditions,

$$T_r = \frac{200 + 273.15}{419.6} = 1.13 \qquad P_r = \frac{70}{40.2} = 1.74$$

we find that

$$Z = Z^0 + \omega Z^1 = 0.476 + (0.187)(0.135) = 0.501$$

Whence

$$V = \frac{(0.501)(83.14)(473.15)}{70} = 281.7 \text{ cm}^3 \text{ mol}^{-1}$$

For H and S, we use a calculational path like that of Fig. 6.14, leading from an initial state of saturated liquid 1-butene at 0°C, where H and S are zero, to the final state of interest. In this case, an initial vaporization step is required, and we have the four-step path shown by Fig. 6.15. The steps are:

(a) Vaporization at T_1 and $P_1 = P^{sat}$.
(b) Transition to the ideal-gas state at (T_1, P_1).
(c) Change to (T_2, P_2) in the ideal-gas state.
(d) Transition to the actual final state at (T_2, P_2).

Step (a). Vaporization of saturated liquid 1-butene at 0°C. The vapor pressure must be estimated, since it is not given. One method is based on Eq. (6.50):

$$\ln P^{sat} = A - \frac{B}{T}$$

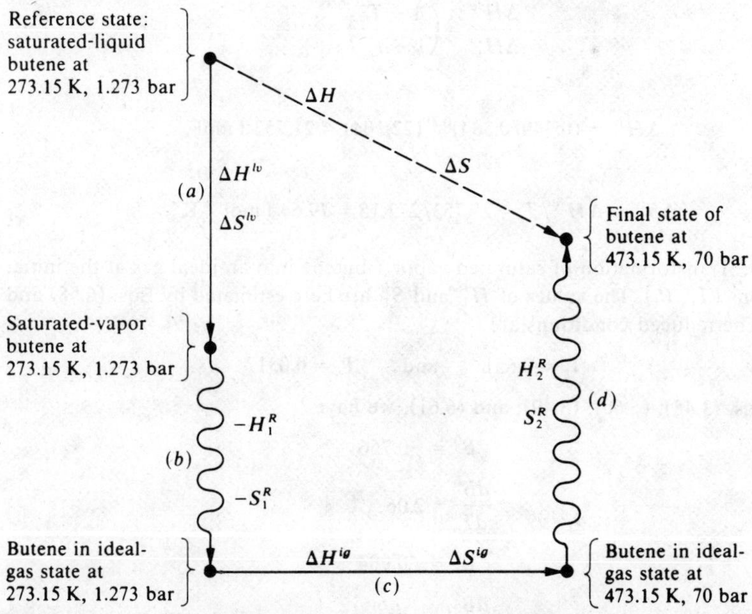

Figure 6.15

We know two points on the vapor-pressure curve: the normal boiling point, for which $P^{sat} = 1.0133$ bar at 267 K, and the critical point, for which $P^{sat} = 40.2$ bar at 419.6 K. For these two points,

$$\ln 1.0133 = A - \frac{B}{267}$$

and

$$\ln 40.2 = A - \frac{B}{419.6}$$

Simultaneous solution of these two equations gives

$$A = 10.134 \qquad \text{and} \qquad B = 2{,}702.21$$

For 0°C or 273.15 K, we then find that $P^{sat} = 1.273$ bar. This result is used in steps (*b*) and (*c*). Here, we need an estimate of the latent heat of vaporization. Equation (4.12) provides the value at the normal boiling point, where $T_{r_n} = 267/419.6 = 0.636$:

$$\frac{\Delta H_n^{lv}}{RT_n} = \frac{1.092(\ln P_c - 1.013)}{0.930 - T_{r_n}} = \frac{1.092(\ln 40.2 - 1.013)}{0.930 - 0.636}$$

$$= 9.958$$

Whence

$$\Delta H_n^{lv} = (9.958)(8.314)(267) = 22{,}104 \text{ J mol}^{-1}$$

Equation (4.13) now yields the latent heat at 273.15 K, where $T_r = 273.15/419.6 = 0.651$:

$$\frac{\Delta H^{lv}}{\Delta H_n^{lv}} = \left(\frac{1 - T_r}{1 - T_{r_n}}\right)^{0.38}$$

or

$$\Delta H^{lv} = (0.349/0.364)^{0.38}(22{,}104) = 21{,}753 \text{ J mol}^{-1}$$

and

$$\Delta S^{lv} = \Delta H^{lv}/T = 21{,}753/273.15 = 79.64 \text{ J mol}^{-1} \text{ K}^{-1}$$

Step (*b*). Transformation of saturated-vapor 1-butene into an ideal gas at the initial conditions (T_1, P_1). The values of H_1^R and S_1^R are here estimated by Eqs. (6.58) and (6.59). The reduced conditions are

$$T_r = 0.651 \qquad \text{and} \qquad P_r = 0.0317$$

From Eqs. (3.48), (6.60), (3.49), and (6.61), we have

$$B^0 = -0.756$$

$$\frac{dB^0}{dT_r} = 2.06$$

$$B^1 = -0.904$$

$$\frac{dB^1}{dT_r} = 6.73$$

Substitution of these values into Eqs. (6.58) and (6.59) gives

$$\frac{H_1^R}{RT_c} = 0.0317[(-0.756 - 0.651 \times 2.06) + 0.187(-0.904 - 0.651 \times 6.73)] = -0.0978$$

and

$$\frac{S_1^R}{R} = -0.0317[2.06 + (0.187)(6.73)] = -0.105$$

Whence

$$H_1^R = (-0.0978)(8.314)(419.6) = -341 \text{ J mol}^{-1}$$

$$S_1^R = (-0.105)(8.314) = -0.87 \text{ J mol}^{-1} \text{ K}^{-1}$$

Step (*c*). Changes in the ideal-gas state from (273.15 K, 1.273 bar) to (473.15 K, 70 bar). Here, ΔH^{ig} and ΔS^{ig} are given by Eqs. (6.64) and (6.65), which require values of $C_{P_{\text{mh}}}^{ig}$ and $C_{P_{\text{ms}}}^{ig}$. These are evaluated by Eqs. (4.8) and (5.17), wherein $T_{\text{am}} = 373.15$ K and $T_{\text{lm}} = 364.04$ K. Moreover, from the given equation for C_P^{ig}, we have

$$A = 1.967 \qquad B = 31.630 \times 10^{-3} \qquad C = -9.837 \times 10^{-6}$$

Whence

$$C_{P_{\text{mh}}}^{ig}/R = 12.367 \qquad \text{and} \qquad C_{P_{\text{ms}}}^{ig}/R = 12.145$$

Substitution of these values into Eqs. (6.64) and (6.65) gives

$$\Delta H^{ig} = (12.367)(8.314)(473.15 - 273.15) = 20{,}564 \text{ J mol}^{-1}$$

and

$$\Delta S^{ig} = (12.145)(8.314) \ln \frac{473.15}{273.15} - 8.314 \ln \frac{70}{1.273} = 22.16 \text{ J mol}^{-1} \text{ K}^{-1}$$

Step (*d*). Transformation of 1-butene from the ideal-gas state to the real-gas state at T_2 and P_2. The final reduced conditions are

$$T_r = 1.13 \qquad \text{and} \qquad P_r = 1.74$$

In this instance, we estimate H_2^R and S_2^R by Eqs. (6.56) and (6.57). Substitution of values taken from Figs. (6.7), (6.9), (6.11), and (6.13) gives:

$$\frac{H_2^R}{RT_c} = -2.34 + (0.187)(-0.62) = -2.46$$

and

$$\frac{S_2^R}{R} = -1.63 + (0.187)(-0.56) = -1.73$$

Whence

$$H_2^R = (-2.46)(8.314)(419.6) = -8{,}582 \text{ J mol}^{-1}$$

$$S_2^R = (-1.73)(8.314) = -14.38 \text{ J mol}^{-1} \text{ K}^{-1}$$

The sums of the enthalpy and entropy changes for the four steps give the total changes for the process leading from the initial reference state (where H and S are set equal to zero) to the final state:

$$H = \Delta H = 21{,}753 + 341 + 20{,}564 - 8{,}582 = 34{,}076 \text{ J mol}^{-1}$$

and

$$S = \Delta S = 79.64 + 0.87 + 22.16 - 14.38 = 88.29 \text{ J mol}^{-1} \text{ K}^{-1}$$

The internal energy is

$$U = H - PV = 34{,}076 - \frac{(70)(280.9)}{10 \text{ cm}^3 \text{ bar J}^{-1}} = 32{,}110 \text{ J mol}^{-1}$$

These results are in far better agreement with experimental values than would have been the case had we assumed 1-butene vapor an ideal gas.

PROBLEMS

6.1 Starting with Eq. (6.8), show that isobars in the vapor region of a Mollier (HS) diagram must have positive slope and positive curvature.

6.2 Making use of the fact that Eq. (6.20) is an exact differential expression, show that

$$(\partial C_P/\partial P)_T = -T(\partial^2 V/\partial T^2)_P$$

What is the result of application of this equation to an ideal gas?

6.3 A frequent assumption is that pressure has a negligible effect on liquid-phase properties, and that the properties of a compressed liquid are essentially those of the saturated liquid at the same temperature. Estimate the errors when the enthalpy and entropy of liquid ammonia at 270 K and 1,500 kPa are assumed equal to the enthalpy and entropy of saturated liquid ammonia at 270 K. For saturated liquid ammonia at 270 K, $P^{sat} = 381$ kPa, $V^l = 1.551 \times 10^{-3}$ m^3 kg^{-1}, and $\beta = 2.095 \times 10^{-3}$ K^{-1}.

6.4 Liquid propane is throttled through a valve from an initial state of 40°C and 3,000 kPa to a final pressure of 2,000 kPa. Estimate the temperature change and the entropy change of the propane. The specific heat of liquid propane at 40°C is 2.84 J g^{-1} °C^{-1}.

6.5 Liquid water at 25°C and 1 bar fills a rigid vessel. If heat is added to the water until its temperature reaches 50°C, what pressure is developed? The average value of β between 25 and 50°C is 36.2×10^{-5} K^{-1}. The value of κ at 1 bar and 50°C is 4.42×10^{-5} bar^{-1}, and may be assumed independent of P. The specific volume of liquid water at 25°C is 1.0030 cm^3 g^{-1}.

6.6 A good estimate of the latent heat of vaporization of 1,3-butadiene at 60°C is required. The vapor pressure of 1,3-butadiene is given by the equation:

$$\ln P^{sat}/\text{kPa} = 13.7578 - \frac{2{,}142.66}{T/\text{K} - 34.30}$$

From this and from an estimate of ΔV^{lv}, calculate ΔH^{lv} by the Clapeyron equation [Eq. (6.49)].

6.7 A thin-walled metal container, filled with saturated steam at 100°C, is tightly capped and allowed to cool slowly. If the container can support a pressure difference of no more than 20 kPa and if the surrounding pressure is 101.33 kPa, at what temperature does the container collapse? If steam were an ideal gas, what would be the temperature?

6.8 The state of 1(lb$_m$) of steam is changed from saturated vapor at 10(psia) to superheated vapor at 30(psia) and 1,200(°F). What are the enthalpy and entropy changes of the steam? What would the enthalpy and entropy changes be if steam were an ideal gas?

6.9 Very pure liquid water can be supercooled at atmospheric pressure to temperatures well below 0°C. Assume that 1 kg has been cooled as a liquid to −6°C. A small ice crystal (of negligible mass) is added to "seed" the supercooled liquid. If the subsequent change occurs adiabatically at atmospheric pressure, what fraction of the system freezes and what is the final temperature? What is ΔS_{total} for the process, and what is its irreversible feature? The latent heat of fusion of water at 0°C = 333.4 J g^{-1}, and the specific heat of supercooled liquid water = 4.226 J g^{-1} °C^{-1}.

6.10 A two-phase system of liquid water and water vapor in equilibrium at 12,000 kPa consists of equal volumes of liquid and vapor. If the total volume $V^t = 0.1$ m^3, what is the total enthalpy H^t and what is the total entropy S^t?

6.11 A vessel contains 1(lb$_m$) of H_2O existing as liquid and vapor in equilibrium at 500 (psia). If the liquid and vapor each occupy half the volume of the vessel, determine H and S for the 1(lb$_m$) of H_2O.

6.12 A pressure vessel contains liquid water and water vapor in equilibrium at 300(°F). The total mass of liquid and vapor is 2(lb$_m$). If the volume of vapor is 100 times the volume of liquid, what is the total enthalpy of the contents of the vessel?

6.13 Wet steam at 230°C has a specific volume of 25.79 cm^3 g^{-1}. Determine x, H, and S.

6.14 A vessel of 0.1-m^3 volume containing saturated-vapor steam at 110°C is cooled to 25°C. Determine the volume and mass of *liquid* water in the vessel.

6.15 Wet steam at 1,800 kPa expands at constant enthalpy (as in a throttling process) to 101.33 kPa, where its temperature is 115°C. What is the quality of the steam in its initial state?

6.16 Steam at 550 kPa and 200°C expands at constant enthalpy (as in a throttling process) to 200 kPa. What is the temperature of the steam in its final state and what is its entropy change? If steam were an ideal gas, what would be its final temperature and its entropy change?

6.17 Steam at 3,000(psia) and 1,000(°F) expands at constant enthalpy (as in a throttling process) to 2,000(psia). What is the temperature of the steam in its final state and what is its entropy change? If steam were an ideal gas, what would be its final temperature and its entropy change?

6.18 Saturated steam at 160(psia) expands at constant enthalpy (as in a throttling process) to 25(psia). What is its final temperature and what is its entropy change?

6.19 A rigid vessel contains 1(lb$_m$) of saturated-vapor steam at 250(°F). Heat is extracted from the vessel until the pressure reaches 15(psia). What is the entropy change of the steam?

6.20 A rigid vessel contains 0.50(ft)^3 of saturated-vapor steam in equilibrium with 0.75(ft)^3 of saturated-liquid water at 212(°F). Heat is transferred to the vessel until one phase just disappears, and a single phase remains. Which phase (liquid or vapor) remains, and what are its temperature and pressure? How much heat is transferred in the process?

6.21 A vessel of 0.3-m^3 capacity is filled with saturated steam at 1,700 kPa. If the vessel is cooled until 35 percent of the steam has condensed, how much heat is transferred and what is the final pressure?

6.22 A vessel of 3-m^3 capacity contains 0.03 m^3 of liquid water and 2.97 m^3 of water vapor at 101.33 kPa. How much heat must be added to the contents of the vessel so that the liquid water is just evaporated?

6.23 A rigid vessel 10(ft)^3 in volume contains saturated-vapor steam at 75(psia). Heat exchange with a single external heat reservoir at 60(°F) reduces the temperature of the contents of the vessel to 60(°F). Determine ΔS_{total}. What is the irreversible feature of this process?

6.24 A rigid vessel of 0.5-m^3 volume is filled with steam at 700 kPa and 325°C. How much heat must be transferred from the steam to bring its temperature to 175°C.

6.25 A rigid, nonconducting vessel is divided in half by a rigid partition. Initially one side of the vessel contains steam at 3,400 kPa and 275°C, and the other side is evacuated. The partition is removed, and the steam expands adiabatically to fill the vessel. What are the final temperature and pressure of the steam?

6.26 One kilogram of steam undergoes the following changes in state. Calculate Q and W for each process.

(*a*) Initially at 350 kPa and 260°C, it is cooled at constant pressure to 150°C.

(*b*) Initially at 350 kPa and 260°C, it is cooled at constant volume to 150°C.

6.27 One kilogram of steam is contained in a piston/cylinder device at 700 kPa and 260°C.

(*a*) If it undergoes a mechanically reversible, isothermal expansion to 250 kPa, how much heat does it absorb?

(*b*) If it undergoes a reversible, adiabatic expansion to 250 kPa, what is its final temperature and how much work is done?

6.28 Steam at 2,600 kPa containing 5 percent moisture is heated at constant pressure to 475°C. How much heat is required per kilogram?

6.29 Steam at 2,100 kPa and with a quality of 0.85 undergoes a reversible, adiabatic expansion in a nonflow process to 350 kPa. It is then heated at constant volume until it is saturated vapor. Determine Q and W for the process.

6.30 Five kilograms of steam in a piston/cylinder device at 150 kPa and 150°C undergoes a mechanically reversible, isothermal compression to a final pressure such that the steam is just saturated. Determine Q and W for the process.

6.31 Steam at 300(°F) and 1(atm) is compressed isothermally in a mechanically reversible, nonflow process until it reaches a final state of saturated liquid. Determine Q and W for the process.

6.32 One kilogram of water in a piston/cylinder device at 25°C and 1 bar is compressed in a mechanically reversible, isothermal process to 1,500 bar. Estimate Q, W, ΔU, ΔH, and ΔS given that $\beta = 250 \times 10^{-6}\ \mathrm{K}^{-1}$ and $\kappa = 45 \times 10^{-6}\ \mathrm{bar}^{-1}$.

6.33 A piston/cylinder device operating in a cycle with steam as the working fluid executes the following steps:

(*a*) Steam at 525 kPa and 175°C is heated at constant volume to a pressure of 750 kPa.

(*b*) The steam then expands, reversibly and adiabatically, to the initial temperature of 175°C.

(*c*) Finally, the steam is compressed in a mechanically reversible, isothermal process to the initial presure of 525 kPa.

What is the thermal efficiency of the cycle?

6.34 A piston/cylinder device operating in a cycle with steam as the working fluid executes the following steps:

(*a*) Saturated-vapor steam at 500(°F) is heated at constant pressure to 1,000(°F).

(*b*) The steam then expands, reversibly and adiabatically, to the initial temperature of 500(°F).

(*c*) Finally, the steam is compressed in a mechanically reversible, isothermal process to the initial state.

What is the thermal efficiency of the cycle?

6.35 Steam with a quality of 0.85 expands in a mechanically reversible, nonflow process at constant quality from 200 to 40°C. Determine Q and W.

6.36 One kilogram of saturated-liquid water at 1,250 kPa expands at constant internal energy in a mechanically reversible, nonflow process until its tempeature falls to 90°C. What is the work?

6.37 Steam expands isentropically in a turbine, entering at 3,800 kPa and 375°C.

(*a*) For what discharge pressure is the exit stream a saturated vapor?

(*b*) For what discharge pressure is the exit stream a wet vapor with quality of 0.90?

6.38 A steam turbine, operating isentropically, takes in superheated steam at 1,800 kPa and discharges at 30 kPa. What is the minimum superheat required so that the exhaust contains no moisture? What is the power output of the turbine if it operates under these conditions and the steam rate is 5 kg s^{-1}.

6.39 A steam turbine operates adiabatically with a steam rate of 30 kg s^{-1}. The steam is supplied at 1,050 kPa and 375°C and discharges at 20 kPa and 75°C. Determine tthe power output of the turbine and the efficiency of its operation in comparison with a turbine that operates isentropically from the same initial conditions to the same final pressure.

6.40 From steam-table data, estimate values for the residual properties V^R, H^R, and S^R for steam at 200°C and 1,400 kPa, and compare with values found by a suitable generalized correlation.

6.41 Estimate V^R, H^R, and S^R for carbon dioxide at 425 K and 350 bar by appropriate generalized correlations.

6.42 Estimate V^R, H^R, and S^R for sulfur dioxide at 500 K and 235 bar by appropriate generalized correlations.

6.43 From data in the steam tables, determine numerical values for the following:
(*a*) G^l and G^v for saturated liquid and vapor at 135(psia). Should these be the same?
(*b*) $\Delta H^{lv}/T$ and ΔS^{lv} for saturation at 135(psia). Should these be the same?
(*c*) V^R, H^R, and S^R for saturated vapor at 135(psia).
From data for P^{sat} at 130 and 140(psia), estimate a value for dP^{sat}/dT at 135(psia) and apply the Clapeyron equation to estimate ΔS^{lv} at 135(psia). How well does this result agree with the steam-table value? Apply appropriate generalized correlations for evaluation of V^R, H^R, and S^R for saturated vapor at 135(psia). How well do these results compare with the values found in (*c*)?

6.44 From data in the steam tables, determine numerical values for the following:
(*a*) G^l and G^v for saturated liquid and vapor at 900 kPa. Should these be the same?
(*b*) $\Delta H^{lv}/T$ and ΔS^{lv} for saturation at 900 kPa. Should these be the same?
(*c*) V^R, H^R, and S^R for saturated vapor at 900 kPa.
From data for P^{sat} at 875 and 925 kPa, estimate a value for dP^{sat}/dT at 900 kPa and apply the Clapeyron equation to estimate ΔS^{lv} at 900 kPa. How well does this result agree with the steam-table value? Apply appropriate generalized correlations for evaluation of V^R, H^R, and S^R for saturated vapor at 900 kPa. How well do these results compare with the values found in (*c*)?

6.45 Steam undergoes a change from an initial state of 475°C and 3,400 kPa to a final state of 150°C and 275 kPa. Determine ΔH and ΔS:
(*a*) From steam-table data.
(*b*) By equations for an ideal gas.
(*c*) By appropriate generalized correlations.

6.46 Propane gas at 1 bar and 50°C is compressed to a final state of 125 bar and 245°C. Estimate the molar volume of the propane in the final state and the enthalpy and entropy changes for the process. In its initial state, propane may be assumed an ideal gas.

6.47 Propane at 320 K and 101.33 kPa is compressed isothermally to 1,603 kPa, its vapor pressure at 320 K. Estimate ΔH and ΔS for the process by suitable generalized correlations.

6.48 Estimate the molar volume, enthalpy, and entropy for propylene as a saturated vapor and as a saturated liquid at 55°C. The enthalpy and entropy are set equal to zero for the ideal-gas state at 101.33 kPa and 0°C. The normal boiling point of propylene is −47.7°C, and its vapor pressure at 55°C is 22.94 bar.

6.49 Estimate the molar volume, enthalpy, and entropy for *n*-butane as a saturated vapor and as a saturated liquid at 370 K. The enthalpy and entropy are set equal to zero for the ideal-gas state at 101.33 kPa and 273.15 K. The normal boiling point of *n*-butane is 272.67 K, and its vapor pressure at 370 K is 14.35 bar.

6.50 A quantity of 5 mol calcium carbide is combined with 10 mol of liquid water in a closed, rigid, high-pressure vessel of 750-cm^3 capacity. Acetylene gas is produced by the reaction:

$$CaC_2(s) + 2H_2O(l) \rightarrow C_2H_2(g) + Ca(OH)_2(s)$$

Initial conditions are 25°C and 1 bar, and the reaction goes to completion. For a final temperature of 125°C, determine:
(*a*) The final pressure.
(*b*) The heat transferred.
At 125°C, the molar volume of $Ca(OH)_2$ is 33.0 $cm^3\ mol^{-1}$. Ignore the effect of any gas present in the tank initially.

6.51 Propylene gas at 134°C and 43 bar is throttled in a steady-state flow process to 1 bar, where it may be assumed an ideal gas. Estimate the final temperature of the propylene and its entropy change.

6.52 Propane gas at 20 bar and 400 K is throttled in a steady-state flow process to 1 bar. Estimate the entropy change of the propane caused by this process. In its final state, propane may be assumed an ideal gas.

6.53 Carbon dioxide expands at constant enthalpy (as in a throttling process) from 1,500 kPa and 30°C to 101.33 kPa. Estimate ΔS for the process.

6.54 A stream of ethylene gas at 260°C and 4,100 kPa expands isentropically in a turbine to 140 kPa. Determine the temperature of the expanded gas and the work produced if the properties of ethylene are calculated by
(*a*) Equations for an ideal gas.
(*b*) Appropriate generalized correlations.

6.55 A stream of ethane gas at 200°C and 25 bar expands isentropically in a turbine to 2 bar. Determine the temperature of the expanded gas and the work produced if the properties of ethane are calculated by
(*a*) Equations for an ideal gas.
(*b*) Appropriate generalized correlations.

6.56 Estimate the final temperature and the work required when 1 mol of 1,3-butadiene is compressed isentropically in a steady-flow process from 1 bar and 60°C to 7 bar.

CHAPTER

SEVEN

THERMODYNAMICS OF FLOW PROCESSES

Most equipment used in the chemical, petroleum, and related industries is designed for the movement of fluids, and an understanding of fluid flow is essential to a chemical engineer. The underlying discipline is fluid mechanics,† which is based on the law of mass conservation, the linear momentum principle (Newton's second law), and the first and second laws of thermodynamics.

The application of thermodynamics to flow processes is also based on conservation of mass and on the first and second laws. The addition of the linear momentum principle makes fluid mechanics a broader field of study. The usual separation between *thermodynamics problems* and *fluid-mechanics problems* depends on whether this principle is required for solution. Those problems whose solutions depend only on conservation of mass and on the laws of thermodynamics are commonly set apart from the study of fluid mechanics and are treated in courses on thermodynamics. Fluid mechanics then deals with the broad spectrum of problems which *require* application of the momentum principle. This division is arbitrary, but it is traditional and convenient.

The applications of thermodynamics to flow processes usually are to finite amounts of fluid undergoing finite changes in state. One might, for example, deal with the flow of gas through a pipeline. If the states and thermodynamic properties of the gas entering and leaving the pipeline are known, then application of the

† Fluid mechanics is treated as an integral part of transport processes by R. B. Bird, W. E. Stewart, and E. N. Lightfoot in *Transport Phenomena*, John Wiley, New York, 1960, by C. O. Bennett and J. E. Myers in *Momentum Heat and Mass Transfer*, 2d ed., McGraw-Hill, New York, 1982, and by R. W. Fahien in *Fundamentals of Transport Phenomena*, McGraw-Hill, New York, 1984.

first law establishes the magnitude of the energy exchange with the surroundings of the pipeline. The mechanism of the process, the details of flow, and the state path actually followed by the fluid between entrance and exit are not pertinent to this calculation.

On the other hand, if one has only incomplete knowledge of the initial or final state of the gas, then more detailed information about the process is needed before any calculations are made. For example, the exit pressure of the gas may not be specified. In this case, one must apply the momentum principle of fluid mechanics, and this requires an empirical or theoretical expression for the shear stress at the pipe wall.

The fundamental equations generally applicable to flow processes are presented in Sec. 7.1, and in later sections these equations are applied to specific processes.

7.1 FUNDAMENTAL EQUATIONS

Two idealizations are imposed from the start to facilitate the application of thermodynamic principles to flow processes:

1. We presume that flow is unidirectional at any cross section of a conduit where thermodynamic, kinetic, and dynamic properties are assigned or evaluated, namely, at entrances to and exits from the equipment under consideration.
2. We also imagine that at such a cross section these same properties do not vary in the direction perpendicular to the direction of flow. Thus properties such as velocity, temperature, and density, assigned or evaluated for the cross section, have values which are appropriate averages over the cross section.

These idealizations are pragmatic in nature, and for most practical purposes they introduce negligible error.

Conservation of Mass

The law of conservation of mass for fluids in flow processes is most conveniently written so as to apply to a *control volume*, which is equivalent to a thermodynamic system as defined in Sec. 2.3. A control volume is an arbitrary volume enclosed by a bounding *control surface*, which may or may not be identified with physical boundaries, but which in the general case is pervious to matter. The flow processes of interest to chemical engineers usually permit identification of almost the entire control surface with actual material surfaces. Only at specifically provided entrances and exits is the control surface subject to arbitrary location, and here it is universal practice to place the control surface perpendicular to the direction of flow, so as to allow direct imposition of idealizations 1 and 2. An example of a control volume with one entrance and one exit is shown in Fig. 7.1. The actual

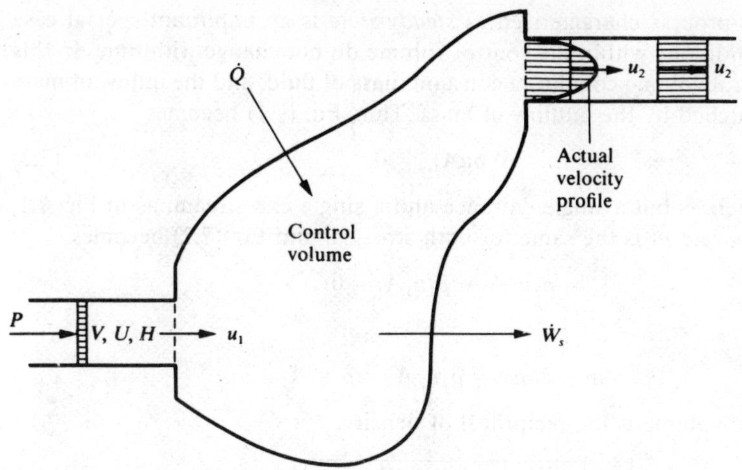

Figure 7.1 Control volume with one entrance and one exit.

velocity profile shown at the exit is equivalent to the uniform velocity profile indicated to the right that provides the same mass flow rate (idealization 2).

The principle of conservation of mass for a flow process may be written in words as:

$$\left\{\begin{matrix}\text{Rate of accumulation}\\ \text{of mass within the}\\ \text{control volume}\end{matrix}\right\} = \left\{\begin{matrix}\text{mass flow}\\ \text{rate in at}\\ \text{entrances}\end{matrix}\right\} - \left\{\begin{matrix}\text{mass flow}\\ \text{rate out}\\ \text{at exits}\end{matrix}\right\}$$

or

$$\left\{\begin{matrix}\text{Rate of accumulation}\\ \text{of mass within the}\\ \text{control volume}\end{matrix}\right\} + \left\{\begin{matrix}\text{net mass flow}\\ \text{rate out by}\\ \text{flowing streams}\end{matrix}\right\} = 0$$

The first term on the left is the rate of change with time of the total mass within the control volume, dm/dt. The mass flow rates of streams at entrances and exits is given by

$$\dot{m} = \text{mass flow rate} = \rho u A$$

where ρ is the average fluid density, u is its average velocity, and A is the cross-sectional area of the entrance or exit duct. The mass-conservation equation (also called the *continuity equation*) is therefore expressed mathematically as:

$$\frac{dm}{dt} + \Delta(\rho u A)_{\text{fs}} = 0 \tag{7.1}$$

where the symbol Δ denotes the difference between exit and entrance streams and the subscript "fs" indicates that the term applies to all flowing streams.

The flow process characterized as *steady-state* is an important special case for which conditions within the control volume do not change with time. In this case the control volume contains a constant mass of fluid, and the inflow of mass is exactly matched by the outflow of mass. Thus Eq. (7.1) becomes

$$\Delta(\rho u A)_{\text{fs}} = 0 \tag{7.2}$$

Further, if there is but a single entrance and a single exit stream, as in Fig. 7.1, the mass flow rate $\dot{m}$ is the same for both streams, and Eq. (7.2) becomes

$$\rho_2 u_2 A_2 - \rho_1 u_1 A_1 = 0$$

or

$$\dot{m} = \text{const} = \rho_2 u_2 A_2 = \rho_1 u_1 A_1$$

Since specific volume is the reciprocal of density,

$$\boxed{\dot{m} = \frac{u_1 A_1}{V_1} = \frac{u_2 A_2}{V_2} = \frac{uA}{V}} \tag{7.3}$$

This form of the continuity equation finds frequent use.

Conservation of Energy

In Chap. 2 the first law of thermodynamics was applied to closed systems (nonflow processes) and to single-stream, steady-state flow processes to provide specific equations of energy conservation for these important applications. Our purpose here is to present a more general equation applicable to an open system or to a control volume.

The basic conservation requirement may be expressed in words:

$$\left\{\begin{matrix}\text{Rate of accumulation}\\ \text{of energy within the}\\ \text{control volume}\end{matrix}\right\} = \left\{\begin{matrix}\text{rate of energy}\\ \text{transport in at}\\ \text{entrances}\end{matrix}\right\} - \left\{\begin{matrix}\text{rate of energy}\\ \text{transport out}\\ \text{at exits}\end{matrix}\right\}$$
$$+ \left\{\begin{matrix}\text{heat flow,}\\ \dot{Q}\text{, in across the}\\ \text{control surface}\end{matrix}\right\} - \left\{\begin{matrix}\text{net power, } \dot{W}\text{,}\\ \text{out across the}\\ \text{control surface}\end{matrix}\right\}$$

or

$$\left\{\begin{matrix}\text{Rate of accumulation}\\ \text{of energy within the}\\ \text{control volume}\end{matrix}\right\} + \left\{\begin{matrix}\text{net rate of energy}\\ \text{transport out by}\\ \text{flowing streams}\end{matrix}\right\} = \dot{Q} - \dot{W}$$

The first term on the left is the rate of change with time of the total internal energy within the control volume, $d(mU)_{\text{cv}}/dt$. Associated with each flowing stream are three forms of energy: internal, kinetic on account of its velocity u, and potential on account of its elevation z above a datum level. Thus on the basis of a unit mass, each stream has a total energy $U + \frac{1}{2}u^2 + zg$ and transports energy at the rate $(U + \frac{1}{2}u^2 + zg)\dot{m}$. The energy-conservation equation is therefore

$$\frac{d(mU)_{\text{cv}}}{dt} + \Delta[(U + \tfrac{1}{2}u^2 + zg)\dot{m}]_{\text{fs}} = \dot{Q} - \dot{W} \tag{7.4}$$

where subscript "cv" denotes the control volume and g is the local acceleration of gravity.

The power or work rate $\dot{W}$ consists of two parts. The first is the shaft-work rate $\dot{W}_s$ shown in Fig. 7.1. Less obvious is the work associated with moving the flowing streams into and out of the control volume at entrances and exits. The fluid at any entrance or exit has a set of average properties, P, V, U, H, etc. We imagine that a unit mass of fluid with these properties exists in a conduit adjacent to the entrance or exit, as shown in Fig. 7.1 at the entrance. This unit mass of fluid is pushed into the control volume by additional fluid, here replaced by a piston which exerts the constant pressure P. The work done by this piston in pushing the unit mass into the control volume is PV, and the work rate is $(PV)\dot{m}$. The net work done at all entrance and exit sections is then $\Delta[(PV)\dot{m}]_{\text{fs}}$. Thus

$$\dot{W} = \dot{W}_s + \Delta[(PV)\dot{m}]_{\text{fs}}$$

Combining this with Eq. (7.4) gives

$$\frac{d(mU)_{\text{cv}}}{dt} + \Delta[(U + PV + \tfrac{1}{2}u^2 + zg)\dot{m}]_{\text{fs}} = \dot{Q} - \dot{W}_s$$

Since $U + PV = H$, this is more conveniently written:

$$\boxed{\frac{d(mU)_{\text{cv}}}{dt} + \Delta[(H + \tfrac{1}{2}u^2 + zg)\dot{m}]_{\text{fs}} = \dot{Q} - \dot{W}_s} \tag{7.5}$$

Although Eq. (7.5) is an energy balance of considerable generality, it has inherent limitations. In particular, it is based on the presumption that the control volume is a constant volume and that it is at rest. This means that kinetic- and potential-energy changes of the fluid in the control volume can be neglected. For virtually all applications of interest to chemical engineers, Eq. (7.5) is adequate. Indeed, for most applications, kinetic- and potential-energy changes in the flowing streams are also negligible, and Eq. (7.5) simplifies to

$$\frac{d(mU)_{\text{cv}}}{dt} + \Delta(H\dot{m})_{\text{fs}} = \dot{Q} - \dot{W}_s \tag{7.6}$$

Since $\dot{m} = dm/dt$, $\dot{Q} = dQ/dt$, and $\dot{W}_s = dW_s/dt$, multiplication of this equation by dt puts it into differential form:

$$d(mU)_{\text{cv}} + \Delta(H\,dm)_{\text{fs}} = dQ - dW_s \tag{7.7}$$

This equation may be applied to a variety of processes of a transient nature, as illustrated in the following examples.

Example 7.1 Consider the filling of an evacuated tank with a gas from a constant-pressure line. What is the relation between the enthalpy of the gas in the entrance line and the internal energy of the gas in the tank? Neglect heat transfer between the gas and the tank. If the gas is ideal and has constant heat capacities, how is the temperature of the gas in the tank related to the temperature in the entrance line?

SOLUTION If the tank is chosen as the control volume, there is but one opening into the tank and it serves as an entrance, because gas flows into the tank. Since there is no shaft work, $dW_s = 0$. In the absence of any specific information, we assume that kinetic- and potential-energy changes are negligible. By Eq. (7.7) we have

$$d(mU)_{\text{tank}} - H'\,dm' = 0$$

where the prime (′) identifies the entrance stream and the minus sign is required because it *is* an entrance stream. Since H' is constant, integration gives

$$\Delta(mU)_{\text{tank}} = m_2U_2 - m_1U_1 = H'm'$$

Since the mass in the tank initially is zero, $m_1 = 0$ and $m_2 = m'$. Therefore the preceding equation reduces to

$$U_2 = H' \qquad (A)$$

This result shows that in the absence of heat transfer the energy of the gas contained within the tank at the end of the process is equal to the enthalpy of the gas added.

If the gas is ideal,

$$H' = U' + P'V' = U' + RT'$$

and Eq. (A) becomes

$$U_2 - U' = RT'$$

For constant heat capacity,

$$U_2 - U' = C_V(T_2 - T')$$

whence

$$C_V(T_2 - T') = RT'$$

or

$$\frac{T_2 - T'}{T'} = \frac{R}{C_V} = \frac{C_P - C_V}{C_V}$$

If C_P/C_V is set equal to γ, this reduces to

$$T_2 = \gamma T'$$

which indicates that the final temperature is independent of the amount of gas admitted to the tank. This result is strongly conditioned by the initial stipulation that heat transfer between the gas and the tank be neglected.

Example 7.2 A 1.5-m³ tank contains 500 kg of liquid water in equilibrium with pure water vapor, which fills the remainder of the tank. The temperature and pressure are 100°C and 101.33 kPa. From a water line at a constant temperature of 70°C and a constant pressure somewhat above 101.33 kPa, 750 kg is bled into the tank. If the temperature and pressure in the tank are not to change as a result of the process, how much energy as heat must be transferred to the tank?

SOLUTION Choose the tank as the control volume. As in Example 7.1, there is no shaft work, and again we assume negligible kinetic- and potential-energy effects. Equation (7.7) therefore is written

$$d(mU)_{\text{tank}} - H'\,dm' = dQ$$

where the prime denotes the state of the inlet stream. Integration of this equation with H' constant gives

$$Q = \Delta(mU)_{\text{tank}} - H'm'$$

The definition of enthalpy may be applied to the entire contents of the tank to give

$$\Delta(mU)_{\text{tank}} = \Delta(mH)_{\text{tank}} - \Delta(PmV)_{\text{tank}}$$

Since the total volume mV of the tank and the pressure are constant, $\Delta(PmV)_{\text{tank}} = 0$. Therefore

$$Q = \Delta(mH)_{\text{tank}} - H'm' = (m_2H_2 - m_1H_1)_{\text{tank}} - H'm'$$

where m' is the mass added in the inlet stream, and m_1 and m_2 are the masses of water in the tank at the beginning and end of the process. At the end of the process the tank still contains saturated liquid and saturated vapor in equilibrium at 100°C and 101.33 kPa. Hence m_1H_1 and m_2H_2 each consist of two terms, one for the liquid phase and one for the vapor phase.

The numerical solution makes use of the following enthalpies taken from the steam tables:

$$\begin{aligned} H' &= 293.0 \text{ kJ kg}^{-1}\text{; saturated liquid at 70°C} \\ H^l_{\text{tank}} &= 419.1 \text{ kJ kg}^{-1}\text{; saturated liquid at 100°C} \\ H^v_{\text{tank}} &= 2{,}676.0 \text{ kJ kg}^{-1}\text{; saturated vapor at 100°C} \end{aligned}$$

The volume of vapor in the tank initially is 1.5 m^3 minus the volume occupied by the 500 kg of liquid water. Thus

$$m^v_1 = \frac{1.5 - (500)(0.001044)}{1.673} = 0.772 \text{ kg}$$

where 0.001044 and 1.673 m^3 kg^{-1} are the specific volumes of saturated liquid and saturated vapor at 100°C from the steam tables. Then

$$\begin{aligned} (m_1H_1)_{\text{tank}} &= m^l_1H^l_1 + m^v_1H^v_1 = 500(419.1) + 0.772(2{,}676.0) \\ &= 211{,}616 \text{ kJ} \end{aligned}$$

At the end of the process, the masses of liquid and vapor are determined by the conservation of mass and by the fact that the tank volume is still 1.5 m^3. These constraints give the equations:

$$m_2 = 500 + 0.772 + 750 = m^v_2 + m^l_2$$

$$1.5 = 1.673m^v_2 + 0.001044\,m^l_2$$

Whence

$$m_2^l = 1{,}250.65 \text{ kg}$$

$$m_2^v = 0.116 \text{ kg}$$

Then since $H_2^l = H_1^l$ and $H_2^v = H_1^v$,

$$(m_2 H_2)_{\text{tank}} = 1{,}250.65(419.1) + 0.116(2{,}676.0) = 524{,}458 \text{ kJ}$$

Finally, substituting the values for $(m_1 H_1)_{\text{tank}}$ and $(m_2 H_2)_{\text{tank}}$ in the equation for Q gives

$$Q = 524{,}458 - 211{,}616 - 750(293.0) = 93{,}092 \text{ kJ}$$

Energy Balances for Steady-State Flow Processes

For a steady-state flow process, the total internal energy of the control volume is constant, and $d(mU)_{\text{cv}}/dt$ is zero. Equation (7.5) therefore becomes

$$\boxed{\Delta[(H + \tfrac{1}{2}u^2 + zg)\dot{m}]_{\text{fs}} = \dot{Q} - \dot{W}_s} \tag{7.8}$$

This equation is widely used, because steady-state flow processes represent the norm in the chemical-process industry.

A further specialization results when there is but one entrance and one exit to the control volume. In this case the mass flow rate $\dot{m}$ is the same for both streams, and Eq. (7.8) reduces to

$$\Delta(H + \tfrac{1}{2}u^2 + zg)\dot{m} = \dot{Q} - \dot{W}_s \tag{7.9}$$

Division by $\dot{m}$ gives

$$\Delta(H + \tfrac{1}{2}u^2 + zg) = \frac{\dot{Q}}{\dot{m}} - \frac{\dot{W}_s}{\dot{m}} = Q - W_s$$

or

$$\boxed{\Delta H + \frac{\Delta u^2}{2} + g\,\Delta z = Q - W_s} \tag{7.10}$$

which is a restatement of Eq. (2.10a). In this equation, each term is based on a unit mass of fluid flowing through the control volume.

The kinetic-energy terms of the various energy balances developed here include the velocity u, which is the bulk-mean velocity as defined by the equation $u = \dot{m}/\rho A$. Fluids flowing in pipes exhibit a velocity profile, as shown in Fig. 7.1, which rises from zero at the wall (the no-slip condition) to a maximum at the center of the pipe. The kinetic energy of a fluid in a pipe depends on the actual velocity profile. For the case of laminar flow, the velocity profile is parabolic, and integration across the pipe shows that the kinetic-energy term should properly be u^2. In fully developed turbulent flow, the more common case in practice, the velocity across the major portion of the pipe is not far from uniform, and the expression $u^2/2$, as used in the energy equations, is more nearly correct.

In all of the equations written here, the energy unit is presumed to be the joule, in accord with the SI system of units. For the English system of units, the kinetic- and potential-energy terms, wherever they appear, require division by the dimensional constant g_c (see Secs. 1.3 and 1.8). However, in many applications, the kinetic- and potential-energy terms are omitted, because they are negligible compared with other terms. Exceptions are applications to nozzles, metering devices, wind tunnels, and hydroelectric power stations.

Mechanical Energy Balance; Bernoulli Equation

Equation (7.10) applies to the steady-state flow of fluid through a control volume to which there is but one entrance and one exit. In addition, we have the fundamental property relation of Eq. (6.8):

$$dH = T\,dS + V\,dP$$

For a reversible change of state, $T\,dS = dQ$. Then

$$dH = dQ + V\,dP$$

Integration gives

$$\Delta H = Q + \int_{P_1}^{P_2} V\,dP$$

Substituting for ΔH in Eq. (7.10), we get

$$-W_s = \int_{P_1}^{P_2} V\,dP + \frac{\Delta u^2}{2} + g\,\Delta z$$

This equation is based on the assumption that the change of state resulting from the process is accomplished *reversibly*. However, the viscous nature of real fluids induces fluid friction that makes changes of state in flow processes inherently irreversible because of the dissipation of mechanical energy into internal energy. In order to correct for this, we add to the equation a friction term F. The *mechanical-energy balance* is then written:

$$-W_s = \int_{P_1}^{P_2} V\,dP + \frac{\Delta u^2}{2} + g\,\Delta z + F \tag{7.11}$$

The determination of numerical values for F is a problem in fluid mechanics. For evaluation of the integral term, one must know or assume a V-vs.-P relation. For liquids, the common assumption is that the specific volume V is constant, independent of pressure.

Bernoulli's famous equation, formulated over a century prior to the development of the first law of thermodynamics, is a special case of the mechanical-energy balance. It applies to a nonviscous, incompressible fluid which does not exchange shaft work with the surroundings. For a nonviscous fluid, F is zero, and for an incompressible fluid

$$\int_{P_1}^{P_2} V\,dP = V\,\Delta P = \frac{\Delta P}{\rho}$$

where ρ is fluid density. Equation (7.11) reduces to

$$\frac{\Delta P}{\rho} + \frac{\Delta u^2}{2} + g\,\Delta z = 0 \tag{7.12}$$

which is Bernoulli's equation. As an alternative expression, we have

$$\Delta\left(\frac{P}{\rho} + \frac{u^2}{2} + gz\right) = 0$$

or

$$\frac{P}{\rho} + \frac{u^2}{2} + gz = \text{const}$$

7.2 FLOW IN PIPES

The quantity of most immediate interest with respect to the steady-state flow of fluid in a straight length of pipe is the pressure change accompanying flow. The appropriate equation for this calculation is Eq. (7.11), the mechanical-energy balance. To allow for the continuous change of properties in a flowing fluid, we write Eq. (7.11) in differential form:

$$V\,dP + u\,du + g\,dz + dF = 0 \tag{7.13}$$

Integration over the length of the pipe requires an empirical expression for the friction term dF. This is usually given by the Fanning equation:

$$dF = \frac{2fu^2}{D}\,dL$$

where D = pipe diameter
L = length along the pipe
$f = f(Du\rho/\mu)$ = dimensionless friction factor
ρ = fluid density
μ = fluid viscosity

Further empirical methods are required to account for the additional friction effects resulting from bends, valves, changes in pipe size, etc. The detailed treatment of friction calculations is beyond the scope of thermodynamics.†

† For evaluation of the friction factor f and other friction effects, see W. L. McCabe, J. C. Smith, and P. Harriott, *Unit Operations of Chemical Engineering*, 4th ed., chap. 5, McGraw-Hill, New York, 1985; R. H. Perry and Don Green, *Perry's Chemical Engineers' Handbook*, 6th ed., sec. 5, McGraw-Hill, New York, 1984.

A topic within the purview of thermodynamics is the maximum velocity attainable in pipe flow. Consider a gas in steady-state adiabatic flow in a horizontal pipe of constant cross-sectional area. Equation (7.10) is the applicable energy balance, and it here becomes:

$$\Delta H + \frac{\Delta u^2}{2} = 0$$

In differential form it is

$$dH = -u\,du \tag{7.14}$$

Equation (7.3) is also applicable. Since $\dot{m}$ is constant, the differential form is

$$d(uA/V) = 0 \tag{7.15}$$

When A is constant, $d(u/V) = 0$; whence

$$\frac{du}{V} - \frac{u\,dV}{V^2} = 0$$

and

$$du = \frac{u\,dV}{V} \qquad \text{(const } A\text{)} \tag{7.16}$$

Substituting this result into Eq. (7.14) gives

$$dH = -\frac{u^2\,dV}{V} \tag{7.17}$$

The fundamental property relation of Eq. (6.8) can be written

$$T\,dS = dH - V\,dP$$

Replacing dH by Eq. (7.17), we get

$$T\,dS = -\frac{u^2\,dV}{V} - V\,dP \tag{7.18}$$

As gas flows along a pipe in the direction of decreasing pressure, its specific volume increases, as does its velocity in accord with Eq. (7.3). Thus in the direction of increasing velocity, dP is negative, dV is positive, and the two terms of the preceding equation contribute in opposite directions to the entropy change. According to the second law, dS must be positive (with a limiting value of zero) for an adiabatic process. This condition is met so long as the final term in the equation makes a sufficiently large positive contribution to overbalance the negative contribution of the preceding term. However, as the pressure decreases, the specific volume increases ever more rapidly. Thus it is possible to reach a pressure such that the negative contribution of the first term on the right becomes equal to the positive contribution of the second. In this case a maximum velocity is reached at that differential length of pipe for which $dS = 0$. An expression for

u_{max} is obtained if we set the right-hand side of Eq. (7.18) equal to zero:

$$\frac{u_{max}^2\,dV}{V} + V\,dP = 0 \qquad (\text{const } S)$$

Rearrangement gives

$$u_{max}^2 = -V^2\left(\frac{\partial P}{\partial V}\right)_S \tag{7.19}$$

This is identical to the equation derived in physics for the speed of sound in the fluid. Therefore, the maximum fluid velocity obtainable in a pipe of constant cross-sectional area is the speed of sound. This does not imply that higher velocities are impossible; they are, in fact, readily obtained in converging/diverging nozzles (Sec. 7.3). However, the speed of sound is the maximum value that can be reached in a conduit of constant cross section, provided the entrance velocity is subsonic. The sonic velocity must be reached at the *exit* of the pipe. If the pipe length is increased, the mass rate of flow decreases so that the sonic velocity is still obtained at the outlet of the lengthened pipe.

7.3 EXPANSION PROCESSES

Flow processes accompanied by sharp reductions in pressure are called expansion processes. They include flow through nozzles, through turbines or expanders, and through throttling devices such as orifices and valves.

Nozzles

A nozzle is a device that causes the interchange of internal and kinetic energy of a fluid as a result of a changing cross-sectional area available for flow. A common example is the converging nozzle designed to produce a high-velocity stream. However, converging and diverging sections are used, separately or combined, for many purposes as in turbines, jet engines, ejectors, and diffusers. The relationship between nozzle length and cross-sectional area is not susceptible to thermodynamic analysis, but is a problem in fluid mechanics. Largely on the basis of experience, nozzles can be tapered to achieve near-isentropic flow.

Since $W_s = 0$ and heat transfer† and potential-energy changes are negligible, the energy equation as given by Eq. (7.14) applies:

$$dH = -u\,du$$

The mechanical-energy balance [Eq. (7.13)] takes the form

$$-V\,dP = u\,du + dF$$

† Flow in nozzles is nearly adiabatic, because the velocity is high (short residence time of fluid) and the area for heat transfer is small.

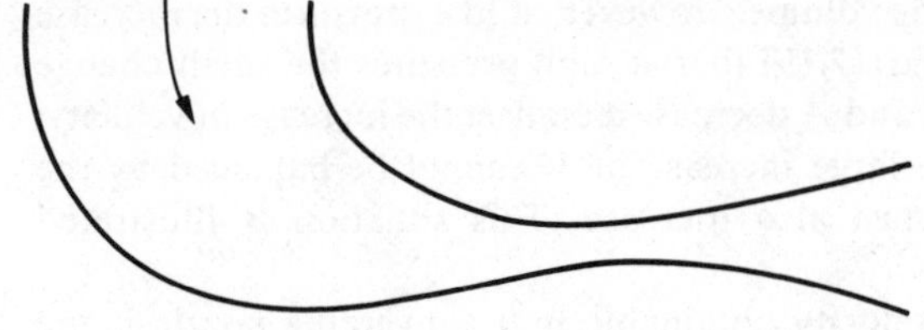

Figure 7.2 Converging/diverging nozzle.

and if the flow is isentropic, this further reduces to

$$-V\,dP = u\,du \qquad (\text{const } S) \tag{7.20}$$

The other relation available for steady flow (constant $\dot{m}$) is Eq. (7.15), $d(uA/V) = 0$.

Equations (7.14), (7.15), and (7.20), combined with the relations between the thermodynamic properties at constant entropy, determine how the velocity varies with cross-sectional area of the nozzle. The variety of results for compressible fluids (e.g., gases), depends in part on whether the velocity is below or above the speed of sound in the fluid. For subsonic flow in a converging nozzle, the velocity increases and pressure decreases as the cross-sectional area diminishes. In a diverging nozzle with supersonic flow, the area increases, but still the velocity increases and the pressure decreases. The various cases are summarized elsewhere.† We limit the rest of this treatment of nozzles to application of the equations to a few specific cases.

The speed of sound is significant in the treatment of nozzles, because this is the velocity at the throat (minimum cross-sectional area) of a converging/diverging nozzle (Fig. 7.2) in which the exit velocity is supersonic. This result follows from the fact that at the throat A is constant, and Eq. (7.16) applies:

$$du = u\frac{dV}{V}$$

Substituting this expression into Eq. (7.20) for isentropic flow gives the throat velocity as

$$u^2_{\text{throat}} = -V^2\left(\frac{\partial P}{\partial V}\right)_S \tag{7.21}$$

Comparison with Eq. (7.19) shows that u_{throat} is equal to the speed of sound.

It is also true that in the *converging section* of a converging/diverging nozzle the maximum obtainable fluid velocity is the speed of sound, reached at the throat. This is because a further decrease in pressure requires an increase in cross-sectional area, i.e., a diverging section. The explanation for this is as follows. At the relatively high pressures in the converging section, a given pressure drop

† M. M. Abbott and H. C. Van Ness, *Theory and Problems of Thermodynamics*, Schaum's Outline Series, pp. 221–224, McGraw-Hill, New York, 1972.

causes a small increase in specific volume. However, at low pressures the increase in V is large. Thus we see by Eq. (7.15) that at high pressures the small change in V does not have much effect, and A decreases to offset the increase in velocity. However, at low pressures, the large increase in V cannot be balanced by the increase in velocity, and A must also increase. This situation is illustrated numerically in Example 7.3.

Since the maximum fluid velocity obtainable in a converging nozzle is the speed of sound, a nozzle of this kind can deliver a constant flow rate into a region of variable pressure. Suppose a compressible fluid enters a converging nozzle at pressure P_1 and discharges from the nozzle into a chamber of variable pressure P_2. If this discharge pressure is P_1, the flow is zero. As P_2 decreases below P_1, the flow rate and velocity increase. Ultimately, the pressure ratio P_2/P_1 reaches a critical value at which the velocity in the throat is sonic. Further reduction in P_2 has no effect on the conditions in the nozzle. The flow remains constant, and the velocity in the throat is that given by Eq. (7.21), regardless of the value of P_2/P_1, provided it is always less than the critical value. For steam, the critical value of this ratio is about 0.55 at moderate temperatures and pressures.

The relation of velocity to pressure in a nozzle can be given analytically if the fluid behaves as an ideal gas. When an ideal gas with constant heat capacities undergoes isentropic expansion, Eq. (3.24) provides a relation between P and V, that is, $PV^\gamma = \text{const}$. Integration of Eq. (7.20) then gives

$$u_2^2 - u_1^2 = -2\int_{P_1}^{P_2} V\,dP = \frac{2\gamma P_1 V_1}{\gamma - 1}\left[1 - \left(\frac{P_2}{P_1}\right)^{(\gamma-1)/\gamma}\right] \tag{7.22}$$

where conditions at the nozzle entrance are denoted by subscript 1. Equation (7.22) together with Eq. (7.21) gives the value of the pressure ratio P_2/P_1 (for $u_1 = 0$) such that the speed of sound is obtained in the throat of a converging nozzle. Evaluation of the derivative $(\partial P/\partial V)_S$ for the isentropic expansion of an ideal gas with constant heat capacities from Eq. (3.24), $PV^\gamma = \text{const}$, reduces Eq. (7.21) to

$$u_{\text{throat}}^2 = \gamma P_2 V_2$$

Substituting this value of the throat velocity for u_2 in Eq. (7.22) and solving for the pressure ratio with $u_1 = 0$ gives

$$\frac{P_2}{P_1} = \left(\frac{2}{\gamma + 1}\right)^{\gamma/(\gamma-1)} \tag{7.23}$$

A general relationship between velocity and cross-sectional area, expressed not in terms of the properties (P, T, V, H) of the fluid but in terms of the speed of sound, results from Eqs. (7.15), (7.19), and (7.20). We start with Eq. (7.15), but express the derivative as

$$\frac{1}{V}(u\,dA + A\,du) - uA\frac{dV}{V^2} = 0$$

or

$$\frac{u\,dA + A\,du}{uA} = \frac{V\,dV}{V^2}$$

Then replacing V in the numerator on the right-hand side by its value from Eq. (7.20), we have for an isentropic process:

$$\frac{dA}{A} + \frac{du}{u} = \frac{u\,du}{-V^2(\partial P/\partial V)_S}$$

By Eq. (7.19) the denominator of the right-hand side is the square of the speed of sound. Hence

$$\frac{dA}{A} = \frac{u\,du}{u_{\text{sonic}}^2} - \frac{du}{u} = \left(\frac{u^2}{u_{\text{sonic}}^2} - 1\right)\frac{du}{u}$$

The ratio of the actual velocity to the speed of sound is called the Mach number **M**. Hence this equation expresses a relation between the cross-sectional area, velocity, and the local Mach number at any axial position in the nozzle, i.e.,

$$\frac{dA}{A} = (\mathbf{M}^2 - 1)\frac{du}{u} \tag{7.24}$$

Depending on whether **M** is greater than unity (supersonic) or less than unity (subsonic), the cross-sectional area increases or decreases with velocity increase. Equation (7.24) is applicable to any type of nozzle, as long as the flow is isentropic.

The speed of sound is attained at the throat of a converging/diverging nozzle only when the pressure at the throat is low enough that the critical value of P_2/P_1 is reached. If insufficient pressure drop is available in the nozzle for the velocity to become sonic, the diverging section of the nozzle acts as a diffuser. That is, after the throat is reached the pressure rises and the velocity decreases; this is the conventional behavior for subsonic flow in diverging sections. The relationships between velocity, area, and pressure in a nozzle are illustrated numerically in Example 7.3.

Example 7.3 A high-velocity nozzle is designed to operate with steam at 700 kPa and 300°C. At the nozzle inlet the velocity is 30 m s^{-1}. Calculate values of the ratio A/A_1 (where A_1 is the cross-sectional area of the nozzle inlet) for the sections where the pressure is 600, 500, 400, 300, and 200 kPa. Assume that the nozzle operates isentropically.

SOLUTION The required area ratios are given by Eq. (7.3):

$$\frac{A}{A_1} = \frac{u_1 V}{V_1 u}$$

The velocity u is found from the integrated form of Eq. (7.14)

$$u^2 = u_1^2 - 2(H - H_1)$$

With units for velocity of m s^{-1}, u^2 has the units of $m^2\,s^{-2}$. Units of J kg^{-1} for H are consistent with these, because 1 J = 1 kg $m^2\,s^{-2}$, whence 1 J kg^{-1} = 1 $m^2\,s^{-2}$.†

From the steam tables, we have initial values for entropy, enthalpy, and specific volume:

$$S_1 = 7.2997 \text{ kJ kg}^{-1}\text{ K}^{-1}$$

$$H_1 = 3{,}059.8 \times 10^3 \text{ J kg}^{-1}$$

$$V_1 = 371.39 \text{ cm}^3\text{ g}^{-1}$$

Thus,

$$\frac{A}{A_1} = \left(\frac{30}{371.39}\right)\frac{V}{u} \tag{A}$$

and

$$u^2 = 900 - 2(H - 3{,}059.8 \times 10^3) \tag{B}$$

Since the expansion process is isentropic, at 600 kPa,

$$S = 7.2997 \text{ kJ kg}^{-1}\text{ K}^{-1}$$

$$H = 3{,}020.4 \times 10^3 \text{ J kg}^{-1}$$

$$V = 418.25 \text{ cm}^3\text{ g}^{-1}$$

From Eq. (*B*)

$$u = 282.3 \text{ m s}^{-1}$$

and by Eq. (*A*),

$$\frac{A}{A_1} = \left(\frac{30}{371.39}\right)\left(\frac{418.25}{282.3}\right) = 0.120$$

Area ratios for other pressures are evaluated the same way, and the results are summarized in the following table. The pressure at the throat of the nozzle is about 380 kPa. At lower pressures, the nozzle clearly diverges.

P/kPa	V/cm^3 g^{-1}	u/m s^{-1}	A/A_1
700	371.39	30	1.0
600	418.25	282.3	0.120
500	481.26	411.2	0.095
400	571.23	523.0	0.088
300	711.93	633.0	0.091
200	970.04	752.2	0.104

Example 7.4 Consider again the nozzle of Example 7.3, assuming now that steam behaves as an ideal gas. Calculate:

(*a*) The critical pressure ratio and the velocity at the throat.

(*b*) The discharge pressure if a Mach number of 2.0 is required at the nozzle exhaust.

† When u is in (ft)(s)$^{-1}$, H in (Btu)(lb$_m$)$^{-1}$ must be multiplied by 778.16(ft lb$_f$)(Btu)$^{-1}$ and by the dimensional constant g_c = 32.174(lb$_m$)(ft)(lb$_f$)$^{-1}$(s)$^{-2}$.

SOLUTION (*a*) The ratio of specific heats for steam is about 1.3. Substituting in Eq. (7.23),

$$\frac{P_2}{P_1} = \left(\frac{2}{1.3+1}\right)^{1.3/(1.3-1)} = 0.55$$

The velocity at the throat, which is equal to the speed of sound, can be found from Eq. (7.22). When P_1 is in Pa (1 Pa = 1 kg m^{-1} s^{-2}) and V_1 is in m^3 kg^{-1}, the product $P_1 V_1$ is in m^2 s^{-2}, the units of velocity squared. Thus

$$u_{\text{throat}}^2 = (30)^2 + \frac{(2)(1.3)(700{,}000)(0.37139)}{1.3-1}[1-(0.55)^{(1.3-1)/1.3}]$$

$$= 900 + 290{,}354 = 291{,}254$$

$$u_{\text{throat}} = 539.7 \text{ m s}^{-1}$$

These results compare favorably with values obtained in Example 7.3, because steam at these conditions closely approximates an ideal gas.

(*b*) For a Mach number of 2.0 (based on conditions at the nozzle throat) the discharge velocity is 1,079.4 m s^{-1}. Substitution of this value in Eq. (7.22) allows calculation of the pressure ratio:

$$(1{,}079.4)^2 = (30)^2 + \frac{(1.3)(700{,}000)(0.37139)}{1.3-1}\left[1-\left(\frac{P_2}{P_1}\right)^{(1.3-1)/1.3}\right]$$

or

$$\left(\frac{P_2}{P_1}\right)^{(1.3-1)/1.3} = 0.483$$

Whence

$$P_2 = (0.0427)(700) = 29.9 \text{ kPa}$$

Turbines or Expanders

The expansion of a gas in a nozzle to produce a high-velocity stream is a process that converts internal energy into kinetic energy. This kinetic energy can in turn be converted into shaft work when the stream impinges on blades attached to a rotating shaft. Thus a turbine (or expander) consists of alternate sets of nozzles and rotating blades through which gas flows in a steady-state expansion process whose overall effect is the efficient conversion of the internal energy of a high-pressure stream into shaft work. When steam provides the motive force as in a power plant, the device is called a turbine; when a high-pressure gas, such as ammonia or ethylene in a chemical or petrochemical plant, is the working fluid, the device is often called an expander. In either case, the process is represented in Fig. 7.3.

Equations (7.9) and (7.10) are appropriate energy relations. However, the potential-energy term can be omitted, because there is little change in elevation. Moreover, in any properly designed turbine, heat transfer is negligible and the inlet and exit pipes are sized to make fluid velocities relatively low. Equations

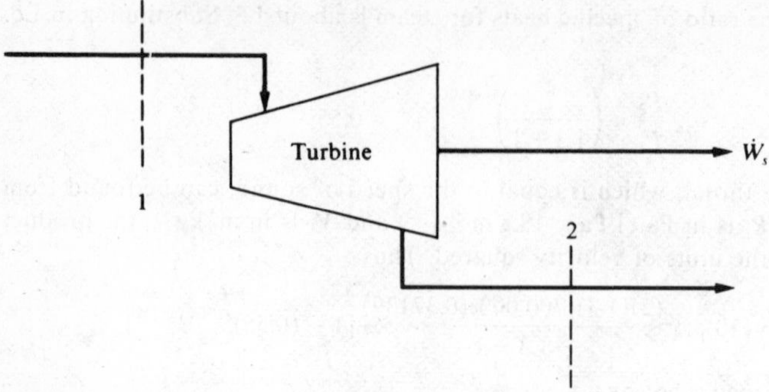

Figure 7.3 Steady-state flow through a turbine or expander.

(7.9) and (7.10) therefore reduce to

$$\dot{W}_s = -\dot{m}\,\Delta H \tag{7.25}$$

and

$$W_s = -\Delta H \tag{7.26}$$

Normally, we know the inlet conditions T_1 and P_1 and the discharge pressure P_2. Thus in Eq. (7.26) we know only H_1, and are left with both H_2 and W_s as unknowns. The energy equation alone does not allow any calculations to be made. However, if the fluid in the turbine undergoes an expansion process that is reversible as well as adiabatic, then the process is isentropic, and $S_2 = S_1$. This second equation allows us to determine the final state of the fluid and hence H_2. For this special case, we can evaluate W_s by Eq. (7.26), written as

$$W_s(\text{isentropic}) = -(\Delta H)_S \tag{7.27}$$

The shaft work given by Eq. (7.27) is the *maximum* that can be obtained from an adiabatic turbine with given inlet conditions and given discharge pressure. Actual turbines produce less work, because the actual expansion process is irreversible. We therefore define a turbine efficiency as

$$\eta = \frac{W_s}{W_s(\text{isentropic})}$$

where W_s is the actual shaft work. By Eqs. (7.26) and (7.27)

$$\eta = \frac{\Delta H}{(\Delta H)_S} \tag{7.28}$$

Values of η for properly designed turbines or expanders are usually in the range of 70 to 80 percent.

Figure 7.4 shows an *HS* diagram on which are compared an actual expansion process in a turbine and the reversible process for the same intake conditions and the same discharge pressure. The reversible path is a vertical line of constant

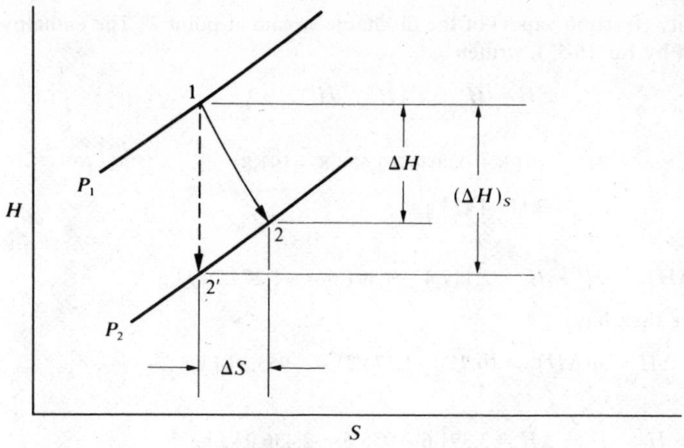

Figure 7.4 Adiabatic expansion process in a turbine or expander.

entropy from point 1 at the intake pressure P_1 to point 2′ at the discharge pressure P_2. The line representing the actual irreversible process starts also from point 1, but is directed downward and to the right, in the direction of increasing entropy. Since the process is adiabatic, irreversibilities cause an increase in entropy of the fluid. The process terminates at point 2 on the isobar for P_2. The more irreversible the process, the further this point lies to the right on the P_2 isobar, and the lower the efficiency η of the process.

Example 7.5 A steam turbine with rated capacity of 56,400 kW operates with steam at inlet conditions of 8,600 kPa and 500°C, and discharges into a condenser at a pressure of 10 kPa. Assuming a turbine efficiency of 75 percent, determine the state of the steam at discharge and the mass rate of flow of the steam.

SOLUTION At the inlet conditions of 8,600 kPa and 500°C, the following values are given in the steam tables:

$$H_1 = 3{,}391.6 \text{ kJ kg}^{-1}$$

$$S_1 = 6.6858 \text{ kJ kg}^{-1} \text{ K}^{-1}$$

If the expansion to 10 kPa is isentropic, then

$$S_2' = S_1 = 6.6858$$

Steam with this entropy at 10 kPa is wet, and we apply Eq. (6.53), with $M = S$:

$$S = (1 - x^v)S^l + x^v S^v = S^l + x^v(S^v - S^l)$$

Whence

$$6.6858 = 0.6493 + x_2'(8.1511 - 0.6493)$$

and

$$x_2' = 0.80467$$

This is the quality (fraction vapor) of the discharge stream at point 2′. The enthalpy H'_2 is also given by Eq. (6.53), written

$$H = H^l + x^v(H^v - H^l)$$

Thus

$$H'_2 = 191.8 + 0.80467(2{,}584.8 - 191.8)$$
$$= 2{,}117.4 \text{ kJ kg}^{-1}$$

and

$$(\Delta H)_S = H'_2 - H_1 = 2{,}117.4 - 3{,}391.6 = -1{,}274.2 \text{ kJ kg}^{-1}$$

By Eq. (7.28) we then have

$$\Delta H = \eta(\Delta H)_S = (0.75)(-1{,}274.2) = -955.6 \text{ kJ kg}^{-1}$$

Whence

$$H_2 = H_1 + \Delta H = 3{,}391.6 - 955.6 = 2{,}436.0 \text{ kJ kg}^{-1}$$

Thus the steam in its actual final state is also wet, and its quality is found from the equation:

$$2{,}436.0 = 191.8 + x_2(2{,}584.8 - 191.8)$$

Solution gives

$$x_2 = 0.93782$$

Finally,

$$S_2 = 0.6493 + (0.93782)(8.1511 - 0.6493)$$
$$= 7.6864 \text{ kJ kg}^{-1} \text{ K}^{-1}$$

This value may be compared with the initial value of $S_1 = 6.6858$.

The steam rate is found from Eq. (7.25). With $\dot{W} = 56{,}400$ kW or 56,400 kJ s^{-1}, we have

$$56{,}400 = -\dot{m}(2{,}436.0 - 3{,}391.6)$$

and

$$\dot{m} = 59.02 \text{ kg s}^{-1}$$

Example 7.5 was worked with the aid of the steam tables. When a comparable set of tables is not available for the motive fluid, the generalized correlations of Sec. 6.6 may be used in conjunction with Eqs. (6.62) and (6.63), as illustrated in the following example.

Example 7.6 A stream of ethylene gas at 300°C and 45 bar is expanded adiabatically in a turbine to 2 bar. Calculate the isentropic work produced. Determine the properties of ethylene by (*a*) equations for an ideal gas, and (*b*) appropriate generalized correlations.

SOLUTION The enthalpy and entropy changes for this process are given by Eqs. (6.62) and (6.63):

$$\Delta H = C^{ig}_{P_{mh}}(T_2 - T_1) + H^R_2 - H^R_1 \tag{6.62}$$

and

$$\Delta S = C^{ig}_{P_{\mathrm{ms}}} \ln \frac{T_2}{T_1} - R \ln \frac{P_2}{P_1} + S^R_2 - S^R_1 \tag{6.63}$$

As given values, we have $P_1 = 45$ bar, $P_2 = 2$ bar, and $T_1 = 300 + 273.15 = 573.15$ K.

(a) If ethylene is assumed an ideal gas, then all residual properties are zero, and the preceding equations reduce to:

$$\Delta H = C^{ig}_{P_{\mathrm{mh}}}(T_2 - T_1)$$

and

$$\Delta S = C^{ig}_{P_{\mathrm{ms}}} \ln \frac{T_2}{T_1} - R \ln \frac{P_2}{P_1}$$

For an isentropic process, $\Delta S = 0$, and the last equation becomes:

$$\frac{C^{ig}_{P_{\mathrm{ms}}}}{R} \ln \frac{T_2}{T_1} = \ln \frac{P_2}{P_1} = \ln \frac{2}{45} = -3.1135$$

or

$$\ln T_2 = \frac{-3.1135}{C^{ig}_{P_{\mathrm{ms}}}/\mathrm{R}} + \ln 573.15$$

Whence

$$T_2 = \exp\left(\frac{-3.1135}{C^{ig}_{P_{\mathrm{ms}}}/R} + 6.3511\right) \tag{A}$$

Equation (5.17) with $D = 0$ (in accord with the heat-capacity data for ethylene given in Table 4.1) is

$$\frac{C^{ig}_{P_{\mathrm{ms}}}}{R} = A + BT_{\mathrm{lm}} + CT_{\mathrm{am}}T_{\mathrm{lm}} \tag{B}$$

where

$$A = 1.424$$

$$B = 14.394 \times 10^{-3}$$

$$C = -4.392 \times 10^{-6}$$

$$T_{\mathrm{am}} = \frac{T_1 + T_2}{2}$$

and

$$T_{\mathrm{lm}} = \frac{T_1 - T_2}{\ln (T_1/T_2)}$$

In these equations T_2 is the only unknown. It is conveniently found by iteration between Eqs. (B) and (A). We assume a value of T_2, calculate $C^{ig}_{P_{\mathrm{ms}}}/R$ by Eq. (B), calculate T_2 by Eq. (A), return to Eq. (B), and repeat to convergence. The result is:

$$T_2 = 370.79 \text{ K}$$

Then

$$W_s(\text{isentropic}) = -(\Delta H)_S = -C^{ig}_{P_{\mathrm{mh}}}(T_2 - T_1)_S$$

By Eq. (4.7),

$$\frac{C_{P_{\text{mh}}}^{ig}}{R} = A + BT_{\text{am}} + \frac{C}{3}(4T_{\text{am}}^2 - T_1 T_2)$$

With

$$T_{\text{am}} = \frac{573.15 + 370.79}{2} = 471.97 \text{ K}$$

this gives

$$\frac{C_{P_{\text{mh}}}^{ig}}{R} = 7.224$$

Whence

$$W_s(\text{isentropic}) = -(7.224)(8.314)(370.79 - 573.15)$$
$$= 12{,}154 \text{ J mol}^{-1}$$

(*b*) For ethylene,

$$T_c = 282.4 \text{ K} \qquad P_c = 50.4 \text{ bar} \qquad \omega = 0.085$$

At the initial state,

$$T_{r_1} = \frac{573.15}{282.4} = 2.032 \qquad P_{r_1} = \frac{45}{50.4} = 0.893$$

According to Fig. 3.16, the generalized correlations based on second virial coefficients should be satisfactory. Application of Eqs. (3.48), (6.60), (3.49), and (6.61) for the initial state yields:

$$B^0 = -0.053 \qquad \frac{dB^0}{dT_r} = 0.107$$

$$B^1 = 0.130 \qquad \frac{dB^1}{dT_r} = 0.018$$

Equations (6.58) and (6.59) then give

$$\frac{H_1^R}{RT_c} = -0.234 \qquad \frac{S_1^R}{R} = -0.097$$

Whence

$$H_1^R = (-0.234)(8.314)(282.4)$$
$$= -550 \text{ J mol}^{-1}$$

and

$$S_1^R = (-0.097)(8.314) = -0.806 \text{ J mol}^{-1} \text{ K}^{-1}$$

For the purpose of getting an initial estimate of S_2^R, we assume that $T_2 = 370.79$ K, the value determined in part (*a*). Then

$$T_{r_2} = \frac{370.79}{282.4} = 1.313 \qquad P_{r_2} = \frac{2}{50.4} = 0.040$$

and by Eqs. (6.60) and (6.61),

$$\frac{dB^0}{dT_r} = 0.332 \qquad \frac{dB^1}{dT_r} = 0.175$$

Equation (6.59) then gives

$$S_2^R = -0.115 \text{ J mol}^{-1} \text{ K}^{-1}$$

If the expansion process is isentropic, Eq. (6.63) gives

$$0 = C_{P_{ms}}^{ig} \ln \frac{T_2}{573.15} - 8.314 \ln \frac{2}{45} - 0.115 + 0.806$$

from which

$$\ln \frac{T_2}{573.15} = \frac{-26.577}{C_{P_{ms}}^{ig}}$$

or

$$T_2 = \exp\left(\frac{-26.577}{C_{P_{ms}}^{ig}} + 6.3511\right)$$

An iteration process exactly like that of part (a) yields the result,

$$T_2 = 365.79 \text{ K}$$

For the recomputation of S_2^R, we now find

$$T_{r_2} = 1.295 \qquad P_{r_2} = 0.040$$

$$\frac{dB^0}{dT_r} = 0.345 \qquad \frac{dB^1}{dT_r} = 0.188$$

and

$$S_2^R = -0.120 \text{ J mol}^{-1} \text{ K}^{-1}$$

This result is so little changed from the initial value that another recalculation of T_2 is unnecessary. We therefore evaluate H_2^R at the reduced conditions already established,

$$B^0 = -0.196 \qquad B^1 = 0.081$$

and by Eq. (6.58)

$$H_2^R = -62 \text{ J mol}^{-1}$$

Equation (6.62) now gives

$$(\Delta H)_S = C_{P_{mh}}^{ig}(365.79 - 573.15) - 62 + 550$$

Evaluation of $C_{P_{mh}}^{ig}$ as in part (a) with $T_{am} = 469.47$ K gives

$$C_{P_{mh}}^{ig} = 59.843 \text{ J mol}^{-1} \text{ K}^{-1}$$

Whence

$$(\Delta H)_S = -11{,}920 \text{ J mol}^{-1}$$

and

$$W_s(\text{isentropic}) = -(\Delta H)_S = 11{,}920 \text{ J mol}^{-1}$$

Throttling Processes

When a fluid flows through a restriction, such as an orifice, a partly closed valve, or a porous plug, without any appreciable change in kinetic energy, the primary result of the process is a pressure drop in the fluid. Such a *throttling process* produces no shaft work and results in negligible change in elevation. In the absence of heat transfer, Eq. (7.10) reduces to

$$\Delta H = 0$$

or

$$H_2 = H_1$$

and the process occurs at constant enthalpy.

Since the enthalpy of an ideal gas depends on temperature only, a throttling process does not change the temperature of an ideal gas. For most real gases at moderate conditions of temperature and pressure, a reduction in pressure at constant enthalpy results in a decrease in temperature. For example, if steam at 1,000 kPa and 300°C is throttled to 101.325 kPa (atmospheric pressure),

$$H_2 = H_1 = 3{,}052.1 \text{ kJ kg}^{-1}$$

Interpolation in the steam tables at 101.325 kPa shows that steam has this enthalpy at a temperature of 288.8°C. The temperature has decreased, but the effect is small. The following example illustrates the use of generalized correlations in calculations for a throttling process.

Example 7.7 Propane gas at 20 bar and 400 K is throttled in a steady-state flow process to 1 bar. Estimate the final temperature of the propane and its entropy change. Properties of propane can be found from suitable generalized correlations.

SOLUTION Applying Eq. (6.62) to this constant-enthalpy process gives:

$$\Delta H = C^{ig}_{P_{\text{mh}}}(T_2 - T_1) + H_2^R - H_1^R = 0$$

If propane in its final state at 1 bar is assumed an ideal gas, then $H_2^R = 0$, and the preceding equation gives

$$T_2 = \frac{H_1^R}{C^{ig}_{P_{\text{mh}}}} + T_1 \qquad (A)$$

For propane,

$$T_c = 369.8 \text{ K} \qquad P_c = 42.5 \text{ bar} \qquad \omega = 0.152$$

and for the initial state

$$T_{r_1} = \frac{400}{369.8} = 1.0817 \qquad P_{r_1} = \frac{20}{42.5} = 0.4706$$

At these conditions the generalized correlation based on second virial coefficients is satisfactory (see Fig. 3.16), and Eqs. (3.48), (6.60), (3.49), and (6.61) yield:

$$B^0 = -0.289 \qquad \frac{dB^0}{dT_r} = 0.550$$

$$B^1 = 0.015 \qquad \frac{dB^1}{dT_r} = 0.480$$

Whence by Eq. (6.58),

$$\frac{H_1^R}{RT_c} = -0.452$$

and

$$H_1^R = (8.314)(369.8)(-0.452) = -1{,}390 \text{ J mol}^{-1}$$

The only remaining quantity in Eq. (A) to be evaluated is $C_{P_{mh}}^{ig}$. Taking data for propane from Table 4.1, we have

$$\frac{C_P^{ig}}{R} = 1.213 + 28.785 \times 10^{-3}T - 8.824 \times 10^{-6}T^2$$

For an initial calculation, we assume that $C_{P_{mh}}^{ig}$ is approximately the value of C_P^{ig} at the initial temperature of 400 K. This provides the value

$$C_{P_{mh}}^{ig} = 94.074 \text{ J mol}^{-1} \text{ K}^{-1}$$

Equation (A) now gives

$$T_2 = \frac{-1{,}390}{94.074} + 400 = 385.2 \text{ K}$$

Clearly, the temperature change is small, and we can reevaluate $C_{P_{mh}}^{ig}$ to an excellent approximation by calculating C_P^{ig} at the arithmetic mean temperature,

$$T_{am} = \frac{400 + 385.2}{2} = 392.6 \text{ K}$$

This gives

$$C_{P_{mh}}^{ig} = 92.734 \text{ J mol}^{-1} \text{ K}^{-1}$$

and recalculation of T_2 by Eq. (A) yields the final value:

$$T_2 = 385.0 \text{ K}$$

The entropy change of the propane is given by Eq. (6.63), which here becomes

$$\Delta S = C_{P_{ms}}^{ig} \ln \frac{T_2}{T_1} - R \ln \frac{P_2}{P_1} - S_1^R$$

Since the temperature change is so small, we can take

$$C_{P_{ms}}^{ig} = C_{P_{mh}}^{ig} = 92.734 \text{ J mol}^{-1} \text{ K}^{-1}$$

Calculation of S_1^R by Eq. (6.59) gives

$$S_1^R = -2.437 \text{ J mol}^{-1} \text{ K}^{-1}$$

Then

$$\Delta S = 92.734 \ln \frac{385.0}{400} - 8.314 \ln \frac{1}{20} + 2.437$$

$$= 23.80 \text{ J mol}^{-1} \text{ K}^{-1}$$

The positive value reflects the irreversibility of throttling processes.

When a wet vapor is throttled to a sufficiently low pressure, the liquid evaporates and the vapor becomes superheated. Thus if wet steam at 1,000 kPa ($t^{sat} = 179.88°C$) with a quality of 0.96 is throttled to 101.325 kPa,

$$H_2 = H_1 = (0.04)(762.6) + (0.96)(2{,}776.2)$$
$$= 2{,}695.7 \text{ kJ kg}^{-1}$$

Steam with this enthalpy at 101.325 kPa has a temperature of 109.8°C, and is superheated. (At this pressure, $t^{sat} = 100°C$.) The considerable temperature drop here results from evaporation of liquid.

If a saturated liquid is throttled to a lower pressure, some of the liquid vaporizes or *flashes*, producing a mixture of saturated liquid and saturated vapor at the lower pressure. Thus if saturated liquid water at 1,000 kPa ($t^{sat} = 179.88°C$) is flashed to 101.325 kPa ($t^{sat} = 100°C$),

$$H_2 = H_1 = 762.605 \text{ kJ kg}^{-1}$$

At 101.325 kPa the quality of the resulting stream is found from:

$$762.605 = (1 - x)(419.064) + x(2{,}676.0)$$
$$= 419.064 + x(2{,}676.0 - 419.1)$$

Whence

$$x = 0.1522$$

Thus 15.22 percent of the original liquid vaporized in the process. Again, the large temperature drop results from evaporation of liquid.

Throttling processes find frequent application in refrigeration systems (Chap. 9).

7.4 COMPRESSION PROCESSES

Just as expansion processes result in pressure reductions in a flowing fluid, so compression processes bring about pressure increases. Compressors, pumps, fans, blowers, and vacuum pumps are all devices designed for this purpose. They are vital for the transport of fluids, for fluidization of particulate solids, for bringing fluids to the proper pressure for reaction or processing, etc. We are here concerned not with the design of such devices, but with specification of energy requirements for the steady-state compression of fluids from one pressure to a higher one.

Compressors

The compression of gases may be accomplished in equipment with rotating blades (like a turbine operating in reverse) or in cylinders with reciprocating pistons. Rotary equipment is used for high-volume flow where the discharge pressure is not too high. For high pressures, reciprocating compressors are required.

The energy equations are independent of the type of equipment; indeed, they are the same as for turbines or expanders, because here too potential- and

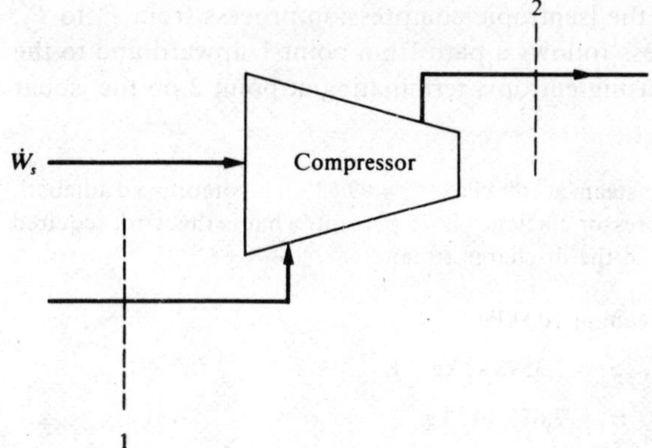

Figure 7.5 Steady-state compression process.

kinetic-energy changes are presumed negligible. Thus Eqs. (7.25) through (7.27) apply to adiabatic compression, a process represented by Fig. 7.5.

In a compression process, the isentropic work, as given by Eq. (7.27), is the *minimum* shaft work required for compression of a gas from a given initial state to a given discharge pressure. Thus we define a compressor efficiency as

$$\eta = \frac{W_s(\text{isentropic})}{W_s}$$

In view of Eqs. (7.26) and (7.27), this is also given by

$$\eta = \frac{(\Delta H)_S}{\Delta H} \tag{7.29}$$

Compressor efficiencies are usually in the range of 70 to 80 percent. The compression process is shown on an *HS* diagram in Fig. 7.6. The vertical path rising from

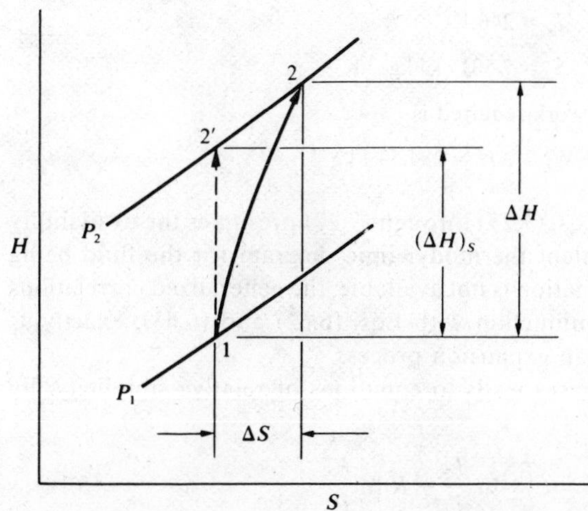

Figure 7.6 Adiabatic compression process.

point 1 to point 2′ represents the isentropic compression process from P_1 to P_2. The actual compression process follows a path from point 1 upward and to the right in the direction of increasing entropy, terminating at point 2 on the isobar for P_2.

Example 7.8 Saturated-vapor steam at 100 kPa ($t^{sat} = 99.63°C$) is compressed adiabatically to 300 kPa. If the compressor efficiency is 75 percent, what is the work required and what are the properties of the discharge stream?

SOLUTION For saturated steam at 100 kPa,

$$S_1 = 7.3598 \text{ kJ kg}^{-1} \text{ K}^{-1}$$

$$H_1 = 2{,}675.4 \text{ kJ kg}^{-1}$$

For isentropic compression to 300 kPa,

$$S_2' = S_1 = 7.3598 \text{ kJ kg}^{-1} \text{ K}^{-1}$$

By interpolation in the tables for superheated steam at 300 kPa, we find that steam with this entropy has an enthalpy of

$$H_2' = 2{,}888.8 \text{ kJ kg}^{-1}$$

Thus

$$(\Delta H)_S = 2{,}888.8 - 2{,}675.4 = 213.4 \text{ kJ kg}^{-1}$$

By Eq. (7.29),

$$\Delta H = \frac{(\Delta H)_S}{\eta} = \frac{213.4}{0.75} = 284.5 \text{ kJ kg}^{-1}$$

Whence

$$H_2 = H_1 + \Delta H = 2{,}675.4 + 284.5 = 2{,}959.9 \text{ kJ kg}^{-1}$$

Again by interpolation, we find that superheated steam with this enthalpy has the additional properties:

$$T_2 = 246.1°\text{C}$$

$$S_2 = 7.5019 \text{ kJ kg}^{-1} \text{ K}^{-1}$$

Moreover, by Eq. (7.26), the work required is

$$-W_s = \Delta H = 284.5 \text{ kJ kg}^{-1}$$

The direct application of Eqs. (7.25) through (7.27) presumes the availability of tables of data or an equivalent thermodynamic diagram for the fluid being compressed. Where such information is not available, the generalized correlations of Sec. 6.6 may be used in conjunction with Eqs. (6.62) and (6.63), exactly as illustrated in Example 7.6 for an expansion process.

The assumption of ideal gases leads to equations of relative simplicity. By Eq. (5.18) for an ideal gas

$$\Delta S = C_{P_{ms}} \ln \frac{T_2}{T_1} - R \ln \frac{P_2}{P_1} \tag{5.18}$$

where for simplicity of notation the superscript "ig" has been omitted from the mean heat capacity. If the compression is isentropic, $\Delta S = 0$, and this equation becomes

$$T_2' = T_1 \left(\frac{P_2}{P_1}\right)^{R/C'_{P_{\mathrm{ms}}}} \tag{7.30}$$

where T_2' is the temperature that results when compression from T_1 and P_1 to P_2 is *isentropic* and where $C'_{P_{\mathrm{ms}}}$ is the mean heat-capacity for the temperature range from T_1 to T_2'.

The enthalpy change for isentropic compression is given by Eq. (4.8), written as

$$(\Delta H)_S = C'_{P_{\mathrm{mh}}}(T_2' - T_1)$$

In accord with Eq. (7.27), we then have

$$W_s(\text{isentropic}) = -C'_{P_{\mathrm{mh}}}(T_2' - T_1) \tag{7.31}$$

This result may be combined with the compressor efficiency to give

$$W_s = \frac{W_s(\text{isentropic})}{\eta} \tag{7.32}$$

The *actual* discharge temperature T_2 resulting from compression is also found from Eq. (4.8), now written

$$\Delta H = C_{P_{\mathrm{mh}}}(T_2 - T_1)$$

Whence

$$T_2 = T_1 + \frac{\Delta H}{C_{P_{\mathrm{mh}}}} \tag{7.33}$$

where by Eq. (7.26) $\Delta H = -W_s$. Here $C_{P_{\mathrm{mh}}}$ is the mean heat-capacity for the temperature range from T_1 to T_2.

For the special case of an ideal gas with constant heat capacities,

$$C'_{P_{\mathrm{mh}}} = C_{P_{\mathrm{mh}}} = C'_{P_{\mathrm{ms}}} = C_P$$

Equations (7.30) and (7.31) therefore become

$$T_2' = T_1 \left(\frac{P_2}{P_1}\right)^{R/C_P}$$

and

$$W_s(\text{isentropic}) = -C_P(T_2' - T_1)$$

Combining these equations gives

$$W_s(\text{isentropic}) = -C_P T_1 \left[\left(\frac{P_2}{P_1} \right)^{R/C_P} - 1 \right] \tag{7.34}†$$

For monatomic gases, such as argon and helium, $R/C_P = 2/5 = 0.4$. For diatomic gases, such as oxygen, nitrogen, and air at moderate temperatures, an approximate value is $R/C_P = 2/7 = 0.2857$. For gases of greater molecular complexity the ideal-gas heat capacity depends more strongly on temperature, and Eq. (7.34) is less likely to be suitable. One can easily show that the assumption of constant heat capacities also leads to the result:

$$T_2 = T_1 + \frac{T_2' - T_1}{\eta} \tag{7.35}$$

Example 7.9 If methane (assumed to be an ideal gas) is compressed adiabatically from 20°C and 140 kPa to 560 kPa, estimate the work requirement and the discharge temperature of the methane. The compressor efficiency is 75 percent.

SOLUTION Application of Eq. (7.30) requires evaluation of the exponent $R/C'_{P_{ms}}$. By Eq. (5.17) with $D = 0$ (in accord with the heat-capacity data for ethylene given in Table 4.1),

$$\frac{C_{P_{ms}}}{R} = A + BT_{lm} + CT_{am}T_{lm}$$

where

$$A = 1.702$$

$$B = 9.081 \times 10^{-3}$$

$$C = -2.164 \times 10^{-6}$$

$$T_{am} = \frac{T_1 + T_2}{2}$$

and

$$T_{lm} = \frac{T_1 - T_2}{\ln (T_1/T_2)}$$

We choose a value for T_2' somewhat higher than the initial temperature $T_1 = 293.15$ K. Evaluation of $C'_{P_{ms}}/R$ then provides a value for the exponent in Eq. (7.30). With

† Since $R = C_P - C_V$ for an ideal gas, we can write

$$\frac{R}{C_P} = \frac{C_P - C_V}{C_P} = \frac{C_P/C_V - 1}{C_P/C_V} = \frac{\gamma - 1}{\gamma}$$

An alternative form of Eq. (7.34) is therefore

$$W_s(\text{isentropic}) = -\frac{\gamma R T_1}{\gamma - 1} \left[\left(\frac{P_2}{P_1} \right)^{(\gamma-1)/\gamma} - 1 \right]$$

Although this form is the one most commonly encountered, Eq. (7.34) is simpler and more easily applied.

$P_2/P_1 = 560/140 = 4.0$ and $T_1 = 293.15$ K, we then calculate T'_2. The procedure is repeated until no further significant change occurs in the value of T'_2. This process results in the values:

$$T'_2 = 397.37 \text{ K} \qquad \text{and} \qquad \frac{C'_{P_{\text{ms}}}}{R} = 4.5574$$

For the same T_1 and T'_2, we evaluate $C'_{P_{\text{mh}}}/R$ by Eq. (4.7):

$$\frac{C'_{P_{\text{mh}}}}{R} = A + BT_{\text{am}} + \frac{C}{3}(4T_{\text{am}}^2 - T_1 T_2)$$

This gives

$$\frac{C'_{P_{\text{mh}}}}{R} = 4.5774 \qquad \text{and} \qquad C'_{P_{\text{mh}}} = 38.056 \text{ J mol}^{-1} \text{ K}^{-1}$$

Then by Eq. (7.31),

$$W_s(\text{isentropic}) = -(38.056)(397.37 - 293.15) = -3{,}966.2 \text{ J mol}^{-1}$$

The actual work is found from Eq. (7.32) as

$$W_s = \frac{-3{,}966.2}{0.75} = -5{,}288.3 \text{ J mol}^{-1}$$

Application of Eq. (7.33) for the calculation of T_2 gives

$$T_2 = 293.15 + \frac{5{,}288.3}{C_{P_{\text{mh}}}}$$

Since $C_{P_{\text{mh}}}$ depends on T_2, we again iterate. With T'_2 as a starting value, this leads to the results:

$$T_2 = 428.65 \text{ K} \qquad \text{or} \qquad t_2 = 155.5°\text{C}$$

and

$$C_{P_{\text{mh}}} = 39.027 \text{ J mol}^{-1} \text{ K}^{-1}$$

Pumps

Liquids are usually moved by pumps, generally rotating equipment. The same equations apply to adiabatic pumps as to adiabatic compressors. Thus, Eqs. (7.25) through (7.27) and (7.29) are valid. However, application of Eq. (7.26) for the calculation of $W_s = -\Delta H$ requires values of the enthalpy of compressed liquids, and these are seldom available. The fundamental property relation, Eq. (6.8), provides an alternative. For an isentropic process,

$$dH = V\,dP \qquad (\text{const } S)$$

Combining this with Eq. (7.27) gives

$$W_s(\text{isentropic}) = -(\Delta H)_S = -\int_{P_1}^{P_2} V\,dP$$

The usual assumption for liquids (at conditions well removed from the critical point) is that V is independent of P. Integration then gives

$$W_s(\text{isentropic}) = -(\Delta H)_S = -V(P_2 - P_1) \tag{7.36}$$

Also useful are the following equations from Chap. 6:

$$dH = C_P\,dT + V(1 - \beta T)\,dP \tag{6.27}$$

and

$$dS = C_P\frac{dT}{T} - \beta V\,dP \tag{6.28}$$

where the volume expansivity β is defined by Eq. (3.2). Since temperature changes in the pumped fluid are very small and since the properties of liquids are insensitive to pressure (again at conditions not close to the critical point), these equations are usually integrated on the assumption that C_P, V, and β are constant, usually at initial values. Thus, to a good approximation:

$$\Delta H = C_P\,\Delta T + V(1 - \beta T)\,\Delta P \tag{7.37}$$

and

$$\Delta S = C_P \ln\frac{T_2}{T_1} - \beta V\,\Delta P \tag{7.38}$$

Example 7.10 Water at 45°C and 10 kPa enters an adiabatic pump and is discharged at a pressure of 8,600 kPa. Assume the pump efficiency to be 75 percent. Calculate the work of the pump, the temperature change of the water, and the entropy change of the water.

SOLUTION The following properties are available for saturated liquid water at 45°C (318.15 K):

$$V = 1{,}010\ \text{cm}^3\ \text{kg}^{-1}$$

$$\beta = 425 \times 10^{-6}\ \text{K}^{-1}$$

$$C_P = 4.178\ \text{kJ kg}^{-1}\ \text{K}^{-1}$$

By Eq. (7.36),

$$W_s(\text{isentropic}) = -(\Delta H)_S = -(1{,}010)(8{,}600 - 10)$$
$$= -8.676 \times 10^6\ \text{kPa cm}^3\ \text{kg}^{-1}$$

Since $1\ \text{kJ} = 10^6\ \text{kPa cm}^3$,

$$W_s(\text{isentropic}) = -(\Delta H)_S = -8.676\ \text{kJ kg}^{-1}$$

By Eq. (7.29),

$$\Delta H = \frac{(\Delta H)_S}{\eta} = \frac{8.676}{0.75} = 11.57\ \text{kJ kg}^{-1}$$

Since $W_s = -\Delta H$,

$$W_s = -11.57\ \text{kJ kg}^{-1}$$

The temperature change of the water during pumping is found from Eq. (7.37):

$$11.57 = 4.178\,\Delta T + 1{,}010[1 - (425 \times 10^{-6})(318.15)]\,\frac{8{,}590}{10^6}$$

Solution for ΔT gives

$$\Delta T = 0.97\text{ K} \qquad \text{or} \qquad 0.97°\text{C}$$

The entropy change of the water is given by Eq. (7.38):

$$\Delta S = 4.178 \ln \frac{319.12}{318.15} - (425 \times 10^{-6})(1{,}010)\,\frac{8{,}590}{10^6}$$

$$= 0.0090\text{ kJ kg}^{-1}\text{ K}^{-1}$$

Ejectors

Ejectors remove gases or vapors from an evacuated space and compress them for discharge at a higher pressure. Where the mixing of the gases or vapors with the driving fluid is allowable, ejectors are usually lower in first cost and maintenance costs than other types of vacuum pumps. As illustrated in Fig. 7.7 an ejector consists of an inner converging-diverging nozzle through which the driving fluid (commonly steam) is fed, and an outer, larger nozzle through which both the extracted gases or vapors and the driving fluid pass. The momentum of the high-speed fluid leaving the driving nozzle is partly transferred to the extracted gases or vapors, and the mixture velocity is therefore less than that of the driving fluid leaving the smaller nozzle. It is nevertheless higher than the speed of sound, and the larger nozzle therefore acts as a converging-diverging *diffuser* in which the pressure rises and the velocity decreases, passing through the speed of sound at the throat. Although the usual energy equations for nozzles apply, the mixing process is complex, and as a result ejector design is empirical.

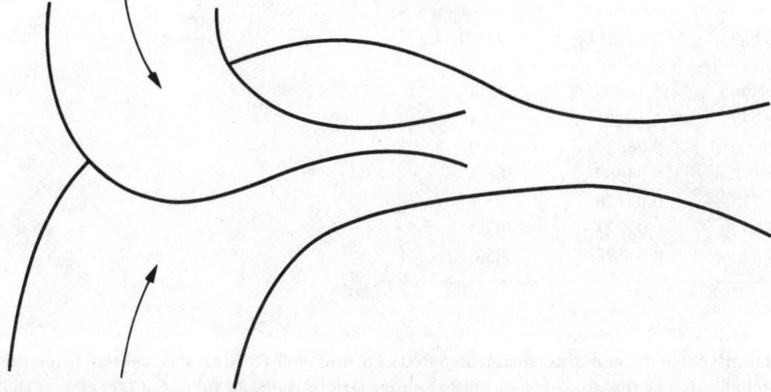

Figure 7.7 Single-stage ejector.

PROBLEMS

7.1 Two boilers discharge equal amounts of steam into the same steam main. The steam from one is at 1,400 kPa and 225°C; from the other, at 1,400 kPa with a quality of 0.94. Determine ΔS_{total} for the process. What is the irreversible feature of the process?

7.2 Two nonconducting tanks of negligible heat capacity and of equal volume initially contain equal quantities of the same ideal gas at the same T and P. Tank A discharges to the atmosphere through a small turbine in which the gas expands isentropically; tank B discharges to the atmosphere through a porous plug. Both devices operate until discharge ceases.

(*a*) When discharge ceases is the temperature in tank A less than, equal to, or greater than the temperature in tank B?

(*b*) When the pressures in both tanks have fallen to half the initial pressure, is the temperature of the gas discharging from the turbine less than, equal to, or greater than the temperature of the gas discharging from the porous plug?

(*c*) During the discharge process, is the temperature of the gas leaving the turbine less than, equal to, or greater than the temperature of the gas leaving tank A at the same instant?

(*d*) During the discharge process, is the temperature of the gas leaving the porous plug less than, equal to, or greater than the temperature of the gas leaving tank B at the same instant?

(*e*) When discharge ceases, is the mass of gas remaining in tank A less than, equal to, or greater than the mass of gas remaining in tank B?

7.3 A rigid tank of $100(\text{ft})^3$ capacity contains $5{,}100(\text{lb}_\text{m})$ of saturated liquid water at 460(°F). This amount of liquid almost completely fills the tank, the small remaining volume being occupied by saturated-vapor steam. Since a bit more vapor space in the tank is wanted, a valve at the top of the tank is opened, and saturated-vapor steam is vented to the atmosphere until the temperature in the tank falls to 450(°F). Assuming no heat transfer to the contents of the tank, determine the mass of steam vented.

7.4 Liquid nitrogen is stored in 0.5-m^3 metal tanks that are thoroughly insulated. Consider the process of filling an evacuated tank, initially at 295 K. It is attached to a line containing liquid nitrogen at its normal boiling point of 77.35 K and at a pressure of several bars. At this condition, its enthalpy is -120.8 kJ kg^{-1}. When a valve in the line is opened, the nitrogen flowing into the tank at first evaporates in the process of cooling the tank. If the tank has a mass of 30 kg and the metal has a specific heat of 0.43 J g^{-1} K^{-1}, what mass of nitrogen must flow into the tank just to cool it to a temperature such that *liquid* nitrogen begins to accumulate in the tank? Assume that the nitrogen and the tank are always at the same temperature.

The properties of saturated nitrogen vapor at several temperatures are given as follows:

T/K	P/bar	V^v/m^3 kg^{-1}	H^v/kJ kg^{-1}
80	1.396	0.1640	78.9
85	2.287	0.1017	82.3
90	3.600	0.06628	85.0
95	5.398	0.04487	86.8
100	7.775	0.03126	87.7
105	10.83	0.02223	87.4
110	14.67	0.01598	85.6

7.5 A tank of 60-m^3 capacity contains steam at 5,000 kPa and 400°C. Steam is vented from the tank through a relief valve to the atmosphere until the pressure in the tank falls to 4,000 kPa. If the venting process is adiabatic, estimate the final temperature of the steam in the tank and the mass of steam vented.

7.6 A tank of $3(\text{ft})^3$ [0.085-m^3] volume contains air at 70(°F) [21°C] and 14.7(psia) [101.33 kPa]. The tank is connected to a compressed-air line which supplies air at the constant conditions of 100(°F) [38°C] and 200(psia) [1,380 kPa]. A valve in the line is cracked so that air flows slowly into the tank until the pressure equals the line pressure. If the process occurs slowly enough that the temperature in the tank remains at 70(°F) [21°C], how much heat is lost from the tank? Assume air an ideal gas for which $C_P = (7/2)R$ and $C_V = (5/2)R$.

7.7 A small adiabatic air compressor is used to pump air into a $700(\text{ft})^3$ [19.8-m^3] insulated tank. The tank initially contains air at 80(°F) [26.7°C] and 1(atm) [101.33 kPa], exactly the conditions at which air enters the compressor. The pumping process continues until the pressure in the tank reaches 8(atm) [810 kPa]. If the process is adiabatic and if compression is isentropic, what is the shaft work of the compressor? Assume air an ideal gas for which $C_P = (7/2)R$ and $C_V = (5/2)R$.

7.8 A tank of 3-m^3 capacity contains 1,200 kg of liquid water at 200°C in equilibrium with its vapor, which fills the rest of the tank. A quantity of 800 kg of water at 60°C is pumped into the tank. How much heat must be added during this process if the temperature in the tank is not to change?

7.9 Gas at constant T and P is contained in a supply line connected through a valve to a closed tank containing the same gas at a lower pressure. The valve is opened to allow flow of gas into the tank, and then is shut again.

(*a*) Develop a general equation relating n_1 and n_2, the moles (or mass) of gas in the tank at the beginning and end of the process, to the properties U_1 and U_2, the internal energy of the gas in the tank at the beginning and end of the process, and H', the enthalpy of the gas in the supply line, and to Q, the heat transferred to the material in the tank during the process.

(*b*) Reduce the general equation to its simplest form for the special case of an ideal gas with constant heat capacities.

(*c*) Further reduce the equation of (*b*) for the case of $n_1 = 0$.

(*d*) Further reduce the equation of (*c*) for the case in which, in addition, $Q = 0$.

(*e*) Apply the appropriate equation to the case in which a steady supply of nitrogen at 25°C and 3 bar flows into an evacuated tank of 4-m^3 volume, and calculate the number of moles of nitrogen that flow into the tank to equalize the pressures if:

1. It is assumed that no heat flows from the gas to the tank or through the tank walls.
2. The tank weighs 400 kg, is perfectly insulated, has a specific heat of 0.46 J g^{-1} K^{-1}, has an initial temperature of 25°C, and is heated by the gas so as always to be at the temperature of the gas in the tank.

Assume nitrogen an ideal gas for which $C_P = (7/2)R$.

7.10 Develop equations which may be solved to give the final temperature of the gas remaining in a tank after the tank has been bled from an initial pressure P_1 to a final pressure P_2. Known quantities are the initial temperature, the tank volume, the heat capacity of the gas, the total heat capacity of the containing tank, P_1, and P_2. Assume the tank to be always at the temperature of the gas remaining in the tank, and the tank to be perfectly insulated.

7.11 A well-insulated tank of 70-m^3 volume initially contains 23,000 kg of water distributed between liquid and vapor phases at 25°C. Saturated steam at 1,100 kPa is admitted to the tank until the pressure reaches 700 kPa. What mass of steam is added?

7.12 An insulated evacuated tank of 1.5-m^3 volume is attached to a line containing steam at 350 kPa and 200°C. Steam flows into the tank until the pressure in the tank reaches 350 kPa. Assuming no heat flow from the steam to the tank, prepare graphs showing the mass of steam in the tank and its temperature as functions of pressure in the tank.

7.13 A 3-m^3 tank initially contains a mixture of saturated-vapor steam and saturated liquid water at 3,400 kPa. Of the total mass, 15 percent is vapor. Saturated-liquid water is bled from the tank through a valve until the total mass in the tank is 40 percent of the initial total mass. If during the process the temperature of the contents of the tank is kept constant, how much heat is transferred?

7.14 A stream of water at 65°C, flowing at the rate of 3 kg s^{-1}, is formed by mixing water at 20°C with saturated steam at 140°C. Assuming adiabatic operation, at what rates are the steam and water fed to the mixer?

7.15 In a desuperheater, water at 2,900 kPa and 40°C is sprayed into a stream of superheated steam at 2,800 kPa and 325°C in an amount such that a single stream of saturated-vapor steam at 2,700 kPa flows from the desuperheater at the rate of 10 kg s^{-1}. Assuming adiabatic operation, what is the mass flow rate of the water? What is ΔS_{total} for the process? What is the irreversible feature of the process?

7.16 Superheated steam at 100(psia) and 500(°F) flowing at the rate of 100(lb$_m$)(s)$^{-1}$ is mixed with liquid water at 100(°F) to produce steam at 100(psia) and 380(°F). Assuming adiabatic operation, at what rate is water supplied to the mixer? What is ΔS_{total} for the process? What is the irreversible feature of the process?

7.17 A stream of air at 10 bar and 800 K is mixed with another stream of air at 1 bar and 300 K with three times the mass flow rate. If this process is accomplished reversibly and adiabatically, what are the temperature and pressure of the resulting air stream? Assume air an ideal gas for which $C_P = (7/2)R$.

7.18 A stream of hot nitrogen gas at 700(°F) and atmospheric pressure, flows into a waste-heat boiler at the rate of 30(lb$_m$) (s)$^{-1}$, and transfers heat to water boiling at 1(atm). The water feed to the boiler is saturated liquid at 1(atm), and it leaves the boiler as superheated steam at 1(atm) and 350(°F). If the nitrogen is cooled to 250(°F) and if heat is lost to the surroundings at a rate of 50(Btu) for each (lb$_m$) of steam generated, what is the steam-generation rate? If the surroundings are at 70(°F), what is ΔS_{total} for the process? Assume nitrogen an ideal gas for which $C_P = (7/2)R$.

7.19 A stream of hot nitrogen gas at 370°C and atmospheric pressure, flows into a waste-heat boiler at the rate of 1 kg s^{-1}, and transfers heat to water boiling at 101.33 kPa. The water feed to the boiler is saturated liquid at 101.33 kPa, and it leaves the boiler as superheated steam at 101.33 kPa and 175°C. If the nitrogen is cooled to 120°C and if heat is lost to the surroundings at a rate of 100 kJ for each kilogram of steam generated, what is the steam-generation rate? If the surroundings are at 20°C, what is ΔS_{total} for the process? Assume nitrogen an ideal gas for which $C_P = (7/2)R$.

7.20 Air expands through a nozzle from a negligible initial velocity to a final velocity of 350 m s^{-1}. What is the temperature drop of the air, if air is assumed an ideal gas for which $C_P = (7/2)R$?

7.21 Steam enters a nozzle at 700 kPa and 280°C at negligible velocity and discharges at a pressure of 475 kPa. Assuming isentropic expansion of the steam in the nozzle, what is the exit velocity and what is the cross-sectional area at the nozzle exit for a flow rate of 0.5 kg s^{-1}?

7.22 Steam enters a converging nozzle at 700 kPa and 260°C with negligible velocity. If expansion is isentropic, what is the minimum pressure that can be reached in such a nozzle and what is the cross-sectional area at the nozzle throat at this pressure for a flow rate of 0.5 kg s^{-1}?

7.23 A gas enters a converging nozzle at pressure P_1 with negligible velocity, expands isentropically in the nozzle, and discharges into a chamber at pressure P_2. Sketch graphs showing the velocity at the throat and the mass flow rate as functions of the pressure ratio P_2/P_1.

7.24 For a converging/diverging nozzle with negligible entrance velocity in which expansion is isentropic, sketch graphs of mass flow rate $\dot{m}$, velocity u, and area ratio A/A_1 vs. the pressure ratio P/P_1. Here, A is the cross-sectional area of the nozzle at the point in the nozzle where the pressure is P, and subscript 1 denotes the nozzle entrance.

7.25 An ideal gas with constant heat capacities enters a converging/diverging nozzle with negligible velocity. If it expands isentropically within the nozzle, show that the throat velocity is given by

$$u_{\text{throat}}^2 = \frac{\gamma R T_1}{M}\left(\frac{2}{\gamma+1}\right)$$

where T_1 is the temperature of the gas entering the nozzle and R is the gas constant in units of J (kg mol)$^{-1}$ K^{-1}.

7.26 Steam expands isentropically in a converging/diverging nozzle from inlet conditions of 200(psia), 600(°F), and negligible velocity to a discharge pressure of 50(psia). At the throat, the cross-sectional area is 1(in)^2. Determine the mass flow rate of the steam and the state of the steam at the exit of the nozzle.

7.27 Steam expands adiabatically in a nozzle from inlet conditions of 100(psia), 400(°F), and a velocity of 200(ft)(s)$^{-1}$ to a discharge pressure of 20(psia) where its velocity is 2,000(ft)(s)$^{-1}$. What is the state of the steam at the nozzle exit, and what is ΔS_{total} for the process.

7.28 Air discharges from an adiabatic nozzle at 40(°F) [4.4°C] with a velocity of 1,800(ft)(s)$^{-1}$ [550 m s^{-1}]. What is the temperature at the entrance of the nozzle if the entrance velocity is negligible? Assume air an ideal gas for which $C_P = (7/2)R$.

7.29 A steam turbine operates adiabatically at a power level of 3,000 kW. Steam enters the turbine at 2,100 kPa and 475°C and exhausts from the turbine as saturated vapor at 30 kPa. What is the steam rate through the turbine, and what is the turbine efficiency?

7.30 A portable power-supply system consists of a 30-liter bottle of compressed nitrogen, connected to a small adiabatic turbine. The bottle is initially charged to 13,800 kPa at 27°C and in operation drives the turbine continuously until the pressure drops to 700 kPa. The turbine exhausts at 101.33 kPa. Neglecting all heat transfer to the gas, calculate the maximum possible work that can be obtained during the process. Assume nitrogen an ideal gas for which $C_P = (7/2)R$.

7.31 A turbine operates adiabatically with superheated steam entering at T_1 and P_1 with a mass flow rate $\dot{m}$. The exhaust pressure is P_2 and the turbine efficiency is η. For one of the following sets of operating conditions, determine the power output of the turbine and the enthalpy and entropy of the exhaust steam.

(*a*) $T_1 = 450°\text{C}$, $P_1 = 8{,}000$ kPa, $\dot{m} = 80$ kg s^{-1}, $P_2 = 30$ kPa, $\eta = 0.80$.
(*b*) $T_1 = 550°\text{C}$, $P_1 = 9{,}000$ kPa, $\dot{m} = 90$ kg s^{-1}, $P_2 = 20$ kPa, $\eta = 0.77$.
(*c*) $T_1 = 600°\text{C}$, $P_1 = 8{,}600$ kPa, $\dot{m} = 70$ kg s^{-1}, $P_2 = 10$ kPa, $\eta = 0.82$.
(*d*) $T_1 = 400°\text{C}$, $P_1 = 7{,}000$ kPa, $\dot{m} = 65$ kg s^{-1}, $P_2 = 50$ kPa, $\eta = 0.75$.
(*e*) $T_1 = 200°\text{C}$, $P_1 = 1{,}400$ kPa, $\dot{m} = 5$ kg s^{-1}, $P_2 = 200$ kPa, $\eta = 0.75$.
(*f*) $T_1 = 900(°\text{F})$, $P_1 = 1{,}200$(psia), $\dot{m} = 150(\text{lb}_\text{m})(\text{s})^{-1}$, $P_2 = 2$(psia), $\eta = 0.80$.
(*g*) $T_1 = 800(°\text{F})$, $P_1 = 1{,}000$(psia), $\dot{m} = 100(\text{lb}_\text{m})(\text{s})^{-1}$, $P_2 = 4$(psia), $\eta = 0.75$.

7.32 The steam rate to a turbine for variable output is controlled by a throttle valve in the inlet line. Steam is supplied to the throttle valve at 240(psia) and 440(°F). During a test run, the pressure at the turbine inlet is 160(psia), the exhaust steam at 1(psia) has a quality of 0.95, the steam flow rate is $1(\text{lb}_\text{m})(\text{s})^{-1}$, and the power output of the turbine is 240(hp).

(*a*) What are the heat losses from the turbine?

(*b*) What would be the power output if the steam supplied to the throttle valve were expanded isentropically to the final pressure?

7.33 Isobutane expands adiabatically in a turbine from 700(psia) [4,826 kPa] and 500(°F) [260°C] to 70(psia) [483 kPa] at the rate of 1.5(lb mol)(s)$^{-1}$ [0.68 kg mol s^{-1}]. If the turbine efficiency is 0.80, what is the power output of the turbine and what is the temperature of the isobutane leaving the turbine?

7.34 Combustion products from a burner enter a gas turbine at 7.5 bar and 900°C and discharge at 1.2 bar. The turbine operates adiabatically with an efficiency of 80 percent. Assuming the combustion products to be an ideal-gas mixture with a heat capacity of 30 J mol^{-1} °C^{-1}, what is the work output of the turbine per mole of gas, and what is the temperature of the gases discharging from the turbine?

7.35 An expander operates adiabatically with nitrogen entering at T_1 and P_1 with a molar flow rate $\dot{n}$. The exhaust pressure is P_2, and the expander efficiency is η. Estimate the power output of the expander and the temperature of the exhaust stream for one of the following sets of operating conditions.

(*a*) $T_1 = 480°\text{C}$, $P_1 = 6$ bar, $\dot{n} = 200$ mol s^{-1}, $P_2 = 1$ bar, $\eta = 0.80$.
(*b*) $T_1 = 400°\text{C}$, $P_1 = 5$ bar, $\dot{n} = 150$ mol s^{-1}, $P_2 = 1$ bar, $\eta = 0.75$.
(*c*) $T_1 = 500°\text{C}$, $P_1 = 7$ bar, $\dot{n} = 175$ mol s^{-1}, $P_2 = 1$ bar, $\eta = 0.78$.
(*d*) $T_1 = 450°\text{C}$, $P_1 = 8$ bar, $\dot{n} = 100$ mol s^{-1}, $P_2 = 2$ bar, $\eta = 0.85$.
(*e*) $T_1 = 900(°\text{F})$, $P_1 = 95$(psia), $\dot{n} = 0.5$(lb mol)(s)$^{-1}$, $P_2 = 15$(psia), $\eta = 0.80$.

7.36 Saturated steam at 175 kPa is compressed adiabatically in a centrifugal compressor to 650 kPa at the rate of 1.5 kg s^{-1}. The compressor efficiency is 75 percent. What is the power requirement of the compressor and what are the enthalpy and entropy of the steam in its final state?

7.37 A compressor operates adiabatically with air entering at T_1 and P_1 with a molar flow rate $\dot{n}$. The discharge pressure is P_2 and the compressor efficiency is η. Estimate the power requirement of the compressor and the temperature of the discharge stream for one of the following sets of operating conditions.

(*a*) $T_1 = 25°\text{C}$, $P_1 = 101.33$ kPa, $\dot{n} = 100$ mol s^{-1}, $P_2 = 375$ kPa, $\eta = 0.75$.
(*b*) $T_1 = 80°\text{C}$, $P_1 = 375$ kPa, $\dot{n} = 100$ mol s^{-1}, $P_2 = 1{,}000$ kPa, $\eta = 0.70$.
(*c*) $T_1 = 30°\text{C}$, $P_1 = 100$ kPa, $\dot{n} = 150$ mol s^{-1}, $P_2 = 500$ kPa, $\eta = 0.80$.
(*d*) $T_1 = 100°\text{C}$, $P_1 = 500$ kPa, $\dot{n} = 50$ mol s^{-1}, $P_2 = 1{,}300$ kPa, $\eta = 0.75$.
(*e*) $T_1 = 80$(°F), $P_1 = 14.7$(psia), $\dot{n} = 0.5$(lb mol)(s)$^{-1}$, $P_2 = 55$(psia), $\eta = 0.75$.
(*f*) $T_1 = 150$(°F), $P_1 = 55$(psia), $\dot{n} = 0.5$(lb mol)(s)$^{-1}$, $P_2 = 135$(psia), $\eta = 0.70$.

7.38 Ammonia gas is compressed from 21°C and 200 kPa to 1,000 kPa in an adiabatic compressor with an efficiency of 0.82. Estimate the work required per mol of ammonia and the enthalpy and entropy changes of the ammonia.

7.39 Propylene is compressed adiabatically from 11.5 bar and 30°C to 18 bar at the rate of 1 kg mol s^{-1}. If the compressor efficiency is 0.8, what is the power requirement of the compressor and what is the discharge temperature of the propylene?

7.40 Methane is compressed adiabatically in a pipeline pumping station from 500(psia) [3,450 kPa] and 77(°F) [25°C] to 725(psia) [5,000 kPa] at the rate of 2.5(lb mol)(s)$^{-1}$ [1.134 kg mol s^{-1}]. If the compressor efficiency is 0.75, what is the power requirement of the compressor and what is the discharge temperature of the methane?

7.41 A pump operates adiabatically with liquid water entering at T_1 and P_1 with a mass flow rate $\dot{m}$. The discharge pressure is P_2, and the pump efficiency is η. For one of the following sets of operating conditions, determine the power requirement of the pump and the temperature of the water stream discharged from the pump.

(*a*) $T_1 = 25°\text{C}$, $P_1 = 1$ bar, $\dot{m} = 20$ kg s^{-1}, $P_2 = 20$ bar, $\eta = 0.75$, $\beta = 257.2 \times 10^{-6}$ K^{-1}.
(*b*) $T_1 = 90°\text{C}$, $P_1 = 2$ bar, $\dot{m} = 30$ kg s^{-1}, $P_2 = 50$ bar, $\eta = 0.70$, $\beta = 696.2 \times 10^{-6}$ K^{-1}.
(*c*) $T_1 = 60°\text{C}$, $P_1 = 20$ kPa, $\dot{m} = 15$ kg s^{-1}, $P_2 = 5{,}000$ kPa, $\eta = 0.75$, $\beta = 523.1 \times 10^{-6}$ K^{-1}.
(*d*) $T_1 = 70$(°F), $P_1 = 1$(atm), $\dot{m} = 50$(lb$_m$)(s)$^{-1}$, $P_2 = 20$(atm), $\eta = 0.70$, $\beta = 217.3 \times 10^{-6}$ K^{-1}.
(*e*) $T_1 = 200$(°F), $P_1 = 15$(psia), $\dot{m} = 80$(lb$_m$)(s)$^{-1}$, $P_2 = 1{,}500$(psia), $\eta = 0.75$, $\beta = 714.3 \times 10^{-6}$ K^{-1}.

CHAPTER
EIGHT
CONVERSION OF HEAT INTO WORK BY POWER CYCLES

Prior to the development of nuclear power, all significant contributions to the mechanical energy used by humankind had the sun as their source. However, economical methods have not been developed as yet for *directly* converting solar radiation into work on a large scale. The total rate at which energy reaches the earth from the sun is staggering, but the rate at which it falls on a square meter of surface is small. The difficulty is therefore to concentrate the heat gathered over a large surface so that it is a practical energy source for the production of work. Research in this area continues, and progress has been made on the direct use of solar energy for heat. For example, solar radiation is used to heat homes, to produce high temperatures for metallurgical operations (solar furnaces), and to concentrate aqueous solutions by evaporation.

The kinetic energy associated with mass movement of air has been used to some extent for the production of work (windmills), especially in rural areas. Variations and uncertainties in wind speed, and the need for large-size equipment to produce significant quantities of work, are problems in this field.

Conceivably, the potential energy of tides could be exploited. Attempts in this direction on a large scale have been made in parts of the world where tides are particularly high. However, total power production from this source is unlikely to be significant in comparison with world demands for energy.

By far the most important sources of power are the chemical (molecular) energy of fuels, nuclear energy, and the potential energy of water. The use of water power involves the conversion of mechanical energy from one form to another, and an efficiency of 100 percent is theoretically possible. On the other hand, all present-day methods for the *large-scale* use of molecular or nuclear energy are based on the evolution of heat and subsequent conversion of part of the heat into useful work. Accordingly, the efficiency of all such processes is destined to be low (values greater than 35 percent are uncommon), despite improvements in the design of equipment. This is, of course, a direct consequence of the second law. When it is possible to convert the energy in fuels into work without the intermediate generation of heat, conversion efficiency is considerably improved. The usual device for the direct conversion of chemical energy into electrical energy is the electrolytic cell. Progress has been made in developing cells which operate on hydrogen and on carbonaceous fuels such as natural gas or coal. Such *fuel cells* are already in use to supply modest power requirements for special purposes. The efficiency of these cells ranges from 65 to 80 percent, about twice the value obtained by the conventional process of first converting the chemical energy into heat.

In a conventional power plant the molecular energy of fuel is released by a combustion process. The function of the work-producing device is to convert part of the heat of combustion into mechanical energy. In a nuclear power plant the fission or fusion process releases the energy of the nucleus of the atom as heat, and then this heat is partially converted into work. Thus, the thermodynamic analysis of heat engines, as presented in this chapter, applies equally well to conventional (fossil-fuel) and nuclear power plants.

In one form of heat engine, the steam power plant, the working fluid (steam) is completely enclosed and goes through a cyclic process, accomplished by vaporization and condensation. Heat is transferred to the fluid from another part of the plant across a physical boundary. In a coal-fired plant the combustion gases are separated from the steam by boiler-tube walls. The *internal*-combustion engine is another form of heat engine, wherein high temperatures are attained by conversion of the chemical energy of a fuel directly into internal energy within the work-producing device. Examples of this type are the Otto engine and the gas turbine.†

To illustrate the calculation of thermal efficiencies, we analyze in this chapter several common heat-engine cycles.

8.1 THE STEAM POWER PLANT

The Carnot-engine cycle, described in Chap. 5, operates reversibly and consists of two isothermal steps connected by two adiabatic steps. In the isothermal step at higher temperature T_H, heat $|Q_H|$ is absorbed by the working fluid of the engine, and in the isothermal step at lower temperature T_C, heat $|Q_C|$ is discarded

† Details of steam power plants and internal-combustion engines can be found in E. B. Woodruff, H. B. Lammers, and T. S. Lammers, *Steam Plant Operation*, 5th ed., McGraw-Hill, New York, 1983; and C. F. Taylor and E. S. Taylor, *The Internal Combustion Engine*, International Textbook, Scranton, Pa., 1962.

by the fluid. The work produced is $W = |Q_H| - |Q_C|$, and the thermal efficiency of the Carnot engine [Eq. (5.8)] is

$$\eta = \frac{W}{|Q_H|} = 1 - \frac{T_C}{T_H}$$

Clearly, η increases as T_H increases and as T_C decreases. Although the efficiencies of practical heat engines are lowered by irreversibilities, it is still true that their efficiencies are increased when the average temperature at which heat is absorbed is increased and when the average temperature at which heat is rejected is decreased.

Figure 8.1 shows a simple steady-state flow process in which steam generated in a boiler is expanded in an adiabatic turbine to produce work. The discharge stream from the turbine passes to a condenser from which it is pumped adiabatically back to the boiler. The power produced by the turbine is much greater than that required by the pump, and the net power output is equal to the difference between the rate of heat input in the boiler $\dot{Q}_H$ and the rate of heat rejection in the condenser $\dot{Q}_C$.

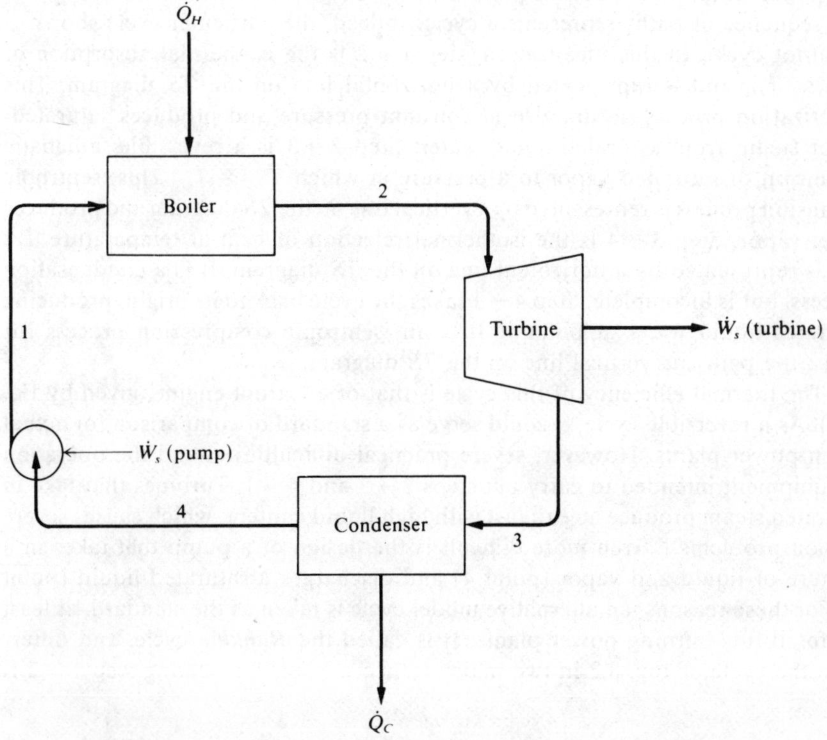

Figure 8.1 Simple steam power plant.

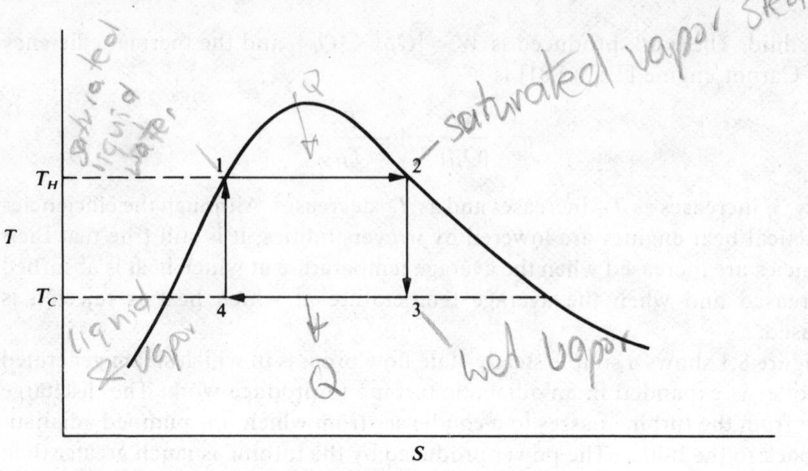

Figure 8.2 Carnot cycle on a *TS* diagram.

The property changes of the fluid as it flows through the individual pieces of equipment may be shown as paths on a *TS* diagram, as illustrated in Fig. 8.2. The sequence of paths represents a cycle. Indeed, the particular cycle shown is a Carnot cycle. In this idealization, step 1 → 2 is the isothermal absorption of heat at T_H, and is represented by a horizontal line on the *TS* diagram. This vaporization process occurs also at constant pressure and produces saturated-vapor steam from saturated-liquid water. Step 2 → 3 is a reversible, adiabatic expansion of saturated vapor to a pressure at which $T^{\text{sat}} = T_C$. This isentropic expansion process is represented by a vertical line on the *TS* diagram and produces a wet vapor. Step 3 → 4 is the isothermal rejection of heat at temperature T_C, and is represented by a horizontal line on the *TS* diagram. It is a condensation process, but is incomplete. Step 4 → 1 takes the cycle back to its origin, producing saturated-liquid water at point 1. It is an isentropic compression process for which the path is a vertical line on the *TS* diagram.

The thermal efficiency of this cycle is that of a Carnot engine, given by Eq. (5.8). As a reversible cycle, it could serve as a standard of comparison for actual steam power plants. However, severe practical difficulties attend the operation of equipment intended to carry out steps 2 → 3 and 4 → 1. Turbines that take in saturated steam produce an exhaust with high liquid content, which causes severe erosion problems.† Even more difficult is the design of a pump that takes in a mixture of liquid and vapor (point 4) and discharges a saturated liquid (point 1). For these reasons, an alternative model cycle is taken as the standard, at least for fossil-fuel-burning power plants. It is called the *Rankine* cycle, and differs from the cycle of Fig. 8.2 in two major respects. First, the heating step 1 → 2 is

† Nevertheless, present-day nuclear power plants generate saturated steam and operate with turbines designed to eject liquid at various stages of expansion.

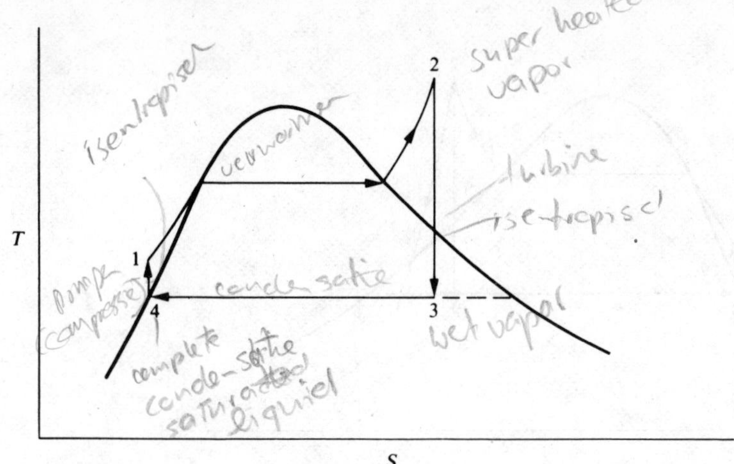

Figure 8.3 The Rankine cycle.

carried well beyond vaporization, so as to produce a superheated vapor, and second, the cooling step 3 → 4 brings about complete condensation, yielding saturated liquid to be pumped to the boiler. The Rankine cycle therefore consists of the four steps shown by Fig. 8.3, and described as follows:

1 → 2 A constant-pressure heating process in a boiler. The path lies along an isobar (the pressure of the boiler), and consists of three sections: heating of liquid water to its saturation temperature, vaporization at constant temperature and pressure, and superheating of the vapor to a temperature well above its saturation temperature.

2 → 3 Reversible, adiabatic (isentropic) expansion of vapor in a turbine to the pressure of the condenser. The path normally crosses the saturation curve, producing a wet exhaust. However, the superheating accomplished in step 1 → 2 shifts the path far enough to the right on Fig. 8.3 that the moisture content is not too large.

3 → 4 A constant-pressure, constant-temperature process in a condenser to produce saturated liquid at point 4.

4 → 1 Reversible, adiabatic (isentropic) pumping of the condensed liquid to the pressure of the boiler. The vertical path (whose length is exaggerated in Fig. 8.3) is very short, because the temperature rise associated with compression of a liquid is small.

Power plants can be built to operate on a cycle that departs from the Rankine cycle only to the extent that the work-producing and work-requiring steps are irreversible. We show in Fig. 8.4 the effects of these irreversibilities on steps 2 → 3 and 4 → 1. The paths are no longer vertical, but tend in the direction of increasing entropy. The turbine exhaust is normally still wet, but as long as the moisture

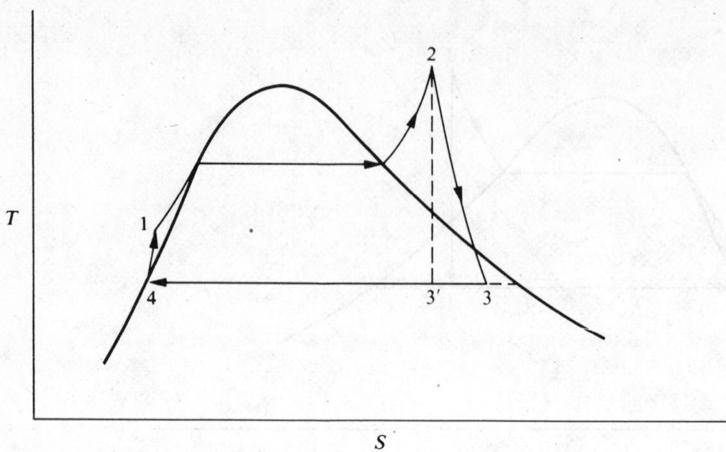

Figure 8.4 Simple practical power cycle.

content is less than about 10 percent, erosion problems are not serious. Slight subcooling of the condensate in the condenser may occur, but the effect is inconsequential.

The boiler serves to transfer heat from a burning fuel to the cycle, and the condenser transfers heat from the cycle to the surroundings. Neglecting kinetic- and potential-energy changes reduces the energy relations, Eqs. (7.9) and (7.10), in either case to

$$\dot{Q} = \dot{m}\,\Delta H \tag{8.1}$$

and

$$Q = \Delta H \tag{8.2}$$

Turbine and pump calculations are treated in detail in Chap. 7.

Example 8.1 Steam generated in a power plant at a pressure of 8,600 kPa and a temperature of 500°C is fed to a turbine. Exhaust from the turbine enters a condenser at 10 kPa, where it is condensed to saturated liquid, which is then pumped to the boiler.

(*a*) Determine the thermal efficiency of a Rankine cycle operating at these conditions.

(*b*) Determine the thermal efficiency of a practical cycle operating at these conditions if the turbine efficiency and pump efficiency are both 75 percent.

(*c*) If the rating of the power cycle of part (*b*) is 80,000 kW, what is the steam rate and what are the heat-transfer rates in the boiler and condenser?

SOLUTION (*a*) The turbine operates under the same conditions as the turbine of Example 7.5, where we found

$$(\Delta H)_S = -1{,}274.2 \text{ kJ kg}^{-1}$$

Thus,

$$W_s(\text{isentropic}) = -(\Delta H)_S = 1{,}274.2 \text{ kJ kg}^{-1}$$

Moreover, we found the enthalpy at the end of isentropic expansion (H'_2 in Example 7.5) to be

$$H'_3 = 2{,}117.4 \text{ kJ kg}^{-1}$$

The enthalpy of saturated liquid at 10 kPa (and $t^{\text{sat}} = 45.83°\text{C}$) is

$$H_4 = 191.8 \text{ kJ kg}^{-1}$$

Thus by Eq. (8.2) applied to the condenser,

$$Q(\text{condenser}) = H_4 - H'_3 = 191.8 - 2{,}117.4$$
$$= -1{,}925.6 \text{ kJ kg}^{-1}$$

where the minus sign signifies that the heat flows out of the system.

The pump operates under essentially the same conditions as the pump of Example 7.10, where we found

$$W_s(\text{isentropic}) = -(\Delta H)_S = -8.7 \text{ kJ kg}^{-1}$$

Thus,

$$H_1 = H_4 + (\Delta H)_S = 191.8 + 8.7 = 200.5 \text{ kJ kg}^{-1}$$

The enthalpy of superheated steam at 8,600 kPa and 500°C is

$$H_2 = 3{,}391.6 \text{ kJ kg}^{-1}$$

By Eq. (8.2) applied to the boiler,

$$Q(\text{boiler}) = H_2 - H_1 = 3{,}391.6 - 200.5$$
$$= 3{,}191.1 \text{ kJ kg}^{-1}$$

The net work of the Rankine cycle is the sum of the turbine work and the pump work:

$$W_s(\text{Rankine}) = 1{,}274.2 - 8.7$$
$$= 1{,}265.5 \text{ kJ kg}^{-1}$$

This result is of course also given by

$$W_s(\text{Rankine}) = Q(\text{boiler}) + Q(\text{condenser})$$
$$= 3{,}191.1 - 1{,}925.6 = 1{,}265.5 \text{ kJ kg}^{-1}$$

The thermal efficiency of the cycle is

$$\eta = \frac{W_s(\text{Rankine})}{Q(\text{boiler})} = \frac{1{,}265.5}{3{,}191.1} = 0.3966$$

(*b*) If the turbine efficiency is 0.75, then we also have from Example 7.5 that

$$W_s(\text{turbine}) = -\Delta H = 995.6 \text{ kJ kg}^{-1}$$

and

$$H_3 = H_2 + \Delta H = 3{,}391.6 - 955.6$$
$$= 2{,}436.0 \text{ kJ kg}^{-1}$$

For the condenser,

$$Q(\text{condenser}) = H_4 - H_3 = 191.8 - 2{,}436.0$$
$$= 2{,}244.2 \text{ kJ kg}^{-1}$$

By Example 7.10 for the pump,

$$W_s(\text{pump}) = -\Delta H = -11.6 \text{ kJ kg}^{-1}$$

Whence

$$H_1 = H_4 + \Delta H = 191.8 + 11.6$$
$$= 203.4 \text{ kJ kg}^{-1}$$

Then

$$Q(\text{boiler}) = H_2 - H_1 = 3{,}391.6 - 203.4$$
$$= 3{,}188.2 \text{ kJ kg}^{-1}$$

The thermal efficiency of the cycle is therefore

$$\eta = \frac{W_s(\text{net})}{Q(\text{boiler})} = \frac{955.6 - 11.6}{3{,}188.2} = 0.2961$$

which may be compared with the result of part (*a*).

(*c*) For a power rating of $\dot{W}_s(\text{net}) = 80{,}000$ kW, we have

$$\dot{W}_s(\text{net}) = \dot{m} W_s(\text{net})$$

or

$$\dot{m} = \frac{\dot{W}_s(\text{net})}{W_s(\text{net})} = \frac{80{,}000 \text{ kJ s}^{-1}}{944.0 \text{ kJ kg}^{-1}} = 84.75 \text{ kg s}^{-1}$$

Whence by Eq. (8.1),

$$\dot{Q}(\text{boiler}) = (84.75)(3{,}188.2) = 270.2 \times 10^3 \text{ kJ s}^{-1}$$

and

$$\dot{Q}(\text{condenser}) = (84.75)(2{,}244.2) = 190.2 \times 10^3 \text{ kJ s}^{-1}$$

Note that

$$\dot{Q}(\text{boiler}) - \dot{Q}(\text{condenser}) = \dot{W}_s(\text{net})$$

The thermal efficiency of a steam power cycle is increased when the pressure and hence the vaporization temperature in the boiler is raised. It is also increased by increased superheating in the boiler. Thus, high boiler pressures and temperatures favor high efficiencies. However, these same conditions increase the capital investment in the plant, because they require heavier construction and more expensive materials of construction. Moreover, these costs increase ever more rapidly as more severe conditions are imposed. Thus, in practice power plants seldom operate at pressures much above 10,000 kPa and temperatures much above 600°C. The thermal efficiency of a power plant increases as the pressure and hence the temperature in the condenser is reduced. However, the condensation temperature must be higher than the temperature of the cooling medium, usually water, and this is controlled by local conditions of climate and geography. Power plants universally operate with the condenser pressure as low as practicable.

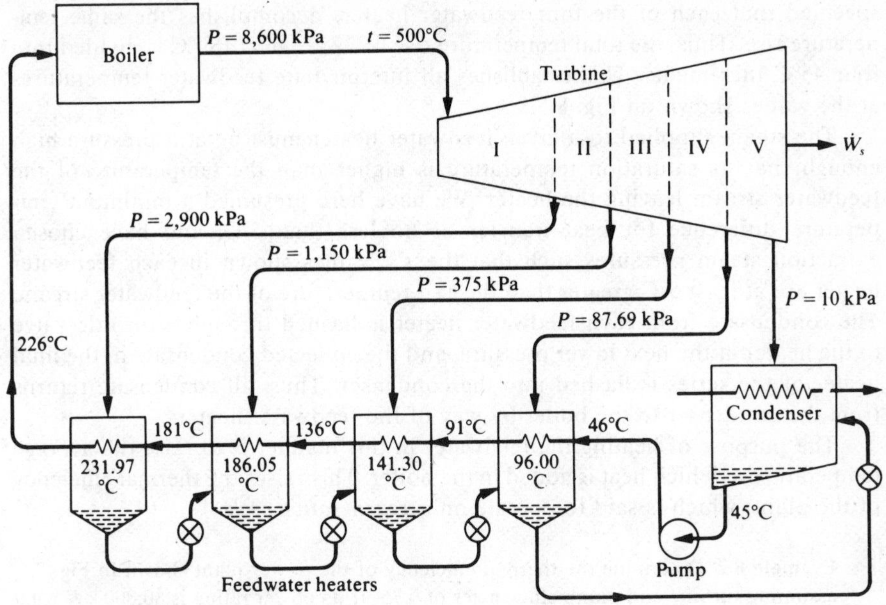

Figure 8.5 Steam power plant with feedwater heating.

Most modern power plants operate on a modification of the Rankine cycle that incorporates feedwater heaters. Water from the condenser, rather than being pumped directly back to the boiler, is first heated by steam extracted from the turbine. This is normally done in several stages, with steam taken from the turbine at several intermediate states of expansion. An arrangement with four feedwater heaters is shown in Fig. 8.5. The operating conditions indicated on this figure and described in the following paragraphs are typical, and are the basis for the illustrative calculations of Example 8.2.

The conditions of steam generation in the boiler are the same as in Example 8.1: 8,600 kPa and 500°C. The exhaust pressure of the turbine, 10 kPa, is also the same. The saturation temperature of the exhaust steam is therefore 45.83°C. Allowing for slight subcooling of the condensate, we fix the temperature of the liquid water from the condenser at 45°C. The feedwater pump, which operates under exactly the conditions of the pump in Example 7.10, causes a temperature rise of about 1°C, making the temperature of the feedwater entering the series of heaters equal to 46°C.

The saturation temperature of steam at the boiler pressure of 8,600 kPa is 300.06°C, and the temperature to which the feedwater can be raised in the heaters is certainly less. This temperature is a design variable, which is ultimately fixed by economic considerations. However, a value must be chosen before any thermodynamic calculations can be made. We have therefore arbitrarily specified a temperature of 226°C for the feedwater stream entering the boiler. We have also specified that each of the four feedwater heaters accomplishes the same temperature rise. Thus, the total temperature rise of 226 − 46 = 180°C is divided into four 45°C increments. This establishes all intermediate feedwater temperatures at the values shown on Fig. 8.5.

The steam supplied to a given feedwater heater must be at a pressure high enough that its saturation temperature is higher than the temperature of the feedwater stream leaving the heater. We have here presumed a minimum temperature difference for heat transfer of no less than 5°C, and have chosen extraction steam pressures such that the t^{sat} values shown in each feedwater heater are at least 5°C greater than the exit temperature of the feedwater stream. The condensate from each feedwater heater is flashed through a throttle valve to the heater at the next lower pressure, and the collected condensate in the final heater of the series is flashed into the condenser. Thus, all condensate returns from the condenser to the boiler by way of the feedwater heaters.

The purpose of heating the feedwater in this manner is to raise the average temperature at which heat is added in the boiler. This raises the thermal efficiency of the plant, which is said to operate on a *regenerative cycle.*

Example 8.2 Determine the thermal efficiency of the power plant shown in Fig. 8.5, assuming turbine and pump efficiencies of 0.75. If its power rating is 80,000 kW what is the steam rate from the boiler and the heat-transfer rates in the boiler and condenser?

SOLUTION Initial calculations are made on the basis of 1 kg of steam entering the turbine from the boiler. The turbine is in effect divided into five sections, as indicated in Fig. 8.5. Because steam is extracted at the end of each section, the flow rate in the turbine decreases from one section to the next. The amounts of steam extracted from the first four sections are determined by energy balances.

For this, we need enthalpies of the compressed feedwater streams. The effect of pressure at constant temperature on a liquid is given by Eq. (7.37) written as

$$\Delta H = V(1 - \beta T)\,\Delta P \qquad (\text{const } T)$$

For saturated liquid water at 226°C (499.15 K), we find from the steam tables:

$$P^{sat} = 2{,}598.2 \text{ kPa}$$

$$H = 971.5 \text{ kJ kg}^{-1}$$

$$V = 1{,}201 \text{ cm}^3 \text{ kg}^{-1}$$

In addition, at this temperature

$$\beta = 1.582 \times 10^{-3} \text{ K}^{-1}$$

Thus, for a pressure change from the saturation pressure to 8,600 kPa,

$$\Delta H = 1{,}201[1 - (1.528 \times 10^{-3})(499.15)]\,\frac{(8{,}600 - 2{,}598.2)}{10^6}$$

$$= 1.5 \text{ kJ kg}^{-1}$$

and

$$H = H(\text{sat liq}) + \Delta H = 971.5 + 1.5 = 973.0 \text{ kJ kg}^{-1}$$

Similar calculations yield the enthalpies of the feedwater at other temperatures. All pertinent values are given as follows:

$t/°C$	226	181	136	91	46
$H/kJ\ kg^{-1}$ for water at t and $P = 8{,}600$ kPa	973.0	771.3	577.4	387.5	200.0

Consider the first section of the turbine and the first feedwater heater, as shown by Fig. 8.6. The enthalpy and entropy of the steam entering the turbine are found from the tables for superheated steam. The assumption of isentropic expansion of steam in section I of the turbine to 2,900 kPa leads to the result:

$$(\Delta H)_S = -320.5 \text{ kJ kg}^{-1}$$

If we assume that the turbine efficiency is independent of the pressure to which the steam expands, then Eq. (7.28) gives:

$$\Delta H = \eta(\Delta H)_S = (0.75)(-320.5) = -240.4 \text{ kJ kg}^{-1}$$

By Eq. (7.26),

$$W_s(\mathrm{I}) = -\Delta H = 240.4 \text{ kJ}$$

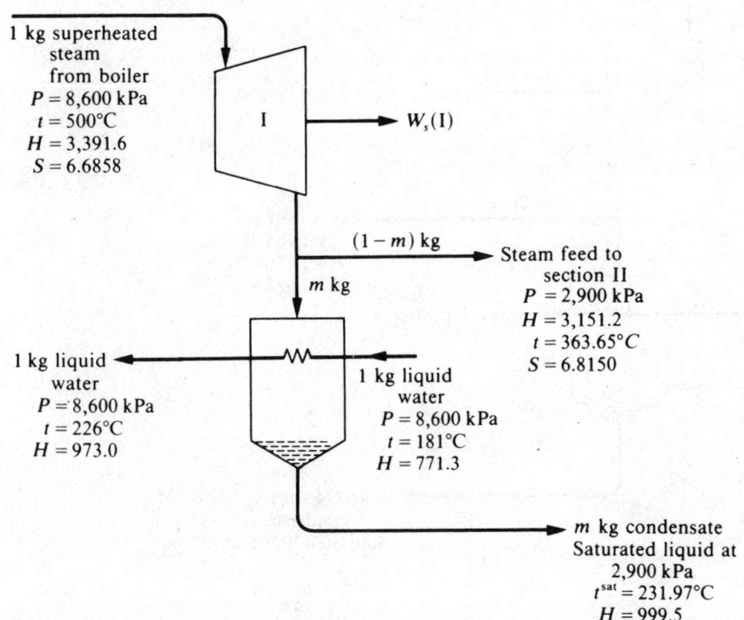

Figure 8.6 Section I of turbine and first feedwater heater. Enthalpies in $kJ\ kg^{-1}$; entropies in $kJ\ kg^{-1}\ K^{-1}$.

In addition, the enthalpy of the steam discharged from this section of the turbine is

$$H = 3{,}391.6 - 240.4 = 3{,}151.2 \text{ kJ kg}^{-1}$$

An energy balance on the feedwater heater requires application of Eq. (7.8). Neglecting kinetic- and potential-energy changes and noting that $\dot{Q} = \dot{W}_s = 0$, we have

$$\Delta(\dot{m}H)_{fs} = 0$$

This equation expresses mathematically the requirement that the total enthalpy change for the process be zero. Thus on the basis of 1 kg of steam entering the turbine (see Fig. 8.6),

$$m(999.5 - 3{,}151.2) + (1)(973.0 - 771.3) = 0$$

Whence

$$m = 0.09374 \text{ kg} \qquad \text{and} \qquad 1 - m = 0.90626 \text{ kg}$$

On the basis of 1 kg of steam entering the turbine, $1 - m$ is the mass of steam flowing into section II of the turbine.

Section II of the turbine and the second feedwater heater are shown in Fig. 8.7. In doing the same calculations as for section I, we assume that each kilogram of steam leaving section II expands from its state *at the turbine entrance* to the exit of

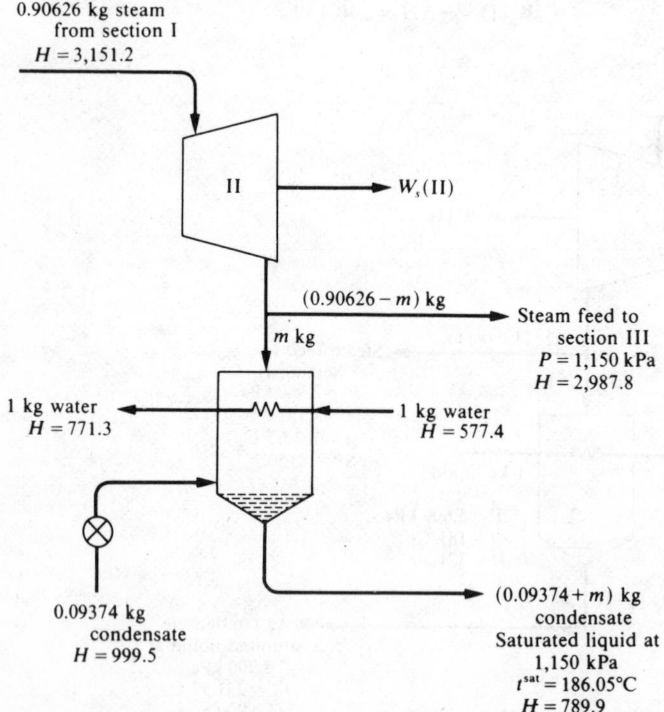

Figure 8.7 Section II of turbine and second feedwater heater. Enthalpies in kJ kg^{-1}; entropies in kJ kg^{-1} K^{-1}.

section II with an efficiency of 75 percent compared with isentropic expansion. The enthalpy of the steam leaving section II found in this way is

$$H = 2{,}987.8 \text{ kJ kg}^{-1}$$

Then on the basis of 1 kg of steam entering the turbine,

$$W_s(\text{II}) = -(2{,}987.8 - 3{,}151.2)(0.90626)$$
$$= 148.08 \text{ kJ}$$

An energy balance on the feedwater heater (see Fig. 8.7) gives:

$$(0.09374 + m)(789.9) - (0.09374)(999.5) - m(2{,}987.8) + (1)(771.3 - 577.4) = 0$$

Whence

$$m = 0.07971 \text{ kg}$$

Note that throttling the condensate stream does not change its enthalpy.

These results and those of similar calculations for the remaining sections of the turbine are listed in the following table:

	H/kJ kg^{-1} at section exit	W_s/kJ for section	t/°C at section exit	State	m/kg of steam extracted
Sec. I	3,151.2	240.40	363.65	Superheated vapor	0.09374
Sec. II	2,987.8	148.08	272.48	Superheated vapor	0.07928
Sec. III	2,827.4	132.65	183.84	Superheated vapor	0.06993
Sec. IV	2,651.3	133.32	96.00	Wet vapor $x = 0.9919$	0.06257
Sec. V	2,435.9	149.59	45.83	Wet vapor $x = 0.9378$	
		$\sum W_s = 804.0$ kJ			$\sum m = 0.3055$

Thus for every kilogram of steam entering the turbine, the work produced is 804.0 kJ and 0.3055 kg of steam is extracted from the turbine for the feedwater heaters. The work required by the pump is exactly the work calculated for the pump in Example 7.10, that is, 11.6 kJ. The net work of the cycle is therefore

$$W_s(\text{net}) = 804.0 - 11.6 = 792.4 \text{ kJ}$$

on the basis of 1 kg of steam generated in the boiler. On the same basis, the heat added in the boiler is

$$Q(\text{boiler}) = \Delta H = 3{,}391.6 - 973.0 = 2{,}418.6 \text{ kJ}$$

The thermal efficiency of the cycle is therefore

$$\eta = \frac{W_s(\text{net})}{Q(\text{boiler})} = \frac{792.4}{2{,}418.6} = 0.3276$$

This is a significant improvement over the value of 0.2961 found in Example 8.1.
Since $\dot{W}_s(\text{net}) = 80{,}000 \text{ kJ s}^{-1}$,

$$\dot{m} = \frac{\dot{W}_s(\text{net})}{W_s(\text{net})} = \frac{80{,}000}{792.4} = 100.96 \text{ kg s}^{-1}$$

This is the steam rate to the turbine, and with it we can calculate the heat-transfer rate in the boiler:

$$\dot{Q}(\text{boiler}) = \dot{m}\Delta H = (100.96)(2{,}418.6)$$
$$= 244.2 \times 10^3 \text{ kJ s}^{-1}$$

The heat-transfer rate to the cooling water in the condenser is

$$\dot{Q}(\text{condenser}) = \dot{Q}(\text{boiler}) - \dot{W}_s(\text{net})$$
$$= 244.2 \times 10^3 - 80.0 \times 10^3$$
$$= 164.2 \times 10^3 \text{ kJ s}^{-1}$$

Although the steam generation rate is higher than was found in Example 8.1, the heat-transfer rates in the boiler and condenser are appreciably less, because their functions are partly taken over by the feedwater heaters.

8.2 INTERNAL-COMBUSTION ENGINES

In a steam power plant, the steam is an inert medium to which heat is transferred from a burning fuel or from a nuclear reactor. It is therefore characterized by large heat-transfer surfaces: (1) for the absorption of heat by the steam at a high temperature in the boiler, and (2) for the rejection of heat from the steam at a relatively low temperature in the condenser. The disadvantage is that when heat must be transferred through walls (as through the metal walls of boiler tubes) the ability of the walls to withstand high temperatures and pressures imposes a limit on the temperature of heat absorption. In an internal-combustion engine, on the other hand, a fuel is burned within the engine itself, and the combustion products serve as the working-medium, acting for example on a piston in a cylinder. High temperatures are internal, and do not involve heat-transfer surfaces.

The burning of fuel within the internal-combustion engine does complicate thermodynamic analysis. Moreover, fuel and air flow steadily into an internal-combustion engine and combustion products flow steadily out of it; there is no working medium that undergoes a cyclic process, as does the steam in a steam power plant. However, for making simple analyses, one imagines cyclic engines with air as the working fluid that are equivalent in performance to actual internal-combustion engines. In addition, the combustion step is replaced by the addition to the air of an equivalent amount of heat. In each of the following sections, we first present a qualitative description of an internal-combustion engine. Quantitative analysis is then made of an ideal cycle in which air, treated as an ideal gas with constant heat capacities, is the working medium.

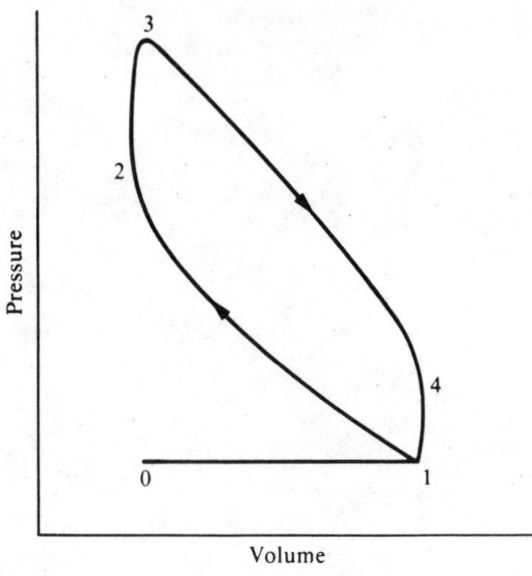

Figure 8.8 Otto internal-combustion-engine cycle.

8.3 THE OTTO ENGINE

The most common internal-combustion engine, because of its use in automobiles, is the Otto engine. Its cycle consists of four strokes, and starts with an intake stroke at essentially constant pressure, during which a piston moving outward draws a fuel/air mixture into a cylinder. This is represented by line $0 \rightarrow 1$ in Fig. 8.8. During the second stroke (line $1 \rightarrow 3$), all valves are closed, and the fuel/air mixture is compressed, approximately adiabatically, along line $1 \rightarrow 2$. The mixture is then ignited, and combustion occurs so rapidly that the volume remains nearly constant while the pressure rises along line $2 \rightarrow 3$. It is during the third stroke (line $3 \rightarrow 1$) that work is produced. The high-temperature, high-pressure products of combustion expand, approximately adiabatically, along line $3 \rightarrow 4$. Then the exhaust valve opens and the pressure falls rapidly at nearly constant volume along line $4 \rightarrow 1$. During the fourth or exhaust stroke (line $1 \rightarrow 0$), the piston pushes the remaining combustion gases (except for the contents of the clearance volume) from the cylinder. The volume plotted in Fig. 8.8 is the total volume of gas contained in the engine between the piston and the cylinder head.

The effect of increasing the compression ratio, defined as the ratio of the volumes at the beginning and end of the compression stroke, is to increase the efficiency of the engine, i.e., to increase the work produced per unit quantity of fuel. We demonstrate this for an idealized cycle, called the air-standard cycle, shown in Fig. 8.9. It consists of two adiabatic and two constant-volume steps, which comprise a heat-engine cycle for which air is the working fluid. In step DA, sufficient heat is absorbed by the air at constant volume to raise its temperature and pressure to the values resulting from combustion in an actual Otto engine. Then the air is expanded adiabatically and reversibly (step AB), cooled

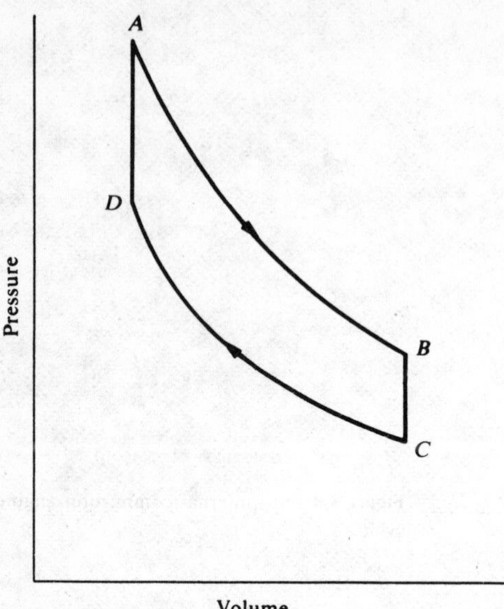

Figure 8.9 Air-standard Otto cycle.

at constant volume (step BC), and finally compressed adiabatically and reversibly to the initial state at D.

The thermal efficiency η of the air-standard cycle shown in Fig. 8.9 is simply

$$\eta = \frac{W_s(\text{net})}{Q_{DA}} = \frac{Q_{DA} + Q_{BC}}{Q_{DA}} \tag{8.3}$$

For 1 mol of air with constant heat capacities,

$$Q_{DA} = C_V(T_A - T_D)$$

$$Q_{BC} = C_V(T_C - T_B)$$

Substituting these expressions in Eq. (8.3) gives

$$\eta = \frac{C_V(T_A - T_D) - C_V(T_B - T_C)}{C_V(T_A - T_D)}$$

or

$$\eta = 1 - \frac{T_B - T_C}{T_A - T_D} \tag{8.4}$$

The thermal efficiency is also related in a simple way to the compression ratio $r = V_C / V_D$. We replace each temperature in Eq. (8.4) by an appropriate group PV/R, in accord with the ideal-gas equation. Thus

$$T_B = \frac{P_B V_B}{R} = \frac{P_B V_C}{R}$$

$$T_C = \frac{P_C V_C}{R}$$

$$T_A = \frac{P_A V_A}{R} = \frac{P_A V_D}{R}$$

$$T_D = \frac{P_D V_D}{R}$$

Substituting into Eq. (8.4) leads to

$$\eta = 1 - \frac{V_C}{V_D}\left(\frac{P_B - P_C}{P_A - P_D}\right) = 1 - r\left(\frac{P_B - P_C}{P_A - P_D}\right) \tag{8.5}$$

For the two adiabatic, reversible steps, we have PV^γ = const; whence

$$P_A V_D^\gamma = P_B V_C^\gamma \qquad \text{(since } V_D = V_A \text{ and } V_C = V_B\text{)}$$

$$P_C V_C^\gamma = P_D V_D^\gamma$$

These expressions are combined to eliminate the volumes:

$$\frac{P_B}{P_C} = \frac{P_A}{P_D}$$

Also

$$\frac{P_C}{P_D} = \left(\frac{V_D}{V_C}\right)^\gamma = \left(\frac{1}{r}\right)^\gamma$$

These equations transform Eq. (8.5) as follows:

$$\eta = 1 - r\frac{(P_B/P_C - 1)P_C}{(P_A/P_D - 1)P_D} = 1 - r\frac{P_C}{P_D}$$

or

$$\eta = 1 - r\left(\frac{1}{r}\right)^\gamma = 1 - \left(\frac{1}{r}\right)^{\gamma - 1} \tag{8.6}$$

This equation shows that the thermal efficiency increases rapidly with the compression ratio r at low values of r, but more slowly at high compression ratios. This agrees with the results of actual tests on Otto engines.

8.4 THE DIESEL ENGINE

The Diesel engine differs from the Otto engine primarily in that the temperature at the end of compression is sufficiently high that combustion is initiated spontaneously. This higher temperature results because of a higher compression ratio that carries the compression step to a higher pressure. The fuel is not injected until the end of the compression step, and then is added slowly enough that the combustion process occurs at approximately constant pressure.

For the same compression ratio, the Otto engine has a higher efficiency than the Diesel engine. However, preignition limits the compression ratio attainable in the Otto engine. The Diesel engine therefore operates at higher compression ratios, and consequently at higher efficiencies.

Example 8.3 Sketch the air-standard Diesel cycle on a *PV* diagram, and derive an equation giving the thermal efficiency of this cycle in relation to the compression ratio r_c (ratio of volumes at the beginning and end of the compression step) and the expansion ratio r_e (ratio of volumes at the end and beginning of the expansion step).

SOLUTION The air-standard Diesel cycle is the same as the air-standard Otto cycle, except that the heat-absorption step (corresponding to the combustion process in the actual engine) is at constant pressure, as indicated by line *DA* in Fig. 8.10.

On the basis of one mol of air, considered to be an ideal gas with constant heat capacities, the heat absorbed in the cycle is

$$Q_{DA} = C_P(T_A - T_D)$$

The heat rejected in step *BC* is

$$Q_{BC} = C_V(T_C - T_B)$$

By an energy balance, $W_s = Q_{DA} + Q_{BC}$, and the thermal efficiency is given by

$$\eta = 1 - \frac{C_V}{C_P}\left(\frac{T_B - T_C}{T_A - T_D}\right) = 1 - \frac{1}{\gamma}\left(\frac{T_B - T_C}{T_A - T_D}\right) \tag{A}$$

For reversible, adiabatic expansion (step *AB*) and reversible, adiabatic compression (step *CD*), Eq. (3.22) applies:

$$T_A(V_A)^{\gamma-1} = T_B(V_B)^{\gamma-1}$$

and

$$T_D(V_D)^{\gamma-1} = T_C(V_C)^{\gamma-1}$$

By definition, the compression ratio is $r = V_C/V_D$; in addition the expansion ratio

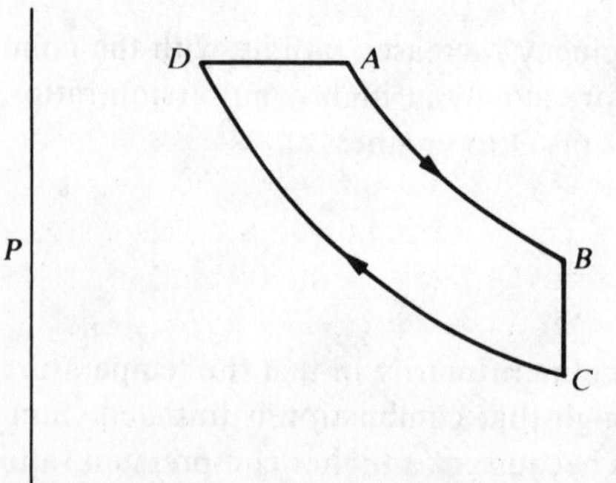

Figure 8.10 Air-standard Diesel cycle.

is defined as $r_e = V_B/V_A$. Thus

$$T_B = T_A\left(\frac{1}{r_e}\right)^{\gamma-1} \tag{B}$$

and

$$T_C = T_D\left(\frac{1}{r}\right)^{\gamma-1} \tag{C}$$

Substituting Eqs. (B) and (C) into Eq. (A) gives

$$\eta = 1 - \frac{1}{\gamma}\left[\frac{T_A(1/r_e)^{\gamma-1} - T_D(1/r)^{\gamma-1}}{T_A - T_D}\right] \tag{D}$$

Also, $P_A = P_D$, and from the ideal-gas equation,

$$P_D V_D = RT_D \qquad \text{and} \qquad P_A V_A = RT_A$$

Hence

$$\frac{T_D}{T_A} = \frac{V_D}{V_A} = \frac{V_D/V_C}{V_A/V_B} = \frac{r_e}{r}$$

This relation combines with Eq. (D) to give:

$$\eta = 1 - \frac{1}{\gamma}\left[\frac{(1/r_e)^{\gamma-1} - (r_e/r)(1/r)^{\gamma-1}}{1 - r_e/r}\right]$$

or

$$\eta = 1 - \frac{1}{\gamma}\left[\frac{(1/r_e)^{\gamma} - (1/r)^{\gamma}}{1/r_e - 1/r}\right] \tag{8.7}$$

8.5 THE GAS-TURBINE POWER PLANT

Consideration of the Otto and Diesel engines has shown that direct use of the energy of high-temperature and high-pressure gases, without transfer of external heat, possesses some advantages in power production. On the other hand, the turbine is more efficient than the reciprocating engine, primarily because of friction between the reciprocating piston and cylinder and because of fluid friction generated by action of the valves. The gas turbine combines in one unit the advantages of internal combustion with the advantages of the turbine.

The gas turbine is driven by high-temperature gases from a combustion space, as indicated in Fig. 8.11. The entering air is compressed (supercharged) to a pressure of several bars before combustion. The centrifugal compressor operates on the same shaft as the turbine, and part of the work of the turbine serves to drive the compressor. The unit shown in Fig. 8.11 is a complete power plant, as are Otto and Diesel engines. The gas turbine is just one part of the assembly and performs the same function as the steam turbine in a steam power plant (Fig. 8.1).

The higher the temperature of the combustion gases entering the turbine, the higher the efficiency of the unit, i.e., the greater the work produced per unit of

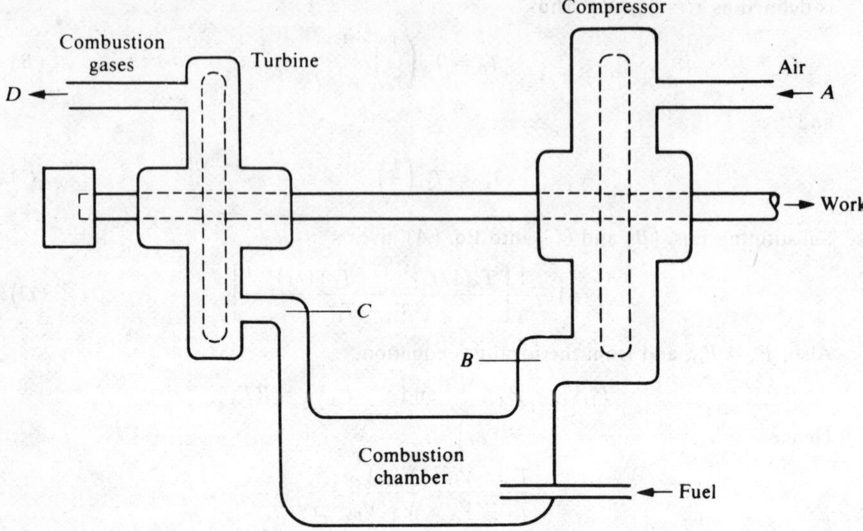

Figure 8.11 Gas-turbine power plant.

fuel burned. The limiting temperature is determined by the strength of the metal turbine blades, and is much lower than the theoretical flame temperature (Sec. 4.7) of the fuel. Sufficient excess air must be supplied to keep the combustion temperature at a safe level.

The idealization of the gas-turbine cycle (based on air, and called the Brayton cycle) is shown on a *PV* diagram in Fig. 8.12. The compression step *AB* is represented by an adiabatic, reversible (isentropic) path in which the pressure increases from P_A (atmospheric pressure) to P_B. The combustion process is replaced by the constant-pressure addition of an amount of heat Q_{BC}. Work is produced in the turbine as the result of isentropic expansion of the air to pressure

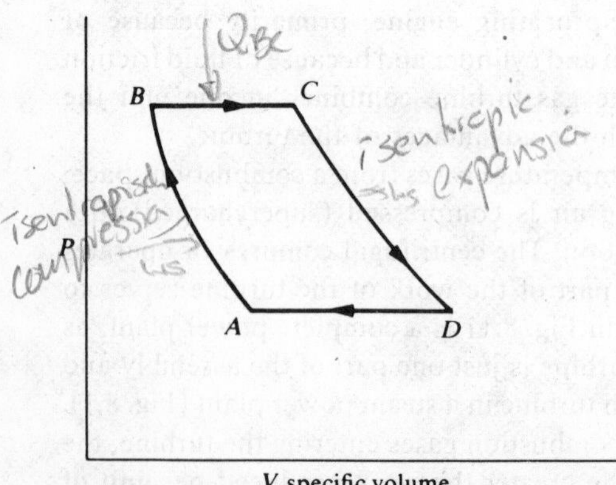

Figure 8.12 Ideal cycle for gas-turbine power plant.

P_D. Since the hot gases from the turbine are exhausted to the atmosphere, $P_D = P_A$. The thermal efficiency of the cycle is given by

$$\eta = \frac{W_s(\text{net})}{Q_{BC}} = \frac{W_{CD} + W_{AB}}{Q_{BC}} \tag{8.8}$$

where each energy quantity is based on 1 mol of air.

The work done as the air passes through the compressor is given by Eq. (7.26):

$$-W_{AB} = H_B - H_A$$

For air as an ideal gas with constant heat capacities,

$$-W_{AB} = H_B - H_A = C_P(T_B - T_A)$$

Similarly, for the combustion and turbine processes,

$$Q_{BC} = C_P(T_C - T_B)$$

$$-W_{CD} = C_P(T_D - T_C)$$

Substituting these equations into Eq. (8.8) and simplifying leads to:

$$\eta = 1 - \frac{T_D - T_A}{T_C - T_B} \tag{8.9}$$

Since processes AB and CD are isentropic, the temperatures and pressures are related as follows [Eq. (3.23)]:

$$\frac{T_B}{T_A} = \left(\frac{P_B}{P_A}\right)^{(\gamma-1)/\gamma} \tag{8.10}$$

and

$$\frac{T_D}{T_C} = \left(\frac{P_D}{P_C}\right)^{(\gamma-1)/\gamma} = \left(\frac{P_A}{P_B}\right)^{(\gamma-1)/\gamma} \tag{8.11}$$

With these equations, T_A and T_D may be eliminated to give:

$$\eta = 1 - \left(\frac{P_A}{P_B}\right)^{(\gamma-1)/\gamma} \tag{8.12}$$

Example 8.4 A gas-turbine power plant operates with a pressure ratio P_B/P_A of 6. The temperature of the air entering the compressor is 25°C, and the maximum permissible temperature in the turbine is 760°C.

(*a*) What is the efficiency of the reversible ideal-gas cycle for these conditions if $\gamma = 1.4$?

(*b*) If the compressor and turbine operate adiabatically but irreversibly with efficiencies $\eta_c = 0.83$ and $\eta_t = 0.86$, what is the thermal efficiency of the power plant for the given conditions?

SOLUTION

(*a*) Direct substitution in Eq. (8.12) gives the ideal-cycle efficiency:

$$\eta = 1 - (1/6)^{(1.4-1)/1.4} = 1 - 0.60 = 0.40$$

(*b*) Irreversibilities in the compressor and turbine greatly reduce the thermal efficiency of the power plant, because the net work is the difference between the work required by the compressor and the work produced by the turbine. The temperature of the air entering the compressor T_A and the temperature of the air entering the turbine, the specified maximum for T_C, are the same as for the ideal cycle. However, the temperature after irreversible compression in the compressor T_B is higher than the temperature after *isentropic* compression T'_B, and the temperature after irreversible expansion in the turbine T_D is higher than the temperature after *isentropic* expansion T'_D.

The work required by the compressor is

$$-W(\text{comp}) = C_P(T_B - T_A) \tag{A}$$

Alternatively, this may be found from the isentropic work:

$$-W(\text{comp}) = \frac{C_P(T'_B - T_A)}{\eta_c} \tag{B}$$

Similarly, the work produced by the turbine is

$$W(\text{turb}) = -C_P(T_D - T_C) = -C_P\eta_t(T'_D - T_C) \tag{C}$$

and the heat absorbed in place of combustion is

$$Q = C_P(T_C - T_B) \tag{D}$$

These equations are combined to give the thermal efficiency of the power plant:

$$\eta = \frac{W(\text{comp}) + W(\text{turb})}{Q} = \frac{-[(T'_B - T_A)/\eta_c] + \eta_t(T_C - T'_D)}{T_C - T_B}$$

Combining Eqs. (*A*) and (*B*) and using the result to eliminate T_B from this equation gives after simplification:

$$\eta = \frac{-(T'_B/T_A - 1) + \eta_t\eta_c(T_C/T_A - T'_D/T_A)}{\eta_c(T_C/T_A - 1) - (T'_B/T_A - 1)} \tag{E}$$

The temperature ratio T'_B/T_A is related to the pressure ratio by Eq. (8.10). The ratio T_C/T_A depends on given conditions. In view of Eq. (8.11), the ratio T'_D/T_A can be written:

$$\frac{T'_D}{T_A} = \frac{T_C T'_D}{T_A T_C} = \frac{T_C}{T_A}\left(\frac{P_A}{P_B}\right)^{(\gamma-1)/\gamma}$$

Substituting these expressions in Eq. (*E*) gives

$$\eta = \frac{\eta_t\eta_c(T_C/T_A)(1 - 1/\alpha) - (\alpha - 1)}{\eta_c(T_C/T_A - 1) - (\alpha - 1)} \tag{8.13}$$

where

$$\alpha = \left(\frac{P_B}{P_A}\right)^{(\gamma-1)/\gamma}$$

It can be shown from Eq. (8.13) that the thermal efficiency of the gas-turbine power plant increases as the temperature of the air entering the turbine (T_C) increases, and also as the compressor and turbine efficiencies η_c and η_t increase.

The given efficiency values are here

$$\eta_t = 0.86 \qquad \text{and} \qquad \eta_c = 0.83$$

Other given data provide:

$$\frac{T_C}{T_A} = \frac{760 + 273.15}{25 + 273.15} = 3.47$$

and

$$\alpha = (6)^{(1.4-1)/1.4} = 1.67$$

Substituting these quantities in Eq. (8.13) gives

$$\eta = \frac{(0.86)(0.83)(3.47)(1 - 1/1.67) - (1.67 - 1)}{(0.83)(3.47 - 1) - (1.67 - 1)} = 0.235$$

This analysis shows that, even with a compressor and turbine of rather high efficiencies, the thermal efficiency (23.5 percent) is considerably reduced from the ideal-cycle value of 40 percent.

8.6 JET ENGINES; ROCKET ENGINES

In the power cycles considered up to this point the high-temperature, high-pressure gas has been expanded in a turbine (steam power plant, gas turbine) or in the cylinder of a reciprocating Otto or Diesel engine. In either case, the power becomes available through a rotating shaft. Another device for expanding the hot gases is a nozzle. Here the power is available as kinetic energy in the jet of exhaust gases leaving the nozzle. The entire power plant, consisting of a compression device and a combustion chamber, as well as a nozzle, is known as a jet engine. Since the kinetic energy of the exhaust gases is directly available for propelling the engine and its attachments, jet engines are most commonly used to power aircraft. There are several types of jet-propulsion engines based on different ways of accomplishing the compression and expansion processes. Since the air striking the engine has kinetic energy (with respect to the engine), its pressure may be increased in a diffuser.

The turbojet engine illustrated in Fig. 8.13 takes advantage of a diffuser to reduce the work of compression. The axial-flow compressor completes the job of compression, and then the fuel is injected and burned in the combustion chamber. The hot combustion-product gases first pass through a turbine where the expansion provides just enough power to drive the compressor. The remainder of the expansion to the exhaust pressure is accomplished in the nozzle. Here, the velocity of the gases with respect to the engine is increased to a level above that of the entering air. This increase in velocity provides a thrust (force) on the engine in the forward direction. If the compression and expansion processes are adiabatic and reversible, the turbojet-engine cycle is identical to the ideal gas-turbine-power-plant cycle shown in Fig. 8.11. The only differences are that, physically, the compression and expansion steps are carried out in devices of different types.

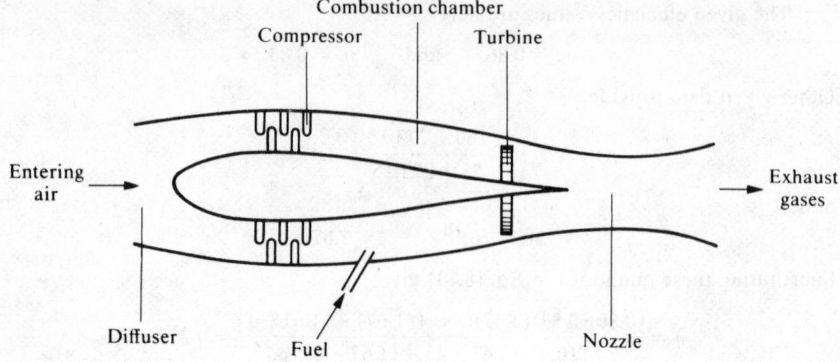

Figure 8.13 The turbojet power plant.

A rocket engine differs from a jet engine in that the oxidizing agent is carried with the engine. Instead of depending on the surrounding air for burning the fuel, the rocket is self-contained. This means that the rocket operates in vacuum such as in space. In fact, the performance is better in a vacuum, because none of the thrust is required to overcome friction forces.

In rockets burning liquid fuels the oxidizing agent (e.g., liquid oxygen) is pumped from tanks into the combustion chamber. Simultaneously, fuel (e.g., kerosene) is pumped into the chamber and burned. The combustion takes place at a constant high pressure and produces high-temperature product gases that are expanded in a nozzle, as indicated in Fig. 8.14.

In rockets burning solid fuels the fuel (organic polymers) and oxidizer (e.g., ammonium perchlorate) are contained together in a solid matrix and stored at the forward end of the combustion chamber.

In an ideal rocket, the combustion and expansion steps are the same as those for an ideal jet engine (Fig. 8.12). A solid-fuel rocket requires no compression

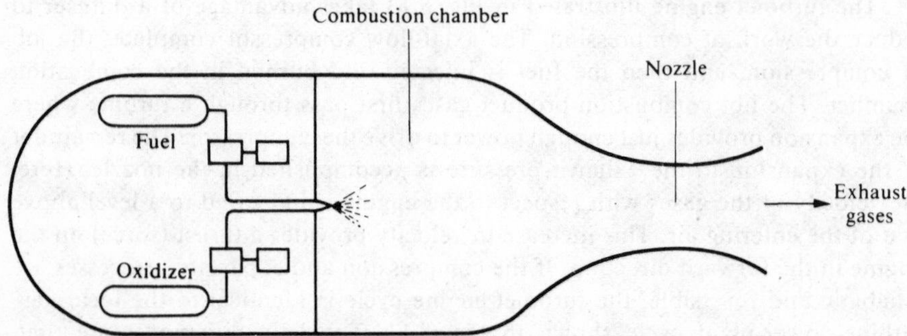

Figure 8.14 Liquid-fuel rocket engine.

work, and in a liquid-fuel rocket the compression energy is small, since the fuel and oxidizer are pumped as liquids.

PROBLEMS

8.1 The basic cycle for a steam power plant is shown by Fig. 8.1. Suppose that the turbine operates adiabatically with inlet steam at 6,500 kPa and 525°C and that the exhaust steam enters the condenser at 100°C with a quality of 0.98. Saturated liquid water leaves the condenser, and is pumped to the boiler. Neglecting pump work and kinetic- and potential-energy changes, determine the thermal efficiency of the cycle and the turbine efficiency.

8.2 A steam power plant operates on the cycle of Fig. 8.4. For one of the following sets of operating conditions, determine the steam rate, the heat-transfer rates in the boiler and condenser, and the thermal efficiency of the plant.

(*a*) $P_1 = P_2 = 10{,}000$ kPa; $T_2 = 600$°C; $P_3 = P_4 = 10$ kPa; η(turbine) = 0.80; η(pump) = 0.75; power rating = 80,000 kW.

(*b*) $P_1 = P_2 = 7{,}000$ kPa; $T_2 = 550$°C; $P_3 = P_4 = 20$ kPa; η(turbine) = 0.75; η(pump) = 0.75; power rating = 100,000 kW.

(*c*) $P_1 = P_2 = 8{,}500$ kPa; $T_2 = 600$°C; $P_3 = P_4 = 10$ kPa; η(turbine) = 0.80; η(pump) = 0.80; power rating = 70,000 kW.

(*d*) $P_1 = P_2 = 6{,}500$ kPa; $T_2 = 525$°C; $P_3 = P_4 = 101.33$ kPa; η(turbine) = 0.78; η(pump) = 0.75; power rating = 50,000 kW.

(*e*) $P_1 = P_2 = 950$(psia); $T_2 = 1{,}000$(°F); $P_3 = P_4 = 14.7$(psia); η(turbine) = 0.78; η(pump) = 0.75; power rating = 50,000 kW.

(*f*) $P_1 = P_2 = 1{,}450$(psia); $T_2 = 1{,}100$(°F); $P_3 = P_4 = 1$(psia); η(turbine) = 0.80; η(pump) = 0.75; power rating = 80,000 kW.

8.3 Steam enters the turbine of a power plant operating on the Rankine cycle (Fig. 8.3) at 3,500 kPa and exhausts at 20 kPa. To show the effect of superheating on the performance of the cycle, calculate the thermal efficiency of the cycle and the quality of the exhaust steam from the turbine for turbine-inlet steam temperatures of 400, 500, and 600°C.

8.4 Steam enters the turbine of a power plant operating on the Rankine cycle (Fig. 8.3) at 450°C and exhausts at 20 kPa. To show the effect of boiler pressure on the performance of the cycle, calculate the thermal efficiency of the cycle and the quality of the exhaust steam from the turbine for boiler pressures of 4,000, 6,000, 8,000 and 10,000 kPa.

8.5 A steam power plant employs two adiabatic turbines in series. Steam enters the first turbine at 600°C and 6,500 kPa and discharges from the second turbine at 10 kPa. The system is designed for equal power outputs from the two turbines, based on a turbine efficiency of 76 percent for *each* turbine. Determine the temperature and pressure of the steam in its intermediate state between the two turbines. What is the overall efficiency of the two turbines together with respect to isentropic expansion of the steam from the initial to the final state?

8.6 A steam power plant operating on a regenerative cycle, as illustrated in Fig. 8.5, includes just one feedwater heater. Steam enters the turbine at 4,000 kPa and 450°C and exhausts at 20 kPa. Steam for the feedwater heater is extracted from the turbine at 300 kPa, and in condensing raises the temperature of the feedwater to within 6°C of its condensation temperature at 300 kPa. If the turbine and pump efficiencies are both 78 percent, what is the thermal efficiency of the cycle and what fraction of the steam entering the turbine is extracted for the feedwater heater?

8.7 A steam power plant operating on a regenerative cycle, as illustrated in Fig. 8.5, includes just one feedwater heater. Steam enters the turbine at 600(psia) and 850(°F) and exhausts at 1(psia). Steam for the feedwater heater is extracted from the turbine at 45(psia), and in condensing raises the temperature of the feedwater to within 11(°F) of its condensation temperature at 45(psia). If the turbine and pump efficiencies are both 78 percent, what is the thermal efficiency of the cycle and what fraction of the steam entering the turbine is extracted for the feedwater heater?

8.8 A steam power plant operating on a regenerative cycle, as illustrated in Fig. 8.5, includes two feedwater heaters. Steam enters the turbine at 6,000 kPa and 500°C and exhausts at 10 kPa. Steam for the feedwater heaters is extracted from the turbine at pressures such that the feedwater is heated to 180°C in two equal increments of temperature rise, with 5-°C approaches to the steam-condensation temperature in each feedwater heater. If the turbine and pump efficiencies are both 80 percent, what is the thermal efficiency of the cycle and what fraction of the steam entering the turbine is extracted for each feedwater heater?

8.9 A power plant operating on heat recovered from the exhaust gases of internal-combustion engines uses isobutane as the working medium in a modified Rankine cycle in which the upper pressure level is above the critical pressure of isobutane. Thus the isobutane does not undergo a change of phase as it absorbs heat prior to its entry into the turbine. Isobutane vapor is heated at 4,800 kPa to 260°C, and enters the turbine as a supercritical fluid at these conditions. Isentropic expansion in the turbine produces superheated vapor at 450 kPa, which is cooled and condensed at constant pressure. The resulting saturated liquid enters the pump for return to the heater. If the power output of the modified Rankine cycle is 1,000 kW, what is the isobutane flow rate, the heat-transfer rates in the heater and condenser, and the thermal efficiency of the cycle?

The vapor pressure of isobutane is given by:

$$\ln P/\text{kPa} = 14.57100 - \frac{2{,}606.775}{t/°\text{C} + 274.068}$$

8.10 A power plant operating on heat from a geothermal source uses isobutane as the working medium in a Rankine cycle. Isobutane is heated at 3,400 kPa (a pressure just a little below its critical pressure) to a temperature of 140°C, at which conditions it enters the turbine. Isentropic expansion in the turbine produces superheated vapor at 450 kPa, which is cooled and condensed at constant pressure. The resulting saturated liquid enters the pump for return to the heater/boiler. If the flow rate of isobutane is 75 kg s^{-1}, what is the power output of the Rankine cycle and what are the heat-transfer rates in the heater/boiler and condenser? What is the thermal efficiency of the cycle?

The vapor pressure of isobutane is given in the preceding problem.

8.11 Show that the thermal efficiency of the air-standard Diesel cycle can be expressed as

$$\eta = 1 - \left(\frac{1}{r}\right)^{\gamma-1} \frac{r_c^{\gamma} - 1}{\gamma(r_c - 1)}$$

where r is the compression ratio and r_c is the *cutoff ratio*, defined as $r_c = V_A/V_D$. (See Fig. 8.10.)

Show that for the same compression ratio the thermal efficiency of the air-standard Otto engine is greater than the thermal efficiency of the air-standard Diesel cycle.

Hint: Show that the fraction which multiplies $(1/r)^{\gamma-1}$ in the above equation for η is greater than unity by expanding r_c^{γ} in a Taylor's series with remainder taken to the first derivative.

If $\gamma = 1.4$, how does the thermal efficiency of an air-standard Otto cycle with a compression ratio of 8 compare with the thermal efficiency of an air-standard Diesel cycle with the same compression ratio and a cutoff ratio of 2? How is the comparison changed if the cutoff ratio is 3?

8.12 An air-standard Diesel cycle absorbs 1,500 J mol^{-1} of heat (step *DA* of Fig. 8.10, which simulates combustion). The pressure and temperature at the beginning of the compression step are 1 bar and 20°C, and the pressure at the end of the compression step is 4 bar. Assuming air to be an ideal gas for which $C_P = (7/2)R$ and $C_V = (5/2)R$, what are the compression ratio and the expansion ratio of the cycle?

8.13 Calculate the efficiency for an air-standard gas-turbine cycle (the Brayton cycle) operating with a pressure ratio of 3. Repeat for pressure ratios of 5, 7, and 9. Take $\gamma = 1.35$.

8.14 An air-standard gas-turbine cycle is modified by installation of a regenerative heat exchanger to transfer energy from the air leaving the turbine to the air leaving the compressor. In an optimum countercurrent exchanger, the temperature of the air leaving the compressor is raised to that of point *D* in Fig. 8.12, and the temperature of the gas leaving the turbine is cooled to that of point *B* in Fig.

8.12. Show that the thermal efficiency of this cycle is given by

$$\eta = 1 - \frac{T_A}{T_C}\left(\frac{P_B}{P_A}\right)^{(\gamma-1)/\gamma}$$

8.15 Consider an air-standard cycle for representing the turbojet power plant shown in Fig. 8.13. The temperature and pressure of the air entering the compressor are 1 bar and 30°C. The pressure ratio in the compressor is 6.5, and the temperature at the turbine inlet is 1,100°C. If expansion in the nozzle is isentropic and if the nozzle exhausts at 1 bar, what is the pressure at the nozzle inlet (turbine exhaust) and what is the velocity of the air leaving the nozzle?

CHAPTER

NINE

REFRIGERATION AND LIQUEFACTION

Refrigeration is best known for its use in the air conditioning of buildings and in the treatment, transportation, and preservation of foods and beverages. It also finds large-scale industrial use, for example, in the manufacture of ice and the dehydration of gases. Applications in the petroleum industry include lubricating-oil purification, low-temperature reactions, and separation of volatile hydrocarbons. A closely related process is gas liquefaction, which has important commercial applications.

The purpose of this chapter is to present a thermodynamic analysis of refrigeration and liquefaction processes. However, the details of equipment design are left to specialized books.†

The word *refrigeration* implies the maintenance of a temperature below that of the surroundings. This requires continuous absorption of heat at a low temperature level, usually accomplished by evaporation of a liquid in a steady-state flow process. The vapor formed may be returned to its original liquid state for reevaporation in either of two ways. Most commonly, it is simply compressed and then condensed. Alternatively, it may be absorbed by a liquid of low volatility, from which it is subsequently evaporated at higher pressure. Before treating these practical refrigeration cycles, we consider the Carnot refrigerator, which provides a standard of comparison.

† *ASHRAE Handbook and Product Directory: Equipment*, 1983; *Fundamentals*, 1981; *Systems*, 1980, American Society of Heating, Refrigerating and Air-Conditioning Engineers, Brown and Briley, Atlanta.

9.1 THE CARNOT REFRIGERATOR

In a continuous refrigeration process, the heat absorbed at a low temperature must be continuously rejected to the surroundings at a higher temperature. Basically, a refrigeration cycle is a reversed heat-engine cycle. Heat is transferred from a low temperature level to a higher one; according to the second law, this cannot be accomplished without the use of external energy. The ideal refrigerator, like the ideal heat engine (Sec. 5.2), operates on a Carnot cycle, consisting in this case of two isothermal steps in which heat $|Q_C|$ is absorbed at the lower temperature T_C and heat $|Q_H|$ is rejected at the higher temperature T_H and two adiabatic steps. The cycle requires the addition of net work $|W|$ to the system. Since ΔU of the working fluid is zero for the cycle, the first law gives

$$|W| = |Q_H| - |Q_C| \tag{9.1}$$

The usual measure of performance of a refrigerator is called the *coefficient of performance* ω, defined as

$$\omega = \frac{\text{heat absorbed at the lower temperature}}{\text{net work}}$$

Thus

$$\omega \equiv \frac{|Q_C|}{|W|} \tag{9.2}$$

Division of Eq. (9.1) by $|Q_C|$ gives

$$\frac{|W|}{|Q_C|} = \frac{|Q_H|}{|Q_C|} - 1$$

But according to Eq. (5.7),

$$\frac{|Q_H|}{|Q_C|} = \frac{T_H}{T_C}$$

whence

$$\frac{|W|}{|Q_C|} = \frac{T_H}{T_C} - 1 = \frac{T_H - T_C}{T_C}$$

and Eq. (9.2) becomes

$$\omega = \frac{T_C}{T_H - T_C} \tag{9.3}$$

This equation applies only to a refrigerator operating on a Carnot cycle, and it gives the maximum possible value of ω for any refrigerator operating between given values of T_H and T_C. It shows clearly that the refrigeration effect per unit of work decreases as the temperature of the refrigerator T_C decreases and as the temperature of heat rejection T_H increases. For refrigeration at a temperature level of 5°C and a surroundings temperature of 30°C, the value of ω for a Carnot refrigerator is

$$\omega = \frac{5 + 273.15}{(30 + 273.15) - (5 + 273.15)} = 11.13$$

9.2 THE VAPOR-COMPRESSION CYCLE

A liquid evaporating at constant pressure provides a means for heat absorption at constant temperature. Likewise, condensation of the vapor, after compression to a higher pressure, provides for the rejection of heat at constant temperature. The liquid from the condenser is returned to its original state by an expansion process. This can be carried out in a turbine from which work is obtained. When compression and expansion are isentropic, this sequence of processes constitutes the cycle of Fig. 9.1*a*. It is equivalent to the Carnot cycle, except that superheated vapor from the compressor (point 3 in Fig. 9.1*a*) must be cooled to its saturation temperature before condensation begins.

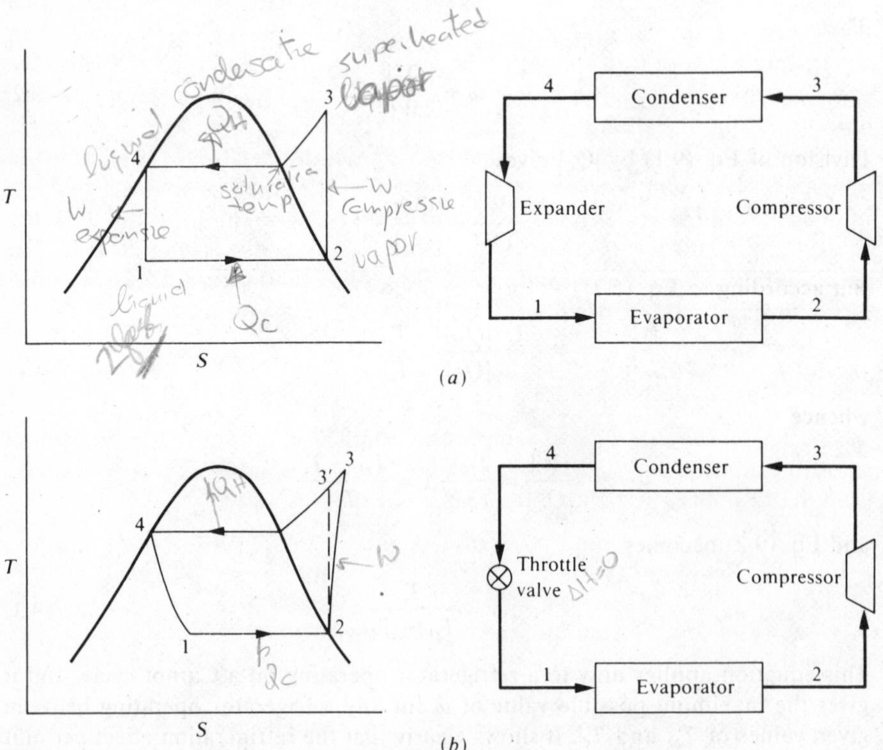

Figure 9.1 Vapor-compression refrigeration cycles.

On the basis of a unit mass of fluid, the heat absorbed in the evaporator is

$$|Q_C| = \Delta H = H_2 - H_1$$

This equation follows from Eq. (7.10) when the small changes in potential and kinetic energy are neglected. Likewise, the heat rejected in the condenser is

$$|Q_H| = H_3 - H_4$$

By Eq. (9.1),

$$|W| = (H_3 - H_4) - (H_2 - H_1)$$

and by Eq. (9.2), the coefficient of performance is

$$\omega = \frac{H_2 - H_1}{(H_3 - H_4) - (H_2 - H_1)} \tag{9.4}$$

The process requires a turbine or expander that operates on a two-phase liquid/vapor mixture. Such a machine is impractical for small units. Therefore, the cycle of Fig. 9.1*a* is used only for large installations. More commonly, expansion is accomplished by throttling the liquid from the condenser through a partly opened valve. The pressure drop in this irreversible process results from fluid friction in the valve. In small units, such as household refrigerators and air conditioners, the simplicity and lower cost of the throttle valve outweigh the energy savings possible with a turbine. As shown in Sec. 7.3, the throttling process occurs at constant enthalpy.

The vapor-compression cycle incorporating an expansion valve is shown in Fig. 9.1*b*, where line 4 → 1 represents the constant-enthalpy throttling process. Line 2 → 3, representing an actual compression process, slopes in the direction of increasing entropy, reflecting the irreversibility inherent in the process. The dashed line 2 → 3′ is the path of isentropic compression (see Fig. 7.6). For this cycle, the coefficient of performance is simply

$$\omega = \frac{H_2 - H_1}{H_3 - H_2} \tag{9.5}$$

Design of the evaporator, compressor, condenser, and auxiliary equipment requires knowledge of the rate of circulation of refrigerant $\dot{m}$. This is determined from the heat absorbed in the evaporator† by the equation:

$$\dot{m} = \frac{|Q_C|}{H_2 - H_1} \tag{9.6}$$

The vapor-compression cycle of Fig. 9.1*b* is shown on a *PH* diagram in Fig. 9.2. Such diagrams are more commonly used in refrigeration work than *TS*

† In the United States refrigeration equipment is commonly rated in *tons of refrigeration*; a ton of refrigeration is defined as heat absorption at the rate of 12,000(Btu) or 12,660 kJ per hour. This corresponds approximately to the rate of heat removal required to freeze 1(ton) of water, initially at 32(°F), per day.

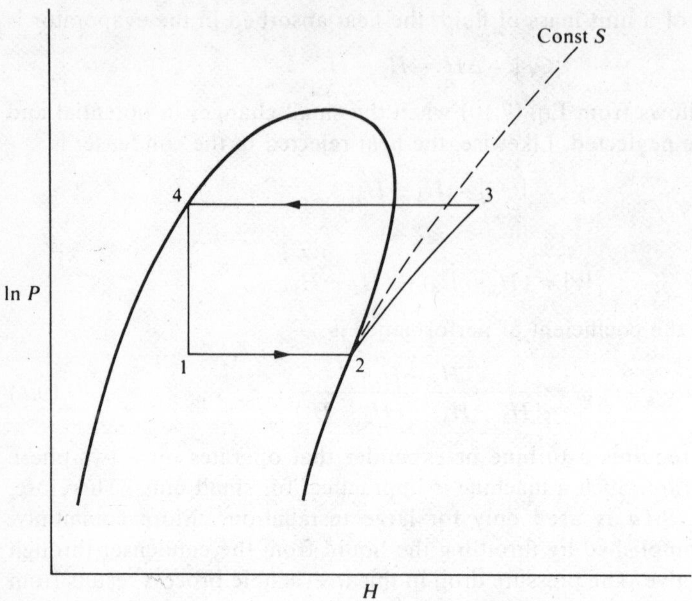

Figure 9.2 Vapor-compression refrigeration cycle on a *PH* diagram.

diagrams, because they show directly the required enthalpies. Although the evaporation and condensation processes are represented by constant-pressure paths, small pressure drops do occur because of fluid friction.

9.3 COMPARISON OF REFRIGERATION CYCLES

The effectiveness of a refrigeration cycle is measured by its coefficient of performance. For given values of T_C and T_H, the highest possible value is attained by the Carnot refrigerator. The vapor-compression cycle with reversible compression and expansion approaches this upper limit. A vapor-compression cycle with expansion in a throttle valve has a somewhat lower value, and this is reduced further when compression is not isentropic. The following example provides an indication of the magnitudes of coefficients of performance.

Example 9.1 A refrigerated space is maintained at 10(°F), and cooling water is available at 70(°F). The evaporator and condenser are of sufficient size that a 10(°F) minimum-temperature difference for heat transfer can be realized in each. The refrigeration capacity is $120{,}000(\text{Btu})(\text{hr})^{-1}$, and the refrigerant is Freon-12.

(*a*) What is the value of ω for a Carnot refrigerator?

(*b*) Calculate ω and $\dot{m}$ for the vapor-compression cycle of Fig. 9.1*a*.

(*c*) Calculate ω and $\dot{m}$ for the vapor-compression cycle of Fig. 9.1*b* if the compressor efficiency is 80 percent.

SOLUTION (*a*) For a Carnot refrigerator, Eq. (9.3) gives

$$\omega = \frac{0 + 459.67}{(80 + 459.67) - (0 + 459.67)} = 5.75$$

(*b*) Since Freon-12 is the refrigerant, the enthalpies for states 1, 2, 3, and 4 of Fig. 9.1*a* are read from Table 9.1 and Fig. 9.3. From the entry at 10 − 10 = 0(°F) in Table 9.1, we see that Freon-12 vaporizes in the evaporator at a pressure of 23.85(psia). Its properties as a saturated vapor at these conditions are:

$$H_2 = 77.27(\text{Btu})(\text{lb}_m)^{-1}$$

$$S_2 = 0.1689(\text{Btu})(\text{lb}_m)^{-1}(\text{R})^{-1}$$

From the entry at 70 + 10 = 80(°F) in Table 9.1, we find that Freon-12 condenses at 98.87(psia); its properties as a saturated liquid at these conditions are:

$$H_4 = 26.37(\text{Btu})(\text{lb}_m)^{-1}$$

$$S_4 = 0.0548(\text{Btu})(\text{lb}_m)^{-1}(\text{R})^{-1}$$

Since $S_3 = S_2 = 0.1689$, the enthalpy from Fig. 9.3 at this entropy and at a pressure of 98.87(psia) is

$$H_3 = 88.3(\text{Btu})(\text{lb}_m)^{-1}$$

State 1 is a two-phase mixture to which Eq. (6.53) applies. Written for the entropy, it is

$$S_1 = (1 - x)S^l + xS^v$$

where x is the quality (mass fraction of the mixture that is vapor). Since $S_1 = S_4 = 0.0548$, this becomes

$$0.0548 = (1 - x)(0.0193) + x(0.1689)$$

Solution for x gives

$$x = 0.2373$$

Similarly,

$$H_1 = (1 - x)H^l + xH^v$$
$$= (0.7627)(8.52) + (0.2373)(77.27) = 24.83(\text{Btu})(\text{lb}_m)^{-1}$$

Evaluation of the coefficient of performance by Eq. (9.4) gives

$$\omega = \frac{77.27 - 24.83}{(88.3 - 26.37) - (77.27 - 24.83)} = 5.53$$

By Eq. (9.6), the Freon-12 circulation rate is

$$\dot{m} = \frac{120{,}000}{77.27 - 24.83} = 2{,}288(\text{lb}_m)(\text{hr})^{-1}$$

(*c*) For the expansion step of the cycle shown in Fig. 9.1*b*, $H_1 = H_4 = 26.37(\text{Btu})(\text{lb}_m)^{-1}$. For the compression step,

$$(\Delta H)_S = (H_3 - H_2)_S = 88.3 - 77.27 = 11.03$$

Table 9.1 Thermodynamic properties of saturated Freon-12†

t (°F)	P (psia)	Volume $(ft)^3(lb_m)^{-1}$ V^l	V^v	Enthalpy $(Btu)(lb_m)^{-1}$ H^l	H^v	Entropy $(Btu)(lb_m)^{-1}(R)^{-1}$ S^l	S^v
−40	9.31	0.01056	3.875	0.00	72.91	0.0000	0.1737
−38	9.80	0.01059	3.692	0.42	73.13	0.0010	0.1734
−36	10.32	0.01067	3.520	0.84	73.35	0.0020	0.1731
−34	10.86	0.01063	3.357	1.27	73.58	0.0030	0.1729
−32	11.42	0.01065	3.204	1.69	73.80	0.0040	0.1726
−30	12.00	0.01067	3.059	2.11	74.02	0.0050	0.1723
−28	12.60	0.01070	2.921	2.54	74.23	0.0059	0.1720
−26	13.23	0.01072	2.792	2.96	74.45	0.0069	0.1718
−24	13.89	0.01074	2.669	3.38	74.67	0.0079	0.1715
−22	14.56	0.01076	2.553	3.81	74.89	0.0089	0.1713
−20	15.27	0.01079	2.443	4.24	75.11	0.0098	0.1710
−18	16.00	0.01081	2.339	4.66	75.33	0.0108	0.1708
−16	16.75	0.01083	2.240	5.09	75.55	0.0118	0.1706
−14	17.54	0.01086	2.146	5.52	75.76	0.0127	0.1703
−12	18.35	0.01088	2.057	5.94	75.98	0.0137	0.1701
−10	19.19	0.01091	1.973	6.37	76.20	0.0146	0.1699
−8	20.06	0.01093	1.892	6.80	76.41	0.0156	0.1697
−6	20.96	0.01096	1.816	7.23	76.63	0.0165	0.1695
−4	21.89	0.01098	1.744	7.66	76.84	0.0174	0.1693
−2	22.85	0.01101	1.675	8.09	77.06	0.0184	0.1691
0	23.85	0.01103	1.609	8.52	77.27	0.0193	0.1689
2	24.88	0.01106	1.546	8.95	77.49	0.0203	0.1687
4	25.94	0.01108	1.487	9.38	77.70	0.0212	0.1685
6	27.04	0.01111	1.430	9.82	77.91	0.0221	0.1683
8	28.17	0.01113	1.376	10.25	78.12	0.0230	0.1682
10	29.34	0.01116	1.324	10.68	78.34	0.0240	0.1680
12	30.54	0.01119	1.275	11.12	78.55	0.0249	0.1678
14	31.78	0.01121	1.228	11.55	78.76	0.0258	0.1677
16	33.06	0.01124	1.183	11.99	78.97	0.0267	0.1675
18	34.38	0.01127	1.140	12.43	79.18	0.0276	0.1673
20	35.74	0.01130	1.099	12.86	79.39	0.0285	0.1672
22	37.14	0.01132	1.060	13.30	79.59	0.0294	0.1670
24	38.57	0.01135	1.022	13.74	79.80	0.0303	0.1669
26	40.06	0.01138	0.986	14.18	80.01	0.0312	0.1668
28	41.58	0.01141	0.952	14.62	80.21	0.0321	0.1666
30	43.15	0.01144	0.919	15.06	80.42	0.0330	0.1665
32	44.76	0.01147	0.887	15.50	80.62	0.0339	0.1664
34	46.42	0.01150	0.857	15.94	80.83	0.0348	0.1662
36	48.12	0.01153	0.828	16.38	81.03	0.0357	0.1661
38	49.87	0.01156	0.800	16.83	81.23	0.0366	0.1660
40	51.67	0.01159	0.774	17.27	81.44	0.0375	0.1659
42	53.51	0.01162	0.748	17.72	81.64	0.0383	0.1657
44	55.41	0.01165	0.723	18.16	81.84	0.0392	0.1656
46	57.35	0.01168	0.700	18.61	82.04	0.0401	0.1655
48	59.35	0.01171	0.677	19.06	82.24	0.0410	0.1654
50	61.39	0.01175	0.655	19.51	82.44	0.0418	0.1653
52	63.49	0.01178	0.634	19.96	82.63	0.0427	0.1652
54	65.65	0.01181	0.614	20.41	82.83	0.0436	0.1651
56	67.85	0.01185	0.595	20.86	83.02	0.0444	0.1650
58	70.12	0.01188	0.576	21.31	83.22	0.0453	0.1649

† Reprinted by permission. Courtesy of E. I. du Pont de Nemours and Co., copyright 1967.

Table 9.1 (*Continued*)

t (°F)	P (psia)	Volume $(ft)^3(lb_m)^{-1}$ V^l	V^v	Enthalpy $(Btu)(lb_m)^{-1}$ H^l	H^v	Entropy $(Btu)(lb_m)^{-1}(R)^{-1}$ S^l	S^v
60	72.75	0.01191	0.5584	21.77	83.41	0.0462	0.1648
62	74.81	0.01195	0.5411	22.22	83.60	0.0470	0.1647
64	77.24	0.01198	0.5245	22.68	83.79	0.0479	0.1646
66	79.73	0.01202	0.5085	23.13	83.98	0.0488	0.1645
68	82.28	0.01205	0.4931	23.59	84.17	0.0496	0.1644
70	84.89	0.01209	0.4782	24.05	84.36	0.0505	0.1643
72	87.56	0.01213	0.4638	24.51	84.55	0.0513	0.1643
74	90.29	0.01216	0.4500	24.97	84.73	0.0522	0.1642
76	93.09	0.01220	0.4367	25.44	84.92	0.0530	0.1641
78	95.95	0.01224	0.4238	25.90	85.10	0.0539	0.1640
80	98.87	0.01228	0.4114	26.37	85.28	0.0548	0.1639
82	101.86	0.01232	0.3994	26.83	85.46	0.0556	0.1638
84	104.92	0.01236	0.3878	27.30	85.64	0.0565	0.1638
86	108.04	0.01240	0.3766	27.77	85.82	0.0573	0.1637
88	111.23	0.01244	0.3658	28.24	86.00	0.0581	0.1636
90	114.49	0.01248	0.3553	28.71	86.17	0.0590	0.1635
92	117.82	0.01252	0.3452	29.19	86.35	0.0598	0.1635
94	121.22	0.01256	0.3354	29.66	86.52	0.0607	0.1634
96	124.70	0.01261	0.3259	30.14	86.69	0.0615	0.1633
98	128.24	0.01265	0.3168	30.62	86.86	0.0624	0.1632
100	131.86	0.01269	0.3079	31.10	87.03	0.0632	0.1631
102	135.56	0.01274	0.2994	31.58	87.20	0.0640	0.1631
104	139.33	0.01278	0.2911	32.07	87.36	0.0649	0.1630
106	143.18	0.01283	0.2830	32.55	87.52	0.0658	0.1629
108	147.11	0.01288	0.2752	33.04	87.68	0.0666	0.1629
110	151.11	0.01292	0.2677	33.53	87.84	0.0675	0.1628
112	155.19	0.01297	0.2604	34.02	88.00	0.0683	0.1627
114	159.36	0.01302	0.2533	34.52	88.16	0.0691	0.1626
116	163.61	0.01307	0.2464	35.01	88.31	0.0700	0.1626
118	167.94	0.01312	0.2397	35.51	88.46	0.0708	0.1625
120	172.35	0.01317	0.2333	36.01	88.61	0.0717	0.1624
122	176.85	0.01323	0.2270	36.52	88.76	0.0725	0.1623
124	181.43	0.01328	0.2209	37.02	88.90	0.0734	0.1623
126	186.10	0.01334	0.2150	37.53	89.04	0.0742	0.1622
128	190.86	0.01339	0.2092	38.04	89.18	0.0751	0.1621
130	195.71	0.01345	0.2036	38.55	89.32	0.0759	0.1620
132	200.64	0.01350	0.1982	39.07	89.46	0.0768	0.1619
134	205.67	0.01356	0.1929	39.59	89.59	0.0776	0.1619
136	210.79	0.01362	0.1878	40.11	89.72	0.0785	0.1618
138	216.01	0.01368	0.1828	40.63	89.84	0.0793	0.1617
140	221.32	0.01375	0.1780	41.16	89.97	0.0802	0.1616
142	226.72	0.01381	0.1733	41.69	90.09	0.0811	0.1615
144	232.22	0.01387	0.1687	42.23	90.20	0.0819	0.1614
146	237.82	0.01394	0.1642	42.77	90.32	0.0828	0.1613
148	243.51	0.01401	0.1599	43.31	90.43	0.0837	0.1612
150	249.31	0.01408	0.1556	43.85	90.53	0.0845	0.1611
152	255.20	0.01415	0.1515	44.40	90.64	0.0854	0.1610
154	261.20	0.01422	0.1475	44.95	90.74	0.0863	0.1609
156	267.30	0.01430	0.1436	45.51	90.83	0.0872	0.1608
158	273.51	0.01437	0.1398	46.07	90.92	0.0880	0.1607

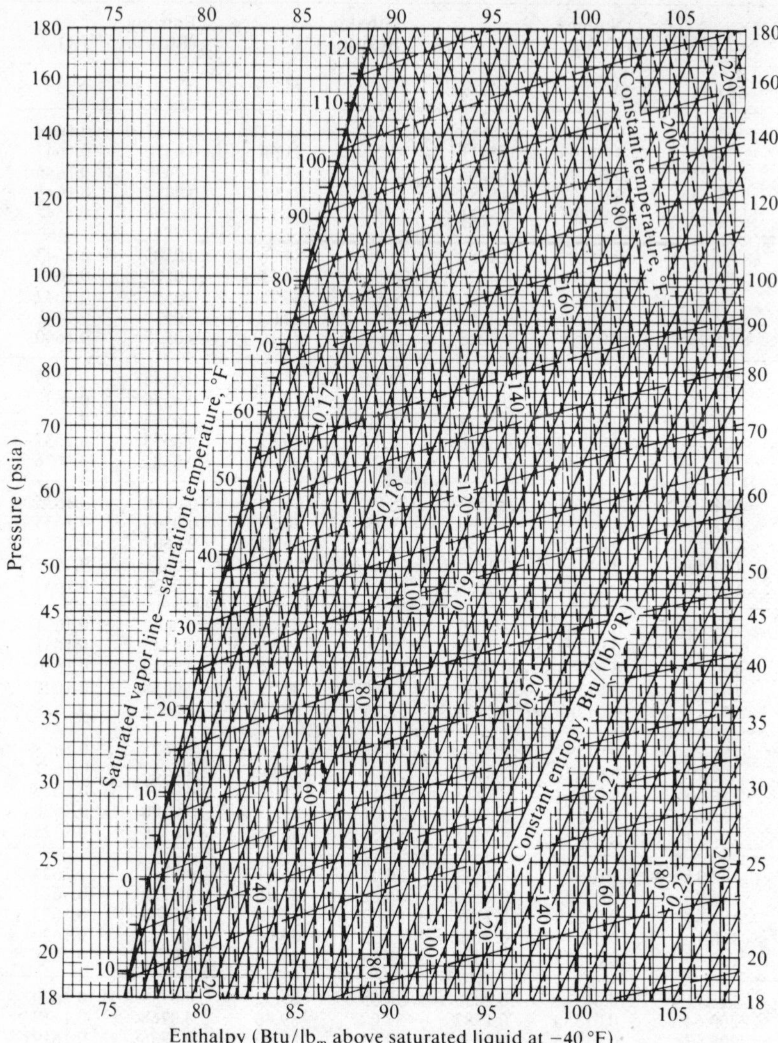

Figure 9.3 Pressure/enthalpy diagram for Freon-12. *(Reprinted by permission. Courtesy of E. I. du Pont de Nemours and Co., Copyright 1967.)*

By Eq. (7.29) for a compressor efficiency of 0.80,

$$\Delta H = H_3 - H_2 = \frac{(\Delta H)_S}{\eta} = \frac{11.03}{0.80} = 13.8(\text{Btu})(\text{lb}_\text{m})^{-1}$$

The coefficient of performance is now found from Eq. (9.5):

$$\omega = \frac{H_2 - H_1}{H_3 - H_2} = \frac{77.27 - 26.37}{13.8} = 3.69$$

The Freon-12 circulation rate is

$$\dot{m} = \frac{120{,}000}{77.27 - 26.37} = 2{,}358(\mathrm{lb_m})(\mathrm{hr})^{-1}$$

Results are summarized as follows:

Cycle	ω	$\dot{m}\ (\mathrm{lb_m})(\mathrm{hr})^{-1}$
(*a*) Carnot	5.75	
(*b*) Fig. 9.1*a*	5.53	2,288
(*c*) Fig. 9.1*b*	3.69	2,358

9.4 THE CHOICE OF REFRIGERANT

As shown in Sec. 5.2, the efficiency of a Carnot heat engine is independent of the working medium of the engine. Similarly, the coefficient of performance of a Carnot refrigerator is independent of the refrigerant. However, the irreversibilities inherent in the vapor-compression cycle cause the coefficient of performance of practical refrigerators to depend to some extent on the refrigerant. Nevertheless, such characteristics as its toxicity, flammability, cost, corrosion properties, and vapor pressure in relation to temperature are of greater importance in the choice of refrigerant. So that air cannot leak into the refrigeration system, the vapor pressure of the refrigerant at the evaporator temperature should be greater than atmospheric pressure. On the other hand, the vapor pressure at the condenser temperature should not be unduly high, because of the high initial cost and operating expense of high-pressure equipment. These two requirements limit the choice of refrigerant to relatively few fluids.† The final selection then depends on the other characteristics mentioned.

Ammonia(R-717),‡ methyl chloride(R-40), carbon dioxide(R-744), propane(R-290) and other hydrocarbons, and various halogenated hydrocarbons are used as refrigerants. Of the last, Freon-12 (dichlorodifluoromethane, also designated R-12), is widely employed in small units. Pressure/enthalpy diagrams for Freon-12 and ammonia are shown in Figs. 9.3 and 9.4, and Tables 9.1 and 9.2 provide saturation data for Freon-12 and ammonia.§ Tables and diagrams for a variety of refrigerants are given by Perry and Green.¶

† R. H. Perry and D. Green, *Perry's Chemical Engineers' Handbook*, 6th ed., table 12-6, p. 12-25, McGraw-Hill, New York, 1984.

‡ The R-designation for refrigerants is standard nomenclature of the American Society of Heating, Refrigerating, and Air-Conditioning Engineers.

§ The data for ammonia in Table 9.2 and Fig. 9.5, from *NBS Circular 142*, 1923, and adequate for instructional purposes, have been superseded by the very extensive tables of L. Haar and J. S. Gallagher, *J. Phys. Chem. Ref. Data*, 7: 635, 1978.

¶ R. H. Perry and D. Green, op. cit., sec. 3.

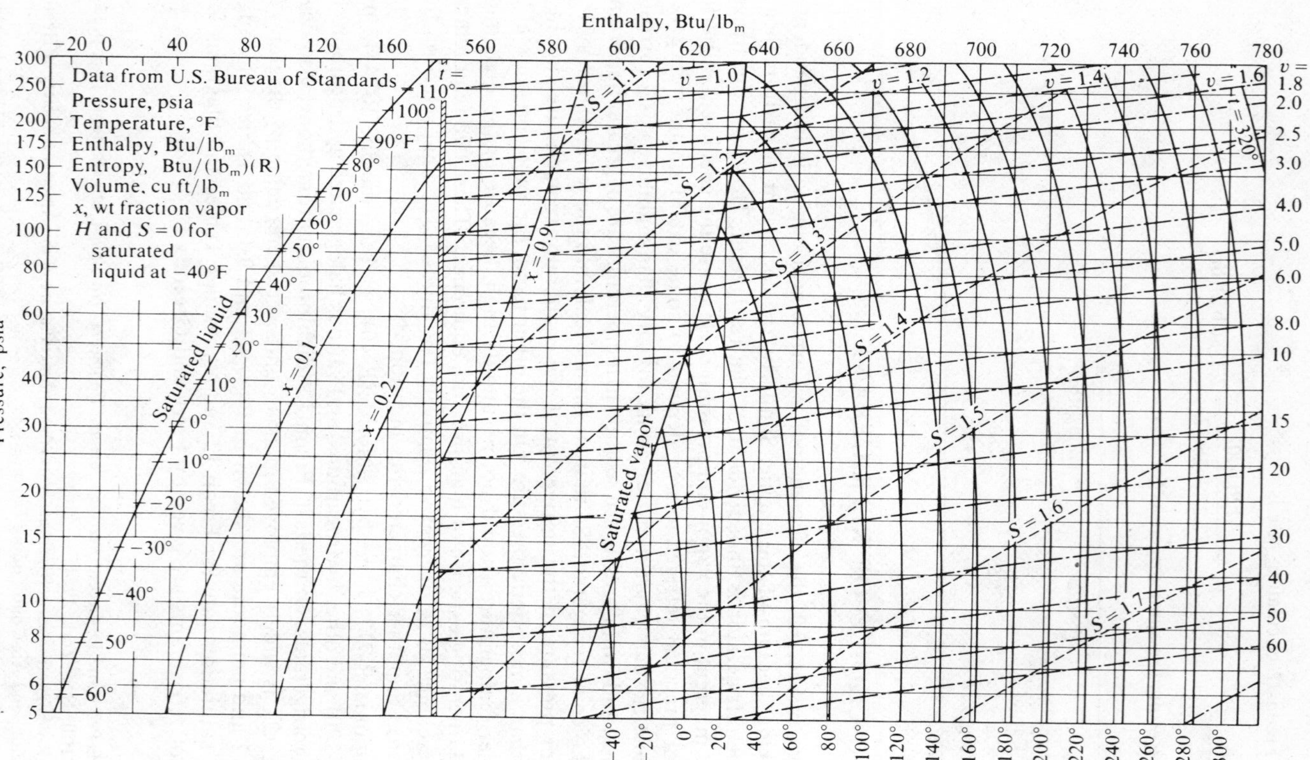

Figure 9.4 Pressure/enthalpy diagram for ammonia.

Table 9.2 Thermodynamic properties of saturated ammonia†

t (°F)	P (psia)	Volume $(ft)^3(lb_m)^{-1}$ V^l	V^v	Enthalpy $(Btu)(lb_m)^{-1}$ H^l	H^v	Entropy $(Btu)(lb_m)^{-1}(R)^{-1}$ S^l	S^v
−40	10.41	0.02322	24.86	0.0	597.6	0.0000	1.4242
−38	11.04	0.02327	23.53	2.1	598.3	0.0051	1.4193
−36	11.71	0.02331	22.27	4.3	599.1	0.0101	1.4144
−34	12.41	0.02336	21.10	6.4	599.9	0.0151	1.4096
−32	13.14	0.02340	20.00	8.5	600.6	0.0201	1.4048
−30	13.90	0.02345	18.97	10.7	601.4	0.0250	1.4001
−28	14.71	0.02349	18.00	12.8	602.1	0.0300	1.3955
−26	15.55	0.02354	17.09	14.9	602.8	0.0350	1.3909
−24	16.42	0.02359	16.24	17.1	603.6	0.0399	1.3863
−22	17.34	0.02364	15.43	19.2	604.3	0.0448	1.3818
−20	18.30	0.02369	14.68	21.4	605.0	0.0497	1.3774
−18	19.30	0.02373	13.97	23.5	605.7	0.0545	1.3729
−16	20.34	0.02378	13.29	25.6	606.4	0.0594	1.3686
−14	21.43	0.02383	12.66	27.8	607.1	0.0642	1.3643
−12	22.56	0.02388	12.06	30.0	607.8	0.0690	1.3600
−10	23.74	0.02393	11.50	32.1	608.5	0.0738	1.3558
−8	24.97	0.02398	10.97	34.3	609.2	0.0786	1.3516
−6	26.26	0.02403	10.47	36.4	609.8	0.0833	1.3474
−4	27.59	0.02408	9.991	38.6	610.5	0.0880	1.3433
−2	28.98	0.02413	9.541	40.7	611.1	0.0928	1.3393
0	30.42	0.02419	9.116	42.9	611.8	0.0975	1.3352
2	31.92	0.02424	8.714	45.1	612.4	0.1022	1.3312
4	33.47	0.02430	8.333	47.2	613.0	0.1069	1.3273
6	35.09	0.02435	7.971	49.4	613.6	0.1115	1.3234
8	36.77	0.02441	7.629	51.6	614.3	0.1162	1.3195
10	38.51	0.02446	7.304	53.8	614.9	0.1208	1.3157
12	40.31	0.02452	6.996	56.0	615.5	0.1254	1.3118
14	42.18	0.02457	6.703	58.2	616.1	0.1300	1.3081
16	44.12	0.02463	6.425	60.3	616.6	0.1346	1.3043
18	46.13	0.02468	6.161	62.5	617.2	0.1392	1.3006
20	48.21	0.02474	5.910	64.7	617.8	0.1437	1.2969
22	50.36	0.02479	5.671	66.9	618.3	0.1483	1.2933
24	52.59	0.02485	5.443	69.1	618.9	0.1528	1.2897
26	54.90	0.02491	5.227	71.3	619.4	0.1573	1.2861
28	57.28	0.02497	5.021	73.5	619.9	0.1618	1.2825
30	59.74	0.02503	4.825	75.7	620.5	0.1663	1.2790
32	62.29	0.02509	4.637	77.9	621.0	0.1708	1.2755
34	64.91	0.02515	4.459	80.1	621.5	0.1753	1.2721
36	67.63	0.02521	4.289	82.3	622.0	0.1797	1.2686
38	70.43	0.02527	4.126	84.6	622.5	0.1841	1.2652
40	73.32	0.02533	3.971	86.8	623.0	0.1885	1.2618
42	76.31	0.02539	3.823	89.0	623.4	0.1930	1.2585
44	79.38	0.02545	3.682	91.2	623.9	0.1974	1.2552
46	82.55	0.02551	3.547	93.5	624.4	0.2018	1.2519
48	85.82	0.02557	3.418	95.7	624.8	0.2062	1.2486
50	89.19	0.02564	3.294	97.9	625.2	0.2105	1.2453
52	92.66	0.02571	3.176	100.2	625.7	0.2149	1.2421
54	96.23	0.02577	3.063	102.4	626.1	0.2192	1.2389
56	99.91	0.02584	2.954	104.7	626.5	0.2236	1.2357
58	103.7	0.02590	2.851	106.9	626.9	0.2279	1.2325

† From *U.S. Natl. Bur. Stand. Circ.* 142 (1923).

(*Continued*)

Table 9.2 (*Continued*)

		Volume $(ft)^3(lb_m)^{-1}$		Enthalpy $(Btu)(lb_m)^{-1}$		Entropy $(Btu)(lb_m)^{-1}(R)^{-1}$	
t (°F)	P (psia)	V^l	V^v	H^l	H^v	S^l	S^v
60	107.6	0.02597	2.751	109.2	627.3	0.2322	1.2294
62	111.6	0.02604	2.656	111.5	627.7	0.2365	1.2262
64	115.7	0.02611	2.565	113.7	628.0	0.2408	1.2231
66	120.0	0.02618	2.477	116.0	628.4	0.2451	1.2201
68	124.3	0.02625	2.393	118.3	628.8	0.2494	1.2170
70	128.8	0.02632	2.312	120.5	629.1	0.2537	1.2140
72	133.4	0.02639	2.235	122.8	629.4	0.2579	1.2110
74	138.1	0.02646	2.161	125.1	629.8	0.2622	1.2080
76	143.0	0.02653	2.089	127.4	630.1	0.2664	1.2050
78	147.9	0.02660	2.021	129.7	630.4	0.2706	1.2020
80	153.0	0.02668	1.955	132.0	630.7	0.2749	1.1991
82	158.3	0.02675	1.892	134.3	631.0	0.2791	1.1962
84	163.7	0.02683	1.831	136.6	631.3	0.2833	1.1933
86	169.2	0.02691	1.772	138.9	631.5	0.2875	1.1904
88	174.8	0.02699	1.716	141.2	631.8	0.2917	1.1875
90	180.6	0.02707	1.661	143.5	632.0	0.2958	1.1846
92	186.6	0.02715	1.609	145.8	632.2	0.3000	1.1818
94	192.7	0.02723	1.559	148.2	632.5	0.3041	1.1789
96	198.9	0.02731	1.510	150.5	632.7	0.3083	1.1761
98	205.3	0.02739	1.464	152.9	632.9	0.3125	1.1733
100	211.9	0.02747	1.419	155.2	633.0	0.3166	1.1705
102	218.6	0.02755	1.375	157.6	633.2	0.3207	1.1677
104	225.4	0.02763	1.334	159.9	633.4	0.3248	1.1649
106	232.5	0.02772	1.293	162.3	633.5	0.3289	1.1621
108	239.7	0.02781	1.254	164.6	633.6	0.3330	1.1593
110	247.0	0.02790	1.217	167.0	633.7	0.3372	1.1566
112	254.5	0.02799	1.180	169.4	633.8	0.3413	1.1538
114	262.2	0.02808	1.145	171.8	633.9	0.3453	1.1510
116	270.1	0.02817	1.112	174.2	634.0	0.3495	1.1483
118	278.2	0.02826	1.079	176.6	634.0	0.3535	1.1455
120	286.4	0.02836	1.047	179.0	634.0	0.3576	1.1427
122	294.8	0.02845	1.017	181.4	634.0	0.3618	1.1400
124	303.4	0.02855	0.987	183.9	634.0	0.3659	1.1372

Limits placed on the operating pressures of the evaporator and condenser of a refrigeration system also limit the temperature difference $T_H - T_C$ over which a simple vapor-compression cycle can operate. With T_H fixed by the temperature of the surroundings, a lower limit is placed on the temperature level of refrigeration. This can be overcome by the operation of two or more refrigeration cycles employing different refrigerants in a *cascade.* A two-stage cascade is shown in Fig. 9.5. Here, the two cycles operate so that the heat absorbed in the interchanger by the refrigerant of the higher-temperature cycle (cycle 2) serves to condense the refrigerant in the lower-temperature cycle (cycle 1). The two refrigerants are so chosen that at the required temperature levels each cycle operates at reasonable

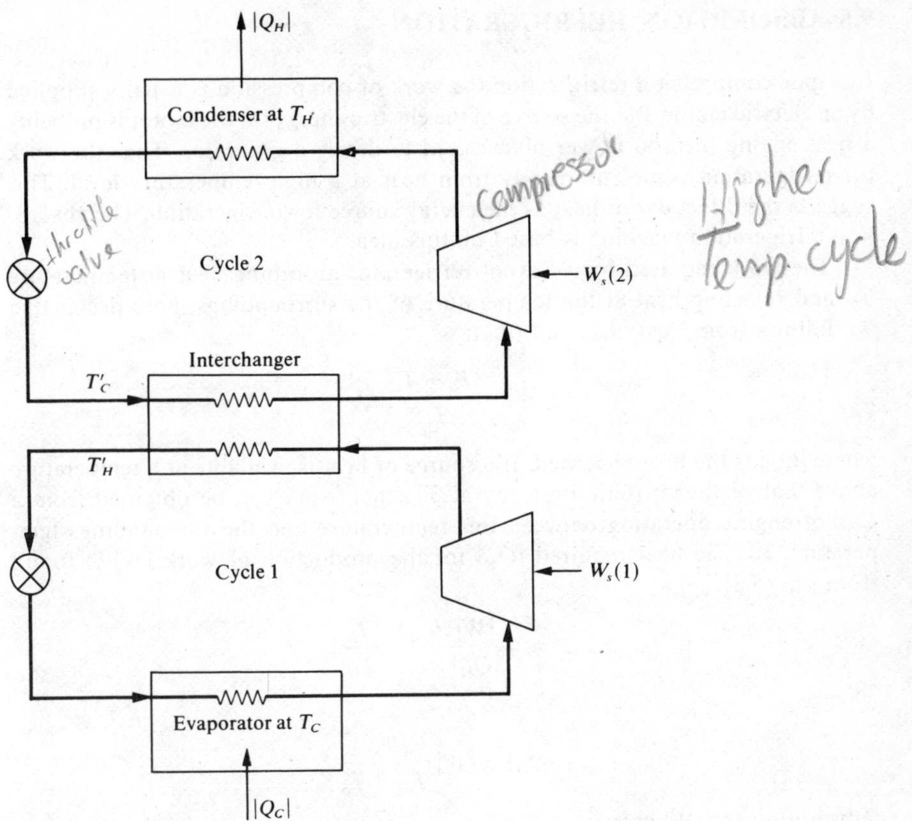

Figure 9.5 A two-stage cascade refrigeration system.

pressures. For example, let us assume the following operating temperatures

$$T_H = 86(°F)$$
$$T'_C = 0(°F)$$
$$T'_H = 10(°F)$$
$$T_C = -50(°F)$$

If Freon-12 is the refrigerant in cycle 1, then the intake and discharge pressures for the compressor are about 24(psia) and 108(psia), and the pressure ratio is about 4.5. If propylene is the refrigerant in cycle 2, these pressures are about 16 and 58(psia), and the pressure ratio is about 3.6. All of these are reasonable values. On the other hand, for a single cycle operating between −50 and 86(°F) with Freon-12 as refrigerant, the intake pressure to the condenser is about 7(psia), well below atmospheric pressure. Moreover, for a discharge pressure of about 108(psia) the pressure ratio is about 15.4, too high a value for a single-stage compressor.

9.5 ABSORPTION REFRIGERATION

In vapor-compression refrigeration the work of compression is usually supplied by an electric motor. But the source of the electric energy for the motor is probably a heat engine (central power plant) used to derive a generator. Thus the work for refrigeration comes ultimately from heat at a high temperature level. This suggests the direct use of heat as the energy source for refrigeration. The absorption-refrigeration machine is based on this idea.

The work required by a Carnot refrigerator absorbing heat at temperature T_C and rejecting heat at the temperature of the surroundings, here designated T_S, follows from Eqs. (9.2) and (9.3):

$$|W| = \frac{T_S - T_C}{T_C}|Q_C|$$

where $|Q_C|$ is the heat absorbed. If a source of heat is available at a temperature above that of the surroundings, say at T_H, then work can be obtained from a Carnot engine operating between this temperature and the surroundings temperature T_S. The heat required $|Q_H|$ for the production of work $|W|$ is found from Eq. (5.8):

$$\eta = \frac{|W|}{|Q_H|} = 1 - \frac{T_S}{T_H}$$

whence

$$|Q_H| = |W|\frac{T_H}{T_H - T_S}$$

Substitution for $|W|$ gives

$$|Q_H| = |Q_C|\frac{T_H}{T_H - T_S}\,\frac{T_S - T_C}{T_C} \tag{9.7}$$

The value of $|Q_H|/|Q_C|$ given by this equation is of course a minimum, because Carnot cycles cannot be achieved in practice.

A schematic diagram for a typical absorption refrigerator is shown in Fig. 9.6. The essential difference between a vapor-compression and an absorption refrigerator is in the different means employed for compression. The section of the absorption unit to the right of the dashed line in Fig. 9.6 is the same as in a vapor-compression refrigerator, but the section to the left accomplishes compression by what amounts to a heat engine. Refrigerant as vapor from the evaporator is absorbed in a relatively nonvolatile liquid solvent at the pressure of the evaporator and at relatively low temperature. The heat given off in the process is discarded to the surroundings at T_S. This is the lower temperature level of the heat engine. The liquid solution from the absorber, which contains a relatively high concentration of refrigerant, passes to a pump, which raises the pressure of the liquid to that of the condenser. Heat from the higher temperature source at T_H is transferred to the compressed liquid solution, raising its temperature and

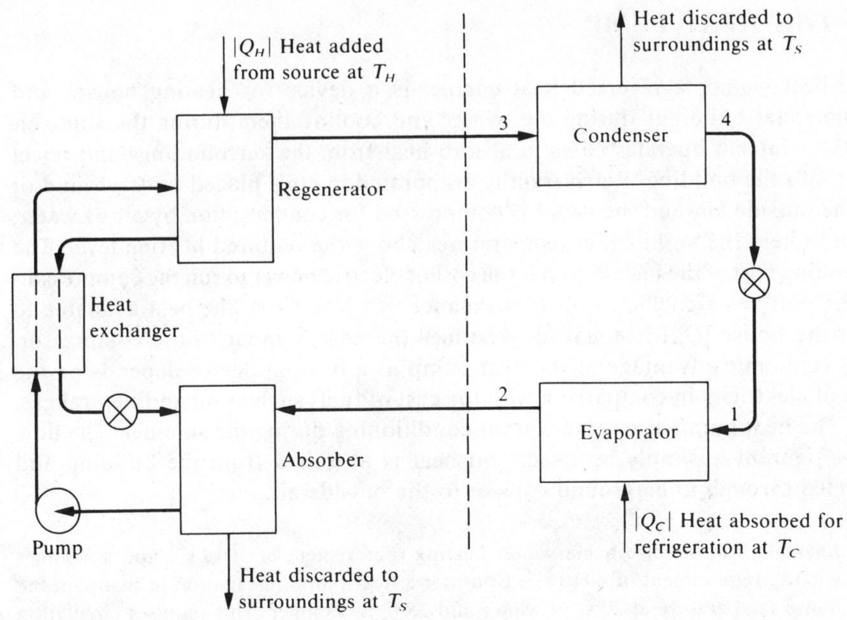

Figure 9.6 Schematic diagram of an absorption-refrigeration unit.

evaporating the refrigerant from the solvent. Vapor passes from the regenerator to the condenser, and solvent (which now contains a relatively low concentration of refrigerant) returns to the absorber. The heat exchanger conserves energy and also adjusts stream temperatures toward proper values. Low-pressure steam is the usual source of heat for the regenerator.

The most commonly used absorption-refrigeration system operates with water as the refrigerant and a lithium bromide solution as the absorbent. This system is obviously limited to refrigeration temperatures above the freezing point of water. It is treated in detail by Perry and Green.† For lower temperatures the usual system operates with ammonia as refrigerant and water as the solvent.

As an example, one might have refrigeration at a temperature level of −10°C ($T_C = 263.15$ K) and a heat source of condensing steam at atmospheric pressure ($T_H = 373.15$ K). For a surroundings temperature of 30°C ($T_S = 303.15$ K), the minimum possible value of $|Q_H|/|Q_C|$ is found from Eq. (9.7):

$$\frac{|Q_H|}{|Q_C|} = \left(\frac{373.15}{373.15 - 303.15}\right)\left(\frac{303.15 - 263.15}{263.15}\right) = 0.81$$

For an actual absorption refrigerator, the value would be on the order of three times this result.

† R. H. Perry and D. Green, op. cit., pp. 12-39–12-41.

9.6 THE HEAT PUMP

The heat pump, a reversed heat engine, is a device for heating houses and commercial buildings during the winter and cooling them during the summer. In the winter it operates so as to absorb heat from the surroundings and reject heat into the building. Refrigerant is evaporated in coils placed underground or in the outside air, and the vapor is compressed for condensation by air or water, used to heat the building, at temperatures above the required heating level. The operating cost of the installation is the cost of electric power to run the compressor. If the unit has a coefficient of performance, $|Q_C|/|W| = 4$, the heat available to heat the house $|Q_H|$ is equal to five times the energy input to the compressor. Any economic advantage of the heat pump as a heating device depends on the cost of electricity in comparison with the cost of fuels such as oil and natural gas.

The heat pump also serves for air conditioning during the summer. The flow of refrigerant is simply reversed, and heat is absorbed from the building and rejected through underground coils or to the outside air.

Example 9.2 A house has a winter heating requirement of 30 kJ s^{-1} and a summer cooling requirement of 60 kJ s^{-1}. Consider a heat-pump installation to maintain the house temperature at 20°C in winter and 25°C in summer. This requires circulation of the refrigerant through interior exchanger coils at 30°C in winter and 5°C in summer. Underground coils provide the heat source in winter and the heat sink in summer. For a year-round ground temperature of 15°C, the heat-transfer characteristics of the coils necessitate refrigerant temperatures of 10°C in winter and 25°C in summer. What are the minimum power requirements for winter heating and summer cooling?

SOLUTION The minimum power requirements are provided by a Carnot heat pump. For winter heating, the house coils are at the higher-temperature level T_H, and we know that $|Q_H| = 30$ kJ s^{-1}. Application of Eq. (5.7) gives

$$|Q_C| = |Q_H|\frac{T_C}{T_H} = 30\left(\frac{10 + 273.15}{30 + 273.15}\right) = 28.02 \text{ kJ s}^{-1}$$

This is the heat absorbed in the ground coils. By Eq. (9.1) we now have

$$|W| = |Q_H| - |Q_C| = 30 - 28.02 = 1.98 \text{ kJ s}^{-1}$$

Thus the power requirement is 1.98 kW.

For summer cooling, $|Q_C| = 60$ kJ s^{-1}, and the house coils are at the lower-temperature level T_C. Combining Eqs. (9.2) and (9.3) and solving for W, we get

$$|W| = |Q_C|\frac{T_H - T_C}{T_C}$$

Whence

$$|W| = 60\left(\frac{25 - 5}{5 + 273.15}\right) = 4.31 \text{ kJ s}^{-1}$$

The power requirement here is therefore 4.31 kW.

9.7 LIQUEFACTION PROCESSES

Liquefied gases are common for a variety of purposes. For example, liquid propane in cylinders is used as a domestic fuel, liquid oxygen is carried in rockets, natural gas is liquefied for ocean transport, and liquid nitrogen is used for low-temperature refrigeration. In addition, gas mixtures (e.g., air) are liquefied for separation into their component species by fractionation.

Liquefaction results when a gas is cooled to a temperature in the two-phase region. This may be accomplished in several ways:

1. By heat exchange at constant pressure.
2. By expansion in a turbine from which work is obtained.
3. By a throttling process.

The first method requires a heat sink at a temperature lower than that to which the gas is cooled, and is most commonly used to precool a gas prior to its liquefaction by the other two methods. An external refrigerator is required for a gas temperature below that of the surroundings.

The three methods are illustrated in Fig. 9.7. The constant-pressure path (1) approaches the two-phase region (and liquefaction) most closely for a given drop in temperature. The throttling process (3) does not result in liquefaction unless the initial state is at a high enough pressure and low enough temperature for the constant-enthalpy path to cut into the two-phase region. This does not occur when the initial state is at *A*. If the initial state is at *A′*, where the temperature is the same but the pressure is higher than at *A*, then isenthalpic expansion by path (3′) does result in the formation of liquid. The change of state from *A* to *A′* is most easily accomplished by compression of the gas to the final pressure at *B*, followed by constant-pressure cooling to *A′*. Liquefaction by isentropic expansion along path (2) may be accomplished from lower pressures (for given

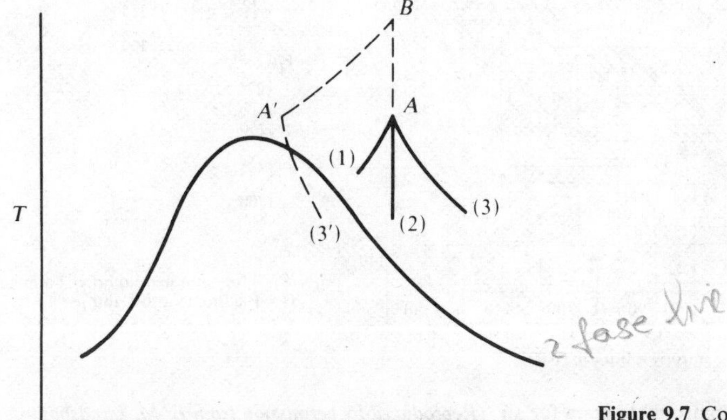

Figure 9.7 Cooling processes on a temperature/entropy diagram.

temperature) than by throttling. For example, continuation of process (2) from initial state *A* ultimately results in liquefaction.

The throttling process (3) is the one commonly employed in small-scale commercial liquefaction plants. The temperature of the gas must of course decrease during expansion. This is indeed what happens with most gases at usual conditions of temperature and pressure. The exceptions are hydrogen and helium, which increase in temperature upon throttling unless the initial temperature is below about 100 K for hydrogen and 20 K for helium. Liquefaction of these gases by throttling requires initial reduction of the temperature to lower values by method 1 or 2.

As already mentioned, the temperature must be low enough and the pressure high enough prior to throttling that the constant-enthalpy path cuts into the two-phase region. For example, reference to the *TS* diagram for air of Fig. 9.8

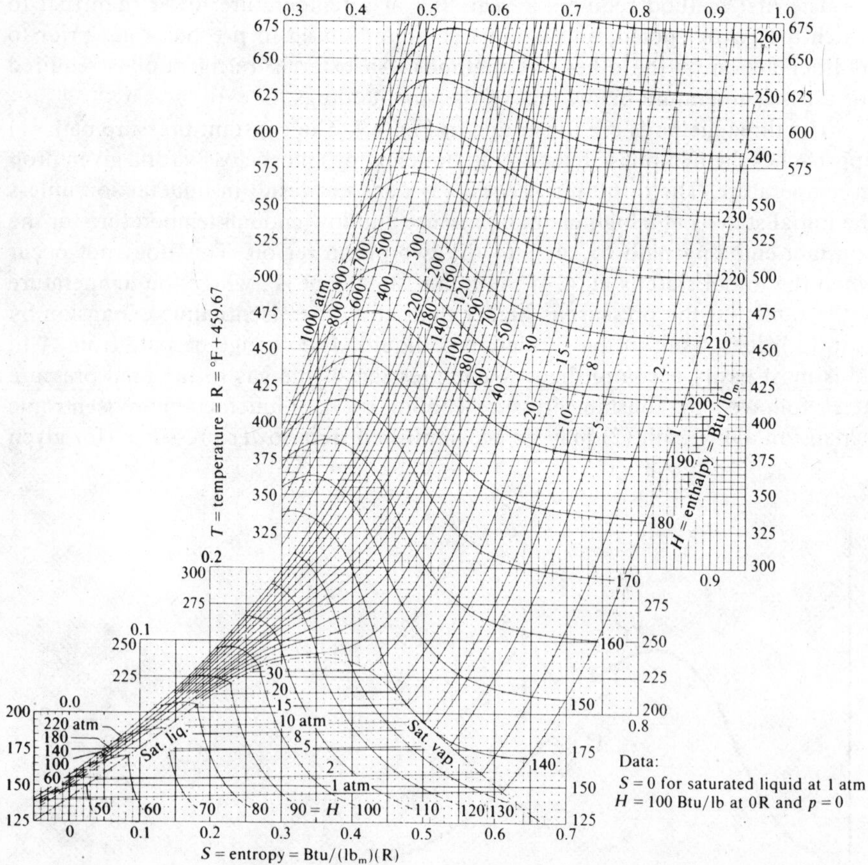

Figure 9.8 Temperature/entropy diagram for air. *(Reproduced by permission from E. M. Landsbaum, W. S. Dodds, W. F. Stevens, B. J. Sollami, and L. F. Stutzman, AIChE J.,* **1**: *392, 1955.)*

shows that at a pressure of 100(atm) the temperature must be less than 305(R) for any liquefaction to occur along a path of constant enthalpy. In other words, if air is compressed to 100(atm) and cooled to below 305(R), it can be partly liquefied by throttling. The most economical way to cool the air is by countercurrent heat exchange with the unliquefied portion of the air from the expansion process.

This simplest kind of liquefaction system, known as the Linde process, is shown in Fig. 9.9. After compression, the gas is precooled to ambient temperature. It may even be further cooled by refrigeration. The lower the temperature of the gas entering the throttle valve, the greater the fraction of gas that is liquefied. For example, evaporating a refrigerant in the precooler at −40(°F) gives a lower temperature into the valve than if water at 70(°F) is the cooling medium.

Under steady-state conditions, an energy balance [Eq. (7.8)] around the separator, valve, and cooler gives $\Delta(\dot{m}H)_{fs} = 0$, or

$$H_6 z + H_8(1 - z) = H_3 \tag{9.8}$$

where the enthalpies are for a unit mass of fluid at the positions indicated in Fig. 9.9. Knowledge of the enthalpies allows solution of Eq. (9.8) for z, the fraction of the gas that is liquefied.

The flow diagram for the Claude process, shown by Fig. 9.10, is the same as for the Linde process, except that an expansion engine or turbine replaces the

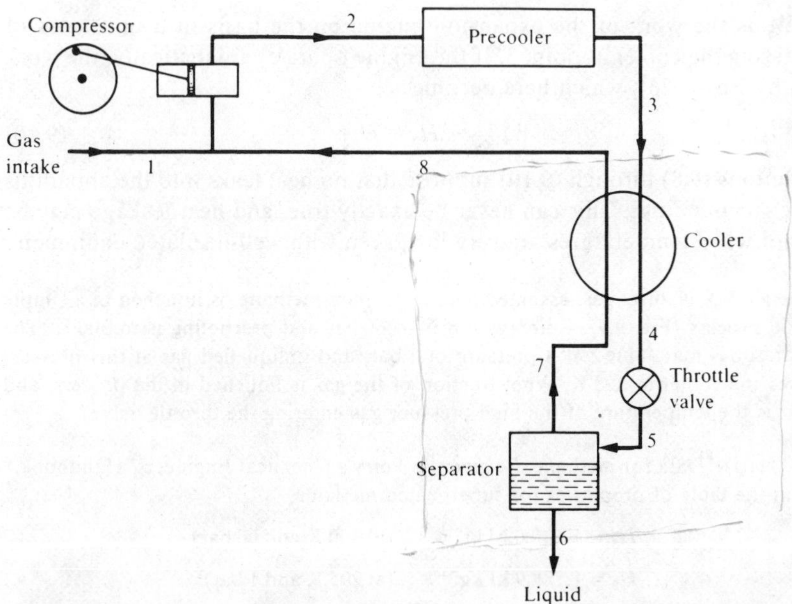

Figure 9.9 Linde liquefaction process.

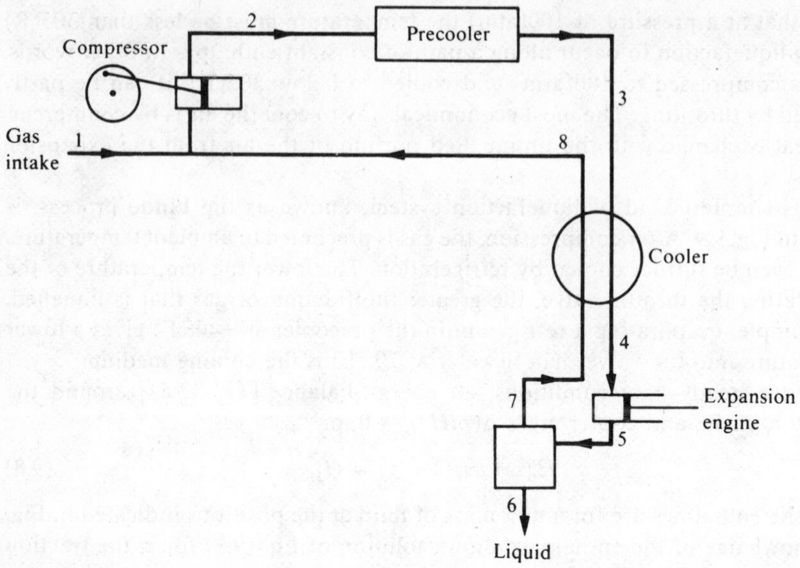

Figure 9.10 Claude liquefaction process.

throttle valve. The energy balance here becomes

$$H_6 z + H_8(1 - z) + W_s = H_3 \tag{9.9}$$

where W_s is the work of the expansion engine on the basis of a unit mass of fluid entering the cooler at point 3. If the engine operates adiabatically, the work is given by Eq. (7.26), which here becomes

$$W_s = -(H_5 - H_4) \tag{9.10}$$

Equations (9.8) through (9.10) suppose that no heat leaks into the apparatus from the surroundings. This can never be exactly true, and heat leakage may be significant when temperatures are very low, even with well-insulated equipment.

Example 9.3 Natural gas, assumed here to be pure methane, is liquefied in a simple Linde process (Fig. 9.9). Compression is to 60 bar and precooling is to 300 K. The separator is maintained at a pressure of 1 bar, and unliquefied gas at this pressure leaves the cooler at 295 K. What fraction of the gas is liquefied in the process, and what is the temperature of the high-pressure gas entering the throttle valve?

SOLUTION Data for methane are given in Perry's Chemical Engineers' Handbook.† From the table of properties for superheated methane,

$$H_3 = 1{,}140.0 \text{ kJ kg}^{-1} \qquad \text{(at 300 K and 60 bar)}$$

$$H_8 = 1{,}188.9 \text{ kJ kg}^{-1} \qquad \text{(at 295 K and 1 bar)}$$

† R. H. Perry and D. Green, op. cit., pp. 3-203.

By interpolation in the table of properties for saturated liquid and vapor, we find for a pressure of 1 bar that

$$T^{\text{sat}} = 111.45 \text{ K}$$

$$H_6 = 285.4 \text{ kJ kg}^{-1} \qquad \text{(saturated liquid)}$$

$$H_7 = 796.9 \text{ kJ kg}^{-1} \qquad \text{(saturated vapor)}$$

Solution of Eq. (9.8) for z gives

$$z = \frac{H_8 - H_3}{H_8 - H_6} = \frac{1{,}188.9 - 1{,}140.0}{1{,}188.9 - 285.4} = 0.0541$$

Thus 5.41 percent of the gas entering the throttle valve emerges as liquid.

The temperature of the gas at point 4 is found from its enthalpy, which is calculated by an energy balance around the cooler:

$$(1)(H_4 - H_3) + (1 - z)(H_8 - H_7) = 0$$

Solution for H_4 and substitution of known values yields

$$H_4 = 1{,}140.0 - (0.9459)(1{,}188.9 - 796.9) = 769.2 \text{ kJ kg}^{-1}$$

Interpolation in the tables for superheated methane at 60 bar gives the temperature of the gas entering the throttle valve as 206.5 K.

PROBLEMS

9.1 A Carnot engine is coupled to a Carnot refrigerator so that all of the work produced by the engine is used by the refrigerator in extraction of heat from a heat reservoir at 270 K at the rate of 4 kJ s^{-1}. The source of energy for the Carnot engine is a heat reservoir at 500 K. If both devices discard heat to the surroundings at 300 K, how much heat does the engine absorb from the 500-K reservoir?

If the actual coefficient of performance of the refrigerator is $\omega = \omega_{\text{Carnot}}/1.5$ and if the thermal efficiency of the engine is $\eta = \eta_{\text{Carnot}}/1.5$, how much heat does the engine absorb from the 500-K reservoir?

9.2 A refrigeration system requires 1 kW of power for a refrigeration rate of 3 kJ s^{-1}.

(*a*) What is the coefficient of performance?

(*b*) How much heat is rejected from the system?

(*c*) If heat rejection is at 35°C, what is the lowest temperature the system can possibly maintain?

9.3 A conventional vapor-compression refrigeration system operates on the cycle of Fig. 9.1*b*. For one of the following sets of operating conditions, determine the circulation rate of the refrigerant, the heat-transfer rate in the condenser, the power requirement, the coefficient of performance of the cycle, and the coefficient of performance of a Carnot refrigeration cycle operating between the same temperature levels.

(*a*) Refrigerant is ammonia; evaporation t = 30(°F);
condensation t = 90(°F); η(compressor) = 0.80;
refrigeration rate = 3,000(Btu)(s)$^{-1}$.

(*b*) Refrigerant is ammonia; evaporation t = 0(°F);
condensation t = 90(°F); η(compressor) = 0.75;
refrigeration rate = 1,500(Btu)(s)$^{-1}$.

(*c*) Refrigerant is Freon-12; evaporation t = 10(°F);
condensation t = 80(°F); η(compressor) = 0.77;
refrigeration rate = 400(Btu)(s)$^{-1}$.

(*d*) Refrigerant is Freon-12; evaporation $t = 20(°F)$;
condensation $t = 85(°F)$; η(compressor) = 0.80;
refrigeration rate = $100(\text{Btu})(\text{s})^{-1}$.

(*e*) Refrigerant is Freon-12; evaporation $t = -20(°F)$;
condensation $t = 85(°F)$; η(compressor) = 0.80;
refrigeration rate = $200(\text{Btu})(\text{s})^{-1}$.

(*f*) Refrigerant is water; evaporation $t = 4°C$;
condensation $t = 30°C$; η(compressor) = 0.76;
refrigeration rate = 1,000 kJ s^{-1}.

9.4 A refrigerator with Freon-12 as refrigerant operates with an evaporation temperature of −14(°F) and a condensation temperature of 76(°F). The saturated liquid Freon-12 from the condenser flows through an expansion valve into the evaporator, from which it emerges as saturated vapor.

(*a*) What is the circulation rate of the Freon-12 for refrigeration at the rate of $5(\text{Btu})(\text{s})^{-1}$?

(*b*) By how much would the circulation rate be reduced if the throttle valve were replaced by a turbine in which the Freon-12 expands isentropically?

(*c*) Suppose the cycle of (*a*) is modified by the inclusion of a countercurrent heat exchanger between the condenser and the throttle valve in which heat is transferred to vapor returning from the evaporator. If liquid from the condenser enters the exchanger at 76(°F) and if vapor from the evaporator enters the exchanger at −14(°F) and leaves at 65(°F), what is the circulation rate of the Freon-12?

(*d*) For each of (*a*), (*b*), and (*c*), determine the coefficient of performance for isentropic compression of the vapor.

9.5 A vapor-compression refrigeration system is conventional except that a countercurrent heat exchanger is installed to subcool the liquid from the condenser by heat exchange with the vapor stream from the evaporator. The minimum temperature difference for heat transfer is 10(°F). Ammonia is the refrigerant, evaporating at 22(°F) and condensing at 80(°F). The heat load on the evaporator is $2{,}000(\text{Btu})(\text{s})^{-1}$. If the compressor efficiency is 75 percent, what is the power requirement?

How does this result compare with the power required by the compressor if the system operates without the heat exchanger? How do the ammonia circulation rates compare for the two cases?

9.6 Consider the vapor-compression refrigeration cycle of Fig. 9.1*b* with Freon-12 as refrigerant. If the evaporation temperature is 10(°F), show the effect of condensation temperature on the coefficient of performance by making calculations for condensation temperatures of 60, 80, and 100(°F).

(*a*) Assume isentropic compression of the vapor.

(*b*) Assume a compressor efficiency of 75 percent.

9.7 A heat pump is used to heat a house in the winter and to cool it in the summer. During the winter, the outside air serves as a low-temperature heat source; during the summer, it acts as a high-temperature heat sink. The heat-transfer rate through the walls and roof of the house is 0.75 kJ s^{-1} for each °C of temperature difference between the inside and outside of the house, summer and winter. The heat-pump motor is rated at 1.5 kW. Determine the minimum outside temperature for which the house can be maintained at 20°C during the winter and the maximum outside temperature for which the house can be maintained at 20°C during the summer.

9.8 Dry air is supplied by a compressor and precooling system to the cooler of a Linde liquid-air system (Fig. 9.9) at 180(atm) and 80(°F). The low-pressure air leaves the cooler at a temperature 10(°F) lower than the temperature of the incoming high-pressure air. The separator operates at 1(atm), and the product is saturated liquid at this pressure. What is the maximum fraction of the air entering the cooler that can be liquefied.

9.9 Rework the preceding problem for air entering at 200(atm), and precooled to −40(°F) by external refrigeration.

CHAPTER

TEN

SYSTEMS OF VARIABLE COMPOSITION. IDEAL BEHAVIOR

In Chap. 6 we treated the thermodynamic properties of constant-composition fluids. However, many applications of chemical-engineering thermodynamics are to systems wherein multicomponent mixtures of gases or liquids undergo composition changes as the result of mixing or separation processes, the transfer of species from one phase to another, or chemical reaction. The properties of such systems depend on composition as well as on temperature and pressure. Our first task in this chapter is therefore to develop a fundamental property relation for homogeneous fluid mixtures of variable composition. We then derive equations applicable to mixtures of ideal gases and ideal solutions. Finally, we treat in detail a particularly simple description of multicomponent vapor/liquid equilibrium known as Raoult's law.

10.1 FUNDAMENTAL PROPERTY RELATION

Equation (6.6) expresses the basic relation connecting the Gibbs energy to the temperature and pressure in any closed system:

$$d(nG) = (nV)\,dP - (nS)\,dT \tag{6.6}$$

We apply this equation to the case of a single-phase fluid that does not undergo chemical reaction. The system is then of constant composition, and we can write immediately that

$$\left[\frac{\partial(nG)}{\partial P}\right]_{T,n} = nV \quad \text{and} \quad \left[\frac{\partial(nG)}{\partial T}\right]_{P,n} = -nS$$

where the subscript n indicates that the numbers of moles of *all* chemical species are held constant.

We are now prepared to treat the more general case of a single-phase, *open* system that can interchange matter with its surroundings. The total Gibbs energy nG is still a function of T and P; since material may be taken from or added to the system, nG is now also a function of the number of moles of each chemical species present. Thus

$$nG = g(P, T, n_1, n_2, \ldots, n_i, \ldots)$$

where the n_i are mole numbers of the species. The total differential of nG is

$$d(nG) = \left[\frac{\partial(nG)}{\partial P}\right]_{T,n} dP + \left[\frac{\partial(nG)}{\partial T}\right]_{P,n} dT + \sum_i \left[\frac{\partial(nG)}{\partial n_i}\right]_{P,T,n_j} dn_i$$

where the summation is over all species present, and subscript n_j indicates that all mole numbers except the ith are held constant. Replacing the first two partial derivatives by (nV) and $-(nS)$, we have

$$d(nG) = (nV)\,dP - (nS)\,dT + \sum_i \left[\frac{\partial(nG)}{\partial n_i}\right]_{P,T,n_j} dn_i$$

The derivative of nG with respect to the number of moles of species i has a special significance, and is given its own symbol and name. Thus, we define the *chemical potential* of species i in the mixture as

$$\mu_i \equiv \left[\frac{\partial(nG)}{\partial n_i}\right]_{P,T,n_j} \tag{10.1}$$

Expressed in terms of μ_i, the general equation for $d(nG)$ is

$$\boxed{d(nG) = (nV)\,dP - (nS)\,dT + \sum \mu_i\,dn_i} \tag{10.2}$$

Equation (10.2) is the fundamental property relation for single-phase fluid systems of constant or variable mass and constant or variable composition. It is the foundation equation upon which the structure of solution thermodynamics is built. It is applied initially in the following section, and will appear again in subsequent chapters.

10.2 THE CHEMICAL POTENTIAL AS A CRITERION OF PHASE EQUILIBRIUM

Consider a closed system consisting of two phases in equilibrium. Within this *closed* system, each of the individual phases is an *open* system, free to transfer mass to the other. Equation (10.2) may therefore be written for each phase:

$$d(nG)^\alpha = (nV)^\alpha\,dP - (nS)^\alpha\,dT + \sum \mu_i^\alpha\,dn_i^\alpha$$

$$d(nG)^\beta = (nV)^\beta\,dP - (nS)^\beta\,dT + \sum \mu_i^\beta\,dn_i^\beta$$

where superscripts α and β identify the phases. In writing these expressions, we have supposed that at equilibrium T and P are uniform throughout the entire system. The total change in the Gibbs energy of the system is the sum of these equations. When each total-system property is expressed by an equation of the form

$$nM = (nM)^\alpha + (nM)^\beta$$

this sum is given by

$$d(nG) = (nV)\,dP - (nS)\,dT + \sum \mu_i^\alpha\,dn_i^\alpha + \sum \mu_i^\beta\,dn_i^\beta$$

Since the two-phase system is closed, Eq. (6.6) must also be valid. Comparison of the two equations shows that at equilibrium

$$\sum \mu_i^\alpha\,dn_i^\alpha + \sum \mu_i^\beta\,dn_i^\beta = 0$$

However, the changes dn_i^α and dn_i^β result from mass transfer between the phases, and mass conservation requires that

$$dn_i^\alpha = -dn_i^\beta$$

Therefore

$$\sum (\mu_i^\alpha - \mu_i^\beta)\,dn_i^\alpha = 0$$

Since the dn_i^α are independent and arbitrary, the left-hand side of this equation can be zero in general only if each term in parentheses is separately zero. Hence

$$\mu_i^\alpha = \mu_i^\beta \qquad (i = 1, 2, 3, \ldots, N)$$

where N is the number of species present in the system. Although not given here, a similar but more comprehensive derivation shows (as we have supposed) that T and P must also be the same in the two phases at equilibrium.

By successively considering pairs of phases, we may readily generalize to more than two phases the equality of chemical potentials; the result for π phases is

$$\boxed{\mu_i^\alpha = \mu_i^\beta = \cdots = \mu_i^\pi} \qquad (i = 1, 2, \ldots, N) \tag{10.3}$$

Thus multiple phases at the same T and P are in equilibrium when the chemical potential of each species is the same in all phases.

The application of Eq. (10.3) to specific phase-equilibrium problems requires use of *models* of solution behavior, which provide expressions for G or for the μ_i as functions of temperature, pressure, and composition. The simplest of such expressions are for mixtures of ideal gases and for mixtures that form ideal solutions. These expressions, developed in this chapter, lead directly to Raoult's law, the simplest realistic relation between the compositions of phases coexisting in vapor/liquid equilibrium. Models of more general validity are treated in Chaps. 11 and 12.

10.3 THE IDEAL-GAS MIXTURE

If n moles of an ideal-gas mixture occupy a total volume V^t at temperature T, the pressure is

$$P = \frac{nRT}{V^t}$$

If the n_k moles of species k in this mixture occupy the same total volume alone at the same temperature, the pressure is

$$p_k = \frac{n_k RT}{V^t}$$

Dividing the latter equation by the former gives

$$\frac{p_k}{P} = \frac{n_k}{n} = y_k$$

or

$$p_k = y_k P \qquad (k = 1, 2, \ldots, N) \tag{10.4}$$

where y_k is the mole fraction of species k in the gas mixture, and p_k is known as the *partial pressure* of species k. The sum of the partial pressures as given by Eq. (10.4) equals the total pressure.

An ideal gas is a model gas comprised of imaginary molecules of zero volume that do not interact. Each chemical species in an ideal-gas mixture therefore has its own private properties, uninfluenced by the presence of other species. This is the basis of *Gibbs's theorem*:

> A total thermodynamic property (nU, nH, nC_P, nS, nA, or nG) of an ideal-gas mixture is the sum of the total properties of the individual species, each evaluated at the mixture temperature but at its own partial pressure.

This is expressed mathematically for general property M by the equation

$$nM^{ig}(T, P) = \sum n_k M_k^{ig}(T, p_k)$$

where the superscript *ig* denotes an ideal-gas property. Division by n gives

$$M^{ig}(T, P) = \sum y_k M_k^{ig}(T, p_k) \tag{10.5}$$

Since the enthalpy of an ideal gas is independent of pressure,

$$H_k^{ig}(T, p_k) = H_k^{ig}(T, P)$$

Therefore, Eq. (10.5) becomes

$$\boxed{H^{ig} = \sum y_k H_k^{ig}} \tag{10.6}$$

where H^{ig} and H_k^{ig} are understood to be values at the *mixture T and P*. Analogous equations apply for U^{ig} and other properties that are independent of pressure. [See Eq. (4.10) for C_P^{ig}.]

When Eq. (10.6) is written

$$H^{ig} - \sum y_k H_k^{ig} = 0$$

the difference on the left is the enthalpy change associated with a process in which appropriate amounts of the pure species at T and P are mixed to form one mole of mixture at the same T and P. For ideal gases, this *enthalpy change of mixing* is zero.

The entropy of an ideal gas does depend on pressure, and by Eq. (6.23),

$$dS_k^{ig} = -R\, d \ln P \qquad (\text{const } T)$$

Integration from p_k to P gives

$$S_k^{ig}(T, P) - S_k^{ig}(T, p_k) = -R \ln \frac{P}{p_k} = -R \ln \frac{P}{y_k P} = R \ln y_k$$

whence

$$S_k^{ig}(T, p_k) = S_k^{ig}(T, P) - R \ln y_k$$

Substituting this result into Eq. (10.5) written for the entropy gives

$$S^{ig}(T, P) = \sum y_k S_k^{ig}(T, P) - R \sum y_k \ln y_k$$

or more simply

$$\boxed{S^{ig} = \sum y_k S_k^{ig} - R \sum y_k \ln y_k} \qquad (10.7)$$

where S^{ig} and the S_k^{ig} are values at the mixture T and P.

When this equation is rearranged as

$$S^{ig} - \sum y_k S_k^{ig} = R \sum y_k \ln \frac{1}{y_k}$$

we have on the left the *entropy change of mixing* for ideal gases. Since $1/y_k > 1$, this quantity is always positive, in agreement with the second law. The mixing process is inherently irreversible, and for ideal gases mixing at constant T and P is not accompanied by heat transfer [Eq. (10.6)].

For the Gibbs energy of an ideal-gas mixture, $G^{ig} = H^{ig} - TS^{ig}$. Substitution for H^{ig} and S^{ig} by Eqs. (10.6) and (10.7) gives

$$G^{ig} = \sum y_k H_k^{ig} - T \sum y_k S_k^{ig} + RT \sum y_k \ln y_k$$

or

$$\boxed{G^{ig} = \sum y_k G_k^{ig} + RT \sum y_k \ln y_k} \qquad (10.8)$$

where G^{ig} and the G_k^{ig} are values at the mixture T and P.

The chemical potential of species i in an ideal-gas mixture is found by application of Eq. (10.1):

$$\mu_i^{ig} = \left[\frac{\partial(nG^{ig})}{\partial n_i}\right]_{P,T,n_j} \tag{A}$$

Multiplication of Eq. (10.8) by n gives

$$nG^{ig} = \sum n_k G_k^{ig} + RT \sum n_k \ln y_k$$

Since $y_k \equiv n_k/n$, where $n = \sum n_k$, this becomes

$$nG^{ig} = \sum n_k G_k^{ig} + RT \sum n_k \ln n_k - RTn \ln n$$

Separating particular species i from the set $\{k\}$ of all species, we are left with set $\{j\}$ of all species except i. Then

$$nG^{ig} = n_i G_i^{ig} + \sum n_j G_j^{ig} + RTn_i \ln n_i + RT \sum n_j \ln n_j - RTn \ln n$$

Since differentiation according to Eq. (A) is at constant T and P, G_i^{ig} and the G_j^{ig} are constant; moreover, all n_j are constant. Differentiation therefore gives

$$\mu_i^{ig} = G_i^{ig} + RT\left[n_i\left(\frac{\partial \ln n_i}{\partial n_i}\right)_{n_j} + \ln n_i\right] - RT\left[n\left(\frac{\partial \ln n}{\partial n_i}\right)_{n_j} + (\ln n)\left(\frac{\partial n}{\partial n_i}\right)_{n_j}\right]$$

Since $n = n_i + \sum n_j$, we also have $(\partial n/\partial n_i)_{n_j} = 1$. The preceding equation therefore reduces to

$$\mu_i^{ig} = G_i^{ig} + RT \ln \frac{n_i}{n}$$

or

$$\boxed{\mu_i^{ig} = G_i^{ig} + RT \ln y_i} \tag{10.9}$$

This equation is applied in the development of Raoult's law in Sec. 10.5.

10.4 THE IDEAL SOLUTION

The equations just derived show that for ideal gases a mixture property depends only on the properties of the pure ideal gases which comprise the mixture. No information about the mixture other than its composition is required. This circumstance is not limited to ideal gases, but extends more generally to any solution wherein all molecules are of the same size and all forces between molecules (like and unlike) are equal. Equations based on these characteristics provide a model of behavior known as the *ideal solution.*

The ideal gas, consisting of molecules with zero volume that do not interact, fulfills the conditions of solution ideality as a special case. When ideal gases are mixed, there is no volume change of mixing, because the molar volume of the mixture V^{ig} and the molar volumes of the pure species V_i^{ig} are all equal to RT/P.

Thus for ideal gases, the equation

$$V^{ig} = \sum y_i V_i^{ig}$$

is a simple identity. However, an analogous equation written for the ideal-*solution* model provides an essential relation:

$$\boxed{V^{id} = \sum x_i V_i} \tag{10.10}$$

where V^{id} is the molar volume of the ideal solution formed from pure species with *actual* molar volumes V_i at the temperature and pressure of the mixture. Thus the volume change of mixing is zero for ideal solutions as well as for ideal gases. In Eq. (10.10), x_i is used for mole fraction, because our immediate application of the ideal-solution model is to liquids. Since the formation of an ideal solution results in no change in molecular energies or volumes, we can write an equation for the enthalpy of an ideal solution analogous to Eq. (10.6):

$$\boxed{H^{id} = \sum x_i H_i} \tag{10.11}$$

where H_i is the enthalpy of pure species i at the mixture T and P.

For solutions comprised of species of equal molecular volume in which all molecular interactions are the same, one can show by the methods of statistical thermodynamics that the lowest possible value of the entropy is given by an equation analogous to Eq. (10.7). Thus we complete the definition of an ideal solution by specifying that its entropy be given by the equation:

$$\boxed{S^{id} = \sum x_i S_i - R \sum x_i \ln x_i} \tag{10.12}$$

The Gibbs energy of an ideal solution then follows from its defining equation, $G^{id} = H^{id} - TS^{id}$:

$$\boxed{G^{id} = \sum x_i G_i + RT \sum x_i \ln x_i} \tag{10.13}$$

Finally, the chemical potential of species i in an ideal solution follows from Eq. (10.13) by a derivation completely analogous to the derivation of Eq. (10.9):

$$\boxed{\mu_i^{id} = G_i + RT \ln x_i} \tag{10.14}$$

In the preceding equations, the quantities S_i and G_i are the properties of pure species i at the mixture T and P.

Ideal-solution behavior is often approximated by solutions comprised of molecules not too different in size and of the same chemical nature. Thus, a mixture of isomers, such as *ortho*-, *meta*-, and *para*-xylene, conforms very closely to ideal-solution behavior. So do mixtures of adjacent members of a homologous series, as for example, *n*-hexane/*n*-heptane, ethanol/propanol, and benzene/toluene. Other examples are acetone/acetonitrile and acetonitrile/nitromethane.

10.5 RAOULT'S LAW

When combined with the ideal-gas and ideal-solution models of phase behavior, the criterion of vapor/liquid equilibrium produces a simple and useful equation known as Raoult's law. Consider a liquid phase and a vapor phase, both comprised of N chemical species, coexisting in equilibrium at temperature T and pressure P, a condition of vapor/liquid equilibrium for which Eq. (10.3) becomes

$$\mu_i^v = \mu_i^l \qquad (i = 1, 2, \ldots, N) \tag{10.15}$$

If the vapor phase is an ideal gas and the liquid phase is an ideal solution, we may replace the chemical potentials in this equality by Eqs. (10.9) and (10.14):

$$G_i^{ig} + RT \ln y_i = G_i^l + RT \ln x_i$$

Rearrangement gives

$$RT \ln \frac{y_i}{x_i} = G_i^l(T, P) - G_i^{ig}(T, P) \tag{A}$$

where we indicate explicitly that the pure-species properties are evaluated at the equilibrium T and P. Assuming a negligible effect of pressure on G_i^l, we write

$$G_i^l(T, P) = G_i^l(T, P_i^{\text{sat}}) \tag{B}$$

where P_i^{sat} is the saturation or vapor pressure of pure species i at temperature T. For pure i as an ideal gas, we have from Eq. (6.10) that

$$dG_i^{ig} = V_i^{ig}\, dP \qquad (\text{const } T)$$

Integration at temperature T from P to P_i^{sat} yields

$$G_i^{ig}(T, P_i^{\text{sat}}) - G_i^{ig}(T, P) = \int_P^{P_i^{\text{sat}}} \frac{RT}{P}\, dP = RT \ln \frac{P_i^{\text{sat}}}{P} \tag{C}$$

Combining Eqs. (A), (B), and (C) gives

$$RT \ln \frac{y_i}{x_i} = G_i^l(T, P_i^{\text{sat}}) - G_i^{ig}(T, P_i^{\text{sat}}) + RT \ln \frac{P_i^{\text{sat}}}{P}$$

But the first two terms on the right are the Gibbs energies of pure liquid i and pure vapor i at the pure-species equilibrium conditions T and P_i^{sat}; according to Eq. (6.47), they are equal. The preceding equation therefore reduces to

$$\boxed{y_i P = x_i P_i^{\text{sat}}} \qquad (i = 1, 2, \ldots, N) \tag{10.16}$$

This equation expresses *Raoult's*† *law.* According to Eq. (10.4), the left-hand

† Francois Marie Raoult (1830–1901), French chemist.

side is the partial pressure of species i in the vapor phase, equal here to the product of the liquid-phase mole fraction of species i and its vapor pressure at temperature T.

Since P_i^{sat} is a function of temperature only, Raoult's law is a set of N equations in the variables T, P, $\{y_i\}$, and $\{x_i\}$. There are, in fact, $N-1$ independent vapor-phase mole fractions (the y_i's), $N-1$ independent liquid-phase mole fractions (the x_i's), and T and P. This makes a total of $2N$ independent variables related by N equations. The specification of N of these variables in the formulation of a vapor/liquid equilibrium problem allows the remaining N variables to be determined by the simultaneous solution of the N equilibrium relations given here by Raoult's law. In practice, one usually specifies either T or P *and* either the liquid-phase or the vapor-phase composition, fixing $1+(N-1)=N$ variables.

Example 10.1 The binary system acetonitrile(1)/nitromethane(2) conforms closely to Raoult's law. Vapor pressures for the pure species are given by the following Antoine equations:

$$\ln P_1^{\text{sat}}/\text{kPa} = 14.2724 - \frac{2{,}945.47}{t/°\text{C} + 224.00}$$

$$\ln P_2^{\text{sat}}/\text{kPa} = 14.2043 - \frac{2{,}972.64}{t/°\text{C} + 209.00}$$

(*a*) Prepare a graph showing P vs. x_1 and P vs. y_1 for a temperature of 75°C.
(*b*) Prepare a graph showing t vs. x_1 and t vs. y_1 for a pressure of 70 kPa.

SOLUTION (*a*) At 75°C, vapor pressures calculated from the given equations are

$$P_1^{\text{sat}} = 83.21 \qquad \text{and} \qquad P_2^{\text{sat}} = 41.98 \text{ kPa}$$

We write Eq. (10.16) for each of the two species:

$$y_1 P = x_1 P_1^{\text{sat}} \tag{A}$$

$$y_2 P = x_2 P_2^{\text{sat}} \tag{B}$$

Since $y_1 + y_2 = 1$, addition gives

$$P = x_1 P_1^{\text{sat}} + x_2 P_2^{\text{sat}} \tag{C}$$

When $1 - x_1$ is substituted for x_2, this becomes

$$P = P_2^{\text{sat}} + (P_1^{\text{sat}} - P_2^{\text{sat}})x_1 \tag{D}$$

Thus a plot of P vs. x_1 is a straight line connecting P_2^{sat} at $x_1 = 0$ with P_1^{sat} at $x_1 = 1$.

We can, of course, calculate P for a single value of x_1. For example, when $x_1 = 0.6$,

$$P = 41.98 + (83.21 - 41.98)(0.6) = 66.72 \text{ kPa}$$

The corresponding value of y_1 is then found from Eq. (*A*):

$$y_1 = \frac{x_1 P_1^{\text{sat}}}{P} = \frac{(0.6)(83.21)}{66.72} = 0.7483$$

These results mean that at 75°C a liquid mixture of 60 mole percent acetonitrile and

40 mole percent nitromethane is in equilibrium with a vapor containing 74.83 mole percent acetonitrile at a pressure of 66.72 kPa. The results of this and similar calculations for 75°C are tabulated as follows:

x_1	y_1	P/kPa
0.0	0.0	41.98
0.2	0.3313	50.23
0.4	0.5692	58.47
0.6	0.7483	66.72
0.8	0.8880	74.96
1.0	1.0	83.21

These same results are shown by the *Pxy* diagram of Fig. 10.1.

This figure is an example of a *phase diagram*, because the lines represent phase boundaries. Thus the line labeled $P - x_1$ represents states of saturated liquid; the subcooled-liquid region lies above this line. The curve labeled $P - y_1$ represents states of saturated vapor; the superheated-vapor region lies below the $P - y_1$ curve. Points lying between the saturated-liquid and saturated-vapor lines are in the two-phase

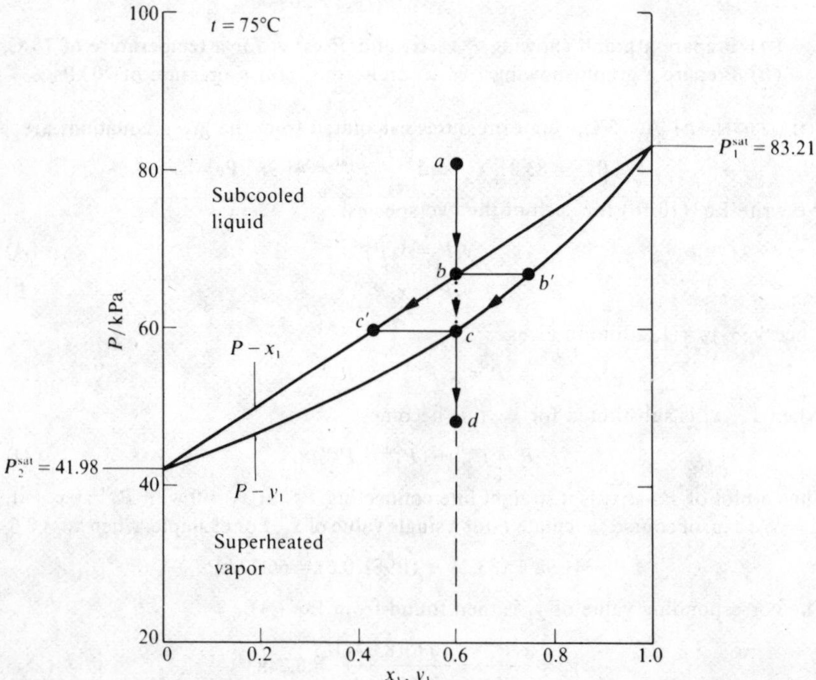

Figure 10.1 *Pxy* diagram for acetonitrile(1)/nitromethane(2) at 75°C as given by Raoult's law.

region, where saturated liquid and saturated vapor coexist in equilibrium. The $P - x_1$ and $P - y_1$ lines meet at the edges of the diagram, where saturated liquid and saturated vapor of the pure species coexist at the vapor pressures P_1^{sat} and P_2^{sat}.

We can illustrate the nature of phase behavior in this binary (two-component) system by following the course of a constant-temperature process on the *Pxy* diagram. Imagine a subcooled liquid mixture of 60 mole percent acetonitrile and 40 mole percent nitromethane existing in a piston/cylinder arrangement at 75°C. Its state is represented by point *a* in Fig. 10.1. The pressure is reduced slowly enough so that the system is always in equilibrium at 75°C. Since the system is closed, the overall composition remains constant during the process, and the states of the system *as a whole* fall on the vertical line descending from point *a*. When the pressure decreases to the state represented by point *b*, the system is saturated liquid on the verge of vaporizing. A minute further decrease in pressure is accompanied by the appearance of a bubble of vapor, represented by point *b'*. The two points *b* and *b'* together represent the equilibrium state at $x_1 = 0.6$, $P = 66.72$ kPa, and $y_1 = 0.7483$ for which calculations were illustrated. This is known as a *BUBL P calculation*, because the bubble (vapor) composition y_1 and the pressure are calculated from given values of x_1 and *t*. Point *b* is called a *bubble* point, and the $P - x_1$ line is the locus of bubble points.

As the pressure is further reduced, the amount of vapor increases and the amount of liquid decreases, with the states of the two phases following paths *b'c* and *bc'*, respectively. The dotted line from *b* to *c* represents the *overall* states of the two-phase system. Finally, as point *c* is approached, the liquid phase, represented by point *c'*, has almost disappeared, with only minute drops (dew) remaining. Point *c* is therefore called a *dew point*, and the $P - y_1$ line is the locus of dew points. Once the dew has evaporated, only saturated vapor at point *c* remains, and further pressure reduction leads to superheated vapor at point *d*.

The composition of the vapor at point *c* is $y_1 = 0.6$, but the composition of the liquid at point *c'* and the pressure must either be read from the graph or calculated. This is a *DEW P calculation*, because the dew (liquid) composition x_1 and the pressure are calculated for given values of y_1 and *t*. We write Eqs. (*A*) and (*B*) as

$$x_1 = \frac{y_1 P}{P_1^{\text{sat}}} \tag{E}$$

$$x_2 = \frac{y_2 P}{P_2^{\text{sat}}} \tag{F}$$

Since $x_1 + x_2 = 1$, addition gives

$$1 = \frac{y_1 P}{P_1^{\text{sat}}} + \frac{y_2 P}{P_2^{\text{sat}}} = P\left(\frac{y_1}{P_1^{\text{sat}}} + \frac{y_2}{P_2^{\text{sat}}}\right)$$

whence

$$P = \frac{1}{y_1/P_1^{\text{sat}} + y_2/P_2^{\text{sat}}} \tag{G}$$

For $y_1 = 0.6$ and $t = 75°\text{C}$,

$$P = \frac{1}{0.6/83.21 + 0.4/41.98} = 59.74 \text{ kPa}$$

By Eq. (E),

$$x_1 = \frac{(0.6)(59.74)}{83.21} = 0.4308$$

This is the liquid-phase composition at point c'.

(b) When pressure P is fixed, the temperature varies along with x_1 and y_1. Since temperature enters calculations based on Raoult's law only indirectly through the vapor pressures, we cannot solve explicitly for t, and an iterative procedure is indicated. For a given pressure, the temperature range is bounded by the saturation temperatures t_1^{sat} and t_2^{sat}, the temperatures at which the pure species exert vapor pressures equal to P. For the present system, these temperatures are calculated from the Antoine equations with $P_i^{sat} = P = 70$ kPa:

$$t_1^{sat} = 69.84 \qquad \text{and} \qquad t_2^{sat} = 89.58°\text{C}$$

When x_1 is known along with P, Eq. (C) is the basis of solution for t:

$$P = x_1 P_1^{sat} + x_2 P_2^{sat} = P_2^{sat}\left(x_1 \frac{P_1^{sat}}{P_2^{sat}} + x_2\right)$$

or

$$P_2^{sat} = \frac{P}{x_1 \alpha_{12} + x_2} \tag{H}$$

where $\alpha_{12} \equiv P_1^{sat}/P_2^{sat}$. Although P_1^{sat} and P_2^{sat} both increase rapidly with increasing temperature, α_{12} is a weak function of t. Values of α_{12} are readily calculated from the Antoine equations. Subtracting $\ln P_2^{sat}$ from $\ln P_1^{sat}$, we get

$$\ln \alpha_{12} = 0.0681 - \frac{2{,}945.47}{t + 224.00} + \frac{2{,}972.64}{t + 209.00} \tag{I}$$

The iteration procedure is as follows:

1. Choosing a value of α_{12} calculated at some intermediate temperature, calculate P_2^{sat} by Eq. (H).
2. Calculate t from the Antoine equation for species 2:

$$t = \frac{2{,}972.64}{14.2043 - \ln P_2^{sat}} - 209.00$$

3. Determine a new value of α_{12} by Eq. (I) and a new value of P_2^{sat} by Eq. (H).
4. Return to step 2, and iterate to convergence.

When y_1 is known along with P, Eq. (G) is the basis of solution:

$$P = \frac{1}{y_1/P_1^{sat} + y_2/P_2^{sat}} = \frac{P_1^{sat}}{y_1 + y_2 \alpha_{12}}$$

or

$$P_1^{sat} = P(y_1 + y_2 \alpha_{12}) \tag{J}$$

The iteration procedure is the same as before, except that

$$t = \frac{2{,}945.47}{14.2724 - \ln P_1^{sat}} - 224.00$$

For the purpose of preparing a *txy* diagram, the simplest procedure is to select values of t between t_1^{sat} and t_2^{sat}, calculate P_1^{sat} and P_2^{sat} for these temperatures, and evaluate x_1 by Eq. (D):

$$x_1 = \frac{P - P_2^{sat}}{P_1^{sat} - P_2^{sat}}$$

For example, at 78°C,

$$P_1^{sat} = 91.76 \qquad \text{and} \qquad P_2^{sat} = 46.84\,\text{kPa}$$

whence

$$x_1 = \frac{70 - 46.84}{91.76 - 46.84} = 0.5156$$

By Eq. (A),

$$y_1 = \frac{x_1 P_1^{sat}}{P} = \frac{(0.5156)(91.76)}{70} = 0.6759$$

The results of this and similar calculations for $P = 70$ kPa are given in the following table:

x_1	y_1	t/°C
0.0	0.0	89.58 (t_2^{sat})
0.1424	0.2401	86
0.3184	0.4742	82
0.5156	0.6759	78
0.7378	0.8484	74
1.0	1.0	69.84 (t_1^{sat})

Figure 10.2 is the *txy* diagram showing these results.

This figure is another example of a phase diagram, drawn here for a constant pressure of 70 kPa. The $t - y_1$ curve represents states of saturated vapor, with states of superheated vapor lying above it. The $t - x_1$ curve represents states of saturated liquid, with states of subcooled liquid lying below it. The two-phase region lies between these curves.

With reference to the *txy* diagram, we describe the course of a constant-pressure heating process leading from a state of subcooled liquid at point a to a state of superheated vapor at point d. The path shown on the figure is for a constant composition of 60 mole percent acetonitrile. The temperature of the liquid increases as the result of heating from point a to point b, where the first bubble of vapor appears. Thus point b is a bubble point, and the $t - x_1$ curve is the locus of bubble points.

We here know $x_1 = 0.6$ and $P = 70$ kPa; t is therefore determined by the iteration scheme described in connection with Eq. (H). The result in this case is $t = 76.42$°C, the temperature of points b and b'. At this temperature, $P_1^{sat} = 87.17$ kPa, and by Eq. (A) we find the composition of point b':

$$y_1 = \frac{x_1 P_1^{sat}}{P} = \frac{(0.6)(87.17)}{70} = 0.7472$$

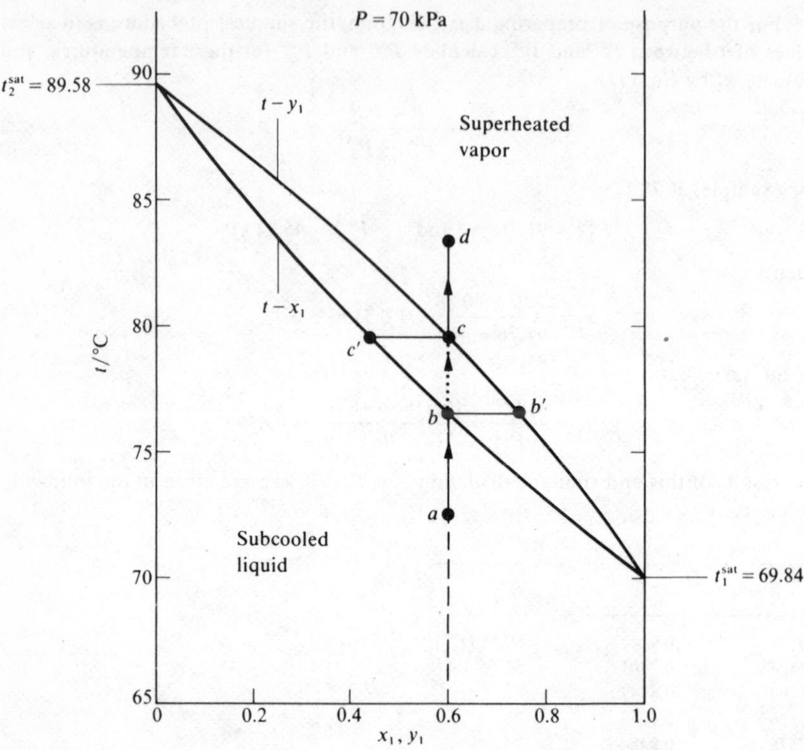

Figure 10.2 Diagram showing t vs. y_1 and t vs. x_1 for acetonitrile(1)/nitromethane(2) at 70 kPa as given by Raoult's law.

This is a *BUBL T calculation*, because the bubble composition y_1 and the temperature are calculated from given values of x_1 and P.

Vaporization of a mixture at constant pressure, unlike vaporization of a pure species, does not in general occur at constant temperature. As the heating process continues beyond point b, the temperature rises, the amount of vapor increases, and the amount of liquid decreases. During this process, the vapor- and liquid-phase compositions change as indicated by paths $b'c$ and bc', until the dew point is reached at point c, where the last droplets of liquid disappear. The $t - y_1$ curve is the locus of dew points.

The vapor composition at point c is $y_1 = 0.6$; since the pressure is also known ($P = 70$ kPa), we may carry out a *DEW T calculation* according to the iteration scheme associated with Eq. (J). The result here is $t = 79.58$°C, the temperature of points c and c'. With $P_1^{sat} = 96.53$ kPa, we find by Eq. (E) that the composition at point c' is

$$x_1 = \frac{y_1 P}{P_1^{sat}} = \frac{(0.6)(70)}{96.53} = 0.4351$$

Thus the temperature rises from 76.42 to 79.58°C during the vaporization step from point b to point c. Continued heating simply superheats the vapor to point d.

Straightforward generalizations of the procedures for binary systems allow application of Raoult's law to multicomponent systems. For a BUBL P calculation, given $\{x_k\}$ and t, we calculate $\{y_k\}$ and P. Since Raoult's law gives

$$y_k P = x_k P_k^{\text{sat}} \qquad (k = 1, 2, \ldots, N)$$

then

$$P = \sum_k x_k P_k^{\text{sat}} \tag{10.17}$$

Once P is calculated by Eq. (10.17), each y_k is found from Raoult's law.

For a DEW P calculation, we know $\{y_k\}$ and t, and calculate $\{x_k\}$ and P. Since

$$x_k = \frac{y_k P}{P_k^{\text{sat}}} \qquad (k = 1, 2, \ldots, N) \tag{10.18}$$

then

$$1 = P \sum_k \frac{y_k}{P_k^{\text{sat}}}$$

and

$$P = \frac{1}{\sum_k (y_k / P_k^{\text{sat}})} \tag{10.19}$$

Once P is calculated by Eq. (10.19), each x_k is given by Eq. (10.18).

A BUBL T calculation of $\{y_k\}$ and t, given $\{x_k\}$ and P, is based on Eq. (10.17), written

$$P = P_i^{\text{sat}} \sum_k x_k \frac{P_k^{\text{sat}}}{P_i^{\text{sat}}}$$

where i is an arbitrarily selected member of set $\{k\}$. Solution for P_i^{sat} gives

$$P_i^{\text{sat}} = \frac{P}{\sum_k x_k \alpha_{ki}} \tag{10.20}$$

where

$$\alpha_{ki} \equiv P_k^{\text{sat}} / P_i^{\text{sat}}$$

When the vapor pressures are given by Antoine equations,

$$\ln \alpha_{ki} = A_k - A_i - \frac{B_k}{t + C_k} + \frac{B_i}{t + C_i} \tag{10.21}$$

An iterative procedure starts with solution of Eq. (10.21) with an initial value of t provided by the equation

$$t_0 = \sum_k x_k t_k^{\text{sat}} \tag{10.22}$$

Equation (10.20) then yields P_i^{sat}, and we get an improved value of t from the Antoine equation:

$$t = \frac{B_i}{A_i - \ln P_i^{sat}} - C_i \tag{10.23}$$

This calculational sequence is repeated until there is no significant change in t from one iteration to the next. Final values of P_k^{sat} are found from the Antoine equations, and final y_k values come from Raoult's law.

The DEW T calculation is similar. Since we know $\{y_k\}$ and P and seek $\{x_k\}$ and t, we write Eq. (10.19) as

$$P = \frac{P_i^{sat}}{\sum_k y_k(P_i^{sat}/P_k^{sat})}$$

or

$$P_i^{sat} = P \sum_k (y_k/\alpha_{ki}) \tag{10.24}$$

Again an iterative process starts with Eq. (10.21), now with an initial value

$$t_0 = \sum_k y_k t_k^{sat} \tag{10.25}$$

Equation (10.24) then yields P_i^{sat}, and Eq. (10.23), an improved value of t with which to repeat the calculations. After convergence, we evaluate the P_k^{sat}, and calculate the final x_k by Eq. (10.18).

Example 10.2 For the acetone(1)/acetonitrile(2)/nitromethane(3) system, we have the following Antoine equations:

$$\ln P_1^{sat} = 14.5463 - \frac{2{,}940.46}{t + 237.22}$$

$$\ln P_2^{sat} = 14.2724 - \frac{2{,}945.47}{t + 224.00}$$

$$\ln P_3^{sat} = 14.2043 - \frac{2{,}972.64}{t + 209.00}$$

where t is in °C and the vapor pressures are in kPa. Assuming that Raoult's law is appropriate to this system, calculate:
(*a*) P and $\{y_k\}$, given that $t = 80$°C, $x_1 = 0.25$, $x_2 = 0.35$, and $x_3 = 0.40$.
(*b*) P and $\{x_k\}$, given that $t = 70$°C, $y_1 = 0.50$, $y_2 = 0.30$, and $y_3 = 0.20$.
(*c*) t and $\{y_k\}$, given that $P = 80$ kPa, $x_1 = 0.30$, $x_2 = 0.45$, and $x_3 = 0.25$.
(*d*) t and $\{x_k\}$, given that $P = 90$ kPa, $y_1 = 0.60$, $y_2 = 0.20$, and $y_3 = 0.20$.

SOLUTION (*a*) A BUBL P calculation. For $t = 80$°C, we calculate the following vapor pressures:

$$P_1^{sat} = 195.75 \qquad P_2^{sat} = 97.84 \qquad P_3^{sat} = 50.32 \text{ kPa}$$

Then by Eq. (10.17) with the given values of x_k,

$$P = (0.25)(195.75) + (0.35)(97.84) + (0.40)(50.32)$$
$$= 103.31 \text{ kPa}$$

From Eq. (10.16) written as $y_k = x_k P_k^{\text{sat}}/P$, we get

$$y_1 = 0.4737 \qquad y_2 = 0.3315 \qquad y_3 = 0.1948$$

The sum of these mole fractions equals unity, as it should.

(*b*) A DEW P calculation. For $t = 70°\text{C}$,

$$P_1^{\text{sat}} = 144.77 \qquad P_2^{\text{sat}} = 70.37 \qquad P_3^{\text{sat}} = 43.80 \text{ kPa}$$

Then by Eq. (10.19) with the given values of y_k,

$$P = 74.27 \text{ kPa}$$

and by Eq. (10.18),

$$x_1 = 0.2565 \qquad x_2 = 0.3166 \qquad x_3 = 0.4269$$

Again, the mole fractions sum to unity.

(*c*) A BUBL T calculation. For a ternary system with $i = 3$, Eq. (10.20) becomes

$$P_3^{\text{sat}} = \frac{P}{x_1\alpha_{13} + x_2\alpha_{23} + x_3\alpha_{33}}$$
$$= \frac{80}{0.30\alpha_{13} + 0.45\alpha_{23} + 0.25}$$

Setting each P_k^{sat} in the Antoine equations equal to 80 kPa, we find:

$$t_1^{\text{sat}} = 52.07 \qquad t_2^{\text{sat}} = 73.81 \qquad t_3^{\text{sat}} = 93.64°\text{C}$$

Equation (10.22) then gives

$$t_0 = 72.25°\text{C}$$

For this initial temperature, we find from Eq. (10.21) that

$$\alpha_{13} = 4.0951 \qquad \text{and} \qquad \alpha_{23} = 2.0037$$

whence

$$P_3^{\text{sat}} = 33.61 \text{ kPa}$$

Then by Eq. (10.23),

$$t = \frac{2{,}972.64}{14.2043 - \ln 33.61} - 209.00 = 69.09°\text{C}$$

This new value of t allows the calculations to be repeated. Further iteration leads to a final value of

$$t = 68.60°\text{C}$$

At this temperature,

$$P_1^{\text{sat}} = 138.56 \qquad P_2^{\text{sat}} = 67.08 \qquad P_3^{\text{sat}} = 32.98 \text{ kPa}$$

and by Eq. (10.16),

$$y_1 = 0.5196 \qquad y_2 = 0.3773 \qquad y_3 = 0.1031$$

(*d*) A DEW T calculation. We again take $i = 3$, and Eq. (10.24) becomes

$$P_3^{\text{sat}} = P\left(\frac{y_1}{\alpha_{13}} + \frac{y_2}{\alpha_{23}} + \frac{y_3}{\alpha_{33}}\right)$$

$$= 90\left(\frac{0.6}{\alpha_{13}} + \frac{0.2}{\alpha_{23}} + 0.2\right)$$

At $P = 90$ kPa, the saturation temperatures are

$$t_1^{\text{sat}} = 55.47 \qquad t_2^{\text{sat}} = 77.40 \qquad t_3^{\text{sat}} = 97.32°\text{C}$$

and by Eq. (10.25),

$$t_0 = 68.23°\text{C}$$

At this temperature we find from Eq. (10.21) that

$$\alpha_{13} = 4.2123 \qquad \alpha_{23} = 2.0370$$

whence

$$P_3^{\text{sat}} = 39.66 \text{ kPa}$$

By Eq. (10.23),

$$t = 73.46°\text{C}$$

This new estimate of t allows the calculations to be repeated; continued iteration leads to a final value of

$$t = 73.95°\text{C}$$

At this temperature,

$$P_1^{\text{sat}} = 163.47 \qquad P_2^{\text{sat}} = 80.37 \qquad P_3^{\text{sat}} = 40.39 \text{ kPa}$$

Equation (10.18) then yields

$$x_1 = 0.3303 \qquad x_2 = 0.2240 \qquad x_3 = 0.4457$$

One further vapor/liquid equilibrium problem is the *flash calculation.* The origin of the name is in the change that occurs when a liquid under pressure passes through a valve to a pressure low enough that some of the liquid vaporizes or "flashes," producing a two-phase stream of vapor and liquid in equilibrium. We consider here only the P,T-flash, which refers to any calculation of the quantities and compositions of the vapor and liquid phases making up a two-phase system in equilibrium at known P, T, and *overall* composition.

Consider such a system containing a total of one mole of chemical species, and having an *overall* composition represented by the set of mole fractions $\{z_i\}$. Let L be the moles of liquid, with mole fractions $\{x_i\}$, and let V be the moles of vapor, with mole fractions $\{y_i\}$. The material-balance equations are

$$L + V = 1$$

$$z_i = x_i L + y_i V \qquad (i = 1, 2, \ldots, N)$$

Choosing to eliminate L from these equations, we get

$$z_i = x_i(1 - V) + y_i V \qquad (i = 1, 2, \ldots, N) \tag{10.26}$$

As a matter of convenience, we write Raoult's law as

$$y_i = K_i x_i \tag{10.27}$$

where K_i is known as a "K-value," given here by

$$K_i = P_i^{\text{sat}}/P \tag{10.28}$$

Since P_i^{sat} is a function of T only, K_i is a function of T and P. Substituting $x_i = y_i/K_i$ in Eq. (10.26) and solving for y_i gives

$$y_i = \frac{z_i K_i}{1 + V(K_i - 1)} \qquad (i = 1, 2, \ldots, N) \tag{10.29}$$

Since $\sum y_i = 1$, the sum of Eqs. (10.29) gives

$$\sum_i \frac{z_i K_i}{1 + V(K_i - 1)} = 1 \tag{10.30}$$

In a flash calculation, T, P, and $\{z_i\}$ are known; the only unknown in Eq. (10.30) is therefore V. Solution is by trial. (Note that there is always a trivial solution at $V = 1$.) The y_i are then found from Eq. (10.29), and the x_i from Eq. (10.27).

Example 10.3 The system acetone(1)/acetonitrile(2)/nitromethane(3) at 80°C and 110 kPa has the overall composition, $z_1 = 0.45$, $z_2 = 0.35$, $z_3 = 0.20$. Determine L, V, $\{x_i\}$, and $\{y_i\}$.

SOLUTION There is no assurance at the outset that at the stated conditions the system is actually in the two-phase region. This should be determined before a flash calculation is attempted. A two-phase system at a given temperature and with given *overall* composition can exist over a range of pressures from the bubble point at P_b, where $V = 0$ and $\{z_i\} = \{x_i\}$, to the dew point at P_d, where $V = 1$ and $\{z_i\} = \{y_i\}$. If the given pressure lies between P_b and P_d, then the system is indeed made up of two phases at the stated conditions.

The vapor pressures of the pure species at 80°C are given in Example 10.2(*a*):

$$P_1^{\text{sat}} = 195.75 \qquad P_2^{\text{sat}} = 97.84 \qquad P_3^{\text{sat}} = 50.32 \text{ kPa}$$

First, we do a BUBL P calculation with $\{z_i\} = \{x_i\}$ to determine P_b. By Eq. (10.17),

$$P_b = x_1 P_1^{\text{sat}} + x_2 P_2^{\text{sat}} + x_3 P_3^{\text{sat}}$$

Numerically,

$$P_b = (0.45)(195.75) + (0.35)(97.84) + (0.20)(50.32)$$
$$= 132.40 \text{ kPa}$$

Second, we do a DEW P calculation with $\{z_i\} = \{y_i\}$ to determine P_d. By Eq. (10.19),

$$P_d = \frac{1}{y_1/P_1^{\text{sat}} + y_2/P_2^{\text{sat}} + y_3/P_3^{\text{sat}}} = \frac{1}{0.45/195.75 + 0.35/97.84 + 0.20/50.32}$$
$$= 101.52 \text{ kPa}$$

Since the given pressure lies between P_b and P_d, we proceed to the flash calculation.

By Eq. (10.28),

$$K_1 = \frac{195.75}{110} = 1.7795$$

Similarly,

$$K_2 = 0.8895$$

$$K_3 = 0.4575$$

Substitution of known values into Eq. (10.30) gives

$$\frac{(0.45)(1.7795)}{1 + 0.7795\,V} + \frac{(0.35)(0.8895)}{1 - 0.1105\,V} + \frac{(0.20)(0.4575)}{1 - 0.5425\,V} = 1$$

Solution for V by trial yields

$$V = 0.7364 \text{ mol}$$

whence

$$L = 1 - V = 0.2636 \text{ mol}$$

By Eq. (10.29),

$$y_1 = \frac{(0.45)(1.7795)}{1 + (0.7795)(0.7364)} = 0.5087$$

Similarly,

$$y_2 = 0.3389$$

$$y_3 = 0.1524$$

By Eq. (10.27),

$$x_1 = \frac{y_1}{K_1} = \frac{0.5087}{1.7795} = 0.2859$$

Similarly,

$$x_2 = 0.3810$$

$$x_3 = 0.3331$$

Obviously, we must have $\sum y_i = \sum x_i = 1$.

PROBLEMS

10.1 What is the change in entropy when 0.8 m^3 of nitrogen and 0.2 m^2 of oxygen, each at 1 bar and 25°C blend to form a homogeneous gas mixture at the same conditions? Assume ideal gases.

10.2 A vessel is divided into two parts by a partition, and contains 2 mol of nitrogen gas at 80°C and 40 bar on one side and 3 mol of argon gas at 150°C and 15 bar on the other. If the partition is removed and the gases mix adiabatically and completely, what is the change in entropy? Assume nitrogen an ideal gas with $C_V = (5/2)R$ and argon an ideal gas with $C_V = (3/2)R$.

10.3 A stream of nitrogen flowing at the rate of $14{,}000(\text{lb}_\text{m})(\text{hr})^{-1}$ and a stream of hydrogen flowing at the rate of $3{,}024(\text{lb}_\text{m})(\text{hr})^{-1}$ mix adiabatically in a steady-flow process. If the gases are ideal and if both are at the same T and P, what is the rate of entropy increase $[(\text{Btu})(\text{hr})^{-1}(\text{R})^{-1}]$ as a result of the process?

10.4 A design for purifying helium consists of an adiabatic process that splits a helium stream containing 30-mole-percent methane into two product streams, one containing 97-mole-percent helium and the other 90-mole-percent methane. The feed enters at 10 bar and 117°C; the methane-rich product leaves at 1 bar and 27°C; the helium-rich product leaves at 50°C and 15 bar. Moreover, work is produced by the process. Assuming helium an ideal gas with $C_P = (5/2)R$ and methane an ideal gas with $C_P = (9/2)R$, calculate the total entropy change of the process on the basis of 1 mol of feed to confirm that the process does not violate the second law.

10.5 A liquid mixture containing 40 mole percent benzene and 60 mole percent toluene is fed to a distillation column. The overhead product is nearly pure benzene and the bottoms product, pure toluene. The reboiler is heated by steam condensing at 140°C at the rate of 80 kg for each kilogram mole of feed. The overhead condenser is cooled by water at the essentially constant temperature of 20°C. Neglecting heat losses and sensible heat effects and assuming that the feed mixture is an ideal solution, calculate the total change in entropy resulting from the separation of 1 kg mol of feed.

10.6 Assuming Raoult's law to be valid for the system acetonitrile(1)/nitromethane(2),
(*a*) Prepare a *Pxy* diagram for a temperature of 100°C.
(*b*) Prepare a *txy* diagram for a pressure of 101.33 kPa.
Vapor pressures of the pure species are given by the following Antoine equations (P_i^{sat} in kPa and t in °C):

$$\ln P_1^{\text{sat}} = 14.2724 - \frac{2{,}945.47}{t + 224.00}$$

$$\ln P_2^{\text{sat}} = 14.2043 - \frac{2{,}972.64}{t + 209.00}$$

10.7 Assuming Raoult's law to be valid for the system benzene(1)/ethylbenzene(2),
(*a*) Prepare a *Pxy* diagram for a temperature of 100°C.
(*b*) Prepare a *txy* diagram for a pressure of 101.33 kPa.
Vapor pressures of the pure species are given by the following Antoine equations (P_i^{sat} in kPa and t in °C):

$$\ln P_1^{\text{sat}} = 13.8858 - \frac{2{,}788.51}{t + 220.79}$$

$$\ln P_2^{\text{sat}} = 14.0045 - \frac{3{,}279.47}{t + 213.20}$$

10.8 Assuming Raoult's law to be valid for the system 1-chlorobutane(1)/chlorobenzene(2),
(*a*) Prepare a *Pxy* diagram for a temperature of 100°C.
(*b*) Prepare a *txy* diagram for a pressure of 101.33 kPa.
Vapor pressures of the pure species are given by the following Antoine equations (P_i^{sat} in kPa and t in °C):

$$\ln P_1^{\text{sat}} = 13.9600 - \frac{2{,}826.26}{t + 224.10}$$

$$\ln P_2^{\text{sat}} = 13.9926 - \frac{3{,}295.12}{t + 217.55}$$

10.9 For the system acetone(1)/acetonitrile(2), the vapor pressures of the pure species are given by

$$\ln P_1^{\text{sat}} = 14.5463 - \frac{2{,}940.46}{t + 237.22}$$

$$\ln P_2^{\text{sat}} = 14.2724 - \frac{2{,}945.47}{t + 224.00}$$

where t is in °C and the vapor pressures are in kPa. Assuming Raoult's law to describe the vapor/liquid

equilibrium states of this system, determine:
(a) x_1 and y_1 for the equilibrium phases at 54°C and 65 kPa.
(b) t and y_1 for $P = 65$ kPa and $x_1 = 0.4$.
(c) P and y_1 for $t = 54$°C and $x_1 = 0.4$.
(d) t and x_1 for $P = 65$ kPa and $y_1 = 0.4$.
(e) P and x_1 for $t = 54$°C and $y_1 = 0.4$.
(f) The fraction of the system that is liquid, x_1, and y_1 at 54°C and 65 kPa, when the overall composition of the system is 70 mole percent acetone.
(g) The fraction of the system that is liquid, x_1, and y_1 at 54°C and 65 kPa, when the overall composition of the system is 60 mole percent acetone.

10.10 For the system n-pentane(1)/n-heptane(2), the vapor pressures of the pure species are given by

$$\ln P_1^{\text{sat}} = 13.8183 - \frac{2{,}477.07}{t + 233.21}$$

$$\ln P_2^{\text{sat}} = 13.8587 - \frac{2{,}911.32}{t + 216.64}$$

where t is in °C and the vapor pressures are in kPa. Assuming Raoult's law to describe the vapor/liquid equilibrium states of this system, determine:
(a) x_1 and y_1 for the equilibrium phases at 63°C and 95 kPa.
(b) t and y_1 for $P = 95$ kPa and $x_1 = 0.34$.
(c) P and y_1 for $t = 60$°C and $x_1 = 0.44$.
(d) t and x_1 for $P = 85$ kPa and $y_1 = 0.86$.
(e) P and x_1 for $t = 70$°C and $y_1 = 0.08$.
(f) The fraction of the system that is liquid, x_1, and y_1 at 60°C and 115 kPa, when the overall composition of the system is equimolar.
(g) The fraction of the system that is liquid, x_1, and y_1 at 60°C and 115 kPa, when the overall composition of the system is 60 mole percent n-pentane.

10.11 For the system benzene(1)/toluene(2)/ethylbenzene(3), the vapor pressures of the pure species are given by

$$\ln P_1^{\text{sat}} = 13.8858 - \frac{2{,}788.51}{t + 220.79}$$

$$\ln P_2^{\text{sat}} = 13.9987 - \frac{3{,}096.52}{t + 219.48}$$

$$\ln P_3^{\text{sat}} = 14.0045 - \frac{3{,}279.47}{t + 213.20}$$

where t is in °C and the vapor pressures are in kPa. Assuming Raoult's law to describe the vapor/liquid equilibrium states of this system, determine:
(a) P and $\{y_k\}$, given that $t = 110$°C, $x_1 = 0.22$, $x_2 = 0.37$, $x_3 = 0.41$.
(b) P and $\{x_k\}$, given that $t = 105$°C, $y_1 = 0.45$, $y_2 = 0.32$, $y_3 = 0.23$.
(c) t and $\{y_k\}$, given that $P = 90$ kPa, $x_1 = 0.47$, $x_2 = 0.18$, $x_3 = 0.35$.
(d) t and $\{x_k\}$, given that $P = 95$ kPa, $y_1 = 0.52$, $y_2 = 0.28$, $y_3 = 0.20$.

10.12 For the system of the preceding problem at a temperature of 100°C and an overall composition $z_1 = 0.41$, $z_2 = 0.34$, and $z_3 = 0.25$, determine:
(a) The bubble-point pressure P_b and the bubble composition.
(b) The dew-point pressure P_d and the dew composition.
(c) L, V, $\{x_i\}$, and $\{y_i\}$ for a pressure equal to $\frac{1}{2}(P_b + P_d)$.

10.13 The system 1-chlorobutane(1)/benzene(2)/chlorobenzene(3) conforms closely to Raoult's law. The vapor pressures of the pure species are given by the following Antoine equations:

$$\ln P_1^{\text{sat}} = 13.9600 - \frac{2{,}826.26}{t + 224.10}$$

$$\ln P_2^{\text{sat}} = 13.8858 - \frac{2{,}788.51}{t + 220.79}$$

$$\ln P_3^{\text{sat}} = 13.9926 - \frac{3{,}295.12}{t + 217.55}$$

where t is in °C and the vapor pressures are in kPa. Determine:
(*a*) P and $\{y_k\}$, given that $t = 90°\text{C}$, $x_1 = 0.16$, $x_2 = 0.22$, $x_3 = 0.62$.
(*b*) P and $\{x_k\}$, given that $t = 95°\text{C}$, $y_1 = 0.39$, $y_2 = 0.27$, $y_3 = 0.34$.
(*c*) t and $\{y_k\}$, given that $P = 101.33$ kPa, $x_1 = 0.24$, $x_2 = 0.52$, $x_3 = 0.24$.
(*d*) t and $\{x_k\}$, given that $P = 101.33$ kPa, $y_1 = 0.68$, $y_2 = 0.12$, $y_3 = 0.20$.

10.14 For the system of the preceding problem at a temperature of 125°C and an overall composition $z_1 = 0.20$, $z_2 = 0.30$, and $z_3 = 0.50$, determine:
(*a*) The bubble-point pressure P_b and the bubble composition.
(*b*) The dew-point pressure P_d and the dew composition.
(*c*) L, V, $\{x_i\}$, and $\{y_i\}$ for a pressure of 175 kPa.

10.15 The system *n*-pentane(1)/*n*-hexane(2)/*n*-heptane(3) conforms closely to Raoult's law. The vapor pressures of the pure species are given by the following Antoine equations:

$$\ln P_1^{\text{sat}} = 13.8183 - \frac{2{,}477.07}{t + 233.21}$$

$$\ln P_2^{\text{sat}} = 13.8216 - \frac{2{,}697.55}{t + 224.37}$$

$$\ln P_3^{\text{sat}} = 13.8587 - \frac{2{,}911.32}{t + 216.64}$$

where t is in °C and the vapor pressures are in kPa. Determine:
(*a*) P and $\{y_k\}$, given that $t = 70°\text{C}$, $x_1 = 0.09$, $x_2 = 0.57$, $x_3 = 0.34$.
(*b*) P and $\{x_k\}$, given that $t = 80°\text{C}$, $y_1 = 0.43$, $y_2 = 0.36$, $y_3 = 0.21$.
(*c*) t and $\{y_k\}$, given that $P = 250$ kPa, $x_1 = 0.48$, $x_2 = 0.28$, $x_3 = 0.24$.
(*d*) t and $\{x_k\}$, given that $P = 300$ kPa, $y_1 = 0.44$, $y_2 = 0.47$, $y_3 = 0.09$.

10.16 For the system of the preceding problem at a temperature of 105°C and an overall composition $z_1 = 0.25$, $z_2 = 0.45$, and $z_3 = 0.30$, determine:
(*a*) The bubble-point pressure P_b and the bubble composition.
(*b*) The dew-point pressure P_d and the dew composition.
(*c*) L, V, $\{x_i\}$, and $\{y_i\}$ for a pressure equal to $\frac{1}{2}(P_b + P_d)$.

CHAPTER

ELEVEN

SYSTEMS OF VARIABLE COMPOSITION. NONIDEAL BEHAVIOR

The properties of mixtures of ideal gases and of ideal solutions depend solely on the properties of the pure constituent species, and are calculated from them by simple equations, as illustrated in Chap. 10. Although these models approximate the behavior of certain fluid mixtures, they do not adequately represent the behavior of most solutions of interest to chemical engineers, and Raoult's law is not in general a realistic relation for vapor/liquid equilibrium. However, these models of ideal behavior—the ideal gas, the ideal solution, and Raoult's law—provide convenient references to which the behavior of nonideal solutions may be compared.

In this chapter we lay the foundation for a general treatment of vapor/liquid equilibrium (Chap. 12) through introduction of two auxiliary thermodynamic properties related to the Gibbs energy, namely, the fugacity coefficient and the activity coefficient. These properties, relating directly to deviations from ideal behavior, will serve in Chap. 12 as correction factors that transform Raoult's law into a valid general expression for vapor/liquid equilibrium. Their definitions depend on development of the concept of *fugacity*, which provides an alternative to the chemical potential as a criterion for phase equilibrium. This treatment first requires the introduction of a new class of thermodynamic properties known as *partial properties.* The mathematical definition of these quantities endows them with all the characteristics of properties of the individual species as they exist in solution.

11.1 PARTIAL PROPERTIES

The definition of the chemical potential by Eq. (10.1) as the mole-number derivative of nG suggests that such derivatives may be of particular use in solution thermodynamics. We can, for example, write

$$\bar{V}_i \equiv \left[\frac{\partial(nV)}{\partial n_i}\right]_{P,T,n_j} \tag{11.1}$$

This equation defines the *partial molar volume* $\bar{V}_i$ of species i in solution. It is simply the volumetric response of the system to the addition at constant T and P of a differential amount of species i. A partial molar property may be defined in like fashion for each extensive thermodynamic property. Letting M represent the molar value of such a property, we write the general defining equation for a partial molar property as

$$\boxed{\bar{M}_i \equiv \left[\frac{\partial(nM)}{\partial n_i}\right]_{P,T,n_j}} \tag{11.2}$$

Here, $\bar{M}_i$ may represent the partial molar internal energy $\bar{U}_i$, the partial molar enthalpy $\bar{H}_i$, the partial molar entropy $\bar{S}_i$, the partial molar Gibbs energy $\bar{G}_i$, etc. Comparison of Eq. (10.1) with Eq. (11.2) written for the Gibbs energy shows that the chemical potential and the partial molar Gibbs energy are identical, that is, $\mu_i \equiv \bar{G}_i$.

Example 11.1 What physical interpretation can be given to the defining expression [Eq. (11.1)] for the partial molar volume?

SOLUTION Consider an open beaker containing an equimolar mixture of alcohol and water. The mixture occupies a total volume nV at room temperature T and atmospheric pressure P. Now add to this solution a small drop of pure water, also at T and P, containing Δn_w moles, and mix it thoroughly into the solution, allowing sufficient time for heat exchange so that the contents of the beaker return to the initial temperature. What is the volume change of the solution in the beaker? One might suppose that the volume increases by an amount equal to the volume of the water added, i.e., by $V_w \, \Delta n_w$, where V_w is the molar volume of pure water at T and P. If this were true, we would have

$$\Delta(nV) = V_w \, \Delta n_w$$

However, we find by experiment that the actual value of $\Delta(nV)$ is somewhat less than that given by this equation. Evidently, the *effective* molar volume of the added water in solution is less than the molar volume of pure water at the same T and P. Designating the effective molar volume in solution by $\tilde{V}_w$, we can write

$$\Delta(nV) = \tilde{V}_w \, \Delta n_w \tag{A}$$

or

$$\tilde{V}_w = \frac{\Delta(nV)}{\Delta n_w} \tag{B}$$

If this effective molar volume is to represent the property of species i in the original equimolar solution, it must be based on data for a solution of this composition. However, in the process described a finite drop of water is added to the equimolar solution, causing a small but finite change in composition. We may, however, consider the limiting case for which $\Delta n_w \to 0$. Then Eq. (B) becomes

$$\tilde{V}_w = \bar{V}_w = \lim_{\Delta n_w \to 0} \frac{\Delta(nV)}{\Delta n_w} = \frac{d(nV)}{dn_w} \tag{C}$$

Since T, P, and n_a (the number of moles of alcohol) are constant, this equation is more appropriately written:

$$\bar{V}_w = \left[\frac{\partial(nV)}{\partial n_w}\right]_{P,T,n_a} \tag{D}$$

which is a particular case of Eq. (11.1). Thus the partial molar volume of the water in solution is the rate of change of the total solution volume with n_w at constant T, P, and n_a.

If we write Eq. (A) for the addition of dn_w moles of water to the solution, it becomes

$$d(nV) = \bar{V}_w \, dn_w \tag{E}$$

When $\bar{V}_w$ is considered the effective molar property of water as it exists in solution, the total volume change $d(nV)$ is merely this molar property multiplied by the number of moles of water added.

If dn_w moles of water is added to a volume of *pure* water, then we have every reason to expect the volume change of the system to be given by

$$d(nV) = V_w \, dn_w \tag{F}$$

where V_w is the molar volume of pure water at T and P. Comparison of Eqs. (E) and (F) indicates that $\bar{V}_w = V_w$ when the "solution" is taken as pure water.

The definition of a partial molar property, Eq. (11.2), provides the means for calculation of partial properties from solution-property data. Implicit in this definition is a second, equally important, equation that allows the calculation of solution properties from knowledge of the partial properties. The derivation of this second equation starts with the observation that the thermodynamic properties of a homogeneous phase are functions of temperature, pressure, and the numbers of moles of the individual species which comprise the phase. For thermodynamic property M we may therefore write

$$mM = M(T, P, n_1, n_2, n_3, \ldots)$$

The total differential of nM is then

$$d(nM) = \left[\frac{\partial(nM)}{\partial P}\right]_{T,n} dP + \left[\frac{\partial(nM)}{\partial T}\right]_{P,n} dT + \sum_i \left[\frac{\partial(nM)}{\partial n_i}\right]_{P,T,n_j} dn_i$$

where subscript n indicates that *all* mole numbers are held constant, and subscript n_j that all mole numbers *except* n_i are held constant. Because the first two partial derivatives on the right are evaluated at constant n and in view of Eq. (11.2), this equation may be written more simply as

$$d(nM) = n\left(\frac{\partial M}{\partial P}\right)_{T,x} dP + n\left(\frac{\partial M}{\partial T}\right)_{P,x} dT + \sum \bar{M}_i \, dn_i \tag{11.3}$$

where subscript x denotes differentiation at constant composition.

Since $n_i = x_i n$,

$$dn_i = x_i \, dn + n \, dx_i$$

Replacing dn_i by this expression and replacing $d(nM)$ by the identity

$$d(nM) \equiv n \, dM + M \, dn$$

we write Eq. (11.3) as

$$n \, dM + M \, dn = n\left(\frac{\partial M}{\partial P}\right)_{T,x} dP + n\left(\frac{\partial M}{\partial T}\right)_{P,x} dT + \sum \bar{M}_i(x_i \, dn + n \, dx_i)$$

When the terms containing n are collected and separated from those containing dn, this equation becomes

$$\left[dM - \left(\frac{\partial M}{\partial P}\right)_{T,x} dP - \left(\frac{\partial M}{\partial T}\right)_{P,x} dT - \sum \bar{M}_i \, dx_i\right] n + \left[M - \sum x_i \bar{M}_i\right] dn = 0$$

In application, one is free to choose a system of any size, as represented by n, and to choose any variation in its size, as represented by dn. Thus, n and dn are independent and arbitrary. The only way that the left-hand side of this equation can then, in general, be zero is for both quantities enclosed by brackets to be zero. We therefore have:

$$dM = \left(\frac{\partial M}{\partial P}\right)_{T,x} dP + \left(\frac{\partial M}{\partial T}\right)_{P,x} dT + \sum \bar{M}_i \, dx_i \tag{11.4}$$

and

$$\boxed{M = \sum x_i \bar{M}_i} \tag{11.5}$$

Multiplication of Eq. (11.5) by n yields the alternative expression

$$\boxed{nM = \sum n_i \bar{M}_i} \tag{11.6}$$

Equation (11.4) is in fact just a special case of Eq. (11.3), obtained by setting $n = 1$, which also makes $n_i = x_i$. Equations (11.5) and (11.6) on the other hand are new and vital. They allow the calculation of mixture properties from partial properties, playing a role opposite to that of Eq. (11.2), which provides for the calculation of partial properties from mixture properties.

One further important equation follows directly from Eqs. (11.4) and (11.5). Since Eq. (11.5) is a general expression for M, differentiation yields a general expression for dM:

$$dM = \sum x_i\, d\bar{M}_i + \sum \bar{M}_i\, dx_i$$

comparison of this equation with Eq. (11.4), another general equation for dM, yields the *Gibbs/Duhem*† *equation*:

$$\left(\frac{\partial M}{\partial P}\right)_{T,x} dP + \left(\frac{\partial M}{\partial T}\right)_{P,x} dT - \sum x_i\, d\bar{M}_i = 0 \tag{11.7}$$

This equation must be satisfied for all changes in P, T, and the $\bar{M}_i$ caused by changes of state in a homogeneous phase. For the important special case of changes at constant T and P, it simplifies to:

$$\sum x_i\, d\bar{M}_i = 0 \qquad \text{(const } T, P) \tag{11.8}$$

Equation (11.5) implies that a molar solution property is given as a sum of its parts and that $\bar{M}_i$ is the molar property of species i as it exists in solution. This is a proper interpretation provided one understands that the defining equation for $\bar{M}_i$, Eq. (11.2), is an apportioning formula which *arbitrarily* assigns to each species i a share of the mixture property, subject to the constraint of Eq. (11.5).‡

The constituents of a solution are in fact intimately intermixed, and owing to molecular interactions cannot have private properties of their own. Nevertheless they can have *assigned* property values, and partial molar properties, as defined by Eq. (11.2), have all the characteristics of properties of the individual species as they exist in solution.

The properties of solutions as represented by the symbol M may be on a unit-mass basis as well as on a mole basis. The equations relating solution properties are unchanged in form; one merely replaces the various n's, representing moles, by m's, representing mass, and speaks of partial *specific* properties rather than of partial *molar* properties. In order to accommodate either, we generally speak simply of partial properties.

Since we are concerned here primarily with the properties of solutions, we represent molar (or unit-mass) properties of the solution by the plain symbol M. Partial properties are denoted by an overbar, and a subscript identifies the species, giving the symbol $\bar{M}_i$. In addition, we need a symbol for the properties of the individual species as they exist in the *pure state at the T and P of the solution.* These molar (or unit-mass) properties are identified by only a subscript, and the symbol is M_i. In summary, three kinds of properties used in solution thermodynamics are distinguished by the following symbolism:

† Pierre-Maurice-Marie Duhem (1861–1916), French physicist.

‡ Other apportioning equations, which make different allocations of the mixture property, are possible and are equally valid.

Solution properties	M,	for example, U, H, S, G
Partial properties	$\bar{M}_i$,	for example, $\bar{U}_i, \bar{H}_i, \bar{S}_i, \bar{G}_i$
Pure-species properties	M_i,	for example, U_i, H_i, S_i, G_i

Example 11.2 The need arises in a laboratory for 2,000 cm^3 of an antifreeze solution consisting of a 30 mole percent solution of methanol in water. What volumes of pure methanol and of pure water at 25°C must be mixed to form the 2,000 cm^3 of antifreeze, also at 25°C? Partial molar volumes for methanol and water in a 30 mole percent methanol solution at 25°C are:

Methanol(1): $\bar{V}_1 = 38.632 \text{ cm}^3 \text{ mol}^{-1}$
Water(2): $\bar{V}_2 = 17.765 \text{ cm}^3 \text{ mol}^{-1}$

For the pure species at 25°C:

Methanol(1): $V_1 = 40.727 \text{ cm}^3 \text{ mol}^{-1}$
Water(2): $V_2 = 18.068 \text{ cm}^3 \text{ mol}^{-1}$

SOLUTION Equation (11.5) written for the volume of a binary solution is

$$V = x_1 \bar{V}_1 + x_2 \bar{V}_2$$

All quantities on the right are known, and we calculate the molar volume of the antifreeze solution:

$$V = (0.3)(38.632) + (0.7)(17.765) = 24.025 \text{ cm}^3 \text{ mol}^{-1}$$

The required total volume of solution is

$$V^t = nV = 2{,}000 \text{ cm}^3$$

Thus the total number of moles required is

$$n = \frac{V^t}{V} = \frac{2{,}000}{24.025} = 83.246 \text{ mol}$$

Of this, 30 percent is methanol, and 70 percent is water:

$$n_1 = (0.3)(83.246) = 24.974 \text{ mol}$$

$$n_2 = (0.7)(83.246) = 58.272 \text{ mol}$$

The volume of each pure species is $V_i^t = n_i V_i$; thus

$$V_1^t = (24.974)(40.727) = 1{,}017 \text{ cm}^3$$

$$V_2^t = (58.272)(18.068) = 1{,}053 \text{ cm}^3$$

Note that the simple sum of the initial volumes gives a total of 2,070 cm^3, a volume more than 3 percent larger than that of the solution formed.

11.2 FUGACITY AND FUGACITY COEFFICIENT

For a constant-composition fluid at constant temperature, Eq. (6.10) becomes

$$dG = V\,dP \qquad (\text{const } T) \tag{11.9}$$

An inherent problem with use of the Gibbs energy has its origin with this equation.

Integration at constant temperature from the state of a gas at a low pressure P^* to the state at higher pressure P gives

$$G^* = G - \int_{P^*}^{P} V\,dP$$

In the limit as P^* approaches zero, V becomes infinite, making the integral infinite as well. Thus

$$\lim_{P^* \to 0} G^* = G - \infty$$

If we are to have finite values of G at positive pressures, then the Gibbs energy must approach the awkward limit of $-\infty$ as P^* approaches zero.

We can, however, define an auxiliary property that is mathematically better behaved. A clue to the nature of such a property is found in Eq. (11.9) written for an ideal gas:

$$dG^{ig} = V^{ig}\,dP = \frac{RT}{P}\,dP \qquad \text{(const } T\text{)}$$

or

$$dG^{ig} = RT\,d\ln P \qquad \text{(const } T\text{)} \tag{11.10}$$

Although correct only for an ideal gas, the simplicity of this equation suggests writing another equation of exactly the same form for a real fluid that defines a new property f that also has dimensions of pressure:

$$\boxed{dG \equiv RT\,d\ln f} \qquad \text{(const } T\text{)} \tag{11.11}$$

Equation (11.11) serves as a partial definition of f, which is called *fugacity*.†

Subtraction of Eq. (11.10) from Eq. (11.11) gives

$$dG - dG^{ig} = RT\,d\ln f - RT\,d\ln P$$

or

$$d(G - G^{ig}) = RT\,d\ln\frac{f}{P} \qquad \text{(const } T\text{)}$$

According to the definition of Eq. (6.32), $G - G^{ig}$ is the *residual Gibbs energy*, G^R; the dimensionless ratio f/P is a mixture property called the *fugacity coefficient* and given the symbol ϕ. Thus,

$$dG^R = RT\,d\ln\phi \qquad \text{(const } T\text{)} \tag{11.12}$$

† Introduced by Gilbert Newton Lewis (1875–1946), American physical chemist, who also developed the concepts of the partial property and the ideal solution.

where

$$\boxed{\phi \equiv \frac{f}{P}} \tag{11.13}$$

Integration of Eq. (11.12) yields the general relation,

$$G^R = RT \ln \phi + C(T) \tag{11.14}$$

where the integration constant is a function of temperature only. We now complete the definition of fugacity by setting the fugacity of an ideal gas equal to its pressure:

$$f^{ig} = P \tag{11.15}$$

Thus for the special case of an ideal gas, $G^R = 0$, $\phi = 1$, and the integration constant in Eq. (11.14) must vanish. Therefore $C(T) = 0$, and Eq. (11.14) may be written

$$\boxed{\frac{G^R}{RT} = \ln \phi} \tag{11.16}$$

This general equation applies to a mixture.

For the special case of pure species i, Eq. (11.11) is written

$$dG_i = RT\, d \ln f_i \qquad \text{(const } T\text{)} \tag{11.17}$$

Equation (11.16) here becomes

$$\frac{G_i^R}{RT} = \ln \phi_i \tag{11.18}$$

where

$$\phi_i = \frac{f_i}{P} \tag{11.19}$$

The identification of $\ln \phi$ with G^R/RT allows Eq. (6.40) to be rewritten as

$$\boxed{\ln \phi = \int_0^P (Z-1) \frac{dP}{P}} \qquad \text{(const } T, x\text{)} \tag{11.20}$$

Fugacity coefficients (and therefore fugacities) are evaluated by this equation from *PVT* data or from an equation of state. For example, when the compressibility factor is given by Eq. (3.31), we have

$$Z - 1 = \frac{BP}{RT}$$

where the second virial coefficient B is a function of temperature only for a constant-composition gas. Substitution into Eq. (11.20) gives

$$\ln \phi = \frac{B}{RT}\int_0^P dP \qquad \text{(const } T, x\text{)}$$

whence

$$\ln \phi = \frac{BP}{RT} \tag{11.21}$$

Equation (11.17), which defines the fugacity of pure species i, may be integrated for the change of state from saturated liquid to saturated vapor, both at temperature T and at the vapor pressure P_i^{sat}:

$$G_i^v - G_i^l = RT \ln \frac{f_i^v}{f_i^l}$$

According to Eq. (6.47), $G_i^v - G_i^l = 0$; therefore

$$f_i^v = f_i^l = f_i^{\text{sat}} \tag{11.22}$$

where f_i^{sat} indicates the value for either saturated liquid or saturated vapor. The corresponding fugacity coefficient is

$$\phi_i^{\text{sat}} = \frac{f_i^{\text{sat}}}{P_i^{\text{sat}}} \tag{11.23}$$

whence

$$\phi_i^v = \phi_i^l = \phi_i^{\text{sat}} \tag{11.24}$$

Since coexisting phases of saturated liquid and saturated vapor are in equilibrium, the equality of fugacities as expressed by Eqs. (11.22) and (11.24) is a criterion of vapor/liquid equilibrium for pure species.

Because of the equality of fugacities of saturated liquid and vapor, the calculation of fugacity for species i as a compressed liquid is done in two steps. First, one calculates the fugacity coefficient of saturated vapor $\phi_i^v = \phi_i^{\text{sat}}$ by an integrated form of Eq. (11.20), evaluated for $P = P_i^{\text{sat}}$. Then by Eqs. (11.22) and (11.23),

$$f_i^l = f_i^{\text{sat}} = \phi_i^{\text{sat}} P_i^{\text{sat}}$$

The second step is the evaluation of the change in fugacity of the liquid with an increase in pressure above P_i^{sat}. The required equation follows directly from Eq. (11.17),

$$dG_i = RT\, d \ln f_i \qquad \text{(const } T\text{)}$$

together with Eq. (11.9) written for pure species i,

$$dG_i = V_i\, dP \qquad \text{(const } T\text{)}$$

Whence

$$d \ln f_i = \frac{V_i}{RT}\, dP \qquad \text{(const } T\text{)} \tag{11.25}$$

Integration from the state of saturated liquid to that of compressed liquid gives

$$\ln \frac{f_i}{f_i^{\text{sat}}} = \frac{1}{RT} \int_{P_i^{\text{sat}}}^{P} V_i \, dP$$

Since V_i, the liquid-phase molar volume, is a very weak function of P at temperatures well below T_c, an excellent approximation is often obtained when evaluation of the integral is based on the assumption that V_i is constant at the value for saturated liquid, V_i^l:

$$\ln \frac{f_i}{f_i^{\text{sat}}} = \frac{V_i^l(P - P_i^{\text{sat}})}{RT}$$

Substituting $f_i^{\text{sat}} = \phi_i^{\text{sat}} P_i^{\text{sat}}$ and solving for f_i gives

$$f_i = \phi_i^{\text{sat}} P_i^{\text{sat}} \exp \frac{V_i^l(P - P_i^{\text{sat}})}{RT} \tag{11.26}$$

The exponential is known as the Poynting† factor.

Example 11.3 For H_2O at a temperature of 300°C and for pressures up to 10,000 kPa (100 bar) plot values of f_i and ϕ_i calculated from data in the steam tables vs. P.

SOLUTION Equation (11.17) may be written:

$$d \ln f_i = \frac{1}{RT} \, dG_i$$

Integration from a low-pressure reference state (designated by *) to a state at pressure P, both at the same temperature T, gives

$$\ln \frac{f_i}{f_i^*} = \frac{1}{RT}(G_i - G_i^*)$$

By the definition of the Gibbs energy,

$$G_i = H_i - TS_i$$

and

$$G_i^* = H_i^* - TS_i^*$$

Whence

$$\ln \frac{f_i}{f_i^*} = \frac{1}{R}\left[\frac{H_i - H_i^*}{T} - (S_i - S_i^*)\right]$$

If the reference-state pressure P^* is low enough that the fluid closely approximates an ideal gas, then $f_i^* = P^*$, and

$$\ln \frac{f_i}{P^*} = \frac{1}{R}\left[\frac{H_i - H_i^*}{T} - (S_i - S_i^*)\right] \tag{A}$$

† John Henry Poynting (1852–1914), British physicist.

The lowest pressure for which data at 300°C are given in the steam tables is 1 kPa, and we assume that steam at these conditions is for practical purposes an ideal gas. Data for this state provide the following reference values:

$$P^* = 1 \text{ kPa}$$

$$H_i^* = 3{,}076.8 \text{ J g}^{-1}$$

$$S_i^* = 10.3450 \text{ J g}^{-1} \text{ K}^{-1}$$

Equation (A) may now be applied to states of superheated steam at 300°C for various values of P from 1 kPa to the saturation pressure of 8,592.7 kPa. For example, at $P = 4{,}000$ kPa and 300°C,

$$H_i = 2{,}962.0 \text{ J g}^{-1}$$

$$S_i = 6.3642 \text{ J g}^{-1} \text{ K}^{-1}$$

These values must be multiplied by the molar mass of water (18.016) to put them on a molar basis for substitution into Eq. (A):

$$\ln \frac{f_i}{P^*} = \frac{18.016}{8.314}\left[\frac{2{,}962.0 - 3{,}076.8}{573.15} - (6.3642 - 10.3450)\right]$$

$$= 8.1922$$

and $f_i/P^* = 3{,}612.5$. Since $P^* = 1$ kPa, $f_i = 3{,}612.5$ kPa. The fugacity coefficient is given by

$$\phi_i = \frac{f_i}{P} = \frac{3{,}612.5}{4{,}000} = 0.9031$$

Similar calculations at other pressures lead to the values plotted in Fig. 11.1 at pressures up to the saturation pressure of 8,592.7 kPa, where $f_i^{\text{sat}} = 6{,}742.2$ kPa and $\phi_i^{\text{sat}} = 0.7846$. According to Eqs. (11.22) and (11.24), the saturation values are unchanged by condensation.

Values of f_i and ϕ_i for liquid water at higher pressures are found by applications of Eq. (11.26). Taking V_i equal to the molar volume of saturated liquid water at 300°C, we have,

$$V_i = (1.404)(18.016) = 25.29 \text{ cm}^3 \text{ mol}^{-1}$$

For a pressure of 10,000 kPa, Eq. (11.26) then gives

$$f_i = 6{,}742.2 \exp \frac{25.29(10{,}000 - 8{,}592.7)}{(8{,}314)(573.15)} = 6{,}792.7 \text{ kPa}$$

The fugacity coefficient for liquid water at these conditions is then

$$\phi_i = f_i/P = 6{,}792.7/10{,}000 = 0.6793$$

Such calculations allow completion of Fig. 11.1, where the solid lines show how f_i and ϕ_i vary with pressure.

The curve for f_i deviates increasingly with increasing pressure from ideal-gas behavior, which is shown by the dashed line, $f_i = P$. At P_i^{sat} there is a sharp break, and the curve then rises very slowly with increasing pressure. Thus the fugacity of liquid water at 300°C is a weak function of pressure. This behavior is characteristic of liquids at temperatures well below the critical temperature. The fugacity coefficient ϕ_i decreases steadily from its zero-pressure value of unity as the pressure rises. Its

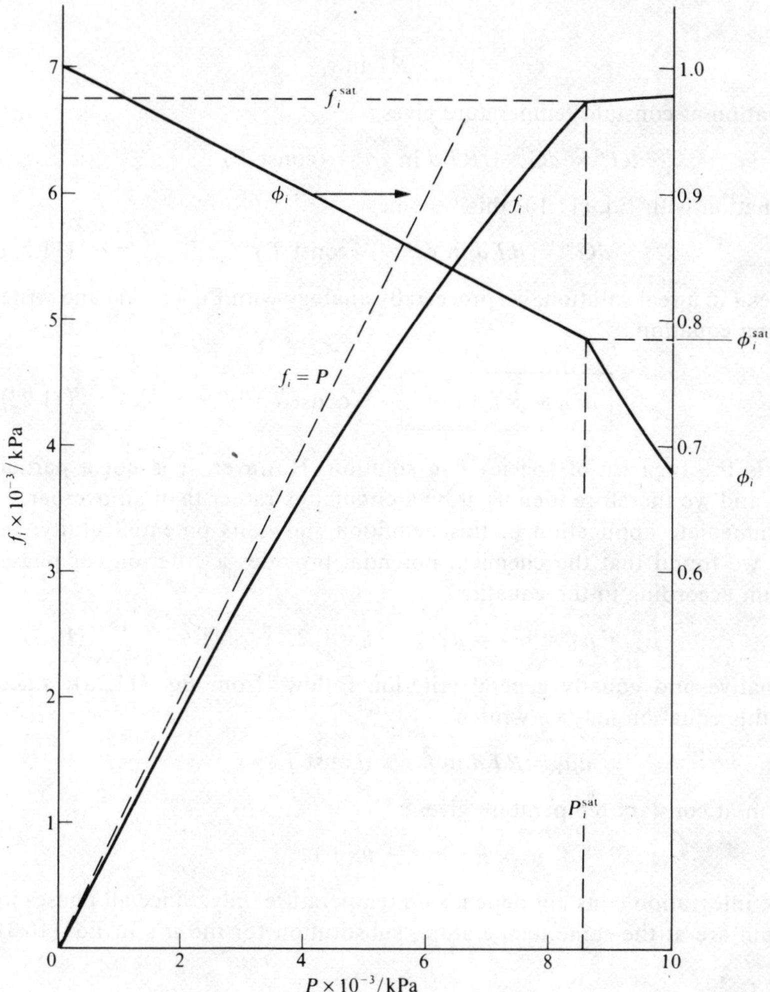

Figure 11.1 Fugacity and fugacity coefficients of pure species i as functions of pressure at constant temperature.

rapid decrease in the liquid region is a consequence of the near constancy of the fugacity itself.

11.3 FUGACITY AND FUGACITY COEFFICIENT FOR SPECIES i IN SOLUTION

For a species in solution, we recall that the chemical potential μ_i is identical with the partial molar Gibbs energy. Therefore, we write Eq. (10.9) for an ideal gas as

$$\bar{G}_i^{ig} = G_i^{ig} + RT \ln y_i$$

Differentiation at constant temperature gives

$$d\bar{G}_i^{ig} = dG_i^{ig} + RT\, d \ln y_i \qquad \text{(const } T\text{)}$$

In combination with Eq. (11.10) this becomes

$$d\bar{G}_i^{ig} = RT\, d \ln y_i P \qquad \text{(const } T\text{)} \tag{11.27}$$

For species i in a real solution, we proceed by analogy with Eq. (11.11) and write the defining equation:

$$\boxed{d\bar{G}_i \equiv RT\, d \ln \hat{f}_i} \qquad \text{(const } T\text{)} \tag{11.28}$$

where $\hat{f}_i$ is the fugacity of species i in solution. However, it is not a partial property, and we therefore identify it by a circumflex rather than an overbar.

An immediate application of this definition shows its potential utility. In Sec. 10.2 we found that the chemical potential provides a criterion for phase equilibrium according to the equation

$$\mu_i^\alpha = \mu_i^\beta = \cdots = \mu_i^\pi \qquad (i = 1, 2, \ldots, N) \tag{10.3}$$

An alternative and equally general criterion follows from Eq. (11.28); since $\mu_i = \bar{G}_i$, this equation may be written

$$d\mu_i = RT\, d \ln \hat{f}_i \qquad \text{(const } T\text{)}$$

Integration at constant temperature gives

$$\mu_i = RT \ln \hat{f}_i + \theta_i(T)$$

where the integration constant depends on temperature only. Since all phases in equilibrium are at the same temperature, substitution for the μ's in Eq. (10.3) leads to

$$\boxed{\hat{f}_i^\alpha = \hat{f}_i^\beta = \cdots = \hat{f}_i^\pi} \qquad (i = 1, 2, \ldots, N) \tag{11.29}$$

Thus multiple phases at the same T and P are in equilibrium when the fugacity of each species is uniform throughout the system. This criterion of equilibrium is the one usually applied by chemical engineers in the solution of phase-equilibrium problems.

For the specific case of multicomponent vapor/liquid equilibrium, Eq. (11.29) becomes

$$\hat{f}_i^v = \hat{f}_i^l \qquad (i = 1, 2, \ldots, N) \tag{11.30}$$

Equation (11.22) results as a special case when this relation is applied to the vapor/liquid equilibrium of *pure* species i.

The definition of the residual Gibbs energy as given by Eq. (6.32) is readily combined with Eq. (11.2), the definition of a partial property, to provide a defining equation for the partial residual Gibbs energy. Thus, upon multiplication by n, Eq. (6.32) becomes

$$nG^R = nG - nG^{ig}$$

This equation applies to n moles of mixture. Differentiation with respect to n_i at constant T, P, and the n_j gives:

$$\left[\frac{\partial(nG^R)}{\partial n_i}\right]_{P,T,n_j} = \left[\frac{\partial(nG)}{\partial n_i}\right]_{P,T,n_j} - \left[\frac{\partial(nG^{ig})}{\partial n_i}\right]_{P,T,n_j}$$

Reference to Eq. (11.2) shows that each term has the form of a partial molar property. Thus,

$$\boxed{\bar{G}_i^R = \bar{G}_i - \bar{G}_i^{ig}} \tag{11.31}$$

an equation which defines the *partial residual Gibbs energy*, $\bar{G}_i^R$.

Subtracting Eq. (11.27) from Eq. (11.28) gives

$$d(\bar{G}_i - \bar{G}_i^{ig}) = RT\,d\ln\frac{\hat{f}_i}{y_i P} \qquad \text{(const } T\text{)}$$

By Eq. (11.31), $\bar{G}_i - \bar{G}_i^{ig}$ is the partial residual Gibbs energy $\bar{G}_i^R$; the dimensionless ratio $\hat{f}_i/y_iP$ is called the *fugacity coefficient of species i in solution*, and is given the symbol $\hat{\phi}_i$. Then

$$d\bar{G}_i^R = RT\,d\ln\hat{\phi}_i \qquad \text{(const } T\text{)} \tag{11.32}$$

where

$$\boxed{\hat{\phi}_i \equiv \frac{\hat{f}_i}{y_iP}} \tag{11.33}$$

Integration of Eq. (11.32) at constant temperature yields the general equation

$$\bar{G}_i^R = RT\ln\hat{\phi}_i + \beta(T)$$

where the integration constant is a function of T. However, if this equation is applied to a pure species, it must reduce to Eq. (11.18). Thus $\beta(T) = 0$, and we have

$$\boxed{\frac{\bar{G}_i^R}{RT} = \ln\hat{\phi}_i} \tag{11.34}$$

This general result is the analog of Eqs. (11.16) and (11.18), which relate ϕ to

G^R and ϕ_i to $\bar{G}_i^R$. For an ideal gas, $\bar{G}_i^R$ is necessarily zero; therefore $\hat{\phi}_i^{ig} = 1$, and

$$\hat{f}_i^{ig} = y_i P \tag{11.35}$$

Thus the fugacity of a species in an ideal-gas mixture is equal to the partial pressure of the species.

Since $\bar{G}_i^R/RT$ is a partial property with respect to G^R/RT, Eqs. (11.34) and (11.16) show that $\ln \hat{\phi}_i$ is a partial property with respect to $\ln \phi$. As a result of Eqs. (11.2) and (11.5) we therefore have the following important relations:

$$\boxed{\ln \hat{\phi}_i = \left[\frac{\partial(n \ln \phi)}{\partial n_i}\right]_{P,T,n_j}} \tag{11.36}$$

and

$$\boxed{\ln \phi = \sum x_i \ln \hat{\phi}_i} \tag{11.37}$$

In addition, the Gibbs/Duhem equation as given by Eq. (11.8) becomes

$$\boxed{\sum x_i \, d \ln \hat{\phi}_i = 0} \qquad \text{(const } T, P) \tag{11.38}$$

Example 11.4 Develop a general equation for calculation of $\ln \hat{\phi}_i$ values from compressibility-factor data.

SOLUTION For n moles of a constant-composition mixture, Eq. (11.20) becomes

$$n \ln \phi = \int_0^P (nZ - n) \frac{dP}{P}$$

Direct application of Eq. (11.36) to this expression gives

$$\ln \hat{\phi}_i = \int_0^P \left[\frac{\partial(nZ - n)}{\partial n_i}\right]_{P,T,n_j} \frac{dP}{P}$$

Since $\partial(nZ)/\partial n_i = \bar{Z}_i$ and $\partial n/\partial n_i = 1$, this becomes

$$\ln \hat{\phi}_i = \int_0^P (\bar{Z}_i - 1) \frac{dP}{P} \tag{11.39}$$

where integration is at constant temperature and composition. This general equation is the partial-property analog of Eq. (11.20). It allows the calculation of $\hat{\phi}_i$ values from PVT data.

11.4 GENERALIZED CORRELATIONS FOR THE FUGACITY COEFFICIENT

The generalized methods developed in Sec. 3.6 for the compressibility factor Z and in Sec. 6.6 for the residual enthalpy and entropy of pure gases are applied here to the fugacity coefficient. Equation (11.20) is put into generalized form by substitution of the relationships

$$P = P_c P_r \qquad dP = P_c\, dP_r$$

Whence

$$\ln \phi = \int_0^{P_r} (Z - 1)\frac{dP_r}{P_r} \tag{11.40}$$

where integration is at constant T_r. Substitution for Z by Eq. (3.45) yields

$$\ln \phi = \int_0^{P_r} (Z^0 - 1)\frac{dP_r}{P_r} + \omega \int_0^{P_r} Z^1 \frac{dP_r}{P_r}$$

Alternatively, we may write

$$\ln \phi = \ln \phi^0 + \omega \ln \phi^1 \tag{11.41}$$

where

$$\ln \phi^0 \equiv \int_0^{P_r} (Z^0 - 1)\frac{dP_r}{P_r}$$

and

$$\ln \phi^1 \equiv \int_0^{P_r} Z^1 \frac{dP_r}{P_r}$$

Calculated values of $\ln \phi^0$ and $\ln \phi^1$ result from evaluation of the integrals for various T_r and P_r from the compressibility-factor data of Figs. 3.12 through 3.15, and we may plot these quantities vs. P_r for selected values of T_r. We also have the option of plotting ϕ^0 and ϕ^1 rather than their logarithms. Equation (11.41) is then written

$$\phi = (\phi^0)(\phi^1)^\omega \tag{11.42}$$

This is the choice made here, and Figs. 11.2 through 11.5 provide a three-parameter generalized correlation for the fugacity coefficient. Figures 11.2 and 11.4 for ϕ^0 can be used alone as a two-parameter correlation which does not incorporate the refinement introduced by the acentric factor.

Example 11.5 Estimate from Eq. (11.42) a value for the fugacity of 1-butene vapor at 200°C and 70 bar.

SOLUTION These are the same conditions given in Example 6.5, where we found

$$T_r = 1.13 \qquad P_r = 1.74 \qquad \omega = 0.187$$

From Figs. 11.3 and 11.5 at these conditions,

$$\phi^0 = 0.620 \qquad \text{and} \qquad \phi^1 = 1.095$$

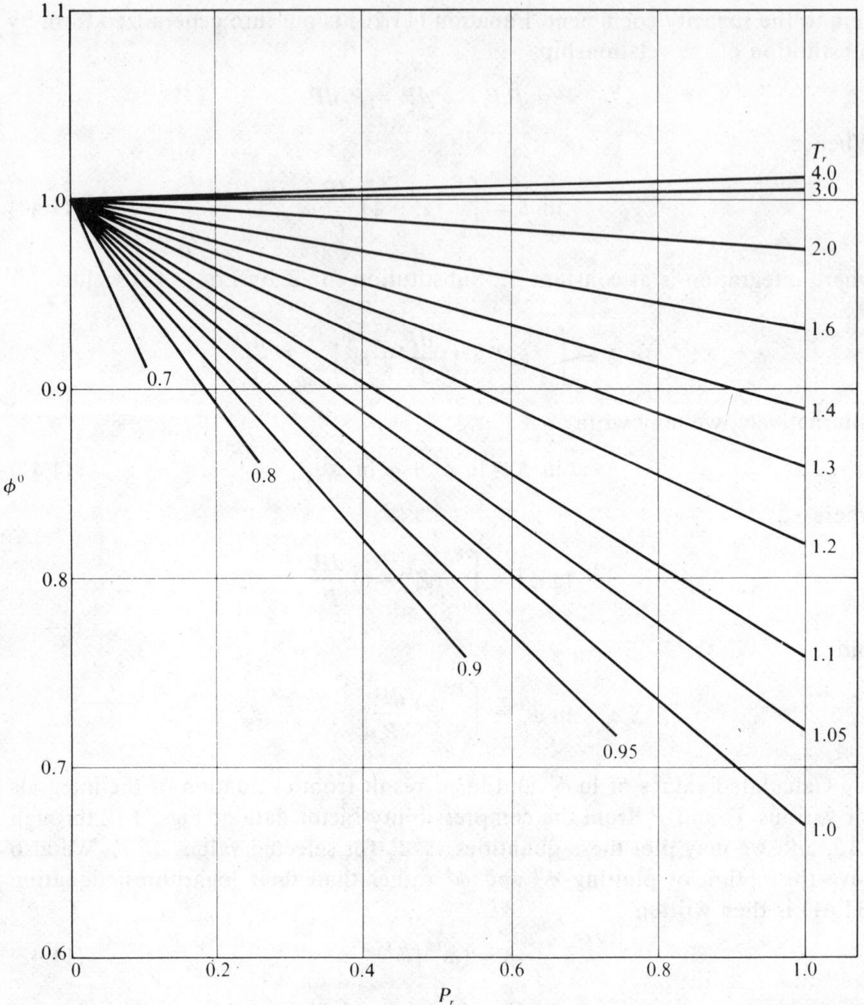

Figure 11.2 Generalized correlation for ϕ^0, $P_r < 1.0$. *(Based on data of B. I. Lee and M. G. Kesler, AIChE J.,* **21**: *510–527, 1975.)*

Equation (11.42) then gives:

$$\phi = (0.620)(1.095)^{0.187} = 0.631$$

and

$$f = \phi P = (0.631)(70) = 44.17 \text{ bar}$$

A particularly simple generalized correlation for $\ln \phi$ results when the simplest form of the virial equation is valid. Equations (3.46) and (3.47) combine

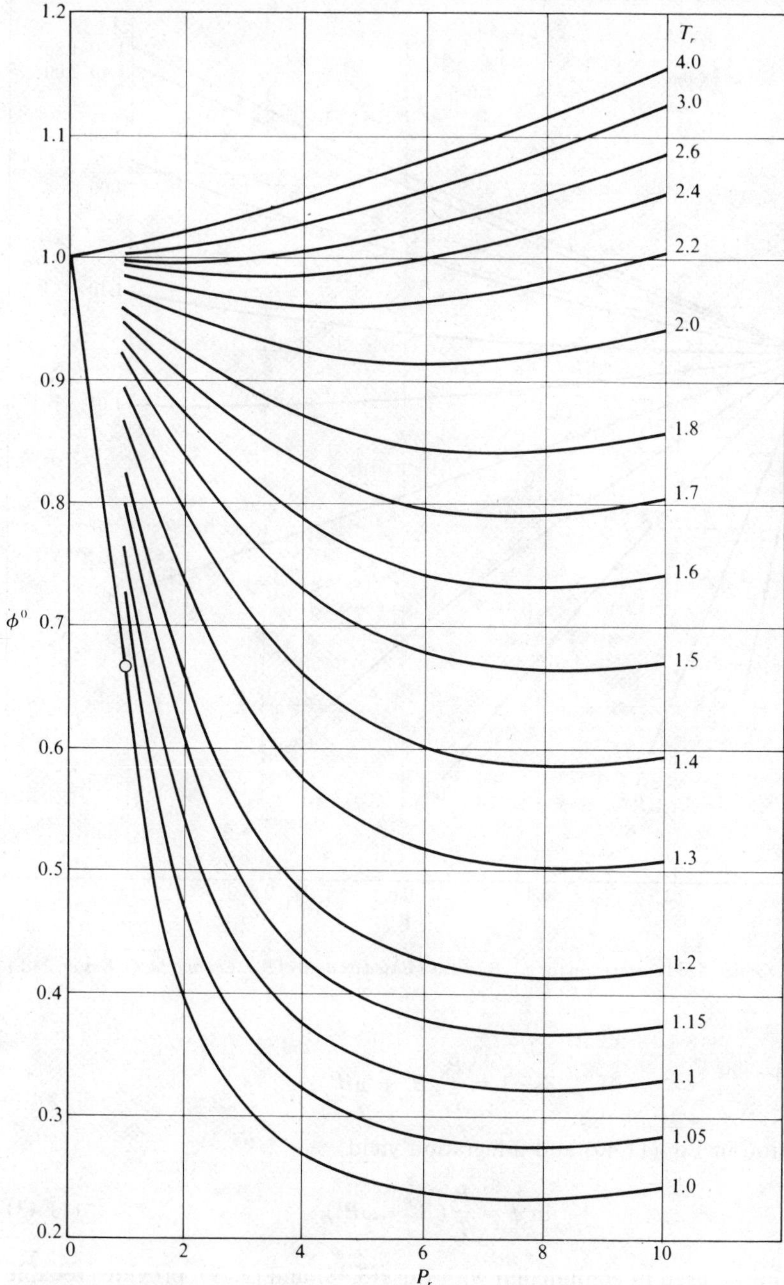

Figure 11.3 Generalized correlation for ϕ^0, $P_r > 1.0$. *(Based on data of B. I. Lee and M. G. Kesler, ibid.)*

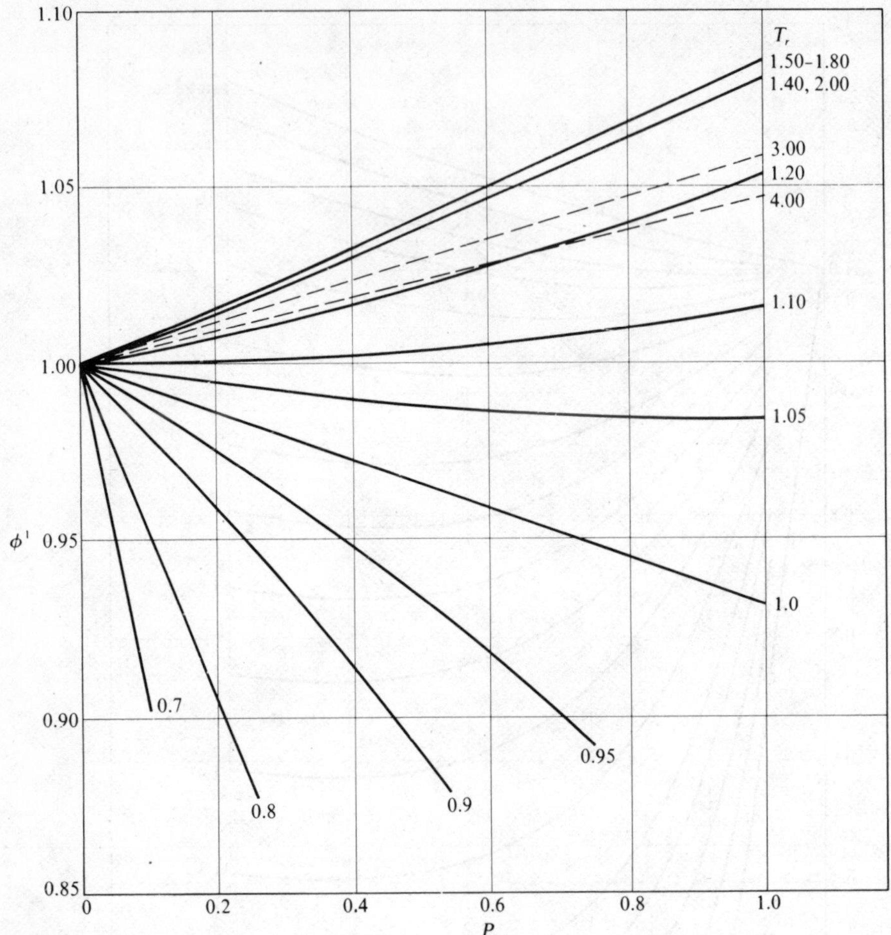

Figure 11.4 Generalized correlation for ϕ^1, $P_r < 1.0$. (*Based on data of B. I. Lee and M. G. Kesler,* ibid.)

to give

$$Z - 1 = \frac{P_r}{T_r}(B^0 + \omega B^1)$$

Substitution in Eq. (11.40) and integration yield

$$\ln \phi = \frac{P_r}{T_r}(B^0 + \omega B^1) \tag{11.43}$$

This equation, used in conjunction with Eqs. (3.48) and (3.49), provides reliable values of ϕ for any nonpolar or slightly polar gas when applied at conditions where Z is linear in pressure. Figure 3.16 again serves as a guide to the validity of this criterion.

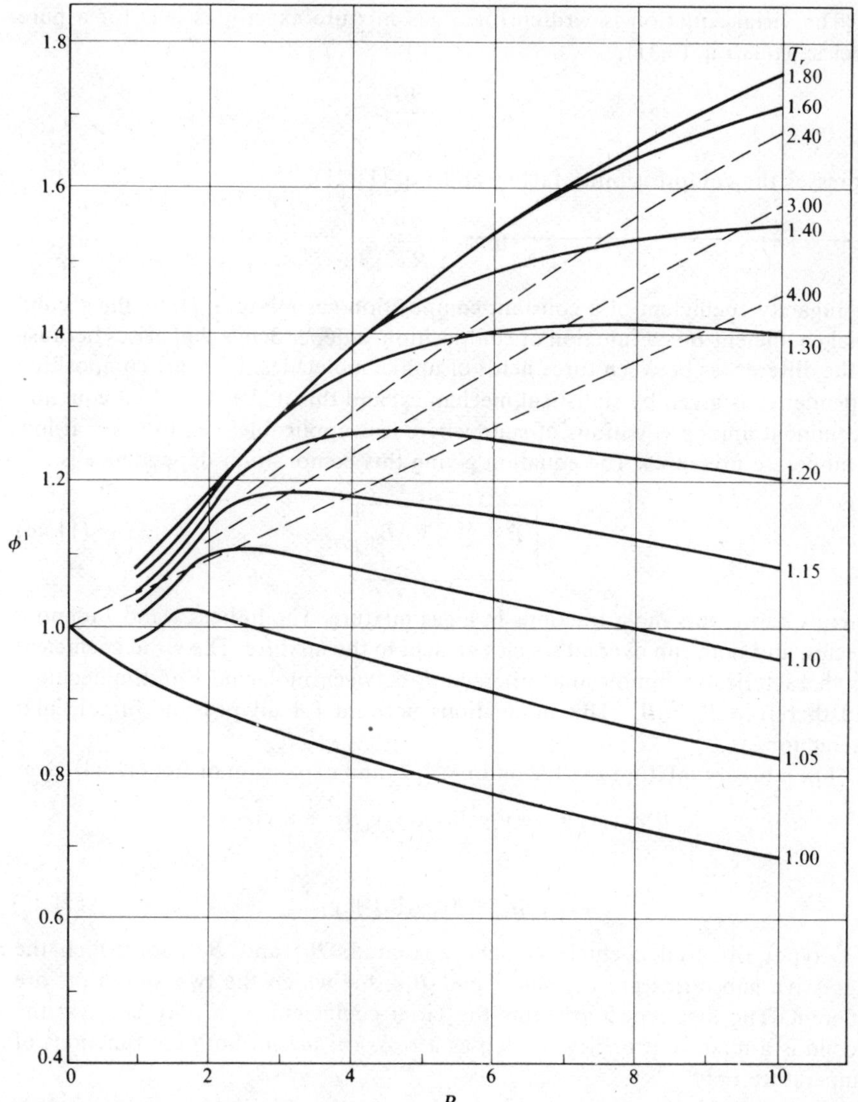

Figure 11.5 Generalized correlation for ϕ^1, $P_r > 1.0$. (*Based on data of B. I. Lee and M. G. Kesler*, ibid.)

Although we have omitted an identifying subscript in the preceding equations, their application so far has been to the development of generalized correlations for *pure* gases only. In the remainder of this section we show how the virial equation may be generalized to allow calculation of fugacity coefficients $\hat{\phi}_i$ of species in gas *mixtures*.

The virial equation is written for a gas mixture exactly as it is for a pure species. Thus Eq. (3.31),

$$Z = 1 + \frac{BP}{RT}$$

expresses the compressibility factor, and Eq. (11.21),

$$\ln \phi = \frac{BP}{RT}$$

the fugacity coefficient of a constant-composition gas mixture. Here, the second virial coefficient B is a function of composition, a dependence that arises because of the differences between force fields of unlike molecules. Its *exact* composition dependence is given by statistical mechanics, and this makes the virial equation preeminent among equations of state where it is applicable, i.e., to gases at low to moderate pressures. The equation giving this composition dependence is

$$\boxed{B = \sum_i \sum_j y_i y_j B_{ij}} \tag{11.44}$$

where y represents mole fractions in a gas mixture. The indices i and j identify species, and both run over all species present in the mixture. The virial coefficient B_{ij} characterizes a bimolecular interaction between molecule i and molecule j, and therefore $B_{ij} = B_{ji}$. The summations account for all possible bimolecular interactions.

For a binary mixture $i = 1, 2$ and $j = 1, 2$, and expansion of Eq. (11.44) gives

$$B = y_1 y_1 B_{11} + y_1 y_2 B_{12} + y_2 y_1 B_{21} + y_2 y_2 B_{22}$$

or

$$B = y_1^2 B_{11} + 2 y_1 y_2 B_{12} + y_2^2 B_{22} \tag{11.45}$$

Two types of virial coefficients have appeared: B_{11} and B_{22}, for which the successive subscripts are the same, and B_{12}, for which the two subscripts are different. The first type represents the virial coefficient of a pure species; the second is a mixture property, known as a *cross coefficient.* Both are functions of temperature only.

Equation (11.45) allows us to find expressions for $\ln \hat{\phi}_1$ and $\ln \hat{\phi}_2$ for a binary gas mixture that obeys Eq. (3.31), the simplest form of the virial equation. Equation (11.21) for the mixture may be multiplied by n:

$$n \ln \phi = \frac{(nB)P}{RT}$$

Differentiation with respect to n_1 gives

$$\left[\frac{\partial(n \ln \phi)}{\partial n_1}\right]_{P,T,n_2} = \frac{P}{RT}\left[\frac{\partial(nB)}{\partial n_1}\right]_{T,n_2}$$

In view of Eq. (11.36) this may be written

$$\ln \hat{\phi}_1 = \frac{P}{RT}\left[\frac{\partial(nB)}{\partial n_1}\right]_{T,n_2}$$

All that remains is evaluation of the derivative.

The second virial coefficient as given by Eq. (11.45) may be written:

$$B = y_1(1 - y_2)B_{11} + 2y_1y_2B_{12} + y_2(1 - y_1)B_{22}$$
$$= y_1B_{11} - y_1y_2B_{11} + 2y_1y_2B_{12} + y_2B_{22} - y_1y_2B_{22}$$

or

$$B = y_1B_{11} + y_2B_{22} + y_1y_2\delta_{12}$$

where

$$\delta_{12} \equiv 2B_{12} - B_{11} - B_{22}$$

Since $y_i = n_i/n$,

$$nB = n_1B_{11} + n_2B_{22} + \frac{n_1n_2}{n}\delta_{12}$$

Differentiation gives

$$\left[\frac{\partial(nB)}{\partial n_1}\right]_{T,n_2} = B_{11} + \left(\frac{1}{n} - \frac{n_1}{n^2}\right)n_2\delta_{12}$$
$$= B_{11} + (1 - y_1)y_2\delta_{12} = B_{11} + y_2^2\delta_{12}$$

Therefore

$$\ln \hat{\phi}_1 = \frac{P}{RT}(B_{11} + y_2^2\delta_{12}) \tag{11.46}$$

and similarly,

$$\ln \hat{\phi}_2 = \frac{P}{RT}(B_{22} + y_1^2\delta_{12}) \tag{11.47}$$

Equations (11.46) and (11.47) are readily extended for application to multicomponent mixtures;† the general equation is:

$$\ln \hat{\phi}_k = \frac{P}{RT}\left[B_{kk} + \frac{1}{2}\sum_i\sum_l y_iy_l(2\delta_{ik} - \delta_{il})\right] \tag{11.48}$$

where the dummy indices i and l run over all species, and

$$\delta_{ik} \equiv 2B_{ik} - B_{ii} - B_{kk}$$
$$\delta_{il} \equiv 2B_{il} - B_{ii} - B_{ll}$$

with $\delta_{ii} = 0$, $\delta_{kk} = 0$, etc., and $\delta_{ki} = \delta_{ik}$, etc.

† H. C. Van Ness and M. M. Abbott, *Classical Thermodynamics of Nonelectrolyte Solutions: With Applications to Phase Equilibria*, pp. 135–140, McGraw-Hill, New York, 1982.

Values of the pure-species virial coefficients B_{kk}, B_{ii}, etc., can be determined from the generalized correlation represented by Eqs. (3.47) through (3.49). The cross coefficients B_{ik}, B_{ij}, etc., are found from an extension of the same correlation. For this purpose, Prausnitz† has rewritten Eq. (3.47) in the more general form:

$$B_{ij} = \frac{RT_{cij}}{P_{cij}}(B^0 + \omega_{ij}B^1) \tag{11.49}$$

where B^0 and B^1 are the same functions of T_r as given by Eqs. (3.48) and (3.49). The combining rules proposed by Prausnitz for calculation of ω_{ij}, T_{cij}, and P_{cij} are:

$$\omega_{ij} = \frac{\omega_i + \omega_j}{2} \tag{11.50}$$

$$T_{cij} = (T_{ci}T_{cj})^{1/2}(1 - k_{ij}) \tag{11.51}$$

and

$$P_{cij} = \frac{Z_{cij}RT_{cij}}{V_{cij}} \tag{11.52}$$

where

$$Z_{cij} = \frac{Z_{ci} + Z_{cj}}{2} \tag{11.53}$$

and

$$V_{cij} = \left(\frac{V_{ci}^{1/3} + V_{cj}^{1/3}}{2}\right)^3 \tag{11.54}$$

In Eq. (11.51), k_{ij} is an empirical interaction parameter specific to an i-j molecular pair. When $i = j$ or when the species are chemically similar, $k_{ij} = 0$. Otherwise, it is a small positive number evaluated from minimal *PVT* data or in the absence of data set equal to zero.

When $i = j$, all equations reduce to the appropriate values for a pure species. When $i \neq j$, these equations define a set of interaction parameters having no physical significance. Reduced temperature is given for each ij pair by $T_{rij} = T/T_{cij}$.

For a mixture, values of B_{ij} from Eq. (11.49) substituted into Eq. (11.44) yield the mixture second virial coefficient B, and substituted into Eq. (11.48) [Eqs. (11.46) and (11.47) for a binary] they yield values of $\ln \hat{\phi}_i$.

The primary virtue of the generalized correlation for second virial coefficients presented here is simplicity; more accurate, but more complex, correlations appear in the literature.‡

† J. M. Prausnitz, *Molecular Thermodynamics of Fluid-Phase Equilibria*, chap. 5, Prentice Hall, Englewood Cliffs, N.J., 1969.

‡ See, for example: J. G. Hayden and J. P. O'Connell, *Ind. Eng. Chem. Proc. Des. Dev.*, **14**: 209, 1975; D. W. McCann and R. P. Danner, *Ind. Eng. Chem. Proc. Des. Dev.*, **23**: 529, 1984.

Example 11.6 Estimate $\hat{\phi}_1$ and $\hat{\phi}_2$ by Eqs. (11.46) and (11.47) for an equimolar mixture of methyl ethyl ketone(1)/toluene(2) at 50°C and 25 kPa. Set all $k_{ij} = 0$.

SOLUTION The required data are as follows:

ij	T_{cij}/K	P_{cij}/bar	V_{cij}/cm³ mol⁻¹	Z_{cij}	ω_{ij}
11	535.6	41.5	267.	0.249	0.329
22	591.7	41.1	316.	0.264	0.257
12	563.0	41.3	291.	0.256	0.293

where values in the last row have been calculated by Eqs. (11.50) through (11.54). The values of T_{rij}, together with B^0, B^1, and B_{ij} calculated for each ij pair by Eqs. (3.48), (3.49), and (11.49), are as follows:

ij	T_{rij}	B^0	B^1	B_{ij}/cm³ mol⁻¹
11	0.603	−0.865	−1.300	−1,387
22	0.546	−1.028	−2.045	−1,860
12	0.574	−0.943	−1.632	−1,611

Calculating δ_{12} according to its definition, we get

$$\delta_{12} = 2B_{12} - B_{11} - B_{22} = (2)(-1{,}611) + 1{,}387 + 1{,}860$$
$$= 25 \text{ cm}^3 \text{ mol}^{-1}$$

Equations (11.46) and (11.47) then yield:

$$\ln \hat{\phi}_1 = \frac{P}{RT}(B_{11} + y_2^2\delta_{12}) = \frac{25}{(8{,}314)(323.15)}[-1{,}387 + (0.5)^2(25)]$$
$$= -0.0128$$

$$\ln \hat{\phi}_2 = \frac{P}{RT}(B_{22} + y_1^2\delta_{12}) = \frac{25}{(8{,}314)(323.15)}[-1{,}860 + (0.5)^2(25)]$$
$$= -0.0172$$

Whence

$$\hat{\phi}_1 = 0.987 \qquad \text{and} \qquad \hat{\phi}_2 = 0.983$$

These results are representative of values obtained for vapor phases at typical conditions of low-pressure vapor/liquid equilibrium.

11.5 THE EXCESS GIBBS ENERGY

The residual Gibbs energy and the fugacity coefficient are directly related to experimental *PVT* data by Eqs. (6.40) and (11.20). Where such data can be adequately correlated by equations of state, thermodynamic-property information is advantageously provided by these and other residual properties. Indeed, if convenient treatment of all fluids by means of equations of state were possible, the thermodynamic-property relations already presented would suffice. However, *liquid* solutions are often more easily dealt with through properties that measure their deviations, not from ideal-gas behavior, but from ideal-solution behavior. Thus the mathematical formalism of *excess* properties is analogous to that of the residual properties.

If M represents the molar value of an extensive thermodynamic property (for example, V, U, H, S, G, etc.), then an excess property M^E is defined as the difference between the actual property value of a solution and the value it would have as an ideal solution (Sec. 10.4) at the same temperature, pressure, and composition. Thus,

$$\boxed{M^E \equiv M - M^{id}} \tag{11.55}$$

where the superscript *id* denotes an ideal-solution value. This definition is analogous to the definition of a residual property as given by Eq. (6.35). However, excess properties have no meaning for pure species, whereas residual properties exist for pure species as well as for mixtures.

The only excess property of immediate interest is the excess Gibbs energy,

$$\boxed{G^E = G - G^{id}} \tag{11.56}$$

Multiplication of this equation by n and differentiation with respect to n_i at constant T, P, and n_j leads to the analog of Eq. (11.31), which was derived in exactly the same way:

$$\boxed{\bar{G}_i^E = \bar{G}_i - \bar{G}_i^{id}} \tag{11.57}$$

Equation (11.57) defines the partial excess Gibbs energy.

Equation (11.28) may be integrated *at constant T and P* for the change of species i from a state of pure i, where $\bar{G}_i = G_i$ and $\hat{f}_i = f_i$, to a state in solution at arbitrary mole fraction x_i:

$$\bar{G}_i - G_i = RT \ln \frac{\hat{f}_i}{f_i} \tag{11.58}$$

Since the chemical potential μ_i and the partial molar Gibbs energy are identical, Eq. (10.14) gives the partial molar Gibbs energy for species i in an ideal solution:

$$\bar{G}_i^{id} - G_i = RT \ln x_i$$

The difference between this expression and Eq. (11.58) is

$$\bar{G}_i - \bar{G}_i^{id} = RT \ln \frac{\hat{f}_i}{x_1 f_i}$$

According to Eq. (11.57), $\bar{G}_i - \bar{G}_i^{id}$ is the partial excess Gibbs energy $\bar{G}_i^E$; the dimensionless ratio $\hat{f}_i / x_i f_i$ is called the *activity coefficient of species i in solution,* and is given the symbol γ_i. Thus, by definition,

$$\boxed{\gamma_i \equiv \frac{\hat{f}_i}{x_i f_i}} \tag{11.59}$$

and

$$\bar{G}_i^E = RT \ln \gamma_i$$

or

$$\boxed{\frac{\bar{G}_i^E}{RT} = \ln \gamma_i} \tag{11.60}$$

Comparison with Eq. (11.34) shows that Eq. (11.60) relates γ_i to $\bar{G}_i^E$ exactly as Eq. (11.34) relates $\hat{\phi}_i$ to $\bar{G}_i^R$.

For an ideal solution, $\bar{G}_i^E = 0$, and therefore $\gamma_i = 1$. For this case, Eq. (11.59) becomes

$$\hat{f}_i^{id} = x_i f_i \tag{11.61}$$

This expression is known as the *Lewis/Randall rule.*

Since $\bar{G}_i^E / RT$ is a partial property with respect to G^E/RT, it follows from Eq. (11.60) that $\ln \gamma_i$ is also a partial property with respect to G^E/RT. As a result of Eqs. (11.2), (11.5), and (11.8) we therefore have the following important relations:

$$\boxed{\ln \gamma_i = \left[\frac{\partial(nG^E/RT)}{\partial n_i}\right]_{P,T,n_j}} \tag{11.62}$$

$$\boxed{\frac{G^E}{RT} = \sum x_i \ln \gamma_i} \tag{11.63}$$

and

$$\boxed{\sum x_i \, d \ln \gamma_i = 0} \qquad \text{(const } T, P) \tag{11.64}$$

The usefulness of these equations derives from the fact that γ_i values are experimentally accessible through vapor/liquid equilibrium (VLE) data, as explained in the following section. Once established, values of the activity coefficients are used in the calculation of phase compositions for systems in vapor/liquid equilibrium, as discussed in Chap. 12.

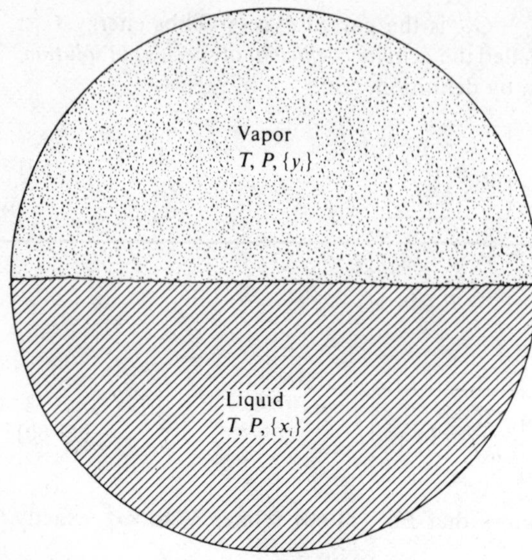

Figure 11.6 A state of vapor/liquid equilibrium represented schematically.

11.6 ACTIVITY COEFFICIENTS FROM VLE DATA

Figure 11.6 shows a vessel in which a vapor mixture and a liquid solution coexist in equilibrium. The temperature T and pressure P are uniform throughout the vessel, and can be measured with appropriate instruments. Samples of the vapor and liquid phases may be withdrawn for analysis, and this provides experimental values for the mole fractions in the vapor $\{y_i\}$ and the mole fractions in the liquid $\{x_i\}$. For species i in the vapor mixtures, Eq. (11.33) is written:

$$\hat{f}_i^v = y_i \hat{\phi}_i P$$

and for species i in the liquid solution, Eq. (11.59) becomes

$$\hat{f}_i^l = x_i \gamma_i f_i$$

According to Eq. (11.30) these two expressions must be equal; whence

$$\boxed{y_i \hat{\phi}_i P = x_i \gamma_i f_i} \qquad (i = 1, 2, \ldots, N) \tag{11.65}$$

Superscripts v and l are omitted here with the understanding that $\hat{\phi}_i$ refers to the vapor phase and that γ_i and f_i are liquid-phase properties. Substituting for f_i by Eq. (11.26) and solving for γ_i gives

$$\gamma_i = \frac{y_i \Phi_i P}{x_i P_i^{sat}} \qquad (i = 1, 2, \ldots, N) \tag{11.66}$$

Table 11.1 VLE Data for methyl ethyl ketone(1)/toluene(2) at 50°C

P/kPa	x_1	y_1	$\ln \gamma_1$	$\ln \gamma_2$	G^E/RT	G^E/x_1x_2RT
12.30†	0.0000	0.0000	......	0.000	0.000	
15.51	0.0895	0.2716	0.266	0.009	0.032	0.389
18.61	0.1981	0.4565	0.172	0.025	0.054	0.342
21.63	0.3193	0.5934	0.108	0.049	0.068	0.312
24.01	0.4232	0.6815	0.069	0.075	0.072	0.297
25.92	0.5119	0.7440	0.043	0.100	0.071	0.283
29.96	0.6096	0.8050	0.023	0.127	0.063	0.267
30.12	0.7135	0.8639	0.010	0.151	0.051	0.248
31.75	0.7934	0.9048	0.003	0.173	0.038	0.234
34.15	0.9102	0.9590	−0.003	0.237	0.019	0.227
36.09‡	1.0000	1.0000	0.000	......	0.000	

† P_2^{sat}
‡ P_1^{sat}

where

$$\Phi_i \equiv \frac{\hat{\phi}_i}{\phi_i^{\text{sat}}} \exp\left[-\frac{V_i(P - P_i^{\text{sat}})}{RT}\right] \tag{11.67}$$

We could of course calculate Φ_i values by Eq. (11.67) for conditions of low-pressure VLE and combine them with experimental values of P, T, x_i, and y_i for the evaluation of activity coefficients by Eq. (11.66). However, at low pressures (up to at least 1 bar), vapor phases usually approximate ideal gases, for which $\hat{\phi}_i = \phi_i^{\text{sat}} = 1$, and the Poynting factor (represented by the exponential) differs from unity by only a few parts per thousand. Moreover, values of $\hat{\phi}_i$ and ϕ_i^{sat} differ significantly less from each other than from unity, and their influence in Eq. (11.67) tends to cancel. Thus the assumption that $\Phi_i = 1$ introduces little error for low-pressure VLE, and it reduces Eq. (11.66) to

$$\gamma_i = \frac{y_i P}{x_i P_i^{\text{sat}}} \qquad (i = 1, 2, \ldots, N) \tag{11.68}$$

This simple equation is adequate to our present purpose, allowing easy calculation of activity coefficients from experimental low-pressure VLE data. For comparison, when a system obeys Raoult's law, $y_i P = x_i P_i^{\text{sat}}$, and $\gamma_i = 1$.

The first three columns of Table 11.1 contain experimental P-x_1-y_1 data for the methyl ethyl ketone(1)/toluene(2) system at 50°C.† These data points are also shown as circles on Fig. 11.7. Values of $\ln \gamma_1$ and $\ln \gamma_2$ calculated for each data point by Eq. (11.68) are listed in columns 4 and 5 of Table 11.1, and are shown by the open squares and triangles in Fig. 11.8. These are combined

† M. Díaz Peña, A. Crespo Colin, and A. Compostizo, *J. Chem. Thermodyn.*, **10**: 337, 1978.

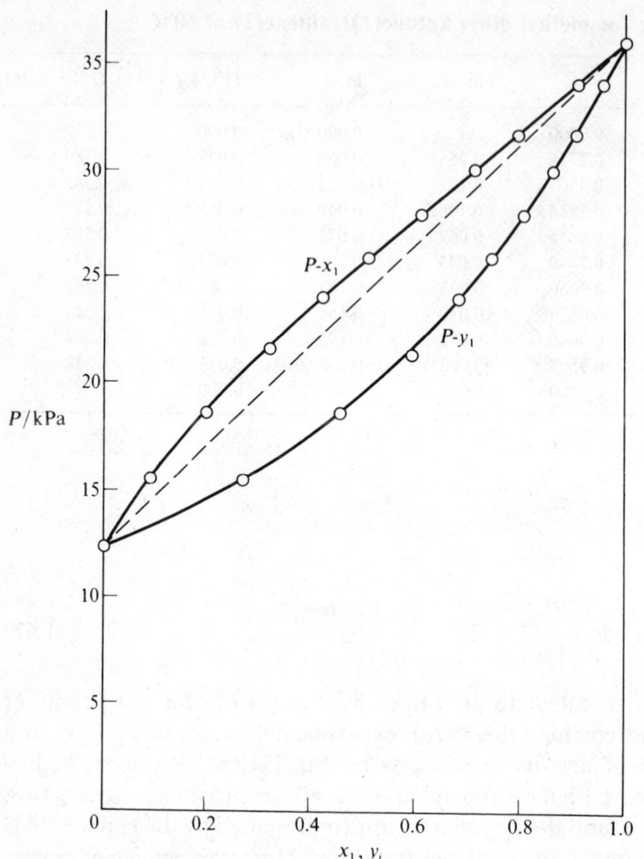

Figure 11.7 *Pxy* data at 50°C for methyl ethyl ketone(1)/toluene(2).

according to Eq. (11.63) written for a binary system:

$$\frac{G^E}{RT} = x_1 \ln \gamma_1 + x_2 \ln \gamma_2 \tag{11.69}$$

The values of G^E/RT so calculated are divided by x_1x_2 to provide in addition values of G^E/x_1x_2RT; the two sets of numbers are listed in columns 6 and 7 of Table 11.1 and appear as solid circles on Fig. 11.8.

The four thermodynamic functions for which we have experimental values, $\ln \gamma_1$, $\ln \gamma_2$, G^E/RT, and G^E/x_1x_2RT, are properties of the liquid phase. Figure 11.8 shows how each varies with composition. This figure is characteristic of systems for which

$$\gamma_i \geqq 1 \qquad \text{and} \qquad \ln \gamma_i \geq 0 \qquad (i = 1, 2)$$

Such systems are said to show *positive deviations from Raoult's law.* This is seen

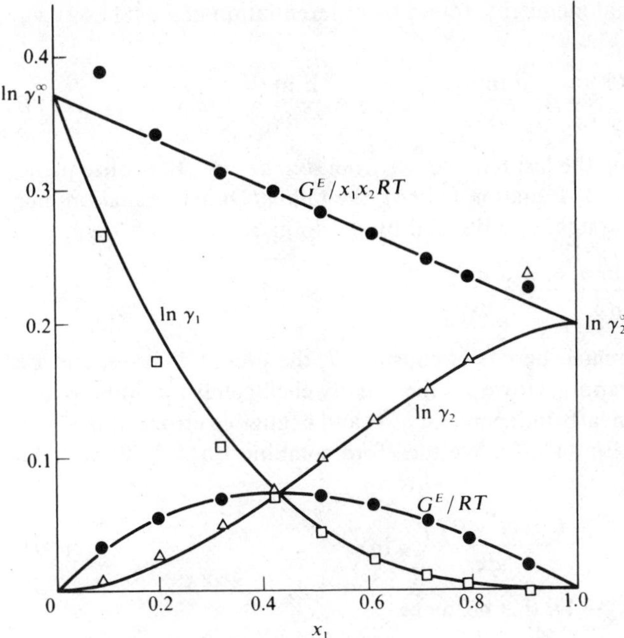

Figure 11.8 Liquid-phase properties from VLE data for methyl ethyl ketone(1)/toluene(2) at 50°C.

also in Fig. 11.7, where the P-x_1 data points all lie above the dashed line representing the linear relation of Raoult's law.

The points on Fig. 11.8 representing ln γ_i ($i = 1, 2$) are seen to tend toward zero as $x_i \to 1$. This is in accord both with Eq. (11.59) and Eq. (11.68); by the latter,

$$\lim_{x_i \to 1} \gamma_i = \lim_{x_i \to 1} \frac{y_i P}{x_i P_i^{\text{sat}}} = \frac{(1)(P_i^{\text{sat}})}{(1)(P_i^{\text{sat}})} = 1$$

Thus the activity coefficient of a species in solution becomes unity as the species becomes pure. At the other limit, where $x_i \to 0$ and species i becomes infinitely dilute, ln γ_i is seen to approach some finite limit, which we represent by ln γ_i^∞.

In the limit as $x_1 \to 0$, the dimensionless excess Gibbs energy G^E/RT as given by Eq. (11.69) becomes

$$\lim_{x_1 \to 0} \frac{G^E}{RT} = (0) \ln \gamma_i^\infty + (1)(0) = 0$$

The same result is obtained for $x_2 \to 0$ ($x_1 \to 1$). Thus the value of G^E/RT goes to zero at both $x_1 = 0$ and $x_1 = 1$.

The quantity G^E/x_1x_2RT becomes indeterminate both at $x_1 = 0$ and $x_1 = 1$, because G^E is zero in both limits, as is the product x_1x_2. Thus for $x_1 \to 0$, we have

$$\lim_{x_1 \to 0} \frac{G^E}{x_1 x_2 RT} = \lim_{x_1 \to 0} \frac{G^E/RT}{x_1} = \lim_{x_1 \to 0} \frac{d(G^E/RT)}{dx_1} \tag{A}$$

The derivative of the final member is found by differentiation of Eq. (11.69) with respect to x_1:

$$\frac{d(G^E/RT)}{dx_1} = x_1 \frac{d \ln \gamma_1}{dx_1} + \ln \gamma_1 + x_2 \frac{d \ln \gamma_2}{dx_1} - \ln \gamma_2 \tag{B}$$

The minus sign preceding the last term comes from $dx_2/dx_1 = -1$, a consequence of the equation $x_1 + x_2 = 1$. Equation (11.64), the Gibbs/Duhem equation, may be written for a binary system and divided by dx_1 to give:

$$x_1 \frac{d \ln \gamma_1}{dx_1} + x_2 \frac{d \ln \gamma_2}{dx_1} = 0 \qquad \text{(const } T, P\text{)} \tag{11.70}$$

Although the data set treated here is at constant T, the pressure varies, and Eq. (11.70) strictly does not apply. However, the activity coefficients for liquid phases at low pressure are very nearly independent of P, and negligible error is introduced through application of Eq. (11.70). We therefore combine Eq. (11.70) with Eq. (B) to get

$$\frac{d(G^E/RT)}{dx_1} = \ln \frac{\gamma_1}{\gamma_2} \tag{11.71}$$

In the limit as $x_1 \to 0$ $(x_2 \to 1)$, this becomes

$$\lim_{x_1 \to 0} \frac{d(G^E/RT)}{dx_1} = \lim_{x_1 \to 0} \ln \frac{\gamma_1}{\gamma_2} = \ln \gamma_1^\infty$$

and by Eq. (A),

$$\lim_{x_1 \to 0} \frac{G^E}{x_1 x_2 RT} = \ln \gamma_1^\infty$$

Similarly, as $x_1 \to 1$ $(x_2 \to 0)$,

$$\lim_{x_1 \to 1} \frac{G^E}{x_1 x_2 RT} = \ln \gamma_2^\infty$$

Thus the limiting values of $G^E/x_1 x_2 RT$ are equal to the infinite-dilution limits of $\ln \gamma_1$ and $\ln \gamma_2$.

Equation (11.70), the Gibbs/Duhem equation, has further influence on the nature of Fig. 11.8. Rewritten as

$$\frac{d \ln \gamma_1}{dx_1} = -\frac{x_2}{x_1} \frac{d \ln \gamma_2}{dx_1}$$

it shows the direct relation required between the slopes of curves drawn through the data points for $\ln \gamma_1$ and $\ln \gamma_2$. Qualitatively, we observe that at every composition the slope of the $\ln \gamma_1$ curve is of opposite sign to the slope of the $\ln \gamma_2$ curve. Furthermore, when $x_2 \to 0$ (and $x_1 \to 1$), the slope of the $\ln \gamma_1$ curve is zero. Similarly, when $x_1 \to 0$, the slope of the $\ln \gamma_2$ curve is zero. Thus, each $\ln \gamma_i$ $(i = 1, 2)$ curve becomes horizontal at $x_i = 1$.

Of the sets of points shown in Fig. 11.8, those for G^E/x_1x_2RT most closely conform to a simple mathematical relation. Thus we draw a straight line as a reasonable approximation to this set of points, and we give mathematical expression to this assumed linear relation by an equation of the form

$$\frac{G^E}{x_1x_2RT} = A_{21}x_1 + A_{12}x_2 \tag{11.72a}$$

where A_{21} and A_{12} are constants in any particular application. Alternatively,

$$\frac{G^E}{RT} = (A_{21}x_1 + A_{12}x_2)x_1x_2 \tag{11.72b}$$

Application of Eq. (11.62) to this expression leads to equations for $\ln \gamma_1$ and $\ln \gamma_2$:

$$\ln \gamma_1 = x_2^2[A_{12} + 2(A_{21} - A_{12})x_1] \tag{11.73a}$$

and

$$\ln \gamma_2 = x_1^2[A_{21} + 2(A_{12} - A_{21})x_2] \tag{11.73b}$$

These are the *Margules† equations*, and they represent a commonly used empirical model of solution behavior. For the limiting conditions of infinite dilution, they show that when $x_1 = 0$, $\ln \gamma_1^\infty = A_{12}$, and when $x_2 = 0$, $\ln \gamma_2^\infty = A_{21}$. For the methyl ethyl ketone/toluene system considered here, the curves of Fig. 11.8 for G^E/RT, $\ln \gamma_1$, and $\ln \gamma_2$ represent Eqs. (11.72*b*), (11.73*a*), and (11.73*b*) with $A_{12} = 0.372$ and $A_{21} = 0.198$, the intercepts at $x_1 = 0$ and $x_1 = 1$ of the straight line drawn to represent the G^E/x_1x_2RT data points.

What we have accomplished is the *reduction* of a set of VLE data to a simple mathematical equation for the dimensionless excess Gibbs energy,

$$\frac{G^E}{RT} = (0.198x_1 + 0.372x_2)x_1x_2$$

which concisely stores the information of the data set. Indeed, with the Margules equations for $\ln \gamma_1$ and $\ln \gamma_2$, we can easily construct a correlation of the original P-x_1-y_1 data.

Rearrangement of Eq. (11.68) provides a *modified Raoult's law*:

$$\boxed{y_iP = x_i\gamma_iP_i^{\text{sat}}} \qquad (i = 1, 2, \ldots, N) \tag{11.74}$$

For species 1 and 2 of a binary system,

$$y_1P = x_1\gamma_1P_1^{\text{sat}}$$

and

$$y_2P = x_2\gamma_2P_2^{\text{sat}}$$

† Max Margules (1856–1920), Austrian meteorologist and physicist.

Table 11.2 VLE Data for chloroform(1)/1,4-dioxane(2) at 50°C

P/kPa	x_1	y_1	$\ln \gamma_1$	$\ln \gamma_2$	G^E/RT	G^E/x_1x_2RT
15.79†	0.0000	0.0000	....	0.000	0.000	
17.51	0.0932	0.1794	−0.722	0.004	−0.064	−0.758
18.15	0.1248	0.2383	−0.694	−0.000	−0.086	−0.790
19.30	0.1757	0.3302	−0.648	−0.007	−0.120	−0.825
19.89	0.2000	0.3691	−0.636	−0.007	−0.133	−0.828
21.37	0.2626	0.4628	−0.611	−0.014	−0.171	−0.882
24.95	0.3615	0.6184	−0.486	−0.057	−0.212	−0.919
29.82	0.4750	0.7552	−0.380	−0.127	−0.248	−0.992
34.80	0.5555	0.8378	−0.279	−0.218	−0.252	−1.019
42.10	0.6718	0.9137	−0.192	−0.355	−0.245	−1.113
60.38	0.8780	0.9860	−0.023	−0.824	−0.120	−1.124
65.39	0.9398	0.9945	−0.002	−0.972	−0.061	−1.074
69.36‡	1.0000	1.0000	0.000	....	0.000	

† P_2^{sat}
‡ P_1^{sat}

Addition gives:

$$\boxed{P = x_1\gamma_1 P_1^{sat} + x_2\gamma_2 P_2^{sat}} \tag{11.75}$$

whence

$$\boxed{y_1 = \frac{x_1\gamma_1 P_1^{sat}}{x_1\gamma_1 P_1^{sat} + x_2\gamma_2 P_2^{sat}}} \tag{11.76}$$

Finding values of γ_1 and γ_2 from Eqs. (11.73) with A_{12} and A_{21} as determined for the methyl ethyl ketone(1)/toluene(2) system and taking P_1^{sat} and P_2^{sat} as the experimental values, we calculate P and y_1 by Eqs. (11.75) and (11.76) at various values of x_1. The results are shown by the P-x_1 and P-y_1 curves of Fig. 11.7, which provide an adequate correlation of the experimental data points.

A second set of P-x_1-y_1 data, for chloroform(1)/1,4-dioxane(2) at 50°C,† is given in Table 11.2, along with values of pertinent thermodynamic functions. Figures 11.9 and 11.10 display as points all of the experimentally determined values. This system shows *negative deviations from Raoult's law*; since γ_1 and γ_2 are less than unity, values of $\ln \gamma_1$, $\ln \gamma_2$, G^E/RT, and G^E/x_1x_2RT are negative. Moreover, the P-x_1 data points in Fig. 11.9 all lie below the dashed line representing the Raoult's-law relation. Again the data points for G^E/x_1x_2RT are reasonably well correlated by Eq. (11.72*a*), and the Margules equations [Eqs. (11.73)] again apply, here with $A_{12} = -0.72$ and $A_{21} = -1.27$. Values of G^E/RT, $\ln \gamma_1$, $\ln \gamma_2$, P,

† M. L. McGlashan and R. P. Rastogi, *Trans. Faraday Soc.*, **54**: 496, 1958.

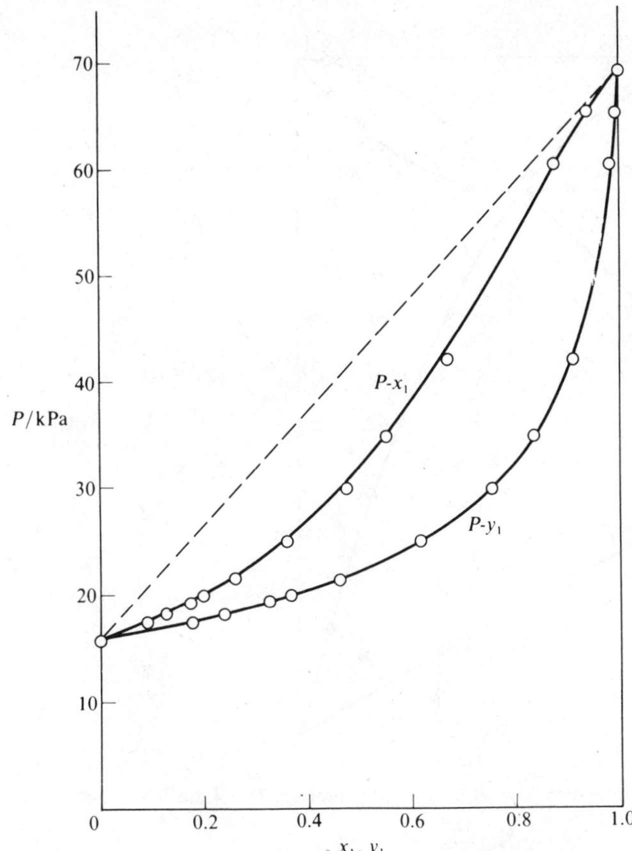

Figure 11.9 *Pxy* data at 50°C for chloroform(1)/1,4-dioxane(2).

and y_1 calculated by Eqs. (11.72*b*), (11.73), (11.75), and (11.76) provide the curves shown for these quantities in Figs. 11.9 and 11.10. Again, the experimental *Pxy* data are adequately correlated.

Although the correlations provided by the Margules equations for the two sets of VLE data presented here are satisfactory, they are not perfect. The two possible reasons are, first, that the Margules equations are not precisely suited to the data set; second, that the data themselves are systematically in error such that they do not conform to the requirements of Gibbs/Duhem equation.

We have presumed in applying the Margules equations that the deviations of the experimental points for G^E/x_1x_2RT from the straight lines drawn to represent them result from random error in the data. Indeed, the straight lines do provide excellent correlations of all but a few data points. Only toward edges of a diagram are there significant deviations, and these have been discounted, because the error bounds widen rapidly as the edges of a diagram are approached.

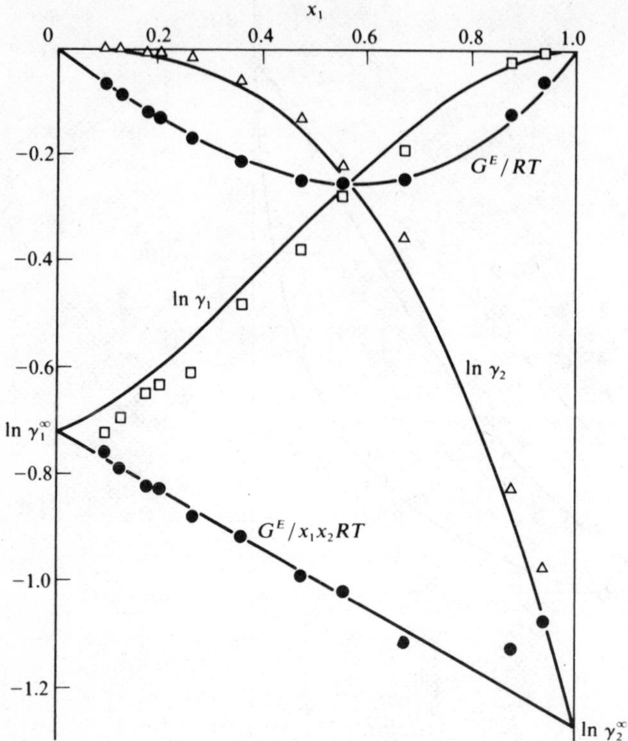

Figure 11.10 Liquid-phase properties from VLE data for chloroform(1)/1,4-dioxane at 50°C.

In the limits as $x_1 \to 0$ and $x_1 \to 1$, G^E/x_1x_2RT becomes indeterminate; experimentally this means that the values are subject to unlimited error and are not measurable. However, we cannot rule out the possibility that the correlation would be improved were the G^E/x_1x_2RT points represented by an appropriate *curve*. Finding the correlation that best represents the data is a trial procedure.

The Gibbs/Duhem equation imposes a constraint on the activity coefficients that may not be satisfied by experimental values that contain systematic error. If this is the case, the experimental values of $\ln \gamma_1$ and $\ln \gamma_2$ used for calculation of G^E/RT by Eq. (11.69), which does not depend on the Gibbs/Duhem equation, will not agree with values of $\ln \gamma_1$ and $\ln \gamma_2$ later calculated by equations derived from Eq. (11.62), which do implicitly contain the Gibbs/Duhem equation. It is then impossible to find a correlating equation that precisely represents the original data. The following example provides an illustration.

Example 11.7 Reduce the VLE data set for diethyl ketone(1)/*n*-hexane(2) at 65°C given by Maripuri and Ratcliff.†

† V. C. Maripuri and G. A. Ratcliff, *J. Appl. Chem. Biotechnol.*, **22**: 899, 1972.

Table 11.3 VLE Data for diethyl ketone(1)/*n*-hexane(2) at 65°C

P/kPa	x_1	y_1	$\ln \gamma_1$	$\ln \gamma_2$	G^E/x_1x_2RT
90.15†	0.000	0.000	......	0.000	
91.78	0.063	0.049	0.901	0.033	1.481
88.01	0.248	0.131	0.472	0.121	1.114
81.67	0.372	0.182	0.321	0.166	0.955
78.89	0.443	0.215	0.278	0.210	0.972
76.82	0.508	0.248	0.257	0.264	1.043
73.39	0.561	0.268	0.190	0.306	0.977
66.45	0.640	0.316	0.123	0.337	0.869
62.95	0.702	0.368	0.129	0.393	0.993
57.70	0.763	0.412	0.072	0.462	0.909
50.16	0.834	0.490	0.016	0.536	0.740
45.70	0.874	0.570	0.027	0.548	0.844
29.00‡	1.000	1.000	0.000		

† P_2^{sat}
‡ P_1^{sat}

SOLUTION The experimental P-x_1-y_1 values for this system are reproduced in the first three columns of Table 11.3. The remaining columns present values of $\ln \gamma_1$, $\ln \gamma_2$, and G^E/x_1x_2RT calculated from the data by Eqs. (11.68) and (11.69). All values are shown as points on Figs. 11.11 and 11.12. The object of data reduction is to arrive at an equation for G^E/RT which provides a suitable correlation of the data.

The data points of Fig. 11.12 for G^E/x_1x_2RT show scatter, but are adequate to define a straight line, drawn here by eye and represented by the equation:

$$\frac{G^E}{x_1x_2RT} = 0.70x_1 + 1.35x_2$$

This is Eq. (11.72*a*) with $A_{21} = 0.70$ and $A_{12} = 1.35$. Equations (11.73) allow calculation of values for $\ln \gamma_1$ and $\ln \gamma_2$ at various values of x_1, and Eqs. (11.75) and (11.76) provide for the calculation of P and y_1 at the same values of x_1. Results of such calculations are plotted as the solid lines of Figs. 11.11 and 11.12. They clearly do not represent a good correlation of the data.

The problem is that the data are not *consistent* with the Gibbs/Duhem equation. That is, the *experimental* values of $\ln \gamma_1$ and $\ln \gamma_2$ do not conform to Eq. (11.70). However, the values of $\ln \gamma_1$ and $\ln \gamma_2$ *found from the correlation* necessarily obey this equation; the two sets of values therefore cannot possibly agree, and the resulting correlation cannot provide a precise representation of the complete set of P-x_1-y_1 data. Although this is true regardless of the means of data reduction, the method just described produces a correlation that is unnecessarily divergent from the experimental values.

An alternative is to process just the P-x_1 data; this is possible because the P-x_1-y_1 data set includes more information than necessary. The procedure requires a computer, but in principle is simple enough. Assuming that the Margules equation is appropriate to the data, one merely searches for values of the parameters A_{12} and A_{21} that yield pressures by Eq. (11.75) that are as close as possible to the measured values. The

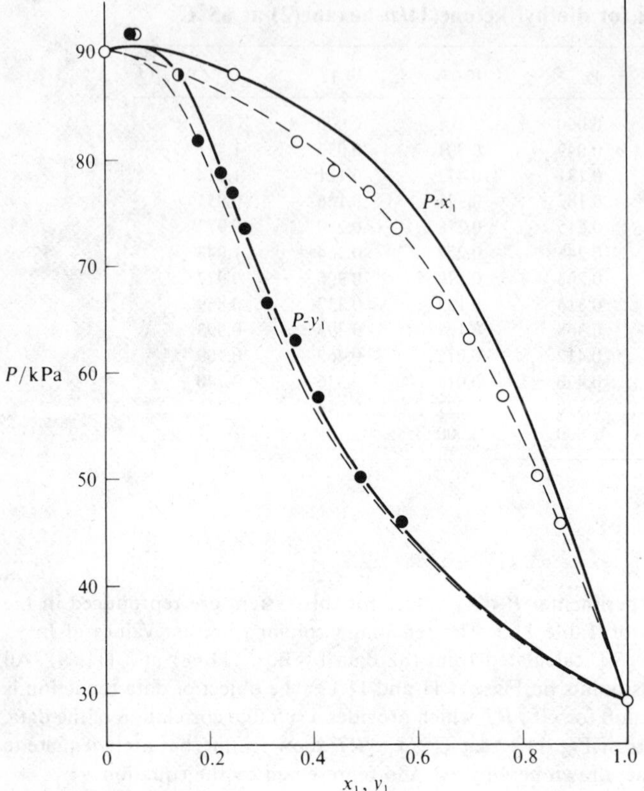

Figure 11.11 *Pxy* data at 65°C for diethyl ketone(1)/*n*-hexane(2).

method is applicable regardless of the correlating equation assumed, and is known as *Barker's method.*† Applied to the present data set, it yields the parameters

$$A_{12} = 1.153 \qquad \text{and} \qquad A_{21} = 0.596$$

Use of these parameters in Eqs. (11.72*a*), (11.73), (11.75), and (11.76) produces the results described by the dashed lines of Figs. 11.11 and 11.12. The correlation cannot be precise, but it clearly provides a better overall representation of the experimental *Pxy* data.

PROBLEMS

11.1 Prove that the "partial molar mass" of a species in solution is equal to its molar mass (molecular weight).

† J. A. Barker, *Austral. J. Chem.*, **6**: 207, 1953.

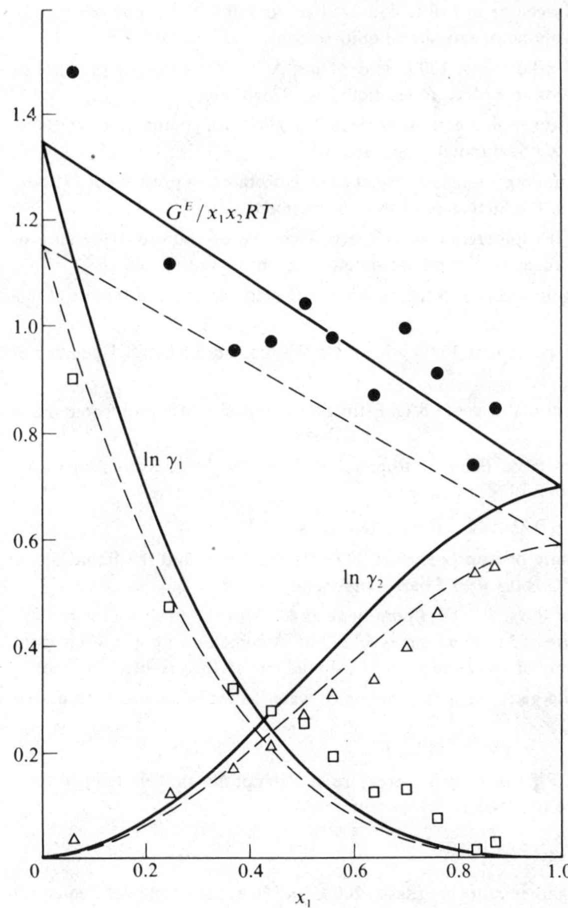

Figure 11.12 Liquid-phase properties from VLE data for diethyl ketone(1)/n-hexane(2) at 65°C.

11.2 From the following compressibility data for hydrogen at 0°C determine the fugacity of hydrogen at 1,000(atm).

P(atm)	Z	P(atm)	Z
100	1.069	600	1.431
200	1.138	700	1.504
300	1.209	800	1.577
400	1.283	900	1.649
500	1.356	1,000	1.720

11.3 For ammonia at 600 K and 300 bar, determine good estimates of the fugacity and of G^R/RT.

11.4 Estimate the fugacity of n-pentane as a gas

(a) At 280°C and 100 bar.

(b) At 280°C and 20 bar.

11.5 Estimate the fugacity of liquid acetone at 110°C and 275 bar. At 110°C the vapor pressure of acetone is 4.360 bar and the molar volume of saturated-liquid acetone is 73 $cm^3 mol^{-1}$.

11.6 Estimate the fugacity of liquid *n*-butane at 120°C and 34 bar. At 120°C the vapor pressure of *n*-butane is 22.38 bar and the molar volume of saturated liquid is 137 $cm^3 mol^{-1}$.

11.7 From data in the steam tables, determine a good estimate for f/f^{sat} for liquid water at 100°C and 100 bar, where f^{sat} is the fugacity of saturated liquid at 100°C.

11.8 Steam at 13,000 kPa and 380°C undergoes an isothermal change of state to a pressure of 275 kPa. Determine the ratio of the fugacity in the final state to that in the initial state.

11.9 Steam at 1,850(psia) and 700(°F) undergoes an isothermal change of state to a pressure of 40(psia). Determine the ratio of the fugacity in the final state to that in the initial state.

11.10 The normal boiling point of *n*-butane is 0.5°C. Estimate the fugacity of liquid *n*-butane at this temperature and 200 bar.

11.11 The normal boiling point of 1-pentene is 30.0°C. Estimate the fugacity of liquid 1-pentene at this temperature and 350 bar.

11.12 The normal boiling point of isobutane is −11.8°C. Estimate the fugacity of liquid isobutane at this temperature and 150 bar.

11.13 Prepare plots of f vs. P and of ϕ vs. P for isopropanol at 200°C for the pressure range from 0 to 50 bar. For the vapor phase, values of Z are given by

$$Z = 1 - 9.86 \times 10^{-3}\, P - 11.41 \times 10^{-5}\, P^2$$

where P is in bars. The vapor pressure of isopropanol at 200°C is 31.92 bar, and the liquid-phase isothermal compressibility κ at 200°C is 0.3×10^{-3} bar^{-1}, independent of P.

11.14 Prepare plots of f vs. P and of ϕ vs. P for 1,3-butadiene at 40°C for the pressure range from 0 to 10 bar. At 40°C the vapor pressure of 1,3-butadiene is 4.287 bar. Assume that Eq. (11.43) is valid for the vapor phase. The molar volume of saturated liquid 1,3-butadiene at 40°C is 90.45 $cm^3 mol^{-1}$.

11.15 The *saturation humidity formula* gives the mole fraction of water vapor in air that is saturated with water vapor:

$$y_{H_2O} = P^{sat}_{H_2O}/P$$

where P is the ambient pressure and $P^{sat}_{H_2O}$ is the vapor pressure of water at the ambient temperature. Derive this formula, starting with the phase-equilibrium criterion

$$\hat{f}^{l}_{H_2O} = \hat{f}^{v}_{H_2O}$$

State and justify any assumptions.

11.16 The fugacity coefficient of a binary mixture of gases at 200°C and 50 bar is given by the equation

$$\ln \phi = (1 + y_2)y_1 y_2$$

where y_1 and y_2 are the mole fractions of species 1 and 2. Multiply this equation through by n, eliminate all remaining mole fractions in favor of mole numbers, and apply Eq. (11.36) to find expressions for $\hat{f}_1$ and $\hat{f}_2$. Then determine values of the fugacities for the species in an equimolar mixture at the given conditions.

11.17 Equation (11.30) is a fundamental criterion of vapor/liquid equilibrium. The question has arisen as to whether at equilibrium it is also true that

$$f^{l} = f^{v}$$

In, words, is it true that the fugacity of a liquid mixture is equal to the fugacity of the vapor mixture with which it is in equilibrium?

11.18 For the system ethylene(1)/propylene(2) as a gas, estimate $\hat{f}_1$, $\hat{f}_2$, $\hat{\phi}_1$, and $\hat{\phi}_2$ at $t = 200$°C, $P = 20$ bar, and $y_1 = 0.25$:

(*a*) Through application of Eqs. (11.46) and (11.47).

(*b*) Assuming that the mixture is an ideal solution.

Apply Eq. (11.37) to the results of parts (*a*) and (*b*) and determine values of ϕ and f for the mixture.

11.19 For the system methane(1)/ethane(2)/propane(3) as a gas, estimate $\hat{f}_1$, $\hat{f}_2$, $\hat{f}_3$, $\hat{\phi}_1$, $\hat{\phi}_2$, and $\hat{\phi}_3$ at $t = 40$°C, $P = 20$ bar, $y_1 = 0.17$, and $y_2 = 0.35$:

(*a*) Through application of Eq. (11.48).
(*b*) Assuming that the mixture is an ideal solution.

Apply Eq. (11.37) to the results of parts (*a*) and (*b*) and determine values of ϕ and f for the mixture.

11.20 Prove that Eqs. (11.73) do indeed follow from Eq. (11.72*b*) by application of Eq. (11.62). To do this, first multiply Eq. (11.72*b*) through by n; then eliminate all remaining mole fractions by the substitution, $x_i = n_i/n$. Finally, apply Eq. (11.62), noting that n cannot be treated as a constant.

11.21 A special case of Eq. (11.72*b*) results when $A_{12} = A_{21} = A$:

$$G^E/RT = Ax_1x_2$$

This is the simplest realistic expression for the excess Gibbs energy, and applies to binary systems comprised of species that are chemically similar.

(*a*) What are the expressions for $\ln \gamma_1$ and $\ln \gamma_2$ that result from this expression?

(*b*) For a particular binary system to which these equations are known to apply, data are available for a single data point:

$$t = 45°\text{C} \qquad P = 37 \text{ kPa} \qquad x_1 = 0.398 \qquad y_1 = 0.428$$

In addition, $P_1^{sat} = 27.78$ kPa and $P_2^{sat} = 29.82$ kPa. From these data, determine the value of A.

(*c*) Using the value of A determined in (*b*) and for $t = 45°$C, calculate P and y_1 for $x_1 = 0.500$.

11.22 Given in what follows are values of infinite-dilution activity coefficients and pure-species vapor pressures for binary systems at specified temperatures. For one of the systems, determine the Margules parameters, and then apply the Margules equation to a sufficient number of VLE calculations to allow construction of a *Pxy* diagram for the given temperature. Base your calculations on the modified Raoult's-law expression, i.e., Eq. (11.74).

(*a*) For diethyl ether(1)/chloroform(2) at 30°C,

$$\gamma_1^\infty = 0.71; \; \gamma_2^\infty = 0.57; \; P_1^{sat} = 33.73 \text{ kPa}; \; P_2^{sat} = 86.59 \text{ kPa}$$

(*b*) For acetone(1)/benzene(2) at 45°C,

$$\gamma_1^\infty = 1.60; \; \gamma_2^\infty = 1.47; \; P_1^{sat} = 68.36 \text{ kPa}; \; P_2^{sat} = 29.82 \text{ kPa}$$

(*c*) For 2-butanone(1)/toluene(2) at 50°C,

$$\gamma_1^\infty = 1.47; \; \gamma_2^\infty = 1.30; \; P_1^{sat} = 36.09 \text{ kPa}; \; P_2^{sat} = 12.30 \text{ kPa}$$

(*d*) For benzene(1)/acetonitrile(2) at 45°C,

$$\gamma_1^\infty = 2.74; \; \gamma_2^\infty = 3.01; \; P_1^{sat} = 29.81 \text{ kPa}; \; P_2^{sat} = 28.12 \text{ kPa}$$

(*e*) For diethyl ether(1)/acetone(2) at 30°C,

$$\gamma_1^\infty = 1.78; \; \gamma_2^\infty = 2.18; \; P_1^{sat} = 85.93 \text{ kPa}; \; P_2^{sat} = 38.01 \text{ kPa}$$

11.23 The following is a set of VLE data for the system carbon disulfide(1)/chloroform(2) at 25°C [N. D. Litvinov, *Zh. Fiz. Khim.*, **26**: 1144, 1952]. Assuming the validity of Eq. (11.74), find parameter values for the Margules equation that provide a suitable correlation of these data, and prepare a *Pxy* diagram that compares the experimental points with curves determined from the correlation.

P/kPa	x_1	y_1
27.30	0.000	0.000
31.61	0.100	0.219
34.98	0.200	0.363
37.74	0.300	0.468
39.84	0.400	0.555
41.64	0.500	0.630
43.16	0.600	0.699
44.46	0.700	0.768
45.49	0.800	0.838
46.30	0.900	0.914
46.85	1.000	1.000

11.24 The following is a set of VLE data for the system acetone(1)/chloroform(2) at 50°C [H. Röck and W. Schröder, *Z. Phys. Chem. (Frankfurt)*, **11**: 41, 1957]. Assuming the validity of Eq. (11.74), find parameter values for the Margules equation that provide a suitable correlation of these data, and prepare a *Pxy* diagram that compares the experimental points with curves determined from the correlation.

P/kPa	x_1	y_1
69.38	0.000	0.000
66.11	0.104	0.066
63.07	0.198	0.153
61.25	0.298	0.269
60.60	0.401	0.414
62.01	0.502	0.562
64.53	0.591	0.676
68.29	0.695	0.793
72.75	0.797	0.879
77.13	0.895	0.946
81.75	1.000	1.000

CHAPTER

TWELVE

PHASE EQUILIBRIA AT LOW TO MODERATE PRESSURES

A number of industrially important processes, such as distillation, absorption, and extraction, bring two phases into contact. When the phases are not in equilibrium, mass transfer occurs between the phases. The rate of transfer of each species depends on the departure of the system from equilibrium. Quantitative treatment of mass-transfer rates requires knowledge of the equilibrium states (T, P, and compositions) of the system.

In most industrial processes coexisting phases are vapor and liquid, although liquid/liquid, vapor/solid, and liquid/solid systems are also encountered. In this chapter we present a general qualitative discussion of vapor/liquid phase behavior (Sec. 12.3) and describe the calculation of temperatures, pressures, and phase compositions for systems in vapor/liquid equilibrium (VLE) at low to moderate pressures (Sec. 12.4).† Comprehensive expositions are given of dew-point, bubble-point, and P, T-flash calculations.

12.1 THE NATURE OF EQUILIBRIUM

Equilibrium is a static condition in which no changes occur in the macroscopic properties of a system with time. This implies a balance of all potentials that may cause change. In engineering practice, the assumption of equilibrium is justified when it leads to results of satisfactory accuracy. For example, in the

† For VLE at high pressures, see chap. 14.

reboiler for a distillation column, equilibrium between vapor and liquid phases is commonly assumed. For finite vaporization rates this is an approximation, but it does not introduce significant error into engineering calculations.

If a system containing fixed amounts of chemical species and consisting of liquid and vapor phases in intimate contact is completely isolated, then in time there is no further tendency for any change to occur within the system. The temperature, pressure, and phase compositions reach final values which thereafter remain fixed. The system is in equilibrium. Nevertheless, at the microscopic level, conditions are not static. The molecules comprising one phase at a given instant are not the same molecules as those in that phase at a later time. Molecules with sufficiently high velocities that are near the interphase boundary overcome surface forces and pass into the other phase. However, the average rate of passage of molecules is the same in both directions, and there is no net transfer of material between the phases.

12.2 THE PHASE RULE. DUHEM'S THEOREM

The phase rule for nonreacting systems, presented without proof in Sec. 2.8, results from application of a rule of algebra. The number of phase-rule variables which must be arbitrarily specified in order to fix the intensive state of a system at equilibrium, called the degrees of freedom F, is the difference between the total number of phase-rule variables and the number of independent equations that can be written connecting these variables.

The *intensive* state of a PVT system containing N chemical species and π phases in equilibrium is characterized by the temperature T, the pressure P, and $N-1$ mole fractions† for each phase. These are the phase-rule variables, and their number is $2+(N-1)(\pi)$. The masses of the phases are not phase-rule variables, because they have no influence on the intensive state of the system.

The phase-equilibrium equations that may be written connecting the phase-rule variables are given by Eqs. (10.3) or Eqs. (11.29):

$$\mu_i^\alpha = \mu_i^\beta = \cdots = \mu_i^\pi \qquad (i = 1, 2, \ldots, N) \tag{10.3}$$

$$\hat{f}_i^\alpha = \hat{f}_i^\beta = \cdots = \hat{f}_i^\pi \qquad (i = 1, 2, \ldots, N) \tag{11.29}$$

Either set contains $(\pi-1)(N)$ independent phase-equilibrium equations. They are equations connecting the phase-rule variables, because the chemical potentials and fugacities are functions of temperature, pressure, and composition. The difference between the number of phase-rule variables and the number of equations connecting them is the degrees of freedom:

$$F = 2 + (N-1)(\pi) - (\pi - 1)(N)$$

† Only $N-1$ mole fractions are required, because $\sum x_i = 1$.

This reduces to Eq. (2.12):

$$\boxed{F = 2 - \pi + N} \tag{2.12}$$

Applications of the phase rule were discussed in Sec. 2.8.

Duhem's theorem is another rule, similar to the phase rule, but less celebrated. It applies to closed systems for which the extensive state as well as the intensive state of the system is fixed. The state of such a system is said to be *completely determined*, and is characterized not only by the $2 + (N - 1)\pi$ intensive phase-rule variables but also by the π extensive variables represented by the masses (or mole numbers) of the phases. Thus the total number of variables is

$$2 + (N - 1)\pi + \pi = 2 + N\pi$$

If the system is closed and formed from specified amounts of the chemical species present, then we can write a material-balance equation for each of the N chemical species. These in addition to the $(\pi - 1)N$ phase-equilibrium equations provide a total number of independent equations equal to

$$(\pi - 1)N + N = \pi N$$

The difference between the number of variables and the number of equations is therefore

$$2 + N\pi - \pi N = 2$$

On the basis of this result, Duhem's theorem is stated as follows:

> For any closed system formed initially from given masses of prescribed chemical species, the equilibrium state is completely determined when any two independent variables are fixed.

The two independent variables subject to specification may in general be either intensive or extensive. However, the number of *independent intensive* variables is given by the phase rule. Thus when $F = 1$, at least one of the two variables must be extensive, and when $F = 0$, both must be extensive.

12.3 PHASE BEHAVIOR FOR VAPOR/LIQUID SYSTEMS

Vapor/liquid equilibrium (VLE) refers to systems in which a single liquid phase is in equilibrium with its vapor. In this qualitative discussion, we limit consideration to systems comprised of two chemical species, because systems of greater complexity cannot be adequately represented graphically.

When $N = 2$, the phase rule becomes $F = 4 - \pi$. Since there must be at least one phase ($\pi = 1$), the maximum number of phase-rule variables which must be specified to fix the intensive state of the system is *three*: namely, P, T, and one mole (or mass) fraction. All equilibrium states of the system can therefore be represented in three-dimensional *P*-*T*-composition space. Within this space, the states of *pairs* of phases coexisting at equilibrium ($F = 4 - 2 = 2$) define surfaces. A schematic three-dimensional diagram illustrating these surfaces for VLE is shown in Fig. 12.1.

This figure shows schematically the *P*-*T*-composition surfaces which represent equilibrium states of saturated vapor and saturated liquid for a binary system. The under surface represents saturated-vapor states; it is the *PTy* surface. The upper surface represents saturated-liquid states; it is the *PTx* surface. These surfaces intersect along the lines $UBHC_1$ and KAC_2, which represent the vapor pressure-vs.-*T* curves for pure species 1 and 2. Moreover, the under and upper surfaces form a continuous rounded surface across the top of the diagram between C_1 and C_2, the critical points of pure species 1 and 2; the critical points of the

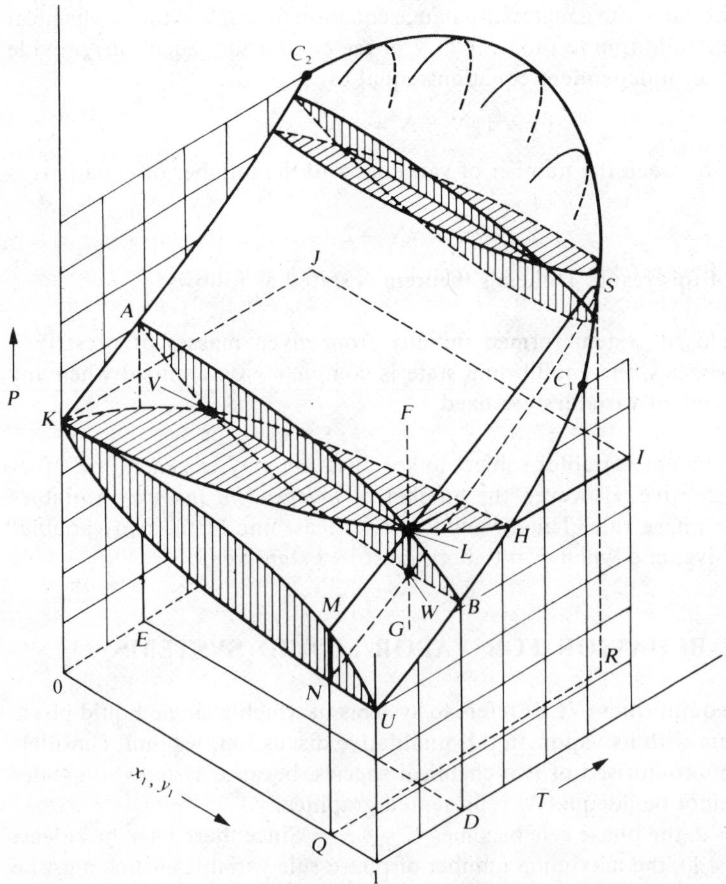

Figure 12.1 *PTxy* diagram for vapor/liquid equilibrium.

various mixtures of 1 and 2 lie along a line on the rounded edge of the surface between C_1 and C_2. This critical locus is defined by the points at which vapor and liquid phases in equilibrium become identical. Further discussion of the critical region is given later.

The region lying above the upper surface of Fig. 12.1 is the subcooled-liquid region; that below the under surface is the superheated-vapor region. The interior space between the two surfaces is the region of coexistence of both liquid and vapor phases. If one starts with a liquid at F and reduces the pressure at constant temperature and composition along vertical line FG, the first bubble of vapor appears at point L, which lies on the upper surface. Thus, L is a *bubble point*, and the upper surface is the bubble-point surface. The state of the vapor bubble in equilibrium with the liquid at L must be represented by a point on the under surface at the temperature and pressure of L. This point is indicated by the letter V. Line VL is an example of a *tie line*, which connects points representing phases in equilibrium.

As the pressure is further reduced along line FG, more and more liquid vaporizes until at W the process is complete. Thus W lies on the under surface and represents a state of saturated vapor having the mixture composition. Since W is the point at which the last drops of liquid (dew) disappear, it is a *dew point*, and the lower surface is the dew-point surface. Continued reduction of pressure merely leads into the superheated vapor region.

Because of the complexity of Fig. 12.1, the detailed characteristics of binary VLE are usually depicted by two-dimensional graphs that display what is seen on various planes that cut the three-dimensional diagram. The three principal planes, each perpendicular to one of the coordinate axes, are illustrated in Fig. 12.1. Thus a vertical plane perpendicular to the temperature axis is outlined as $ALBDEA$. The lines on this plane represent a Pxy phase diagram at constant T, of which we have already seen examples in Figs. 10.1, 11.7, 11.9, and 11.11. If the lines from several such planes are projected on a single parallel plane, a diagram like Fig. 12.2 is obtained. It shows Pxy plots for three different temperatures. The one for T_a represents the section of Fig. 12.1 indicated by $ALBDEA$. The horizontal lines are tie lines connecting the compositions of phases in equilibrium. The temperature T_b lies between the two pure-species critical temperatures identified by C_1 and C_2 in Fig. 12.1, and temperature T_d is above both critical temperatures. The curves for these two temperatures therefore do not extend all the way across the diagram. However, the first passes through one mixture critical point, and the second through two such points. All three of these critical points are denoted by the letter C. Each is a tangent point at which a horizontal line touches the curve. This is so because all tie lines connecting phases in equilibrium are horizontal, and the tie line connecting *identical* phases (the definition of a critical point) must therefore be the last such line to cut the diagram.

A horizontal plane passed through Fig. 12.1 perpendicular to the P axis is identified by $HIJKLH$. Viewed from the top, the lines on this plane represent a Txy diagram similar to that of Fig. 10.2. When lines for several pressures are projected on a parallel plane, the resulting diagram appears as in Fig. 12.3. This

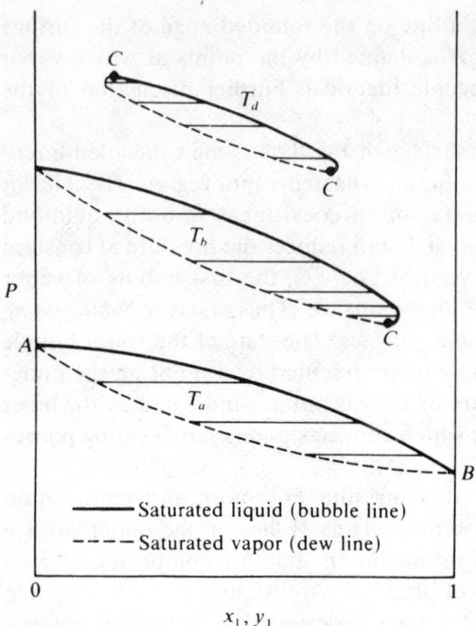

Figure 12.2 *Pxy* diagram for three constant temperatures.

figure is analogous to Fig. 12.2, except that it represents values for three constant pressures, P_a, P_b, and P_d.

It is also possible to plot the vapor mole fraction y_1 vs. the liquid mole fraction x_1 for either the constant-temperature conditions of Fig. 12.2 or the constant-pressure conditions of Fig. 12.3. Examples of such *xy* diagrams are shown later.

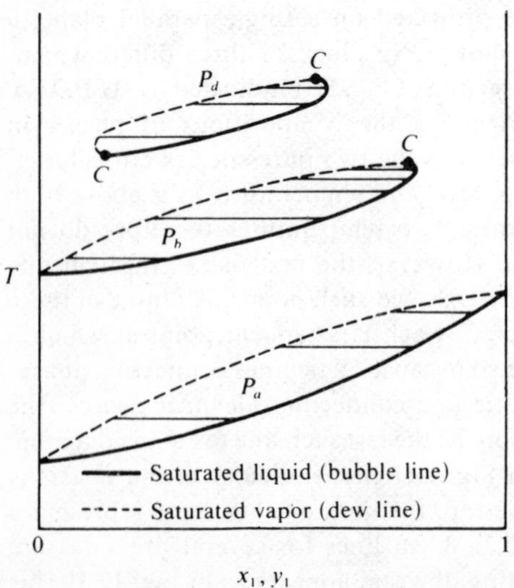

Figure 12.3 *Txy* diagram for three constant pressures.

The third plane identified in Fig. 12.1 is the vertical one perpendicular to the composition axis and indicated by *MNQRSLM*. When projected on a parallel plane, the lines from several such planes present a diagram such as that shown by Fig. 12.4. This is the *PT* diagram; lines UC_1 and KC_2 are vapor-pressure curves for the pure species, identified by the same letters as in Fig. 12.1. Each interior loop represents the *PT* behavior of saturated liquid and of saturated vapor for a *mixture of fixed composition*; the different loops are for different compositions. Clearly, the *PT* relation for saturated liquid is different from that for saturated vapor of the same composition. This is in contrast with the behavior of a pure species, for which the bubble line and the dew line coincide. At points *A* and *B* in Fig. 12.4 saturated-liquid and saturated-vapor lines intersect. At such points a saturated liquid of one composition and a saturated vapor of another composition have the same *T* and *P*, and the two phases are therefore in equilibrium. The tie lines connecting the coinciding points at *A* and at *B* are perpendicular to the *PT* plane, as illustrated by the tie line *VL* in Fig. 12.1.

The critical point of a binary mixture occurs where the nose of a loop in Fig. 12.4 is tangent to the envelope curve. Put another way, the envelope curve is the

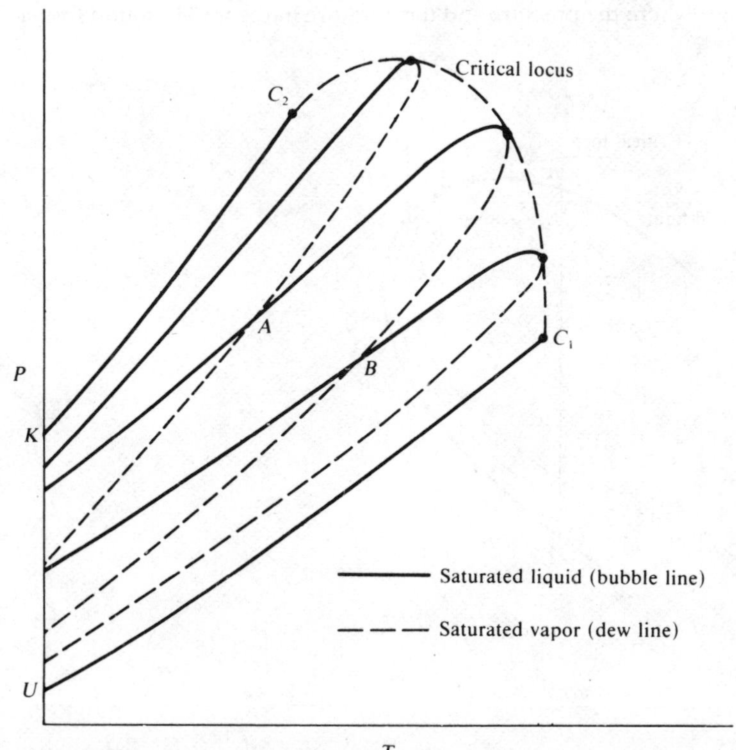

Figure 12.4 *PT* diagram for several compositions.

critical locus. One can verify this by considering two closely adjacent loops and noting what happens to the point of intersection as their separation becomes infinitesimal. Figure 12.4 illustrates that the location of the critical point on the nose of the loop varies from one composition to another. For a pure species the critical point is the highest temperature and highest pressure at which vapor and liquid phases can coexist, but for a mixture it is, in general, neither. Therefore under certain conditions a condensation process occurs as the result of a *reduction* in pressure.

Consider the enlarged nose section of a single *PT* loop shown in Fig. 12.5. The critical point is at *C*. The points of maximum pressure and maximum temperature are identified as M_P and M_T. The dashed curves of Fig. 12.5 indicate the fraction of the overall system that is liquid in a two-phase mixture of liquid and vapor. To the left of the critical point *C* a reduction in pressure along a line such as *BD* is accompanied by vaporization from the bubble point to the dew point, as would be expected. However, if the original condition corresponds to point *F*, a state of saturated *vapor*, liquefaction occurs upon reduction of the pressure and reaches a maximum at *G*, after which vaporization takes place until the dew point is reached at *H*. This phenomenon is called *retrograde condensation*. It is of considerable importance in the operation of certain deep natural-gas wells where the pressure and temperature in the underground forma-

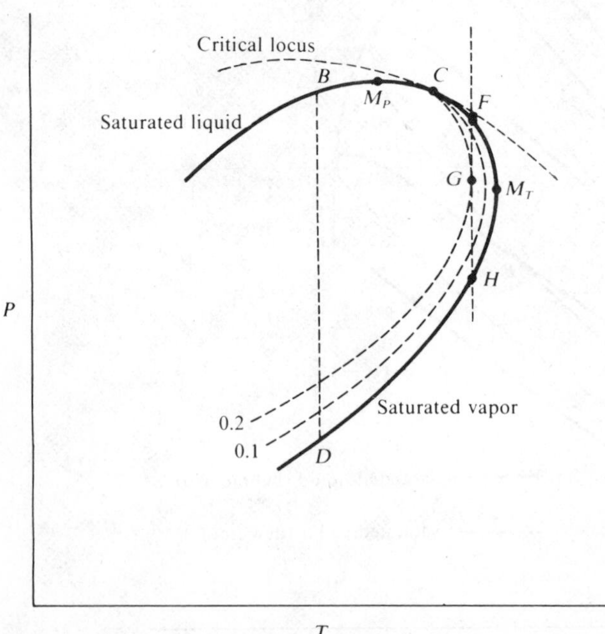

Figure 12.5 Portion of a *PT* diagram showing phase behavior in the critical region.

tion are approximately the conditions represented by point *F*. If one then maintains the pressure at the wellhead at a value near that of point *G*, considerable liquefaction of the product stream is accomplished along with partial separation of the heavier species of the mixture. Within the underground formation itself, the pressure tends to drop as the gas supply is depleted. If not prevented, this leads to the formation of a liquid phase and a consequent reduction in the production of the well. Repressuring is therefore a common practice; i.e., lean gas (gas from which the heavier species have been removed) is returned to the underground reservoir to maintain an elevated pressure.

A *PT* diagram for the ethane/heptane system is shown in Fig. 12.6, and a *yx* diagram for several pressures for the same system appears in Fig. 12.7. According to convention, one plots as *y* and *x* the mole fractions of the more volatile species in the mixture. The maximum and minimum concentrations of the more volatile species obtainable by distillation at a given pressure are indicated by the points of intersection of the appropriate *yx* curve with the diagonal, for at these points the vapor and liquid have the same composition. They are in fact mixture critical points, unless $y = x = 0$ or $y = x = 1$. Point *A* in Fig. 12.7

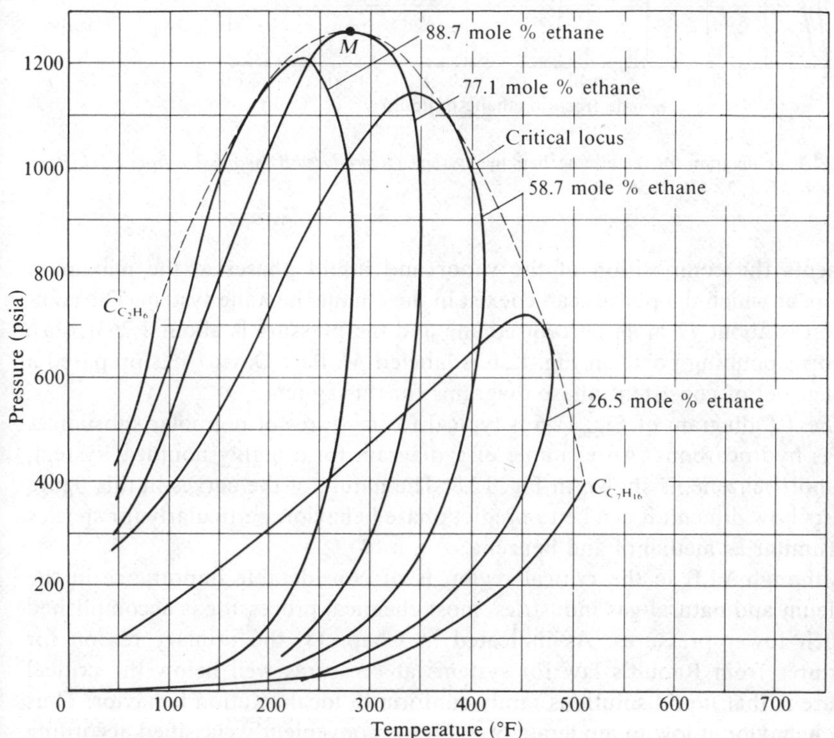

Figure 12.6 Pressure/temperature diagram for the ethane/heptane system. *(Reproduced by permission from F. H. Barr-David, AIChE J., 2: 426, 1956.)*

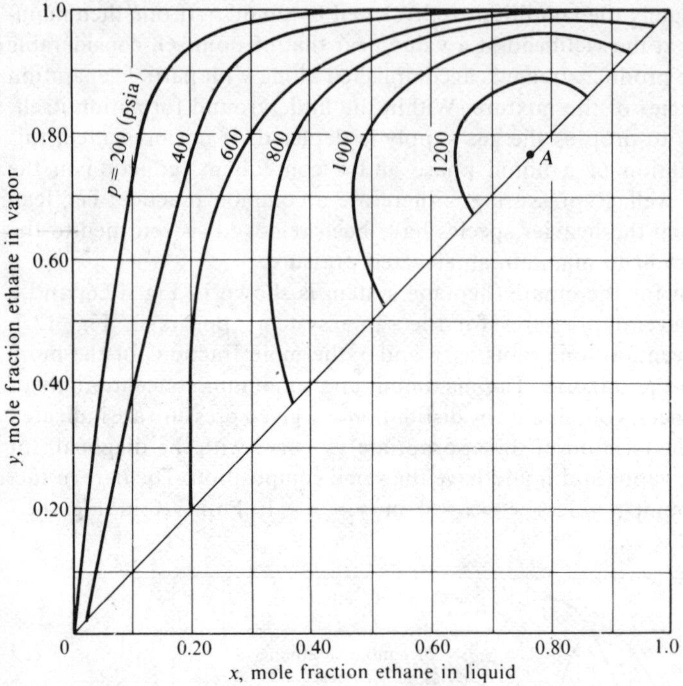

Figure 12.7 *yx* diagram for the ethane/heptane system. *(Reproduced by permission from F. H. Barr-David, AIChE J., 2: 426, 1956.)*

represents the composition of the vapor and liquid phases at the maximum pressure at which the phases can coexist in the ethane/heptane system. The composition is about 77 mole percent ethane and the pressure is about 1,263(psia). The corresponding point on Fig. 12.6 is labeled *M*. Barr-David† has prepared a complete set of consistent phase diagrams for this system.

The *PT* diagram of Fig. 12.6 is typical for mixtures of nonpolar substances such as hydrocarbons. An example of a diagram for a highly nonideal system, methanol/benzene, is shown in Fig. 12.8. The nature of the curves in this figure suggests how difficult it can be to predict phase behavior, particularly for species so dissimilar as methanol and benzene.

Although VLE in the critical region is of considerable importance in the petroleum and natural-gas industries, most chemical processing is accomplished at much lower pressures. As indicated in Chap. 11, the primary reason for departures from Raoult's law for systems at pressures well below the critical pressure is that liquid solutions rarely conform to ideal-solution behavior. Thus phase behavior at low to moderate pressures is conveniently classified according

† F. H. Barr-David, *AIChE J.*, **2**: 426, 1956.

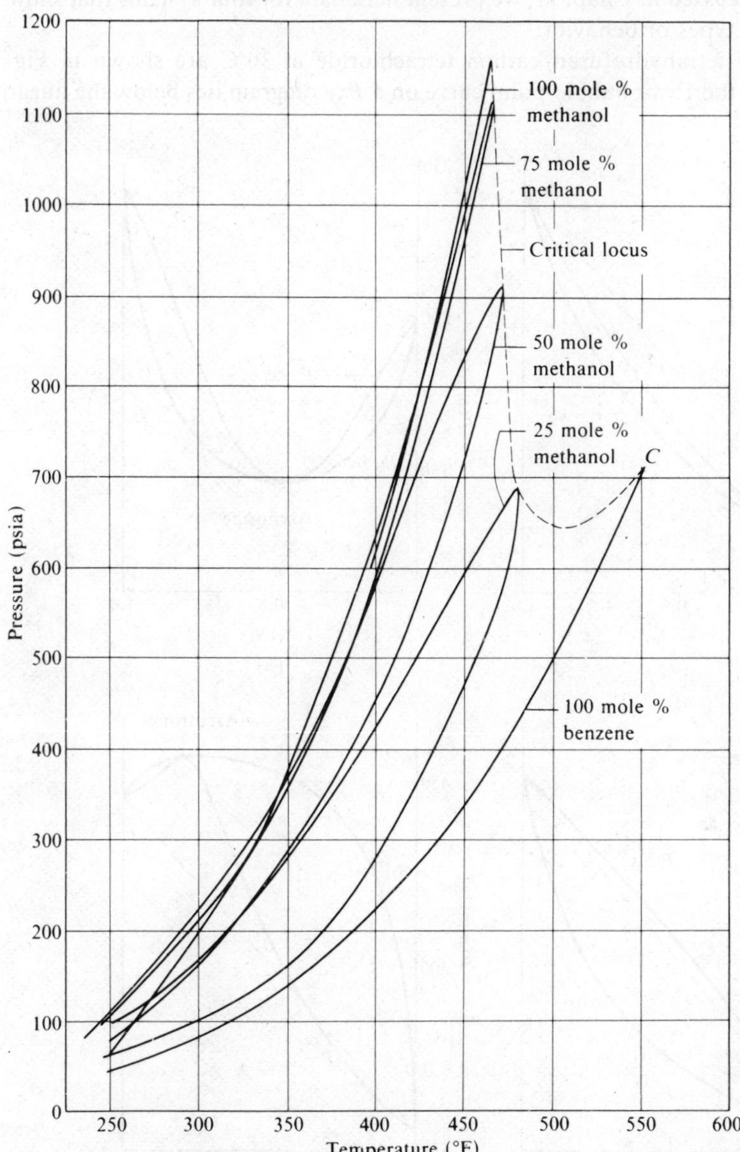

Figure 12.8 Pressure/temperature diagram for the methanol/benzene system. *(Reproduced by permission from P. G. McCracken and J. M. Smith, AIChE J., 2: 498, 1956.)*

to the sign and magnitude of deviations from Raoult's law. In addition to the examples discussed in Chap. 11, we present here data for four systems that show the common types of behavior.

Data for tetrahydrofuran/carbon tetrachloride at 30°C are shown in Fig. 12.9*a*. Here, the *Px* or bubble-point curve on a *Pxy* diagram lies below the linear

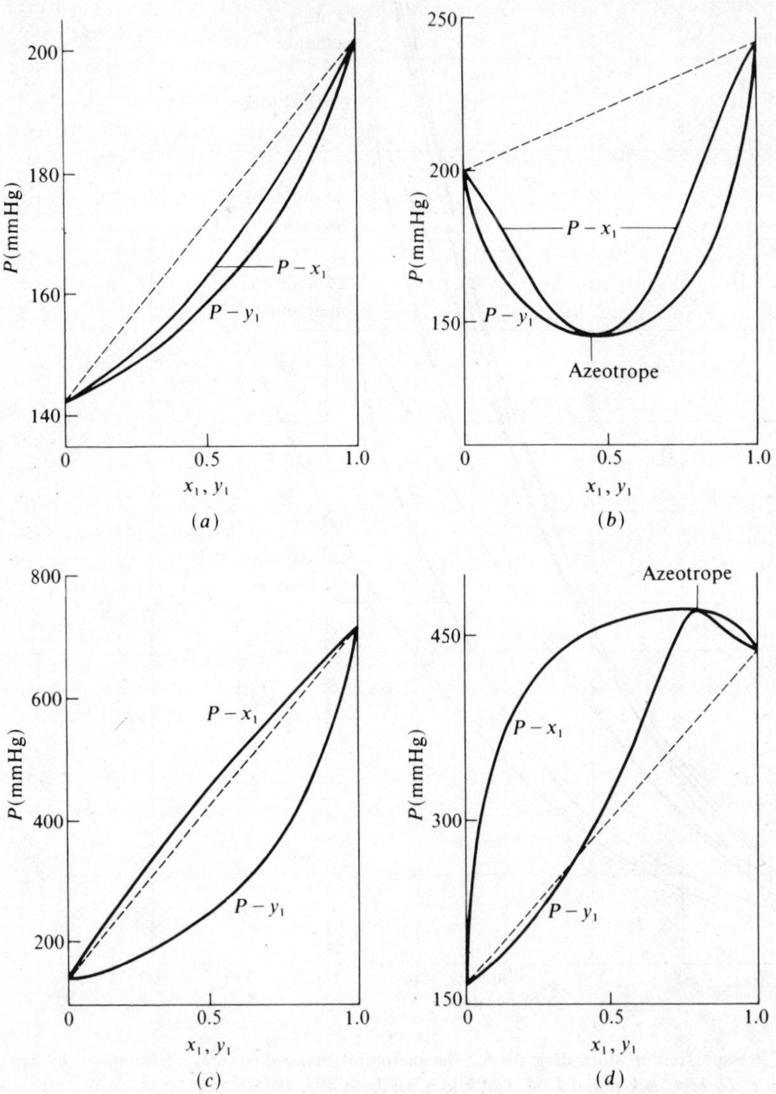

Figure 12.9 *Pxy* diagrams at constant temperature. (*a*) Tetrahydrofuran(1)/carbon tetrachloride(2) at 30°C; (*b*) chloroform(1)/tetrahydrofuran(2) at 30°C; (*c*) furan(1)/carbon tetrachloride(2) at 30°C; (*d*) ethanol(1)/toluene(2) at 65°C. Dashed lines: *Px* relation for Raoult's law.

Px relation of Raoult's law, and the system therefore exhibits negative deviations. When the deviations become sufficiently large relative to the difference between the two pure-species vapor pressures, the *Px* curve exhibits a minimum, as illustrated in Fig. 12.9*b* for the chloroform/tetrahydrofuran system at 30°C. This figure shows that the *Py* curve also has a minimum at the same point. Thus at this point where $x = y$ the dew-point and bubble-point curves are tangent to the same horizontal line. A boiling liquid of this composition produces a vapor of exactly the same composition, and the liquid therefore does not change in composition as it evaporates. No separation of such a constant-boiling solution is possible by distillation. The term *azeotrope* is used to describe this state.

The data for furan/carbon tetrachloride at 30°C shown by Fig. 12.9*c* provide an example of a system that exhibits small positive deviations from Raoult's law. Ethanol/toluene is a system for which the positive deviations are sufficiently large to lead to a maximum in the *Px* curve, as shown for 65°C by Fig. 12.9*d*. Just as a mimimum on the *Px* curve represents an azeotrope, so does a maximum. Thus there are minimum-pressure and maximum-pressure azeotropes. In either case the vapor and liquid phases at the azeotropic state are of identical composition.

At the molecular level, appreciable negative deviations from Raoult's law reflect stronger forces of intermolecular attraction in the liquid phase between unlike than between like pairs of molecules. Conversely, appreciable positive deviations result for solutions in which intermolecular forces between like molecules are stronger than between unlike. In this latter case the forces between like molecules may be so strong as to prevent complete miscibility, and the system then forms two separate liquid phases over a range of compositions. Systems of limited miscibility are treated in Sec. 13.9.

Since distillation processes are carried out more nearly at constant pressure than at constant temperature, *txy* diagrams of data at constant *P* are in common use. The four such diagrams corresponding to those of Fig. 12.9 are shown for atmospheric pressure in Fig. 12.10. Note that the dew-point (*ty*) curves lie above the bubble-point (*tx*) curves. Moreover, the minimum-pressure azeotrope of Fig. 12.9*b* corresponds to the maximum-temperature (or maximum-boiling) azeotrope of Fig. 12.10*b*. There is an analogous correspondence between Figs. 12.9*d* and 12.10*d*. The *yx* diagrams at constant *P* for the same four systems are shown in Fig. 12.11. The point at which a curve crosses the diagonal line of the diagram represents an azeotrope, for at such a point $y_1 = x_1$.

12.4 LOW-PRESSURE VLE FROM CORRELATIONS OF DATA

In Sec. 10.5 we treated dew- and bubble-point calculations for multicomponent systems that obey Raoult's law [Eq. (10.16)], an equation valid for low-pressure VLE when an ideal-liquid solution is in equilibrium with an ideal gas. Calculations for the general case are carried out in exactly the same way as for Raoult's law,

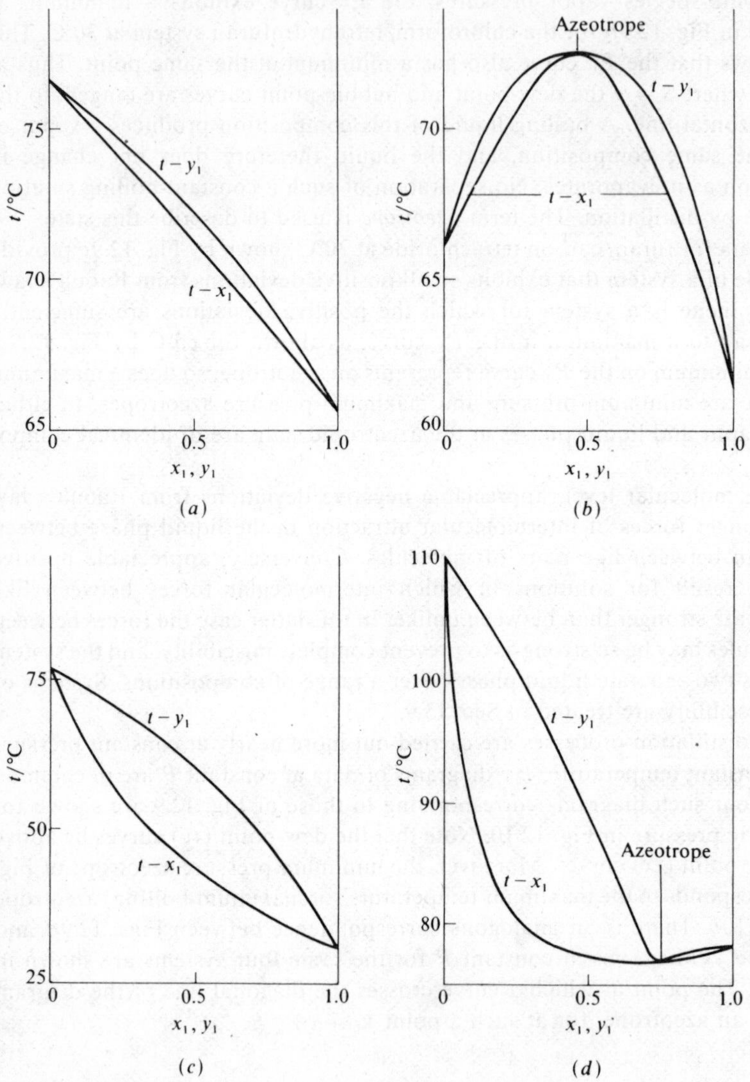

Figure 12.10 *txy* diagrams at a constant pressure of 1(atm). (*a*) Tetrahydrofuran(1)/carbon tetrachloride(2); (*b*) chloroform(1)/tetrahydrofuran(2); (*c*) furan(1)/carbon tetrachloride(2); (*d*) ethanol(1)/toluene(2).

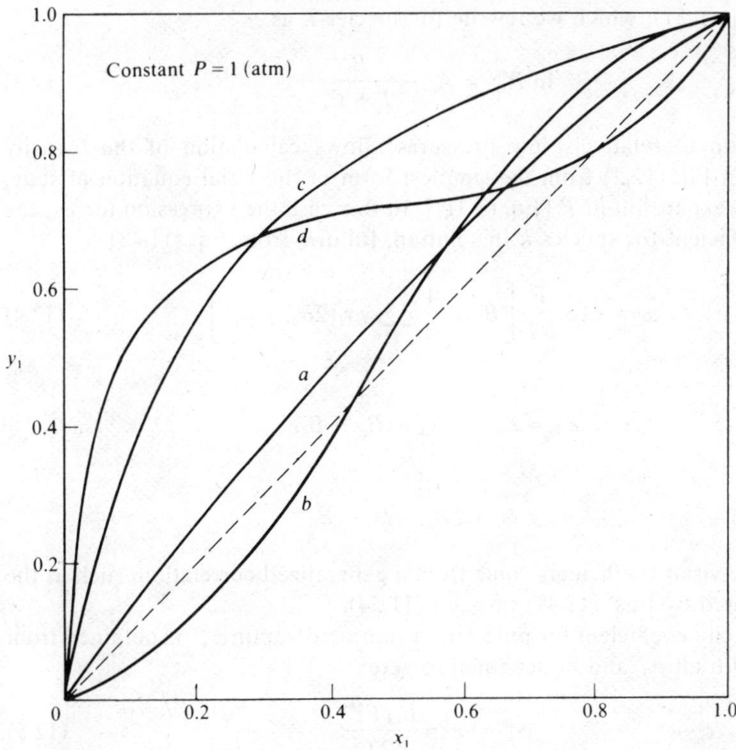

Figure 12.11 Equilibrium yx diagrams at a constant pressure of 1(atm). (a) Tetrahydrofuran(1)/carbon tetrachloride(2); (b) chloroform(1)/tetrahydrofuran(2); (c) furan(1)/carbon tetrachloride(2); (d) ethanol(1)/toluene(2).

but with equations of greater complexity. The equilibrium relation provided by Eq. (11.66) may be written

$$\boxed{y_k \Phi_k P = x_k \gamma_k P_k^{\text{sat}}} \qquad (k = 1, 2, \ldots, N) \tag{12.1}$$

where Φ_k is defined by Eq. (11.67). At low to moderate pressures, the Poynting factor is negligible, and Eq. (11.67) simplifies to

$$\Phi_k = \frac{\hat{\phi}_k}{\phi_k^{\text{sat}}} \tag{12.2}$$

Systematic application of Eqs. (12.1) and (12.2) depends on the availability of correlations of data from which values may be obtained for P_k^{sat}, Φ_k, and γ_k. We consider each of these in turn.

The vapor pressures of the pure species are usually calculated from equations that give P_k^{sat} as a function of temperature. Most commonly used is the Antoine equation, Eq. (6.51), which we rewrite for species k as

$$\ln P_k^{\text{sat}} = A_k - \frac{B_k}{T + C_k} \tag{12.3}$$

Restriction to relatively low pressures allows calculation of the fugacity coefficients in Eq. (12.2) from the simplest form of the virial equation of state, the two-term expansion in P [Eq. (3.31)]. In this case the expression for $\hat{\phi}_k$, the fugacity coefficient for species k in solution, follows from Eq. (11.48):

$$\hat{\phi}_k = \exp \frac{P}{RT}\left[B_{kk} + \frac{1}{2}\sum_i \sum_l y_i y_l (2\delta_{ik} - \delta_{il})\right] \tag{12.4}$$

where

$$\delta_{ik} = \delta_{ki} = 2B_{ik} - B_{ii} - B_{kk}$$

and

$$\delta_{il} = \delta_{li} = 2B_{il} - B_{ii} - B_{ll}$$

Values of the virial coefficients come from a generalized correlation, such as the one represented by Eqs. (11.49) through (11.54).

The fugacity coefficient for pure k as a saturated vapor ϕ_k^{sat} is obtained from Eq. (12.4) with all δ_{ik} and δ_{il} set equal to zero:

$$\phi_k^{\text{sat}} = \exp \frac{B_{kk} P_k^{\text{sat}}}{RT} \tag{12.5}$$

This result also follows from Eq. (11.21).

Combination of Eqs. (12.2), (12.4), and (12.5) gives:

$$\Phi_k = \exp \frac{B_{kk}(P - P_k^{\text{sat}}) + \frac{1}{2}\sum_i \sum_l y_i y_l (2\delta_{ik} - \delta_{il})}{RT} \tag{12.6}$$

For a binary system comprised of species 1 and 2, this becomes:

$$\Phi_1 = \exp \frac{B_{11}(P - P_1^{\text{sat}}) + P y_2^2 \delta_{12}}{RT} \tag{12.7}$$

and

$$\Phi_2 = \exp \frac{B_{22}(P - P_2^{\text{sat}}) + P y_1^2 \delta_{12}}{RT} \tag{12.8}$$

Activity coefficients γ_k have traditionally been calculated from correlating equations for G^E/RT by application of Eq. (11.62). The excess Gibbs energy is a function of T, P, and composition, but for liquids at low to moderate pressures it is a very weak function of P. Under these conditions, its pressure dependence and therefore the pressure dependence of the activity coefficients are usually neglected. This is consistent with our earlier omission of the Poynting factor from the evaluation of Φ_k. Thus we have for data *at constant T*:

$$\frac{G^E}{RT} = g(x_1, x_2, \ldots, x_N) \quad \text{(const } T\text{)}$$

The Margules equations [Eq. (11.72)] provide an example of this functionality.

Other equations are also in common use for the correlation of activity coefficients. For binary systems the function often most conveniently represented by an equation is G^E/x_1x_2RT, and one procedure is to express this function as a power series in x_1:

$$\frac{G^E}{x_1x_2RT} = a + bx_1 + cx_1^2 + \cdots \qquad \text{(const } T\text{)}$$

Since $x_2 = 1 - x_1$ for a binary system of species 1 and 2, x_1 can be taken as the single independent variable. An equivalent power series with certain advantages is known as the Redlich/Kister expansion:†

$$\frac{G^E}{x_1x_2RT} = B + C(x_1 - x_2) + D(x_1 - x_2)^2 + \cdots$$

In application, different truncations of this series are appropriate. For each particular expression representing G^E/x_1x_2RT, specific expressions for $\ln \gamma_1$ and $\ln \gamma_2$ result from application of Eq. (11.62). Thus, when $B = C = D = \cdots = 0$, $G^E/RT = 0$, $\ln \gamma_2 = 0$, and $\ln \gamma_2 = 0$. In this event $\gamma_1 = \gamma_2 = 1$, and the solution is ideal.

If $C = D = \cdots = 0$, then

$$\frac{G^E}{x_1x_2RT} = B$$

where B is a constant for a given temperature. The corresponding equations for $\ln \gamma_1$ and $\ln \gamma_2$ are

$$\ln \gamma_1 = Bx_2^2 \tag{12.9}$$

and

$$\ln \gamma_2 = Bx_1^2 \tag{12.10}$$

The symmetrical nature of these relations is evident. The infinite-dilution values of the activity coefficients are given by $\ln \gamma_1^\infty = \ln \gamma_2^\infty = B$.

If $D = \cdots = 0$, then

$$\frac{G^E}{x_1x_2RT} = B + C(x_1 - x_2)$$

and in this case G^E/x_1x_2RT is linear in x_1. Multiplication of B by $x_1 + x_2$ (= 1)

† O. Redlich, A. T. Kister, and C. E. Turnquist, *Chem. Eng. Progr. Symp. Ser.*, **48**(2): 49, 1952.

gives

$$\frac{G^E}{x_1 x_2 RT} = B(x_1 + x_2) + C(x_1 - x_2) = (B + C)x_1 + (B - C)x_2$$

Letting $B + C = A_{21}$ and $B - C = A_{12}$, we have

$$\frac{G^E}{x_1 x_2 RT} = A_{21}x_1 + A_{12}x_2 \tag{12.11}$$

The corresponding equations for the activity coefficients are

$$\ln \gamma_1 = x_2^2[A_{12} + 2(A_{21} - A_{12})x_1] \tag{12.12}$$

$$\ln \gamma_2 = x_1^2[A_{21} + 2(A_{12} - A_{21})x_2] \tag{12.13}$$

These are the Margules equations, written earlier as Eqs. (11.73). Note that when $x_1 = 0$, $\ln \gamma_1^\infty = A_{12}$; when $x_2 = 0$, $\ln \gamma_2^\infty = A_{21}$.

Another well-known equation is obtained when we write the reciprocal expression $x_1 x_2 RT/G^E$ as a linear function of x_1:

$$\frac{x_1 x_2}{G^E/RT} = B' + C'(x_1 - x_2)$$

This may also be written:

$$\frac{x_1 x_2}{G^E/RT} = B'(x_1 + x_2) + C'(x_1 - x_2) = (B' + C')x_1 + (B' - C')x_2$$

We now let $B' + C' = 1/A'_{21}$ and $B' - C' = 1/A'_{12}$. Then

$$\frac{x_1 x_2}{G^E/RT} = \frac{x_1}{A'_{21}} + \frac{x_2}{A'_{12}} = \frac{A'_{12}x_1 + A'_{21}x_2}{A'_{12}A'_{21}}$$

or

$$\frac{G^E}{x_1 x_2 RT} = \frac{A'_{12}A'_{21}}{A'_{12}x_1 + A'_{21}x_2} \tag{12.14}$$

The activity coefficients implied by this equation are given by

$$\ln \gamma_1 = A'_{12}\left(1 + \frac{A'_{12}x_1}{A'_{21}x_2}\right)^{-2} \tag{12.15}$$

$$\ln \gamma_2 = A'_{21}\left(1 + \frac{A'_{21}x_2}{A'_{12}x_1}\right)^{-2} \tag{12.16}$$

These are known as the van Laar† equations. When $x_1 = 0$, $\ln \gamma_1^\infty = A'_{12}$; when $x_2 = 0$, $\ln \gamma_2^\infty = A'_{21}$.

The Redlich/Kister expansion, the Margules equations, and the van Laar equations are all special cases of a very general treatment based on rational functions, i.e., on equations for G^E given by ratios of polynomials. These are

† Johannes Jacobus van Laar (1860–1938), Dutch physical chemist.

presented in detail by Van Ness and Abbott.† They provide great flexibility in the fitting of VLE data for binary systems. However, they have scant theoretical foundation, and as a result there is no rational basis for their extension to multicomponent systems. Moreover, they do not incorporate an explicit temperature dependence for the parameters, though this can be supplied on an ad hoc basis.

Modern theoretical developments in the molecular thermodynamics of liquid-solution behavior are based on the concept of *local composition.* Within a liquid solution, local compositions, different from the overall mixture composition, are presumed to account for the short-range order and nonrandom molecular orientations that result from differences in molecular size and intermolecular forces. The concept was introduced by G. M. Wilson in 1964 with the publication of a model of solution behavior since known as the Wilson equation.‡ The success of this equation in the correlation of VLE data prompted the development of alternative local-composition models, most notably the NRTL (Non-Random-Two-Liquid) equation of Renon and Prausnitz§ and the UNIQUAC (UNIversal QUAsi-Chemical) equation of Abrams and Prausnitz.¶ A further significant development, based on the UNIQUAC equation, is the UNIFAC method,†† in which activity coefficients are calculated from contributions of the various groups making up the molecules of a solution.

The Wilson equation, like the Margules and van Laar equations, contains just two parameters for a binary system (Λ_{12} and Λ_{21}), and is written:

$$\frac{G^E}{RT} = -x_1 \ln (x_1 + x_2\Lambda_{12}) - x_2 \ln (x_2 + x_1\Lambda_{21}) \tag{12.17}$$

$$\ln \gamma_1 = -\ln (x_1 + x_2\Lambda_{12}) + x_2\left(\frac{\Lambda_{12}}{x_1 + x_2\Lambda_{12}} - \frac{\Lambda_{21}}{x_2 + x_1\Lambda_{21}}\right) \tag{12.18}$$

$$\ln \gamma_2 = -\ln (x_2 + x_1\Lambda_{21}) - x_1\left(\frac{\Lambda_{12}}{x_1 + x_2\Lambda_{12}} - \frac{\Lambda_{21}}{x_2 + x_1\Lambda_{21}}\right) \tag{12.19}$$

For infinite dilution, these equations become

$$\ln \gamma_1^\infty = -\ln \Lambda_{12} + 1 - \Lambda_{21}$$

and

$$\ln \gamma_2^\infty = -\ln \Lambda_{21} + 1 - \Lambda_{12}$$

We note that Λ_{12} and Λ_{21} must always be positive numbers.

† H. C. Van Ness and M. M. Abbott, *Classical Thermodynamics of Nonelectrolyte Solutions: With Applications to Phase Equilibria,* sec. 5-7, McGraw-Hill, New York, 1982.

‡ G. M. Wilson, *J. Am. Chem. Soc.,* **86**: 127, 1964.

§ H. Renon and J. M. Prausnitz, *AIChE J.,* **14**: 135, 1968.

¶ D. S. Abrams and J. M. Prausnitz, *AIChE J.,* **21**: 116, 1975.

†† *UNI*QUAC *F*unctional-group *A*ctivity *C*oefficients; proposed by Aa. Fredenslund, R. L. Jones, and J. M. Prausnitz, *AIChE J.,* **21**: 1086, 1975; given detailed treatment in the monograph: Aa. Fredenslund, J. Gmehling, and P. Rasmussen, *Vapor-Liquid Equilibrium using UNIFAC,* Elsevier, Amsterdam, 1977.

The NRTL equation contains three parameters for a binary system and is written:

$$\frac{G^E}{x_1 x_2 RT} = \frac{G_{21}\tau_{21}}{x_1 + x_2 G_{21}} + \frac{G_{12}\tau_{12}}{x_2 + x_1 G_{12}} \tag{12.20}$$

$$\ln \gamma_1 = x_2^2\left[\tau_{21}\left(\frac{G_{21}}{x_1 + x_2 G_{21}}\right)^2 + \frac{G_{12}\tau_{12}}{(x_2 + x_1 G_{12})^2}\right] \tag{12.21}$$

$$\ln \gamma_2 = x_1^2\left[\tau_{12}\left(\frac{G_{12}}{x_2 + x_1 G_{12}}\right)^2 + \frac{G_{21}\tau_{21}}{(x_1 + x_2 G_{21})^2}\right] \tag{12.22}$$

Here

$$G_{12} = \exp(-\alpha\tau_{12}) \qquad G_{21} = \exp(-\alpha\tau_{21})$$

and

$$\tau_{12} = \frac{b_{12}}{RT} \qquad \tau_{21} = \frac{b_{21}}{RT}$$

where α, b_{12}, and b_{21}, parameters specific to a particular pair of species, are independent of composition and temperature. The infinite-dilution values of the activity coefficients are given by the equations:

$$\ln \gamma_1^\infty = \tau_{21} + \tau_{12}\exp(-\alpha\tau_{12})$$

$$\ln \gamma_2^\infty = \tau_{12} + \tau_{21}\exp(-\alpha\tau_{21})$$

The UNIQUAC equation and the UNIFAC method are models of greater complexity and are treated in App. D.

The local-composition models have limited flexibility in the fitting of data, but they are adequate for most engineering purposes. Moreover, they are implicitly generalizable to multicomponent systems without the introduction of any parameters beyond those required to describe the constituent binary systems. For example, the Wilson equation for multicomponent systems is written:

$$\frac{G^E}{RT} = -\sum_i x_i \ln \sum_j x_j \Lambda_{ij} \tag{12.23}$$

and

$$\ln \gamma_i = 1 - \ln \sum_j x_j \Lambda_{ij} - \sum_k \frac{x_k \Lambda_{ki}}{\sum_j x_j \Lambda_{kj}} \tag{12.24}$$

where $\Lambda_{ij} = 1$ for $i = j$, etc. All indices in these equations refer to the same species, and all summations are over *all* species. For each ij pair there are two parameters, because $\Lambda_{ij} \neq \Lambda_{ji}$. For example, in a ternary system the three possible ij pairs are associated with the parameters Λ_{12}, Λ_{21}; Λ_{13}, Λ_{31}; and Λ_{23}, Λ_{32}.

The temperature dependence of the parameters is given by:

$$\Lambda_{ij} = \frac{V_j}{V_i}\exp\frac{-a_{ij}}{RT} \qquad (i \neq j) \tag{12.25}$$

where V_j and V_i are the molar volumes at temperature T of pure liquids j and i, and a_{ij} is a constant independent of composition and temperature. Thus the Wilson equation, like all other local-composition models, has built into it an approximate temperature dependence for the parameters. Moreover, all parameters are found from data for binary (in contrast to multicomponent) systems. This makes parameter determination for the local-composition models a task of manageable proportions.

12.5 DEW-POINT AND BUBBLE-POINT CALCULATIONS

Although VLE problems with other combinations of variables are possible, those of engineering interest are usually dew-point or bubble-point calculations; there are four classes:

BUBL P: Calculate $\{y_k\}$ and P, given $\{x_k\}$ and T
DEW P: Calculate $\{x_k\}$ and P, given $\{y_k\}$ and T
BUBL T: Calculate $\{y_k\}$ and T, given $\{x_k\}$ and P
DEW T: Calculate $\{x_k\}$ and T, given $\{y_k\}$ and P

Thus, one specifies either T or P *and* either the liquid-phase or the vapor-phase composition, fixing $1 + (N - 1)$ or N phase-rule variables, exactly the number required by the phase rule for vapor/liquid equilibrium. All of these calculations require iterative schemes because of the complex functionality implicit in Eqs. (12.1) and (12.2). In particular, we have the following functional relationships for low-pressure VLE:

$$\Phi_k = \Phi(T, P, y_1, y_2, \ldots, y_{N-1})$$

$$\gamma_k = \gamma(T, x_1, x_2, \ldots, x_{N-1})$$

$$P_k^{\text{sat}} = f(T)$$

For example, when solving for $\{y_k\}$ and P, we do not have values necessary for calculation of the Φ_k, and when solving for $\{x_k\}$ and T, we can evaluate neither the P_k^{sat} nor the γ_k. Simple iterative procedures, described in the following paragraphs, allow efficient solution of each of the four types of problem.

In all cases Eq. (12.1) provides the basis of calculation. This equation, valid for each species k in a multicomponent system, may be written either as

$$y_k = \frac{x_k \gamma_k P_k^{\text{sat}}}{\Phi_k P} \tag{12.26}$$

or as

$$x_k = \frac{y_k \Phi_k P}{\gamma_k P_k^{\text{sat}}} \tag{12.27}$$

Since $\sum y_k = 1$ and $\sum x_k = 1$, we also have

$$1 = \sum_k \frac{x_k \gamma_k P_k^{sat}}{\Phi_k P}$$

or

$$P = \sum_k \frac{x_k \gamma_k P_k^{sat}}{\Phi_k} \tag{12.28}$$

and

$$1 = \sum_k \frac{y_k \Phi_k P}{\gamma_k P_k^{sat}}$$

or

$$P = \frac{1}{\sum_k y_k \Phi_k / \gamma_k P_k^{sat}} \tag{12.29}$$

When $\gamma_k = \Phi_k = 1$, Eqs. (12.28) and (12.29) reduce to the Raoult's-law expressions, Eqs. (10.17) and (10.19).

BUBL P. The iteration scheme for this simple and direct bubble-point calculation is shown in Fig. 12.12. With reference to a computer program for carrying it out, one reads and stores the given values of T and $\{x_k\}$, along with all constants required in evaluation of the P_k^{sat}, γ_k, and Φ_k. Since $\{y_k\}$ is not given, we cannot

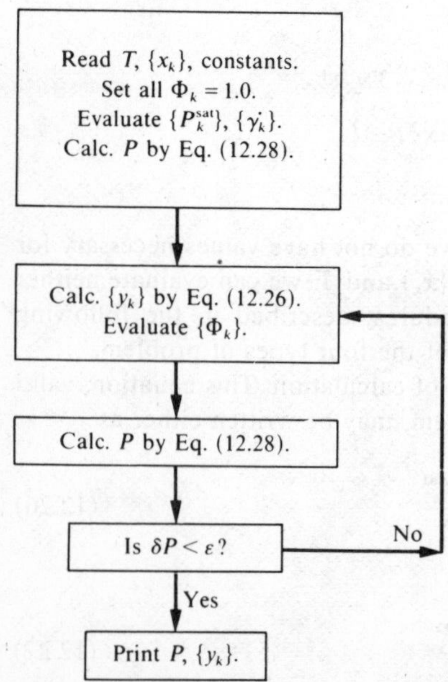

Figure 12.12 Block diagram for the calculation BUBL P.

yet determine values for the Φ_k, and each is set equal to unity. Values for $\{P_k^{sat}\}$ are found from the Antoine equation [Eq. (12.3)] and values of $\{\gamma_k\}$ come from an activity-coefficient correlation. Equations (12.28) and (12.26) are now solved for P and $\{y_k\}$. Values of Φ_k from Eq. (12.6) allow recalculation of P by Eq. (12.28). Iteration leads to final values for P and $\{y_k\}$.

DEW P. The calculational scheme here is shown in Fig. 12.13. We read and store T and $\{y_k\}$, along with appropriate constants. Since we can calculate neither the Φ_k nor the γ_k, all values of each are set equal to unity. Values of $\{P_k^{sat}\}$ are found from the Antoine equation, and Eqs. (12.29) and (12.27) are then solved for P and $\{x_k\}$. Evaluation of $\{\gamma_k\}$ now allows recalculation of P by Eq. (12.29). With this rather good estimate of P, we evaluate $\{\Phi_k\}$ and enter an inner iteration loop that converges on values for $\{x_k\}$ and $\{\gamma_k\}$. Subsequent recalculation of P by Eq. (12.29) leads to the outer iteration loop that establishes the final value of P. Since the x_k calculated within the inner loop are not constrained to sum to

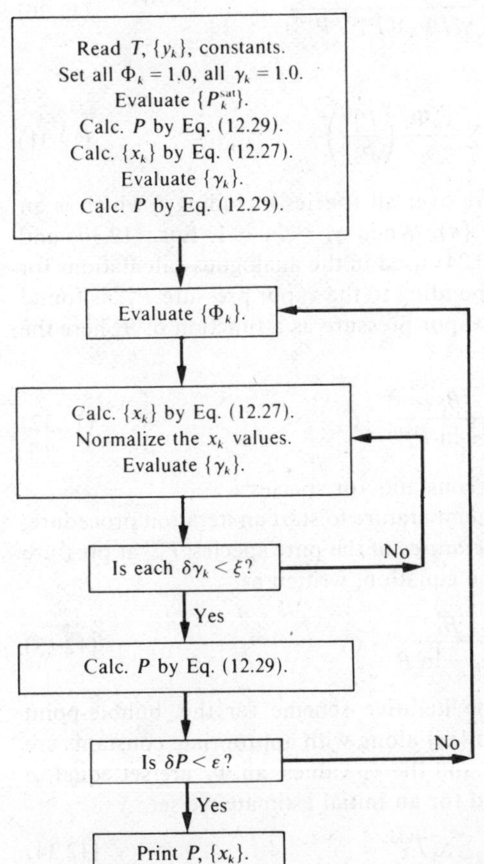

Figure 12.13 Block diagram for the calculation DEW P.

unity, each value is divided by $\sum x_k$:

$$x_k = \frac{x_k}{\sum x_k}$$

This yields a set of *normalized* x_k values, which do sum to unity. Actually, the inner loop can be omitted; it is included simply to make the calculational procedure more efficient.

In the BUBL P and DEW P calculations, the temperature is known initially, and this allows immediate calculation of the key quantities P_k^{sat}. This is not the case for the two remaining procedures, BUBL T and DEW T, where the temperature is to be found. Here, as with the analogous Raoult's law calculations, we deal with vapor-pressure *ratios*, because they are weak functions of temperature. To introduce these ratios on the right-hand sides of Eqs. (12.28) and (12.29), we multiply by P_i^{sat} (outside the summation) and divide by P_i^{sat} (inside the summation). Solution for the P_i^{sat} outside the summation then gives:

$$P_i^{\text{sat}} = \frac{P}{\sum_k (x_k \gamma_k / \Phi_k)(P_k^{\text{sat}}/P_i^{\text{sat}})} \tag{12.30}$$

and

$$P_i^{\text{sat}} = P \sum_k \frac{y_k \Phi_k}{\gamma_k} \left(\frac{P_i^{\text{sat}}}{P_k^{\text{sat}}} \right) \tag{12.31}$$

In these equations the summations are over all species including i, which is an arbitrarily selected species of the set $\{k\}$. When $\gamma_k = \Phi_k = 1$, Eqs. (12.30) and (12.31) reduce to Eqs. (10.20) and (10.24) used in the analogous calculations for Raoult's law. The temperature corresponding to the vapor pressure P_i^{sat} is found from an appropriate equation giving vapor pressure as a function of T, here the Antoine equation:

$$T = \frac{B_i}{A_i - \ln P_i^{\text{sat}}} - C_i \tag{12.32}$$

where A_i, B_i, and C_i are the Antoine constants for species i.

For purposes of finding an initial temperature to start an iteration procedure, we need values of the saturation *temperatures* of the pure species T_k^{sat} at pressure P. These are also given by the Antoine equation, written as:

$$T_k^{\text{sat}} = \frac{B_k}{A_k - \ln P} - C_k \tag{12.33}$$

BUBL T. Figure 12.14 shows the iterative scheme for this bubble-point calculation. The given values of P and $\{x_k\}$ along with appropriate constants are read and stored. In the absence of T and the y_k values, all Φ_k are set equal to unity. Iteration is controlled by T, and for an initial estimate we set

$$T = \sum_k x_k T_k^{\text{sat}} \tag{12.34}$$

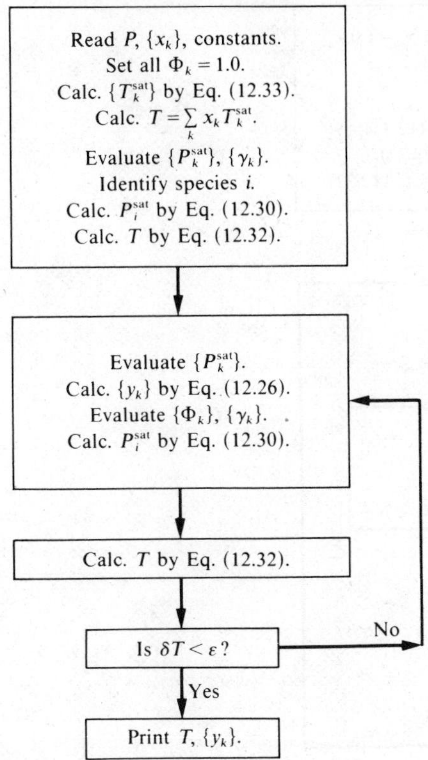

Figure 12.14 Block diagram for the calculation BUBL T.

where the T_k^{sat} are found from Eq. (12.33). With this initial value of T, we find values for $\{P_k^{sat}\}$ from the Antoine equations and values of $\{\gamma_k\}$ from the activity-coefficient correlation. Species i is identified, P_i^{sat} is calculated by Eq. (12.30), and a new value of T is found from Eq. (12.32). The P_k^{sat} are immediately reevaluated, and the y_k are calculated by Eq. (12.26). Values can now be found for both $\{\Phi_k\}$ and $\{\gamma_k\}$, allowing a revised value of P_i^{sat} to be calculated by Eq. (12.30) and a better estimate of T to be found from Eq. (12.32). Iteration then leads to final values of T and $\{y_k\}$.

DEW T. The scheme for this dew-point calculation is shown in Fig. 12.15. Since we know neither the x_k values nor the temperature, all values of both Φ_k and γ_k are set equal to unity. Iteration is again controlled by T, and here we find an initial value by

$$T = \sum_k y_k T_k^{sat} \tag{12.35}$$

With this value of T, we determine $\{P_k^{sat}\}$ from the Antoine equations. All quantities on the right-hand side of Eq. (12.31) are now fixed; we identify species i and solve for P_i^{sat}, from which we get a new value for T by Eq. (12.32). We immediately

DEW T

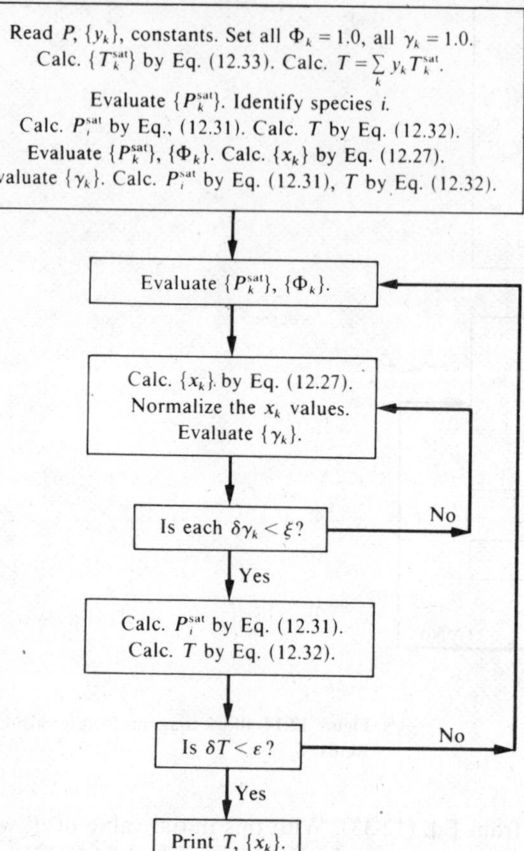

Figure 12.15 Block diagram for the calculation DEW T.

reevaluate $\{P_k^{sat}\}$ which, together with $\{\Phi_k\}$, permits calculation of the x_k by Eq. (12.27). This allows recalculation of P_i^{sat} by Eq. (12.31) and of T by Eq. (12.32). With this rather good estimate of T, we again evaluate $\{P_k^{sat}\}$ and $\{\Phi_k\}$, and enter an inner iteration loop that converges on values of $\{x_k\}$ and $\{\gamma_k\}$. Subsequent recalculation of P_i^{sat} and T then leads to the outer iteration loop that produces a final value of T. As in the DEW P procedure, the x_k calculated within the inner loop are not constrained to sum to unity, and each value is divided by $\sum x_k$:

$$x_k = \frac{x_k}{\sum x_k}$$

This set of normalized x_k values does sum to unity. Again, the inner loop is included simply to make the calculational procedure more efficient.

Table 12.1 shows the results of a BUBL T calculation for the system n-hexane(1)/ethanol(2)/methylcyclopentane(3)/benzene(4). The given pressure P

Table 12.1 Results of BUBL T calculations for the system, *n*-hexane/ethanol/methylcyclopentane(MCP)/benzene at 1(atm)

Species k	x_k	y_k (calc)	y_k (exp)	P_k^{sat} (atm)	Φ_k	γ_k
n-Hexane(1)	0.162	0.139	0.140	0.797	0.993	1.073
Ethanol(2)	0.068	0.279	0.274	0.498	0.999	8.241
MCP(3)	0.656	0.500	0.503	0.725	0.990	1.042
Benzene(4)	0.114	0.082	0.083	0.547	0.983	1.289

T(calc) = 334.82 K T(exp) = 334.85 K Iterations = 4

is 1(atm), and the given liquid-phase mole fractions x_k are listed in the second column of Table 12.1. Parameters for the Antoine equations† [T in kelvins, P in (atm)], supplied as input data, are:

$$A_1 = 9.2033 \qquad B_1 = 2{,}697.55 \qquad C_1 = -48.78$$

$$A_2 = 12.2786 \qquad B_2 = 3{,}803.98 \qquad C_2 = -41.68$$

$$A_3 = 9.1690 \qquad B_3 = 2{,}731.00 \qquad C_3 = -47.11$$

$$A_4 = 9.2675 \qquad B_4 = 2{,}788.51 \qquad C_4 = -52.36$$

As additional input information, we supply the following virial coefficients‡ (in $\mathrm{cm^3\ mol^{-1}}$):

$$B_{11} = -1{,}360.1 \qquad B_{12} = -657.0 \qquad B_{13} = -1{,}274.2 \qquad B_{14} = -1{,}218.8$$

$$B_{22} = -1{,}174.7 \qquad B_{23} = -621.8 \qquad B_{24} = -589.7$$

$$B_{33} = -1{,}191.9 \qquad B_{34} = -1{,}137.9$$

$$B_{44} = -1{,}086.9$$

Finally, the input information includes parameters for the UNIFAC method (App. D). The calculated values of T and the vapor-phase mole fractions y_k compare favorably with experimental values.§ Also listed in Table 12.1 are final computed values of P_k^{sat}, Φ_k, and γ_k.

The BUBL T calculation for which results are given in Table 12.1 is for a pressure of 1(atm), a pressure for which vapor phases are often assumed to be ideal gases. With this assumption, Φ_k is unity for each species. In fact, these values lie between 0.98 and 1.00. Thus in this example, and usually at pressures of 1(atm) and less, the assumption of ideal gases introduces little error. When

† R. C. Reid, J. M. Prausnitz, and T. K. Sherwood, *The Properties of Gases and Liquids*, 3d ed., app. A, McGraw-Hill, New York, 1977.

‡ From the correlation of J. G. Hayden and J. P. O'Connell, *Ind. Eng. Chem. Proc. Des. Dev.*, **14**: 209, 1975.

§ J. E. Sinor and J. H. Weber, *J. Chem. Eng. Data*, **4**: 243, 1960.

this assumption is made, Eq. (12.1) reduces to Eq. (11.74):

$$y_k P = x_k \gamma_k P_k^{\text{sat}} \qquad (k = 1, 2, \ldots, N) \tag{11.74}$$

This modified Raoult's law was used for data reduction in Sec. 11.6. Bubble- and dew-point calculations made with Eq. (11.74) are, of course, somewhat simpler than those shown by Figs. 12.12 through 12.15. Indeed, the BUBL P calculation yields final results in a single step, without iteration. The additional assumption of liquid-phase ideality ($\gamma_k = 1$), on the other hand, is justified only infrequently. We note that γ_k for ethanol in Table 12.1 is greater than 8.

Values of parameters for the Margules, van Laar, Wilson, NRTL, and UNIQUAC equations are given for many binary pairs by Gmehling et al.† in a summary collection of the world's published VLE data for low to moderate pressures. These values are based on reduction of data through application of Eq. (11.74). On the other hand, data reduction for determination of parameters in the UNIFAC method (App. D) is carried out with Eq. (12.1).

Example 12.1 For the system 2-propanol(1)/water(2), the following parameter values are recommended for the Wilson equation:

$$a_{12} = 437.98 \qquad a_{21} = 1{,}238.00 \text{ cal mol}^{-1}$$

$$V_1 = 76.92 \qquad V_2 = 18.07 \text{ cm}^3 \text{ mol}^{-1}$$

In addition, we have the following Antoine equations:

$$\ln P_1^{\text{sat}} = 16.6780 - \frac{3{,}640.20}{T - 53.54}$$

$$\ln P_2^{\text{sat}} = 16.2887 - \frac{3{,}816.44}{T - 46.13}$$

where T is in kelvins and the vapor pressures are in kPa. Assuming the validity of Eq. (11.74), calculate:

(*a*) P and $\{y_k\}$, for $T = 353.15$ K (80°C) and $x_1 = 0.25$.
(*b*) P and $\{x_k\}$, for $T = 353.15$ K (80°C) and $y_1 = 0.60$.
(*c*) T and $\{y_k\}$, for $P = 101.33$ kPa [1(atm)] and $x_1 = 0.85$.
(*d*) T and $\{x_k\}$, for $P = 101.33$ kPa [1(atm)] and $y_1 = 0.40$.
(*e*) P^{az}, the azeotropic pressure, and $x_1^{az} = y_1^{az}$, the azeotropic composition, for $T = 353.15$ K (80°C).

SOLUTION Since we have assumed the validity of Eq. (11.74), $\Phi_k = 1.0$ throughout this problem. This, together with the fact that we are considering a binary system, makes the solution simple enough that the steps can be explained as though carried out by hand calculations.

(*a*) A BUBL P calculation. For $T = 353.15$ K, the Antoine equations yield the following vapor pressures:

$$P_1^{\text{sat}} = 92.59 \qquad P_2^{\text{sat}} = 47.38 \text{ kPa}$$

† J. Gmehling, U. Onken, and W. Arlt, "Vapor-Liquid Equilibrium Data Collection," Chemistry Data Series, vol. 1, Parts 1-8, DECHEMA, Frankfurt/Main, 1977-1984.

The activity coefficients as given by the Wilson equation are calculated by Eqs. (12.18) and (12.19). First, however, we must find the values of Λ_{12} and Λ_{21} by Eq. (12.25). Thus

$$\Lambda_{12} = \frac{V_2}{V_1} \exp \frac{-a_{12}}{RT} = \frac{18.07}{76.92} \exp \frac{-437.98}{(1.987)(353.15)}$$
$$= 0.1258$$

and

$$\Lambda_{21} = \frac{V_1}{V_2} \exp \frac{-a_{21}}{RT} = \frac{76.92}{18.07} \exp \frac{-1{,}238.00}{(1.987)(353.15)}$$
$$= 0.7292$$

Substituting known values into Eqs. (12.18) and (12.19) gives:

$$\ln \gamma_1 = -\ln (0.25 + 0.75 \times 0.1258)$$
$$+ 0.75\left(\frac{0.1258}{0.25 + 0.75 \times 0.1258} - \frac{0.7292}{0.75 + 0.25 \times 0.7292}\right)$$

or

$$\ln \gamma_1 = 1.0661 + 0.75(-0.4168) = 0.7535$$

and

$$\ln \gamma_2 = -\ln (0.75 + 0.25 \times 0.7292) - 0.25(-0.4168)$$
$$= 0.0701 + 0.1042 = 0.1743$$

whence

$$\gamma_1 = 2.1244 \qquad \gamma_2 = 1.1904$$

By Eq. (12.28) with $\Phi = 1.0$,

$$P = (0.25)(2.1244)(92.59) + (0.75)(1.1904)(47.38)$$
$$= 91.48 \text{ kPa}$$

From Eq. (11.74), written as $y_k = x_k \gamma_k P_k^{sat}/P$, we get

$$y_1 = 0.538 \qquad y_2 = 0.462$$

(*b*) A DEW P calculation. With T unchanged from part (*a*), the values of P_1^{sat}, P_2^{sat}, Λ_{12}, and Λ_{21} are the same as already calculated. However, here the liquid-phase composition is unknown. We therefore set $\gamma_k = 1.0$, and Eq. (12.29) reduces to its Raoult's law counterpart:

$$P = \frac{1}{y_1/P_1^{sat} + y_2/P_2^{sat}}$$

From this we find $P = 67.01$ kPa. Equation (12.27), written $x_1 = y_1 P/P_1^{sat}$ now gives:

$$x_1 = \frac{(0.6)(67.01)}{92.59} = 0.434$$

Whence $x_2 = 1 - x_1 = 0.566$. The resulting values of γ_1 and γ_2, calculated by Eqs. (12.18) and (12.19) are:

$$\gamma_1 = 1.4277 \qquad \gamma_2 = 1.4558$$

We recompute P by Eq. (12.29), now written

$$P = \frac{1}{y_1/\gamma_1 P_1^{\text{sat}} + y_2/\gamma_2 P_2^{\text{sat}}}$$

This gives $P = 96.73$kPa. Recalculation of x_1 by Eq. (12.27) gives

$$x_1 = \frac{y_1 P}{\gamma_1 P_1^{\text{sat}}} = \frac{(0.60)(96.73)}{(1.4277)(92.59)} = 0.439$$

Similarly, $x_2 = 0.561$. Equations (12.18) and (12.19) now yield new values of the activity coefficients:

$$\gamma_1 = 1.4167 \qquad \gamma_2 = 1.4646$$

Iteration within the inner loop of Fig. (12.13) leads to the values:

$$x_1 = 0.449 \qquad \gamma_1 = 1.3957 \qquad \gamma_2 = 1.4821$$

Equation (12.29) now gives $P = 96.72$ kPa. Since the Φ_k are fixed at unity, no further iteration is required, and we have for final values:

$$P = 96.72 \text{ kPa} \qquad x_1 = 0.449 \qquad x_2 = 0.551$$

(*c*) A BUBL T calculation. Application of Eq. (12.33) with the given Antoine constants and $P = 101.33$ kPa leads to the values:

$$T_1^{\text{sat}} = 355.39 \qquad T_2^{\text{sat}} = 373.15 \text{ K}$$

An initial value for T is then given by Eq. (12.34):

$$T = (0.85)(355.39) + (0.15)(373.15) = 358.05 \text{ K}$$

Evaluation of the P_k^{sat} values at this temperature by the given Antoine equations gives

$$P_1^{\text{sat}} = 112.60 \qquad P_2^{\text{sat}} = 57.60 \text{ kPa}$$

The activity coefficients at this temperature are calculated by the Wilson equations after evaluation of Λ_{12} and Λ_{21} by Eq. (12.25):

$$\Lambda_{12} = 0.1269 \qquad \Lambda_{21} = 0.7471$$

Then by Eqs. (12.18) and (12.19),

$$\gamma_1 = 1.0197 \qquad \gamma_2 = 2.5265$$

Substitution of values into Eq. (12.30), with $i = 1$ and each $\Phi_k = 1$ gives:

$$P_1^{\text{sat}} = \frac{101.33}{(0.85)(1.0197) + (0.15)(2.5265)(57.60/112.60)}$$

$$= 95.54 \text{ kPa}$$

Equation (12.32) written for species 1 then gives a new value for the temperature, $T = 353.924$ K. The sequence of calculations is now repeated for this temperature, yielding:

$$P_2^{\text{sat}} = 48.88 \text{ kPa} \qquad \Lambda_{12} = 0.1260 \qquad \Lambda_{21} = 0.7320$$

$$\gamma_1 = 1.0197 \qquad \gamma_2 = 2.5287$$

$$P_1^{\text{sat}} = 95.52 \text{ kPa} \qquad T = 353.920 \text{ K}$$

The change in T is negligible, and additional iteration leads to no significant further change in values. We therefore calculate y_1 by Eq. (12.26):

$$y_1 = \frac{(0.85)(1.0197)(95.52)}{(1)(101.33)} = 0.817$$

Thus for final results we have:

$$T = 353.92 \text{ K} \qquad y_1 = 0.817 \qquad y_2 = 0.183$$

(d) A DEW T calculation. Since $P = 101.33$ kPa, the saturation temperatures are the same as those of part (c), but the initial T is given by Eq. (12.35):

$$T = (0.40)(355.39) + (0.60)(373.15) = 366.05 \text{ K}$$

The P_k^{sat} values at this temperature found from the Antoine equations are:

$$P_1^{\text{sat}} = 152.89 \qquad P_2^{\text{sat}} = 78.19 \text{ kPa}$$

For $i = 1$ and $\gamma_k = \Phi_k = 1.0$, we evaluate P_1^{sat} by Eq. (12.31):

$$P_1^{\text{sat}} = 101.33\left[0.40 + 0.60\left(\frac{152.89}{78.19}\right)\right] = 159.41 \text{ kPa}$$

Writing Eq. (12.32) for species 1 gives the new estimate, $T = 367.17$ K. At this temperature, $P_2^{\text{sat}} = 81.54$ kPa, and Λ_{12} and Λ_{21} by Eq. (12.25) are:

$$\Lambda_{12} = 0.1289 \qquad \Lambda_{21} = 0.7801$$

Application of the Wilson equation for evaluation of activity coefficients requires knowledge of the liquid-phase composition. We therefore calculate x_1 by Eq. (12.27):

$$x_1 = \frac{(0.40)(1)(101.33)}{(1)(159.41)} = 0.254$$

Whence $x_2 = 1 - x_1 = 0.746$. Equations (12.18) and (12.19) then give:

$$\gamma_1 = 2.0276 \qquad \gamma_2 = 1.1902$$

We now recalculate P_1^{sat} by Eq. (12.31):

$$P_1^{\text{sat}} = 101.33\left[\frac{0.40}{2.0276} + \frac{0.60}{1.1902}\left(\frac{159.41}{81.54}\right)\right] = 119.86 \text{ kPa}$$

Reevaluation of T by Eq. (12.32) gives $T = 359.65$ K. At this temperature,

$$P_2^{\text{sat}} = 61.31 \text{ kPa} \qquad \Lambda_{12} = 0.1273 \qquad \Lambda_{21} = 0.7529$$

These values remain fixed while the iterations of the inner loop of Fig. 12.15 are carried out. Calculation of x_1 by Eq. (12.27) gives

$$x_1 = \frac{y_1 P}{\gamma_1 P_1^{\text{sat}}} = \frac{(0.40)(101.33)}{(2.0276)(119.86)} = 0.167$$

Similarly, $x_2 = 0.833$. By Eqs. (12.18) and (12.19),

$$\gamma_1 = 2.8103 \qquad \gamma_2 = 1.0999$$

Equation (12.27) yields new values of x_1 and x_2, which are then normalized, and γ_1 and γ_2 are again calculated by Eqs. (12.18) and (12.19). The process is repeated until the γ_1 and γ_2 values do not change appreciably in successive iterations. The results of this procedure are:

$$x_1 = 0.0658 \qquad \gamma_1 = 5.1369 \qquad \gamma_2 = 1.0203$$

Leaving the inner loop, we calculate P_1^{sat} by Eq. (12.31):

$$P_1^{\text{sat}} = 101.33\left[\frac{0.40}{5.1369} + \frac{0.60}{1.0203}\left(\frac{119.86}{61.31}\right)\right] = 124.384 \text{ kPa}$$

By Eq. (12.32), written for species 1, we find $T = 360.61$ K. At this temperature:

$$P_2^{\text{sat}} = 63.62 \text{ kPa} \qquad \Lambda_{12} = 0.1275 \qquad \Lambda_{21} = 0.7563$$

We now return to the inner loop, and iteration for x_1, γ_1, and γ_2 leads to the values:

$$x_1 = 0.0639 \qquad \gamma_1 = 5.0999 \qquad \gamma_2 = 1.0205$$

A return to the outer loop produces no significant change in these results. Thus we find:

$$T = 360.61 \text{ K} \qquad x_1 = 0.0639 \qquad x_2 = 0.9361$$

(*e*) First we determine whether or not an azeotrope exists at the given temperature. This calculation is facilitated by the definition of a quantity called the *relative volatility* α_{12}:

$$\boxed{\alpha_{12} \equiv \frac{y_1/x_1}{y_2/x_2}} \tag{12.36}$$

This quantity becomes unity at an azeotrope. By Eq. (11.74),

$$\frac{y_k}{x_k} = \frac{\gamma_k P_k^{\text{sat}}}{P}$$

Therefore

$$\alpha_{12} = \frac{\gamma_1 P_1^{\text{sat}}}{\gamma_2 P_2^{\text{sat}}} \tag{12.37}$$

At the limits $x_1 = 0$ and $x_1 = 1$, this quantity is given by:

$$(\alpha_{12})_{x_1=0} = \frac{\gamma_1^\infty P_1^{\text{sat}}}{P_2^{\text{sat}}}$$

and

$$(\alpha_{12})_{x_1=1} = \frac{P_1^{\text{sat}}}{\gamma_2^\infty P_2^{\text{sat}}}$$

These values are readily calculated from the given information. If one of them is less than 1 and the other is greater than 1, then an azeotrope exists, because α_{12} is a continuous function of x_1 and must then pass through the value of 1.0 at some intermediate composition.

Values of P_1^{sat} and P_2^{sat} and values of Λ_{12} and Λ_{21} for the Wilson equation are given in part (*a*) for the temperature of interest here. Expressions for the infinite-dilution activity coefficients appear following Eqs. (12.18) and (12.19). Thus

$$\ln \gamma_1^\infty = -\ln \Lambda_{12} + 1 - \Lambda_{21} = -\ln 0.1258 + 1 - 0.7292$$
$$= 2.3439$$

and

$$\ln \gamma_2^\infty = -\ln \Lambda_{21} + 1 - \Lambda_{12} = -\ln 0.7292 + 1 - 0.1258$$
$$= 1.1900$$

Whence

$$\gamma_1^\infty = 10.422 \qquad \gamma_2^\infty = 3.287$$

$$(\alpha_{12})_{x_1=0} = \frac{(10.422)(92.59)}{47.38} = 20.37$$

and

$$(\alpha_{12})_{x_1=1} = \frac{(92.59)}{(3.287)(47.38)} = 0.595$$

From these results, we conclude that an azeotrope does indeed exist.

For $\alpha_{12} = 1$, Eq. (12.37) becomes

$$\frac{\gamma_1^{\mathrm{az}}}{\gamma_2^{\mathrm{az}}} = \frac{P_2^{\mathrm{sat}}}{P_1^{\mathrm{sat}}} = \frac{47.38}{92.59} = 0.5117$$

The difference between Eqs. (12.19) and (12.18), the Wilson equations for γ_2 and γ_1, gives the general expression:

$$\ln \frac{\gamma_1}{\gamma_2} = \ln \frac{x_2 + x_1\Lambda_{21}}{x_1 + x_2\Lambda_{12}} + \frac{\Lambda_{12}}{x_1 + x_2\Lambda_{12}} - \frac{\Lambda_{21}}{x_2 + x_1\Lambda_{21}}$$

Thus the azeotropic composition is the value of x_1 (with $x_2 = 1 - x_1$) for which this equation is satisfied when

$$\ln \frac{\gamma_1}{\gamma_2} = \ln 0.5117 = -0.6700$$

and

$$\Lambda_{12} = 0.1258 \qquad \Lambda_{21} = 0.7292$$

Solution by trial for x_1 gives $x_1^{\mathrm{az}} = 0.7173$. For this value of x_1, we find from Eq. (12.18) that $\gamma_1^{\mathrm{az}} = 1.0787$. With $x_1^{\mathrm{az}} = y_1^{\mathrm{az}}$, Eq. (11.74) becomes

$$P^{\mathrm{az}} = \gamma_1^{\mathrm{az}} P_1^{\mathrm{sat}} = (1.0787)(92.59)$$

Thus

$$P^{\mathrm{az}} = 99.83 \text{ kPa} \qquad x_1^{\mathrm{az}} = y_1^{\mathrm{az}} = 0.7173$$

12.6 FLASH CALCULATIONS

The *P*, *T*-flash calculation was discussed in Sec. 10.5 in connection with Raoult's law. The problem is to calculate for a system of known *overall* composition $\{z_i\}$

at given T and P the fraction of the system that is vapor V and the compositions of both the vapor phase $\{y_i\}$ and the liquid phase $\{x_i\}$. This problem is known to be determinate on the basis of Duhem's theorem, because two independent variables (T and P) are specified for a system made up of fixed quantities of its constituent species.

The flash calculation illustrated by Example 10.3 for a system obeying Raoult's law was solved by a very simple trial procedure. This was possible because K-values ($K_i \equiv y_i/x_i$) could be calculated from knowledge of T and P alone. When the K-values depend not only on T and P but also on the phase compositions, their calculation is inherently more difficult. Moreover, since the phase compositions are not initially known, they are most conveniently found by an iterative computation scheme.

On the basis of material balances and the definition of a K-value, we derived in Sec. 10.5 the equation,

$$y_i = \frac{z_i K_i}{1 + V(K_i - 1)} \qquad (i = 1, 2, \ldots, N) \tag{10.29}$$

Since $x_i = y_i / K_i$, an alternative equation is

$$x_i = \frac{z_i}{1 + V(K_i - 1)} \qquad (i = 1, 2, \ldots, N) \tag{12.38}$$

Since both sets of mole fractions must sum to unity, $\sum x_i = \sum y_i = 1$. Thus, if we sum Eq. (10.29) over all species and subtract unity from this sum, the difference F_y must be zero; that is,

$$F_y = \sum_i \frac{z_i K_i}{1 + V(K_i - 1)} - 1 = 0 \tag{12.39}$$

Similar treatment of Eq. (12.38) yields the difference F_x, which must also be zero:

$$F_x = \sum_i \frac{z_i}{1 + V(K_i - 1)} - 1 = 0 \tag{12.40}$$

Solution to a P, T-flash problem is accomplished when a value of V is found that makes *either* the function F_y *or* F_x equal to zero. However, a more convenient function for use in a *general* solution procedure† is the difference $F_y - F_x = F$:

$$F = \sum_i \frac{z_i(K_i - 1)}{1 + V(K_i - 1)} = 0 \tag{12.41}$$

The advantage of this function is apparent from its derivative:

$$\frac{dF}{dV} = -\sum_i \frac{z_i(K_i - 1)^2}{[1 + V(K_i - 1)]^2} \tag{12.42}$$

† H. H. Rachford, Jr., and J. D. Rice, *J. Petrol. Technol.*, **4**(10): sec. 1, p. 19 and sec. 2, p. 3, October, 1952.

Since dF/dV is always negative, the F vs. V relation is monotonic, and this makes Newton's method (App. E), a rapidly converging iteration procedure, well suited to solution for V. Newton's method here gives

$$V_{j+1} = V_j - \frac{F_j}{(dF/dV)_j} \tag{12.43}$$

where j is the iteration index, and F_j and $(dF/dV)_j$ are found by Eqs. (12.41) and (12.42). In these equations the K-values come from Eq. (11.66) written

$$K_i = \frac{y_i}{x_i} = \frac{\gamma_i P_i^{\text{sat}}}{\Phi_i P} \qquad (i = 1, 2, \ldots, N) \tag{12.44}$$

where Eq. (11.67) without the Poynting factor gives

$$\Phi_i = \frac{\hat{\phi}_i}{\phi_i^{\text{sat}}}$$

The K-values contain all of the thermodynamic information, and are related in a complex way to T, P, $\{y_i\}$, and $\{x_i\}$. Since we are solving for $\{y_i\}$ and $\{x_i\}$, the P, T-flash calculation inevitably requires iteration.

A general solution scheme is shown by the block diagram of Fig. 12.16. The given information is read and stored. Since we do not know in advance whether the system of stated composition at the stated T and P is in fact a mixture of saturated liquid and saturated vapor and not entirely liquid or entirely vapor, we do preliminary calculations to establish the nature of the system. At the given T and overall composition, the system exists as a superheated vapor if its pressure is less than the dew-point pressure P_{dew}. On the other hand, it exists as a subcooled liquid if its pressure is greater than the bubble-point pressure P_{bubl}. Only for pressures between P_{dew} and P_{bubl} is the system an equilibrium mixture of vapor and liquid. We therefore determine P_{dew} by a DEW P calculation (see Fig. 12.13) at the given T and for $\{y_i\} = \{z_i\}$ and P_{bubl} by a BUBL P calculation (see Fig. 12.12) at the given T and for $\{x_i\} = \{z_i\}$. The P, T-flash calculation is performed only if the given pressure P lies between P_{dew} and P_{bubl}. If this is the case, then we make use of the results of the preliminary DEW P and BUBL P calculations to provide initial estimates of $\{\gamma_i\}$, $\{\hat{\phi}_i\}$, and V. For the dew point, we have calculated values of P_{dew}, $\gamma_{i,\text{dew}}$, $\hat{\phi}_{i,\text{dew}}$, and $V_{\text{dew}} = 1$; for the bubble point, we have calculated values of P_{bubl}, $\gamma_{i,\text{bubl}}$, $\hat{\phi}_{i,\text{bubl}}$, and $V_{\text{bubl}} = 0$. The simplest procedure is to interpolate between dew- and bubble-point values in relation to the location of P between P_{dew} and P_{bubl}:

$$\frac{\gamma_i - \gamma_{i,\text{dew}}}{\gamma_{i,\text{bubl}} - \gamma_{i,\text{dew}}} = \frac{\hat{\phi}_i - \hat{\phi}_{i,\text{dew}}}{\hat{\phi}_{i,\text{bubl}} - \hat{\phi}_{i,\text{dew}}} = \frac{P - P_{\text{dew}}}{P_{\text{bubl}} - P_{\text{dew}}}$$

and

$$\frac{V - 1}{0 - 1} = \frac{P - P_{\text{dew}}}{P_{\text{bubl}} - P_{\text{dew}}} \quad \text{or} \quad V = \frac{P_{\text{bubl}} - P}{P_{\text{bubl}} - P_{\text{dew}}}$$

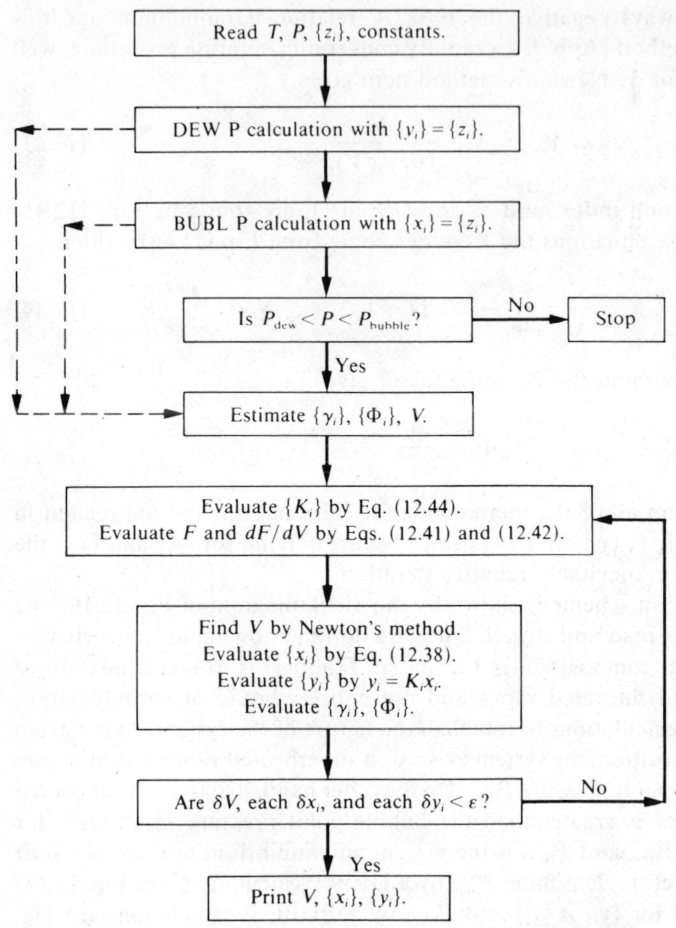

Figure 12.16 Block diagram for a P, T-flash calculation.

With these initial values of the γ_i and $\hat{\phi}_i$, initial values of the K_i can be calculated by Eq. (12.44). The P_i^{sat} and ϕ_i^{sat} values are already available from the preliminary DEW P and BUBL P calculations. Equations (12.41) and (12.42) now provide initial values of $F = F_0$ and $dF/dV = (dF/dV)_0$. The initial value of $V = V_0$ comes from the preceding step. These values are substituted into Eq. (12.43), which represents Newton's method, and repeated application of this equation leads to the value of V for which Eq. (12.41) is satisfied for the present estimates of the K_i. The remaining calculations serve to provide new estimates of the γ_i and Φ_i from which to reevaluate the K_i. The sequence of steps is repeated until there is no significant change in results from one iteration to the next. After the first application of Newton's method, the starting value V_0 in subsequent iterations is simply the most recently calculated value. Once the value of V is established,

Table 12.2 Results of a *P*, *T*-flash calculation for the system, *n*-hexane/ethanol/methylcyclopentane(MCP)/benzene

Species(i)	z_i	x_i	y_i	K_i
n-Hexane(1)	0.250	0.160	0.270	1.694
Ethanol(2)	0.400	0.569	0.362	0.636
MCP(3)	0.200	0.129	0.216	1.668
Benzene(4)	0.150	0.142	0.152	1.070
$P = 1$(atm)	$T = 334.15$ K	$V = 0.8166$		

the x_i values are calculated by Eq. (12.38) and the y_i values by the equation $y_i = K_i x_i$.

Table 12.2 shows the results of a P, T-flash calculation for the system n-hexane(1)/ethanol(2)/methylcyclopentane(3)/benzene(4). This is the same system for which the results of a BUBL T calculation were presented in Table 12.1, and the same correlations and parameter values have been used here. The given P and T are 1(atm) and 334.15 K. The given overall mole fractions for the system $\{z_i\}$ are listed in the table along with the calculated values of the liquid-phase and vapor-phase mole fractions and the K-values. The molar fraction of the system that is vapor is here found to be $V = 0.8166$.

12.7 COMPOSITION DEPENDENCE OF $\hat{f}_i$

Numerical values for the fugacities of species in liquid mixtures are readily calculated from experimental VLE data. According to Eq. (11.30),

$$\hat{f}_i^l = \hat{f}_i^v$$

for each species. If we assume the equilibrium vapor phase to be an ideal gas, then $\hat{f}_i^v = y_i P$, the partial pressure of species i in the vapor, and

$$\hat{f}_i^l = y_i P$$

In the limit of pure species i, where $x_i = y_i = 1$, this becomes $\hat{f}_i^l = f_i^l = P_i^{\text{sat}}$. Thus, for example, we can calculate the fugacities of species 1 and 2 in the liquid mixture methyl ethyl ketone(1)/toluene(2) at 50°C to a good approximation from the $y_1 - P$ data listed in Table 11.1. Specifically, when $P = 25.92$ kPa, $y_1 = 0.744$, and $y_2 = 1 - y_1 = 0.256$, then

$$\hat{f}_1 = (0.744)(25.92) = 19.28 \text{ kPa}$$

and

$$\hat{f}_2 = (0.256)(25.92) = 6.64 \text{ kPa}$$

where superscript l has for simplicity been dropped. The values of $\hat{f}_1$ and $\hat{f}_2$ so calculated from the data of Table 11.1 are plotted in Fig. 12.17 as the solid lines.

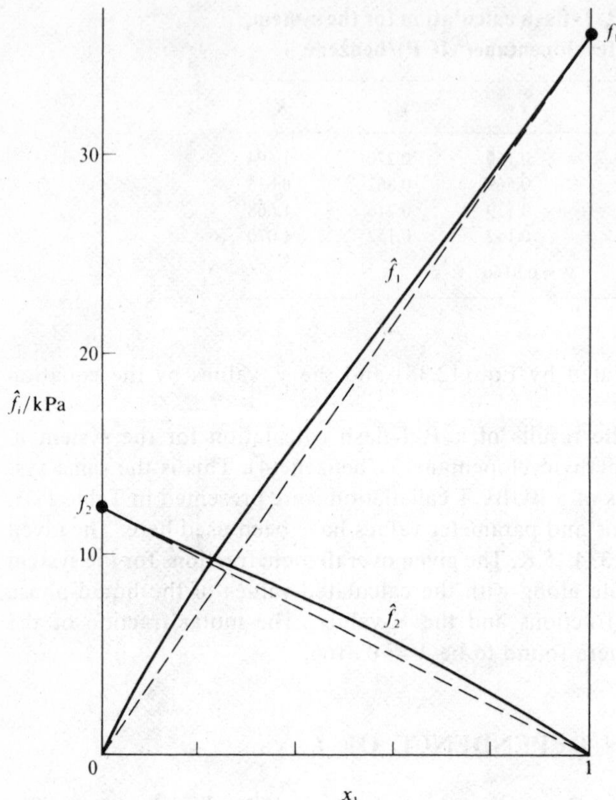

Figure 12.17 Fugacities $\hat{f}_1$ and $\hat{f}_2$ for the system methyl ethyl ketone(1)/toluene(2) at 50°C. The dashed lines represent the Lewis/Randall rule.

The straight dashed lines represent Eq. (11.61), the Lewis/Randall rule, which expresses the composition dependence of the component fugacities in an ideal solution:

$$\hat{f}_i^{id} = x_i f_i \tag{11.61}$$

Figure 12.17, derived from a specific set of data, illustrates the general characteristics of the $\hat{f}_1$ and $\hat{f}_2$ vs. x_1 relationships for a binary liquid solution at constant T. Although P varies, its influence on the $\hat{f}_i$ is very small, and a plot at constant T and P would look the same. Thus in Fig. 12.18 we show a schematic diagram of the $\hat{f}_i$-vs.-x_i relation for species i ($i = 1, 2$) in a binary solution at constant T and P.

The straight dashed line in Fig. 12.18 that represents the Lewis/Randall rule is the only model of ideal-solution behavior so far considered. Alternative models also express the direct proportionality between $\hat{f}_i$ and x_i represented by Eq. (11.61), but with different proportionality constants. We may express this direct

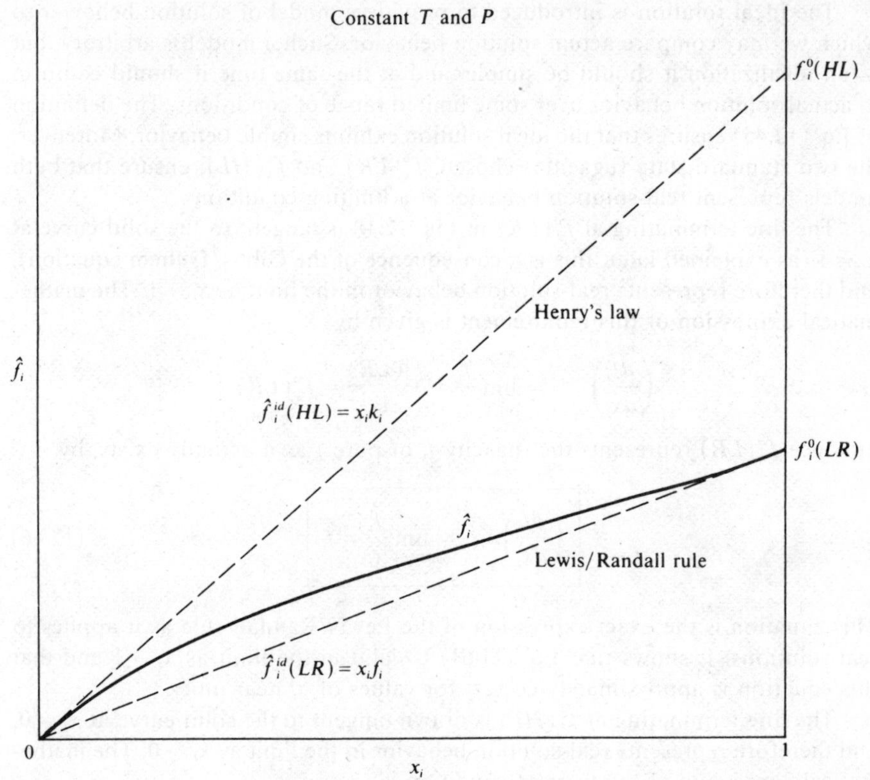

Figure 12.18 Composition dependence of $\hat{f}_i$, showing relation to Henry's law and the Lewis/Randall rule.

proportionality quite generally by writing:

$$\hat{f}_i^{id} = x_i f_i^\circ \tag{12.45}$$

When $x_i = 1$, $\hat{f}_i^{id}$ is equal to the fugacity of pure species i in some state at the mixture T and P. Such states are called *standard states*, and they may be either real or imaginary. When $f_i^\circ = f_i$, Eqs. (11.61) and (12.45) are identical; thus the standard state associated with the Lewis/Randall rule is the *real* state of species i at the T and P of the mixture.

The nature of imaginary (or ficticious or hypothetical) standard states is most easily explained by reference to Fig. 12.18. The two dashed lines shown both conform to ideal-solution behavior as prescribed by Eq. (12.45). The points labeled $f_i^\circ(LR)$ and $f_i^\circ(HL)$ are both fugacities of pure i, but only $f_i^\circ(LR)$ is the fugacity of pure i as it actually exists at the given T and P. The other point $f_i^\circ(HL)$ represents an imaginary state of pure i in which its imaginary properties are fixed at values other than those of the real fluid. Either choice of value for f_i° fixes the entire line which represents $\hat{f}_i^{id} = x_i f_i^\circ$.

The ideal solution is introduced to provide a model of solution behavior to which we may compare actual solution behavior. Such a model is arbitrary, but as an idealization it should be simple, and at the same time it should conform to actual solution behavior over some limited range of conditions. The definition of Eq. (12.45) ensures that the ideal solution exhibits simple behavior. Moreover, the two standard-state fugacities chosen, $f_i^\circ(LR)$ and $f_i^\circ(HL)$, ensure that both models represent real-solution behavior at a limiting condition.

The line terminating at $f_i^\circ(LR)$ in Fig. 12.18 is tangent to the solid curve at $x_i = 1$ (as explained later, this is a consequence of the Gibbs/Duhem equation), and therefore represents real-solution behavior in the limit as $x_i \to 1$. The mathematical expression of this requirement is given by

$$\left(\frac{d\hat{f}_i}{dx_i}\right)_{x_i=1} = \lim_{x_i \to 1} \frac{\hat{f}_i}{x_i} = \frac{f_i^\circ(LR)}{1} = f_i^\circ(LR)$$

or, since $f_i^\circ(LR)$ represents the fugacity f_i of pure i as it actually exists, by

$$\left(\frac{d\hat{f}_i}{dx_i}\right)_{x_i=1} = \lim_{x_i \to 1} \frac{\hat{f}_i}{x_i} = f_i \tag{12.46}$$

This equation is the exact expression of the Lewis/Randall rule as it applies to real solutions. It shows that Eq. (11.61) is valid in the limit as $x_i \to 1$ and that this equation is approximately correct for values of x_i near unity.

The line terminating at $f_i^\circ(HL)$ is drawn tangent to the solid curve at $x_i = 0$, and therefore represents real-solution behavior in the limit as $x_i \to 0$. The mathematical expression of the tangent condition is

$$\lim_{x_i \to 0} \frac{\hat{f}_i}{x_i} = \left(\frac{d\hat{f}_i}{dx_i}\right)_{x_i=0} = f_i^\circ(HL)$$

or more commonly

$$\lim_{x_i \to 0} \frac{\hat{f}_i}{x_i} = k_i \tag{12.47}$$

Equation (12.47) is a statement of *Henry's law* (*HL*) as it applies to real solutions. It shows that the equation $\hat{f}_i = x_i k_i$ applies in the limit as $x_i \to 0$, and that this relation is of approximate validity for small values of x_i. The proportionality factor k_i is called Henry's constant.

Equations (12.46) and (12.47) imply two models of solution ideality. The first is based on the Lewis/Randall rule, for which the standard-state fugacity is

$$f_i^\circ(LR) = f_i$$

and the other is based on Henry's law, for which the standard-state fugacity is

$$f_i^\circ(HL) = k_i$$

Thus in practice the direct proportionality of Eq. (12.45) takes two forms:

$$\hat{f}_i^{id}(LR) = x_i f_i \tag{12.48}$$

and

$$\hat{f}_i^{id}(HL) = x_i k_i \tag{12.49}$$

Both models of ideality are shown in Fig. 12.18 in relation to the curve representing the actual $\hat{f}_i$-vs.-x_i behavior. These equations have two uses. First, they provide approximate values for $\hat{f}_i$ when applied to appropriate composition ranges. Second, they provide reference values to which actual values of $\hat{f}_i$ may be compared. This use is formalized through the activity coefficient, which is defined by

$$\gamma_i = \frac{\hat{f}_i}{\hat{f}_i^{id}} \tag{12.50}$$

For ideality in the sense of the Lewis/Randall rule, this equation is identical with Eq. (11.59). For ideality in the sense of Henry's law, it becomes

$$\gamma_i(HL) = \frac{\hat{f}_i}{x_i k_i} \tag{12.51}$$

Use of activity coefficients based on Henry's law is treated in the following section.

The Gibbs/Duhem equation provides a relation between the Lewis/Randall rule and Henry's law. Substituting $d\bar{G}_i$ from Eq. (11.28) for $d\bar{M}_i$ in Eq. (11.8) gives, for a binary solution at constant T and P,

$$x_1\, d \ln \hat{f}_1 + x_2\, d \ln \hat{f}_2 = 0$$

In the region where Henry's law is valid for component 1, Eq. (12.47) is written $\hat{f}_1 = x_1 k_1$, in which case the foregoing equation becomes

$$x_1\, d \ln (x_1 k_1) + x_2\, d \ln \hat{f}_2 = 0$$

or

$$d \ln \hat{f}_2 = -\frac{x_1}{x_2}\, d \ln (k_1 x_1) = -\frac{x_1}{x_2}\frac{d(k_1 x_1)}{k_1 x_1} = \frac{-dx_1}{x_2}$$

Since $dx_1 + dx_2 = 0$ for composition changes in a binary system,

$$d \ln \hat{f}_2 = \frac{dx_2}{x_2} = d \ln x_2$$

Integration from $x_2 = 1$, where $\hat{f}_2 = f_2$, to arbitrary mole fraction x_2 gives

$$\ln \frac{\hat{f}_2}{f_2} = \ln \frac{x_2}{1}$$

or

$$\hat{f}_2 = f_2 x_2$$

This is the Lewis/Randall rule for species 2, and the derivation shows that it holds whenever Henry's law is valid for species 1. Similarly, $\hat{f}_1 = f_1 x_1$ whenever $\hat{f}_2 = k_2 x_2$.

Figure 12.18 is drawn for a species that shows positive deviations from ideality in the sense of the Lewis/Randall rule. Negative deviations from ideality are also common, and in this case the $\hat{f}_i$-vs.-x_i curve lies below the Lewis/Randall line. In Fig. 12.19 we show the composition dependence of the fugacity of acetone in two different binary solutions at 50°C. When the second component is methanol, acetone shows positive deviations from ideality. On the other hand, when the second component is chloroform, acetone shows negative deviations from ideality. The fugacity of pure acetone f_{acetone} is of course the same regardless of the second component. However, Henry's constants, represented by the slopes of the two dotted lines, are very different for the two cases.

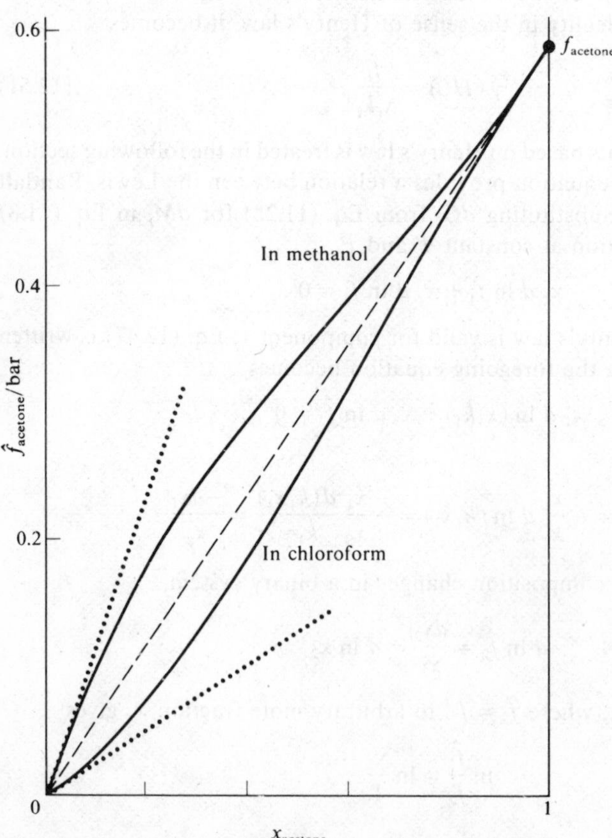

Figure 12.19 Composition dependence of the fugacity of acetone in two binary liquid solutions at 50°C.

12.8 HENRY'S LAW AS A MODEL FOR IDEAL BEHAVIOR OF A SOLUTE

Application of the Lewis/Randall rule, Eq. (11.61),

$$\hat{f}_i^{id} = x_i f_i$$

to species i in a liquid solution requires knowledge of f_i, the fugacity of pure liquid i at the mixture T and P. We have presumed in the preceding discussion that the liquid phase being considered is stable throughout the entire composition range at the given T and P. Where this is true, as for mixtures of subcooled liquids, the ideal-solution model based on the Lewis/Randall rule provides the most convenient values of $\hat{f}_i^{id}$ for reference purposes. However, there is always a range of conditions of T and P for which the full curve of Fig. 12.18 for a given liquid phase cannot be determined, because the phase becomes unstable in some composition range. This is most obvious when gases or solids of limited solubility dissolve in liquids. What, then, is done when pure species i does not *exist* as a liquid at the mixture T and P?

Consider a binary liquid solution of species 1 and 2, wherein species 1 dissolves up to some solubility limit at a specified T and P. Data for the solution can therefore exist only up to this limit, and a plot like Fig. 12.18 is necessarily truncated, as indicated by Fig. 12.20. Clearly, the Lewis/Randall line for species 2, representing the relation

$$\hat{f}_2^{id} = x_2 f_2$$

is readily constructed. However, f_1 does not appear on the figure, and the corresponding Lewis/Randall line for species 1 cannot be drawn. We can, however, construct an alternative line for species 1, representing the alternative model of ideal behavior provided by Henry's law, as shown in Fig. 12.20. Henry's constant, the standard-state fugacity, is the fugacity that pure species 1 would have if species 1 obeyed Henry's law over the full range of mole fractions from $x_1 = 0$ to $x_1 = 1$.

We write Eq. (11.28) for species 1 in solution:

$$d\bar{G}_1 = RT \ln \hat{f}_1 \qquad \text{(const } T\text{)}$$

Integration of this equation at constant T, P, and x_1 for a change from the state of species 1 in an ideal solution in the sense of Henry's law, where $\bar{G}_1 = G_1^{id}(HL)$ and $\hat{f}_1 = x_1 k_1$, to its actual state in solution gives

$$\bar{G}_1 - \bar{G}_1^{id}(HL) = RT \ln \frac{\hat{f}_1}{x_1 k_1}$$

The difference on the left is just an alternative partial excess Gibbs energy, $\bar{G}_1^E(HL)$, and the argument of the logarithm by Eq. (12.51) is:

$$\gamma_1(HL) = \frac{\hat{f}_1}{x_1 k_1} \tag{12.52}$$

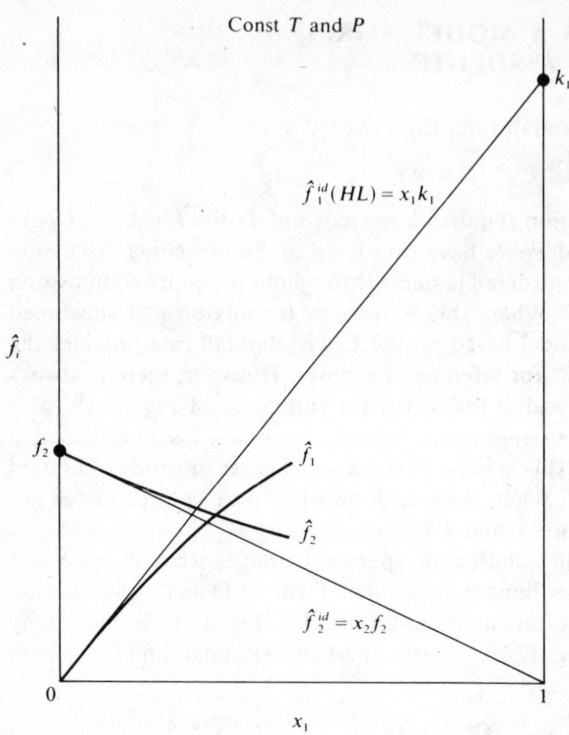

Figure 12.20 Plots of $\hat{f}_1$ and $\hat{f}_2$ vs. x_1 for a binary liquid system wherein species 1 is of limited solubility in species 2.

Therefore

$$\bar{G}_1^E(HL) = RT \ln \gamma_1(HL) \tag{12.53}$$

Analytical representation of the excess Gibbs energy of a system implies knowledge of the standard-state fugacities f_i° and of the $\hat{f}_i$-vs.-x_1 relationships. Since an equation expressing $\hat{f}_1$ as a function of x_1 cannot recognize a solubility limit, it implies an extrapolation of the $\hat{f}_1$-vs.-x_1 curve from the solubility limit to $x_1 = 1$, at which point $\hat{f}_1 = f_1$. This provides a fictitious or hypothetical value for the fugacity of pure species 1 that serves to establish a Lewis/Randall line for this species, as shown by Fig. 12.21. It is also the basis for calculation of the activity coefficient of species 1:

$$\gamma_1 = \frac{\hat{f}_1}{x_1 f_1} \tag{12.54}$$

This equation may be written

$$\frac{\hat{f}_1}{x_1} = \gamma_1 f_1$$

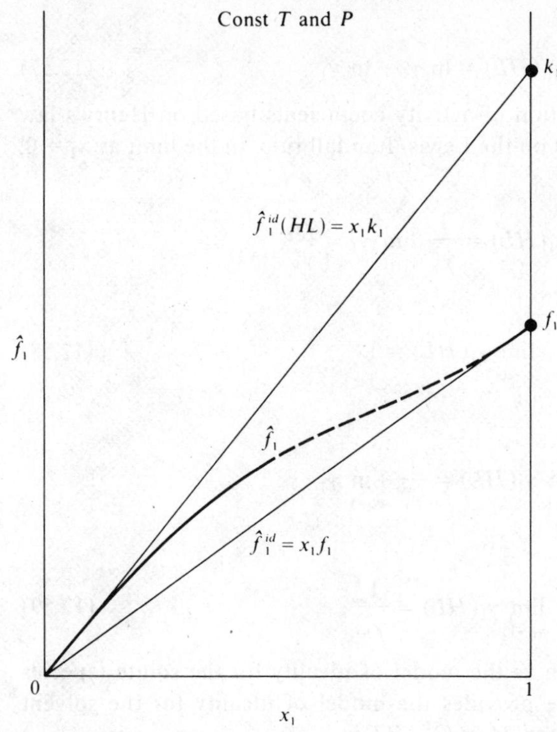

Figure 12.21 Plot of $\hat{f}_1$ vs. x_1 showing extrapolation to $x_1 = 1$. The straight lines represent ideal-solution models based on Henry's law and the Lewis/Randall rule.

whence

$$\lim_{x_1 \to 0} \frac{\hat{f}_1}{x_1} = \lim_{x_1 \to 0} (\gamma_1 f_1) = \gamma_1^\infty f_1$$

In view of Eq. (12.47), this becomes

$$k_1 = \gamma_1^\infty f_1 \tag{12.55}$$

a direct relation between k_1 and f_1, the two fictitious standard-state fugacities of pure species 1.

Solving Eqs. (12.52) and (12.54) for $\hat{f}_1$, we get two expressions for $\hat{f}_1$ which may be equated to give:

$$x_1 k_1 \gamma_1(HL) = x_1 f_1 \gamma_1$$

or

$$\gamma_1(HL) = \frac{\gamma_1}{k_1/f_1}$$

In view of Eq. (12.55) this becomes

$$\gamma_1(HL) = \frac{\gamma_1}{\gamma_1^\infty} \tag{12.56}$$

or

$$\ln \gamma_1(HL) = \ln \gamma_1 - \ln \gamma_1^\infty \tag{12.57}$$

These equations allow calculation of activity coefficients based on Henry's law from activity coefficients based on the Lewis/Randall rule. In the limit as $x_1 \to 0$, Eq. (12.56) yields:

$$\lim_{x_1\to 0} \gamma_1(HL) = \frac{1}{\gamma_1^\infty} \lim_{x_1\to 0} \gamma_1 = \frac{\gamma_1^\infty}{\gamma_1^\infty}$$

or

$$\lim_{x_1\to 0} \gamma_1(HL) = 1 \tag{12.58}$$

In the limit as $x_1 \to 1$

$$\lim_{x_1\to 1} \gamma_1(HL) = \frac{1}{\gamma_1^\infty} \lim_{x_1\to 1} \gamma_1$$

or

$$\lim_{x_1\to 1} \gamma_1(HL) = \frac{1}{\gamma_1^\infty} \tag{12.59}$$

When Henry's law is taken as the model of ideality for the solute (species 1) and the Lewis/Randall rule provides the model of ideality for the solvent (species 2), Eq. (11.5) written for $M = G^E/RT$ is

$$\left(\frac{G^E}{RT}\right)^* = x_1 \frac{\bar{G}_1^E(HL)}{RT} + x_2 \frac{\bar{G}_2^E}{RT}$$

where the asterisk (*) denotes a value based on this asymmetric treatment of solution ideality. As a result of Eqs. (11.60) and (12.53), the preceding equation becomes

$$\left(\frac{G^E}{RT}\right)^* = x_1 \ln \gamma_1(HL) + x_2 \ln \gamma_2 \tag{12.60}$$

Substitution for $\ln \gamma_1(HL)$ by Eq. (12.57) gives

$$\left(\frac{G^E}{RT}\right)^* = x_1 \ln \gamma_1 - x_1 \ln \gamma_1^\infty + x_2 \ln \gamma_2$$

In view of Eq. (11.63) this can be written

$$\left(\frac{G^E}{RT}\right)^* = \frac{G^E}{RT} - x_1 \ln \gamma_1^\infty \tag{12.61}$$

This equation relates the excess Gibbs energy based on the asymmetric treatment of solution ideality to the excess Gibbs energy based entirely on the Lewis/Randall rule.

Example 12.2 Given a binary solution for which the composition dependence at constant T and P of the excess Gibbs energy is expressed by

$$\frac{G^E}{RT} = Bx_1x_2$$

find the corresponding equations for $\ln \gamma_1(HL)$ and $(G^E/RT)^*$.

SOLUTION Equations (12.9) and (12.10) for $\ln \gamma_1$ and $\ln \gamma_2$ are associated with the given equation for G^E/RT:

$$\ln \gamma_1 = Bx_2^2 \tag{A}$$

and

$$\ln \gamma_2 = Bx_1^2 \tag{B}$$

When $x_1 = 0$, $x_2 = 1$, and Eq. (A) becomes

$$\ln \gamma_1^\infty = B$$

Equation (12.57) then yields

$$\ln \gamma_1(HL) = Bx_2^2 - B = B(x_2^2 - 1) = B[(1 - x_1)^2 - 1]$$

or

$$\ln \gamma_1(HL) = -Bx_1(2 - x_1) \tag{C}$$

By Eq. (12.61)

$$\left(\frac{G^E}{RT}\right)^* = Bx_1x_2 - x_1B = Bx_1(x_2 - 1)$$

or

$$\left(\frac{G^E}{RT}\right)^* = -Bx_1^2 \tag{D}$$

We should be able to regenerate Eqs. (B) and (C) by application of Eq. (11.62). Multiplying Eq. (D) by n and substituting $x_1 = n_1/n$ gives

$$\left(\frac{nG^E}{RT}\right)^* = \frac{-Bn_1^2}{n}$$

With the understanding that T and P are held constant, differentiation with respect to n_1 at constant n_2 gives:

$$\ln \gamma_1(HL) = -B\left[\frac{2n_1}{n} - \frac{n_1^2}{n^2}\left(\frac{\partial n}{\partial n_1}\right)_{n_2}\right]$$

Since $(\partial n/\partial n_1) = 1$, this becomes

$$\ln \gamma_1(HL) = -B(2x_1 - x_1^2) = -Bx_1(2 - x_1)$$

in agreement with Eq. (C). Similarly, differentiation with respect to n_2 at const n_1 gives

$$\ln \gamma_2 = -Bn_1^2\left(\frac{-1}{n^2}\right)\left(\frac{\partial n}{\partial n_2}\right)_{n_1}$$

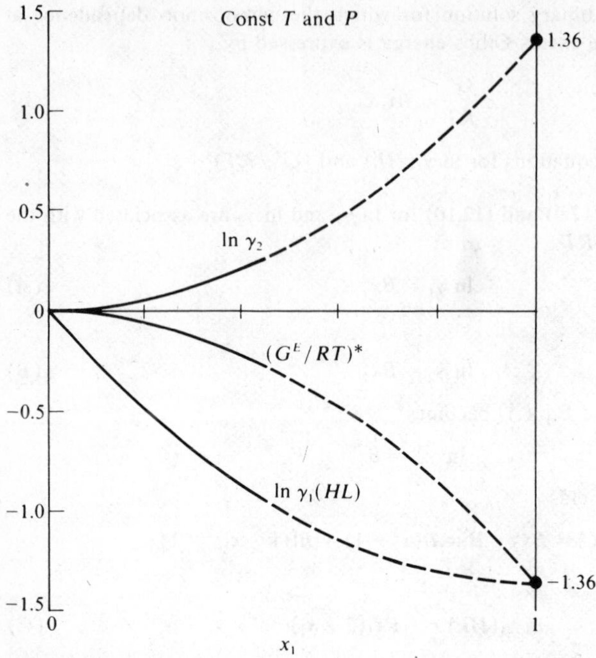

Figure 12.22 Plots showing the excess Gibbs energy and activity coefficients based on the asymmetric treatment of solution ideality.

or

$$\ln \gamma_2 = -Bx_1^2$$

in agreement with Eq. (*B*). Figure 12.22 shows plots of $\ln \gamma_1(HL)$, $\ln \gamma_2$, and $(G^E/RT)^*$ for $B = 1.36$.

PROBLEMS

12.1 To a very good approximation, the excess Gibbs energy for the system acetone(1)/methanol(2) is given by

$$G^E/RT = Bx_1x_2$$

The vapor pressures of acetone and methanol are given by Antoine equations:

$$\ln P_1^{\text{sat}}/\text{kPa} = 14.39155 - \frac{2{,}795.817}{t/°\text{C} + 230.002}$$

$$\ln P_2^{\text{sat}}/\text{kPa} = 16.59381 - \frac{3{,}644.297}{t/°\text{C} + 239.765}$$

(*a*) If $B = 0.64$, independent of T and P, and if the vapor phase is assumed an ideal gas, prepare a Pxy diagram for this system at 50°C and a txy diagram at $P = 75$ kPa.

(*b*) If $B = 0.64$ at 50°C, if $dB/dT = -0.014$, and if the vapor phase is assumed an ideal gas prepare a *txy* diagram for this system at $P = 75$ kPa.

(*c*) If $B = 0.64$ at 50°C and if the virial coefficients are $B_{11} = -1{,}425$, $B_{22} = -1{,}200$, and $B_{12} = -1{,}030$ cm^3 mol^{-1}, prepare a *Pxy* diagram at 50°C.

12.2 The following table gives a set of VLE data for the benzene(1)/acetonitrile(2) system at 45°C:

P/kPa	x_1	y_1
27.78	0.000	0.000
30.04	0.043	0.108
32.33	0.103	0.213
34.37	0.186	0.309
35.79	0.279	0.384
36.78	0.405	0.463
36.98	0.454	0.490
37.07	0.494	0.512
37.00	0.602	0.573
36.46	0.709	0.639
35.29	0.817	0.722
33.55	0.906	0.818
31.96	0.954	0.894
29.82	1.000	1.000

These data can be reasonably well correlated by an equation of the form $G^E/RT = Bx_1x_2$. Making the usual assumptions for low-pressure VLE, determine a suitable value for B and calculate values of the deviations δy_1 and δP between values calculated from the correlation and experimental values, basing the correlation on:

(*a*) Both the P-x_1 and the y_1-x_1 data.
(*b*) Just the P-x_1 data.
(*c*) Just the y_1-x_1 data.

What values are predicted by each correlation for x_1^{az} and P^{az}?

12.3 A liquid mixture of cyclohexanone(1)/phenol(2) for which $x_1 = 0.6$ is in equilibrium with its vapor at 144°C. Determine the equilibrium pressure P and vapor composition y_1 from the following information:

(*a*) Because of the nature of the system, we assume that the composition dependence of G^E is given by an equation of the form $G^E/RT = Bx_1x_2$, where B is a function of temperature only.

(*b*) At 144°C, $P_1^{sat} = 75.20$ and $P_2^{sat} = 31.66$ kPa.

(*c*) The system forms an azeotrope at 144°C for which $x_1^{az} = y_1^{az} = 0.294$.

12.4 Only the three data points given below are available for a particular binary system of interest at temperature *T*. Determine whether these data are better represented by the Margules or van Laar equation at temperature *T*, where $P_1^{sat} = 21$(psia) and $P_2^{sat} = 47$(psia).

P(psia)	x_1	y_1
43.77	0.25	0.188
40.14	0.50	0.378
36.07	0.75	0.545

12.5 The excess Gibbs energy for the system chloroform(1)/ethanol(2) at 55°C is well represented by the Margules equation, written:

$$G^E/RT = (1.42x_1 + 0.59x_2)x_1x_2$$

The vapor pressures of chloroform and ethanol at 55°C are

$$P_1^{sat} = 82.37 \quad \text{and} \quad P_2^{sat} = 37.31 \text{ kPa}$$

(*a*) Prepare a *Pxy* diagram for this system at 55°C, assuming the vapor an ideal gas. What are the pressure and composition of the azeotrope? What are Henry's constants for each species? Over what composition range can Henry's law be used to calculate fugacity values for ethanol if errors are to be no more than 5 percent?

(*b*) Repeat part (*a*) given the virial coefficients: $B_{11} = -963$, $B_{22} = -1{,}523$, and $\delta_{12} = 52 \text{ cm}^3 \text{ mol}^{-1}$.

12.6 For the system acetone(1)/water(2), the following are recommended values for the Wilson parameters:

$$a_{12} = 292.66 \qquad a_{21} = 1{,}445.26 \text{ cal mol}^{-1}$$

$$V_1 = 74.05 \qquad V_2 = 18.07 \text{ cm}^3 \text{ mol}^{-1}$$

The vapor pressures of the pure species are given by:

$$\ln P_1^{sat}/\text{kPa} = 14.39155 - \frac{2{,}795.817}{t/°\text{C} + 230.002}$$

$$\ln P_2^{sat}/\text{kPa} = 16.26205 - \frac{3{,}799.887}{t/°\text{C} + 226.346}$$

Assuming the validity of Eq. (11.74), make the following calculations:
(*a*) BUBL P, given $x_1 = 0.43$ and $t = 76°\text{C}$.
(*b*) DEW P, given $y_1 = 0.43$ and $t = 76°\text{C}$.
(*c*) BUBL T, given $x_1 = 0.32$ and $P = 101.33$ kPa.
(*d*) DEW T, given $y_1 = 0.57$ and $P = 101.33$ kPa.
(*e*) A *P*, *T*-flash for $z_1 = 0.43$, $t = 76°\text{C}$, and $P = \frac{1}{2}(P_b + P_d)$, where P_b and P_d are the bubble- and dew-point pressures determined in (*a*) and (*b*).

12.7 For the system 1-propanol(1)/water(2), the following are recommended values for the Wilson parameters:

$$a_{12} = 775.48 \qquad a_{21} = 1{,}351.90 \text{ cal mol}^{-1}$$

$$V_1 = 75.14 \qquad V_2 = 18.07 \text{ cm}^3 \text{ mol}^{-1}$$

The vapor pressures of the pure species are given by:

$$\ln P_1^{sat}/\text{kPa} = 16.06923 - \frac{3{,}448.660}{t/°\text{C} + 204.094}$$

$$\ln P_2^{sat}/\text{kPa} = 16.26205 - \frac{3{,}799.887}{t/°\text{C} + 226.346}$$

Assuming the validity of Eq. (11.74), make the following calculations:
(*a*) BUBL P, given $x_1 = 0.62$ and $t = 93°\text{C}$.
(*b*) DEW P, given $y_1 = 0.62$ and $t = 93°\text{C}$.
(*c*) BUBL T, given $x_1 = 0.73$ and $P = 101.33$ kPa.
(*d*) DEW T, given $y_1 = 0.38$ and $P = 101.33$ kPa.
(*e*) A *P*, *T*-flash for $z_1 = 0.62$, $t = 93°\text{C}$, and $P = \frac{1}{2}(P_b + P_d)$, where P_b and P_d are the bubble- and dew-point pressures determined in (*a*) and (*b*).

12.8 For the system water(1)/1,4-dioxane(2), the following are recommended values for the Wilson parameters:

$$a_{12} = 1{,}696.98 \qquad a_{21} = -219.39 \text{ cal mol}^{-1}$$

$$V_1 = 18.07 \qquad V_2 = 85.71 \text{ cm}^3 \text{ mol}^{-1}$$

The vapor pressures of the pure species are given by:

$$\ln P_1^{sat}/\text{kPa} = 16.26205 - \frac{3{,}799.887}{t/°\text{C} + 226.346}$$

$$\ln P_2^{sat}/\text{kPa} = 14.1177 - \frac{2{,}966.88}{t/°\text{C} + 210.00}$$

Assuming the validity of Eq. (11.74), make the following calculations:
(*a*) BUBL P, given $x_1 = 0.43$ and $t = 85°\text{C}$.
(*b*) DEW P, given $y_1 = 0.43$ and $t = 85°\text{C}$.
(*c*) BUBL T, given $x_1 = 0.17$ and $P = 101.33$ kPa.
(*d*) DEW T, given $y_1 = 0.82$ and $P = 101.33$ kPa.
(*e*) A *P*, *T*-flash for $z_1 = 0.43$, $t = 85°\text{C}$, and $P = \frac{1}{2}(P_b + P_d)$, where P_b and P_d are the bubble- and dew-point pressures determined in (*a*) and (*b*).

12.9 For the system methanol(1)/acetonitrile(2), the following are recommended values for the Wilson parameters:

$$a_{12} = 504.31 \qquad a_{21} = 196.75 \text{ cal mol}^{-1}$$

$$V_1 = 40.73 \qquad V_2 = 66.30 \text{ cm}^3 \text{ mol}^{-1}$$

The vapor pressures of the pure species are given by:

$$\ln P_1^{sat}/\text{kPa} = 16.59381 - \frac{3{,}644.297}{t/°\text{C} + 239.765}$$

$$\ln P_2^{sat}/\text{kPa} = 14.72577 - \frac{3{,}271.241}{t/°\text{C} + 241.852}$$

Assuming the validity of Eq. (11.74), make the following calculations:
(*a*) BUBL P, given $x_1 = 0.73$ and $t = 70°\text{C}$.
(*b*) DEW P, given $y_1 = 0.73$ and $t = 70°\text{C}$.
(*c*) BUBL T, given $x_1 = 0.79$ and $P = 101.33$ kPa.
(*d*) DEW T, given $y_1 = 0.63$ and $P = 101.33$ kPa.
(*e*) A *P*, *T*-flash for $z_1 = 0.73$, $t = 70°\text{C}$, and $P = \frac{1}{2}(P_b + P_d)$, where P_b and P_d are the bubble- and dew-point pressures determined in (*a*) and (*b*).

12.10 For the system acetone(1)/methanol(2), the following are recommended values for the Wilson parameters:

$$a_{12} = -170.18 \qquad a_{21} = 594.18 \text{ cal mol}^{-1}$$

$$V_1 = 74.05 \qquad V_2 = 40.73 \text{ cm}^3 \text{ mol}^{-1}$$

The vapor pressures of the pure species are given by:

$$\ln P_1^{sat}/\text{kPa} = 14.39155 - \frac{2{,}795.817}{t/°\text{C} + 230.002}$$

$$\ln P_2^{sat}/\text{kPa} = 16.59381 - \frac{3{,}644.297}{t/°\text{C} + 239.765}$$

Assuming the validity of Eq. (11.74), make the following calculations:
(*a*) BUBL P, given $x_1 = 0.31$ and $t = 60°\text{C}$.
(*b*) DEW P, given $y_1 = 0.31$ and $t = 60°\text{C}$.
(*c*) BUBL T, given $x_1 = 0.72$ and $P = 101.33$ kPa.
(*d*) DEW T, given $y_1 = 0.43$ and $P = 101.33$ kPa.
(*e*) A *P*, *T*-flash for $z_1 = 0.31$, $t = 60°\text{C}$, and $P = \frac{1}{2}(P_b + P_d)$, where P_b and P_d are the bubble- and dew-point pressures determined in (*a*) and (*b*).

12.11 For the system methyl acetate(1)/methanol(2), the following are recommended values for the Wilson parameters:

$$a_{12} = -31.19 \qquad a_{21} = 813.18 \text{ cal mol}^{-1}$$

$$V_1 = 79.84 \qquad V_2 = 40.73 \text{ cm}^3 \text{ mol}^{-1}$$

The vapor pressures of the pure species are given by:

$$\ln P_1^{\text{sat}}/\text{kPa} = 14.40150 - \frac{2{,}739.174}{t/°\text{C} + 223.115}$$

$$\ln P_2^{\text{sat}}/\text{kPa} = 16.59381 - \frac{3{,}644.297}{t/°\text{C} + 239.765}$$

Assuming the validity of Eq. (11.74), make the following calculations:
(*a*) BUBL P, given $x_1 = 0.31$ and $t = 55°\text{C}$.
(*b*) DEW P, given $y_1 = 0.31$ and $t = 55°\text{C}$.
(*c*) BUBL T, given $x_1 = 0.86$ and $P = 101.33$ kPa.
(*d*) DEW T, given $y_1 = 0.17$ and $P = 101.33$ kPa.
(*e*) A *P*, *T*-flash for $z_1 = 0.31$, $t = 55°\text{C}$, and $P = \frac{1}{2}(P_b + P_d)$, where P_b and P_d are the bubble- and dew-point pressures determined in (*a*) and (*b*).

12.12 For the system methanol(1)/benzene(2), the following are recommended values for the Wilson parameters:

$$a_{12} = 1{,}713.20 \qquad a_{21} = 187.13 \text{ cal mol}^{-1}$$

$$V_1 = 40.73 \qquad V_2 = 89.41 \text{ cm}^3 \text{ mol}^{-1}$$

The vapor pressures of the pure species are given by:

$$\ln P_2^{\text{sat}}/\text{kPa} = 16.59381 - \frac{3{,}644.297}{t/°\text{C} + 239.765}$$

$$\ln P_1^{\text{sat}}/\text{kPa} = 13.85937 - \frac{2{,}773.779}{t/°\text{C} + 220.069}$$

Assuming the validity of Eq. (11.74), make the following calculations:
(*a*) BUBL P, given $x_1 = 0.82$ and $t = 68°\text{C}$.
(*b*) DEW P, given $y_1 = 0.82$ and $t = 68°\text{C}$.
(*c*) BUBL T, given $x_1 = 0.21$ and $P = 101.33$ kPa.
(*d*) DEW T, given $y_1 = 0.38$ and $P = 101.33$ kPa.
(*e*) A *P*, *T*-flash for $z_1 = 0.82$, $t = 68°\text{C}$, and $P = \frac{1}{2}(P_b + P_d)$, where P_b and P_d are the bubble- and dew-point pressures determined in (*a*) and (*b*).

12.13 For the system ethanol(1)/toluene(2), the following are recommended values for the Wilson parameters:

$$a_{12} = 1{,}556.45 \qquad a_{21} = 210.52 \text{ cal mol}^{-1}$$

$$V_1 = 58.68 \qquad V_2 = 106.85 \text{ cm}^3 \text{ mol}^{-1}$$

The vapor pressures of the pure species are given by:

$$\ln P_1^{\text{sat}}/\text{kPa} = 16.67583 - \frac{3{,}674.491}{t/°\text{C} + 226.448}$$

$$\ln P_2^{\text{sat}}/\text{kPa} = 14.00976 - \frac{3{,}103.010}{t/°\text{C} + 219.787}$$

Assuming the validity of Eq. (11.74), make the following calculations:
(*a*) BUBL P, given $x_1 = 0.31$ and $t = 105°\text{C}$.
(*b*) DEW P, given $y_1 = 0.31$ and $t = 105°\text{C}$.

(*c*) BUBL T, given $x_1 = 0.68$ and $P = 101.33$ kPa.
(*d*) DEW T, given $y_1 = 0.79$ and $P = 101.33$ kPa.
(*e*) A *P*, *T*-flash for $z_1 = 0.31$, $t = 105$°C, and $P = \frac{1}{2}(P_b + P_d)$, where P_b and P_d are the bubble- and dew-point pressures determined in (*a*) and (*b*).

12.14 Determine the azeotropic pressure and composition for one of the following:
(*a*) The system of Prob. 12.7 at a temperature of 93°C.
(*b*) The system of Prob. 12.8 at a temperature of 85°C.
(*c*) The system of Prob. 12.9 at a temperature of 70°C.
(*d*) The system of Prob. 12.10 at a temperature of 60°C.
(*e*) The system of Prob. 12.11 at a temperature of 55°C.
(*f*) The system of Prob. 12.12 at a temperature of 68°C.
(*g*) The system of Prob. 12.13 at a temperature of 105°C.

12.15 For the system ethanol(1)/toluene(2), the following are recommended values for the NRTL parameters:

$$b_{12} = 713.57 \qquad b_{21} = 1{,}147.86 \text{ cal mol}^{-1} \qquad \alpha = 0.529$$

$$V_1 = 58.68 \qquad V_2 = 106.85 \text{ cm}^3 \text{ mol}^{-1}$$

The vapor pressures of the pure species are given by:

$$\ln P_1^{\text{sat}}/\text{kPa} = 16.67583 - \frac{3{,}674.491}{t/°\text{C} + 226.448}$$

$$\ln P_2^{\text{sat}}/\text{kPa} = 14.00976 - \frac{3{,}103.010}{t/°\text{C} + 219.787}$$

Assuming the validity of Eq. (11.74), make the following calculations:
(*a*) BUBL P, given $x_1 = 0.31$ and $t = 105$°C.
(*b*) DEW P, given $y_1 = 0.31$ and $t = 105$°C.
(*c*) BUBL T, given $x_1 = 0.68$ and $P = 101.33$ kPa.
(*d*) DEW T, given $y_1 = 0.79$ and $P = 101.33$ kPa.
(*e*) A *P*, *T*-flash for $z_1 = 0.31$, $t = 105$°C, and $P = \frac{1}{2}(P_b + P_d)$, where P_b and P_d are the bubble- and dew-point pressures determined in (*a*) and (*b*).

12.16 For a binary system the excess Gibbs energy of the liquid phase is given by an equation of the form $G^E/RT = Bx_1x_2$, where *B* is a function of temperature only. Making the usual assumptions for low-pressure VLE, show that

(*a*) The relative volatility of species 1 to species 2 at infinite dilution of species 1 is given by

$$\alpha_{12}(x_1 = 0) = \frac{P_1^{\text{sat}}}{P_2^{\text{sat}}}(\exp B)$$

(*b*) Henry's constant for species 1 is given by

$$k_1 = P_1^{\text{sat}}(\exp B)$$

12.17 The table of Prob. 11.24 provides *Pxy* data for VLE in the system acetone(1)/chloroform(2) at 50°C.

(*a*) Assuming the vapor phase an ideal gas, calculate $\hat{f}_1$ and $\hat{f}_2$ for each data point, and plot the results vs. x_1. Show also by dotted lines the relations given by the Lewis/Randall rule.

(*b*) Plot $\hat{f}_1/x_1$ and $\hat{f}_2/x_2$ vs. x_1. What are the values of Henry's constants k_1 and k_2 indicated by this plot? What are the values of γ_1^∞ and γ_1^∞?

Repeat (*a*) and (*b*) given the virial coefficients:

$$B_{11} = -1{,}425 \qquad B_{22} = -1{,}030 \qquad B_{12} = -785 \text{ cm}^3 \text{ mol}^{-1}$$

12.18 The gas phase in a corked bottle of champagne is largely CO_2 in equilibrium with the liquid of interest. Measurements (perhaps of the elevations attained by popping corks) indicate that at the serving temperature of 5°C the pressure in the unopened bottle is about 5 bar. If Henry's constant at this temperature is 1,000 bar, estimate the mole fraction of CO_2 in the champagne.

12.19 The excess Gibbs energy for binary systems consisting of liquids not too dissimilar in chemical nature is represented to a reasonable approximation by the equation

$$G^E/RT = Bx_1x_2$$

where B is a function of temperature only. For such systems, it is often observed that the ratio of the vapor pressures of the pure species is nearly constant over a considerable temperature range. Let this ratio be r, and determine the range of values of B, expressed as a function of r, for which no azeotrope can exist. Assume the vapor phase an ideal gas.

12.20 The excess Gibbs energy for a particular system is represented by

$$G^E/RT = Bx_1x_2$$

where B is a function of temperature only. Assuming the validity of Eq. (11.74), show that, at every temperature for which an azeotrope exists, the azeotropic composition x^{az} and azeotropic pressure P^{az} are related by

$$\frac{1}{x_1^{az}} = 1 + \left[\frac{\ln(P^{az}/P_1^{sat})}{\ln(P^{az}/P_2^{sat})}\right]^{1/2}$$

12.21 A concentrated binary liquid solution containing mostly species 2 (but $x_2 \neq 1$) is in equilibrium with a vapor phase containing both species 1 and 2. The pressure of this two-phase system is 1 bar; the temperature is 25°C. Starting with Eq. (11.30), determine from the following data good estimates of x_1 and y_1.

$$k_1 = 200 \text{ bar} \qquad P_2^{sat} = 0.10 \text{ bar}$$

State and justify all assumptions.

12.22 A vapor stream for which $z_1 = 0.75$ and $z_2 = 0.25$ is cooled to temperature T in the two-phase region and flows into a separation chamber at a pressure of 1 bar. If the composition of the liquid product is to be $x_1 = 0.50$, what is the required value of T, and what is the value of y_1? For liquid mixtures of species 1 and 2

$$G^E/RT = 1.2x_1x_2$$

The vapor pressures of the pure species are given by

$$\ln P_1^{sat}/\text{bar} = 10.00 - \frac{2{,}950}{T/\text{K} - 36.0}$$

$$\ln P_2^{sat}/\text{bar} = 11.70 - \frac{3{,}840}{T/\text{K} - 44.8}$$

12.23 A stream of isopropanol(1)/water(2) is flashed into a separation chamber at the conditions $t = 80$°C and $P = 91.2$ kPa. A particular analysis of the liquid product shows an isopropanol content of 4.7 mole percent, a value which deviates from the norm. The question arises as to whether an air leak into the separator could be the cause. Is this possible? The following laboratory data on the liquid phase at 80°C are available:

$$P_1^{sat} = 91.11 \text{ kPa} \qquad P_2^{sat} = 47.36 \text{ kPa}$$

G^E/RT is give by the van Laar equation with $A'_{12} = 2.470$ and $A'_{21} = 1.094$.

12.24 Vapor/liquid equilibrium data for the system 1,2-dichloromethane(1)/methanol(2) at 50°C are as follows:

P/kPa	x_1	y_1
55.55	0.000	0.000
58.79	0.042	0.093
61.76	0.097	0.174
64.59	0.189	0.265
65.66	0.292	0.324
65.76 (azeotrope)	0.349	0.349
65.59	0.415	0.367
65.15	0.493	0.386
63.86	0.632	0.418
62.36	0.720	0.438
59.03	0.835	0.484
54.92	0.893	0.537
48.41	0.945	0.620
31.10	1.000	1.000

For these data, assume the vapor phase an ideal gas and plot P vs. x_1, P vs. y_1, $y_1 P$ vs. x_1, and $y_2 P$ vs. x_1. Determine Henry's constant for each species from the partial-pressure curves. For each species, over what composition range does Henry's law predict partial pressures within 5 percent of the true values?

12.25 From the data of the preceding problem, calculate values of $\ln \gamma_1$, $\ln \gamma_2$, and $G^E/x_1 x_2 RT$, and plot these values vs. x_1.

(*a*) Determine from the plot values of $\ln \gamma_1^\infty$ and $\ln \gamma_2^\infty$, and use them to find values of A_{12} and A_{21} in the Margules equation. Draw in the line for $G^E/x_1 x_2 RT$ vs. x_1 that represents the Margules equation with these parameters. Determine the values of A_{12} and A_{21} for the Margules equation from just the azeotrope data, and draw in the line for this pair of constants.

(*b*) Use the values of γ_1^∞ and γ_2^∞ in Eq. (12.55) to determine values for Henry's constants. How do these results compare with the values found in Prob. 12.24?

12.26 Rework part (*a*) of the preceding problem for

(*a*) The van Laar equation, determining the corresponding values of A'_{12} and A'_{21}.

(*b*) The Wilson equation, determining the corresponding values of Λ_{12} and Λ_{21}.

CHAPTER

THIRTEEN

SOLUTION THERMODYNAMICS

We turn in this chapter to a detailed study of the properties of solutions. All the fundamental equations and necessary definitions have been given in preceding chapters. However, the development there is concentrated on the Gibbs energy and related properties, with the specific goal of application to vapor/liquid equilibrium. Here we present general treatments of partial properties, ideal solutions, residual properties, and excess properties. Closely related to excess properties are property changes of mixing, treated in Sec. 13.6. In particular, the enthalpy change of mixing, called the heat of mixing, is applied to practical problems in Sec. 13.7. In Sec. 13.8 we give a general exposition of thermodynamic equilibrium and an elementary discussion of phase stability. This leads finally to an introductory description of binary systems comprised of liquids that are not completely miscible with one another.

13.1 RELATIONS AMONG PARTIAL PROPERTIES FOR CONSTANT-COMPOSITION SOLUTIONS

Partial molar properties were defined and discussed briefly in Sec. 11.1. Here we show how they are related to one another. Recalling that $\mu_i = \bar{G}_i$, we may write Eq. (10.2) as

$$d(nG) = (nV)\,dP - (nS)\,dT + \sum \bar{G}_i\,dn_i \tag{10.2}$$

Application of the criterion of exactness, Eq. (6.12), to this equation yields the

Maxwell relation,

$$\left(\frac{\partial V}{\partial T}\right)_{P,n} = -\left(\frac{\partial S}{\partial P}\right)_{T,n} \tag{6.16}$$

plus the two additional equations:

$$\left(\frac{\partial \bar{G}_i}{\partial T}\right)_{P,n} = -\left[\frac{\partial(nS)}{\partial n_i}\right]_{P,T,n_j}$$

and

$$\left(\frac{\partial \bar{G}_i}{\partial P}\right)_{T,n} = \left[\frac{\partial(nV)}{\partial n_i}\right]_{P,T,n_j}$$

where subscript n indicates constancy of all n_i and therefore of composition. In view of Eq. (11.2), these last two equations are most simply written as:

$$\left(\frac{\partial \bar{G}_i}{\partial T}\right)_{P,x} = -\bar{S}_i \tag{13.1}$$

and

$$\left(\frac{\partial \bar{G}_i}{\partial P}\right)_{T,x} = \bar{V}_i \tag{13.2}$$

These equations allow calculation of the effect of temperature and pressure on the partial Gibbs energy (or chemical potential). They are the partial-property analogs of two equations that follow by inspection from Eq. (10.2):

$$\left[\frac{\partial(nG)}{\partial T}\right]_{P,n} = -nS \qquad \text{or} \qquad \left(\frac{\partial G}{\partial T}\right)_{P,x} = -S$$

and

$$\left[\frac{\partial(nG)}{\partial P}\right]_{T,n} = nV \qquad \text{or} \qquad \left(\frac{\partial G}{\partial P}\right)_{T,x} = V$$

Indeed, for every equation providing a *linear* relation among the thermodynamic properties of a *constant-composition* solution there exists a corresponding equation connecting the corresponding partial properties of each species in the solution. We demonstrate this by example.

Consider the equation that defines the enthalpy

$$H = U + PV \tag{2.6}$$

For n moles,

$$nH = nU + P(nV)$$

Differentiation with respect to n_i at constant T, P, and n_j yields

$$\left[\frac{\partial(nH)}{\partial n_i}\right]_{P,T,n_j} = \left[\frac{\partial(nU)}{\partial n_i}\right]_{P,T,n_j} + P\left[\frac{\partial(nV)}{\partial n_i}\right]_{P,T,n_j}$$

By Eq. (11.2) this becomes

$$\bar{H}_i = \bar{U}_i + P\bar{V}_i$$

which is the partial-property analog of Eq. (2.6).

In a constant-composition solution, $\bar{G}_i$ is a function of P and T. We may therefore write

$$d\bar{G}_i = \left(\frac{\partial \bar{G}_i}{\partial P}\right)_{T,x} dP + \left(\frac{\partial \bar{G}_i}{\partial T}\right)_{P,x} dT$$

As a result of Eqs. (13.1) and (13.2) this becomes

$$d\bar{G}_i = \bar{V}_i \, dP - \bar{S}_i \, dT$$

which may be compared with Eq. (6.10).

These examples are sufficient illustration of the parallelism that exists between equations for a constant-composition solution and the corresponding equations for the partial properties of the species in solution. We can therefore write simply by analogy many equations that relate partial properties.

13.2 THE IDEAL SOLUTION

In Sec. 10.4 we wrote down equations for an ideal solution by analogy to those for an ideal gas. We wish here to formalize development of the equations for an ideal solution. We *define* an ideal solution as a fluid which obeys Eq. (11.61), the Lewis/Randall rule,

$$\boxed{\hat{f}_i^{id} = x_i f_i} \qquad (11.61)$$

where f_i is a function of T and P. Thus, an ideal solution (in the sense of the Lewis/Randall rule) is a model fluid for which the fugacity of each constituent species is given by Eq. (11.61) at all conditions of temperature, pressure, and composition. Combination of Eq. (11.58) with the Lewis/Randall rule gives

$$\boxed{\bar{G}_i^{id} = G_i + RT \ln x_i} \qquad (13.3)$$

Since $\mu_i^{id} = \bar{G}_i^{id}$, this equation is identical with Eq. (10.14). When Eq. (13.3) is differentiated with respect to temperature at constant pressure and composition and then combined with Eq. (13.1) written for an ideal solution, we get

$$\bar{S}_i^{id} = -\left(\frac{\partial \bar{G}_i^{id}}{\partial T}\right)_{P,x} = -\left(\frac{\partial G_i}{\partial T}\right)_P - R \ln x_i$$

Since $(\partial G_i/\partial T)_P$ is simply $-S_i$, this becomes

$$\boxed{\bar{S}_i^{id} = S_i - R \ln x_i} \qquad (13.4)$$

Similarly, as a result of Eq. (13.2),

$$\bar{V}_i^{id} = \left(\frac{\partial \bar{G}_i^{id}}{\partial P}\right)_{T,x} = \left(\frac{\partial G_i}{\partial P}\right)_T$$

or

$$\boxed{\bar{V}_i^{id} = V_i} \tag{13.5}$$

Since $\bar{H}_i^{id} = \bar{G}_i^{id} + T\bar{S}_i^{id}$,

$$\bar{H}_i^{id} = G_i + RT \ln x_i + TS_i - RT \ln x_i$$

or

$$\boxed{\bar{H}_i^{id} = H_i} \tag{13.6}$$

As a special case of Eq. (11.5), we write:

$$M^{id} = \sum x_i \bar{M}_i^{id}$$

Application of this relation to Eqs. (13.3) through (13.6) yields:

$$G^{id} = \sum x_i G_i + RT \sum x_i \ln x_i \tag{13.7}$$

$$S^{id} = \sum x_i S_i - R \sum x_i \ln x_i \tag{13.8}$$

$$V^{id} = \sum x_i V_i \tag{13.9}$$

$$H^{id} = \sum x_i H_i \tag{13.10}$$

A mixture of ideal gases is a special case of an ideal solution for which the Lewis/Randall rule [Eq. (11.61)] simplifies to $\hat{f}_i^{ig} = y_i P$. Equation (11.58) then reduces to

$$\bar{G}_i^{ig} = G_i^{ig} + RT \ln y_i$$

which is the particular form of Eq. (13.3) valid for species i in a mixture of ideal gases. In this case Eq. (13.7) becomes

$$G^{ig} = \sum y_i G_i^{ig} + RT \sum y_i \ln y_i$$

Similarly, Eqs. (13.4) through (13.6) and (13.8) through (13.10) for ideal gases become:

$$\bar{S}_i^{ig} = S_i^{ig} - R \ln y_i \quad \text{and} \quad S^{ig} = \sum y_i S_i^{ig} - R \sum y_i \ln y_i$$

$$\bar{V}_i^{ig} = V_i^{ig} \quad \text{and} \quad V^{ig} = \sum y_i V_i^{ig}$$

$$\bar{H}_i^{ig} = H_i^{ig} \quad \text{and} \quad H^{ig} = \sum y_i H_i^{ig}$$

These equations give the base values from which residual properties are measured.

13.3 THE FUNDAMENTAL RESIDUAL-PROPERTY RELATION

The definition of a residual property is given by Eq. (6.35),

$$M^R \equiv M - M^{ig} \tag{6.35}$$

where M is the molar (or unit-mass) value of a thermodynamic property of a fluid and M^{ig} is the value that the property would have if the fluid were an ideal gas of the same composition at the same T and P. From this we have immediately [see the development of Eq. (11.31)]:

$$\bar{M}_i^R = \bar{M}_i - \bar{M}_i^{ig} \tag{13.11}$$

These equations are the basis for extension of the fundamental property relation, given by Eq. (10.2), to residual properties.

We first develop an alternative form of Eq. (10.2), just as was done in Sec. 6.2, where the fundamental property relation was restricted to phases of constant composition. We make use of the same mathematical identity:

$$d\left(\frac{nG}{RT}\right) \equiv \frac{1}{RT}\,d(nG) - \frac{nG}{RT^2}\,dT$$

Substitution for $d(nG)$ by Eq. (10.2) and for G by Eq. (6.3) gives, after algebraic reduction,

$$\boxed{d\left(\frac{nG}{RT}\right) = \frac{nV}{RT}\,dP - \frac{nH}{RT^2}\,dT + \sum_i \frac{\bar{G}_i}{RT}\,dn_i} \tag{13.12}$$

We note with respect to this equation that all terms have the units of moles; moreover, in contrast to Eq. (10.2), the enthalpy rather than the entropy appears on the right-hand side. Equation (13.12) is a general relation expressing G/RT as a function of *all* of its canonical variables, T, P, and the mole numbers. It reduces to Eq. (6.29) for the special case of 1 mole of a constant-composition phase. Equations (6.30) and (6.31) follow from either equation, and equations for the other thermodynamic properties then come from appropriate defining equations. Knowledge of G/RT as a function of its canonical variables allows evaluation of all other thermodynamic properties, and therefore implicitly contains complete property information. However, we cannot directly exploit this characteristic, and in practice we deal with related properties, the residual and excess Gibbs energies.

Since Eq. (13.12) is general, it may be written for the special case of an ideal gas:

$$d\left(\frac{nG^{ig}}{RT}\right) = \frac{nV^{ig}}{RT}\,dP - \frac{nH^{ig}}{RT^2}\,dT + \sum_i \frac{\bar{G}_i^{ig}}{RT}\,dn_i$$

In view of Eqs. (6.35) and (13.11), the difference between this equation and Eq.

(13.12) is

$$d\left(\frac{nG^R}{RT}\right) = \frac{nV^R}{RT}\,dP - \frac{nH^R}{RT^2}\,dT + \sum_i \frac{\bar{G}_i^R}{RT}\,dn_i \tag{13.13}$$

This equation is the *fundamental residual-property relation*. Its derivation from Eq. (10.2) parallels the derivation in Chap. 6 that led from Eq. (6.10) to Eq. (6.36). Indeed Eqs. (6.10) and (6.36) are special cases of Eqs. (10.2) and (13.13) valid for one mole of a constant-composition fluid. An alternative form of Eq. (13.13) follows by introduction of the fugacity coefficients as given by Eqs. (11.16) and (11.34):

$$d(n \ln \phi) = \frac{nV^R}{RT}\,dP - \frac{nH^R}{RT^2}\,dT + \sum \ln \hat{\phi}_i\,dn_i \tag{13.14}$$

Equations so general as Eqs. (13.13) and (13.14) are useful for practical application only in their restricted forms. Division of Eqs. (13.13) and (13.14) by dP and restriction to constant T and composition leads to:

$$\frac{V^R}{RT} = \left[\frac{\partial(G^R/RT)}{\partial P}\right]_{T,x} = \left(\frac{\partial \ln \phi}{\partial P}\right)_{T,x} \tag{13.15}$$

Similarly, division by dT and restriction to constant P and composition gives:

$$\frac{H^R}{RT} = -T\left[\frac{\partial(G^R/RT)}{\partial T}\right]_{P,x} = -T\left(\frac{\partial \ln \phi}{\partial T}\right)_{P,x} \tag{13.16}$$

These equations are restatements of Eqs. (6.37) and (6.38) wherein the restriction of the derivatives to constant composition is shown explicitly. They lead to Eqs. (6.40), (6.41), (6.42), and (11.20), which allow calculation of residual properties and fugacity coefficients from PVT data and equations of state. It is through the residual properties that this kind of experimental information enters into the practical application of thermodynamics.

In addition, from Eqs. (13.13) and (13.14) we have

$$\ln \hat{\phi}_i = \left[\frac{\partial(n \ln \phi)}{\partial n_i}\right]_{T,P,n_j} = \left[\frac{\partial(nG^R/RT)}{\partial n_i}\right]_{T,P,n_j} \tag{13.17}$$

The first equality is Eq. (11.36), which demonstrates that $\ln \hat{\phi}_i$ is a partial property with respect to $\ln \phi$. It is also a partial property with respect to G^R/RT. The partial-property analogs of Eqs. (13.15) and (13.16) are therefore:

$$\left(\frac{\partial \ln \hat{\phi}_i}{\partial P}\right)_{T,x} = \frac{\bar{V}_i^R}{RT} \tag{13.18}$$

and

$$\left(\frac{\partial \ln \hat{\phi}_i}{\partial T}\right)_{P,x} = -\frac{\bar{H}_i^R}{RT^2} \tag{13.19}$$

Equation (11.39) follows directly from Eq. (13.18).

13.4 THE FUNDAMENTAL EXCESS-PROPERTY RELATION

The definition of an excess property is given by Eq. (11.55):

$$M^E \equiv M - M^{id} \tag{11.55}$$

where M is the molar (or unit-mass) value of a solution property and M^{id} is the property value the solution would have if it were an *ideal* solution of the same composition at the same T and P. This definition is analogous to the definition of a residual property; in addition, we have analogous to Eq. (13.11) the partial-property relation

$$\bar{M}_i^E = \bar{M}_i - \bar{M}_i^{id} \tag{13.20}$$

where $\bar{M}_i^E$ is a partial excess property.

The fundamental excess-property relation is derived in exactly the same way as the fundamental residual-property relation and leads to analogous results. Equation (13.12), written for the special case of an ideal solution, is subtracted from Eq. (13.12) itself, yielding:

$$d\left(\frac{nG^E}{RT}\right) = \frac{nV^E}{RT}\,dP - \frac{nH^E}{RT^2}\,dT + \sum_i \frac{\bar{G}_i^E}{RT}\,dn_i \tag{13.21}$$

This is the *fundamental excess-property relation.* As a result of Eq. (11.60), it may be written in the alternative form:

$$\boxed{d\left(\frac{nG^E}{RT}\right) = \frac{nV^E}{RT}\,dP - \frac{nH^E}{RT^2}\,dT + \sum \ln \gamma_i\,dn_i} \tag{13.22}$$

Again, the generality of these equations precludes their direct practical application. Rather, we make use of restricted forms, which are written by inspection:

$$\frac{V^E}{RT} = \left[\frac{\partial(G^E/RT)}{\partial P}\right]_{T,x} \tag{13.23}$$

$$\frac{H^E}{RT} = -T\left[\frac{\partial(G^E/RT)}{\partial T}\right]_{P,x} \tag{13.24}$$

and

$$\ln \gamma_i = \left[\frac{\partial(nG^E/RT)}{\partial n_i}\right]_{T,P,n_j}$$

The last relation is Eq. (11.62), which demonstrates the partial property relationship that $\ln \gamma_i$ bears to G^E/RT. These equations are analogous to Eqs. (13.15) through (13.17). Whereas the fundamental *residual*-property relation derives its usefulness from its direct relation to experimental *PVT* data and equations of state, the *excess*-property formulation is useful because V^E, H^E, and γ_i are all experimentally accessible. Activity coefficients are found from *VLE* data as discussed earlier, and V^E and H^E values come from mixing experiments as described in Sec. 13.6.

Equations (13.23) and (13.24) allow direct calculation of the effects of pressure and temperature on the excess Gibbs energy. For example, an equimolar mixture of benzene and cyclohexane at 25°C and 1 bar has an excess volume of about 0.65 $cm^3\ mol^{-1}$ and an excess enthalpy of about 800 $J\ mol^{-1}$. Thus at these conditions,

$$\left[\frac{\partial(G^E/RT)}{\partial P}\right]_{T,x} = \frac{0.65}{(83.14)(298.15)} = 2.62 \times 10^{-5}\ \text{bar}^{-1}$$

and

$$\left[\frac{\partial(G^E/RT)}{\partial T}\right]_{P,x} = \frac{-800}{(8.314)(298.15)^2} = -1.08 \times 10^{-3}\ \text{K}^{-1}$$

The most striking observation about these results is that it takes a pressure change of more than 40 bar to have an effect on the excess Gibbs energy equivalent to that of a temperature change of 1 K. This is the reason that for liquids at low pressures the effect of pressure on the excess Gibbs energy (and therefore on the activity coefficients) is usually neglected.

The partial-property analogs of Eqs. (13.23) and (13.24) are:

$$\left(\frac{\partial \ln \gamma_i}{\partial P}\right)_{T,x} = \frac{\bar{V}_i^E}{RT} \tag{13.25}$$

and

$$\left(\frac{\partial \ln \gamma_i}{\partial T}\right)_{P,x} = -\frac{\bar{H}_i^E}{RT^2} \tag{13.26}$$

Just as the fundamental property relation of Eq. (13.12) provides complete property information from a canonical equation of state expressing G/RT as a function of T, P, and composition, so the fundamental *residual*-property relation, Eq. (13.13) or (13.14), provides complete *residual*-property information from a *PVT* equation of state, from *PVT* data, or from generalized *PVT* correlations. However, for complete *property* information, one needs in addition to *PVT* data the ideal-gas-state heat capacities of the species that comprise the system.

Given an equation for G^E/RT as a function of T, P, and composition, the fundamental *excess*-property relation, Eq. (13.21) or (13.22), provides complete *excess*-property information. However, this formulation represents less-complete property information than does the residual-property formulation, because it tells us nothing about the properties of the pure constituent chemical species.

13.5 EVALUATION OF PARTIAL PROPERTIES

The definition of a partial property,

$$\bar{M}_i = \left[\frac{\partial(nM)}{\partial n_i}\right]_{P,T,n_j} \tag{11.2}$$

also applies to residual and to excess properties:

$$\bar{M}_i^R = \left[\frac{\partial(nM^R)}{\partial n_i}\right]_{P,T,n_j} \tag{13.27}$$

and

$$\bar{M}_i^E = \left[\frac{\partial(nM^E)}{\partial n_i}\right]_{P,T,n_j} \tag{13.28}$$

Equations for partial properties can always be derived from an equation for the solution property as a function of composition by direct application of Eq. (11.2), (13.27), or (13.28). For binary systems, however, an alternative procedure may be more convenient.

Written for a binary solution Eq. (11.5) becomes

$$M = x_1\bar{M}_1 + x_2\bar{M}_2 \tag{A}$$

whence

$$dM = x_1\, d\bar{M}_1 + \bar{M}_1\, dx_1 + x_2\, d\bar{M}_2 + \bar{M}_2\, dx_2 \tag{B}$$

However, when M is given as a function of composition at constant P and T, the Gibbs/Duhem equation, Eq. (11.8), is

$$x_1\, d\bar{M}_1 + x_2\, d\bar{M}_2 = 0 \tag{C}$$

Since $x_1 + x_2 = 1$, we also have $dx_2 = -dx_1$. Combining Eqs. (B) and (C) and eliminating dx_2 gives

$$dM = \bar{M}_1\, dx_1 - \bar{M}_2\, dx_1$$

or

$$\frac{dM}{dx_1} = \bar{M}_1 - \bar{M}_2 \tag{D}$$

Eliminating $\bar{M}_2$ from Eqs. (A) and (D), and solving for $\bar{M}_1$, we get

$$\boxed{\bar{M}_1 = M + x_2\frac{dM}{dx_1}} \tag{13.29}$$

Similarly, elimination of $\bar{M}_1$ and solution for $\bar{M}_2$ gives

$$\boxed{\bar{M}_2 = M - x_1\frac{dM}{dx_1}} \tag{13.30}$$

For residual and excess properties, these are written:

$$\bar{M}_1^R = M^R + x_2\frac{dM^R}{dx_1} \tag{13.31}$$

$$\bar{M}_2^R = M^R - x_1\frac{dM^R}{dx_1} \tag{13.32}$$

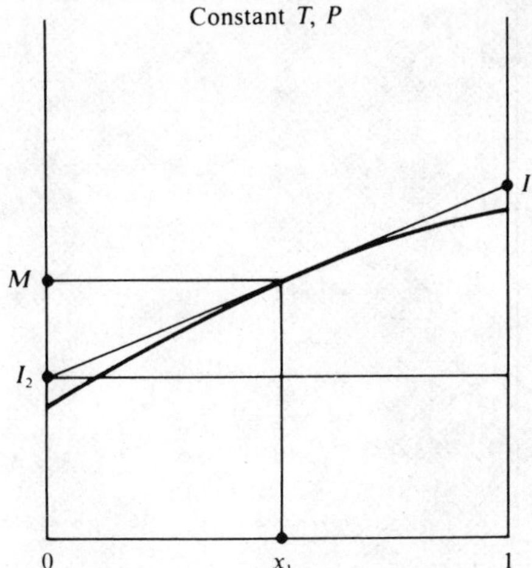

Figure 13.1

and

$$\bar{M}_1^E = M^E + x_2 \frac{dM^E}{dx_1} \tag{13.33}$$

$$\bar{M}_2^E = M^E - x_1 \frac{dM^E}{dx_1} \tag{13.34}$$

Thus for binary systems, the partial properties are readily calculated directly from an expression for the solution property as a function of composition at constant T and P. The corresponding equations for multicomponent systems are much more complex, and are given in detail by Van Ness and Abbott.†

Example 13.1 Describe a graphical interpretation of Eqs. (13.29) and (13.30).

SOLUTION Figure 13.1 shows a representative plot of M vs. x_1 for a binary system. Values of the derivative dM/dx_1 are given by the slopes of lines drawn tangent to the curve of M vs. x_1. One such line drawn tangent at a particular value of x_1 is shown in Fig. 13.1. Its intercepts with the boundaries of the figure at $x_1 = 1$ and $x_1 = 0$ are labeled I_1 and I_2. As is evident from the figure, two equivalent expressions can be written for the slope of this line:

$$\frac{dM}{dx_1} = \frac{M - I_2}{x_1} \qquad \text{and} \qquad \frac{dM}{dx_1} = \frac{I_1 - I_2}{1 - 0} = I_1 - I_2$$

Solving the first equation for I_2 and the second for I_1 (with elimination of I_2) gives

$$I_2 = M - x_1 \frac{dM}{dx_1} \qquad \text{and} \qquad I_1 = M + (1 - x_1) \frac{dM}{dx_1}$$

† H. C. Van Ness and M. M. Abbott, *Classical Thermodynamics of Nonelectrolyte Solutions: With Applications to Phase Equilibria*, pp. 46–54, McGraw-Hill, New York, 1982.

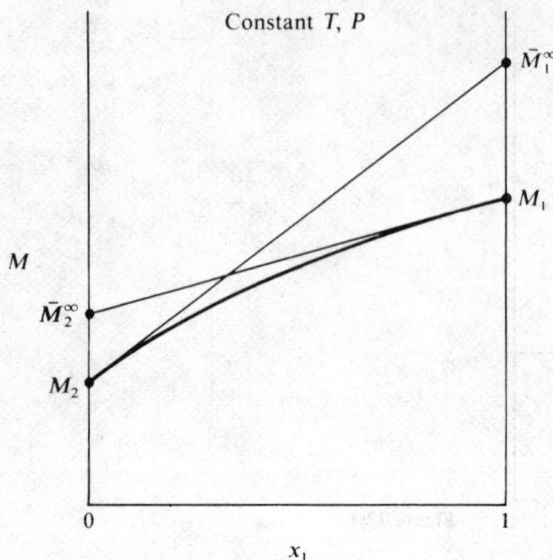

Figure 13.2

Comparison of these expressions with Eqs. (13.29) and (13.30) shows that

$$I_1 = \bar{M}_1 \qquad \text{and} \qquad I_2 = \bar{M}_2$$

Thus the tangent intercepts give directly the values of the two partial properties. These intercepts of course shift as the point of tangency moves along the curve, and the limiting values are indicated by the constructions shown in Fig. 13.2. The tangent drawn at $x_1 = 0$ (pure species 2) gives $\bar{M}_2 = M_2$, consistent with the conclusion reached in Example 11.1 regarding the partial property of a pure species. The opposite intercept gives $\bar{M}_1 = \bar{M}_1^\infty$, the partial property of species 1 when it is present at *infinite dilution* ($x_1 = 0$). Similar comments apply to the tangent drawn at $x_1 = 1$ (pure species 1). In this case $\bar{M}_1 = M_1$ and $\bar{M}_2 = \bar{M}_2^\infty$, since it is species 2 that is present at infinite dilution ($x_1 = 1, x_2 = 0$).

Example 13.2 The enthalpy of a binary liquid system of species 1 and 2 at fixed T and P is represented by the equation

$$H = 400x_1 + 600x_2 + x_1x_2(40x_1 + 20x_2)$$

where H is in J mol^{-1}. Determine expressions for $\bar{H}_1$ and $\bar{H}_2$ as functions of x_1, numerical values for the pure-species enthalpies H_1 and H_2, and numerical values for the partial enthalpies at infinite dilution $\bar{H}_1^\infty$ and $\bar{H}_2^\infty$.

SOLUTION Elimination of x_2 in the given equation for H in favor of x_1 yields

$$H = 600 - 180x_1 - 20x_1^3 \tag{A}$$

whence

$$\frac{dH}{dx_1} = -180 - 60x_1^2$$

By Eq. (13.29),

$$\bar{H}_1 = H + x_2 \frac{dH}{dx_1}$$

Substitution for H and dH/dx_1 gives

$$\bar{H}_1 = 600 - 180x_1 - 20x_1^3 - 180x_2 - 60x_1^2x_2$$

Replacing x_2 by $1 - x_1$ and simplifying, we get

$$\bar{H}_1 = 420 - 60x_1^2 + 40x_1^3 \qquad (B)$$

Similarly, by Eq. (13.30)

$$\bar{H}_2 = H - x_1 \frac{dH}{dx_1}$$

whence

$$\bar{H}_2 = 600 - 180x_1 - 20x_1^3 + 180x_1 + 60x_1^3$$

or

$$\bar{H}_2 = 600 + 40x_1^3 \qquad (C)$$

We could equally well have started with the given equation for H. Since dH/dx_1 is a *total* derivative, x_2 cannot be treated as a constant. In fact, $x_2 = 1 - x_1$, and $dx_2/dx_1 = -1$. Differentiation of the given equation for H therefore gives:

$$\frac{dH}{dx_1} = 400 - 600 + x_1x_2(40 - 20) + (40x_1 + 20x_2)(-x_1 + x_2)$$

When x_2 is replaced by $1 - x_1$, this reduces to the expression previously obtained.

A numerical value for H_1 results when we substitute $x_1 = 1$ in either Eq. (A) or (B). Both equations yield $H_1 = 400\ \text{J mol}^{-1}$. Similarly H_2 is found from either Eq. (A) or (C) when $x_1 = 0$. The result is $H_2 = 600\ \text{J mol}^{-1}$. The infinite-dilution values $\bar{H}_1^\infty$ and $\bar{H}_2^\infty$ are found from Eqs. (B) and (C) when $x_1 = 0$ in Eq. (B) and $x_1 = 1$ in Eq. (C). The results are:

$$\bar{H}_1^\infty = 420 \qquad \text{and} \qquad \bar{H}_2^\infty = 640\ \text{J mol}^{-1}$$

The actual molar volumes of the binary solution methanol(1)/water(2) at 25°C and 1 bar are shown in Fig. 13.3. In addition the values of $\bar{V}_1$ and $\bar{V}_2$ are plotted as functions of x_1. The line drawn tangent to the V-vs.-x_1 curve at $x_1 = 0.3$ illustrates the procedure by which values of $\bar{V}_1$ and $\bar{V}_2$ are obtained. The particular numerical values shown on the graph are those given with Example 11.2.

We note that the curve for $\bar{V}_1$ becomes horizontal $(d\bar{V}_1/dx_1 = 0)$ at $x_1 = 1$ and the curve for $\bar{V}_2$ becomes horizontal at $x_1 = 0$ or $x_2 = 1$. This is a requirement of Eq. (11.8), the Gibbs/Duhem equation, which here becomes

$$x_1\, d\bar{V}_1 + x_2\, d\bar{V}_2 = 0$$

Division of this equation by dx_1 and rearrangement gives:

$$\frac{d\bar{V}_1}{dx_1} = -\frac{x_2}{x_1}\frac{d\bar{V}_2}{dx_1}$$

This result shows that the slopes $d\bar{V}_1/dx_1$ and $d\bar{V}_2/dx_1$ must be of opposite sign. When $x_1 = 1$, $x_2 = 0$ and $d\bar{V}_1/dx_1 = 0$, provided $d\bar{V}_2/dx_1$ remains finite. When $x_1 = 0$, $x_2 = 1$ and $d\bar{V}_2/dx_1 = 0$. The curves for $\bar{V}_1$ and $\bar{V}_2$ in Fig. 13.3 appear to be horizontal at *both* ends; this is a peculiarity of the system considered.

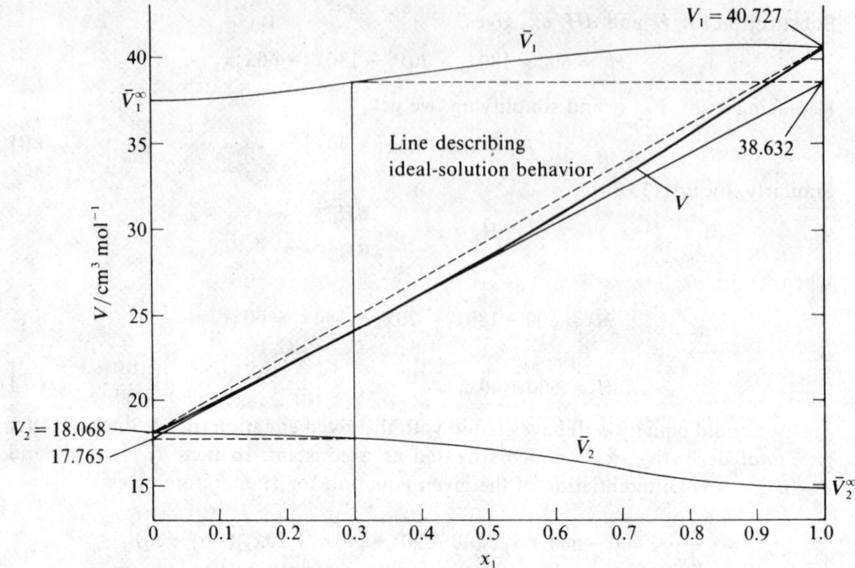

Figure 13.3 Molar volumes for methanol(1)/water(2) at 25°C and 1(atm).

If the methanol/water system is assumed an ideal solution, its volume is given by Eq. (13.9), written here as:

$$V^{id} = x_1 V_1 + x_2 V_2$$

This implies a linear relation between V^{id} and x_1:

$$V^{id} = (V_1 - V_2)x_1 + V_2$$

Thus, for the methanol/water system the straight dashed line shown in Fig. 13.3 connecting the pure-species volumes (V_1 at $x_1 = 1$ and V_2 at $x_1 = 0$) represents the V-vs.-x_1 relation that would result if this system formed an ideal solution.

If in solving Example 11.2 we assume that the solution is ideal, we then use values for V_1 and V_2 in place of the values for $\bar{V}_1$ and $\bar{V}_2$. Otherwise the problem is worked in exactly the same way, and the results are

$$V_1^t = 983 \qquad V_2^t = 1{,}017 \text{ cm}^3$$

Both values are about 3.4 percent low.

13.6 PROPERTY CHANGES OF MIXING

Equations (13.7) through (13.10) are expressions for the properties of ideal solutions. Each may be combined with the defining equation for an excess property, Eq. (11.55), to yield:

$$G^E = G - \sum x_i G_i - RT \sum x_i \ln x_i \tag{13.35}$$

$$S^E = S - \left(\sum x_i S_i - R \sum x_i \ln x_i\right) \tag{13.36}$$

$$V^E = V - \sum x_i V_i \tag{13.37}$$

$$H^E = H - \sum x_i H_i \tag{13.38}$$

In each of these equations there appears to the right of the equals sign a difference that is expressed in general as $M - \sum x_i M_i$. We call this quantity a *property change of mixing* and give it the symbol ΔM. Thus by definition,

$$\Delta M \equiv M - \sum x_i M_i \tag{13.39}$$

where M is a molar (or unit-mass) property of a solution and the M_i are molar (or unit-mass) properties of the pure species, all at the same T and P. Equations (13.35) through (13.38) are now rewritten

$$G^E = \Delta G - RT \sum x_i \ln x_i \tag{13.40}$$

$$S^E = \Delta S + R \sum x_i \ln x_i \tag{13.41}$$

$$V^E = \Delta V \tag{13.42}$$

$$H^E = \Delta H \tag{13.43}$$

where ΔG, ΔS, ΔV, and ΔH are the Gibbs energy change of mixing, the entropy change of mixing, the volume change of mixing, and the enthalpy change of mixing. For an ideal solution, each excess property is zero, and for this special case Eqs. (13.40) through (13.43) become

$$\Delta G^{id} = RT \sum x_i \ln x_i \tag{13.44}$$

$$\Delta S^{id} = -R \sum x_i \ln x_i \tag{13.45}$$

$$\Delta V^{id} = 0 \tag{13.46}$$

$$\Delta H^{id} = 0 \tag{13.47}$$

These equations are just restatements of Eqs. (13.7) through (13.10), and apply to mixtures of ideal gases as a special case.

Equations (13.40) through (13.43) show that excess properties and property changes of mixing are readily calculated one from the other. Although historically the property changes of mixing were introduced first, because of their direct relation to experiment, it is the excess properties that more readily fit into the theoretical framework of solution thermodynamics. The property changes of mixing of major interest, because of their direct measurability, are ΔV and ΔH, and these two properties are identical to the corresponding excess properties.

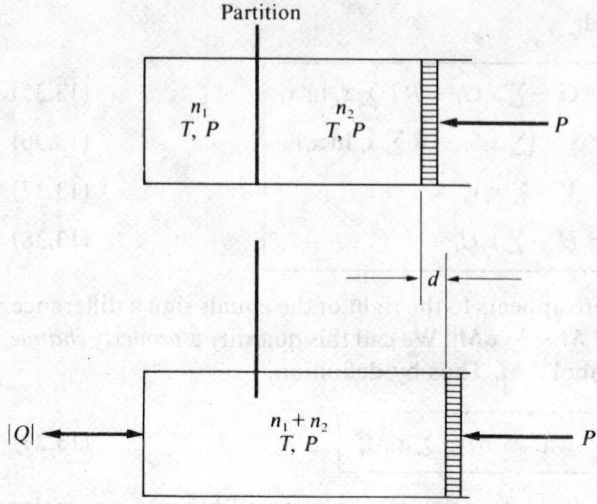

Figure 13.4 Schematic diagram of experimental mixing process.

An experimental mixing process for a binary system is represented schematically in Fig. 13.4. The two pure species, both at T and P, are initially separated by a partition, withdrawal of which allows mixing. As mixing occurs, expansion or contraction of the system is accompanied by movement of the piston so that the pressure is constant. In addition, heat is added or extracted to maintain a constant temperature. When mixing is complete, the total volume change of the system (as indicated by piston displacement d) is

$$\Delta V^t = (n_1 + n_2)V - n_1 V_1 - n_2 V_2$$

Since the process occurs at constant pressure, the total heat transfer Q is equal to the total enthalpy change of the system:

$$Q = \Delta H^t = (n_1 + n_2)H - n_1 H_1 - n_2 H_2$$

Division of these equations by $n_1 + n_2$ gives

$$\Delta V \equiv V - x_1 V_1 - x_2 V_2 = \frac{\Delta V^t}{n_1 + n_2}$$

and

$$\Delta H \equiv H - x_1 H_1 - x_2 H_2 = \frac{Q}{n_1 + n_2}$$

Thus the *volume change of mixing* ΔV and the *enthalpy change of mixing* ΔH are found from the measured quantities ΔV^t and Q. Because of its association with Q, ΔH is usually called the *heat of mixing.*

Figure 13.5 shows experimental heats of mixing ΔH (or excess enthalpies H^E) for the ethanol/water system as a function of composition for several temperatures between 30 and 110°C. This figure illustrates much of the variety of behavior found for $H^E = \Delta H$ and $V^E = \Delta V$ data for binary liquid systems.

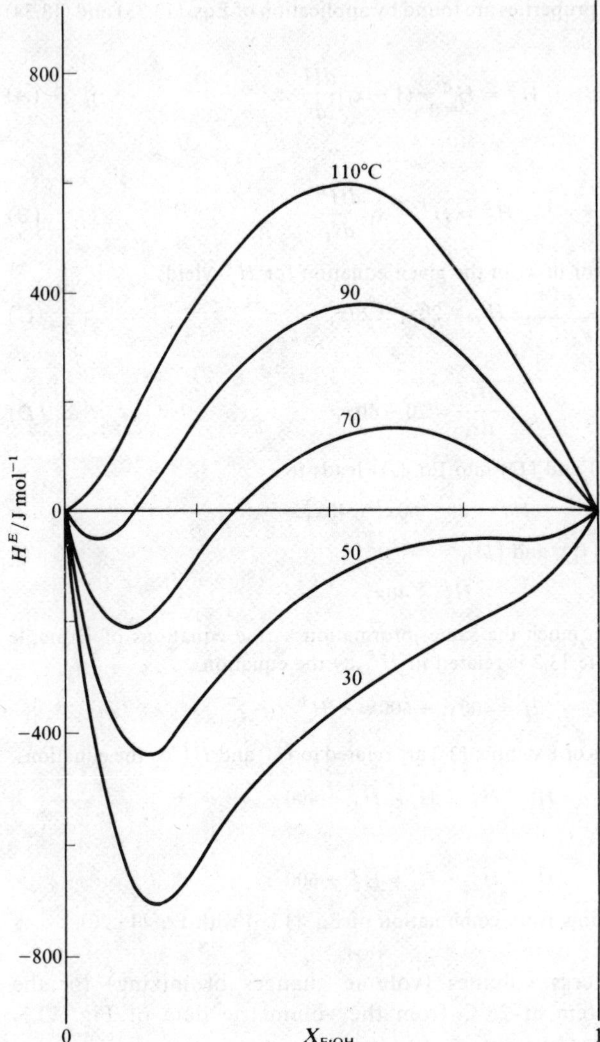

Figure 13.5 Excess enthalpies for ethanol/water.

Such data are also often represented by equations similar to those used for G^E data, in particular by the Redlich/Kister expansion (Sec. 12.4).

Example 13.3 The excess enthalpy (heat of mixing) for a liquid mixture of species 1 and 2 at fixed T and P is represented by the equation:

$$H^E = x_1 x_2(40x_1 + 20x_2)$$

where H^E is in J mol^{-1}. Determine expressions for $\bar{H}_1^E$ and $\bar{H}_2^E$ as functions of x_1.

SOLUTION The partial properties are found by application of Eqs. (13.33) and (13.34) with $M^E = H^E$. Thus,

$$\bar{H}_1^E = H^E + (1 - x_1)\frac{dH^E}{dx_1} \tag{A}$$

and

$$\bar{H}_2^E = H^E - x_1 \frac{dH^E}{dx_1} \tag{B}$$

Elimination of x_2 in favor of x_1 in the given equation for H^E yields

$$H^E = 20x_1 - 20x_1^3 \tag{C}$$

whence

$$\frac{dH^E}{dx_1} = 20 - 60x_1^2 \tag{D}$$

Substitution of Eqs. (C) and (D) into Eq. (A) leads to

$$\bar{H}_1^E = 20 - 60x_1^2 + 40x_1^3$$

Similarly, by Eqs. (B), (C), and (D),

$$\bar{H}_2^E = 40x_1^3$$

These equations contain much the same information as the equations of Example 13.2. Thus H of Example 13.2 is related to H^E by the equation,

$$H = 400x_1 + 600x_2 + H^E$$

and the partial properties of Example 13.2 are related to $\bar{H}_1^E$ and $\bar{H}_2^E$ by the equations,

$$\bar{H}_1 = \bar{H}_1^E + H_1 = \bar{H}_1^E + 400$$

and

$$\bar{H}_2 = \bar{H}_2^E + H_2 = \bar{H}_2^E + 600$$

These two equations follow from combination of Eq. (13.6) with Eq. (13.20).

We can calculate excess volumes (volume changes of mixing) for the methanol(1)/water(2) system at 25°C from the volumetric data of Fig. 13.3. Equation (13.20) specializes to

$$\bar{V}_i^E = \bar{V}_i - \bar{V}_i^{id}$$

According to Eq. (13.5), $\bar{V}_i^{id} = V_i$. Therefore

$$\bar{V}_1^E = \bar{V}_1 - V_1 \qquad \text{and} \qquad \bar{V}_2^E = \bar{V}_2 - V_2$$

Equation (11.5) written for the excess volume of a binary system becomes

$$V^E = x_1 \bar{V}_1^E + x_2 \bar{V}_2^E$$

The results are shown in Fig. 13.6. The values on the figure for $x_1 = 0.3$ come

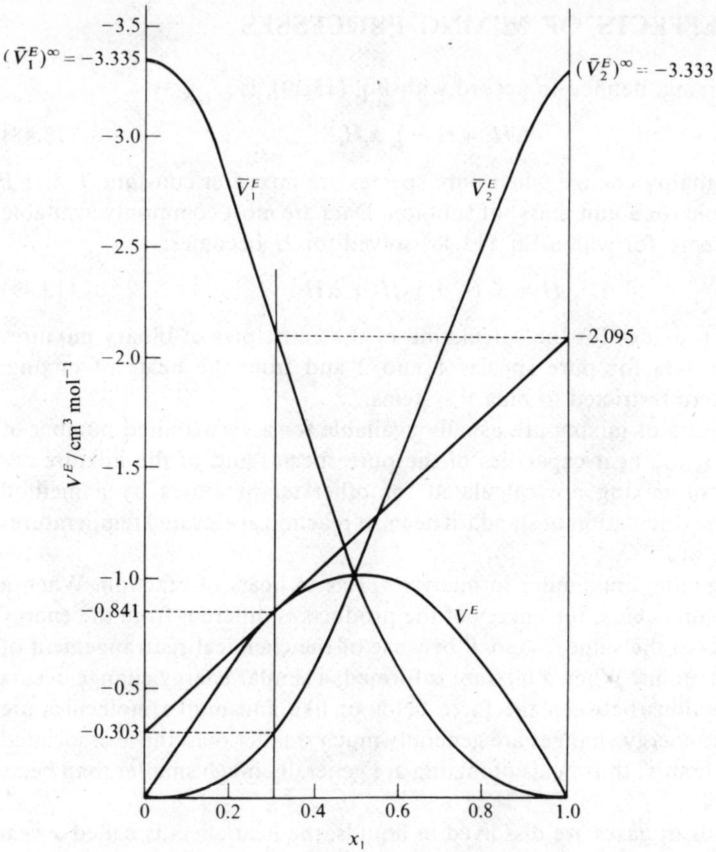

Figure 13.6 Excess volumes for methanol(1)/water(2) at 25°C.

from Example 11.2. Thus

$$\bar{V}_1^E = 38.632 - 40.727 = -2.095 \text{ cm}^3 \text{ mol}^{-1}$$

$$\bar{V}_2^E = 17.765 - 18.068 = -0.303 \text{ cm}^3 \text{ mol}^{-1}$$

and

$$V^E = (0.3)(-2.095) + (0.7)(-0.303) = -0.841 \text{ cm}^3 \text{ mol}^{-1}$$

The tangent line drawn at $x_1 = 0.3$ illustrates the determination of partial excess volumes by the method of tangent intercepts. Whereas the values of V in Fig. 13.3 range from 18.068 to 40.727 cm^3 mol^{-1}, the values of $V^E = \Delta V$ go from zero at $x_1 = 0$ and at $x_1 = 1$ to a value of about -1 cm^3 mol^{-1} at a mole fraction of about 0.5. The curves showing $\bar{V}_1^E$ and $\bar{V}_2^E$ are nearly symmetrical for the methanol/water system, but this is by no means so for all systems.

13.7 HEAT EFFECTS OF MIXING PROCESSES

The heat of mixing, defined in accord with Eq. (13.39), is

$$\Delta H = H - \sum x_i H_i \tag{13.48}$$

It gives the enthalpy change when pure species are mixed at constant T and P to form one mole (or a unit mass) of solution. Data are most commonly available for binary systems, for which Eq. (13.48) solved for H becomes:

$$H = x_1 H_1 + x_2 H_2 + \Delta H \tag{13.49}$$

This equation provides for the calculation of the enthalpies of binary mixtures from enthalpy data for pure species 1 and 2 and from the heats of mixing. Treatment is here restricted to binary systems.

Data for heats of mixing are usually available for a very limited number of temperatures. If the heat capacities of the pure species and of the mixture are known, heats of mixing are calculated for other temperatures by a method analogous to the calculation of standard heats of reaction at elevated temperatures from the value at 25°C.

Heats of mixing are similar in many respects to heats of reaction. When a chemical reaction occurs, the energy of the products is different from the energy of the reactants at the same T and P because of the chemical rearrangement of the constituent atoms. When a mixture is formed, a similar energy change occurs because interactions between the force fields of like and unlike molecules are different. These energy changes are generally much smaller than those associated with chemical bonds; thus heats of mixing are generally much smaller than heats of reaction.

When solids or gases are dissolved in liquids, the heat effect is called a *heat of solution*, and is based on the dissolution of *1 mole of solute.* If we take species 1 as the solute, then x_1 is the moles of solute per mole of solution. Since ΔH is the heat effect per mole of solution, $\Delta H/x_1$ is the heat effect per mole of solute. Thus

$$\widetilde{\Delta H} = \frac{\Delta H}{x_1}$$

where $\widetilde{\Delta H}$ is the heat of solution on the basis of a mole of *solute.*

Solution processes are conveniently represented by *physical-change* equations analogous to chemical-reaction equations. Thus if 1 mole of LiCl is dissolved in 12 moles of H_2O, the process is represented as

$$\text{LiCl}(s) + 12\text{H}_2\text{O}(l) \rightarrow \text{LiCl}(12\text{H}_2\text{O})$$

The designation $\text{LiCl}(12\text{H}_2\text{O})$ means that the product is a solution of 1 mole of LiCl in 12 moles of H_2O. The enthalpy change accompanying this process at 25°C and 1 bar is $\widetilde{\Delta H} = -33{,}614$ J. That is, a solution of 1 mole of LiCl in 12 moles of H_2O has an enthalpy 33,614 J less than that of 1 mole of pure LiCl(s) and 12 moles of pure $H_2O(l)$. Equations for physical changes such as this are readily combined with equations for chemical reactions. This is illustrated in the following example.

Example 13.4 Calculate the heat of formation of LiCl in 12 moles of H_2O at 25°C.

SOLUTION The process implied by the problem statement results in the formation from its constituent elements of 1 mole of LiCl *in solution* in 12 moles of H_2O. The equation representing this process is obtained as follows:

$$\begin{array}{rl} \mathrm{Li} + \tfrac{1}{2}\mathrm{Cl}_2 \rightarrow \mathrm{LiCl}(s) & \Delta H^\circ_{298} = -408{,}610\ \mathrm{J} \\ \mathrm{LiCl}(s) + 12\mathrm{H_2O}(l) \rightarrow \mathrm{LiCl}(12\mathrm{H_2O}) & \widetilde{\Delta H}_{298} = -33{,}614\ \mathrm{J} \\ \hline \mathrm{Li} + \tfrac{1}{2}\mathrm{Cl}_2 + 12\mathrm{H_2O}(l) \rightarrow \mathrm{LiCl}(12\mathrm{H_2O}) & \Delta H^\circ_{298} = -442{,}224\ \mathrm{J} \end{array}$$

The first reaction describes a chemical change resulting in the formation of LiCl(s) from its elements, and the enthalpy change accompanying this reaction is the standard heat of formation of LiCl(s) at 25°C. The second reaction represents the physical change resulting in the solution of 1 mole of LiCl(s) in 12 moles of $H_2O(l)$. The enthalpy change accompanying this reaction is a heat of solution. The enthalpy change of −442,224 J for the overall process is known as the heat of formation of LiCl *in* 12 moles of H_2O. This figure does *not* include the heat of formation of the H_2O.

Often heats of solution are not reported directly and must be calculated from heats of formation by the reverse of the calculation just illustrated. The data given by the Bureau of Standards† for the heats of formation of 1 mole of LiCl are:

LiCl(s)	−408,610 J
LiCl·H_2O(s)	−712,580 J
LiCl·2H_2O(s)	−1,012,650 J
LiCl·3H_2O(s)	−1,311,300 J
LiCl in 3 moles H_2O	−429,366 J
LiCl in 5 moles H_2O	−436,805 J
LiCl in 8 moles H_2O	−440,529 J
LiCl in 10 moles H_2O	−441,579 J
LiCl in 12 moles H_2O	−442,224 J
LiCl in 15 moles H_2O	−442,835 J

From these data heats of solution are readily calculated. Take the case of the solution of 1 mole of LiCl in 5 moles of H_2O. The reaction representing this process is obtained as follows:

$$\begin{array}{rl} \mathrm{Li} + \tfrac{1}{2}\mathrm{Cl}_2 + 5\mathrm{H_2O}(l) \rightarrow \mathrm{LiCl}(5\mathrm{H_2O}) & \Delta H^\circ_{298} = -436{,}805\ \mathrm{J} \\ \mathrm{LiCl}(s) \rightarrow \mathrm{Li} + \tfrac{1}{2}\mathrm{Cl}_2 & \Delta H^\circ_{298} = 408{,}610\ \mathrm{J} \\ \hline \mathrm{LiCl}(s) + 5\mathrm{H_2O}(l) \rightarrow \mathrm{LiCl}(5\mathrm{H_2O}) & \widetilde{\Delta H}_{298} = -28{,}194\ \mathrm{J} \end{array}$$

† "The NBS Tables of Chemical Thermodynamic Properties," *J. Phys. Chem. Ref. Data*, vol. 11, suppl. 2, 1982.

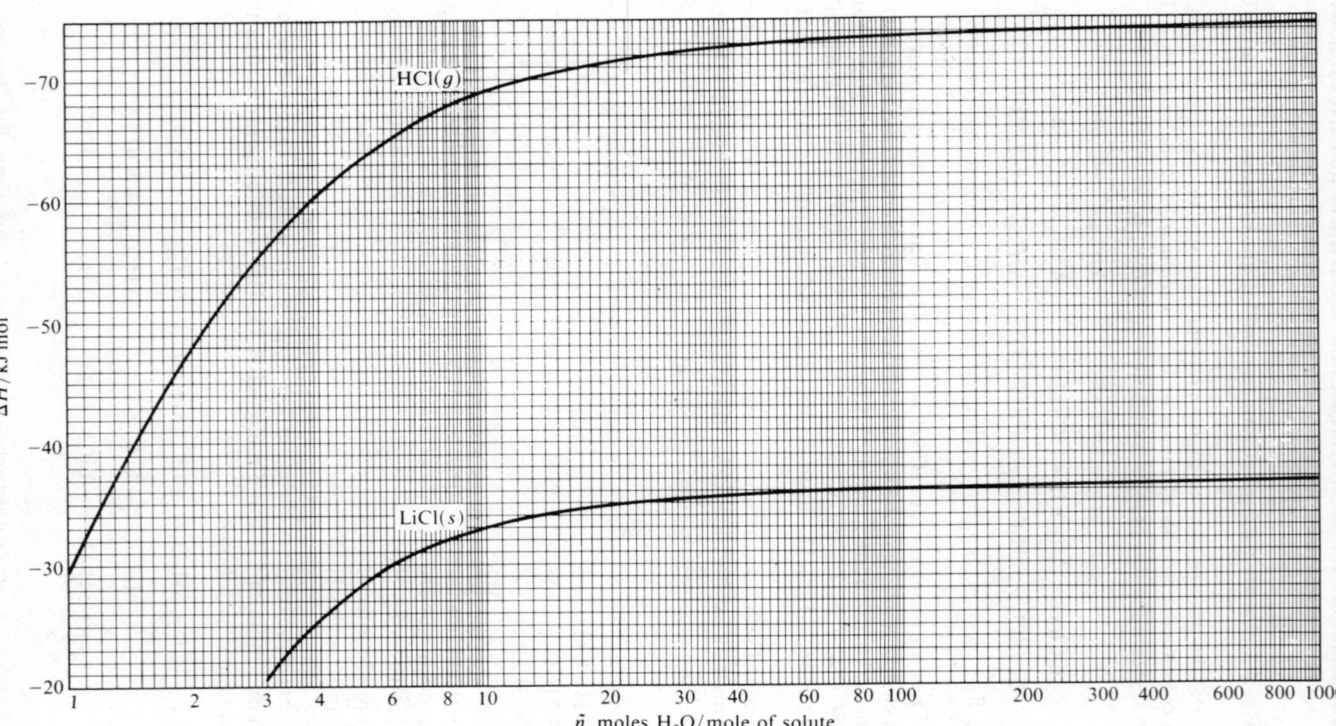

Figure 13.7 Heats of solution at 25°C. *(Based on data from "The NBS Tables of Chemical Thermodynamic Properties," J. Phys. Chem. Ref. Data, vol. 11, suppl. 2, 1982.)*

This calculation can be carried out for each quantity of H_2O for which data are given. The results are then conveniently represented graphically by a plot of $\widetilde{\Delta H}$, the heat of solution per mole of solute, vs. $\tilde{n}$, the moles of solvent per mole of solute. The composition variable, $\tilde{n} = n_2/n_1$, is related to x_1:

$$\tilde{n} = \frac{x_2}{x_1} = \frac{1 - x_1}{x_1}$$

whence

$$x_1 = \frac{1}{1 + \tilde{n}}$$

We therefore have the following relations between ΔH, the heat of mixing based on 1 mole of solution, and $\widetilde{\Delta H}$, the heat of solution based on 1 mole of solute:

$$\widetilde{\Delta H} = \frac{\Delta H}{x_1} = \Delta H(1 + \tilde{n})$$

or

$$\Delta H = \frac{\widetilde{\Delta H}}{1 + \tilde{n}}$$

Figure 13.7 shows plots of $\widetilde{\Delta H}$ vs. $\tilde{n}$ for LiCl(s) and HCl(g) dissolved in water at 25°C. Data in this form are readily applied to the solution of practical problems.

Example 13.5 A single-effect evaporator operating at atmospheric pressure concentrates a 15% (by weight) LiCl solution to 40%. The feed enters the evaporator at the rate of 2 kg s^{-1} at 25°C. The normal boiling point of a 40% LiCl solution is about 132°C, and its specific heat is estimated as 2.72 kJ kg^{-1} $°C^{-1}$. What is the heat-transfer rate in the evaporator?

SOLUTION The 2 kg of 15% LiCl solution entering the evaporator each second consists of 0.30 kg LiCl and 1.70 kg H_2O. A material balance shows that 1.25 kg of H_2O is evaporated and that 0.75 kg of 40% LiCl solution is produced. The process is indicated schematically in Fig. 13.8.

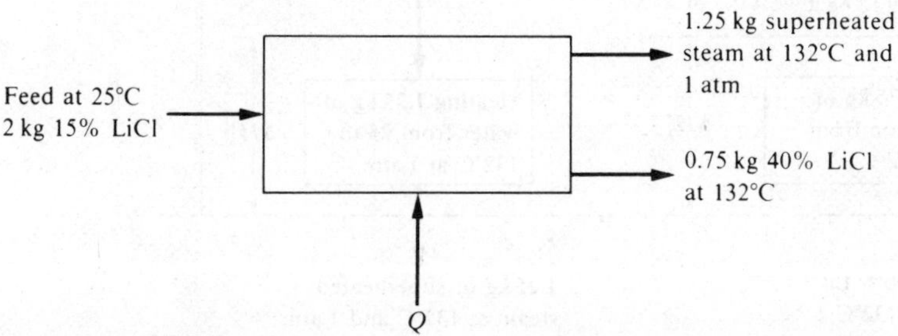

Figure 13.8

The energy balance for this flow process gives $\Delta H^t = Q$, where ΔH^t is the total enthalpy of the product streams minus the total enthalpy of the feed stream. Thus the problem reduces to finding ΔH^t from the available data. Since enthalpy is a state function, the path used for the calculation of ΔH^t is immaterial and may be selected as convenience dictates and without reference to the actual path followed in the evaporator. The data available are heats of solution of LiCl in H_2O at 25°C (see Fig. 13.7), and the calculational path, shown in Fig. 13.9, allows their direct use.

The enthalpy changes for the individual steps shown in this figure must add up to the total enthalpy change:

$$\Delta H^t = \Delta H_a^t + \Delta H_b^t + \Delta H_c^t + \Delta H_d^t$$

The individual enthalpy changes are determined as follows.

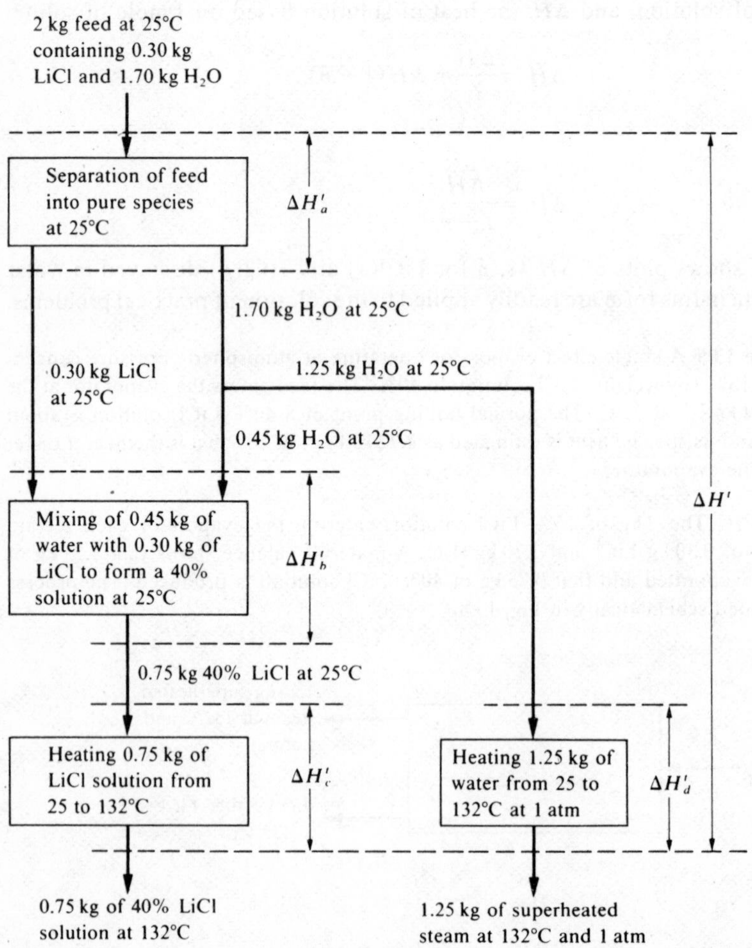

Figure 13.9

ΔH_a^t: This step involves the separation of 2 kg of a 15% LiCl solution into its pure constituents at 25°C. This is an "unmixing" process, and the heat effect is the same as for the corresponding mixing process, but is of opposite sign. For 2 kg of 15% LiCl solution, the moles of material entering are

$$\frac{(0.30)(1{,}000)}{42.39} = 7.077 \text{ mol LiCl}$$

and

$$\frac{(1.70)(1{,}000)}{18.016} = 94.361 \text{ mol } H_2O$$

Thus the solution contains 13.33 moles of H_2O per mole of LiCl. From Fig. 13.7 the heat of solution per mole of LiCl for $\tilde{n} = 13.33$ is −33,800 J. For the "unmixing" of 2 kg of solution,

$$\Delta H_a^t = (+33{,}800)(7.077) = 239{,}250 \text{ J}$$

ΔH_b^t: This step results in the mixing of 0.45 kg of water with 0.30 kg of LiCl to form a 40% solution at 25°C. This solution is made up of

$$0.30 \text{ kg} \quad \text{or} \quad 7.077 \text{ mol LiCl}$$

and

$$0.45 \text{ kg} \quad \text{or} \quad 24.978 \text{ mol } H_2O$$

Thus the final solution contains 3.53 moles of H_2O per mole of LiCl. From Fig. 13.7 the heat of solution per mole of LiCl at this value of $\tilde{n}$ is −23,260 J. Therefore

$$\Delta H_b^t = (-23{,}260)(7.077) = -164{,}630 \text{ J}$$

ΔH_c^t: For this step 0.75 kg of 40% LiCl solution is heated from 25 to 132°C. Since $\Delta H_c^t = mC_P\,\Delta T$,

$$\Delta H_c^t = (0.75)(2.72)(132 - 25) = 218.28 \text{ kJ}$$

or

$$\Delta H_c^t = 218{,}280 \text{ J}$$

ΔH_d^t: In this step liquid water is vaporized and heated to 132°C. The enthalpy change is obtained from the steam tables:

$$\Delta H_d^t = (1.25)(2{,}740.3 - 104.8) = 3{,}294.4 \text{ kJ}$$

or

$$\Delta H_d^t = 3{,}294{,}400 \text{ J}$$

Adding the individual enthalpy changes gives:

$$\begin{aligned}\Delta H^t &= \Delta H_a^t + \Delta H_b^t + \Delta H_c^t + \Delta H_d^t \\ &= 239{,}250 - 164{,}630 + 218{,}280 + 3{,}294{,}400 \\ &= 3{,}587{,}300 \text{ J}\end{aligned}$$

The required heat-transfer rate is therefore 3,587.3 kJ s^{-1}.

The most convenient method for representation of enthalpy data for binary solutions is by *enthalpy/concentration* (Hx) *diagrams.* These diagrams are graphs of the enthalpy plotted as a function of composition (mole fraction or mass fraction of one species) with temperature as parameter. The pressure is a constant and is usually 1 atmosphere. Figure 13.10 shows a partial diagram for the H_2SO_4/H_2O system.

The enthalpy values are based on a mole or a unit mass of *solution*, and Eq. (13.49) is directly applicable. Values of H for the solution depend not only on the heats of mixing, but also on the enthalpies H_1 and H_2 of the pure species. Once H_1 and H_2 are known for a given T and P, H is fixed for all solutions at the same T and P, because ΔH has a unique and measurable value at each composition. Since absolute enthalpies are unknown, arbitrary zero points are chosen for the enthalpies of the pure species. Thus, the *basis* of an enthalpy/concentration diagram is $H_1 = 0$ for some specified state of species 1 and $H_2 = 0$ for some specified state of species 2. The same temperature need not be selected for these states for both species. In the case of the H_2SO_4/H_2O diagram shown in Fig. 13.10, $H_{H_2O} = 0$ for pure liquid H_2O at the triple point [$\approx$32(°F)], and $H_{H_2SO_4} = 0$ for pure liquid H_2SO_4 at 25°C [77(°F)]. In this case the 32(°F) isotherm terminates at $H = 0$ at the end of the diagram representing pure liquid H_2O, and the 77(°F) isotherm terminates at $H = 0$ at the other end of the diagram representing pure liquid H_2SO_4.

The advantage of taking $H = 0$ for pure liquid water at its triple point is that this is the base of the steam tables. Enthalpy values from the steam tables can then be used in conjunction with values taken from the enthalpy/concentration diagram. Were some other base used for the diagram, one would have to apply a correction to the steam-table values to put them on the same basis as the diagram.

For an ideal solution, isotherms on an enthalpy/concentration diagram are straight lines connecting the enthalpy of pure species 2 at $x_1 = 0$ with the enthalpy of pure species 1 at $x_1 = 1$. This follows immediately from Eq. (13.10),

$$H^{id} = x_1 H_1 + (1 - x_1) H_2 = x_1(H_1 - H_2) + H_2$$

and is illustrated for a single isotherm in Fig. 13.11 by the dashed line. The solid curve shows how the isotherm might appear for a real solution. Also shown is a tangent line from which partial enthalpies may be determined. Comparison of Eq. (13.10) with Eq. (13.49) shows that $\Delta H = H - H^{id}$; that is, ΔH is the vertical distance between the curve and the dashed line of Fig. 13.11. The actual isotherm is displaced vertically from the ideal-solution isotherm at a given composition by the value of ΔH at that composition. In the case illustrated ΔH is everywhere negative. This means that heat is evolved whenever the pure species at the given temperature are mixed to form a solution at the same temperature. Such a system is said to be *exothermic.* The H_2SO_4/H_2O system is an example. An *endothermic* system is one for which the heats of solution are positive; in this case heat is absorbed to keep the temperature constant. An example is the methanol/benzene system.

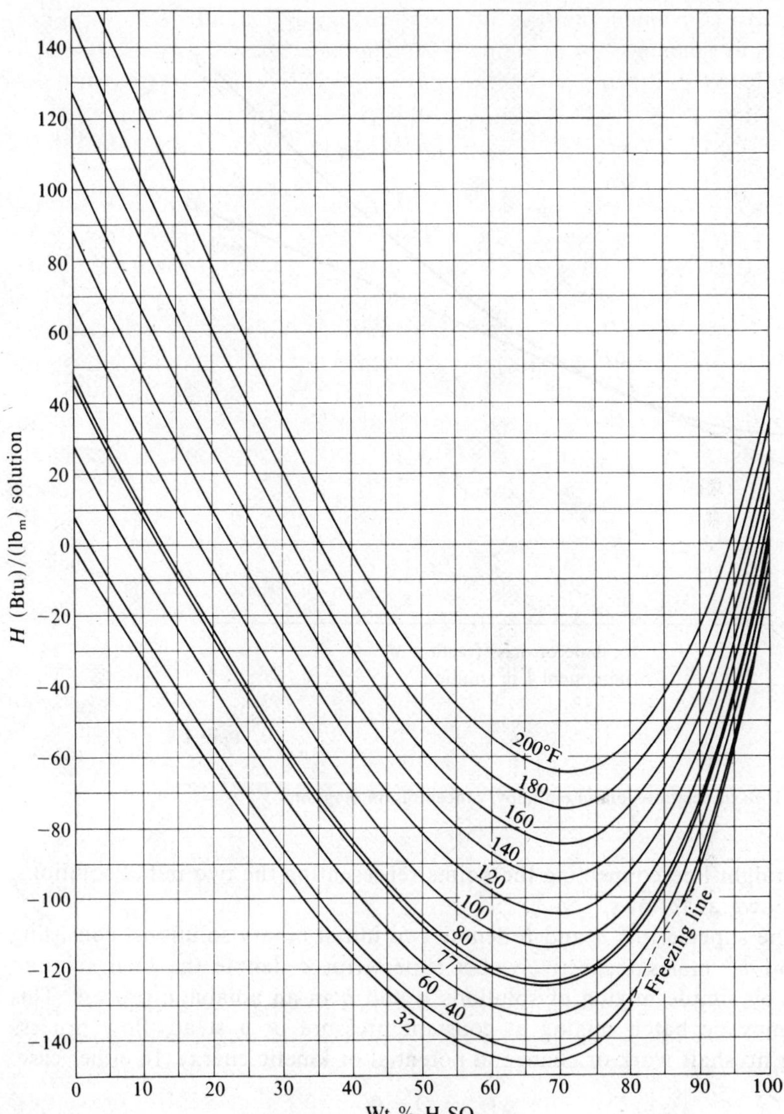

Figure 13.10 Enthalpy/concentration diagram for H_2SO_4/H_2O. *(Redrawn from the data of W. D. Ross, Chem. Eng. Prog.,* **48**: *314, 1952. By permission.)*

One feature of an enthalpy/concentration diagram which makes it particularly useful is the ease with which problems involving adiabatic mixing may be solved. This results from the fact that adiabatic mixing may be represented by a straight line on the *Hx* diagram. More precisely, the point on an *Hx* diagram which represents a solution formed by adiabatic mixing of two other solutions must lie

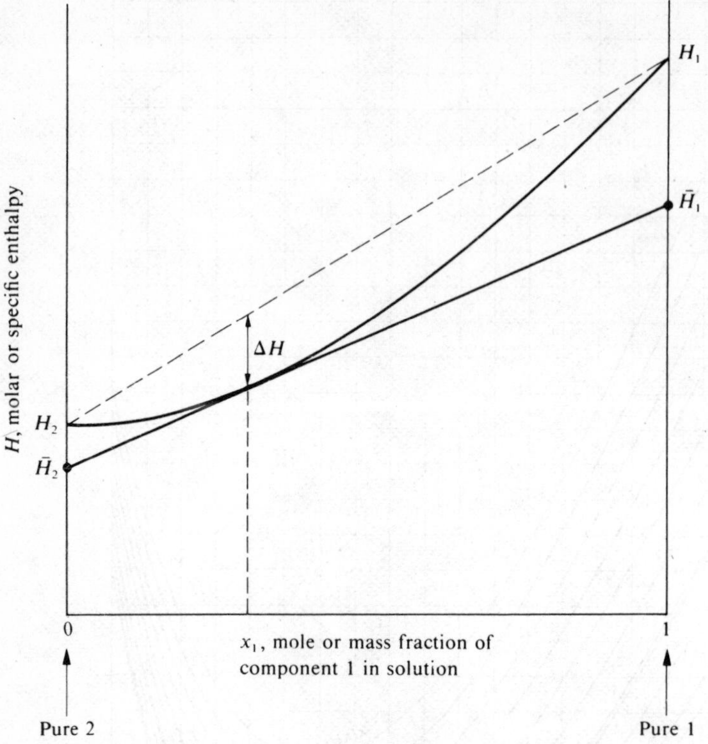

Figure 13.11 Basic relations on an enthalpy/concentration diagram.

on the straight line connecting the points representing the two initial solutions. This is shown as follows.

Let the superscripts a and b denote two initial binary solutions, consisting of n^a and n^b moles respectively. Let superscript c denote the final solution obtained by simple mixing of solutions a and b in an adiabatic process. This process may be batch mixing at constant pressure or a steady-flow process involving no shaft work or change in potential or kinetic energy. In either case,

$$\Delta H^t = Q = 0$$

We may therefore write for the overall change in state:

$$(n^a + n^b)H^c = n^a H^a + n^b H^b$$

In addition, a material balance for species 1 gives

$$(n^a + n^b)x_1^c = n^a x_1^a + n^b x_1^b$$

These two equations may be written:

$$n^a(H^c - H^a) = -n^b(H^c - H^b)$$

and

$$n^a(x_1^c - x_1^a) = -n^b(x_1^c - x_1^b)$$

Division of the first equation by the second gives

$$\frac{H^c + H^a}{x_1^c - x_1^a} = \frac{H^c - H^b}{x_1^c - x_1^b} \tag{A}$$

We now show that the three points *c*, *a*, and *b* represented by (H^c, x_1^c), (H^a, x_1^a), and (H^b, x_1^b) lie along a straight line on an *Hx* diagram. The general equation for a straight line in these coordinates is

$$H = mx_1 + k \tag{B}$$

Assuming that this line passes through points *a* and *b*, we can write:

$$H^a = mx_1^a + k \tag{C}$$

and

$$H^b = mx_1^b + k \tag{D}$$

Subtraction of first Eq. (C) and then Eq. (D) from Eq. (B) gives

$$H - H^a = m(x_1 - x_1^a)$$

and

$$H - H^b = m(x_1 - x_1^b)$$

Dividing the first of these by the second, we obtain

$$\frac{H - H^a}{H - H^b} = \frac{x_1 - x_1^a}{x_1 - x_1^b}$$

or

$$\frac{H - H^a}{x_1 - x_1^a} = \frac{H - H^b}{x_1 - x_1^b}$$

Any point with the coordinates (H, x_1) which satisfies this equation lies on the straight line connecting points *a* and *b*. Equation (A) clearly shows that the point (H^c, x_1^c) satisfies this requirement.

The use of enthalpy/concentration diagrams is illustrated in the following examples for the NaOH/H_2O system, for which an *Hx* diagram is shown in Fig. 13.12.

Example 13.6 A single-effect evaporator concentrates $10{,}000(\text{lb}_m)(\text{hr})^{-1}$ of a 10% (by weight) aqueous solution of NaOH to 50%. The feed enters at 70(°F). The evaporator operates at an absolute pressure of 3(inHg), and under these conditions the boiling point of a 50% solution of NaOH is 190(°F). What is the heat-transfer rate in the evaporator?

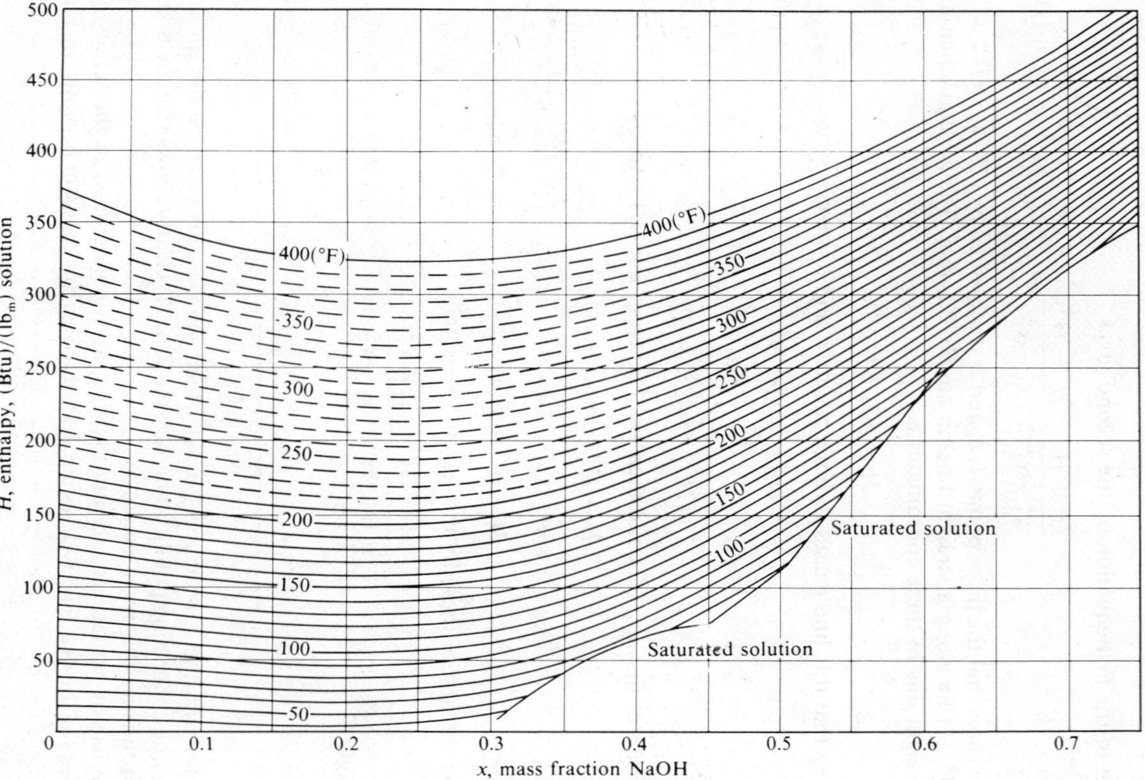

Figure 13.12 Enthalpy/concentration diagram for $NaOH/H_2O$. *(Reproduced by permission. W. L. McCabe, Trans. AIChE.,* **31***: 129, 1935; R. H. Wilson and W. L. McCabe, Ind. Eng. Chem.,* **34***: 558, 1942.)*

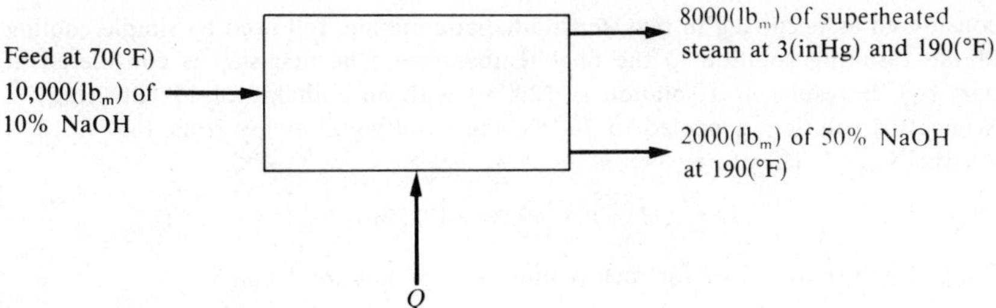

Figure 13.13

SOLUTION On the basis of 10,000(lb$_m$) of 10% NaOH fed to the evaporator, a material balance shows that the product stream consists of 8,000(lb$_m$) of superheated steam at 3(inHg) and 190(°F), and 2,000(lb$_m$) of 50% NaOH at 190(°F). The process is indicated schematically in Fig. 13.13. The energy balance for this flow process is

$$\Delta H^t = Q$$

In this case ΔH^t is easily determined from enthalpy values taken from the Hx diagram of Fig. 13.12 and from the steam tables:

$$\text{Enthalpy of superheated steam at 3(inHg) and 190(°F)} = 1{,}146(\text{Btu})(\text{lb}_m)^{-1}$$

$$\text{Enthalpy of 10\% NaOH solution at 70(°F)} = 34(\text{Btu})(\text{lb}_m)^{-1}$$

$$\text{Enthalpy of 50\% NaOH solution at 190(°F)} = 215(\text{Btu})(\text{lb}_m)^{-1}$$

Thus

$$Q = \Delta H^t = (8{,}000)(1{,}146) + (2{,}000)(215) - (10{,}000)(34)$$

$$= 9{,}260{,}000(\text{Btu})(\text{hr})^{-1}$$

A comparison of this example with Example 13.5 shows the simplification introduced by use of an enthalpy/concentration diagram.

Example 13.7 A 10% aqueous NaOH solution at 70(°F) is mixed with a 70% aqueous NaOH solution at 200(°F) to form a solution containing 40% NaOH.

(*a*) If the mixing is done adiabatically, what is the final temperature of the solution?

(*b*) If the final temperature is brought to 70(°F), how much heat must be removed during the process?

SOLUTION (*a*) A straight line drawn on Fig. 13.12 connecting the points representing the two initial solutions must contain the point representing the final solution obtained by adiabatic mixing. The particular solution represented by a point on this line at a concentration of 40% NaOH has an enthalpy of 192(Btu)(lb$_m$)$^{-1}$. Moreover, the isotherm for 220(°F) passes through this point. Thus the final temperature, obtained graphically, is 220(°F).

(*b*) The overall process cannot be represented by a single straight line on Fig. 13.12. However, we may select any convenient path for calculating ΔH of the process and hence Q, since the energy balance gives $Q = \Delta H$. Thus the process may be considered as occurring in two steps: adiabatic mixing, followed by simple cooling of the resulting solution to the final temperature. The first step is considered in part (*a*). It results in a solution at 220(°F) with an enthalpy of $192(\text{Btu})(\text{lb}_m)^{-1}$. When this solution is cooled to 70(°F), the resulting enthalpy from Fig. 13.12 is $70(\text{Btu})(\text{lb}_m)^{-1}$. Therefore

$$Q = \Delta H = 70 - 192 = -122(\text{Btu})(\text{lb}_m)^{-1}$$

Thus 122(Btu) is *evolved* for each pound *mass* of solution formed.

Example 13.8 Determine the enthalpy of solid NaOH at 68(°F) on the basis used for the NaOH/H_2O enthalpy/concentration diagram of Fig. 13.12.

SOLUTION The isotherms on an *Hx* diagram for a system such as NaOH/H_2O terminate at points where the limit of solubility of the solid in water is reached. Thus the isotherms in Fig. 13.12 do not extend to a mass fraction representing pure NaOH. How, then, is the basis of the diagram with respect to NaOH selected? In the case of the water the basis is $H_{H_2O} = 0$ for liquid water at 32(°F), consistent with the base of the steam tables. For NaOH the basis is $\bar{H}_{\text{NaOH}} = 0$ for NaOH in an infinitely dilute solution at 68(°F).

This means that the partial specific enthalpy of NaOH at infinite dilution (i.e., at $x_{\text{NaOH}} \to 0$) is arbirarily set equal to zero at 68(°F). The graphical interpretation is that the diagram is constructed in such a way that a tangent drawn to the 68(°F) isotherm at $x_{\text{NaOH}} = 0$ intersects the $x_{\text{NaOH}} = 1$ ordinate (not shown) at an enthalpy of zero. The selection of $\bar{H}^{\infty}_{\text{NaOH}}$ as zero at 68(°F) automatically fixes the values of the enthalpy of NaOH in all other states.

In particular, the enthalpy of solid NaOH at 68(°F) can be calculated for the basis selected. If $1(\text{lb}_m)$ of solid NaOH at 68(°F) is dissolved in an infinite amount of water at 68(°F), and if the temperature is held constant by extraction of the heat of solution, the result is an infinitely dilute solution at 68(°F). Since the water is pure in both the initial and final states, its enthalpy does not change. The heat of solution is therefore

$$\widetilde{\Delta H}^{\infty}_{\text{NaOH}} = \bar{H}^{\infty}_{\text{NaOH}} - H_{\text{NaOH}} \qquad [68(°\text{F})]$$

Since

$$\bar{H}^{\infty}_{\text{NaOH}} = 0 \qquad [68(°\text{F})]$$

$$\widetilde{\Delta H}^{\infty}_{\text{NaOH}} = -H_{\text{NaOH}} \qquad [68(°\text{F})]$$

The enthalpy of solid NaOH at 68(°F), H_{NaOH}, is therefore equal to the negative of the heat of solution of NaOH in an infinite amount of water at 68(°F). A literature value† of the *heat evolved* when 1 mole of NaOH is dissolved in water to form an infinitely dilute solution at 18°C is 10,180(cal). Since heat *evolved* is defined as negative,

$$\widetilde{\Delta H}^{\infty}_{\text{NaOH}} = -10{,}180(\text{cal}) \qquad [18°\text{C}]$$

If the difference in temperature between 18°C [64(°F)] and 68(°F) is neglected, the

† R. H. Perry and D. Green, *Perry's Chemical Engineers' Handbook*, 6th ed., p. 3-159, McGraw-Hill, New York, 1984.

enthalpy of solid NaOH at 68(°F) is

$$H_{\mathrm{NaOH}} = -\widetilde{\Delta H}^{\infty}_{\mathrm{NaOH}} = \frac{-(-10{,}180)(1.8)}{40.00}$$

$$= 458(\mathrm{Btu})(\mathrm{lb_m})^{-1}$$

This figure represents the enthalpy of solid NaOH at 68(°F) on the same basis as was selected for the NaOH/H_2O enthalpy/concentration diagram of Fig. 13.12.

Example 13.9 Solid NaOH at 70(°F) is mixed with H_2O at 70(°F) to produce a solution containing 45% NaOH at 70(°F). How much heat must be transferred per pound *mass* of solution formed?

SOLUTION On the basis of 1($\mathrm{lb_m}$) of 45% NaOH solution, 0.45($\mathrm{lb_m}$) of solid NaOH must be dissolved in 0.55($\mathrm{lb_m}$) of H_2O. The energy balance is $\Delta H = Q$.

The enthalpy of H_2O at 70(°F) may be taken from the steam tables, or it may be read from Fig. 13.12 at $x = 0$. In either case, $H_{\mathrm{H_2O}} = 38(\mathrm{Btu})(\mathrm{lb_m})^{-1}$. The enthalpy of 45% NaOH at 70(°F) is read from Fig. 13.12 as $H = 93(\mathrm{Btu})(\mathrm{lb_m})^{-1}$. We assume that the enthalpy of solid NaOH at 70(°F) is essentially the same as the value calculated in the preceding example for 68(°F): $H_{\mathrm{NaOH}} = 458(\mathrm{Btu})(\mathrm{lb_m})^{-1}$. Therefore

$$Q = \Delta H = (1)(93) - (0.55)(38) - (0.45)(458) = -134(\mathrm{Btu})$$

Thus, 134(Btu) is *evolved* for each pound *mass* of solution formed.

13.8 EQUILIBRIUM AND STABILITY

Consider a closed system containing an arbitrary number of species and comprised of an arbitrary number of phases in which the temperature and pressure are uniform (though not necessarily constant). The system is assumed to be initially in a nonequilibrium state with respect to mass transfer between phases and chemical reaction. Any changes which occur in the system are necessarily irreversible, and they take the system ever closer to an equilibrium state. We may imagine that the system is placed in surroundings such that the system and surroundings are always in thermal and mechanical equilibrium. Heat exchange and expansion work are then accomplished reversibly. Under these circumstances the entropy change of the surroundings is given by

$$dS_{\mathrm{surr}} = \frac{dQ_{\mathrm{surr}}}{T_{\mathrm{surr}}} = \frac{-dQ}{T}$$

Here the heat transfer dQ with respect to the system has a sign opposite that of dQ_{surr}, and the temperature of the system T replaces T_{surr}, because both must have the same value for reversible heat transfer. The second law requires that

$$dS^t + dS_{\mathrm{surr}} \geq 0$$

where S^t is the total entropy of the system. Combination of these expressions yields, upon rearrangement:

$$dQ \leq T\,dS^t \tag{13.50}$$

Application of the first law provides

$$dU^t = dQ - dW = dQ - P\,dV^t$$

or

$$dQ = dU^t + P\,dV^t$$

Combining this equation with Eq. (13.50) gives

$$dU^t + P\,dV^t \leq T\,dS^t$$

or

$$\boxed{dU^t + P\,dV^t - T\,dS^t \leq 0} \tag{13.51}$$

Since this relation involves properties only, it must be satisfied for changes in state of *any* closed system of uniform T and P, without restriction to the conditions of mechanical and thermal reversibility assumed in its derivation. The inequality applies to every incremental change of the system between non-equilibrium states, and it dictates the direction of change that leads toward equilibrium. The equality holds for changes between equilibrium states (reversible processes). Thus Eq. (6.1) is just a special case of Eq. (13.51).

Equation (13.51) is so general that application to practical problems is difficult; restricted versions are much more useful. For example, by inspection we see that

$$(dU^t)_{S^t,\,V^t} \leq 0$$

where the subscripts specify properties held constant. Similarly, for processes that occur at constant U^t and V^t,

$$(dS^t)_{U^t,\,V^t} \geq 0$$

An *isolated* system is necessarily constrained to constant internal energy and volume, and for such a system it follows directly from the second law that the latter equation is valid.

If a process is restricted to occur at constant T and P, then Eq. (13.51) may be written:

$$dU^t_{T,P} + d(PV^t)_{T,P} - d(TS^t)_{T,P} \leq 0$$

or

$$d(U^t + PV^t - TS^t)_{T,P} \leq 0$$

From the definition of the Gibbs energy [Eq. (6.3)],

$$G^t = H^t - TS^t = U^t + PV^t - TS^t$$

Therefore

$$\boxed{(dG^t)_{T,P} \leq 0} \tag{13.52}$$

Of the possible specializations of Eq. (13.51), this is the most useful, because T and P are more conveniently treated as constants than are other pairs of variables, such as U^t and V^t.

Equation (13.52) indicates that all irreversible processes occurring at constant T and P proceed in such a direction as to cause a decrease in the Gibbs energy of the system. Therefore:

> The equilibrium state of a closed system is that state for which the total Gibbs energy is a minimum with respect to all possible changes at the given T and P.

This criterion of equilibrium provides a general method for determination of equilibrium states. One writes an expression for G^t as a function of the numbers of moles (mole numbers) of the species in the several phases, and then finds the set of values for the mole numbers that minimizes G^t, subject to the constraints of mass conservation. This procedure can be applied to problems of phase, chemical-reaction, or combined phase and chemical-reaction equilibrium; it is most useful for complex equilibrium problems, and is illustrated for chemical-reaction equilibrium in Sec. 15.9.

At the equilibrium state differential variations can occur in the system at constant T and P without producing any change in G^t. This is the meaning of the equality in Eq. (13.52). Thus another general criterion of equilibrium is

$$\boxed{(dG^t)_{T,P} = 0} \tag{13.53}$$

To apply this criterion, one develops an expression for dG^t as a function of the mole numbers of the species in the various phases, and sets it equal to zero. The resulting equation along with those representing the conservation of mass provide working equations for the solution of equilibrium problems. Equation (13.53) leads directly to Eq. (10.3) for phase equilibrium and it is applied to chemical-reaction equilibrium in Chap. 15.

Equation (13.52) provides a criterion that must be satisfied by any liquid phase that is *stable* with respect to the alternative that it split into two liquid phases. It requires that the Gibbs energy of an equilibrium state be the minimum value with respect to all possible changes at the given T and P. Thus when mixing of two liquids occurs at constant T and P, the total Gibbs energy must decrease, because the mixed state must be the one of lower Gibbs energy with respect to the unmixed state. We can write:

$$G^t \equiv nG < \sum n_i G_i$$

from which

$$G < \sum x_i G_i$$

or

$$G - \sum x_i G_i < 0 \qquad \text{(const } T \text{ and } P\text{)}$$

According to the definition of Eq. (13.39), the quantity on the left is the Gibbs energy change of mixing. Therefore

$$\Delta G < 0$$

Thus the Gibbs energy change of mixing must always be negative, and a plot of ΔG vs. x_1 for a binary system must appear as shown by one of the curves of Fig. 13.14. With respect to curve II, however, there is a further consideration. If, when mixing occurs, a system can achieve a lower value of the Gibbs energy by forming *two* phases than by forming a single phase, then the system splits into two phases. This is in fact the situation represented between points a and b on curve II of Fig. 13.14, because the straight dashed line connecting points a and b represents the ΔG that would obtain for the range of states consisting of two phases of compositions x_1^a and x_1^b in various proportions. Thus the solid curve shown between points a and b cannot represent a stable phase with respect to phase splitting. The equilibrium states between a and b consist of two phases.

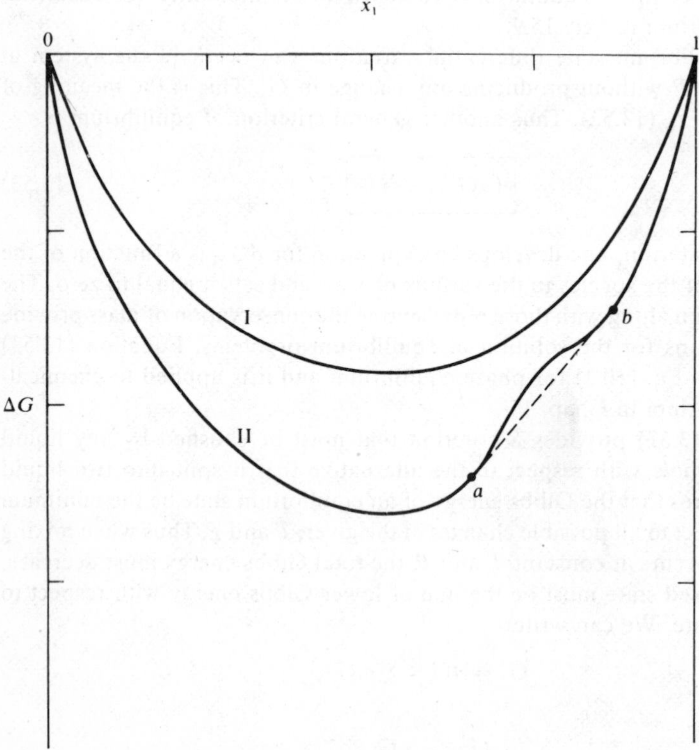

Figure 13.14 Gibbs energy change of mixing. Curve I, complete miscibility; curve II, two phases exist between A and B.

These considerations lead to the following criterion of stability for a single-phase binary system. At constant temperature and pressure, ΔG and its first and second derivatives must be continuous functions of x_1, and the second derivative must everywhere satisfy the inequality

$$\frac{d^2 \Delta G}{dx_1^2} > 0 \qquad \text{(const } T, P\text{)}$$

Since T is constant, this may equally well be expressed by

$$\frac{d^2(\Delta G/RT)}{dx_1^2} > 0 \qquad \text{(const } T, P\text{)} \tag{13.54}$$

This requirement has a number of consequences. Equation (13.40), rearranged and written for a binary system, becomes

$$\frac{\Delta G}{RT} = x_1 \ln x_1 + x_2 \ln x_2 + \frac{G^E}{RT}$$

Substituting for the last term on the right by Eq. (11.69) gives

$$\frac{\Delta G}{RT} = x_1 \ln x_1 + x_2 \ln x_2 + x_1 \ln \gamma_1 + x_2 \ln \gamma_2$$

Differentiation yields

$$\frac{d(\Delta G/RT)}{dx_1} = x_1 \frac{d \ln x_1}{dx_1} + \ln x_1 + x_2 \frac{d \ln x_2}{dx_1} - \ln x_2$$
$$+ x_1 \frac{d \ln \gamma_1}{dx_1} + \ln \gamma_1 + x_2 \frac{d \ln \gamma_2}{dx_1} - \ln \gamma_2$$

where we have made use of the fact that $dx_2/dx_1 = -1$. Simplification is effected as follows. First, we have the mathematical identity:

$$x_1 \frac{d \ln x_1}{dx_1} + x_2 \frac{d \ln x_2}{dx_1} = 0$$

and second, Eq. (11.70), the Gibbs/Duhem equation, provides the relation:

$$x_1 \frac{d \ln \gamma_1}{dx_1} + x_2 \frac{d \ln \gamma_2}{dx_1} = 0$$

Therefore

$$\frac{d(\Delta G/RT)}{dx_1} = \ln x_1 - \ln x_2 + \ln \gamma_1 - \ln \gamma_2$$
$$= \ln x_1 \gamma_1 - \ln x_2 \gamma_2$$

which, by Eq. (11.59), becomes

$$\frac{d(\Delta G/RT)}{dx_1} = \ln \frac{\hat{f}_1}{f_1} - \ln \frac{\hat{f}_2}{f_2}$$

Since f_1 and f_2 are constant at constant T and P, a second differentiation now yields

$$\frac{d^2(\Delta G/RT)}{dx_1^2} = \frac{d \ln \hat{f}_1}{dx_1} - \frac{d \ln \hat{f}_2}{dx_1} \tag{13.55}$$

Another form of the Gibbs/Duhem equation follows from Eq. (11.8) with $\bar{M}_i = \bar{G}_i$ when we substitute for $d\bar{G}_i$ by Eq. (11.28). For a binary solution the result is

$$x_1 \, d \ln \hat{f}_1 + x_2 \, d \ln \hat{f}_2 = 0 \tag{13.56}$$

Division by dx_1 yields:

$$x_1 \frac{d \ln \hat{f}_1}{dx_1} + x_2 \frac{d \ln \hat{f}_2}{dx_1} = 0$$

This equation may be combined with Eq. (13.55) to eliminate either $d \ln \hat{f}_1/dx_1$ or $d \ln \hat{f}_2/dx_1$. The two equations that result are

$$\frac{d^2(\Delta G/RT)}{dx_1^2} = \frac{-1}{x_1} \frac{d \ln \hat{f}_2}{dx_1} = \frac{1}{x_1} \frac{d \ln \hat{f}_2}{dx_2}$$

and

$$\frac{d^2(\Delta G/RT)}{dx_1^2} = \frac{1}{x_2} \frac{d \ln \hat{f}_1}{dx_1}$$

These equations, in conjunction with the criterion of Eq. (13.54), show that for a stable phase the fugacity of each species in a binary solution always increases as its mole fraction increases at constant T and P.

When the preceding equations are applied to a liquid phase in equilibrium with its vapor at constant temperature and sufficiently low pressure, we can assume the vapor to be an ideal gas and replace $\hat{f}_1$ and $\hat{f}_2$ by partial pressures, $\hat{f}_i = y_i P$. Moreover, the constraint to constant pressure can be disregarded, because under these conditions the effect of P on liquid-phase properties is negligible. Thus Eq. (13.55) becomes

$$\begin{aligned} \frac{d^2(\Delta G/RT)}{dx_1^2} &= \frac{d \ln y_1 P}{dx_1} - \frac{d \ln y_2 P}{dx_1} \\ &= \frac{d \ln y_1}{dx_1} - \frac{d \ln y_2}{dx_1} = \frac{1}{y_1} \frac{dy_1}{dx_1} - \frac{1}{y_2} \frac{dy_2}{dx_1} \\ &= \frac{1}{y_1} \frac{dy_1}{dx_1} + \frac{1}{y_2} \frac{dy_1}{dx_1} = \frac{y_1 + y_2}{y_1 y_2} \frac{dy_1}{dx_1} \end{aligned}$$

or since $y_1 + y_2 = 1$,

$$\frac{d^2(\Delta G/RT)}{dx_1^2} = \frac{1}{y_1 y_2} \frac{dy_1}{dx_1} \tag{13.57}$$

In view of Eq. (13.54), we may write the inequality

$$\frac{1}{y_1 y_2}\frac{dy_1}{dx_1} > 0$$

from which it is evident that $dy_1/dx_1 > 0$.

Making the same substitutions in Eq. (13.56), we obtain

$$x_1\, d\ln y_1 + x_1\, d\ln P + x_2\, d\ln y_2 + x_2\, d\ln P = 0$$

This reduces to

$$(x_1 + x_2)\, d\ln P + \frac{x_1}{y_1}\, dy_1 + \frac{x_2}{y_2}\, dy_2 = 0$$

or

$$d\ln P = -\frac{x_1}{y_1}\, dy_1 - \frac{x_2}{y_2}\, dy_2 = -\frac{x_1}{y_1}\, dy_1 + \frac{x_2}{y_2}\, dy_1$$

or

$$\frac{dP}{P} = \left(\frac{x_2}{y_2} - \frac{x_1}{y_1}\right) dy_1 = \frac{x_2 y_1 - x_1 y_2}{y_1 y_2}\, dy_1$$

$$= \frac{(1 - x_1)y_1 - x_1(1 - y_1)}{y_1 y_2}\, dy_1$$

Finally, division by dx_1, further reduction, and rearrangement give

$$\frac{1}{P(y_1 - x_1)}\frac{dP}{dx_1} = \frac{1}{y_1 y_2}\frac{dy_1}{dx_1}$$

In combination with Eqs. (13.57) and (13.54), this result shows that

$$\frac{1}{P(y_1 - x_1)}\frac{dP}{dx_1} > 0$$

from which we conclude that dP/dx_1 and $(y_1 - x_1)$ must have the same sign. Since

$$\frac{dP}{dy_1} = \frac{dP/dx_1}{dy_1/dx_1}$$

it follows that dP/dy_1 and dP/dx_1 also have the same sign.

In summary, the stability requirement implies the following for VLE in binary systems at constant temperature:

$$\boxed{\frac{dy_1}{dx_1} > 0 \qquad \frac{dP}{dx_1}, \frac{dP}{dy_1}, \text{ and } (y_1 - x_1) \text{ have the same sign}}$$

At an azeotrope,

$$\frac{dP}{dx_1} = \frac{dP}{dy_1} = (y_1 - x_1) = 0$$

Although derived for conditions of low pressure, these results are of general validity, as illustrated by the VLE data shown in Fig. 12.9.

13.9 SYSTEMS OF LIMITED LIQUID-PHASE MISCIBILITY

There are many pairs of chemical species which, if they mixed to form a single liquid phase, would not satisfy the stability criterion of Eq. (13.54). Such systems therefore split into two liquid phases, and are important industrially in operations such as solvent extraction.

For a binary system consisting of two liquid phases and one vapor phase in equilibrium, there is (according to the phase rule) but one degree of freedom. For a given pressure, the temperature and the compositions of all three phases are therefore fixed. On a temperature-composition diagram the points representing the states of the three phases in equilibrium fall on a horizontal line at T^*. In Fig. 13.15, points C and D represent the two liquid phases, and point E represents the vapor phase. If more of either species is added to a system whose overall composition lies between points C and D, and if the three-phase equilibrium pressure is maintained, the temperature and the compositions of the phases are unchanged. However, the relative amounts of the phases adjust themselves to reflect the change in overall composition of the system.

At temperatures above T^* in Fig. 13.15, the system may be a single liquid phase, two phases (liquid and vapor), or a single vapor phase, depending on the overall composition. In region α the system is a single liquid rich in species 2; in region β it is again a single liquid, but rich in species 1. In region α-V, liquid and vapor are in equilibrium. The states of the individual phases fall on lines AC and AE. In region β-V, liquid and vapor phases, described by lines BD and BE, also exist at equilibrium. Finally, in the region designated V, the system is a single vapor phase. Below the three-phase temperature T^*, the system is entirely liquid. Single liquid phases exist to the left of line CG and to the right of line DH. Mixtures having overall compositions within region α-β consist of two liquid phases of compositions given by the intersections of horizontal tie lines with lines CG and DH. As indicated in each section of the diagram, horizontal tie lines connect the compositions of phases in equilibrium.

When a vapor is cooled at constant pressure, it follows a path represented on Fig. 13.15 by a vertical line. Several such lines are shown. If one starts at point k, the vapor first reaches its dew point at line BE and then its bubble point at line BD, where condensation into single liquid phase β is complete. This is the same process that takes place when the species are completely miscible. If one starts at point n, no condensation of the vapor occurs until temperature T^* is reached. Then condensation occurs entirely at this temperature, producing the two liquid phases represented by points C and D. If one starts at an intermediate point m, the process is a combination of the two just described. After the dew point is reached the vapor, tracing a path along line BE, is in equilibrium with a

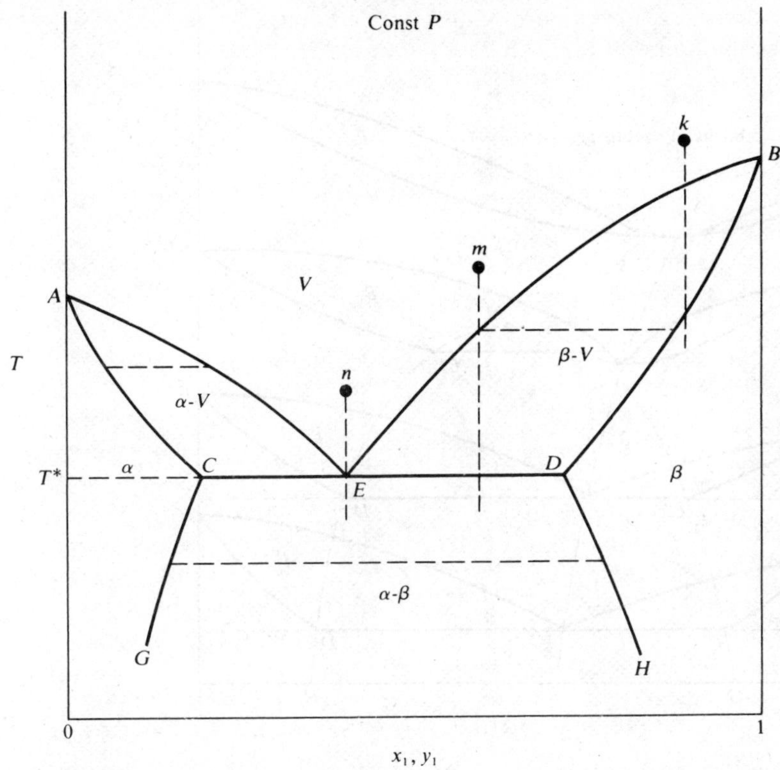

Figure 13.15 Temperature/composition diagram for a binary system of partially miscible liquids at constant pressure.

liquid tracing a path along line *BD*. However, at temperature T^* the vapor phase is at point *E*. All remaining condensation therefore occurs at this temperature, producing the two liquids of points *C* and *D*.

Figure 13.15 is drawn for a single constant pressure; equilibrium phase compositions, and hence the locations of the lines, change with pressure, but the general nature of the diagram is the same over a range of pressures. For the majority of systems the species become more soluble in one another as the temperature increases, as indicated by lines *CG* and *DH* of Fig. 13.15. If this diagram is drawn for successively higher pressures, the corresponding three-phase equilibrium temperatures increase, and lines *CG* and *DH* extend further and further until they meet at the liquid/liquid critical point *M*, as shown by Fig. 13.16. The temperature at which this occurs is known as the upper critical solution temperature, and at this temperature the two liquid phases become identical and merge into a single phase.

As the pressure increases, line *CD* becomes shorter and shorter (as indicated in Fig. 13.16 by lines $C'D'$ and $C''D''$), until at point *M* it diminishes to a

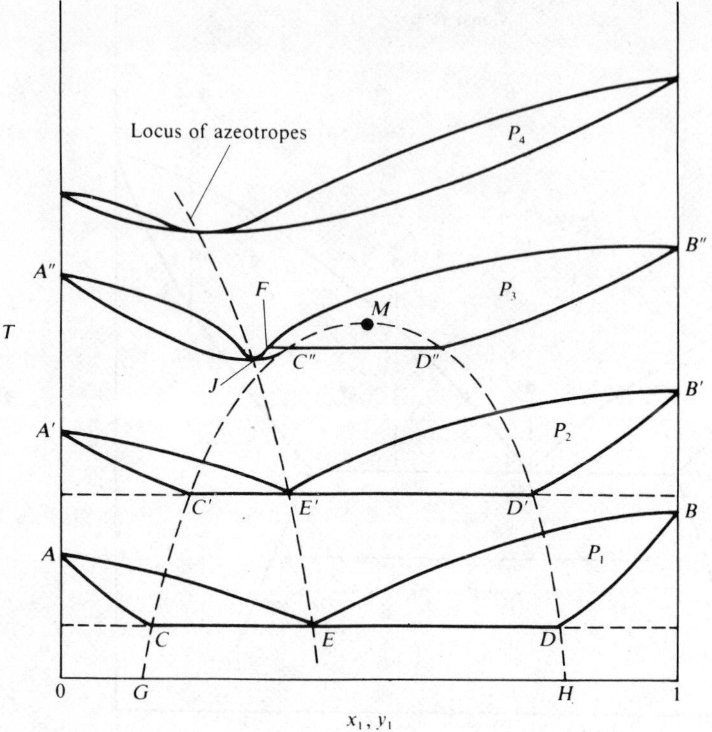

Figure 13.16 Temperature/composition diagram for several pressures.

differential length. For still higher pressures (P_4) the temperature is above the critical solution temperature, and there is but a single liquid phase. The diagram then represents two-phase VLE, and it has the form of Fig. 12.10*d*, exhibiting a minimum-boiling azeotrope.

There is an intermediate range of pressures for which the vapor phase in equilibrium with the two liquid phases has a composition that does not lie between the compositions of the two liquids. This is illustrated in Fig. 13.16 by the curves for P_3, which terminate at A'' and B''. The vapor in equilibrium with the two liquids at C'' and D'' is at point F. In addition the system exhibits an azeotrope, as indicated at point J.

Not all systems behave as described in the preceding paragraphs. Sometimes the upper critical solution temperature is never attained, because a vapor/liquid critical temperature is reached first. In other cases the liquid solubilities increase with a decrease in temperature. In this event a lower critical solution temperature exists, unless solid phases appear first. There are also systems which exhibit both upper and lower critical solution temperatures.

Figure 13.17 is the phase diagram drawn at *constant* T that corresponds to the constant-P diagram of Fig. 13.15. On it we identify the three-phase equilibrium

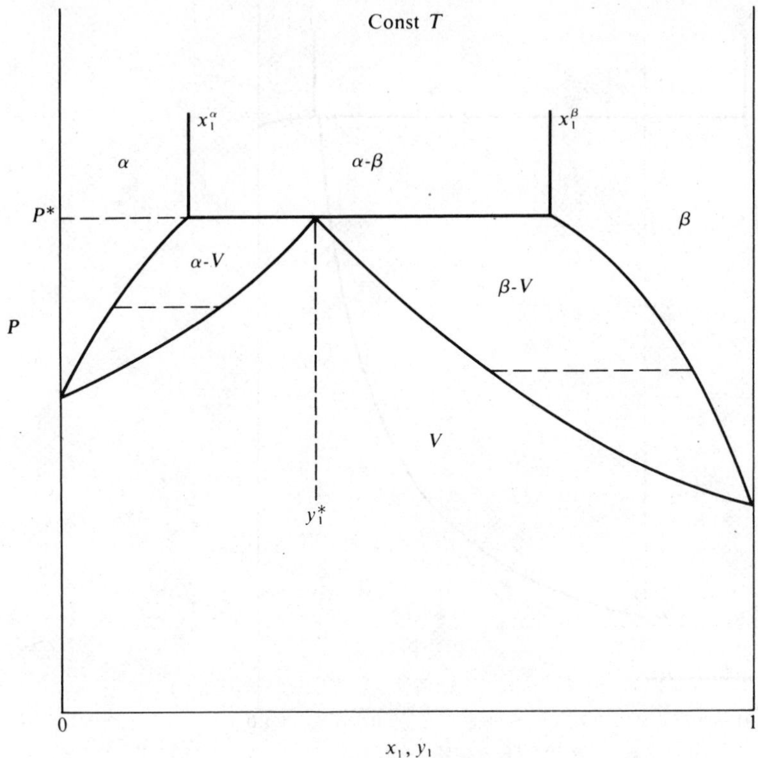

Figure 13.17 *Pxy* diagram for a system of partially miscible liquids at constant T.

pressure as P^*, the three-phase equilibrium vapor composition as y_1^*, and the compositions of the two liquid phases that contribute to the vapor/liquid/liquid equilibrium state as x_1^α and x_1^β. The phase boundaries separating the three liquid-phase regions are nearly vertical, because pressure has only a weak influence on liquid solubilities.

The compositions of the vapor and liquid phases in equilibrium for partially miscible systems are calculated in the same way as for miscible systems. In the regions where a single liquid is in equilibrium with its vapor, the general nature of Fig. 13.17 is not different in any essential way from that of Fig. 12.9*d*. Since limited miscibility implies highly nonideal behavior, any general assumption of liquid-phase ideality is excluded. Even a combination of Henry's law, valid for a species at infinite dilution, and Raoult's law, valid for a species as it approaches purity, is not very useful, because each approximates real behavior only for a very small composition range. Thus G^E is large, and its composition dependence is often not adequately represented by simple equations. However, the UNIFAC method (App. D) is suitable for estimation of activity coefficients.

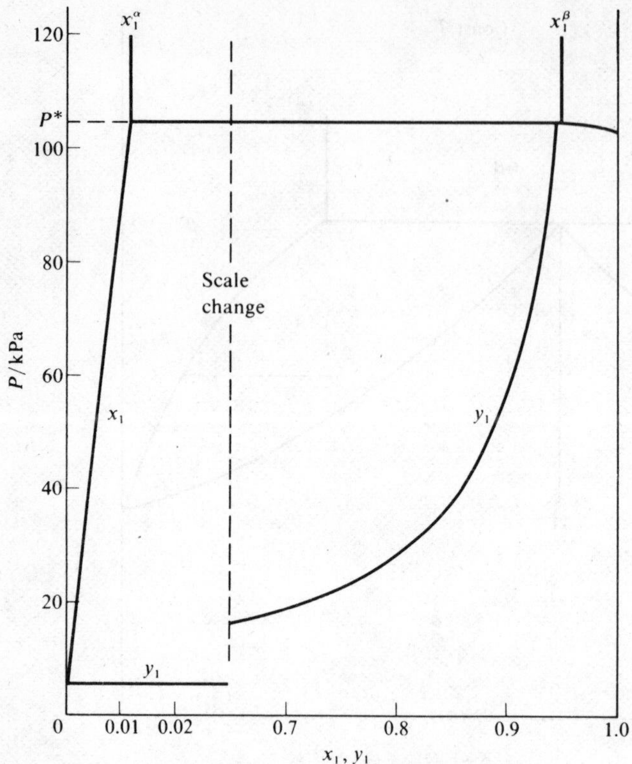

Figure 13.18 *Pxy* diagram for diethyl ether(1)/water(2) at 35°C.

As an example we consider the diethyl ether(1)/water(2) system at 35°C, for which careful measurements have been made.† The *Pxy* behavior of this system is shown by Fig. 13.18, where the very rapid rise in pressure with increasing liquid-phase ether concentration in the dilute-ether region is apparent. The three-phase pressure of $P^* = 104.6$ kPa is reached at an ether mole fraction of only 0.0117. Here, y_1 also increases very rapidly to its three-phase value of $y_1^* = 0.946$. In the dilute-water region, on the other hand, rates of change are quite small, as shown to an expanded scale in Fig. 13.19.

The curves in Figs. 13.18 and 13.19 provide an excellent correlation of the VLE data. They result from BUBL P calculations carried out as indicated in Fig. 12.12. The excess Gibbs energy and activity coefficients are here expressed as functions of liquid-phase composition by the 4-parameter modified Margules

† M. A. Villamañán, A. J. Allawi, and H. C. Van Ness, *J. Chem. Eng. Data*, **29**: 431, 1984.

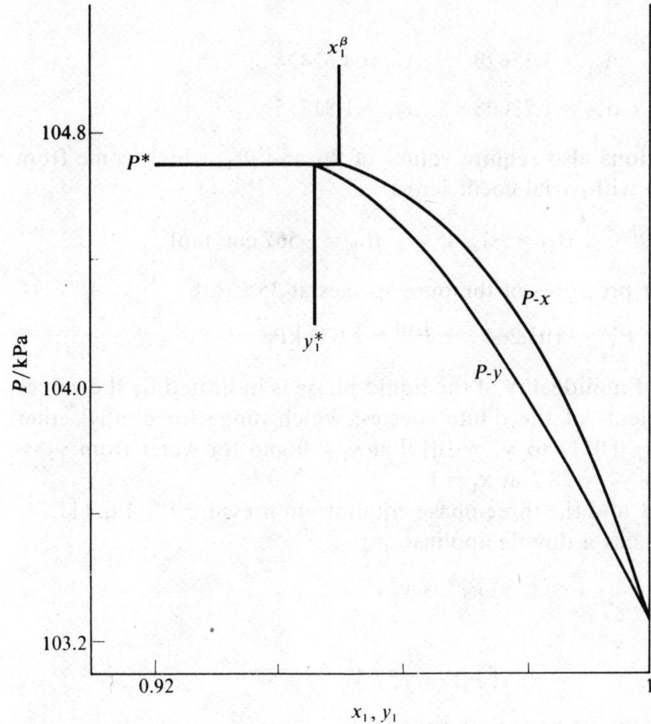

Figure 13.19 *Pxy* diagram for diethyl ether(1)/water(2) for the ether-rich region.

equations [see Eqs. (12.11) through (12.13)]:

$$\frac{G^E}{RT} = A_{12}x_1 + A_{21}x_2 - Q$$

$$\ln \gamma_1 = x_2^2\left[A_{12} + 2(A_{21} - A_{12})x_1 - Q - x_1\frac{dQ}{dx_1}\right]$$

$$\ln \gamma_2 = x_1^2\left[A_{21} + 2(A_{12} - A_{21})\,x_2 - Q + x_2\frac{dQ}{dx_1}\right]$$

where

$$Q = \frac{\alpha_{12}x_1\alpha_{21}x_2}{\alpha_{12}x_1 + \alpha_{21}x_2}$$

$$\frac{dQ}{dx_1} = \frac{\alpha_{12}\alpha_{21}(\alpha_{21}x_2^2 - \alpha_{12}x_1^2)}{(\alpha_{12}x_1 + \alpha_{21}x_2)^2}$$

and

$$A_{21} = 3.35629 \qquad A_{12} = 4.62424$$

$$\alpha_{12} = 3.78608 \qquad \alpha_{21} = 1.81775$$

The BUBL P calculations also require values of Φ_1 and Φ_2, which come from Eqs. (12.7) and (12.8) with virial coefficients:

$$B_{11} = -996 \qquad B_{22} = -1{,}245 \qquad B_{12} = -567 \text{ cm}^3 \text{ mol}^{-1}$$

In addition, the vapor pressures of the pure species at 35°C are

$$P_1^{\text{sat}} = 103.264 \qquad P_2^{\text{sat}} = 5.633 \text{ kPa}$$

The high degree of nonideality of the liquid phase is indicated by the values of the activity coefficients of the dilute species, which range for diethyl ether from $\gamma_1 = 81.8$ at $x_1^\alpha = 0.0117$ to $\gamma_1^\infty = 101.9$ at $x_1 = 0$ and for water from $\gamma_2 = 19.8$ at $x_1^\beta = 0.9500$ to $\gamma_2^\infty = 28.7$ at $x_1 = 1$.

For temperature T and the three-phase equilibrium pressure P^*, Eq. (11.74) for low-pressure VLE has a double application:

$$x_i^\alpha \gamma_i^\alpha P_i^{\text{sat}} = y_i^* P^*$$

and

$$x_i^\beta \gamma_i^\beta P_i^{\text{sat}} = y_i^* P^*$$

Thus for a binary system we have four equations:

$$x_1^\alpha \gamma_1^\alpha P_1^{\text{sat}} = y_1^* P^* \qquad (A)$$

$$x_1^\beta \gamma_1^\beta P_1^{\text{sat}} = y_1^* P^* \qquad (B)$$

$$x_2^\alpha \gamma_2^\alpha P_2^{\text{sat}} = y_2^* P^* \qquad (C)$$

$$x_2^\beta \gamma_2^\beta P_2^{\text{sat}} = y_2^* P^* \qquad (D)$$

All of these equations are correct, but two of them are preferred over the others. Consider the expressions for $y_1^* P^*$:

$$x_1^\alpha \gamma_1^\alpha P_1^{\text{sat}} = x_1^\beta \gamma_1^\beta P_1^{\text{sat}} = y_1^* P^*$$

For the case of two species that approach complete immiscibility,

$$x_1^\alpha \to 0 \qquad \gamma_1^\alpha \to (\gamma_1^\alpha)^\infty \qquad x_1^\beta \to 1 \qquad \gamma_1^\beta \to 1$$

Thus

$$(0)(\gamma_1^\alpha)^\infty P_1^{\text{sat}} = P_1^{\text{sat}} = y_1^* P^*$$

This equation implies that $(\gamma_1^\alpha)^\infty \to \infty$; a similar derivation shows that $(\gamma_2^\beta)^\infty \to \infty$. Thus Eqs. (B) and (C), which include neither of these quantities, are chosen as the more useful expressions. They may be added to give the three-phase pressure,

$$P^* = x_1^\beta \gamma_1^\beta P_1^{\text{sat}} + x_2^\alpha \gamma_2^\alpha P_2^{\text{sat}} \qquad (13.58)$$

In addition, the three-phase vapor composition is

$$y_1^* = \frac{x_1^\beta \gamma_1^\beta P_1^{sat}}{P^*} \tag{13.59}$$

Thus in the case of the diethyl ether(1)/water(2) system,

$$\gamma_1^\beta = 1.0095 \qquad \text{and} \qquad \gamma_2^\alpha = 1.0013$$

These values are in marked contrast to those cited earlier for the dilute species. They allow calculation of P^* and y_1^* by Eqs. (13.58) and (13.59):

$$P^* = (0.9500)(1.0095)(103.264) + (0.9883)(1.0013)(5.633)$$
$$= 104.6 \text{ kPa}$$

and

$$y_1^* = \frac{(0.9500)(1.0095)(103.264)}{104.6} = 0.946$$

Although no two liquids are totally immiscible, this condition is so closely approached in some instances that the assumption of complete immiscibility does not lead to appreciable error. The phase characteristics of an immiscible system are illustrated by the temperature/composition diagram of Fig. 13.20. This

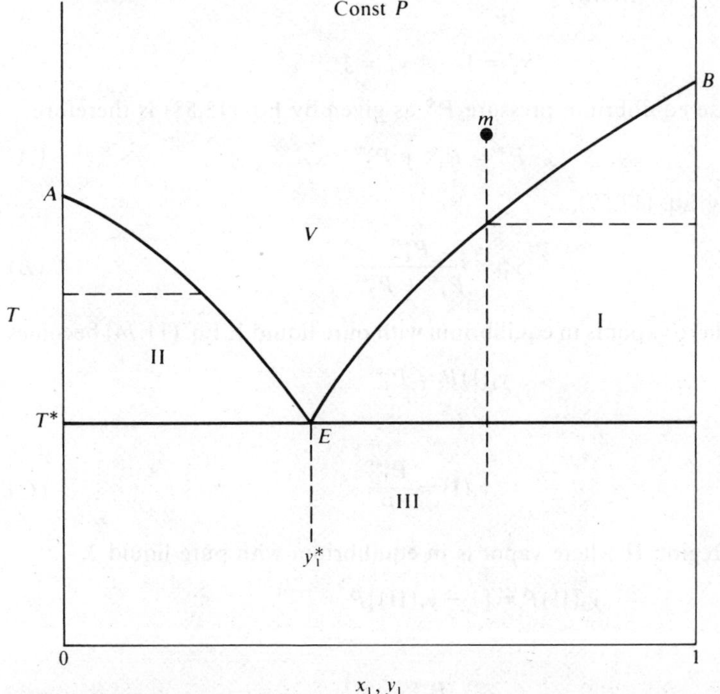

Figure 13.20 Temperature/composition diagram for a binary system of immiscible liquids.

diagram is a special case of Fig. 13.15 wherein phase α is pure species 2 and phase β is pure species 1. Thus lines *ACG* and *BDH* of Fig. 13.15 have in Fig. 13.20 become vertical lines at $x_1 = 0$ and $x_1 = 1$.

In region I, vapor phases with compositions represented by line *BE* are in equilibrium with pure liquid 1. Similarly, in region II, vapor phases whose compositions lie along line *AE* are in equilibrium with pure liquid 2. Liquid/liquid equilibrium exists in region III, where the two phases are pure liquid 1 and pure liquid 2. If one cools a vapor mixture starting at point *m*, the constant-composition path is represented by the vertical line shown in the figure. At the dew point, where this line crosses line *BE*, pure liquid 1 begins to condense. Further reduction in temperature toward T^* causes continued condensation of pure liquid 1; the vapor-phase composition progresses along line *BE* until it reaches point *E*. Here, the remaining vapor condenses at temperature T^*, producing two liquid phases, one of pure species 1 and the other of pure species 2. A similar process, carried out to the left of point *E*, is the same, except that pure liquid 2 condenses initially. The constant-temperature phase diagram for an immiscible system is represented by Fig. 13.21.

Numerical calculations for immiscible systems are particularly simple, because of the following equalities:

$$x_1^\alpha = 0 \qquad \gamma_2^\alpha = 1$$

and

$$x_1^\beta = 1 \qquad \gamma_1^\beta = 1$$

The three-phase equilibrium pressure P^* as given by Eq. (13.58) is therefore:

$$P^* = P_1^{\text{sat}} + P_2^{\text{sat}} \tag{A}$$

from which, by Eq. (13.59),

$$y_1^* = \frac{P_1^{\text{sat}}}{P_1^{\text{sat}} + P_2^{\text{sat}}} \tag{B}$$

For region I where vapor is in equilibrium with pure liquid 1, Eq. (11.74) becomes

$$y_1(\text{I})P = P_1^{\text{sat}}$$

or

$$y_1(\text{I}) = \frac{P_1^{\text{sat}}}{P} \tag{C}$$

Similarly, for region II where vapor is in equilibrium with pure liquid 2,

$$y_2(\text{II})P = [1 - y_1(\text{II})]P = P_2^{\text{sat}}$$

or

$$y_1(\text{II}) = 1 - \frac{P_2^{\text{sat}}}{P} \tag{D}$$

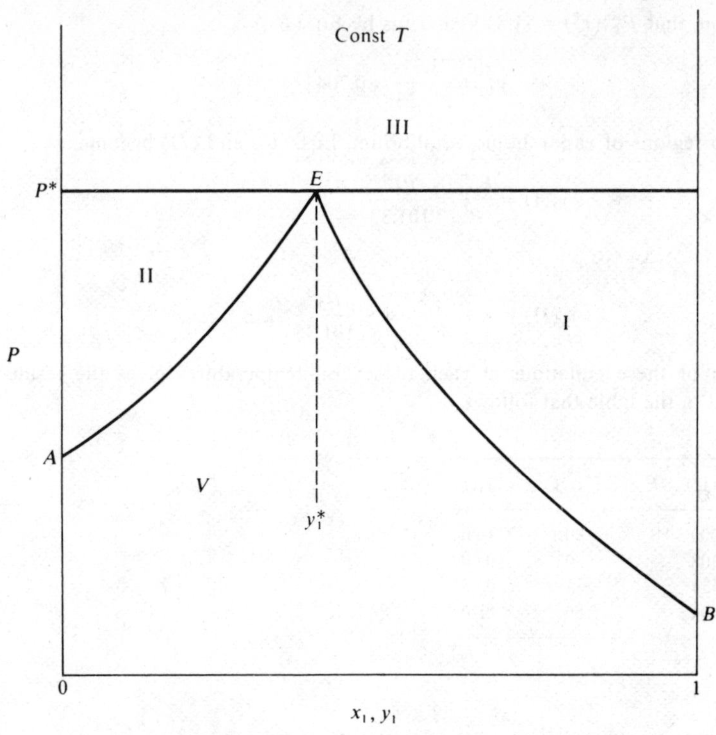

Figure 13.21 *Pxy* diagram for a binary system of immiscible liquids.

Example 13.10 Prepare a table of temperature/composition data for the benzene(1)/water(2) system at a pressure of 101.33 kPa (1 atm) from the following vapor-pressure data:

$t/°C$	P_1^{sat}/kPa	P_2^{sat}/kPa	$P_1^{sat} + P_2^{sat}/\text{kPa}$
60	52.22	19.92	72.14
70	73.47	31.16	104.63
75	86.40		
80	101.05	47.36	
90	136.14	70.11	
100	180.04	101.33	

SOLUTION The three-phase equilibrium temperature t^* is determined from Eq. (*A*), here written as:

$$P = P_1^{sat} + P_2^{sat} = 101.33 \text{ kPa}$$

The last column of the preceding table shows that t^* lies between 60 and 70°C. By interpolation, we find that $t^* = 69.0°C$, and at this temperature we find, again by interpolation, that $P_1^{sat}(t^*) = 71.31$ kPa. Thus by Eq. (B),

$$y_1^* = \frac{71.31}{101.33} = 0.704$$

For the two regions of vapor/liquid equilibrium, Eqs. (C) and (D) become

$$y_1(\mathrm{I}) = \frac{P_1^{sat}}{P} = \frac{P_1^{sat}}{101.33}$$

and

$$y_1(\mathrm{II}) = 1 - \frac{P_2^{sat}}{P} = 1 - \frac{P_2^{sat}}{101.33}$$

Application of these equations at each of several temperatures gives the results summarized in the table that follows.

t/°C	y_1(II)	t/°C	y_1(I)
100	0.000	80.1	1.000
90	0.308	80	0.997
80	0.533	75	0.853
69	0.704	69	0.704

PROBLEMS

13.1 With reference to Example 13.2,

(a) Apply Eq. (11.2) to Eq. (A) to verify Eqs. (B) and (C).

(b) Show that Eqs. (B) and (C), when combined in accord with Eq. (11.5), regenerate Eq. (A).

(c) Show that Eqs. (B) and (C) satisfy Eq. (11.8), the Gibbs/Duhem equation.

(d) Show that at constant T and P

$$(d\bar{H}_1/dx_1)_{x_1=1} = (d\bar{H}_2/dx_1)_{x_1=0} = 0.$$

(e) Plot values of H, $\bar{H}_1$, and $\bar{H}_2$, calculated by Eqs. (A), (B), and (C), vs. x_1. Label points H_1, H_2, $\bar{H}_1^\infty$, and $\bar{H}_2^\infty$, and show their values.

13.2 The molar volume (cm^3 mol^{-1}) of a binary liquid mixture at T and P is given by

$$V = 90x_1 + 50x_2 + (6x_1 + 9x_2)x_1x_2$$

For the given T and P,

(a) Find expressions for the partial molar volumes of species 1 and 2.

(b) Show that when these expressions are combined in accord with Eq. (11.5) the given equation for V is recovered.

(c) Show that these expressions satisfy Eq. (11.8), the Gibbs/Duhem equation.

(d) Show that $(d\bar{V}_1/dx_1)_{x_1=1} = (d\bar{V}_2/dx_1)_{x_1=0} = 0$.

(e) Plot values of V, $\bar{V}_1$, and $\bar{V}_2$ calculated by the given equation for V and by the equations developed in (a) vs. x_1. Label points V_1, V_2, $\bar{V}_1^\infty$, and $\bar{V}_2^\infty$, and show their values.

13.3 Given that

$$M^E = x_1x_2(A_0 + A_1z + A_2z^2)$$

where $z \equiv x_1 - x_2$, derive expressions for $\bar{M}_1^E$ and $\bar{M}_2^E$. Combine the resulting expressions to show that the given equation is recovered. What are the limiting values of $\bar{M}_1^E$, $\bar{M}_2^E$, and M^E/x_1x_2 as $x_1 \to 0$ and as $x_1 \to 1$?

13.4 The excess Gibbs energy of a binary liquid mixture at T and P is given by

$$G^E/RT = (-1.2x_1 - 1.5x_2)x_1x_2$$

For the given T and P,
(*a*) Find expressions for $\ln \gamma_1$ and $\ln \gamma_2$.
(*b*) Show that when these expressions are combined in accord with Eq. (11.69) the given equation for G^E/RT is recovered.
(*c*) Show that these expressions satisfy Eq. (11.70), the Gibbs/Duhem equation.
(*d*) Show that $(d \ln \gamma_1/dx_1)_{x_1=1} = (d \ln \gamma_2/dx_1)_{x_1=0} = 0$.
(*e*) Plot values of G^E/RT, $\ln \gamma_1$, and $\ln \gamma_2$ calculated by the given equation for G^E/RT and by the equations developed in (*a*) vs. x_1. Label points $\ln \gamma_1^\infty$, and $\ln \gamma_2^\infty$, and show their values.

13.5 With reference to Example 13.3,
(*a*) Apply Eq. (11.2) [with $M = H^E$ and $\bar{M}_i = \bar{H}_i^E$] to Eq. (*C*) to verify the expressions for $\bar{H}_1^E$ and $\bar{H}_2^E$.
(*b*) Show that these expressions, when combined in accord with Eq. (11.5), regenerate Eq. (*C*).
(*c*) Show that these expressions satisfy Eq. (11.8), the Gibbs/Duhem equation.
(*d*) Show that at constant T and P $(d\bar{H}_1^E/dx_1)_{x_1=1} = (d\bar{H}_2^E/dx_1)_{x_1=0} = 0$.
(*e*) Plot values of H^E, $\bar{H}_1^E$, and $\bar{H}_2^E$, calculated by the given equation for H^E and by the equations developed in (*a*) vs. x_1. Label points $(\bar{H}_1^E)^\infty$, and $(\bar{H}_2^E)^\infty$, and show their values.

13.6 If the partial volume of species 1 in a binary solution at constant T and P is given by

$$\bar{V}_1 = V_1 + \alpha x_2^2$$

find the corresponding equation for $\bar{V}_2$. What equation for V is consistent with these equations for the partial volumes? What are the corresponding equations for V^E, $\bar{V}_1^E$, and $\bar{V}_2^E$?

13.7 The following equations have been proposed to represent activity-coefficient data for a system at fixed T and P:

$$\ln \gamma_1 = x_2^2(0.5 + 2x_1)$$
$$\ln \gamma_2 = x_1^2(1.5 - 2x_2)$$

Do these equations satisfy the Gibbs/Duhem equation? Determine an expression for G^E/RT for the system.

13.8 The following equations have been proposed to represent activity-coefficient data for a system at fixed T and P:

$$\ln \gamma_1 = Ax_2^2 + Bx_2^2(3x_1 - x_2)$$
$$\ln \gamma_2 = Ax_1^2 + Bx_1^2(x_1 - 3x_2)$$

Do these equations satisfy the Gibbs/Duhem equation? Determine an expression for G^E/RT for the system.

13.9 The following equations have been proposed to represent activity-coefficient data for a system at fixed T and P:

$$\ln \gamma_1 = x_2(a + bx_2)$$
$$\ln \gamma_2 = x_1(a + bx_1)$$

(*a*) Combine these equations in accord with Eq. (11.63) to determine an expression for G^E/RT.
(*b*) Apply Eq. (11.62) to the equation of part (*a*) to develop equations for $\ln \gamma_1$ and $\ln \gamma_2$. Are the given equations for these quantities regenerated?
(*c*) Do the given equations satisfy the Gibbs/Duhem equation?
(*d*) How are the results of parts (*b*) and (*c*) connected?

13.10 For a ternary solution at constant T and P, the composition dependence of molar property M is given by

$$M = x_1M_1 + x_2M_2 + x_3M_3 + x_1x_2x_3C$$

where M_1, M_2, and M_3 are the values of M for pure species 1, 2, and 3, and C is a parameter independent of composition. Determine an expression for $\bar{M}_1$ by application of Eq. (11.2).

13.11 For a particular binary system at constant T and P,

$$H^E = \sum_i x_iA_i(1 - x_i) \qquad (i = 1, 2)$$

Derive expressions for $\bar{H}_1^E$ and $\bar{H}_2^E$. Combine these two equations to show that the original equation is recovered. [*Note:* $\bar{H}_i^E$ cannot be found "by inspection;" it is not equal to $A_i(1 - x_i)$.]

13.12 For a particular binary system at constant T and P, the molar enthalpies of mixtures are represented by the equation

$$H = x_1(a_1 + b_1x_1) + x_2(a_2 + b_2x_2)$$

where the a_i and b_i are constants. Determine an expression for $\bar{H}_1$.

(*Note:* $\bar{H}_1$ cannot be found "by inspection"; it is not equal to $a_1 + b_1x_1$.)

13.13 At 25°C and atmospheric pressure the excess volumes of binary liquid mixtures of species 1 and 2 are given by the equation

$$V^E = x_1x_2(30x_1 + 50x_2)$$

where V^E is in $cm^3\ mol^{-1}$. At the same conditions, $V_1 = 120\ cm^3\ mol^{-1}$ and $V_2 = 150\ cm^3\ mol^{-1}$. Determine the partial molar volumes $\bar{V}_1$ and $\bar{V}_2$ for an equimolar mixture of species 1 and 2 at the given conditions.

13.14 Excess volumes ($cm^3\ mol^{-1}$) for the system ethanol(1)/methyl butyl ether(2) at 25°C are given by the equation

$$V^E = x_1x_2[-1.026 + 0.220(x_1 - x_2)]$$

If $V_1 = 58.63$ and $V_2 = 118.46\ cm^3\ mol^{-1}$, what volume of mixture is formed when 1,000 cm^3 each of pure species 1 and 2 are mixed at 25°C?

13.15 A vapor phase containing pure species 1 is in equilibrium with a binary liquid mixture containing both species 1 and a nonvolatile species 2. At $P = 1.5$ bar and $T = 425$ K, the liquid phase is found to contain 12 mol % of species 1. At 425 K, $P_1^{sat} = 10.6$ bar, and the second virial coefficient is $B_{11} = -450\ cm^3\ mol^{-1}$.

(*a*) From the given data, determine a good estimate of the activity coefficient γ_1 of species 1 in the liquid phase.

(*b*) Why is it not possible from the given information to determine a value for the activity coefficient γ_2?

(*c*) Show that γ_2 can be determined, given the additional information that the excess Gibbs energy of the liquid phase is represented by an equation of the form $G^E/RT = Bx_1x_2$.

13.16 If $LiCl{\cdot}3H_2O(s)$ and $H_2O(l)$ are mixed isothermally at 25°C to form a solution containing 8 moles of water for each mole of LiCl, what is the heat effect per mole of solution?

13.17 If a liquid solution of HCl in water, containing 1 mol of HCl and 3 mol of H_2O, absorbs an additional 1 mol of HCl(g) at the constant temperature of 25°C, what is the heat effect?

13.18 What is the heat effect when 30 kg of LiCl(s) is added to 150 kg of an aqueous solution containing 15-wt-% LiCl in an isothermal process at 25°C?

13.19 A mass of 18 kg s^{-1} of $Cu(NO_3)_2{\cdot}6H_2O$ along with 15 kg s^{-1} of water, both at 25°C, are fed to a tank where mixing takes place. The resulting solution passes through a heat exchanger which adjusts its temperature to 25°C. What is the rate of heat transfer in the exchanger?

Data: For $Cu(NO_3)_2$, $\Delta H^\circ_{f_{298}} = -302.9$ kJ

For $Cu(NO_3)_2{\cdot}6H_2O$, $\Delta H^\circ_{f_{298}} = -2{,}110.8$ kJ

The heat of solution of 1 mol of $Cu(NO_3)_2$ in water at 25°C is -47.84 kJ, independent of $\tilde{n}$ for values of interest here.

13.20 If a liquid solution of HCl in water, containing 1 mol of HCl and 10 mol of H_2O, absorbs an additional 1 mol of HCl(g) at the constant temperature of 25°C, what is the heat effect?

13.21 A liquid solution of LiCl in water at 25°C contains 1 mol of LiCl and 12 mol of water. If an additional 1 mol of LiCl(s) is dissolved isothermally in this solution, what is the heat effect?

13.22 It is required to produce an aqueous LiCl solution by mixing $LiCl \cdot 2H_2O(s)$ with water. The mixing is to occur *both* adiabatically and without change in temperature at 25°C. Determine the mole fraction of LiCl in the final solution.

13.23 Data from the Bureau of Standards (*J. Phys. Chem. Ref. Data*, Vol. 11, suppl. 2, 1982) include the following heats of formation for 1 mol of $CaCl_2$ in water at 25°C:

$CaCl_2$ in 10 mol H_2O	−862.74 kJ
$CaCl_2$ in 15 mol H_2O	−867.85 kJ
$CaCl_2$ in 20 mol H_2O	−870.06 kJ
$CaCl_2$ in 25 mol H_2O	−871.07 kJ
$CaCl_2$ in 50 mol H_2O	−872.91 kJ
$CaCl_2$ in 100 mol H_2O	−873.82 kJ
$CaCl_2$ in 300 mol H_2O	−874.79 kJ
$CaCl_2$ in 500 mol H_2O	−875.13 kJ
$CaCl_2$ in 1,000 mol H_2O	−875.54 kJ

From these data prepare a plot of $\Delta\widetilde{H}$, the heat of solution at 25°C of $CaCl_2$ in water, vs. $\tilde{n}$, the mole ratio of water to $CaCl_2$.

13.24 Solid $CaCl_2 \cdot 6H_2O$ is mixed with water in a continuous process to form a solution containing 20-wt-% $CaCl_2$. What is the heat effect per kilogram of solution formed if the temperature is constant at 25°C? (*Note:* Data for $CaCl_2$ solutions are given in the preceding problem.)

13.25 Consider a plot of $\Delta\widetilde{H}$, the heat of solution based on one mole of solute (species 1), vs. $\tilde{n}$, the moles of solvent per mole of solute, at constant T and P. Figure 13.7 is an example of such a plot, except that the plot considered here has a linear rather than logarithmic scale along the abscissa. Let a tangent drawn to the $\Delta\widetilde{H}$ vs. $\tilde{n}$ curve intercept the ordinate at point I.

(*a*) Prove that the slope of the tangent at a particular point is equal to the partial excess enthalpy of the solvent in a solution with the composition represented by $\tilde{n}$; that is, prove that

$$\frac{d\Delta\widetilde{H}}{d\tilde{n}} = \bar{H}_2^E$$

(*b*) Prove that the intercept I equals the partial excess enthalpy of the solute in the same solution; i.e., prove that

$$I = \bar{H}_1^E$$

13.26 If the heat of mixing at temperature t_0 is ΔH_0 and if the heat of mixing of the same solution at temperature t is ΔH, show that the two heats of mixing are related by

$$\Delta H = \Delta H_0 + \int_{t_0}^{t} \Delta C_P \, dt$$

where ΔC_P is the heat-capacity change of mixing, defined in accord with Eq. (13.39).

13.27 What is the heat effect when 175(lb$_m$) of H_2SO_4 is mixed with 400(lb$_m$) of an aqueous solution containing 30-wt-% H_2SO_4 in an isothermal process at 120(°F)?

13.28 For a 60-wt-% aqueous solution of H_2SO_4 at 100(°F), what is the excess enthalpy H^E in (Btu)(lb$_m$)$^{-1}$?

13.29 A mass of 500(lb$_m$) of 40-wt-% aqueous NaOH solution at 150(°F) is mixed with 250(lb$_m$) of 15-wt-% solution at 200(°F).

(*a*) What is the heat effect if the final temperature is 100(°F)?

(*b*) If the mixing is adiabatic, what is the final temperature?

13.30 A single-effect evaporator concentrates a 15-wt-% aqueous solution of H_2SO_4 to 60%. The feed rate is $20(\text{lb}_m)(\text{s})^{-1}$, and the feed temperature is 100(°F). The evaporator is maintained at an absolute pressure of 1.5(psia), at which pressure the boiling point of 60% H_2SO_4 is 176(°F). What is the heat-transfer rate in the evaporator?

13.31 What temperature results when sufficient NaOH(*s*) at 68(°F) is dissolved adiabatically in a 15-wt-% aqueous NaOH solution, originally at 100(°F), to bring the concentration up to 45%?

13.32 What is the heat effect when sufficient $SO_3(l)$ at 25°C is reacted with H_2O at 25°C to give a 60-wt-% H_2SO_4 solution at 80°C?

13.33 A mass of $100(\text{lb}_m)$ of 10-wt-% solution of H_2SO_4 in water at 180(°F) is mixed at atmospheric pressure with $200(\text{lb}_m)$ of 85-wt-% H_2SO_4 at 77(°F). During the process heat in the amount of 15,000(Btu) is transferred from the system. Determine the temperature of the product solution.

13.34 An insulated tank, open to the atmosphere, contains $1{,}000(\text{lb}_m)$ of 50-wt-% sulfuric acid at 80(°F). It is heated to 200(°F) by injection of live saturated steam at 1(atm), which fully condenses in the process. How much steam is required, and what is the final concentration of H_2SO_4 in the tank?

13.35 For a 20-wt-% aqueous solution of H_2SO_4 at 80(°F), what is the heat of mixing ΔH in $(\text{Btu})(\text{lb}_m)^{-1}$?

13.36 If pure liquid H_2SO_4 at 80(°F) is added adiabatically to pure liquid water at 80(°F) to form a 20-wt-% solution, what is the final temperature of the solution?

13.37 A liquid solution containing 1(lb mol) H_2SO_4 and 10(lb mol) H_2O at 77(°F) absorbs 1(lb mol) of $SO_3(g)$, also at 77(°F), forming a more concentrated sulfuric acid solution. If the process occurs isothermally, determine the heat transferred.

13.38 Determine the heat of mixing ΔH of sulfuric acid in water and the partial specific enthalpies of H_2SO_4 and H_2O for a solution containing 70-wt-% H_2SO_4 at 140(°F).

13.39 It is proposed to cool a stream of 80-wt-% sulfuric acid solution at 160(°F) by diluting it with chilled water at 40(°F). Determine the amount of water that must be added to $1(\text{lb}_m)$ of 80-% acid before cooling below 160(°F) actually occurs.

13.40 The following liquids, all at atmospheric pressure and 100(°F), are mixed: $20(\text{lb}_m)$ of pure water, $30(\text{lb}_m)$ of pure sulfuric acid, and $50(\text{lb}_m)$ of 20-wt-% sulfuric acid.

(*a*) How much heat is liberated if mixing is isothermal at 100(°F)?

(*b*) The mixing process is carried out in two steps: First, the pure sulfuric acid and the 20-% solution are mixed, and the total heat of part (*a*) is extracted; second, the pure water is added adiabatically. What is the enthalpy of the intermediate solution formed in the first step?

13.41 Saturated steam at 30(psia) is throttled to 1(atm) and mixed adiabatically with (and condensed by) 40-wt-% sulfuric acid at 100(°F) in a flow process that raises the temperature of the acid to 180(°F). How much steam is required for each pound *mass* of entering acid, and what is the concentration of the hot acid?

13.42 A batch of 35-wt-% NaOH solution in water at atmospheric pressure and 100(°F) is heated in an insulated tank by injection of live steam drawn through a valve from a line containing saturated steam at 30(psia). The process is stopped when the NaOH solution reaches a concentration of 32.5 wt %. At what temperature does this occur?

13.43 A large quantity of very dilute aqueous NaOH solution is neutralized by addition of the stoichiometric amount of a 20-mol-% aqueous HCl solution. The neutralization reaction, which goes to completion and which yields NaCl as a product, occurs at the constant temperature of 25°C. Determine a good estimate of the heat effect per mole of NaOH. The heat of solution of 1 mol of NaCl(*s*) in an infinite amount of water is 3.88 kJ, and the heat of solution of 1 mol of NaOH in an infinite amount of water is −44.50 kJ.

13.44 A large quantity of very dilute aqueous HCl solution is neutralized by addition of the stoichiometric amount of a 15-wt-% aqueous NaOH solution. The neutralization reaction, which goes to completion and yields NaCl as product, occurs at the constant temperature of 25°C. Determine a good estimate of the heat effect per mole of HCl. The heat of solution of 1 mol of NaCl(*s*) in an infinite amount of water is 3.88 kJ.

13.45 A vapor mixture of methanol(1) and water(2) containing 56-mol-% methanol enters a condenser at 101.33 kPa at its dew point of 82.85°C. It is completely condensed and leaves the condenser at its bubble point of 72.05°C. How much heat must be transferred in the condenser for each mole of mixture condensed? The latent heat of vaporization of methanol at its normal boiling point of 64.7°C is 35,228 J mol^{-1}. The heat of mixing of a liquid mixture containing 56-mol-% methanol at 72°C is estimated as -500 J mol^{-1}.

13.46 Toluene(1) and water(2) are essentially immiscible as liquids. Determine the dew-point temperatures and the compositions of the first drops of liquid formed when vapor mixtures of these species with mole fraction $z_1 = 0.23$ and $z_1 = 0.77$ are cooled at the constant pressure of 101.33 kPa. What is the bubble-point temperature and the composition of the last drop of vapor in each case? The vapor pressure of toluene is given by the Antoine equation:

$$\ln P_1^{\text{sat}}/\text{kPa} = 14.00976 - \frac{3{,}103.010}{t/°\text{C} + 219.787}$$

13.47 n-Heptane(1) and water(2) are essentially immiscible as liquids. A vapor mixture containing 65-mol-% water at 100°C and 101.33 kPa is cooled slowly at constant pressure until condensation is complete. Construct a plot for the process showing temperature vs. the equilibrium mole fraction of heptane in the residual vapor. For n-heptane,

$$\ln P_1^{\text{sat}}/\text{kPa} = 13.87770 - \frac{2{,}918.738}{t/°\text{C} + 216.796}$$

13.48 A 90-wt-% aqueous H_2SO_2 solution at 25°C is added continuously to a tank containing 4,000 kg of pure water also at 25°C over a period of 6 hours. The final concentration of acid in the tank is 50%. The contents of the tank are cooled continuously to maintain a constant temperature of 25°C. However, the cooling system is designed for a constant rate of heat transfer, and this requires the addition of the acid at a variable rate. Determine the instantaneous 90-%-acid rate as a function of time, and plot this rate (kg s^{-1}) vs. time. Heat-of-mixing data for the $H_2SO_4(1)/H_2O(2)$ system at 25°C are as follows (x_1 = mass fraction H_2SO_4):

x_1	$\Delta H/\text{J g}^{-1}$
0.10	−73.27
0.20	−144.21
0.30	−208.64
0.40	−262.83
0.50	−302.84
0.60	−323.31
0.70	−320.98
0.80	−279.58
0.85	−237.25
0.90	−178.87
0.95	−100.71

13.49 Consider VLE in systems for which the excess Gibbs energy of the liquid phase at a particular temperature is represented by $G^E/RT = Bx_1x_2$. For what values of B is phase splitting predicted? That is, for what values of B is the stability requirement of Eq. (13.54) violated for some value of x_1 in the range $0 < x_1 < 1$? Make the usual assumptions for low-pressure VLE.

13.50 Consider a binary system of species 1 and 2 in which the liquid phase exhibits partial miscibility. In the region of miscibility, the excess Gibbs energy at a particular temperature is expressed by the equation,

$$G^E/RT = 2.25x_1x_2$$

In addition, the vapor pressures of the pure species are

$$P_1^{\text{sat}} = 75 \text{ kPa} \qquad \text{and} \qquad P_2^{\text{sat}} = 110 \text{ kPa}$$

Making the usual assumptions for low-pressure VLE, prepare a Pxy diagram for this system at the given temperature.

13.51 The system water(1)/n-pentane(2)/n-hexane(3) exists as a vapor at 101.33 kPa and 100°C with mole fractions $z_1 = 0.45$, $z_2 = 0.25$, $z_3 = 0.30$. The system is slowly cooled at constant pressure until it is completely condensed into a water phase and a hydrocarbon phase. Assuming that the two liquid phases are immiscible, that the vapor phase is an ideal gas, and that the hydrocarbons obey Raoult's law, determine:

(a) The dew-point temperature of the mixture and the composition of the first condensate.

(b) The temperature at which the second liquid phase first appears and its initial composition.

(c) The bubble-point temperature and the composition of the last bubble of vapor.

Vapor pressures of the hydrocarbons are given by Antoine equations

$$\ln P_2^{\text{sat}}/\text{kPa} = 13.8183 - \frac{2{,}477.07}{t/°\text{C} + 233.21}$$

$$\ln P_3^{\text{sat}}/\text{kPa} = 13.8216 - \frac{2{,}697.55}{t/°\text{C} + 224.37}$$

13.52 Repeat the preceding problem for a system composition of $z_1 = 0.32$, $z_2 = 0.41$, $z_3 = 0.27$.

CHAPTER

FOURTEEN

THERMODYNAMIC PROPERTIES AND VLE FROM EQUATIONS OF STATE

As discussed in Chap. 3, equations of state provide concise descriptions of the *PVT* behavior for pure fluids. The only equation of state that we have used extensively is the two-term virial equation,

$$Z = 1 + \frac{BP}{RT} \tag{3.31}$$

suited to gases at low pressures. When put into reduced form for pure gases, this equation leads to generalized correlations for Z [Eqs. (3.46) and (3.47)], H^R [Eq. (6.58)], S^R [Eq. (6.59)], and $\ln \phi$ [Eq. (11.43)]. Moreover, when extended to gas mixtures, it yields a general expression for $\ln \hat{\phi}_k$ [Eq. (11.48)], which is useful for low-pressure VLE calculations [Eq. (12.4)].

Equations of state have a much wider application. In this chapter we first present a general treatment of the calculation of thermodynamic properties of fluids and fluid mixtures from equations of state. Then the use of an equation of state for VLE calculations is described. For this, the fugacity of each species in both liquid and vapor phases must be determined. These calculations are illustrated with the Redlich/Kwong equation. Provided that the equation of state is suitable, such calculations can extend to high pressures.

14.1 PROPERTIES OF FLUIDS FROM THE VIRIAL EQUATIONS OF STATE

Equations of state written for fluid mixtures are exactly the same as the equations of state presented for pure fluids in Secs. 3.4 and 3.5. The additional information needed for application of an equation of state to mixtures is the composition dependence of the parameters. For the virial equations, which apply only to gases, this dependence is given by exact equations arising out of statistical mechanics. The expression for B, the second virial coefficient, is given by

$$B = \sum_i \sum_j y_i y_j B_{ij} \tag{11.44}$$

As indicated in Sec. 11.4, generalized methods are available for evaluation of the B_{ij}. For a binary mixture, Eq. (11.44) reduces to

$$B = y_1^2 B_{11} + 2y_1 y_2 B_{12} + y_2^2 B_{22} \tag{11.45}$$

The third virial coefficient C is expressed as

$$C = \sum_i \sum_j \sum_k y_i y_j y_k C_{ijk} \tag{14.1}$$

where C's with the same subscripts, regardless of order, are equal. For a binary mixture, Eq. (14.1) becomes

$$C = y_1^3 C_{111} + 3y_1^2 y_2 C_{112} + 3y_1 y_2^2 C_{122} + y_2^3 C_{222} \tag{14.2}$$

Here C_{111} and C_{222} are the third virial coefficients for pure species 1 and 2, whereas C_{112} and C_{122} are cross-coefficients. Published generalized correlations for third virial coefficients† are based on a very limited supply of experimental data. Consistent with the mixing rules of Eq. (11.44) and (14.1), the temperature derivatives of B and C are given exactly by

$$\frac{dB}{dT} = \sum_i \sum_j y_i y_j \frac{dB_{ij}}{dT} \tag{14.3}$$

and

$$\frac{dC}{dT} = \sum_i \sum_j \sum_k y_i y_j y_k \frac{dC_{ijk}}{dT} \tag{14.4}$$

As explained in Sec. 13.3, residual properties and fugacity coefficients are readily calculated from equations of state. By Eq. (11.20), applicable to constant-composition fluids,

$$\ln \phi = \int_0^P (Z - 1) \frac{dP}{P} \qquad (\text{const } T, x) \tag{14.5}$$

When the compressibility factor is given by the two-term virial equation,

$$Z - 1 = \frac{BP}{RT} \tag{3.31}$$

Eq. (14.5) yields

$$\ln \phi = \frac{BP}{RT} \tag{14.6}$$

† R. deSantis and B. Grande, *AIChE J*, **25**: 931, 1979; H. Orbey and J. H. Vera, ibid., **29**: 107, 1983.

a restatement of Eq. (11.21). By Eq. (13.16),

$$\frac{H^R}{RT} = -T\left(\frac{\partial \ln \phi}{\partial T}\right)_{P,x} = -T\left(\frac{P}{R}\right)\left(\frac{1}{T}\frac{dB}{dT} - \frac{B}{T^2}\right)$$

or

$$\frac{H^R}{RT} = \frac{P}{R}\left(\frac{B}{T} - \frac{dB}{dT}\right) \tag{14.7}$$

Combination of Eqs. (6.39) and (11.16) gives

$$\frac{S^R}{R} = \frac{H^R}{RT} - \ln \phi \tag{14.8}$$

Whence by Eqs. (14.6) and (14.7),

$$\frac{S^R}{R} = -\frac{P}{R}\frac{dB}{dT} \tag{14.9}$$

The evaluation of residual enthalpies and residual entropies by Eqs. (14.7) and (14.9) is straightforward for given values of T, P, and composition, provided one has sufficient data to evaluate B and dB/dT by Eqs. (11.44) and (14.3). The range of applicability of these equations is the same as for Eq. (3.31), as discussed in Sec. 3.4.

The required values of B_{ij} in Eq. (11.44) can be determined from the generalized correlation for second virial coefficients according to the equation,

$$B_{ij} = \frac{RT_{cij}}{P_{cij}}(B^0 + \omega_{ij}B^1) \tag{11.49}$$

where B^0 and B^1 are given by Eqs. (3.48) and (3.49), and ω_{ij}, T_{cij}, and P_{cij} come from the combining rules of Eqs. (11.50) through (11.54). An equation for dB_{ij}/dT, from which to determine values required in Eq. (14.3), results from differentiation of Eq. (11.49):

$$\frac{dB_{ij}}{dT} = \frac{RT_{cij}}{P_{cij}}\left(\frac{dB^0}{dT} + \omega_{ij}\frac{dB^1}{dT}\right)$$

or

$$\frac{dB_{ij}}{dT} = \frac{R}{P_{cij}}\left(\frac{dB^0}{dT_{rij}} + \omega_{ij}\frac{dB^1}{dT_{rij}}\right) \tag{14.10}$$

where $T_{rij} = T/T_{cij}$. The derivatives dB^0/dT_{rij} and dB^1/dT_{rij} are given as functions of reduced temperature by Eqs. (6.60) and (6.61).

Example 14.1 Estimate V, $\ln \phi$, H^R, and S^R for an equimolar mixture of methyl ethyl ketone(1) and toluene(2) at 50°C and 25 kPa.

SOLUTION The required data are given with Example 11.6, along with calculated values of the B_{ij}. We here need in addition values of dB_{ij}/dT. The values of T_{rij}, together with dB^0/dT_{rij}, dB^1/dT_{rij}, and dB_{ij}/dT calculated for each ij pair by Eqs. (6.60), (6.61), and (14.10), are as follows (note that all $k_{ij} = 0$):

ij	T_{rij}	dB^0/dT_{rij}	dB^1/dT_{rij}	dB_{ij}/dT cm^3 mol^{-1} K^{-1}
11	0.603	2.515	10.020	11.643
22	0.546	3.255	16.793	15.315
12	0.574	2.858	12.948	13.391

With values of B_{ij} calculated in Example 11.6 and values of dB_{ij}/dT calculated here, Eqs. (11.45) and (14.3) yield

$$B = (0.5)^2(-1{,}387) + (2)(0.5)(0.5)(-1{,}611) + (0.5)^2(-1{,}860)$$

$$= -1{,}617 \text{ cm}^3 \text{ mol}^{-1}$$

and

$$dB/dT = (0.5)^2(11.643) + (2)(0.5)(0.5)(13.391) + (0.5)^2(15.315)$$

$$= 13.435 \text{ cm}^3 \text{ mol}^{-1} \text{ K}^{-1}$$

Substitution of these values in Eqs. (3.31), (14.6), (14.7), and (14.9) gives for $T =$ 323.15 K and $P = 25$ kPa:

$$Z = 1 + \frac{BP}{RT} = 1 + \frac{(-1{,}617)(25)}{(8{,}314)(323.15)} = 0.9850$$

$$\ln \phi = \frac{BP}{RT} = \frac{(-1{,}617)(25)}{(8{,}314)(323.15)} = -0.01505$$

$$\frac{H^R}{RT} = \frac{P}{R}\left(\frac{B}{T} - \frac{dB}{dT}\right) = \frac{25}{8{,}314}\left(\frac{-1{,}617}{323.15} - 13.435\right) = -0.05545$$

$$\frac{S^R}{R} = -\frac{P}{R}\frac{dB}{dT} = \frac{-25}{8{,}314}(13.435) = -0.04040$$

From these values we find

$$V = \frac{ZRT}{P} = \frac{(0.9850)(8{,}314)(323.15)}{25} = 105{,}850 \text{ cm}^3 \text{ mol}^{-1}$$

$$\phi = 0.9851$$

$$H^R = (-0.05545)(8.314)(323.15) = -149.0 \text{ J mol}^{-1}$$

$$S^R = (-0.04040)(8.314) = -0.3359 \text{ J mol}^{-1} \text{ K}^{-1}$$

Since Eq. (3.31) expresses Z as a function of P and T, the mathematical operations of Eqs. (14.5) and (13.16) are readily carried out. However, when the equation of state expresses Z as a function of V and T, as is most often the case, Eqs. (14.5) and (13.16) are inappropriate, and must be transformed so that

V rather than P is the independent variable. Since the derivations are tedious,† we here simply present the results:

$$\boxed{\ln \phi = Z - 1 - \ln Z - \int_{\infty}^{V} (Z - 1) \frac{dV}{V}} \tag{14.11}$$

and

$$\boxed{\frac{H^R}{RT} = Z - 1 + T \int_{\infty}^{V} \left(\frac{\partial Z}{\partial T}\right)_V \frac{dV}{V}} \tag{14.12}$$

The residual entropy is found from Eq. (14.8).

When Z is given by the three-term virial equation,

$$Z - 1 = \frac{B}{V} + \frac{C}{V^2} \tag{3.33}$$

Eqs. (14.11) and (14.12) become

$$\ln \phi = \frac{2B}{V} + \frac{(3/2)C}{V^2} - \ln Z \tag{14.13}$$

and

$$\frac{H^R}{RT} = T\left[\left(\frac{B}{T} - \frac{dB}{dT}\right)\frac{1}{V} + \left(\frac{C}{T} - \frac{1}{2}\frac{dC}{dT}\right)\frac{1}{V^2}\right] \tag{14.14}$$

For application of these equations, useful for gases up to moderate pressures, we need values of all B_{ij}, C_{ijk}, and their temperature derivatives for substitution into Eqs. (11.44), (14.1), (14.3), and (14.4).

14.2 PROPERTIES OF FLUIDS FROM CUBIC EQUATIONS OF STATE

As discussed in Sec. 3.5 and illustrated by Fig. 3.10, *PVT* equations of state that are cubic in molar volume are capable of describing the behavior of both liquid and vapor phases of pure fluids.

The application of cubic equations of state to mixtures requires that the equation-of-state parameters be expressed as functions of composition. No exact theory like that for the virial equations prescribes this composition dependence, and we rely instead on empirical *mixing rules* to provide approximate relation-

† H. C. Van Ness and M. M. Abbott, *Classical Thermodynamics of Nonelectrolyte Solutions: With Applications to Phase Equilibria*, app. C, McGraw-Hill, New York, 1982.

ships. For the Redlich/Kwong equation,

$$P = \frac{RT}{V - b} - \frac{a}{T^{1/2}V(V + b)} \tag{3.35}$$

the mixing rules that have found greatest favor are:

$$a = \sum_i \sum_j y_i y_j a_{ij} \tag{14.15}$$

with $a_{ij} = a_{ji}$, and

$$b = \sum_i y_i b_i \tag{14.16}$$

The a_{ij} are of two types: pure-species parameters (like subscripts) and interaction parameters (unlike subscripts). The b_i are parameters for the pure species.

One procedure for evaluation of parameters is a generalization of Eqs. (3.40) and (3.41):

$$a_{ij} = \frac{0.42748R^2T_{cij}^{2.5}}{P_{cij}} \tag{14.17}$$

and

$$b_i = \frac{0.08664RT_{ci}}{P_{ci}} \tag{14.18}$$

where Eqs. (11.51) through (11.54) provide for the calculation of the T_{cij} and P_{cij}.

Multiplication of the Redlich/Kwong equation [Eq. (3.35)] by V/RT leads to its expression in alternative form:

$$Z = \frac{1}{1 - h} - \frac{a}{bRT^{1.5}}\left(\frac{h}{1 + h}\right) \tag{14.19}$$

Whence

$$Z - 1 = \frac{h}{1 - h} - \frac{a}{bRT^{1.5}}\left(\frac{h}{1 + h}\right) \tag{14.20}$$

where

$$h = \frac{bP}{ZRT} \tag{14.21}$$

Equations (14.11) and (14.12) in combination with Eq. (14.20) lead to

$$\ln \phi = Z - 1 - \ln (1 - h)Z - \left(\frac{a}{bRT^{1.5}}\right) \ln (1 + h) \tag{14.22}$$

and

$$\frac{H^R}{RT} = Z - 1 - \left(\frac{3a}{2bRT^{1.5}}\right) \ln (1 + h) \tag{14.23}$$

Once a and b for the mixture are determined by Eqs. (14.15) through (14.18), then for given T and P we find Z, $\ln \phi$, and H^R/RT by Eqs. (14.19) through (14.23) and S^R/R by Eq. (14.8). The procedure requires initial solution of Eqs. (14.19) and (14.21), usually by an iterative scheme, as described in connection with Eq. (3.43) for a gas or vapor phase.

The generalized correlations of Pitzer provide an alternative to the use of a cubic equation of state for the calculation of thermodynamic properties. However, no adequate general method is yet known for the extension of the Pitzer correlations based on the compressibility factor to mixtures. Nevertheless, Z, as given by

$$Z = Z^0 + \omega Z^1 \tag{3.45}$$

depends on T_r, P_r, and ω. Approximate results for mixtures can often be obtained with critical parameters for the mixture and a simple linear mixing rule for the acentric factor. Since values for the actual critical properties T_c and P_c for mixtures are rarely known, use is made of the pseudoparameters T_{pc} and P_{pc}, determined again by a simple linear mixing rule. Thus, by definition,

$$T_{pc} = \sum y_i T_{c_i} \tag{14.24}$$

$$P_{pc} = \sum y_i P_{c_i} \tag{14.25}$$

and

$$\omega = \sum y_i \omega_i \tag{14.26}$$

The pseudoreduced temperature and pseudoreduced pressure, which replace T_r and P_r, are determined by

$$T_{pr} = \frac{T}{T_{pc}} \tag{14.27}$$

and

$$P_{pr} = \frac{P}{P_{pc}} \tag{14.28}$$

Thus, for a mixture at given T_{pr} and P_{pr} we may determine a value of Z by Eq. (3.45) and Figs. 3.12 through 3.15, of H^R/RT_{pc} by Eq. (6.56) and Figs. 6.6 through 6.9, of S^R/R by Eq. (6.57) and Figs. 6.10 through 6.13, and of ϕ by Eq. (11.42) and Figs. 11.2 through 11.5.

Example 14.2 Estimate V, ϕ, H^R, and S^R for an equimolar mixture of carbon dioxide(1) and propane(2) at 450 K and 140 bar by (a) the Redlich/Kwong equation and (b) the generalized correlations of Pitzer.

SOLUTION (a) The required data are given as follows:

ij	T_{cij}/K	P_{cij}/bar	V_{cij}/cm^3 mol^{-1}	Z_{cij}	ω_{ij}
11	304.2	73.8	94.0	0.274	0.225
22	369.8	42.5	203.0	0.281	0.152
12	335.4	54.7	141.6	0.278	0.188

where the values in the last row are calculated by Eqs. (11.50) through (11.54) with $k_{12} = 0$. Substitution of appropriate values into Eq. (14.17) and (14.18) gives:

ij	a_{ij}/bar cm^6 K$^{1/2}$ mol^{-2}	b_i/cm^3 mol^{-1}
11	64.622×10^6	29.69
22	182.837×10^6	62.68
12	111.290×10^6	

Parameters a and b for the mixture are given by Eqs. (14.15) and (14.16):

$$a = y_1^2 a_{11} + 2y_1 y_2 a_{12} + y_2^2 a_{22}$$

$$= (0.5)^2(64.622 \times 10^6) + (2)(0.5)(0.5)(111.290 \times 10^6) + (0.5)^2(182.837 \times 10^6)$$

$$a = 117.51 \times 10^6 \text{ bar cm}^6 \text{ K}^{1/2} \text{ mol}^{-2}$$

$$b = y_1 b_1 + y_2 b_2 = (0.5)(29.69) + (0.5)(62.68)$$

$$b = 46.185 \text{ cm}^3 \text{ mol}^{-1}$$

The dimensionless quantity $a/bRT^{1.5}$ is evaluated as

$$\frac{a}{bRT^{1.5}} = \frac{117.51 \times 10^6}{(46.185)(83.14)(450)^{1.5}} = 3.2059$$

Similarly,

$$\frac{bP}{RT} = \frac{(45.185)(140)}{(83.41)(450)} = 0.17282$$

Therefore Eq. (14.19) becomes

$$Z = \frac{1}{1-h} - 3.2059\left(\frac{h}{1+h}\right)$$

and Eq. (14.21) gives

$$h = \frac{0.17282}{Z}$$

Solution for Z and h yields

$$Z = 0.6922 \qquad \text{and} \qquad h = 0.2496$$

The molar volume is therefore

$$V = \frac{ZRT}{P} = \frac{(0.6922)(83.14)(450)}{140} = 185.0 \text{ cm}^3 \text{ mol}^{-1}$$

By Eq. (14.22),

$$\ln \phi = 0.6922 - 1 - \ln[(0.7504)(0.6922)] - 3.2059 \ln 1.2496$$

$$= -0.3671$$

whence

$$\phi = 0.693$$

By Eq. (14.23),

$$\frac{H^R}{RT} = 0.6922 - 1 - (1.5)(3.2059) \ln 1.2496 = -1.379$$

whence

$$H^R = (-1.379)(8.314)(450) = -5{,}160 \text{ J mol}^{-1}$$

By Eq. (14.8),

$$\frac{S^R}{R} = \frac{H^R}{RT} - \ln \phi = -1.379 + 0.367 = -1.012$$

whence

$$S^R = (-1.012)(8.314) = -8.41 \text{ J mol}^{-1} \text{ K}^{-1}$$

(*b*) The pseudocritical constants are found by Eqs. (14.24) and (14.25):

$$T_{pc} = y_1 T_{c11} + y_2 T_{c22} = (0.5)(304.2) + (0.5)(369.8)$$
$$= 337.0 \text{ K}$$

and

$$P_{pc} = y_1 P_{c11} + y_2 P_{c22} = (0.5)(73.8) + (0.5)(42.5)$$
$$= 58.15 \text{ bar}$$

Whence

$$T_{pr} = \frac{450}{337.0} = 1.335$$

and

$$P_{pr} = \frac{140}{58.15} = 2.41$$

Values of Z^0 and Z^1 from Figs. 3.13 and 3.15 at these reduced conditions are:

$$Z^0 = 0.682 \qquad \text{and} \qquad Z^1 = 0.205$$

With ω given by

$$\omega = y_1\omega_1 + y_2\omega_2 = (0.5)(0.255) + (0.5)(0.152) = 0.188$$

we apply Eq. (3.45):

$$Z = Z^0 + \omega Z^1 = 0.682 + (0.189)(0.205) = 0.721$$

from which

$$V = \frac{ZRT}{P} = \frac{(0.721)(83.14)(450)}{140} = 192.7 \text{ cm}^3 \text{ mol}^{-1}$$

Similarly, from Figs. 6.7 and 6.9,

$$\left(\frac{H^R}{RT_{pc}}\right)^0 = -1.77 \qquad \left(\frac{H^R}{RT_{pc}}\right)^1 = -0.15$$

Substitution into Eq. (6.56) gives

$$\frac{H^R}{RT_{pc}} = -1.77 + (0.188)(-0.15) = -1.80$$

whence

$$H^R = (8.314)(337.0)(-1.80) = -5{,}040 \text{ J mol}^{-1}$$

By Figs. 6.11 and 6.13 and Eq. (6.57),

$$\frac{S^R}{R} = -0.98 + (0.188)(-0.28) = -1.03$$

whence

$$S^R = (8.314)(-1.03) = -8.59 \text{ J mol}^{-1} \text{ K}^{-1}$$

Finally, by Figs. 11.3 and 11.5 and Eq. (11.42),

$$\phi = (0.715)(1.225)^{0.188} = 0.743$$

14.3 VAPOR/LIQUID EQUILIBRIUM FROM CUBIC EQUATIONS OF STATE

In Sec. 11.3 we showed that phases at the same T and P are in equilibrium when the fugacity of each species is the same in all phases. For vapor/liquid equilibrium, this requirement is written

$$\hat{f}_i^v = \hat{f}_i^l \qquad (i = 1, 2, \ldots, N) \tag{11.30}$$

An alternative form of Eq. (11.30) results from introduction of the fugacity coefficient, as defined by Eq. (11.33):

$$y_i P \hat{\phi}_i^v = x_i P \hat{\phi}_i^l$$

or

$$\boxed{y_i \hat{\phi}_i^v = x_i \hat{\phi}_i^l} \qquad (i = 1, 2, \ldots, N) \tag{14.29}$$

For the special case of pure species i, this becomes

$$\phi_i^v = \phi_i^l \tag{14.30}$$

a relation already expressed by Eq. (11.24). We consider first the use of an equation of state with Eq. (14.30) for the calculation of the equilibrium or saturation pressure of pure species i at given temperature T.

As discussed in Sec. 3.5 with respect to cubic equations of state for pure species, a subcritical isotherm on a PVT diagram exhibits a smooth transition from the liquid to the vapor region, shown by the curve labeled $T_2 < T_c$ on Fig. 3.10. We tacitly assumed in that discussion independent knowledge of the vapor pressure at this temperature. In fact, this value is implicit in the equation of state. We reproduce in Fig. 14.1 the subcritical isotherm of Fig. 3.10, without any

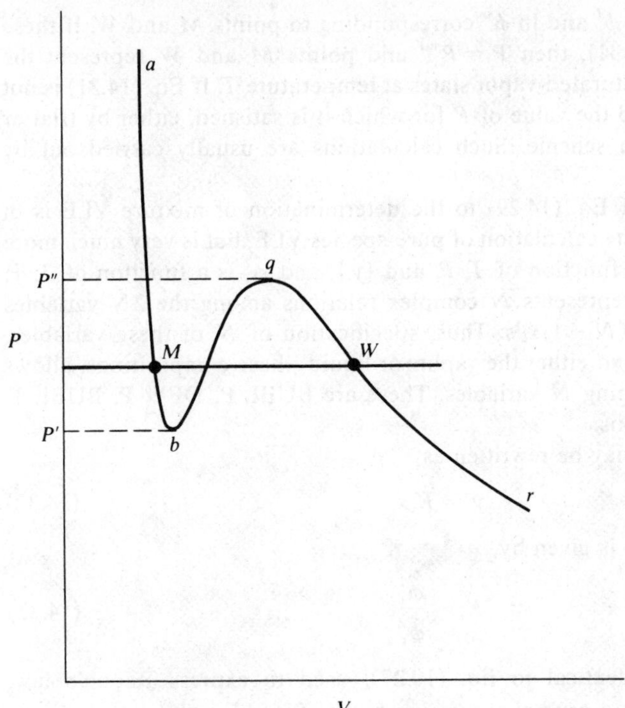

Figure 14.1 Isotherm for $T < T_c$ on PV diagram for a pure fluid.

indication of the location of the equilibrium pressure P^{sat}. However, it clearly must lie between the pressures P' and P'' shown on the figure.

The equilibrium criterion expressed by Eq. (14.30) may be written

$$\ln \phi^l - \ln \phi^v = 0 \tag{14.31}$$

where for economy of notation subscript i is suppressed. The fugacity coefficient of any pure liquid or vapor is a function of its temperature and pressure. For a *saturated* liquid or vapor, the pressure is P^{sat}. This equation therefore implicitly expresses the functional relation,

$$F(T, P^{\text{sat}}) = 0$$

or

$$P^{\text{sat}} = f(T)$$

Application of a cubic equation of state to the isotherm of Fig. 14.1 at a specific P between P' and P'' allows calculation of both a liquid-like volume on branch ab of the isotherm and a vapor-like volume on branch qr, represented for example by points M and W. Since the equation of state [for example, Eq. (14.19)] implies an expression for $\ln \phi$ [for example, Eq. (14.22)], we may calculate the values $\ln \phi^l$ and $\ln \phi^v$ corresponding to points M and W. If these values satisfy Eq. (14.31), then $P = P^{sat}$ and points M and W represent the saturated-liquid and saturated-vapor states at temperature T. If Eq. (14.31) is not satisfied, one must find the value of P for which it is satisfied, either by trial or by a suitable iteration scheme. Such calculations are usually carried out by computer.

The application of Eq. (14.29) to the determination of mixture VLE is in principle the same as the calculation of pure-species VLE, but is very much more difficult. Since $\hat{\phi}_i^v$ is a function of T, P, and $\{y_i\}$, and $\hat{\phi}_i^l$ is a function of T, P, and $\{x_i\}$, Eq. (14.29) represents N complex relations among the $2N$ variables T, P, $(N-1)y_i$'s and $(N-1)x_i$'s. Thus, specification of N of these variables, usually either T or P *and* either the vapor- or liquid-phase compositions, allows solution for the remaining N variables. These are BUBL P, DEW P, BUBL T, and DEW T calculations.

Equation (14.29) may be rewritten as

$$y_i = K_i x_i \tag{14.32}$$

where K_i, the K-value, is given by

$$K_i = \frac{\hat{\phi}_i^l}{\hat{\phi}_i^v} \tag{14.33}$$

Equation (14.32) is identical to Eq. (10.27), used to express Raoult's law. However, Eq. (14.33) is a *general* expression for K_i. Since $\sum y_i = 1$, we can write as a result of Eq. (14.32) that

$$\sum K_i x_i = 1 \tag{14.34}$$

Thus for bubble-point calculations, where the x_i are known, the problem is to find the set of K-values that satisfies Eq. (14.34).

Alternatively, Eq. (14.32) may be written $x_i = y_i/K_i$. Since $\sum x_i = 1$, it follows that

$$\sum \frac{y_i}{K_i} = 1 \tag{14.35}$$

Thus for dew-point calculations, where the y_i are known, the problem is to find the set of K-values that satisfies Eq. (14.35).

Because of the complex functionality of the K-values, these calculations in general require iterative procedures suited only to computer solution. However, in the case of mixtures of light hydrocarbons, in which the molecular force fields are relatively weak and uncomplicated, we may assume as a reasonable approximation that both the liquid and the vapor phases are ideal solutions. By definition of the fugacity coefficient of a species in solution, $\hat{\phi}_i^{id} = \hat{f}_i^{id}/x_i P$. But by Eq. (11.61), $\hat{f}_i^{id} = x_i f_i$. Therefore

$$\hat{\phi}_i^{id} = \frac{x_i f_i}{x_i P} = \frac{f_i}{P} = \phi_i$$

Thus, for an ideal solution the fugacity coefficient of a species in solution is equal to the fugacity coefficient of the pure species at the mixture T and P and in the same physical state (liquid or gas).

The assumption of ideal solutions reduces Eq. (14.33) to

$$K_i = \frac{\phi_i^l(T, P)}{\phi_i^v(T, P)} = \frac{f_i^l(T, P)}{P\phi_i^v(T, P)}$$

The fugacity $f_i^l(T, P)$ is given by Eq. (11.26), which here becomes

$$f_i^l(T, P) = P_i^{\text{sat}} \phi_i^{\text{sat}}(T, P_i^{\text{sat}}) \exp \frac{V_i^l(P - P_i^{\text{sat}})}{RT}$$

where V_i^l is the molar volume of pure species i as a saturated liquid. Thus the K-value is given by

$$K_i = \frac{P_i^{\text{sat}} \phi_i^{\text{sat}}(T, P_i^{\text{sat}})}{P\phi_i^v(T, P)} \exp \frac{V_i^l(P - P_i^{\text{sat}})}{RT} \tag{14.36}$$

The great attraction of this equation is that it contains just properties of the *pure* species and therefore expresses K-values as functions of T and P, independent of the compositions of the liquid and vapor phases. Moreover, ϕ_i^{sat} and ϕ_i^v can be evaluated from equations of state for the pure species or from generalized correlations. This allows K-values for light hydrocarbons to be calculated and correlated as functions of T and P. However, the method is limited for any species to subcritical temperatures, because the vapor-pressure curve terminates at the critical point.

In Figs. 14.2 and 14.3, we present nomographs that give K-values for the light hydrocarbons as functions of T and P. They were prepared by DePriester† on the basis of earlier equation-of-state calculations, and allow for an *average* effect of composition. They give K-values for the light hydrocarbons as functions of T and P. They are suitable for approximate calculations, and provide easy application of K-values to practical problems, as shown in the following example.

Example 14.3 Determine (a) the dew-point pressure and (b) the bubble-point pressure of a mixture of 10 mol % methane, 20 mol % ethane, and 70 mol % propane at 50(°F) if the K-values are given by Fig. 14.2.

SOLUTION (a) When the system is at its dew point, only a minute amount of liquid is present, and the given mole fractions are values of y_i. Since the temperature is specified, the K-values depend on the choice of P, and by trial we find the value for which Eq. (14.35) is satisfied. Results for several values of P are given in the

† C. L. DePriester, *Chem. Eng. Progr. Symp. Ser. 7*, **49**: 1, 1953. They have been published in modified form for direct use with SI units (°C and kPa) by D. B. Dadyburjor, *Chem. Eng. Progr.*, **74**(4): 85, April, 1978.

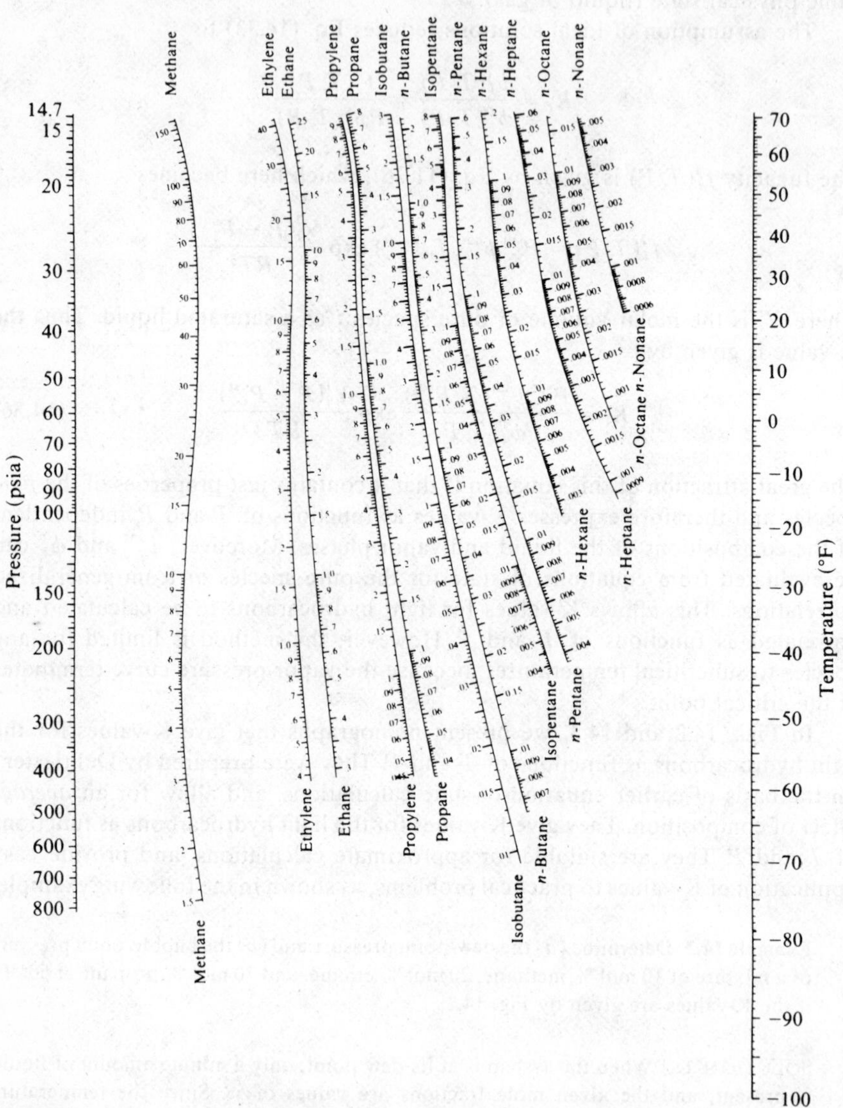

Figure 14.2 Equilibrium constants in light-hydrocarbon systems. Low-temperature range. *(Reproduced by permission from C. L. DePriester, Chem. Eng. Prog. Symp. Ser., 7: 49, 1953.)*

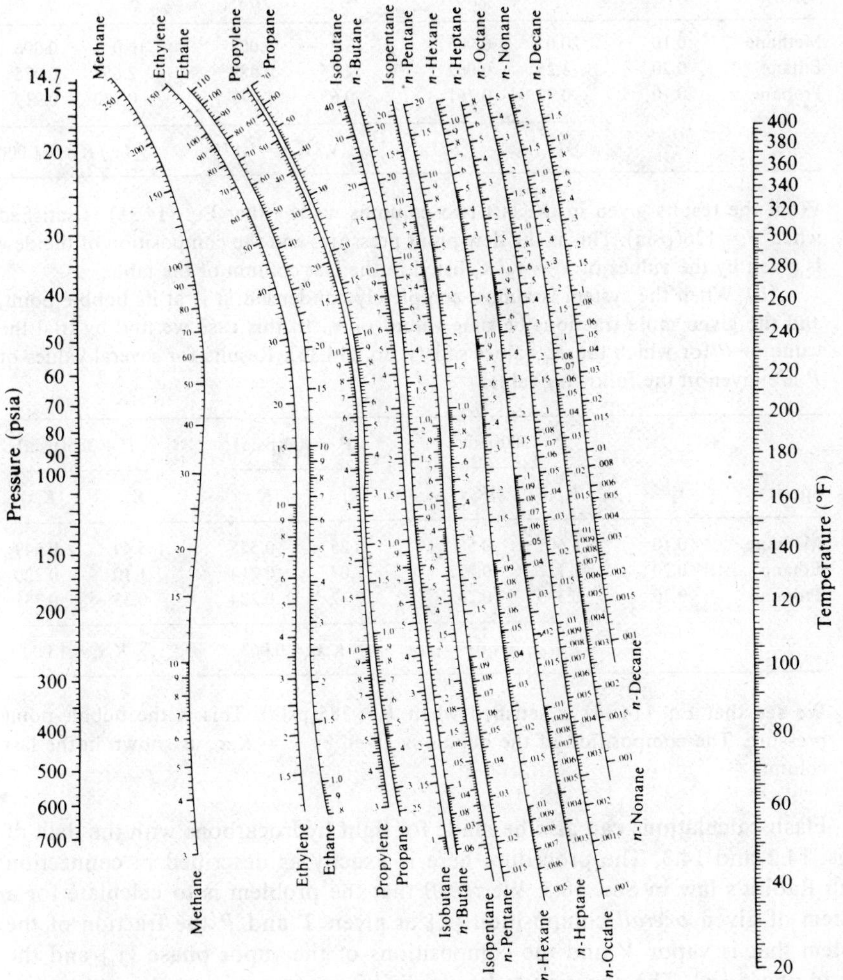

Figure 14.3 Equilibrium constants in light-hydrocarbon systems. High-temperature range. *(Reproduced by permission from C. L. DePriester, Chem. Eng. Prog. Symp. Ser., 7: 49, 1953.)*

following table:

Species	y_i	$P = 100$(psia) K_i	y_i/K_i	$P = 150$(psia) K_i	y_i/K_i	$P = 126$(psia) K_i	y_i/K_i
Methane	0.10	20.0	0.005	13.2	0.008	16.0	0.006
Ethane	0.20	3.25	0.062	2.25	0.089	2.65	0.075
Propane	0.70	0.92	0.761	0.65	1.077	0.762	0.919
		$\sum(y_i/K_i) = 0.827$		$\sum(y_i/K_i) = 1.173$		$\sum(y_i/K_i) = 1.000$	

From the results given in the last two columns we see that Eq. (14.35) is satisfied when $P = 126$(psia). This is the dew-point pressure, and the composition of the dew is given by the values of $x_i = y_i/K_i$ listed in the last column of the table.

(*b*) When the system is almost completely condensed, it is at its bubble point, and the given mole fractions become values of x_i. In this case we find by trial the value of P for which the K_i values satisfy Eq. (14.34). Results for several values of P are given in the following table:

Species	x_i	$P = 380$(psia) K_i	$K_i x_i$	$P = 400$(psia) K_i	$K_i x_i$	$P = 385$(psia) K_i	$K_i x_i$
Methane	0.10	5.60	0.560	5.25	0.525	5.49	0.549
Ethane	0.20	1.11	0.222	1.07	0.214	1.10	0.220
Propane	0.70	0.335	0.235	0.32	0.224	0.33	0.231
		$\sum K_i x_i = 1.017$		$\sum K_i x_i = 0.963$		$\sum K_i x_i = 1.000$	

We see that Eq. (14.34) is satisfied when $P = 385$(psia). This is the bubble-point pressure. The composition of the bubble is given by $y_i = K_i x_i$, as shown in the last column.

Flash calculations can also be made for light hydrocarbons with the data of Figs. 14.2 and 14.3. The procedure here is exactly as described in connection with Raoult's law in Sec. 10.5. We recall that the problem is to calculate for a system of given *overall* composition $\{z_i\}$ at given T and P the fraction of the system that is vapor V and the compositions of the vapor phase $\{y_i\}$ and the liquid phase $\{x_i\}$. The equation to be satisfied is

$$\sum_i \frac{z_i K_i}{1 + V(K_i - 1)} = 1 \tag{10.30}$$

Since T and P are specified, the K_i for light hydrocarbons as given by Figs. 14.2 and 14.3 are known, and V, the only unknown in Eq. (10.30), is found by trial.

Example 14.4 For the system described in Example 14.3, what fraction of the system is vapor when the pressure is 200(psia) and what are the compositions of the equilibrium vapor and liquid phases?

SOLUTION The given pressure lies between the dew- and bubble-point pressures established for this system in Example 14.3. The system therefore surely consists of two phases. The procedure is to find by trial that value of V for which Eq. (10.30) is satisfied. We recall that there is always a trivial solution for $V = 1$. The results of several trials are shown in the following table. The columns headed y_i give values of the terms in the sum of Eq. (10.30), because each such term is in fact a y_i value, as shown by Eq. (10.29).

Species	z_i	K_i	y_i for $V = 0.35$	y_i for $V = 0.25$	y_i for $V = 0.273$	$x_i = y_i/K_i$ for $V = 0.273$
Methane	0.10	10.0	0.241	0.308	0.289	0.029
Ethane	0.20	1.76	0.278	0.296	0.292	0.166
Propane	0.70	0.52	0.438	0.414	0.419	0.805
			$\sum y_i = 0.957$	$\sum y_i = 1.018$	$\sum y_i = 1.000$	$\sum x_i = 1.000$

Thus Eq. (10.30) is satisfied when $V = 0.273$. The phase compositions are given in the last two columns of the table.

When the assumption of ideal solutions is not appropriate, K-values must be calculated by Eq. (14.33), and this requires values of $\hat{\phi}_i^l$ and $\hat{\phi}_i^v$. These come from equations of state that are at least cubic in volume. Just as Eq. (14.5) is inappropriate to the calculation of ln ϕ from equations of state that express Z as a function of T and V, so Eq. (11.39) is not suited to the calculation of ln $\hat{\phi}_i$ from such equations. The transformation of Eq. (11.39) into an equation in which the independent variable is V rather than P gives

$$\ln \hat{\phi}_i = Z - 1 - \ln Z - \int_{\infty}^{V} \left[\left(\frac{\partial (nZ)}{\partial n_i} \right)_{T, V, n_j} - 1 \right] \frac{dV}{V} \tag{14.37}$$

Equation (14.11) is the corresponding expression for ln ϕ.

A generic form of the Redlich/Kwong equation may be written:

$$P = \frac{RT}{V - b} - \frac{\theta}{V(V + b)}$$

or

$$Z = \frac{1}{1 - h} - \left(\frac{\theta}{bRT} \right) \frac{h}{1 + h} \tag{14.38}$$

where

$$h = \frac{b}{V} = \frac{bP}{ZRT}$$

This equation encompasses the original Redlich/Kwong equation and many of its modifications. For the original equation, $\theta = a/T^{1/2}$, where parameter a is a function of composition only. Application of Eq. (14.37) to the generic

Redlich/Kwong equation gives:

$$\ln \hat{\phi}_i = \frac{b_i}{b}(Z-1) - \ln Z(1-h) + \frac{\theta}{bRT}\left(\frac{\bar{b}_i}{b} - \frac{\bar{\theta}_i}{\theta} - 1\right)\ln(1+h) \tag{14.39}$$

where

$$\bar{\theta}_i = \left[\frac{\partial(n\theta)}{\partial n_i}\right]_{T,n_j} \tag{14.40}$$

and

$$\bar{b}_i = \left[\frac{\partial(nb)}{\partial n_i}\right]_{T,n_j} \tag{14.41}$$

These are general equations that do not depend on the particular *mixing rules* adopted for the composition dependence of θ and b.

A commonly used pair of mixing rules is:

$$\theta = \sum_i \sum_j x_i x_j \theta_{ij} \tag{14.42}$$

and

$$b = \sum_i x_i b_i \tag{14.43}$$

In this case, Eqs. (14.40) and (14.41) give

$$\bar{\theta}_i = -\theta + 2\sum_k x_k \theta_{ki} \tag{14.44}$$

and

$$\bar{b}_i = b_i \tag{14.45}$$

Equation (14.39) now becomes

$$\ln \hat{\phi}_i = \frac{b_i}{b}(Z-1) - \ln Z(1-h) + \frac{\theta}{bRT}\left(\frac{b_i}{b} - \frac{2\sum_k x_k \theta_{ki}}{\theta}\right)\ln(1+h) \tag{14.46}$$

For the original Redlich/Kwong equation,

$$\theta = \frac{a}{T^{1/2}} \qquad \text{and} \qquad \theta_{ij} = \frac{a_{ij}}{T^{1/2}}$$

These reduce Eq. (14.42) to

$$a = \sum_i \sum_j x_i x_j a_{ij}$$

Which is equivalent to Eq. (14.15), and Eq. (14.46) becomes

$$\ln \hat{\phi}_i = \frac{b_i}{b}(Z-1) - \ln Z(1-h) + \frac{a}{bRT^{1.5}}\left(\frac{b_i}{b} - \frac{2\sum_k x_k a_{ki}}{a}\right)\ln(1+h) \tag{14.47}$$

Here Z is given by Eq. (14.19), which is Eq. (14.38) with $\theta = a/T^{1/2}$. Parameters

a_{ki} and b_i can be evaluated by Eqs. (14.17) and (14.18). Similar parameter-evaluation procedures are readily formulated for modifications of the Redlich/Kwong equation.

Example 14.5 Estimate $\hat{\phi}_1$ and $\hat{\phi}_2$ for an equimolar vapor mixture of carbon dioxide(1) and propane(2) at 450 K and 140 bar by Eq. (14.47).

SOLUTION For the two species of a binary mixture, Eq. (14.47) reduces to:

$$\ln \hat{\phi}_1 = \frac{b_1}{b}(Z-1) - \ln Z(1-h) + \frac{a}{bRT^{1.5}}\left(\frac{b_1}{b} - \frac{2\sum_k x_k a_{1k}}{a}\right)\ln(1+h)$$

and

$$\ln \hat{\phi}_2 = \frac{b_2}{b}(Z-1) - \ln Z(1-h) + \frac{a}{bRT^{1.5}}\left(\frac{b_2}{b} - \frac{2\sum_k x_k a_{2K}}{a}\right)\ln(1+h)$$

All required values for substitution into these equations are available from Example 14.2. Thus,

$$\frac{b_1}{b} = \frac{29.69}{46.185} = 0.6428$$

$$Z = 0.6922 \qquad h = 0.2496 \qquad a/bRT^{1.5} = 3.2059$$

$$\begin{aligned} 2\sum_k x_k a_{1k} &= 2(x_1 a_{11} + x_2 a_{12}) \\ &= 2[(0.5)(64.622 \times 10^6) + (0.5)(111.290 \times 10^6)] \\ &= 175.91 \times 10^6 \end{aligned}$$

$$\frac{2\sum_k x_k a_{1k}}{a} = \frac{175.91 \times 10^6}{117.51 \times 10^6} = 1.4970$$

and

$$\begin{aligned} \ln \hat{\phi}_1 &= (0.6428)(0.6922 - 1) - \ln[(0.6922)(1 - 0.2496)] \\ &\quad + 3.2059(0.6428 - 1.4970)\ln 1.2496 \end{aligned}$$

$$\ln \hat{\phi}_1 = -0.1530 \qquad \text{and} \qquad \hat{\phi}_1 = 0.8581$$

Similarly,

$$\frac{b_2}{b} = \frac{62.68}{46.185} = 1.3572$$

$$\begin{aligned} 2\sum_k x_k a_{2k} &= 2(x_1 a_{21} + x_2 a_{22}) \\ &= 2[(0.5)(111.290 \times 10^6) + (0.5)(182.837 \times 10^6)] \\ &= 294.127 \times 10^6 \end{aligned}$$

$$\frac{2\sum_k x_k a_{2k}}{a} = \frac{294.127 \times 10^6}{117.51 \times 10^6} = 2.5030$$

and

$$\ln \hat{\phi}_2 = (1.3572)(0.6922 - 1) - \ln [(0.6922)(1 - 0.2496)]$$
$$+ 3.2059(1.3572 - 2.5030) \ln 1.2496$$

$$\ln \hat{\phi}_2 = -0.5812 \quad \text{and} \quad \hat{\phi}_2 = 0.5592$$

Given the means to calculate values of $\hat{\phi}_i^l$ and $\hat{\phi}_i^v$, one can devise computation schemes for solving dew-point, bubble-point, and flash problems. We illustrate here only the BUBL P calculation, for which a block diagram of a computer program is shown by Fig. 14.4.

The input information consists of the given values of T and $\{x_i\}$ and the physical-property data necessary for evaluation of all equation-of-state parameters. We also read in estimates of P and $\{y_i\}$. These values are needed for initial calculation of $\{\hat{\phi}_i^l\}$ and $\{\hat{\phi}_i^v\}$, and can be obtained from a preliminary solution of the problem based on the assumption of ideal solutions.

Application of an equation such as Eq. (14.47) first with liquid-phase mole fractions and then with vapor-phase mole fractions provides initial values for

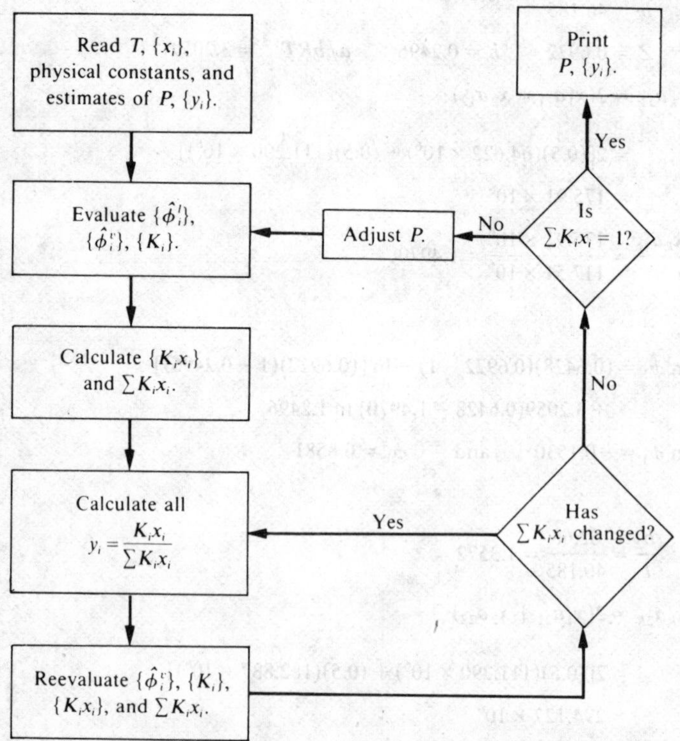

Figure 14.4 Block diagram for BUBL P calculation.

$\{\hat{\phi}_i^l\}$ and $\{\hat{\phi}_i^v\}$. Values for $\{K_i\}$ then come from Eq. (14.33). These allow calculation of $\{K_i x_i\}$; according to Eq. (14.32) this set should be identical to $\{y_i\}$. However, the constraint $\sum y_i = 1$ has not yet been imposed, and it is likely that $\sum K_i x_i \neq 1$. We therefore calculate the y_i by

$$y_i = \frac{K_i x_i}{\sum K_i x_i}$$

and this ensures that the set of y_i's used in subsequent calculations does sum to unity.

This new set of y_i's allows reevaluation of $\{\hat{\phi}_i^v\}$, $\{K_i\}$, $\{K_i x_i\}$, and hence of $\sum K_i x_i$. If the value of $\sum K_i x_i$ has changed, we again calculate the y_i and repeat the sequence of calculations. Iteration leads to a stable value of $\sum K_i x_i$, and we then ask whether $\sum K_i x_i$ is unity. If not, then the value of P is adjusted according to some rational scheme. When $\sum K_i x_i > 1$, P is too low; when $\sum K_i x_i < 1$, P is too high. The entire iterative procedure is then repeated with a new pressure P. The last calculated values of y_i are used as the initial estimate of $\{y_i\}$.

The scheme of Fig. 14.4 illustrates a rational approach to the solution of a BUBL P problem through the use of an equation of state. However, convergence problems sometimes arise, and in this case a solution may not be obtained, even with very good initial estimates of P and $\{y_i\}$. Discussions of such problems and of algorithms for circumventing them are found in the literature.†

Because of its relative simplicity, the original Redlich/Kwong equation was used in Example 14.5 to illustrate the calculation of fugacity coefficients. However, this equation in its original form is rarely satisfactory for VLE calculations, and many modifications have been proposed to make it more suitable. In particular, Soave‡ introduced the acentric factor into the Redlich/Kwong equation by setting θ equal to a function not only of temperature but also of the acentric factor ω. Thus the widely used Soave/Redlich/Kwong (SRK) equation is written:

$$P = \frac{RT}{V - b} - \frac{\theta_{\mathrm{SRK}}}{V(V + b)} \tag{14.48}$$

where

$$\theta_{\mathrm{SRK}} = a'[1 + (0.480 + 1.574\omega - 0.176\omega^2)(1 - T_r^{1/2})]^2$$

$$a' = \frac{0.42748 R^2 T_c^2}{P_c}$$

and

$$b = \frac{0.08664 R T_c}{P_c}$$

† A comprehensive treatment of multiphase equilibrium calculations with an equation of state, including an extensive bibliography, is given by L. X. Nghiem and Yau-Kun Li, *Fluid Phase Equilibria*, **17**: 77, 1984.

‡ G. Soave, *Chem. Eng. Sci.*, **27**: 1197, 1972.

Many modifications of the original Redlich/Kwong equation that appear in the literature are intended for special-purpose applications. The SRK equation, developed for vapor/liquid equilibrium calculations, is designed specifically to yield reasonable vapor pressures for pure fluids. Thus, there is no assurance that molar volumes calculated by the SRK equation are more accurate than values given by the original Redlich/Kwong equation.

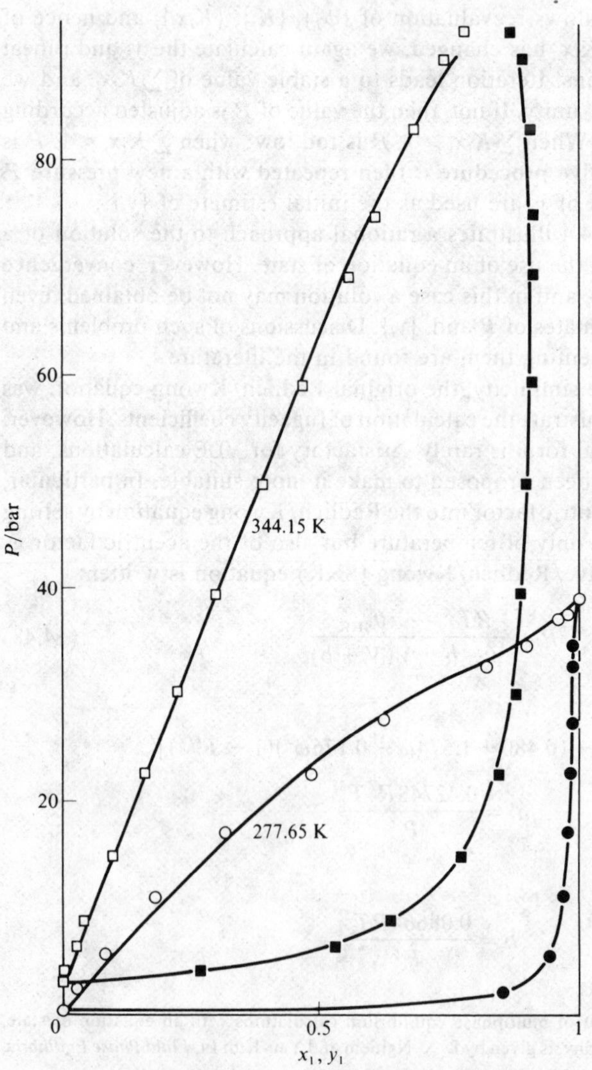

Figure 14.5 *Pxy* diagram for carbon dioxide(1)/*n*-pentane(2).

The Peng/Robinson equation,† also developed specifically for vapor/liquid equilibrium, is an alternative to the SRK equation:

$$P = \frac{RT}{V-b} - \frac{\theta_{PR}}{V^2 + 2bV - b^2} \tag{14.49}$$

where

$$\theta_{PR} = a''[1 + (0.37464 + 1.54226\omega - 0.26992\omega^2)(1 - T_r^{1/2})]^2$$

$$a'' = \frac{0.45724R^2T_c^2}{P_c}$$

and

$$b = \frac{0.07780RT_c}{P_c}$$

As an example of the use of an equation of state for VLE calculations, we have applied the SRK equation together with the BUBL P program of Fig. 14.4 to the system carbon dioxide(1)/n-pentane(2) at 277.65 and 344.15 K. Results are shown in Fig. 14.5, where the lines represent calculated values and the points are the data of Besserer and Robinson.‡ A suitable correlation of the data requires use of an appropriate value of k_{ij} in Eq. (11.51). While one might hope that a single value would here serve for both temperatures, better results are obtained with the values, $k_{12} = 0.12$ at 277.65 K and $k_{12} = 0.14$ at 344.15 K. Calculations such as these can be done routinely with the aid of a computer, but their accuracy depends on knowledge of proper values for k_{ij}, or more generally on the use of appropriate mixing rules. This is an area of active research, and to be informed of progress, one must have recourse to current literature.

PROBLEMS

14.1 Estimate Z, H^R, and S^R at 300 K and 6 bar for propane given the following values of the second virial coefficient for propane.

T/K	B/cm^3 mol^{-1}
250	−584
300	−382
350	−276

14.2 The second virial coefficient for acetonitrile is given approximately by the equation,

$$B/\text{cm}^3\,\text{mol}^{-1} = -8.55\left(\frac{10^3}{T/\text{K}}\right)^{5.5}$$

Determine values for H^R and S^R for acetonitrile vapor at 80°C and 80 kPa.

† D.-Y. Peng and D. B. Robinson, *Ind. Eng. Chem. Fundam.*, **15**: 59, 1976.

‡ G. J. Besserer and D. B. Robinson, *J. Chem. Eng. Data*, **18**: 416, 1973.

14.3 An equimolar mixture of methane and propane is discharged from a compressor at 5,500 kPa and 90°C at the rate of 1.4 kg s^{-1}. If the velocity in the discharge line is not to exceed 30 m s^{-1}, what is the minimum diameter of the discharge line?

14.4 Estimate V, ϕ, H^R, and S^R for one of the following binary vapor mixtures:

(*a*) Acetone(1)/1,3-butadiene(2) with mole fractions $y_1 = 0.28$ and $y_2 = 0.72$ at $t = 60$°C and $P =$ 170 kPa.

(*b*) Acetonitrile(1)/diethyl ether(2) with mole fractions $y_1 = 0.37$ and $y_2 = 0.63$ at $t = 50$°C and $P = 120$ kPa.

(*c*) Methyl chloride(1)/dichlorodifluoromethane(2) with mole fractions $y_1 = 0.43$ and $y_2 = 0.57$ at $t = 25$°C and $P = 150$ kPa.

(*d*) Nitrogen(1)/ammonia(2) with mole fractions $y_1 = 0.17$ and $y_2 = 0.83$ at $t = 20$°C and $P =$ 300 kPa.

(*e*) Ethylene oxide(1)/ethylene(2) with mole fractions $y_1 = 0.68$ and $y_2 = 0.32$ at $t = 25$°C and $P = 420$ kPa.

14.5 Calculate V, ϕ, H^R, and S^R for the binary vapor mixture nitrogen(1)/isobutane(2) with $y_1 = 0.35$, $y_2 = 0.65$, $t = 150$°C, and $P = 60$ bar by the following methods:

(*a*) Assume the mixture an ideal solution with properties of the pure species given by the Pitzer correlations based on the compressibility factor.

(*b*) Apply the Pitzer correlations directly to the mixture.

(*c*) Use the Redlich/Kwong equation with Eqs. (14.17) and (14.18). In Eq. (11.51), set $k_{12} = 0.11$.

14.6 Calculate V, ϕ, H^R, and S^R for the binary vapor mixture hydrogen sulfide(1)/ethane(2) with $y_1 = 0.20$, $y_2 = 0.80$, $t = 140$°C, and $P = 80$ bar by the following methods:

(*a*) Assume the mixture an ideal solution with properties of the pure species given by the Pitzer correlations based on the compressibility factor.

(*b*) Apply the Pitzer correlations directly to the mixture.

(*c*) Use the Redlich/Kwong equation with Eqs. (14.17) and (14.18). In Eq. (11.51), set $k_{12} = 0.06$.

14.7 Using the parameter values calculated in part (*c*) of Prob. 14.5, estimate $\hat{\phi}_1$ and $\hat{\phi}_2$ for the nitrogen(1)/isobutane(2) mixture of Prob. 14.5.

14.8 Using the parameter values calculated in part (*c*) of Prob. 14.6, estimate $\hat{\phi}_1$ and $\hat{\phi}_2$ for the hydrogen sulfide(1)/ethane(2) mixture of Prob. 14.6.

14.9 Assuming the validity of the De Priester charts, make the following VLE calculations for the methane(1)/ethylene(2)/ethane(3) system:

(*a*) BUBL P, given $x_1 = 0.10$, $x_2 = 0.50$, and $t = -60$(°F).

(*b*) DEW P, given $y_1 = 0.50$, $y_2 = 0.25$, and $t = -60$(°F).

(*c*) BUBL T, given $x_1 = 0.12$, $x_2 = 0.40$, and $P = 250$(psia).

(*d*) DEW T, given $y_1 = 0.43$, $y_2 = 0.36$, and $P = 250$(psia).

14.10 Assuming the validity of the De Priester charts, make the following VLE calculations for the ethane(1)/propane(2)/isobutane(3)/isopentane(4) system:

(*a*) BUBL P, given $x_1 = 0.10$, $x_2 = 0.20$, $x_3 = 0.30$, and $t = 140$(°F).

(*b*) DEW P, given $y_1 = 0.48$, $y_2 = 0.25$, $y_3 = 0.15$, and $t = 140$(°F).

(*c*) BULB T, given $x_1 = 0.14$, $x_2 = 0.13$, $x_3 = 0.25$, and $P = 200$(psia).

(*d*) DEW T, given $y_1 = 0.42$, $y_2 = 0.30$, $y_3 = 0.15$, and $P = 200$(psia).

14.11 The stream from a gas well consists of 50-mol-% methane, 10-mol-% ethane, 20-mol-% propane, and 20-mol-% *n*-butane. This stream is fed into a partial condenser maintained at a pressure of 250(psia), where its temperature is brought to 80(°F). Determine the molar fraction of the gas that condenses and the compositions of the liquid and vapor phases leaving the condenser.

14.12 An equimolar mixture of *n*-butane and *n*-hexane at pressure P is brought to a temperature of 200(°F), where it exists as a vapor/liquid mixture in equilibrium. If the mole fraction of *n*-hexane in the liquid phase is 0.75, what is pressure P, what is the molar fraction of the system that is liquid, and what is the composition of the vapor phase?

14.13 A mixture of 25-mol-% *n*-pentane, 45-mol-% *n*-hexane, and 30-mol-% *n*-heptane is brought to a condition of 155(°F) and 1(atm). What molar fraction of the system is liquid, and what are the phase compositions?

14.14 A mixture containing 15-mol-% ethane, 35-mol-% propane, and 50-mol-% *n*-butane is brought to a condition of 100(°F) at pressure *P*. If the molar fraction of liquid in the system is 0.40, what is pressure *P* and what are the compositions of the liquid and vapor phases?

14.15 A mixture containing 1-mol-% ethane, 5-mol-% propane, 44-mol-% *n*-butane, and 50-mol-% isobutane is brought to a condition of 80(°F) at pressure *P*. If the molar fraction of the system that is vapor is 0.2, what is pressure *P*, and what are the compositions of the vapor and liquid phases?

14.16 A mixture of 30-mol-% methane, 10-mol-% ethane, 30-mol-% propane, and 30-mol-% *n*-butane is brought to a condition of −40(°F) at pressure *P*, where it exists as a vapor/liquid mixture in equilibrium. If the mole fraction of the methane in the vapor phase is 0.90, what is pressure *P*?

14.17 The top tray of a distillation column and the condenser are at a pressure of 20(psia). The liquid on the top tray is an equimolar mixture of *n*-butane and *n*-pentane. The vapor from the top tray, assumed to be in equilibrium with the liquid, goes to the condenser where 50 mole percent of the vapor is condensed. What are the temperature and composition of the vapor leaving the condenser?

14.18 *n*-Butane is separated from an equimolar methane/*n*-butane gas mixture by compression of the gas to pressure *P* at 100(°F). If 40 percent of the feed on a mole basis is condensed, what is pressure *P* and what are the compositions of the resulting vapor and liquid phases?

CHAPTER

FIFTEEN

CHEMICAL-REACTION EQUILIBRIA

The transformation of raw materials into products of greater value by means of chemical reaction is a major industry, and a vast number of commercial products is obtained by chemical synthesis. Sulfuric acid, ammonia, ethylene, propylene, phosphoric acid, chlorine, nitric acid, urea, benzene, methanol, ethanol, and ethylene glycol are examples of chemicals produced in the United States in billions of kilograms each year. These in turn are used in the large-scale manufacture of fibers, paints, detergents, plastics, rubber, fertilizers, insecticides, etc. Clearly, the chemical engineer must be familiar with chemical-reactor design and operation.

The rate and maximum possible (or equilibrium) conversion of a chemical reaction are of primary concern in its commercial development. Both depend on the temperature, pressure, and composition of reactants. For a specific example, consider the effect of temperature on the oxidation of sulfur dioxide to sulfur trioxide. This reaction requires a catalyst for a reasonable reaction rate, and the rate becomes appreciable with a vanadium pentoxide catalyst at about 300°C and increases rapidly at higher temperatures. On the basis of rate alone, one would operate the reactor at a high temperature. Although the equilibrium conversion of sulfur trioxide is greater than 90 percent at temperatures below 520°C, it falls off rapidly at higher temperatures, declining to 50 percent at about 680°C. This is the maximum possible conversion at this temperature regardless of catalyst or reaction rate. The evident conclusion from this example is that both equilibrium and rate must be considered in the development of a commercial process for a chemical reaction. Reaction rates are not susceptible to thermodynamic treatment, but equilibrium conversions are found by thermodynamic calculations. Therefore, the purpose of this chapter is to determine the effect of temperature, pressure, and ratio of reactants on the equilibrium conversions of chemical reactions.

Many industrial reactions are not carried to equilibrium. In this circumstance the reactor design is based primarily on reaction rate. However, the choice of operating conditions may still be determined by equilibrium considerations, as already illustrated with respect to the oxidation of sulfur dioxide. In addition, the equilibrium conversion of a reaction provides a goal by which to measure improvements in the process. Similarly, it may determine whether or not an experimental investigation of a new process is worthwhile. For example, if the thermodynamic analysis indicates that a yield of only 20 percent is possible at equilibrium and a 50 percent yield is necessary for the process to be economically attractive, there is no purpose to an experimental study. On the other hand, if the equilibrium yield is 80 percent, an experimental program to determine the reaction rate for various conditions of operation (catalyst, temperature, pressure, etc.) may be warranted.

Calculation of equilibrium conversions is based on the fundamental equations of chemical-reaction equilibrium, which in application require data for the standard Gibbs energy of reaction. The basic equations are developed in Secs. 15.1 through 15.4. These provide the relationship between the standard Gibbs energy change of reaction and the equilibrium constant. Evaluation of the equilibrium constant from thermodynamic data is considered in Sec. 15.5. Application of this information to the calculation of equilibrium conversions for single reactions is taken up in Sec. 15.7. In Sec. 15.8, the phase rule is reconsidered; finally, multireaction equilibrium is treated in Sec. 15.9.†

15.1 THE REACTION COORDINATE

The general chemical reaction of Sec. 4.6 is rewritten here as

$$|\nu_1|A_1 + |\nu_2|A_2 + \cdots \to |\nu_3|A_3 + |\nu_4|A_4 + \cdots \tag{15.1}$$

where the $|\nu_i|$ are stoichiometric coefficients and the A_i stand for chemical formulas. The ν_i themselves are called stoichiometric numbers, and we recall the sign convention that makes them positive for products and negative for reactants. Thus for the reaction

$$CH_4 + H_2O \to CO + 3H_2$$

the stoichiometric numbers are

$$\nu_{CH_4} = -1 \qquad \nu_{H_2O} = -1 \qquad \nu_{CO} = 1 \qquad \nu_{H_2} = 3$$

The stoichiometric number for any inert species present is zero.

† For a comprehensive treatment of chemical-reaction equilibria, see W. R. Smith and R. W. Missen, *Chemical Reaction Equilibrium Analysis*, John Wiley & Sons, New York, 1982.

For the reaction represented by Eq. (15.1), the *changes* in the numbers of moles of the species present are in direct proportion to the stoichiometric numbers. Thus for the preceding reaction, if 0.5 mol of CH_4 disappears by reaction, 0.5 mol of H_2O must also disappear; simultaneously 0.5 mol of CO and 1.5 mol of H_2 are formed by reaction. Applying this principle to a differential amount of reaction, we can write

$$\frac{dn_2}{\nu_2} = \frac{dn_1}{\nu_1} \qquad \frac{dn_3}{\nu_3} = \frac{dn_1}{\nu_1} \qquad \text{etc.}$$

The list continues to include all species. Comparison of these equations shows that

$$\frac{dn_1}{\nu_1} = \frac{dn_2}{\nu_2} = \frac{dn_3}{\nu_3} = \frac{dn_4}{\nu_4} = \cdots$$

Each term is related to an amount of reaction as represented by a change in the number of moles of a chemical species. Since all terms are equal, they can be identified collectively with a single quantity $d\varepsilon$, arbitrarily defined to represent the amount of reaction. Thus a *definition* of $d\varepsilon$ is provided by the equation

$$\boxed{\frac{dn_1}{\nu_1} = \frac{dn_2}{\nu_2} = \frac{dn_3}{\nu_3} = \frac{dn_4}{\nu_4} = \cdots = d\varepsilon} \tag{15.2}$$

The general relation between a differential change dn_i in the number of moles of a reacting species and $d\varepsilon$ is therefore

$$\boxed{dn_i = \nu_i \, d\varepsilon} \qquad (i = 1, 2, \ldots, N) \tag{15.3}$$

This new variable ε, called the *reaction coordinate*, characterizes the extent or degree to which a reaction has taken place.† Equations (15.2) and (15.3) define *changes* in ε with respect to changes in the numbers of moles of the reacting species. The definition of ε itself is completed for each application by the specification that it be *zero* for the initial state of the system prior to reaction. Thus, integration of Eq. (15.3) from an initial unreacted state where $\varepsilon = 0$ and $n_i = n_{i_0}$ to a state reached after an arbitrary amount of reaction gives

$$\int_{n_{i_0}}^{n_i} dn_i = \nu_i \int_0^{\varepsilon} d\varepsilon$$

or

$$n_i = n_{i_0} + \nu_i \varepsilon \qquad (i = 1, 2, \ldots, N) \tag{15.4}$$

Summing over all species gives

$$n = \sum n_i = \sum n_{i_0} + \varepsilon \sum \nu_i$$

† The reaction coordinate ε has been given various other names, such as: degree of advancement, degree of reaction, extent of reaction, and progress variable.

or

$$n = n_0 + \nu\varepsilon$$

where

$$n \equiv \sum n_i \qquad n_0 \equiv \sum n_{i_0} \qquad \nu \equiv \sum \nu_i$$

Thus the mole fractions y_i of the species present are related to ε by

$$\boxed{y_i = \frac{n_i}{n} = \frac{n_{i_0} + \nu_i\varepsilon}{n_0 + \nu\varepsilon}} \tag{15.5}$$

Application of this equation is illustrated in the following examples.

Example 15.1 For a system in which the following reaction occurs,

$$CH_4 + H_2O \rightarrow CO + 3H_2$$

assume there are present initially 2 mol CH_4, 1 mol H_2O, 1 mol CO, and 4 mol H_2. Determine expressions for the mole fractions y_i as functions of ε.

SOLUTION For the given reaction,

$$\nu = \sum \nu_i = -1 - 1 + 1 + 3 = 2$$

For the given numbers of moles of species initially present,

$$n_0 = \sum n_{i_0} = 2 + 1 + 1 + 4 = 8$$

Application of Eq. (15.5) now gives

$$y_{CH_4} = \frac{2 - \varepsilon}{8 + 2\varepsilon} \qquad y_{H_2O} = \frac{1 - \varepsilon}{8 + 2\varepsilon}$$

$$y_{CO} = \frac{1 + \varepsilon}{8 + 2\varepsilon} \qquad y_{H_2} = \frac{4 + 3\varepsilon}{8 + 2\varepsilon}$$

The mole fractions of the species in the reacting mixture are seen to be functions of the single variable ε.

Example 15.2 Consider a vessel which initially contains only n_0 moles of water vapor. If decomposition occurs according to the reaction

$$H_2O \rightarrow H_2 + \tfrac{1}{2}O_2$$

find expressions which relate the number of moles and the mole fraction of each chemical species to the reaction coordinate ε.

SOLUTION For the given reaction, $\nu = -1 + 1 + \frac{1}{2} = \frac{1}{2}$. Application of Eqs. (15.4) and (15.5) gives

$$n_{H_2O} = n_0 - \varepsilon \qquad y_{H_2O} = \frac{n_0 - \varepsilon}{n_0 + \frac{1}{2}\varepsilon}$$

$$n_{H_2} = \varepsilon \qquad y_{H_2} = \frac{\varepsilon}{n_0 + \frac{1}{2}\varepsilon}$$

$$n_{O_2} = \tfrac{1}{2}\varepsilon \qquad y_{O_2} = \frac{\frac{1}{2}\varepsilon}{n_0 + \frac{1}{2}\varepsilon}$$

The fractional decomposition of water vapor is

$$\frac{n_0 - n_{H_2O}}{n_0} = \frac{n_0 - (n_0 - \varepsilon)}{n_0} = \frac{\varepsilon}{n_0}$$

Thus when $n_0 = 1$, ε can be identified with the fractional decomposition of the water vapor.

We see from Eq. (15.3) that either the ν_i's or ε must be expressed in moles and that the other quantity must be a pure number. As a matter of convenience we choose to express the reaction coordinate ε in moles. This allows one to speak of a *mole of reaction*, meaning that ε has changed by a unit amount, i.e., by one mole. When $\Delta\varepsilon = 1$ mol, the reaction proceeds to such an extent that the change in mole number of each reactant and product is equal to its stoichiometric number.

When two or more independent reactions proceed simultaneously, we let subscript j be the reaction index, and associate a separate reaction coordinate ε_j with each reaction. The stoichiometric numbers are doubly subscripted to identify their association with both a species and a reaction. Thus $\nu_{i,j}$ designates the stoichiometric number of species i in reaction j. Since the number of moles of a species n_i may change because of several reactions, the general equation analogous to Eq. (15.3) includes a sum:

$$dn_i = \sum_j \nu_{i,j}\, d\varepsilon_j \qquad (i = 1, 2, \ldots, N)$$

Integration from $n_i = n_{i_0}$ and $\varepsilon_j = 0$ to arbitrary n_i and ε_j gives

$$n_i = n_{i_0} + \sum_j \nu_{i,j}\varepsilon_j \qquad (i = 1, 2, \ldots, N) \tag{15.6}$$

Summing over all species yields

$$n = \sum_i n_{i_0} + \sum_i \sum_j \nu_{i,j}\varepsilon_j$$

This may also be written

$$n = n_0 + \sum_j \left(\sum_i \nu_{i,j} \right) \varepsilon_j$$

Analogous to the definition ν for a single reaction, we here adopt the definition:

$$\nu_j \equiv \sum_i \nu_{i,j}$$

Then

$$n = n_0 + \sum_j \nu_j \varepsilon_j$$

Combination of this equation with Eq. (15.6) gives the mole fraction:

$$\boxed{y_i = \frac{n_{i_0} + \sum_j \nu_{i,j}\varepsilon_j}{n_0 + \sum_j \nu_j \varepsilon_j}} \qquad (i = 1, 2, \ldots, N) \tag{15.7}$$

Example 15.3 Consider a system in which the following reactions occur:

$$CH_4 + H_2O \rightarrow CO + 3H_2 \tag{1}$$

$$CH_4 + 2H_2O \rightarrow CO_2 + 4H_2 \tag{2}$$

where the numbers (1) and (2) indicate the value of j, the reaction index. If there are present initially 2 mol CH_4 and 3 mol H_2O, determine expressions for the y_i as functions of ε_1 and ε_2.

SOLUTION The stoichiometric numbers $\nu_{i,j}$ can be arrayed as follows:

j	CH_4	H_2O	CO	CO_2	H_2	ν_j
1	−1	−1	1	0	3	2
2	−1	−2	0	1	4	2

Application of Eq. (15.7) now gives

$$y_{CH_4} = \frac{2 - \varepsilon_1 - \varepsilon_2}{5 + 2\varepsilon_1 + 2\varepsilon_2}$$

$$y_{H_2O} = \frac{3 - \varepsilon_1 - 2\varepsilon_2}{5 + 2\varepsilon_1 + 2\varepsilon_2}$$

$$y_{CO} = \frac{\varepsilon_1}{5 + 2\varepsilon_1 + 2\varepsilon_2}$$

$$y_{CO_2} = \frac{\varepsilon_2}{5 + 2\varepsilon_1 + 2\varepsilon_2}$$

$$y_{H_2} = \frac{3\varepsilon_1 + 4\varepsilon_2}{5 + 2\varepsilon_1 + 2\varepsilon_2}$$

The composition of the system is a function of the two independent variables ε_1 and ε_2.

15.2 APPLICATION OF EQUILIBRIUM CRITERIA TO CHEMICAL REACTIONS

In Sec. 13.8 it is shown that the total Gibbs energy of a closed system at constant T and P must decrease during an irreversible process and that the condition for equilibrium is reached when

$$(dG^t)_{T,P} = 0 \tag{13.53}$$

Thus if a mixture of chemical species is not in chemical equilibrium, any reaction that occurs must be irreversible and, if the system is maintained at constant T and P, the total Gibbs energy of the system must decrease. The significance of this for a single chemical reaction is seen in Fig. 15.1, which shows a schematic diagram of G^t vs. ε, the reaction coordinate. Since ε is the single variable that

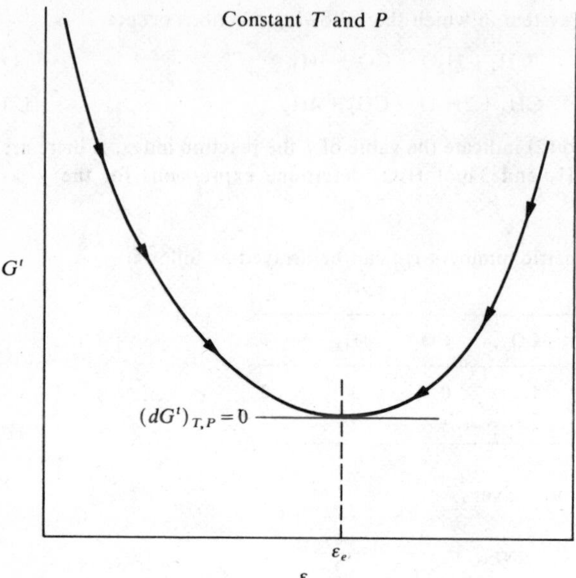

Figure 15.1 The total Gibbs energy in relation to the reaction coordinate.

characterizes the progress of the reaction, and therefore the composition of the system, the total Gibbs energy at constant T and P is determined by ε. The arrows along the curve in Fig. 15.1 indicate the directions of changes in $(G^t)_{T,P}$ that are possible on account of reaction. The reaction coordinate has its equilibrium value ε_e at the minimum of the curve. The meaning of Eq. (13.53) is that differential displacements of the chemical reaction can occur at the equilibrium state without causing changes in the total Gibbs energy of the system. Our purpose is to use this criterion for the calculation of values of ε_e, and hence of the compositions of systems in chemical equilibrium.

The alternative criterion of equilibrium, also discussed in Sec. 13.8, is that the equilibrium state of a closed system at constant T and P is that state for which the total Gibbs energy is a minimum with respect to all possible changes. Figure 15.1 illustrates this criterion for the special case of a single reaction. The equilibrium states of systems in which two or more simultaneous chemical reactions occur is often most conveniently found by application of this criterion. The procedure is to write an expression for the total Gibbs energy of the system and then to find the composition which minimizes G^t for a given T and P, subject to the constraints of the material balances. This method is considered in Sec. 15.9.

These two criteria of equilibrium, which are stated for closed systems at constant T and P, are *not* restricted in application to systems that are actually closed and reach equilibrium states along paths of constant T and P. Once an equilibrium state is reached, no further changes occur, and the system continues to exist in this state at fixed T and P. How this state was *actually* attained does not matter. Once it is known that an equilibrium state exists at given T and P, the criteria apply.

15.3 THE STANDARD GIBBS ENERGY CHANGE AND THE EQUILIBRIUM CONSTANT

Equation (10.2), the fundamental property relation for single-phase systems, provides an expression for the total differential of the Gibbs energy:

$$d(nG) = (nV)\,dP - (nS)\,dT + \sum \mu_i\,dn_i \tag{10.2}$$

If changes in the mole numbers n_i occur as the result of a single chemical reaction in a closed system, then by Eq. (15.3) each dn_i may be replaced by the product $\nu_i\,d\varepsilon$. Equation (10.2) then becomes

$$d(nG) = (nV)\,dP - (nS)\,dT + \sum \nu_i\mu_i\,d\varepsilon$$

Since nG is a state function, the right-hand side of this equation is an exact differential expression; it follows that

$$\sum \nu_i\mu_i = \left[\frac{\partial(nG)}{\partial\varepsilon}\right]_{T,P}$$

Thus the quantity $\sum \nu_i\mu_i$ represents, in general, the rate of change of the total Gibbs energy of the system with the reaction coordinate at constant T and P. Figure 15.1 shows that this quantity is zero at the equilibrium state. Therefore a criterion of chemical-reaction equilibrium is

$$\boxed{\sum \nu_i\mu_i = 0} \tag{15.8}$$

Since the chemical potential μ_i of species i in solution is identically equal to $\bar{G}_i$, Eq. (11.28) may be written:

$$d\mu_i = d\bar{G}_i = RT\,d\ln\hat{f}_i \qquad \text{(const } T\text{)}$$

Integration of this equation at constant T from the standard state of species i (see Sec. 4.3) to a state of species i in solution gives

$$\mu_i - G_i^\circ = RT\ln\frac{\hat{f}_i}{f_i^\circ} \tag{15.9}$$

The ratio $\hat{f}_i/f_i^\circ$ is called the *activity* $\hat{a}_i$ of species i in solution. Thus by definition,

$$\hat{a}_i \equiv \frac{\hat{f}_i}{f_i^\circ} \tag{15.10}$$

and the preceding equation becomes

$$\mu_i = G_i^\circ + RT\ln\hat{a}_i \tag{15.11}$$

Combining Eq. (15.8) with Eq. (15.11) to eliminate μ_i gives for the equilibrium state of a chemical reaction:

$$\sum \nu_i(G_i^\circ + RT \ln \hat{a}_i) = 0$$

or

$$\sum \nu_i G_i^\circ + RT \sum \ln (\hat{a}_i)^{\nu_i} = 0$$

or

$$\ln \prod (\hat{a}_i)^{\nu_i} = \frac{-\sum \nu_i G_i^\circ}{RT} \tag{15.12}$$

where $\prod$ signifies the product over all species i. In exponential form, Eq. (15.12) becomes

$$\boxed{\prod (\hat{a}_i)^{\nu_i} = \exp \frac{-\sum \nu_i G_i^\circ}{RT} \equiv K} \tag{15.13}$$

Included in this equation is the definition of K. Since G_i° is a property of pure species i in its standard state at fixed pressure, it depends only on temperature. It follows from Eq. (15.13) that K is also a function of temperature only. In spite of its dependence on temperature, K is called the equilibrium *constant* for the reaction. Equation (15.12) may now be written

$$\boxed{-RT \ln K = \sum \nu_i G_i^\circ \equiv \Delta G^\circ} \tag{15.14}$$

The final term ΔG° is the conventional way of representing the quantity $\sum \nu_i G_i^\circ$. It is called the *standard Gibbs energy change of reaction.*

The activities $\hat{a}_i$ in Eq. (15.13) provide the connection between the *equilibrium* state of interest and the *standard* states of the individual species, for which data are presumed available, as discussed in Sec. 15.5. The standard states are arbitrary, but must always be at the equilibrium temperature T. The standard states selected need not be the same for all species taking part in a reaction. However, for a *particular* species the standard state represented by G_i° must be the same state represented by the f_i° upon which the activity $\hat{a}_i$ is based.

For a gas the standard state is the ideal-gas state of pure i at a pressure of 1 bar [or 1(atm)]. Since the fugacity of an ideal gas is equal to the pressure, $f_i^\circ = 1$ bar [or $f_i^\circ = 1$(atm)] for each species of a gase-phase reaction. Thus for gas-phase reactions, $\hat{a}_i = \hat{f}_i/f_i^\circ = \hat{f}_i$ and Eq. (15.13) becomes

$$\boxed{K = \prod (\hat{f}_i)^{\nu_i}} \tag{15.15}$$

The fugacities $\hat{f}_i$ must be in bars [or (atm)] because each $\hat{f}_i$ is implicitly divided by 1 bar [or 1(atm)], and K must be dimensionless.

For solids and liquids the usual standard state is the pure solid or liquid at 1 bar [or 1(atm)] and at the temperature of the system. The value of f_i° for such a state is not likely to be 1 bar or 1(atm), and Eq. (15.15) is not valid; the general expression which relates K to activities, given by Eq. (15.13), must therefore be used for equilibrium calculations.

The function $\Delta G^\circ \equiv \sum \nu_i G_i^\circ$ in Eq. (15.14) is the weighted difference (recall that the ν_i's are positive for products and negative for reactants) between the Gibbs energies of the products and reactants when each is in its standard state as a pure substance at the system temperature and at a fixed pressure. Thus the value of ΔG° is fixed for a given reaction once the temperature is established, and is independent of the equilibrium pressure and composition. Other *standard property changes of reaction* are similarly defined. Thus, for the general property M, we write

$$\Delta M^\circ = \sum \nu_i M_i^\circ$$

In accord with this, ΔH° is defined by Eq. (4.14) and ΔC_P° by Eq. (4.16). For the standard entropy change of reaction ΔM° becomes ΔS°. These quantities are all functions of temperature only for a given reaction, and are related to one another by equations analogous to property relations for pure species.

As an example we develop the relation between the standard heat of reaction and the standard Gibbs energy change of reaction. Equation (6.31) written for species i in its standard state becomes

$$H_i^\circ = -RT^2 \frac{d(G_i^\circ/RT)}{dT}$$

Total derivatives are appropriate here because the properties in the standard state are functions of temperature only. Multiplication of both sides of this equation by ν_i and summation over all species gives

$$\sum \nu_i H_i^\circ = -RT^2 \frac{d(\sum \nu_i G_i^\circ/RT)}{dT}$$

In view of the definitions of Eqs. (4.14) and (15.14), this may be written

$$\Delta H^\circ = -RT^2 \frac{d(\Delta G^\circ/RT)}{dT} \tag{15.16}$$

Example 15.4 Devise a gas-phase process for the reversible conversion of reactant species A and B in their standard states into product species L and M in their standard states in accord with the reaction

$$aA + bB \rightarrow lL + mM$$

and show that ΔG° for the process is consistent with Eqs. (15.14) and (15.15).

SOLUTION Figure 15.2 shows a large box containing the reactant and product species in equilibrium at temperature T and pressure P. It is known as a van't Hoff equilibrium box. Material is added and withdrawn through semipermeable membranes that separate the four piston/cylinder assemblies from the box. Each semipermeable membrane permits the passage of only the pure species in its adjacent cylinder.

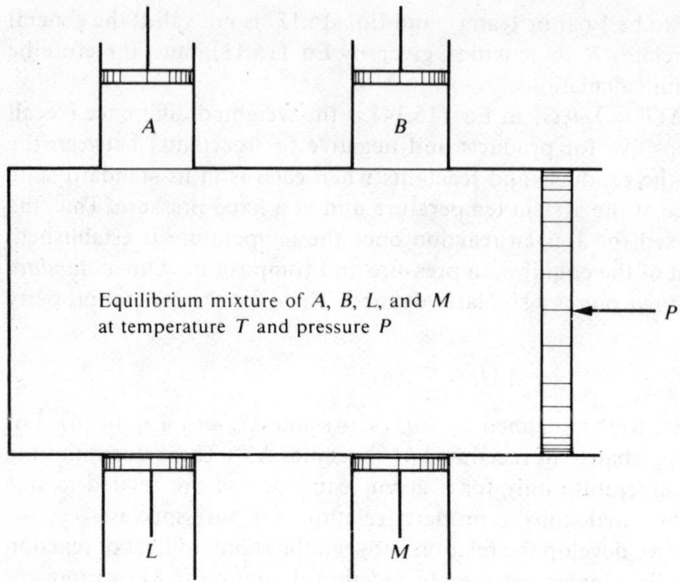

Figure 15.2 Apparatus in which a gas-phase reaction occurs at equilibrium (van't Hoff equilibrium box).

Initially, a moles of species A and b moles of species B are contained in the cylinders shown at the top of the box in Fig. 15.2. Each is stored in its cylinder as a pure gas at temperature T and at a fugacity of 1 bar, i.e., in its standard state. The following series of steps transforms these reactants into l moles of L and m moles of M, the pure product species in their standard states at temperature T and a fugacity of 1 bar. They are collected in the lower cylinders shown in Fig. 15.2.

1. The pure species A and B are isothermally compressed (or expanded, depending on the pressure P) to their equilibrium fugacities in the box. The change in the Gibbs energy for this process is given by Eq. (15.9), here written for one mole as

$$\Delta G_i = RT \ln \frac{\hat{f}_i}{f_i^\circ} = RT \ln \hat{f}_i$$

The total Gibbs energy change for step 1 is therefore

$$\Delta G_1 = RT(a \ln \hat{f}_A + b \ln \hat{f}_B) = RT \ln (\hat{f}_A^a \hat{f}_B^b)$$

2. The a moles of A and b moles of B are added to the box through the semipermeable membranes. Since the fugacities in the cylinders are the same as in the box, the process occurs at equilibrium, and there is no change in the total Gibbs energy of the system:

$$\Delta G_2 = 0$$

3. Once in the box, the reactants are converted into l moles of L and m moles of M under conditions of equilibrium. For this change also the change in total Gibbs

energy is zero, in accord with Eq. (13.53):

$$\Delta G_3 = 0$$

4. In the reverse of step 2, the l moles of L and m moles of M are transferred at constant fugacity into the two product cylinders. Again

$$\Delta G_4 = 0$$

5. Finally, the products are isothermally expanded (or compressed) from their respective equilibrium fugacities to their standard-state fugacities of 1 bar. The Gibbs energy change is calculated as in step 1:

$$\Delta G_5 = RT\left(l \ln \frac{1}{\hat{f}_L} + m \ln \frac{1}{\hat{f}_M}\right) = -RT \ln (\hat{f}_L^l \hat{f}_M^m)$$

The overall change in the Gibbs energy for the entire process, i.e., the sum of the changes for the five steps, is also the *standard* Gibbs energy change of reaction, because the overall result of the process is the conversion of reactants to products, all in their standard states. Therefore

$$\Delta G^\circ = \Delta G_1 + \Delta G_2 + \Delta G_3 + \Delta G_4 + \Delta G_5 = -RT \ln \frac{\hat{f}_L^l \hat{f}_M^m}{\hat{f}_A^a \hat{f}_B^b}$$

If this result is rewritten with the stoichiometric numbers ν_i substituted for the stoichiometric coefficients a, b, l, and m, all the fugacities appear in the numerator of the logarithm, because the ν_i's for the reactants are defined as negative numbers. We may therefore write

$$\Delta G^\circ = \sum \nu_i G_i^\circ = -RT \ln \prod (\hat{f}_i)^{\nu_i} = -RT \ln K$$

which the same result obtained when Eqs. (15.14) and (15.15) are combined.

Use has here been made in steps 2 through 4 of the fact that there is no change in the Gibbs energy for processes carried out under conditions of membrane and chemical-reaction equilibrium. This explains why the value of ΔG° is related directly to the ratios of the equilibrium-state and standard-state fugacities ($f_i^\circ = 1$).

15.4 EFFECT OF TEMPERATURE ON THE EQUILIBRIUM CONSTANT

Since the standard-state temperature is that of the equilibrium mixture, the standard property changes of reaction, such as ΔG° and ΔH°, vary with the equilibrium temperature. The dependence of ΔG° on T is given by Eq. (15.16), which may be rewritten as

$$\frac{d(\Delta G^\circ/RT)}{dT} = \frac{-\Delta H^\circ}{RT^2}$$

According to Eq. (15.14),

$$\frac{\Delta G^\circ}{RT} = -\ln K$$

Therefore

$$\boxed{\frac{d \ln K}{dT} = \frac{\Delta H^\circ}{RT^2}} \tag{15.17}$$

Equation (15.17) gives the effect of temperature on the equilibrium constant, and hence on the equilibrium yield. If ΔH° is negative, i.e., if the reaction is exothermic, the equilibrium constant decreases as the temperature increases. Conversely, K increases with T for an endothermic reaction.

If ΔH°, the standard enthalpy change (heat) of reaction, is assumed independent of T, integration of Eq. (15.17) leads to the simple result,

$$\ln \frac{K}{K_1} = -\frac{\Delta H^\circ}{R}\left(\frac{1}{T} - \frac{1}{T_1}\right) \tag{15.18}$$

This approximate equation implies that a plot of ln K vs. the reciprocal of absolute temperature is a straight line. Figure 15.3, a plot of ln K vs. $1/T$ for a number of common reactions, illustrates this near linearity. Thus, Eq. (15.18) provides a reasonably accurate relation for the interpolation and extrapolation of equilibrium-constant data.

If the standard heat of reaction is known as a function of T, Eq. (15.17) can be integrated rigorously, as indicated by the equation

$$\ln K = \int \frac{\Delta H^\circ}{RT^2}\, dT + I \tag{15.19}$$

where I is a constant of integration. The general expression for ΔH° is given by Eq. (4.20),

$$\Delta H^\circ = J + \int \Delta C_P^\circ \, dT$$

where J is another integration constant. When each $C_{P_i}^\circ$ is given by Eq. (4.4), the expression that results is Eq. (4.22), here written:

$$\frac{\Delta H^\circ}{R} = \frac{J}{R} + (\Delta A)T + \frac{\Delta B}{2}T^2 + \frac{\Delta C}{3}T^3 - \frac{\Delta D}{T} \tag{15.20}$$

Substitution of this result in Eq. (15.19) and integration give

$$\ln K = \frac{-J}{RT} + \Delta A \ln T + \frac{\Delta B}{2}T + \frac{\Delta C}{6}T^2 + \frac{\Delta D}{2T^2} + I \tag{15.21}$$

Since, by Eq. (15.14), $\Delta G^\circ = -RT \ln K$, multiplication of Eq. (15.21) by $-RT$ yields

$$\Delta G^\circ = J - RT\left(\Delta A \ln T + \frac{\Delta B}{2}T + \frac{\Delta C}{6}T^2 + \frac{\Delta D}{2T^2} + I\right) \tag{15.22}$$

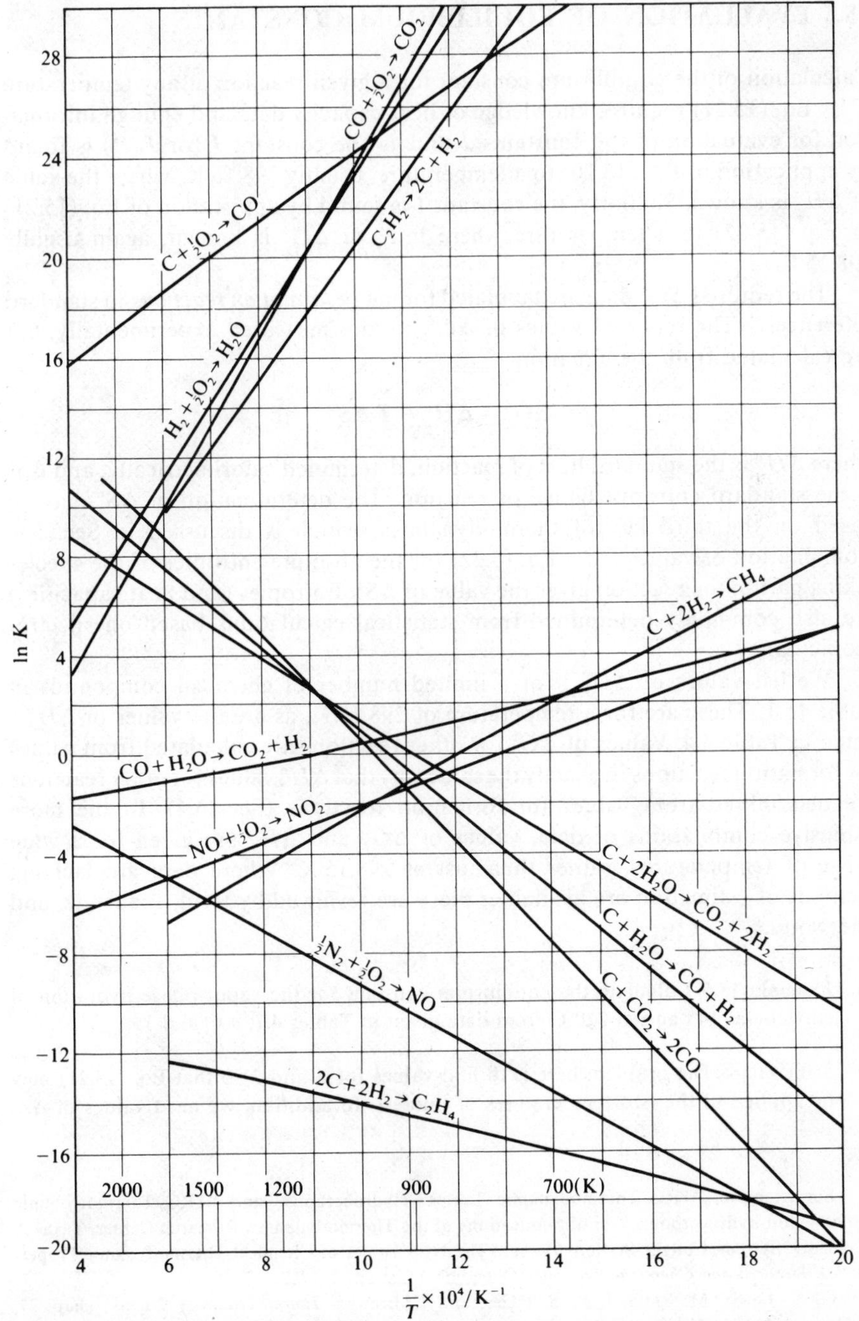

Figure 15.3 Equilibrium constants as a function of temperature for some common reactions.

15.5 EVALUATION OF EQUILIBRIUM CONSTANTS

Calculation of the equilibrium constant for a given reaction at any temperature T by Eq. (15.21) requires knowledge of heat-capacity data and enough information for evaluation of the constants J and I. The constant J (or J/R) is found by application of Eq. (15.20) to a temperature, usually 298.15 K, where the value of $\Delta H°$ is known. Similarly, the constant I is found by application of Eq. (15.21) or Eq. (15.22) to a temperature where $\ln K$ or $\Delta G°$ is known, again usually 298.15 K.

The required $\Delta G°$ data are tabulated for many *formation reactions* in standard references.† The reported values of $\Delta G°$ are not measured experimentally, but are calculated from the equation

$$\Delta G° = \Delta H° - T\,\Delta S°$$

where $\Delta H°$ is the standard heat of reaction, determined calorimetrically, and $\Delta S°$ is the standard entropy change of reaction. The determination of $\Delta S°$ may be based on the third law of thermodynamics, which is discussed in Sec. 5.9. Combination of values from Eq. (5.22) for the absolute entropies of the species taking part in the reaction gives the value of $\Delta S°$. Entropies (and heat capacities) are also commonly determined from statistical calculations based on spectroscopic data.‡

We list values of $\Delta G°_{f_{298}}$ for a limited number of chemical compounds in Table 15.1. These are for a temperature of 298.15 K, as are the values of $\Delta H°_{f_{298}}$ listed in Table 4.4. Values of $\Delta G°$ for other reactions are calculated from values for formation reactions in exactly the same way that $\Delta H°$ values for other reactions are determined from values for formation reactions (Sec. 4.4). In the more extensive compilations of data, values of $\Delta G°_f$ and $\Delta H°_f$ are given for a wide range of temperatures, rather than just at 298.15 K. Where data are lacking, methods of estimation are available; these are reviewed by Reid, Prausnitz, and Sherwood.§

Example 15.5 Calculate the equilibrium constant for the vapor-phase hydration of ethylene at 145 and at 320°C from data given in Tables 4.1, 4.4, and 15.1.

SOLUTION The problem here is to find values for J and I so that Eq. (15.21) may be applied at the two temperatures of interest. In addition, we need values of ΔA,

† For example, "TRC Thermodynamic Tables—Hydrocarbons" and "TRC Thermodynamic Tables—Non-hydrocarbons," serial publications of the Thermodynamics Research Center, Texas A & M Univ. System, College Station, Texas; "The NBS Tables of Chemical Thermodynamic Properties," *J. Physical and Chemical Reference Data*, vol. 11, supp. 2, 1982.

‡ G. N. Lewis, M. Randall, K. S. Pitzer, and L. Brewer, *Thermodynamics*, 2d ed., chap. 27, McGraw-Hill, New York, 1961.

§ R. C. Reid, J. M. Prausnitz, and T. K. Sherwood, *The Properties of Gases and Liquids*, 3d ed., chap. 7, McGraw-Hill, New York, 1977.

ΔB, ΔC, and ΔD, and these come from heat-capacity data. For the reaction

$$C_2H_4(g) + H_2O(g) \rightarrow C_2H_5OH(g)$$

the meaning of Δ is indicated by

$$\Delta = (C_2H_5OH) - (C_2H_4) - (H_2O)$$

Thus, from the heat-capacity data of Table 4.1 we have:

$$\Delta A = 3.518 - 1.424 - 3.470 = -1.376$$

$$\Delta B = (20.001 - 14.394 - 1.450) \times 10^{-3} = 4.157 \times 10^{-3}$$

$$\Delta C = (-6.002 + 4.392 - 0.000) \times 10^{-6} = -1.610 \times 10^{-6}$$

$$\Delta D = (0.000 - 0.000 - 0.121) \times 10^{5} = -0.121 \times 10^{5}$$

Evaluation of the constants J and I by application of Eqs. (15.20) and (15.21) at 298.15 K requires values of ΔH°_{298} and ΔG°_{298} for the hydration reaction. These are found from the heat-of-formation data of Table 4.4 and the Gibbs-energy-of-formation data of Table 15.1:

$$\Delta H^\circ_{298} = -235{,}100 - 52{,}510 - (-241{,}818) = -45{,}792 \text{ J mol}^{-1}$$

and

$$\Delta G^\circ_{298} = -168{,}490 - 68{,}430 - (-228{,}572) = -8{,}348 \text{ J mol}^{-1}$$

By Eq. (15.14) applied at 298.15 K,

$$\ln K = \frac{-\Delta G^\circ}{RT} = \frac{8{,}348}{(8.314)(298.15)} = 3.3677$$

Substitution of known values into Eq. (15.20) for $T = 298.15$ gives

$$\frac{-45{,}792}{8.314} = \frac{J}{R} - (1.376)(298.15) + (2.0785 \times 10^{-3})(298.15)^2 - (0.5367 \times 10^{-6})(298.15)^3 + \frac{12{,}100}{298.15}$$

Whence

$$\frac{J}{R} = -5{,}308.7$$

Substitution of known values into Eq. (15.21) for $T = 298.15$ K gives

$$3.3677 = \frac{5{,}308.7}{298.15} - 1.376 \ln 298.15 + (2.0785 \times 10^{-3})(298.15) - (0.2683 \times 10^{-6})(298.15)^2 - \frac{12{,}100}{(2)(298.15)^2} + I$$

Whence

$$I = -7.125$$

Table 15.1 Standard Gibbs energies of formation at 298.15 K (25°C)†

Joules per mole of the substance formed

Chemical species		State (Note 2)	$\Delta G^\circ_{f_{298}}$
Paraffins:			
Methane	CH_4	*g*	−50,460
Ethane	C_2H_6	*g*	−31,855
Propane	C_3H_8	*g*	−24,290
n-Butane	C_4H_{10}	*g*	−16,570
n-Pentane	C_5H_{12}	*g*	−8,650
n-Hexane	C_6H_{14}	*g*	150
n-Heptane	C_7H_{16}	*g*	8,260
n-Octane	C_8H_{18}	*g*	16,260
1-Alkenes:			
Ethylene	C_2H_4	*g*	68,460
Propylene	C_3H_6	*g*	62,205
1-Butene	C_4H_8	*g*	70,340
1-Pentene	C_5H_{10}	*g*	78,410
1-Hexene	C_6H_{12}	*g*	86,830
Miscellaneous organics:			
Acetaldehyde	C_2H_4O	*g*	−128,860
Acetic acid	$C_2H_4O_2$	*l*	−389,900
Acetylene	C_2H_2	*g*	209,970
Benzene	C_6H_6	*g*	129,665
Benzene	C_6H_6	*l*	124,520
1,3-Butadiene	C_4H_6	*g*	149,795
Cyclohexane	C_6H_{12}	*g*	31,920
Cyclohexane	C_6H_{12}	*l*	26,850
1,2-Ethanediol	$C_2H_6O_2$	*l*	−323,080
Ethanol	C_2H_6O	*g*	−168,490
Ethanol	C_2H_6O	*l*	−174,780
Ethylbenzene	C_8H_{10}	*g*	130,890
Ethylene oxide	C_2H_4O	*g*	−13,010
Formaldehyde	CH_2O	*g*	−102,530
Methanol	CH_4O	*g*	−161,960
Methanol	CH_4O	*l*	−166,270
Methylcyclohexane	C_7H_{14}	*g*	27,480
Methylcyclohexane	C_7H_{14}	*l*	20,560
Styrene	C_8H_8	*g*	213,900
Toluene	C_7H_8	*g*	122,050
Toluene	C_7H_8	*l*	113,630

Chemical species		State (Note 2)	$\Delta G^\circ_{f_{298}}$
Miscellaneous inorganics:			
Ammonia	NH_3	*g*	−16,450
Ammonia	NH_3	*aq*	−26,500
Calcium carbide	CaC_2	*s*	−64,900
Calcium carbonate	$CaCO_3$	*s*	−1,128,790
Calcium chloride	$CaCl_2$	*s*	−748,100
Calcium chloride	$CaCl_2$	*aq*	−816,010
Calcium hydroxide	$Ca(OH)_2$	*s*	−898,490
Calcium hydroxide	$Ca(OH)_2$	*aq*	−868,070
Calcium oxide	CaO	*s*	−604,030
Carbon dioxide	CO_2	*g*	−394,359
Carbon monoxide	CO	*g*	−137,169
Hydrochloric acid	HCl	*g*	−95,299
Hydrogen cyanide	HCN	*g*	124,700
Hydrogen sulfide	H_2S	*g*	−33,560
Iron oxide (hematite)	Fe_2O_3	*s*	−742,200
Iron oxide (magnetite)	Fe_3O_4	*s*	−1,015,400
Iron sulfide (pyrite)	FeS_2	*s*	−166,900
Nitric acid	HNO_3	*l*	−80,710
Nitric acid	HNO_3	*aq*	−111,250
Nitrogen oxides	NO	*g*	86,550
	NO_2	*g*	51,310
	N_2O	*g*	104,200
	N_2O_4	*g*	97,540
Sodium carbonate	Na_2CO_3	*s*	−1,044,440
Sodium chloride	$NaCl$	*s*	−384,138
Sodium chloride	$NaCl$	*aq*	−393,133
Sodium hydroxide	$NaOH$	*s*	−379,494
Sodium hydroxide	$NaOH$	*aq*	−419,150
Sulfur dioxide	SO_2	*g*	−300,194
Sulfur trioxide	SO_3	*g*	−371,060
Sulfuric acid	H_2SO_4	*l*	−690,003
Sulfuric acid	H_2SO_4	*aq*	−744,530
Water	H_2O	*g*	−228,572
Water	H_2O	*l*	−237,129

† Taken from "TRC Thermodynamic Tables—Hydrocarbons", Thermodynamics Research Center, Texas A & M Univ. System, College Station, Texas; "The NBS Tables of Chemical Thermodynamic Properties," *J. Physical and Chemical Reference Data*, vol. 11, supp. 2, 1982.

Notes

1. The standard Gibbs energy of formation $\Delta G^\circ_{f_{298}}$ is the change in the Gibbs energy when 1 mol of the listed compound is formed from its elements with each substance in its standard state at 298.15 K (25°C).
2. Standard states: (*a*) Gases (*g*): the pure ideal gas at 1 bar and 25°C. (*b*) Liquids (*l*) and solids (*s*): the pure substance at 1 bar and 25°C. (*c*) Solutes in aqueous solution (*aq*): The hypothetical ideal 1 molal solution of the solute in water at 1 bar and 25°C.

The general expression for ln K is therefore

$$\ln K = \frac{5{,}308.7}{T} - 1.376 \ln T + 2.0785 \times 10^{-3} T - 0.2683 \times 10^{-6} T^2 - \frac{12{,}100}{2T^2} - 7.125$$

Application of this equation for $T = 145 + 273.15 = 418.1$ K and for $T = 320 + 273.15 = 593.15$ K gives:

$$\text{At } 418.15 \text{ K: } \ln K = -1.948 \quad \text{and} \quad K = 14.26 \times 10^{-2}$$

$$\text{At } 593.15 \text{ K: } \ln K = -5.840 \quad \text{and} \quad K = 2.91 \times 10^{-3}$$

15.6 RELATIONS BETWEEN EQUILIBRIUM CONSTANTS AND COMPOSITION

Gas-phase reactions

Although equilibrium constants for gas-phase reactions are evaluated by Eq. (15.14) with data for ideal-gas standard states, they are related by Eq. (15.15)

$$K = \prod (\hat{f}_i)^{\nu_i} \tag{15.15}$$

to fugacities of the species in the real equilibrium mixture. These fugacities reflect the nonidealities of the equilibrium mixture and are functions of temperature, pressure, and composition. On the other hand, K is a function of temperature only. This means that for a fixed temperature the composition at equilibrium must change with pressure in such a way that $\prod (\hat{f}_i)^{\nu_i}$ remains constant. The fugacity is related to the fugacity coefficient by Eq. (11.33), here written

$$\hat{f}_i = \hat{\phi}_i y_i P$$

Substitution of this equation into Eq. (15.15) provides an equilibrium expression that includes the pressure and the composition:

$$\boxed{\prod (y_i \hat{\phi}_i)^{\nu_i} = P^{-\nu} K} \tag{15.23}$$

where $\nu \equiv \sum \nu_i$ and P must be expressed in bars when the standard-state pressure is 1 bar and in (atm) when the standard-state pressure is 1(atm). The y_i's may be eliminated in favor of the equilibrium value of the reaction coordinate ε_e. Then, for a fixed temperature Eq. (15.23) relates ε_e to P. In principle, specification of the pressure allows solution for ε_e. However, the problem may be complicated by the dependence of the $\hat{\phi}_i$'s on composition, i.e., on ε_e. The methods of Secs. 11.4 and 14.3 can be applied to the calculation of $\hat{\phi}_i$ values, for example, by Eq. (11.48) or (14.47). Because of the complexity of the calculations, an iterative procedure, initiated by setting $\hat{\phi}_i = 1$ and formulated for computer solution, is indicated. Once an initial set of y_i's is calculated, the $\hat{\phi}_i$'s are determined, and the procedure is repeated to convergence.

If the assumption is justified that the equilibrium mixture is an *ideal solution*, then each $\hat{\phi}_i$ becomes ϕ_i, the fugacity coefficient of pure i at T and P. In this case, Eq. (15.23) becomes

$$\boxed{\prod (y_i \phi_i)^{\nu_i} = P^{-\nu} K} \qquad (15.24)$$

Since the ϕ_i's are independent of composition, they can be evaluated from a generalized correlation once the equilibrium T and P are specified.

When the pressure is sufficiently low or the temperature sufficiently high, the equilibrium mixture behaves essentially as an ideal gas. In this event, each $\hat{\phi}_i = 1$, and Eq. (15.23) reduces to

$$\boxed{\prod (y_i)^{\nu_i} = P^{-\nu} K} \qquad (15.25)$$

In this equation the temperature-, pressure-, and composition-dependent terms are distinct and separate, and solution for any one of ε_e, T, or P, given the other two, is straightforward.

Although Eq. (15.25) holds only for an ideal-gas reaction, we can base some conclusions on it that are true in general.

1. According to Eq. (15.17), the effect of temperature on the equilibrium constant K is determined by the sign of ΔH°. Thus when ΔH° is positive, i.e., when the standard reaction is *endothermic*, an increase in T results in an increase in K. Equation (15.25) shows that an increase in K at constant P results in an increase in $\prod (y_i)^{\nu_i}$; this implies a shift of the reaction to the right and an increase in ε_e. Conversely, when ΔH° is negative, i.e., when the standard reaction is *exothermic*, an increase in T causes a decrease in K and a decrease in $\prod (y_i)^{\nu_i}$ at constant P. This implies a shift of the reaction to the left and a decrease in ε_e.
2. If the total stoichiometric number ν $(=\sum \nu_i)$ is negative, Eq. (15.25) shows that an increase in P at constant T causes an increase in $\prod (y_i)^{\nu_i}$, implying a shift of the reaction to the right and an increase in ε_e. If ν is positive, an increase in P at constant T causes a decrease in $\prod (y_i)^{\nu_i}$, a shift of the reaction to the left, and a decrease in ε_e.

Liquid-phase Reactions

For a reaction occurring in the liquid phase, we return to Eq. (15.13), which relates K to activities:

$$K = \prod (\hat{a}_i)^{\nu_i} \qquad (15.26)$$

The most common standard state for liquids is the state of the pure liquid at the system temperature and at 1 bar or 1(atm).† The activities are then given by

$$\hat{a}_i = \frac{\hat{f}_i}{f_i^\circ}$$

where f_i° is the fugacity of pure liquid i at the temperature of the system and at 1 bar.

According to Eq. (11.59), which defines the activity coefficient,

$$\hat{f}_i = \gamma_i x_i f_i$$

where f_i is the fugacity of pure liquid i at the temperature *and pressure* of the equilibrium mixture. The activity can now be expressed as

$$\hat{a}_i = \frac{\gamma_i x_i f_i}{f_i^\circ} = \gamma_i x_i \left(\frac{f_i}{f_i^\circ} \right) \tag{15.27}$$

Since the fugacities of liquids are weak functions of pressure, the ratio f_i/f_i° is often taken as unity. However, it is readily evaluated by means of Eq. (11.25),

$$d \ln f_i = \frac{V_i}{RT}\, dP \qquad \text{(const } T\text{)}$$

Since V_i changes little with pressure for liquids (and solids), integration from the standard-state pressure of 1 bar to pressure P (in bars) gives

$$\ln \frac{f_i}{f_i^\circ} \simeq \frac{V_i(P-1)}{RT}$$

Equation (15.26) may now be written:

$$\boxed{K = \left[\prod (x_i \gamma_i)^{\nu_i} \right] \exp \left[\frac{(P-1)}{RT} \sum (\nu_i V_i) \right]} \tag{15.28}$$

Except for high pressures, the exponential term is close to unity and may be omitted. In this case,

$$K = \prod (x_i \gamma_i)^{\nu_i} \tag{15.29}$$

and the only problem is determination of the activity coefficients. An equation such as the Wilson equation [Eq. (12.24)] or the UNIFAC method can in principle be applied, and the compositions can be found from Eq. (15.29) by a complex iterative computer program. However, the relative ease of experimental investigation for liquid mixtures has worked against the application of Eq. (15.29).

If the equilibrium mixture is an ideal solution, then all the γ_i's are unity, and Eq. (15.29) becomes

$$K = \prod (x_i)^{\nu_i} \tag{15.30}$$

† For liquids and solids, the difference is inconsequential.

This simple relation is known as the *law of mass action.* Since liquids that react are likely to form nonideal solutions, Eq. (15.30) can be expected in most instances to yield poor results.

For species known to be present in high concentration, the equation $\hat{a}_i = x_i$ is usually nearly correct, because the Lewis/Randall rule always becomes valid for a species as its concentration approaches $x_i = 1$, as discussed in Sec. 12.7.

For species at low concentration in aqueous solution, a different procedure has been widely adopted, because in this case the equality of $\hat{a}_i$ and x_i is usually far from correct. The method is based on the use of a fictitious or hypothetical standard state for the solute, taken as the state that would exist if the solute obeyed Henry's law up to a *molality m* of unity. In this application, Henry's law is expressed as

$$\hat{f}_i = k_i m_i \tag{15.31}$$

and it is always valid for a species whose concentration approaches zero. This hypothetical state is illustrated in Fig. 15.4. The dashed line drawn tangent to the curve at the origin represents Henry's law, and is valid in the case shown to a molality much less than unity. However, one can calculate the properties the solute would have if it obeyed Henry's law to a concentration of 1 m, and this hypothetical state often serves as a convenient standard state for solutes.

The standard-state fugacity is

$$\hat{f}_i^\circ = k_i m_i^\circ = k_i(1) = k_i$$

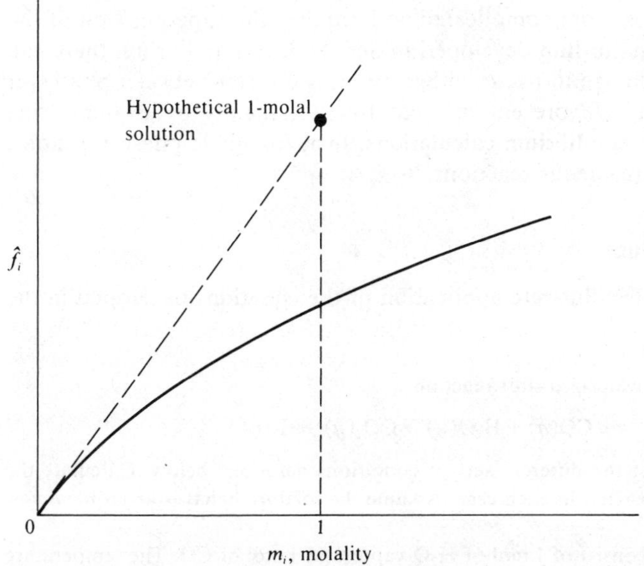

Figure 15.4 Standard state for dilute aqueous solutions.

Hence, for any species at a concentration low enough for Henry's law to hold,

$$\hat{f}_i = k_i m_i = \hat{f}_i^\circ m_i$$

and

$$\hat{a}_i = \frac{\hat{f}_i}{\hat{f}_i^\circ} = m_i \tag{15.32}$$

The advantage of this standard state is that it provides a very simple relation between activity and concentration for cases in which Henry's law is at least approximately valid. Its range does not commonly extend to a concentration of 1 *m*. In the rare case where it does, the standard state is a real state of the solute. This standard state is useful only where ΔG° data are available for the ideal (in the sense of Henry's law) 1-molal standard state, for otherwise the equilibrium constant cannot be evaluated by Eq. (15.14).

15.7 CALCULATION OF EQUILIBRIUM CONVERSIONS FOR SINGLE REACTIONS

Suppose a single reaction occurs in a *homogeneous* system, and suppose the equilibrium constant is known. In this event, the calculation of the phase composition at equilibrium is straightforward if the phase is assumed an ideal gas or an ideal solution. When no assumption of ideality is reasonable, the problem is still tractable for gas-phase reactions through application of an equation of state and solution by computer. For *heterogeneous* systems, where more than one phase is present, the problem is more complicated and requires the superposition of the criterion for phase equilibrium developed in Sec. 11.3. At equilibrium there can be no tendency for change to occur, either by mass transfer between phases or by chemical reaction. We present in what follows, mainly by example, the procedures in use for equilibrium calculations, first, for single-phase reactions, and second, for heterogeneous reactions.

Single-Phase Reactions

The following examples illustrate application of the equations developed in the preceding section.

Example 15.6 The water-gas-shift reaction

$$CO(g) + H_2O(g) \rightarrow CO_2(g) + H_2(g)$$

is carried out under the different sets of conditions described below. Calculate the fraction of steam reacted in each case. Assume the mixture behaves as an ideal gas.

(*a*) The reactants consist of 1 mol of H_2O vapor and 1 mol of CO. The temperature is 1,100 K and the pressure is 1 bar.

(*b*) Same as (*a*) except that the pressure is 10 bar.

(*c*) Same as (*a*) except that 2 mol of N_2 is included in the reactants.
(*d*) The reactants are 2 mol of H_2O and 1 mol of CO. Other conditions are the same as in (*a*).
(*e*) The reactants are 1 mol of H_2O and 2 mol of CO. Other conditions are the same as in (*a*).
(*f*) The initial mixture consists of 1 mol of H_2O, 1 mol of CO, and 1 mol of CO_2. Other conditions are the same as in (*a*).
(*g*) Same as (*a*) except that the temperature is 1,650 K.

SOLUTION (*a*) For the given reaction at 1,100 K, $10^4/T = 9.05$, and Fig. 15.3 provides the value, $\ln K = 0$ or $K = 1$. For this reaction $\nu = \sum \nu_i = 1 + 1 - 1 - 1 = 0$. Since the reaction mixture is an ideal gas, Eq. (15.25) applies, and here becomes:

$$\frac{y_{H_2} y_{CO_2}}{y_{CO} y_{H_2O}} = K = 1 \qquad (A)$$

By Eq. (15.5), we have:

$$y_{CO} = \frac{1 - \varepsilon_e}{2} \qquad y_{H_2O} = \frac{1 - \varepsilon_e}{2}$$

$$y_{CO_2} = \frac{\varepsilon_e}{2} \qquad y_{H_2} = \frac{\varepsilon_e}{2}$$

Substitution of these values into Eq. (*A*) gives

$$\frac{\varepsilon_e^2}{(1 - \varepsilon_e)^2} = 1 \qquad \text{or} \qquad \varepsilon_e = 0.5$$

Therefore the fraction of the steam that reacts is 0.5.

(*b*) Since $\nu = 0$, the increase in pressure has no effect on the ideal-gas reaction, and ε_e is still 0.5.

(*c*) The N_2 does not take part in the reaction, and serves only as a diluent. It does increase the initial number of moles n_0 from 2 to 4, and the mole fractions are all reduced by a factor of 2. However, Eq. (*A*) is unchanged and reduces to the same expression as before. Therefore, ε_e is again 0.5.

(*d*) In this case the mole fractions at equilibrium are:

$$y_{CO} = \frac{1 - \varepsilon_e}{3} \qquad y_{H_2O} = \frac{2 - \varepsilon_e}{3}$$

$$y_{CO_2} = \frac{\varepsilon_e}{3} \qquad y_{H_2} = \frac{\varepsilon_e}{3}$$

and Eq. (*A*) becomes

$$\frac{\varepsilon_e^2}{(1 - \varepsilon_e)(2 - \varepsilon_e)} = 1 \qquad \text{or} \qquad \varepsilon_e = 0.667$$

The fraction of steam that reacts is then $0.667/2 = 0.333$.

(*e*) Here the expressions for y_{CO} and y_{H_2O} are interchanged, but this leaves the equilibrium equation the same as in (*d*). Therefore $\varepsilon_e = 0.667$, and the fraction of steam that reacts is 0.667.

(*f*) In this case Eq. (*A*) becomes

$$\frac{\varepsilon_e(1+\varepsilon_e)}{(1-\varepsilon_e)^2} = 1 \qquad \text{or} \qquad \varepsilon_e = 0.333$$

The fraction of steam reacted is 0.333.

(*g*) At 1,650 K, $10^4/T = 6.06$, and from Fig. 15.3 we have $\ln K = -1.15$ and $K = 0.316$. Therefore Eq. (*A*) becomes

$$\frac{\varepsilon_e^2}{(1-\varepsilon_e)^2} = 0.316 \qquad \text{or} \qquad \varepsilon_e = 0.36$$

Since the reaction is exothermic, the conversion decreases with increasing temperature.

Example 15.7 Estimate the maximum conversion of ethylene to ethanol by vapor-phase hydration at 250°C and 35 bars for an initial steam-to-ethylene ratio of 5.

SOLUTION The general equation for ln *K* as a function of *T* is developed in Example 15.5. For a temperature of 250°C or 523.15 K this equation yields:

$$\ln K = 9.841 \times 10^{-3}$$

The appropriate expression for the equilibrium equation is Eq. (15.23). This equation requires evaluation of the fugacity coefficients of the species present at equilibrium. Although the generalized correlation of Sec. 11.4 is applicable, the calculations involve iteration, because the fugacity coefficients are functions of composition. For purposes of illustration, we carry out only the first iteration, based on the assumption that the reaction mixture is an ideal solution. In this case Eq. (15.23) reduces to Eq. (15.24), which requires fugacity coefficients of the *pure* reacting gases at the equilibrium *T* and *P*. Since $\nu = \sum \nu_i = -1$, this equation becomes

$$\frac{y_{\text{EtOH}}\phi_{\text{EtOH}}}{y_{C_2H_4}\phi_{C_2H_4}y_{H_2O}\phi_{H_2O}} = P(9.841 \times 10^{-3}) \qquad (A)$$

where *P* is in bars.

Equation (11.43) in conjunction with Eqs. (3.48) and (3.49) is suitable for the calculation of values for the ϕ_i's:

$$\ln \phi_i = \frac{P_{r_i}}{T_{r_i}}(B^0 + \omega B^1) \qquad (11.43)$$

where for each species *i*,

$$B^0 = 0.083 - \frac{0.422}{T_{r_i}^{1.6}} \qquad (3.48)$$

$$B^1 = 0.139 - \frac{0.172}{T_{r_i}^{4.2}} \qquad (3.49)$$

The results of these calculations are summarized in the following table:

	T_{c_i}/K	P_{c_i}/bar	ω_i	T_{r_i}	P_{r_i}	B^0	B^1	ϕ_i
C_2H_4	282.4	50.4	0.085	1.853	0.694	−0.074	0.126	0.977
H_2O	647.3	220.5	0.344	0.808	0.159	−0.511	−0.282	0.896
EtOH	516.2	63.8	0.635	1.013	0.548	−0.330	−0.024	0.837

The critical data and ω_i's are from App. B. The temperature and pressure in all cases are 523.15 K and 35 bar. Substitution of values for the ϕ_i's and for P into Eq. (A) gives

$$\frac{y_{\text{EtOH}}}{y_{\text{C}_2\text{H}_4}y_{\text{H}_2\text{O}}} = \frac{(0.977)(0.896)}{(0.837)}(35)(9.841 \times 10^{-3}) = 0.360 \qquad (B)$$

By Eq. (15.5),

$$y_{\text{C}_2\text{H}_4} = \frac{1-\varepsilon_e}{6-\varepsilon_e} \qquad y_{\text{H}_2\text{O}} = \frac{5-\varepsilon_e}{6-\varepsilon_e} \qquad y_{\text{EtOH}} = \frac{\varepsilon_e}{6-\varepsilon_e}$$

Substituting these into Eq. (B) gives

$$\frac{\varepsilon_e(6-\varepsilon_e)}{(5-\varepsilon_e)(1-\varepsilon_e)} = 0.360$$

This reduces to

$$\varepsilon_e^2 - 6.001\varepsilon_e + 1.339 = 0$$

and application of the quadratic formula gives

$$\varepsilon_e = 0.232$$

for the smaller root. Since the larger root is larger than unity, it does not represent a physically possible result. The maximum conversion of ethylene to ethanol under the stated conditions is therefore 23.2 percent.

In this reaction, increasing the temperature decreases K and hence the conversion. Increasing the pressure increases the conversion. Equilibrium considerations therefore suggest that the operating pressure be as high as possible (limited by condensation), and the temperature as low as possible. However, even with the best catalyst known, the minimum temperature for a reasonable reaction rate is about 150°C. This is an instance where both equilibrium and reaction rate influence the commercialization of a reaction process.

The equilibrium conversion is a function of temperature, pressure, and the steam-to-ethylene ratio in the feed. The effects of all three variables are shown in Fig. 15.5. The curves in this figure come from calculations just like those illustrated in this example, except that a less precise equation for K as a function of T was used.

Example 15.8 In a laboratory investigation, acetylene is catalytically hydrogenated to ethylene at 1,120°C and 1 bar. If the feed is an equimolar ratio of acetylene and hydrogen, what is the composition of the product stream at equilibrium?

SOLUTION The required reaction is obtained by addition of the two formation reactions written as follows:

$$C_2H_2 \rightarrow 2C + H_2 \qquad (1)$$

$$2C + 2H_2 \rightarrow C_2H_4 \qquad (2)$$

The sum of reactions (1) and (2) is the hydrogenation reaction

$$C_2H_2 + H_2 \rightarrow C_2H_4$$

Also

$$\Delta G^\circ = \Delta G_1^\circ + \Delta G_2^\circ$$

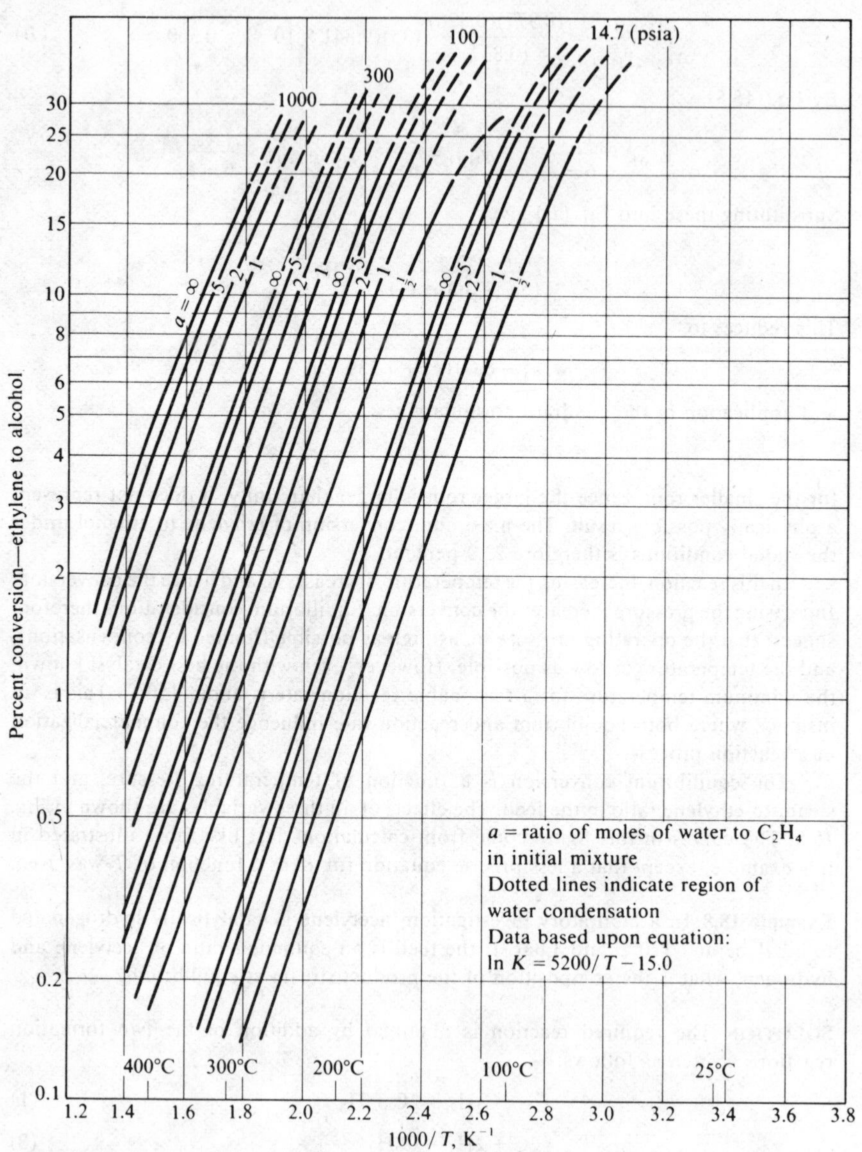

Figure 15.5 Equilibrium conversion of ethylene to ethyl alcohol in the vapor phase.

By Eq. (15.14),

$$-RT \ln K = -RT \ln K_1 - RT \ln K_2$$

or

$$K = K_1 K_2$$

Data for both reactions (1) and (2) are given by Fig. 15.3. At 1,120°C [1,393 K], $10^4/T = 7.18$, and the following values are read from the graph:

$$\ln K_1 = 12.9 \qquad K_1 = 4.0 \times 10^5$$

$$\ln K_2 = -12.9 \qquad K_2 = 2.5 \times 10^{-6}$$

Therefore

$$K = K_1 K_2 = 1.0$$

At this elevated temperature and for a pressure of 1 bar, we can safely assume ideal gases. Application of Eq. (15.25) then leads to the expression

$$\frac{y_{C_2H_4}}{y_{H_2} y_{C_2H_2}} = 1$$

On the basis of one mole initially of each reactant, Eq. (15.5) gives

$$y_{H_2} = y_{C_2H_2} = \frac{1 - \varepsilon_e}{2 - \varepsilon_e} \qquad \text{and} \qquad y_{C_2H_4} = \frac{\varepsilon_e}{2 - \varepsilon_e}$$

Therefore

$$\frac{\varepsilon_e(2 - \varepsilon_e)}{(1 - \varepsilon_e)^2} = 1$$

The smaller root of this quadratic expression (the larger is greater than 1) is

$$\varepsilon_e = 0.293$$

The equilibrium composition of the product gas is then

$$y_{H_2} = y_{C_2H_2} = \frac{1 - 0.293}{2 - 0.293} = 0.414$$

$$y_{C_2H_4} = \frac{0.293}{2 - 0.293} = 0.172$$

Example 15.9 Acetic acid is esterified in the liquid phase with ethanol at 100°C and atmospheric pressure to produce ethyl acetate and water according to the reaction

$$CH_3COOH(l) + C_2H_5OH(l) \rightarrow CH_3COOC_2H_5(l) + H_2O(l)$$

If initially there is one mole each of acetic acid and ethanol, estimate the mole fraction of ethyl acetate in the reacting mixture at equilibrium.

SOLUTION Data for ΔH°_{298} and ΔG°_{298} are given for liquid acetic acid, ethanol, and water in Tables 4.4 and 15.1. For liquid ethyl acetate, the corresponding values are

$$\Delta H^\circ_{f_{298}} = -463{,}250 \text{ J} \qquad \text{and} \qquad \Delta G^\circ_{f_{298}} = -318{,}280 \text{ J}$$

The values of ΔH°_{298} and ΔG°_{298} for the reaction are therefore

$$\Delta H^\circ_{298} = -463{,}250 - 285{,}830 + 484{,}500 + 277{,}690 = 13{,}110 \text{ J}$$

$$\Delta G^\circ_{298} = -318{,}280 - 237{,}130 + 389{,}900 + 174{,}780 = 9{,}270 \text{ J}$$

By Eq. (15.14),

$$\ln K_{298} = \frac{-\Delta G^\circ_{298}}{RT} = \frac{-9{,}270}{(8.314)(298.15)} = -3.740$$

$$K_{298} = 0.0238$$

For the small temperature change from 298.15 to 373.15 K, Eq. (15.18) is adequate for estimation of K_{373}. Thus

$$\ln \frac{K_{373}}{K_{298}} = \frac{-\Delta H^\circ_{298}}{R}\left(\frac{1}{373.15} - \frac{1}{298.15}\right)$$

or

$$\ln \frac{K_{373}}{0.0238} = \frac{-13{,}110}{8.314}\left(\frac{1}{373.15} - \frac{1}{298.15}\right) = 1.0630$$

$$K_{373} = (0.0238)(2.895) = 0.0689$$

For the given reaction Eq. (15.5), with x replacing y, yields

$$x_{\text{AcH}} = x_{\text{EtOH}} = \frac{1 - \varepsilon_e}{2}$$

$$x_{\text{EtAc}} = x_{\text{H}_2\text{O}} = \frac{\varepsilon_e}{2}$$

Since the pressure is low, Eq. (15.29) is applicable. However, in the absence of data for the activity coefficients in this complex system, we assume that the reacting species form an ideal solution. In this case Eq. (15.30) is employed, giving

$$K = \frac{x_{\text{EtAc}}x_{\text{H}_2\text{O}}}{x_{\text{AcH}}x_{\text{EtOH}}}$$

Whence

$$0.0689 = \left(\frac{\varepsilon_e}{1 - \varepsilon_e}\right)^2$$

from which

$$\varepsilon_e = 0.208$$

and

$$x_{\text{EtAc}} = 0.208/2 = 0.104$$

This result is not in very good agreement with experiment. When this reaction is carried out in the laboratory, one finds that the mole fraction of ethyl acetate at equilibrium is about 0.33. The assumption of an ideal solution is unrealistic; if there were no preferential interactions among the species, there would be no reaction.

Reactions in Heterogeneous Systems

When liquid and gas phases are both present in an equilibrium mixture of reacting species, Eq. (11.30), a criterion of vapor/liquid equilibrium, must be satisfied along with the equation of chemical-reaction equilibrium. There is considerable choice in the method of treatment of such cases. For example, consider a reaction of gas A and water B to form an aqueous solution C. The reaction may be assumed to occur entirely in the gas phase with simultaneous transfer of material between phases to maintain phase equilibrium. In this case, the equilibrium constant is evaluated from $\Delta G°$ data based on standard states for the species as gases, i.e., the ideal-gas states at 1 bar and the reaction temperature. On the other hand, the reaction may be assumed to occur in the liquid phase, in which case $\Delta G°$ is based on standard states for the species as liquids. Alternatively, the reaction may be written

$$A(g) + B(l) \rightarrow C(aq)$$

in which case the $\Delta G°$ value is for mixed standard states: C as a solute in an ideal 1-molal aqueous solution, B as a pure liquid at 1 bar, and A as a pure ideal gas at 1 bar. For this choice of standard states, the equilibrium constant as given by Eq. (15.13) becomes

$$K = \frac{\hat{a}_C}{\hat{a}_B \hat{a}_A} = \frac{m_C}{(\gamma_B x_B)(\hat{f}_A)}$$

The last term arises from Eq. (15.32) applied to species C, Eq. (15.27) applied to B, and the fact that $\hat{a}_A = \hat{f}_A$ for species A in the gas phase. Since K depends on the standard states, the value of K is not the same as that obtained when the standard state for each species is chosen as the ideal-gas state at 1 bar. However, all methods theoretically lead to the same equilibrium composition, provided Henry's law as applied to species C in solution is valid. In practice, a particular choice of standard states may simplify calculations or yield more accurate results, because it makes better use of the limited data normally available. The nature of the calculations required for heterogeneous reactions is illustrated in the following example.

Example 15.10 Estimate the compositions of the liquid and vapor phases when ethylene reacts with water to form ethanol at 200°C and 34.5 bar, conditions which assure the presence of both liquid and vapor phases. The reaction vessel is maintained at 34.5 bar by connection to a source of ethylene at this pressure. Assume no other reactions to occur.

SOLUTION According to the phase rule (see Sec. 15.8), the system has two degrees of freedom. Specification of both the temperature and the pressure leaves no other degrees of freedom, and fixes the intensive state of the system, independent of the initial amounts of reactants. Therefore, material-balance equations do not enter into the solution of this problem, and we can make no use of equations that relate compositions to the reaction coordinate. Instead, phase equilibrium relations must be employed to provide a sufficient number of equations to allow solution for the unknown compositions.

The most convenient approach to this problem is to regard the chemical reaction as occurring in the vapor phase. Thus

$$C_2H_4(g) + H_2O(g) \rightarrow C_2H_5OH(g)$$

and the standard states are those of the pure ideal gases at 1 bar. For these standard states, the equilibrium equation is Eq. (15.15), which in this case becomes

$$K = \frac{\hat{f}_{\text{EtOH}}}{\hat{f}_{C_2H_4}\hat{f}_{H_2O}} \tag{A}$$

Furthermore, a general expression for ln K as a function of T is provided by the results of Example 15.5. For 200°C [473.15 K], this equation yields

$$\ln K = -3.511 \qquad K = 0.0299$$

The task now is to incorporate the phase-equilibrium equations,

$$\hat{f}_i^v = \hat{f}_i^l$$

into Eq. (A) and to relate the fugacities to the compositions in such a way that the equations can be readily solved. Equation (A) may be written

$$K = \frac{\hat{f}^v_{\text{EtOH}}}{\hat{f}^v_{C_2H_4}\hat{f}^v_{H_2O}} = \frac{\hat{f}^l_{\text{EtOH}}}{\hat{f}^v_{C_2H_4}\hat{f}^l_{H_2O}} \tag{B}$$

The liquid-phase fugacities are related to activity coefficients by Eq. (11.59):

$$\hat{f}_i^l = x_i \gamma_i f_i^l \tag{C}$$

and the vapor-phase fugacity is related to the fugacity coefficient by Eq. (11.33):

$$\hat{f}_i^v = y_i \hat{\phi}_i P \tag{D}$$

Elimination of the fugacities in Eq. (B) by Eqs. (C) and (D) gives

$$K = \frac{x_{\text{EtOH}}\gamma_{\text{EtOH}} f^l_{\text{EtOH}}}{(y_{C_2H_4}\hat{\phi}_{C_2H_4}P)(x_{H_2O}\gamma_{H_2O}f^l_{H_2O})} \tag{E}$$

The fugacity f_i^l is for pure liquid i at the temperature and pressure of the system. However, pressure has small effect on the fugacity of a liquid, and to a good approximation we can write:

$$f_i^l = f_i^{\text{sat}}$$

and therefore by Eqs. (11.22) and (11.23),

$$f_i^l = \phi_i^{\text{sat}} P_i^{\text{sat}} \tag{F}$$

In this equation ϕ_i^{sat} is the fugacity coefficient of pure saturated i (either liquid or vapor) evaluated at the temperature of the system and at P_i^{sat}, the vapor pressure of pure i. The assumption that the vapor phase is an ideal solution allows substitution of $\phi_{C_2H_4}$ for $\hat{\phi}_{C_2H_4}$, where $\phi_{C_2H_4}$ is the fugacity coefficient of pure ethylene at the system T and P. With this substitution and that of Eq. (F), Eq. (E) becomes

$$K = \frac{x_{\text{EtOH}}\gamma_{\text{EtOH}}\phi^{\text{sat}}_{\text{EtOH}}P^{\text{sat}}_{\text{EtOH}}}{(y_{C_2H_4}\phi_{C_2H_4}P)(x_{H_2O}\gamma_{H_2O}\phi^{\text{sat}}_{H_2O}P^{\text{sat}}_{H_2O})} \tag{G}$$

In addition to Eq. (G) the following expressions can be written. Since $\sum y_i = 1$,

$$y_{C_2H_4} = 1 - y_{EtOH} - y_{H_2O} \tag{H}$$

We can eliminate y_{EtOH} and y_{H_2O} in this equation in favor of x_{EtOH} and x_{H_2O} by the vapor/liquid equilibrium relation:

$$\hat{f}_i^v = \hat{f}_i^l$$

Combining this with Eqs. (C), (D), and (F), we obtain

$$y_i = \frac{\gamma_i x_i \phi_i^{sat} P_i^{sat}}{\phi_i P} \tag{I}$$

where ϕ_i has replaced $\hat{\phi}_i$ because of the assumption that the vapor phase is an ideal solution. Equations (H) and (I) yield

$$y_{C_2H_4} = 1 - \frac{\gamma_{EtOH} x_{EtOH} \phi_{EtOH}^{sat} P_{EtOH}^{sat}}{\phi_{EtOH} P} - \frac{\gamma_{H_2O} x_{H_2O} \phi_{H_2O}^{sat} P_{H_2O}^{sat}}{\phi_{H_2O} P} \tag{J}$$

Since ethylene is far more volatile than ethanol or water, we assume that $x_{C_2H_4} = 0$. Then

$$x_{H_2O} = 1 - x_{EtOH} \tag{K}$$

Equations (G), (J), and (K) form the basis for solution of the problem. The three primary variables in these equations are x_{H_2O}, x_{EtOH}, and $y_{C_2H_4}$, and all other quantities are either given or are determined from correlations of data. The values of P_i^{sat} are readily available. At 200°C they are

$$P_{H_2O}^{sat} = 15.55 \qquad P_{EtOH}^{sat} = 30.22 \text{ bar}$$

The quantities ϕ_i^{sat} and ϕ_i are found from the generalized correlation represented by Eq. (11.43):

$$\ln \phi_i = \frac{P_{r_i}}{T_{r_i}}(B^0 + \omega B^1) \tag{11.43}$$

where B^0 and B^1 are given by Eqs. (3.48) and (3.49). With critical data and the ω_i's from App. B, evaluation of the various fugacity coefficients leads to the following values ($T = 473.15$ K, $P = 34.5$ bar):

	T_c/K	P_c/bar	ω_i	T_{r_i}	P_{r_i}	$P_{r_i}^{sat}$	B^0	B^1	ϕ_i	ϕ_i^{sat}
EtOH	516.2	63.8	0.635	0.917	0.541	0.474	−0.40	−0.11	0.76	0.78
H_2O	647.3	220.5	0.344	0.731	0.156	0.071	−0.61	−0.50	0.85	0.93
C_2H_4	282.4	50.4	0.085	1.675	0.685	.....	−0.10	0.12	0.96	

Substitution of all values so far determined into Eqs. (G), (J), and (K) reduces these three equations to the following:

$$K = \frac{0.0492 x_{EtOH} \gamma_{EtOH}}{y_{C_2H_4} x_{H_2O} \gamma_{H_2O}} \tag{L}$$

$$y_{C_2H_4} = 1 - 0.899 \gamma_{EtOH} x_{EtOH} - 0.493 \gamma_{H_2O} x_{H_2O} \tag{M}$$

$$x_{H_2O} = 1 - x_{EtOH} \tag{K}$$

The only remaining undetermined thermodynamic properties are γ_{H_2O} and γ_{EtOH}. Because of the highly nonideal behavior of a liquid solution of ethanol and water, these must be determined from experimental data. The required data, found from VLE measurements, are given by Otsuki and Williams.† From their results for the ethanol/water system one can estimate values of γ_{H_2O} and γ_{EtOH} at 200°C. (Pressure has little effect on the activity coefficients of liquids.)

A procedure for solution of the foregoing three equations is as follows.

1. Assume a value for x_{EtOH} and calculate x_{H_2O} by Eq. (K).
2. Determine γ_{H_2O} and γ_{EtOH} from data in the reference cited.
3. Calculate $y_{C_2H_4}$ by Eq. (M).
4. Calculate K by Eq. (L) and compare with the value of 0.0299 determined from standard-reaction data.
5. If the two values agree, the assumed value of x_{EtOH} is correct. If they do not agree, assume a new value of x_{EtOH} and repeat the procedure.

If we take $x_{EtOH} = 0.06$, then by Eq. (K), $x_{H_2O} = 0.94$, and from the reference cited,

$$\gamma_{EtOH} \simeq 3.34 \qquad \text{and} \qquad \gamma_{H_2O} \simeq 1.00$$

By Eq. (M),

$$y_{C_2H_4} = 1 - (0.899)(3.34)(0.06) - (0.493)(1.00)(0.94) = 0.356$$

The value of K given by Eq. (L) is then

$$K = \frac{(0.0492)(0.06)(3.34)}{(0.356)(0.94)(1.00)} = 0.0295$$

This result is in essential agreement with the value (0.0299) found from standard-reaction data, and we therefore take $x_{EtOH} = 0.06$ and $x_{H_2O} = 0.94$ as the liquid-phase compositions. The remaining vapor-phase compositions ($y_{C_2H_4}$ has already been determined as 0.356) are found by solution of Eq. (I) for y_{H_2O} or y_{EtOH}. All results are summarized in the following table.

	x_i	y_i
EtOH	0.060	0.180
H_2O	0.940	0.464
C_2H_4	0.000	0.356
	$\sum x_i = 1.000$	$\sum y_i = 1.000$

These results are probably reasonable estimates of actual values, provided no other reactions take place.

† H. Otsuki and F. C. Williams, *Chem. Engr. Progr. Symp. Series No. 6*, **49**: 55, 1953.

15.8 THE PHASE RULE AND DUHEM'S THEOREM FOR REACTING SYSTEMS

The phase rule (applicable to intensive properties) as discussed in Secs. 2.8 and 12.2 for nonreacting systems of π phases and N chemical species is

$$F = 2 - \pi + N$$

It must be modified for application to systems in which chemical reactions occur. The phase-rule variables are the same in either case, namely, temperature, pressure, and $N - 1$ mole fractions in each phase. The total number of these variables is $2 + (N - 1)(\pi)$. The same phase-equilibrium equations apply as before, and they number $(\pi - 1)(N)$. However, Eq. (15.8) provides for each independent reaction an additional relation that must be satisfied at equilibrium. Since the μ_i's are functions of temperature, pressure, and the phase compositions, Eq. (15.8) represents a relation connecting the phase-rule variables. If there are r independent chemical reactions at equilibrium within the system, then there is a total of $(\pi - 1)(N) + r$ independent equations relating the phase-rule variables. Taking the difference between the number of variables and the number of equations, we obtain

$$F = 2 + (N - 1)(\pi) - (\pi - 1)(N) - r$$

or

$$\boxed{F = 2 - \pi + N - r} \tag{15.33}$$

This is the basic equation expressing the phase rule for reacting systems.

The only remaining problem in application is to determine the number of independent chemical reactions. This can be done systematically as follows.

1. Write chemical equations for the formation, from the *constituent elements*, of each chemical compound considered present in the system.
2. Combine these equations so as to eliminate from them all elements not considered present *as elements* in the system. A systematic procedure is to select one equation and combine it with each of the others of the set to eliminate a particular element. Then the process is repeated to eliminate another element from the new set of equations. This is done for each element eliminated (see Example 15.11*d*), and usually reduces the set by one equation for each element eliminated. However, the simultaneous elimination of two or more elements may occur.

The set of r equations resulting from this reduction procedure is a complete set of independent reactions for the N species considered present in the system. However, more than one such set is possible, depending on how the reduction procedure is carried out, but all sets number r and are equivalent.

The reduction procedure also ensures the following relation:

$r \geq$ number of compounds present in the system

− number of constituent elements *not* present *as elements*

The phase-equilibrium and chemical-reaction-equilibrium equations are the only ones considered in the foregoing treatment as interrelating the phase-rule variables. However, in certain situations *special constraints* may be placed on the system that allow additional equations to be written over and above those considered in the development of Eq. (15.33). If the number of equations resulting from special constraints is s, then Eq. (15.33) must be modified to take account of these s additional equations. The still more general form of the phase rule that results is

$$\boxed{F = 2 - \pi + N - r - s} \tag{15.34}$$

Example 15.11 shows how Eqs. (15.33) and (15.34) may be applied to specific systems.

Example 15.11 Determine the number of degrees of freedom F for each of the following systems.

(*a*) A system of two miscible nonreacting species which exists as an azeotrope in vapor/liquid equilibrium.
(*b*) A system prepared by partially decomposing $CaCO_3$ into an evacuated space.
(*c*) A system prepared by partially decomposing NH_4Cl into an evacuated space.
(*d*) A system consisting of the gases CO, CO_2, H_2, H_2O, and CH_4 in chemical equilibrium.

SOLUTION (*a*) The system consists of two nonreacting species in two phases, and application of Eq. (15.33) yields

$$F = 2 - \pi + N - r = 2 - 2 + 2 - 0 = 2$$

This result is in general valid for such a system. However, a special constraint is imposed on the system; it is an azeotrope. This provides an equation, $x_1 = y_1$, not considered in the development of Eq. (15.33). Thus, we apply Eq. (15.34) with $s = 1$. The result is that $F = 1$. If the system is to be an azeotrope, then just one phase-rule variable—T, P, or $x_1 = y_1$—may be arbitrarily specified.

(*b*) Here there is a single chemical reaction:

$$CaCO_3(s) \rightarrow CaO(s) + CO_2(g)$$

and $r = 1$. There are three chemical species and three phases—solid $CaCO_3$, solid CaO, and gaseous CO_2. One might think a special constraint has been imposed by the requirement that the system be prepared in a special way—by decomposing $CaCO_3$. This is not the case, because no equation connecting the phase-rule variables can be written as a result of this requirement. Therefore

$$F = 2 - \pi + N - r - s = 2 - 3 + 3 - 1 - 0 = 1$$

and there is a single degree of freedom. This is the reason that $CaCO_3$ exerts a fixed decomposition pressure at fixed T.

(*c*) The chemical reaction here is

$$NH_4Cl(s) \rightarrow NH_3(g) + HCl(g)$$

Three species, but only two phases, are present in this case, solid NH_4Cl and a gas mixture of NH_3 and HCl. In addition, there is a special constraint, because the requirement that the system be formed by the decomposition of NH_4Cl means that the gas phase is equimolar in NH_3 and HCl. Thus a special equation $y_{NH_3} = y_{HCl}$ (=0.5), connecting the phase-rule variables can be written. Application of Eq. (15.34) gives

$$F = 2 - \pi + N - r - s = 2 - 2 + 3 - 1 - 1 = 1$$

and the system has but one degree of freedom. This result is the same as that for part (*b*), and it is a matter of experience that NH_4Cl has a given decomposition pressure at a given temperature. This conclusion is reached quite differently in the two cases.

(*d*) This system contains five species, all in a single gas phase. There are no special constraints. Only r remains to be determined. The formation reactions for the compounds present are:

$$C + \tfrac{1}{2}O_2 \rightarrow CO \qquad (A)$$

$$C + O_2 \rightarrow CO_2 \qquad (B)$$

$$H_2 + \tfrac{1}{2}O_2 \rightarrow H_2O \qquad (C)$$

$$C + 2H_2 \rightarrow CH_4 \qquad (D)$$

Systematic elimination of C and O_2, the elements not present in the system, leads to two equations. One such pair of equations is obtained in the following way. We eliminate C from this set of equations by combining Eq. (*B*), first with Eq. (*A*) and then with Eq. (*D*). The two resulting reactions are

$$\text{From } (B) \text{ and } (A)\text{:} \quad CO + \tfrac{1}{2}O_2 \rightarrow CO_2 \qquad (E)$$

$$\text{From } (B) \text{ and } (D)\text{:} \quad CH_4 + O_2 \rightarrow 2H_2 + CO_2 \qquad (F)$$

Equations (*C*), (*E*), and (*F*) are the new set, and we now eliminate O_2 by combining Eq. (*C*), first with Eq. (*E*) and then with Eq. (*F*). This gives

$$\text{From } (C) \text{ and } (E)\text{:} \quad CO_2 + H_2 \rightarrow CO + H_2O \qquad (G)$$

$$\text{From } (C) \text{ and } (F)\text{:} \quad CH_4 + 2H_2O \rightarrow CO_2 + 4H_2 \qquad (H)$$

Equations (*G*) and (*H*) are an independent set and indicate that $r = 2$. The use of different elimination procedures produces other pairs of equations, but always just two equations.

Application of Eq. (15.34) yields

$$F = 2 - \pi + N - r - s = 2 - 1 + 5 - 2 - 0 = 4$$

This result means that one is free to specify four phase-rule variables, for example, T, P, and two mole fractions, in an equilibrium mixture of these five chemical species, provided that nothing else is arbitrarily set. In other words, there can be no special constraints, such as the specification that the system be prepared from given amounts of CH_4 and H_2O. This imposes special constraints through material balances that reduce the degrees of freedom to two. (Duhem's theorem; see the following paragraphs.)

Duhem's theorem states that, for any closed system formed initially from given masses of particular chemical species, the equilibrium state is *completely determined* (extensive as well as intensive properties) by specification of any two independent variables. This theorem was developed in Sec. 12.2 for nonreacting systems. It was shown there that the difference between the number of independent variables that completely determine the state of the system and the number of independent equations that can be written connecting these variables is

$$[2 + (N - 1)(\pi) + \pi] - [(\pi - 1)(N) + N] = 2$$

If chemical reactions occur, then we must introduce a new variable, the reaction coordinate ε_j for each independent reaction, in order to formulate the material-balance equations. Furthermore, we are able to write a new equilibrium relation [Eq. (15.8)] for each independent reaction. Therefore, when chemical-reaction equilibrium is superimposed on phase equilibrium, r new variables appear and r new equations can be written. The difference between the number of variables and number of equations therefore is unchanged, and Duhem's theorem as originally stated holds for reacting systems as well as for nonreacting systems.

Most chemical-reaction equilibrium problems are so posed that it is Duhem's theorem that makes them determinate. The usual problem is to find the composition of a system that reaches equilibrium from an initial state of *fixed amounts of reacting species* when the *two* variables T and P are specified.

15.9 MULTIREACTION EQUILIBRIA

When the equilibrium state in a reacting system depends on two or more simultaneous chemical reactions, the equilibrium composition can be found by a direct extension of the methods developed for single reactions. One first determines a set of independent reactions as discussed in Sec. 15.8. With each independent reaction there is associated a reaction coordinate in accord with the treatment of Sec. 15.1. In addition, a separate equilibrium constant is evaluated for each reaction, and Eq. (15.13) becomes

$$K_j = \prod (\hat{a}_i)^{\nu_{i,j}} \tag{15.35}$$

where j is the reaction index. For a gas-phase reaction Eq. (15.35) takes the form

$$K_j = \prod (\hat{f}_i)^{\nu_{i,j}} \tag{15.36}$$

If the equilibrium mixture is an ideal gas, we may write

$$\prod (y_i)^{\nu_{i,j}} = P^{-\nu_j} K_j \tag{15.37}$$

For r independent reactions there are r separate equations of this kind, and the y_i's can be eliminated by Eq. (15.7) in favor of the r reaction coordinates ε_j. The set of equations is then solved simultaneously for the r reaction coordinates. The procedure is illustrated by the following example.

Example 15.12 A bed of coal (assume pure carbon) in a coal gasifier is fed with steam and air and produces a gas stream containing H_2, CO, O_2, H_2O, CO_2, and N_2. If the feed to the gasifier consists of 1 mol of steam and 2.38 mol of air, calculate the equilibrium composition of the gas stream at $P = 20$ bar for temperatures of 1,000, 1,100, 1,200, 1,300, 1,400, and 1,500 K. The following data are available:

	$\Delta G_f^\circ/\text{J mol}^{-1}$		
T/K	H_2O	CO	CO_2
1,000	−192,420	−200,240	−395,790
1,100	−187,000	−209,110	−395,960
1,200	−181,380	−217,830	−396,020
1,300	−175,720	−226,530	−396,080
1,400	−170,020	−235,130	−396,130
1,500	−164,310	−243,740	−396,160

SOLUTION The feed stream to the coal bed consists of 1 mol of steam and 2.38 mol of air, containing

$$O_2\text{: } (0.21)(2.38) = 0.5 \text{ mol}$$

$$N_2\text{: } (0.79)(2.38) = 1.88 \text{ mol}$$

The species present at equilibrium are C, H_2, O_2, N_2, H_2O, CO, and CO_2. The formation reactions for the compounds present are:

$$H_2 + \tfrac{1}{2}O_2 \rightarrow H_2O \tag{1}$$

$$C + \tfrac{1}{2}O_2 \rightarrow CO \tag{2}$$

$$C + O_2 \rightarrow CO_2 \tag{3}$$

Since the elements hydrogen, oxygen, and carbon are themselves presumed present in the system, this set of three reactions is a complete set of independent reactions.

All species are present as gases except carbon, which is present as a pure solid phase. In the basic expression for the equilibrium constant, Eq. (15.35), the activity of the pure carbon is $\hat{a}_C = a_C = f_C/f_C^\circ$. The fugacity ratio is the fugacity of carbon at 20 bar divided by the fugacity of carbon at 1 bar. Since the effect of pressure on the fugacity of a solid is very small, negligible error is introduced by the assumption that this ratio is unity. The activity of the carbon is then $\hat{a}_C \simeq 1$, and it may be omitted from the equilibrium expression. With the assumption that the remaining species are ideal gases, Eq. (15.37) is written for the gas phase only, and it provides the following equilibrium expressions for reactions (1) through (3):

$$K_1 = \frac{y_{H_2O}}{y_{O_2}^{1/2}\,y_{H_2}}\,P^{-1/2}$$

$$K_2 = \frac{y_{CO}}{y_{O_2}^{1/2}}\,P^{1/2}$$

$$K_3 = \frac{y_{CO_2}}{y_{O_2}}$$

The reaction coordinates for the three reactions are designated ε_1, ε_2, and ε_3, and they are here taken to be the equilibrium values. For the initial state, $n_{H_2} = n_{CO} = n_{CO_2} = 0$, $n_{H_2O} = 1$, $n_{O_2} = 0.5$, and $n_{N_2} = 1.88$. Moreover, since only the gas-phase species are considered, $\nu_1 = -\frac{1}{2}$, $\nu_2 = \frac{1}{2}$ and $\nu_3 = 0$. Applying Eq. (15.7) to each species gives:

$$y_{H_2} = \frac{-\varepsilon_1}{3.38 + (\varepsilon_2 - \varepsilon_1)/2} \qquad y_{CO} = \frac{\varepsilon_2}{3.38 + (\varepsilon_2 - \varepsilon_1)/2}$$

$$y_{O_2} = \frac{\frac{1}{2}(1 - \varepsilon_1 - \varepsilon_2) - \varepsilon_3}{3.38 + (\varepsilon_2 - \varepsilon_1)/2} \qquad y_{H_2O} = \frac{1 + \varepsilon_1}{3.38 + (\varepsilon_2 - \varepsilon_1)/2}$$

$$y_{CO_2} = \frac{\varepsilon_3}{3.38 + (\varepsilon_2 - \varepsilon_1)/2} \qquad y_{N_2} = \frac{1.88}{3.38 + (\varepsilon_2 - \varepsilon_1)/2}$$

Substitution of these expressions for y_i into the equilibrium equations gives

$$K_1 = \frac{(1 + \varepsilon_1)(2n)^{1/2}P^{-1/2}}{(1 - \varepsilon_1 - \varepsilon_2 - 2\varepsilon_3)^{1/2}(-\varepsilon_1)}$$

$$K_2 = \frac{\sqrt{2}\varepsilon_2 P^{1/2}}{(1 - \varepsilon_1 - \varepsilon_2 - 2\varepsilon_3)^{1/2}n^{1/2}}$$

$$K_3 = \frac{2\varepsilon_3}{(1 - \varepsilon_1 - \varepsilon_2 - 2\varepsilon_3)}$$

where

$$n \equiv 3.38 + \frac{\varepsilon_2 - \varepsilon_1}{2}$$

Numerical values for the K_i calculated by Eq. (15.14) are found to be very large. For example, at 1,500 K,

$$\ln K_1 = \frac{-\Delta G_1^\circ}{RT} = \frac{164{,}310}{(8.314)(1{,}500)} = 13.2 \qquad K_1 \sim 10^6$$

$$\ln K_2 = \frac{-\Delta G_2^\circ}{RT} = \frac{243{,}740}{(8.314)(1{,}500)} = 19.6 \qquad K_2 \sim 10^8$$

$$\ln K_3 = \frac{-\Delta G_3^\circ}{RT} = \frac{396{,}160}{(8.314)(1{,}500)} = 31.8 \qquad K_3 \sim 10^{14}$$

The only way these K_i's can be so large is for the quantity $1 - \varepsilon_1 - \varepsilon_2 - 2\varepsilon_3$, which appears in the denominator of the expression for each K_i, to be nearly zero. This means that the mole fraction of oxygen in the equilibrium mixture is very small. For practical purposes, no oxygen is present.

We therefore reformulate the problem by eliminating O_2 from the formation reactions. For this, we combine Eq. (1), first with Eq. (2), and then with Eq. (3). This provides the two equations

$$C + CO_2 \rightarrow 2CO \qquad (a)$$

$$H_2O + C \rightarrow H_2 + CO \qquad (b)$$

The corresponding equilibrium equations are

$$K_a = \frac{y_{CO}^2 P}{y_{CO_2}}$$

and

$$K_b = \frac{y_{H_2} y_{CO} P}{y_{H_2O}}$$

The input stream is specified to contain 1 mol H_2, 0.5 mol O_2, and 1.88 mol N_2. Since O_2 has been eliminated from the set of reaction equations, we replace the 0.5 mol of O_2 in the feed by 0.5 mol of CO_2. The presumption is that this amount of CO_2 has been formed by prior reaction of the 0.5 mol O_2 with carbon. Thus the equivalent feed stream contains 1 mol H_2, 0.5 mol CO_2, and 1.88 mol N_2, and application of Eq. (15.7) to Eqs. (*a*) and (*b*) gives

$$y_{H_2} = \frac{\varepsilon_b}{3.38 + \varepsilon_a + \varepsilon_b}$$

$$y_{CO} = \frac{2\varepsilon_a + \varepsilon_b}{3.38 + \varepsilon_a + \varepsilon_b}$$

$$y_{H_2O} = \frac{1 - \varepsilon_b}{3.38 + \varepsilon_a + \varepsilon_b}$$

$$y_{CO_2} = \frac{0.5 - \varepsilon_a}{3.38 + \varepsilon_a + \varepsilon_b}$$

$$y_{N_2} = \frac{1.88}{3.38 + \varepsilon_a + \varepsilon_b}$$

Since values of y_i must lie between zero and unity, we see from the first and third of these expressions that

$$0 \le \varepsilon_b \le 1$$

and from the second and fourth that

$$-0.5 \le \varepsilon_a \le 0.5$$

Combining the expressions for the y_i with the equilibrium equations, we get

$$K_a = \frac{(2\varepsilon_a + \varepsilon_b)^2 P}{(0.5 - \varepsilon_a)(3.38 + \varepsilon_a + \varepsilon_b)} \tag{A}$$

and

$$K_b = \frac{\varepsilon_b(2\varepsilon_a + \varepsilon_b) P}{(1 - \varepsilon_b)(3.38 + \varepsilon_a + \varepsilon_b)} \tag{B}$$

A simpler equation is obtained upon division of Eq. (*A*) by Eq. (*B*):

$$\frac{K_a}{K_b} = \frac{(2\varepsilon_a + \varepsilon_b)(1 - \varepsilon_b)}{(0.5 - \varepsilon_a)\varepsilon_b} \tag{C}$$

Any pair from these three equations is an independent set. We choose to work with Eqs. (*A*) and (*C*), and our problem is to solve them simultaneously for ε_a and ε_b.

We first define a new variable q as

$$q \equiv \frac{2\varepsilon_a + \varepsilon_b}{0.5 - \varepsilon_a} \tag{D}$$

whence

$$\varepsilon_a = \frac{0.5q - \varepsilon_b}{2 + q} \tag{E}$$

and by Eq. (C)

$$\varepsilon_b = \frac{q(K_b/K_a)}{1 + q(K_b/K_a)} \tag{F}$$

Also, by Eq. (A)

$$q = \frac{(K_a/P)(3.38 + \varepsilon_a + \varepsilon_b)}{2\varepsilon_a + \varepsilon_b} \tag{G}$$

The following iteration scheme allows solution for q:

1. Choose an initial value for q.
2. Solve Eqs. (E) and (F) for ε_a and ε_b.
3. Solve Eq. (G) for q.
4. Return to step 2 and iterate to convergence.

For reaction (a),

$$\Delta G^\circ_{1000} = 2(-200{,}240) - (-395{,}790) = -4{,}690$$

and

$$\ln K_a = \frac{4{,}690}{(8.314)(1{,}000)} = 0.5641 \qquad K_a = 1.758$$

Similarly, for reaction (b),

$$\Delta G^\circ_{1000} = -200{,}240 - (-192{,}420) = -7{,}820$$

and

$$\ln K_b = \frac{7{,}820}{(8.314)(1{,}000)} = 0.9406 \qquad K_b = 2.561$$

The same calculations at each temperature for which data are given produce the values of K_a and K_b listed in the following table. Also given are the results of all iterative calculations.

T/K	K_a	K_b	ε_a	ε_b
1,000	1.758	2.561	−0.0506	0.5336
1,100	11.405	11.219	0.1210	0.7124
1,200	53.155	38.609	0.3168	0.8551
1,300	194.430	110.064	0.4301	0.9357
1,400	584.85	268.76	0.4739	0.9713
1,500	1,514.12	583.58	0.4896	0.9863

Values for the mole fractions y_i of the species in the equilibrium mixture are calculated by the equations given earlier. For example, at 1,000 K.

$$y_{H_2} = \frac{0.5336}{3.38 - 0.0506 + 0.5336} = 0.138$$

$$y_{CO} = \frac{(2)(-0.0506) + 0.5336}{3.38 - 0.0506 + 0.5336} = 0.112$$

etc.

The results of all such calculations are given in the following table and are shown graphically in Fig. 15.6.

T/K	y_{H_2}	y_{CO}	y_{H_2O}	y_{CO_2}	y_{N_2}
1,000	0.138	0.112	0.121	0.143	0.486
1,100	0.169	0.226	0.068	0.090	0.447
1,200	0.188	0.327	0.032	0.040	0.413
1,300	0.197	0.378	0.014	0.015	0.396
1,400	0.201	0.398	0.006	0.005	0.390
1,500	0.203	0.405	0.003	0.002	0.387

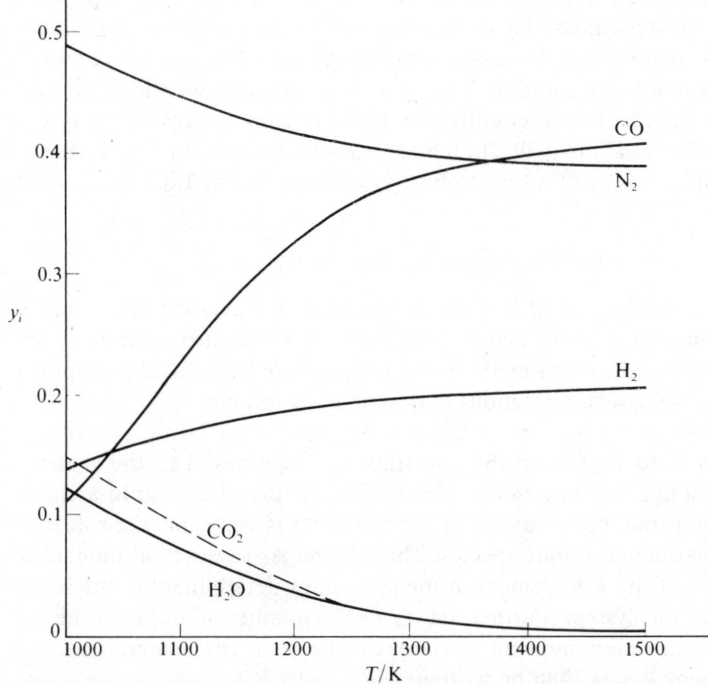

Figure 15.6 Equilibrium compositions of the product gases from a coal gasifier as a function of temperature.

At the higher temperatures the values of ε_a and ε_b are approaching their upper limiting values of 0.5 and 1.0, indicating that reactions (a) and (b) are proceeding nearly to completion. In this limit, which is approached even more closely at still higher temperatures, the mole fractions of CO_2 and H_2O approach zero, and for the product species,

$$y_{H_2} = \frac{1}{3.38 + 0.5 + 1.0} = 0.205$$

$$y_{CO} = \frac{1 + 1}{3.38 + 0.5 + 1.0} = 0.410$$

$$y_{N_2} = \frac{1.88}{3.38 + 0.5 + 1.0} = 0.385$$

In this example we have assumed a sufficient depth for the coal bed that equilibrium is approached by the gases while they are in contact with the incandescent carbon. This need not be the case; if oxygen and steam are supplied at too high a rate, the reactions may not attain equilibrium or may reach equilibrium after they have left the coal bed. In this event, carbon is not present at equilibrium, and the problem must again be reformulated.

The calculations of the preceding example illustrate the complexity of the equations that must be solved simultaneously (even for simple reactions) when the equilibrium-constant method is applied to multireaction equilibria. Moreover, the method does not lend itself to standardization so as to allow a general program to be written for computer solution. The alternative method, mentioned in Sec. 15.2, is based on the fact that at equilibrium the total Gibbs energy of the system has its minimum value. This is illustrated for a single reaction in Fig. 15.1.

The total Gibbs energy of a single-phase system is given by Eq. (10.2), which shows that

$$(G^t)_{T,P} = G(n_1, n_2, n_3, \ldots, n_N)$$

The problem is to find the set of n_i's which minimizes G^t for specified T and P, subject to the constraints of the material balances. The standard solution to this type of problem is based on the method of Lagrange's undetermined multipliers. The procedure for gas-phase reactions is described as follows.

1. The first step is to formulate the constraining equations, i.e., the material balances. Although reacting molecular species are not conserved in a closed system, the total number of atoms of each *element* is constant. Let subscript k identify a particular atomic species. Then define A_k as the total number of atomic masses of the kth element in the system, as determined by the initial constitution of the system. Further, let a_{ik} be the number of atoms of the kth element present in each molecule of chemical species i. The material balance on each element k may then be written:

$$\sum_i n_i a_{ik} = A_k \qquad (k = 1, 2, \ldots, w) \tag{15.38}$$

or

$$\sum_i n_i a_{ik} - A_k = 0 \qquad (k = 1, 2, \ldots, w)$$

2. Next, we introduce the Lagrange multipliers λ_k, one for each element, by multiplying each element balance by its λ_k:

$$\lambda_k\left(\sum_i n_i a_{ik} - A_k\right) = 0 \qquad (k = 1, 2, \ldots, w)$$

These equations are summed over k, giving

$$\sum_k \lambda_k\left(\sum_i n_i a_{ik} - A_k\right) = 0$$

3. Then a new function F is formed by addition of this last sum to G^t. Thus,

$$F = G^t + \sum_k \lambda_k\left(\sum_i n_i a_{ik} - A_k\right)$$

This new function is identical with G^t, because the summation term is zero. However, the partial derivatives of F and G^t with respect to n_i are different, because the function F incorporates the constraints of the material balances.

4. The minimum value of both F and G^t occurs when the partial derivatives of F with respect to n_i are zero. Therefore, we set the expression for these derivatives equal to zero:

$$\left(\frac{\partial F}{\partial n_i}\right)_{T,P,n_j} = \left(\frac{\partial G^t}{\partial n_i}\right)_{T,P,n_j} + \sum_k \lambda_k a_{ik} = 0$$

Since the first term on the right is the definition of the chemical potential [see Eq. (10.1)], this equation can be written:

$$\mu_i + \sum_k \lambda_k a_{ik} = 0 \qquad (i = 1, 2, \ldots, N) \tag{15.39}$$

However, the chemical potential is given by Eq. (15.11):

$$\mu_i = G_i^\circ + RT \ln \hat{a}_i$$

For gas-phase reactions and standard states as the pure gases at 1 bar [or 1(atm)], this becomes

$$\mu_i = G_i^\circ + RT \ln \hat{f}_i$$

If G_i° is arbitrarily set equal to zero for all *elements* in their standard states, then for compounds $G_i^\circ = \Delta G_{f_i}^\circ$, the standard Gibbs-energy change of formation for species i. In addition, the fugacity is eliminated in favor of the fugacity coefficient by Eq. (11.33), $\hat{f}_i = y_i \hat{\phi}_i P$. With these substitutions, the equation for μ_i becomes

$$\mu_i = \Delta G_{f_i}^\circ + RT \ln (y_i \hat{\phi}_i P)$$

Combination with Eq. (15.39) gives

$$\Delta G^\circ_{f_i} + RT \ln (y_i \hat{\phi}_i P) + \sum_k \lambda_k a_{ik} = 0 \qquad (i = 1, 2, \ldots, N) \qquad (15.40)$$

If species i is an element, $\Delta G^\circ_{f_i}$ is zero. The pressure P must be in bars or (atm), depending on whether the standard-state pressure is 1 bar or 1(atm).

There are N equilibrium equations [Eq. (15.40)], one for each species, and there are w material-balance equations [Eq. (15.38)], one for each element—a total of $N + w$ equations. The unknowns in these equations are the n_i's (note that $y_i = n_i/\sum n_i$), of which there are N, and the λ_k's, of which there are w—a total of $N + w$ unknowns. Thus the number of equations is sufficient for the determination of all unknowns.

The foregoing discussion has presumed that the $\hat{\phi}_i$'s are known. If the phase is an ideal gas, then each $\hat{\phi}_i$ is unity. If the phase is an ideal solution, each $\hat{\phi}_i$ becomes ϕ_i, and can at least be estimated. For real gases, each $\hat{\phi}_i$ is a function of the y_i's, the quantities being calculated. Thus an iterative procedure is indicated. The calculations are initiated with each $\hat{\phi}_i$ set equal to unity. Solution of the equations then provides a preliminary set of y_i's. For low pressures or high temperatures this result is usually adequate. Where it is not satisfactory, an equation of state is used together with the calculated y_i's to give a new and more nearly correct set of $\hat{\phi}_i$'s for use in Eq. (15.40). Then a new set of y_i's is determined. The process is repeated until successive iterations produce no significant change in the y_i's. All calculations are well suited to computer solution, including the calculation of the $\hat{\phi}_i$'s by equations such as Eq. (11.48) or (14.47).

In the procedure just described, the question of what chemical reactions are involved never enters directly into any of the equations. However, the choice of a set of species is entirely equivalent to the choice of a set of independent reactions among the species. In any event, a set of species or an equivalent set of independent reactions must always be assumed, and different assumptions produce different results.

Example 15.13 Calculate the equilibrium compositions at 1,000 K and 1 bar of a gas-phase system containing the species CH_4, H_2O, CO, CO_2, and H_2. In the initial unreacted state there are present 2 mol of CH_4 and 3 mol of H_2O. Values of ΔG°_f at 1,000 K are

$$\Delta G^\circ_{f\,CH_4} = 19{,}720 \text{ J mol}^{-1}$$

$$\Delta G^\circ_{f\,H_2O} = -192{,}420 \text{ J mol}^{-1}$$

$$\Delta G^\circ_{f\,CO} = -200{,}240 \text{ J mol}^{-1}$$

$$\Delta G^\circ_{f\,CO_2} = -395{,}790 \text{ J mol}^{-1}$$

SOLUTION The required values of A_k are determined from the initial numbers of moles, and the values of a_{ik} come directly from the chemical formulas of the species. These are shown in the accompanying table.

	Element k		
	Carbon	Oxygen	Hydrogen
	A_k = no. of atomic masses of k in the system		
	$A_C = 2$	$A_O = 3$	$A_H = 14$
Species i	a_{ik} = no. of atoms of k per molecule of i		
CH_4	$a_{CH_4,C} = 1$	$a_{CH_4,O} = 0$	$a_{CH_4,H} = 4$
H_2O	$a_{H_2O,C} = 0$	$a_{H_2O,O} = 1$	$a_{H_2O,H} = 2$
CO	$a_{CO,C} = 1$	$a_{CO,O} = 1$	$a_{CO,H} = 0$
CO_2	$a_{CO_2,C} = 1$	$a_{CO_2,O} = 2$	$a_{CO_2,H} = 0$
H_2	$a_{H_2,C} = 0$	$a_{H_2,O} = 0$	$a_{H_2,H} = 2$

At 1 bar and 1,000 K the assumption of ideal gases is justified and the $\hat{\phi}_i$'s are all unity. Since $P = 1$ bar, Eq. (15.40) is written:

$$\frac{\Delta G^\circ_{fi}}{RT} + \ln \frac{n_i}{\sum n_i} + \sum_k \frac{\lambda_k}{RT} a_{ik} = 0$$

The five equations for the five species then become

$$CH_4: \quad \frac{19{,}720}{RT} + \ln \frac{n_{CH_4}}{\sum n_i} + \frac{\lambda_C}{RT} + \frac{4\lambda_H}{RT} = 0$$

$$H_2O: \quad \frac{-192{,}420}{RT} + \ln \frac{n_{H_2O}}{\sum n_i} + \frac{2\lambda_H}{RT} + \frac{\lambda_O}{RT} = 0$$

$$CO: \quad \frac{-200{,}240}{RT} + \ln \frac{n_{CO}}{\sum n_i} + \frac{\lambda_C}{RT} + \frac{\lambda_O}{RT} = 0$$

$$CO_2: \quad \frac{-395{,}790}{RT} + \ln \frac{n_{CO_2}}{\sum n_i} + \frac{\lambda_C}{RT} + \frac{2\lambda_O}{RT} = 0$$

$$H_2: \quad \ln \frac{n_{H_2}}{\sum n_i} + \frac{2\lambda_H}{RT} = 0$$

The three material-balance equations [Eq. (15.38)] are:

$$C: \quad n_{CH_4} + n_{CO} + n_{CO_2} = 2$$

$$H: \quad 4n_{CH_4} + 2n_{H_2O} + 2n_{H_2} = 14$$

$$O: \quad n_{H_2O} + n_{CO} + 2n_{CO_2} = 3$$

Simultaneous computer solution of these eight equations, with

$$RT = 8{,}314 \text{ J mol}^{-1}$$

and

$$\sum n_i = n_{CH_4} + n_{H_2O} + n_{CO} + n_{CO_2} + n_{H_2}$$

produces the following results ($y_i = n_i/\sum n_i$):

$$y_{CH_4} = 0.0200 \qquad \frac{\lambda_C}{RT} = 0.757$$

$$y_{H_2O} = 0.0983$$

$$y_{CO} = 0.1740 \qquad \frac{\lambda_O}{RT} = 25.06$$

$$y_{CO_2} = 0.0372$$

$$y_{H_2} = 0.6705 \qquad \frac{\lambda_H}{RT} = 0.193$$

$$\sum y_i = 1.0000$$

The values of λ_k/RT are of no significance, but are included for the sake of completeness.

PROBLEMS

15.1 Develop expressions for the mole fractions of the reacting species as functions of the reaction coordinate for:

(a) A system initially containing 1 mol NH_3 and 2 mol O_2 and undergoing the reaction

$$4NH_3(g) + 5O_2(g) \rightarrow 4NO(g) + 6H_2O(g)$$

(b) A system initially containing 3 mol H_2S and 4 mol O_2 and undergoing the reaction

$$2H_2S(g) + 3O_2(g) \rightarrow 2H_2O(g) + 2SO_2(g)$$

(c) A system initially containing 3 mol NH_3, 3 mol NO, and 1 mol H_2O and undergoing the reaction

$$2NH_3(g) + 3NO(g) \rightarrow 3H_2O(g) + 2.5N_2(g)$$

15.2 A system initially containing 3 mol CO_2, 5 mol H_2, and 1 mol H_2O undergoes the following reactions:

$$CO_2(g) + 3H_2(g) \rightarrow CH_3OH(g) + H_2O(g)$$

$$CO_2(g) + H_2(g) \rightarrow CO(g) + H_2O(g)$$

Develop expressions for the mole fractions of the reacting species as functions of the reaction coordinates for the two reactions.

15.3 A system initially containing 3 mol C_2H_4 and 2 mol O_2 undergoes the following reactions:

$$C_2H_4(g) + \tfrac{1}{2}O_2(g) \rightarrow \langle(CH_2)_2\rangle O(g)$$

$$C_2H_4(g) + 3O_2(g) \rightarrow 2CO_2(g) + 2H_2O(g)$$

Develop expressions for the mole fractions of the reacting species as functions of the reaction coordinates for the two reactions.

15.4 Consider the water-gas-shift reaction,

$$H_2(g) + CO_2(g) \rightarrow H_2O(g) + CO(g)$$

At high temperatures and low to moderate pressures the reacting species form an ideal-gas mixture for which we may write Eq. (10.8) as:

$$G = \sum y_i G_i + RT \sum y_i \ln y_i$$

If the Gibbs energies of the elements in their standard states are set equal to zero, $G_i = \Delta G^\circ_{f_i}$ for each species, and then

$$G = \sum y_i \, \Delta G^\circ_{f_i} + RT \sum y_i \ln y_i \qquad (A)$$

With the understanding that T and P are constant, we may write the equilibrium criterion of Eq. (13.53) for this reacting system as:

$$dG^t = d(nG) = n\,dG + G\,dn = 0$$

or

$$n\frac{dG}{d\varepsilon} + G\frac{dn}{d\varepsilon} = 0$$

But for the water-gas-shift reaction, $dn/d\varepsilon = 0$. The equilibrium criterion therefore becomes:

$$\frac{dG}{d\varepsilon} = 0 \qquad (B)$$

Once the y_i are eliminated in favor of ε, Eq. (A) relates G to ε. Data for $\Delta G^\circ_{f_i}$ for the compounds of interest are given with Example 15.12. For a temperature of 1,000 K (the reaction is unaffected by P) and for a feed of 1 mol H_2 and 1 mol CO_2,

(a) Determine the equilibrium value of ε by application of Eq. (B).

(b) Plot G vs. ε, indicating the location of the equilibrium value of ε determined in (a).

15.5 Repeat Prob. 15.4 for a temperature of 1,100 K.

15.6 Repeat Prob. 15.4 for a temperature of 1,200 K.

15.7 Repeat Prob. 15.4 for a temperature of 1,300 K.

15.8 Verify the answer to Prob. 15.4, part (a), by the method of equilibrium constants.

15.9 Verify the answer to Prob. 15.5, part (a), by the method of equilibrium constants.

15.10 Verify the answer to Prob. 15.6, part (a), by the method of equilibrium constants.

15.11 Verify the answer to Prob. 15.7, part (a), by the method of equilibrium constants.

15.12 Develop a general equation for the standard Gibbs energy change of reaction ΔG° as a function of temperature for one of the reactions given in parts (a), (f), (i), (n), (r), (t), (u), (x), and (y) of Prob. 4.20.

15.13 For ideal gases, exact mathematical expressions can be developed for the effect of T and P on ε_e. For conciseness we let $\prod (y_i)^{\nu_i} \equiv K_y$. Then we can write the mathematical relations:

$$\left(\frac{\partial \varepsilon_e}{\partial T}\right)_P = \left(\frac{\partial K_y}{\partial T}\right)_P \frac{d\varepsilon_e}{dK_y} \quad \text{and} \quad \left(\frac{\partial \varepsilon_e}{\partial P}\right)_T = \left(\frac{\partial K_y}{\partial P}\right)_T \frac{d\varepsilon_e}{dK_y}$$

Using Eqs. (15.25) and (15.17), show that

(a) $\left(\frac{\partial \varepsilon_e}{\partial T}\right)_P = \frac{K_y}{RT^2}\frac{d\varepsilon_e}{dK_y}\Delta H^\circ$

(b) $\left(\frac{\partial \varepsilon_e}{\partial P}\right)_T = \frac{K_y}{P}\frac{d\varepsilon_e}{dK_y}(-\nu)$

(c) $d\varepsilon_e/dK_y$ is always positive. (*Note:* It is equally valid and perhaps easier to show that the reciprocal is positive.)

15.14 For the ammonia synthesis reaction written

$$\tfrac{1}{2}N_2(g) + \tfrac{3}{2}H_2(g) \rightarrow NH_3(g)$$

with 0.5 mol N_2 and 1.5 mol H_2 as the initial amounts of reactants and with the assumption that the equilibrium mixture is an ideal gas, show that

$$\varepsilon_e = 1 - (1 + 1.299\,KP)^{-1/2}$$

15.15 Tom, Dick, and Harry, members of a thermodynamics class, are asked to find the equilibrium composition at a particular T and P and for given initial amounts of reactants for the following gas-phase reaction:

$$2HCl + \tfrac{1}{2}O_2 \rightarrow H_2O + Cl_2 \qquad (A)$$

Each solves the problem correctly in a different way. Tom bases his solution on reaction (A) as written. Dick, who prefers whole numbers, multiplies reaction (A) by 2:

$$4HCl + O_2 \rightarrow 2H_2O + 2Cl_2 \qquad (B)$$

Harry, who usually does things backward, deals with the reaction:

$$H_2O + Cl_2 \rightarrow 2HCl + \tfrac{1}{2}O_2 \qquad (C)$$

Write the chemical-equilibrium equations for the three reactions, indicate how the equilibrium constants are related, and show why Tom, Dick, and Harry all obtain the same result.

15.16 The following reaction reaches equilibrium at 550°C and atmospheric pressure:

$$4HCl(g) + O_2(g) \rightarrow 2H_2O(g) + 2Cl_2(g)$$

If the system initially contains 6 mol HCl for each mole of oxygen, what is the composition of the system at equilibrium? Assume ideal gases.

15.17 The following reaction reaches equilibrium at 600°C and atmospheric pressure:

$$N_2(g) + C_2H_2(g) \rightarrow 2HCN(g)$$

If the system initially is an equimolar mixture of nitrogen and acetylene, what is the composition of the system at equilibrium? Assume ideal gases.

15.18 The following reaction reaches equilibrium at 400°C and atmospheric pressure:

$$CH_3CHO(g) + H_2(g) \rightarrow C_2H_5OH(g)$$

If the system initially contains 2 mol H_2 for each mole of acetaldehyde, what is the composition of the system at equilibrium? Assume ideal gases.

15.19 The following reaction reaches equilibrium at 600°C and atmospheric pressure:

$$C_6H_5CH{:}CH_2(g) + H_2(g) \rightarrow C_6H_5.C_2H_5(g)$$

If the system initially contains 2 mol H_2 for each mole of styrene, what is the composition of the system at equilibrium? Assume ideal gases.

15.20 The gas stream from a sulfur burner is composed of 15-mol-% SO_2, 20-mol-% O_2, and 65-mol-% N_2. This gas stream at 1 bar and 480°C enters a catalytic converter, where the SO_2 is further oxidized to SO_3. Assuming that the reaction reaches equilibrium, how much heat must be removed from the converter to maintain isothermal conditions? Base your answer on 1 mol of entering gas.

15.21 For the cracking reaction,

$$C_3H_8(g) \rightarrow C_2H_4(g) + CH_4(g)$$

the equilibrium conversion is negligible at 300 K, but becomes appreciable at temperatuires above 500 K. For a pressure of 1 bar, determine

(a) The fractional conversion of propane at 600 K.

(b) The temperature at which the fractional conversion is 80 percent.

15.22 Ethylene is produced by the dehydrogenation of ethane. If the feed includes 0.4 mol of steam (an inert diluent) per mole of ethane and if the reaction reaches equilibrium at 1,100 K and 2 bar, what is the composition of the product gas on a water-free basis?

15.23 The production of 1,3-butadiene can be carried out by the dehydrogenation of 1-butene:

$$C_2H_5CH{:}CH_2(g) \rightarrow CH_2{:}CHCH{:}CH_2(g) + H_2(g)$$

Side reactions are suppressed by the introduction of steam. If equilibrium is attained at 950 K and

1 bar and if the reactor product contains 10-mol-% 1,3-butadiene, determine
(*a*) The mole fractions of the other species in the product gas.
(*b*) The mole fraction of steam required in the feed.

15.24 The production of 1,3-butadiene can be carried out by the dehydrogenation of *n*-butane:

$$C_4H_{10}(g) \rightarrow CH_2{:}CHCH{:}CH_2(g) + 2H_2(g)$$

Side reactions are suppressed by the introduction of steam. If equilibrium is attained at 925 K and 1 bar and if the reactor product contains 12-mol-% 1,3-butadiene, determine
(*a*) The mole fractions of the other species in the product gas.
(*b*) The mole fraction of steam required in the feed.

15.25 For the ammonia synthesis reaction,

$$\tfrac{1}{2}N_2(g) + \tfrac{3}{2}H_2(g) \rightarrow NH_3(g)$$

the equilibrium conversion to ammonia is large at 300 K, but decreases rapidly with increasing *T*. However, reaction rates become appreciable only at higher temperatures. For a feed mixture of hydrogen and nitrogen in the stoichiometric proportions,

(*a*) Determine the mole fraction of ammonia in the equilibrium mixture at 1 bar and 300 K.

(*b*) At what temperature does the equilibrium mole fraction of ammonia decrease to 0.50 for a pressure of 1 bar?

(*c*) At what temperature does the equilibrium mole fraction of ammonia decrease to 0.50 for a pressure of 100 bar, assuming the equilibrium mixture an ideal gas?

(*d*) At what temperature does the equilibrium mole fraction of ammonia decrease to 0.50 for a pressure of 100 bar, assuming the equilibrium mixture an ideal solution of gases?

15.26 For the methanol synthesis reaction,

$$CO(g) + 2H_2(g) \rightarrow CH_3OH(g)$$

the equilibrium conversion to methanol is large at 300 K, but decreases rapidly with increasing *T*. However, reaction rates become appreciable only at higher temperatures. For a feed mixture of carbon monoxide and hydrogen in the stoichiometric proportions,
(*a*) Determine the mole fraction of methanol in the equilibrium mixture at 1 bar and 300K.
(*b*) At what temperature does the equilibrium mole fraction of methanol decrease to 0.50 for a pressure of 1 bar?
(*c*) At what temperature does the equilibrium mole fraction of methanol decrease to 0.50 for a pressure of 100 bar, assuming the equilibrium mixture an ideal gas?
(*d*) At what temperature does the equilibrium mole fraction of methanol decrease to 0.50 for a pressure of 100 bar, assuming the equilibrium mixture an ideal solution of gases?

15.27 Limestone ($CaCO_3$) decomposes upon heating to yield quicklime (CaO) and carbon dioxide. At what temperature does limestone exert a decomposition pressure of 1(atm)?

15.28 Ammonium chloride [$NH_4Cl(s)$] decomposes upon heating to yield a gas mixture of ammonia and hydrochloric acid. At what temperature does ammonium chloride exert a decomposition pressure of 1(atm)? For $NH_4Cl(s)$, $\Delta H^\circ_{f_{298}} = -314{,}430$ J and $\Delta G^\circ_{f_{298}} = -202{,}870$ J.

15.29 A chemically reactive system contains the following species in the gas phase: NH_3, NO, NO_2, O_2, and H_2O. Determine a complete set of independent reactions for this system. How many degrees of freedom does the system have?

15.30 The relative compositions of the pollutants NO and NO_2 in air are governed by the reaction,

$$NO + \tfrac{1}{2}O_2 \rightarrow NO_2$$

For air containing 21-mol-% O_2 at 25°C and 1.0133 bar, what is the concentration of NO in parts per million if the total concentration of both nitrogen oxides is 5 ppm?

15.31 For the reaction,

$$SO_2(g) + \tfrac{1}{2}O_2(g) \rightarrow SO_3(g)$$

in equilibrium at 900 K what pressure is required for a 90 percent conversion of SO_2 if the initial mixture is equimolar in the reactants? Assume ideal gases.

15.32 Carbon black is produced by the decomposition of methane:

$$CH_4(g) \rightarrow C(s) + 2H_2(g)$$

For equilibrium at 700°C and 1 bar,
(*a*) What is the gas-phase composition if pure methane enters the reactor, and what fraction of the methane decomposes?
(*b*) Repeat part (*a*) if the feed is an equimolar mixture of methane and nitrogen.

15.33 Consider the reactions,

$$\tfrac{1}{2}N_2(g) + \tfrac{1}{2}O_2(g) \rightarrow NO(g)$$

$$\tfrac{1}{2}N_2(g) + O_2(g) \rightarrow NO_2(g)$$

If these reactions come to equilibrium after combustion in an internal combustion engine at 2,000 K and 200 bar, estimate the mole fractions of NO and NO_2 present for mole fractions of nitrogen and oxygen in the combustion products of 0.70 and 0.05.

15.34 Oil refineries frequently have both H_2S and SO_2 to dispose of. The following reaction suggests a means of getting rid of both at once:

$$2H_2S(g) + SO_2(g) \rightarrow 3S(g) + 2H_2O(g)$$

For reactants present in the stoichiometric proportion, estimate the percent conversion of each reactant if the reaction comes to equilibrium at 500°C and 10 bar.

15.35 The species $N_2O_4(a)$ and $NO_2(b)$ as gases attain rapid equilibrium by the reaction:

$$N_2O_4 \rightarrow 2NO_2$$

(*a*) For $T = 350$ K and $P = 5$ bar, calculate y_a, the mole fraction of species *a* in the equilibrium mixture. Assume ideal gases.

(*b*) If an equilibrium mixture of N_2O_4 and NO_2 at the conditions of part (*a*) flows through a throttle valve to a pressure of 1 bar and through a heat exchanger that restores its initial temperature, how much heat must be exchanged, assuming chemical equilibrium is again attained in the final state? Base your answers on an amount of mixture equivalent to 1 mol of N_2O_4, that is, as though all the NO_2 were present as N_2O_4.

15.36 The following isomerization reaction occurs in the *liquid* phase:

$$A \rightarrow B$$

where *A* and *B* are miscible liquids for which

$$G^E/RT = 0.1x_Ax_B$$

If $\Delta G^\circ_{298} = -1{,}000$ J, what is the equilibrium composition of the mixture at 25°C? How much error is introduced if one assumes that *A* and *B* form an ideal solution?

15.37 The feed gas to a methanol synthesis reactor is composed of 75-mol-% H_2, 12-mol-% CO, 8-mol-% CO_2, and 5-mol-% N_2. The system comes to equilibrium at 550 K and 100 bar with respect to the following reactions:

$$2H_2(g) + CO(g) \rightarrow CH_3OH(g)$$

$$H_2(g) + CO_2(g) \rightarrow CO(g) + H_2O(g)$$

Assuming ideal gases, determine the composition of the equilibrium mixture.

15.38 Hydrogen gas is produced by the reaction of steam with "water gas," an equimolar mixture of H_2 and CO obtained by the reaction of steam with coal. A stream of "water gas" mixed with steam is passed over a catalyst to convert CO to CO_2 by the reaction:

$$H_2O(g) + CO(g) \rightarrow H_2(g) + CO_2(g)$$

Subsequently, unreacted water is condensed and carbon dioxide is absorbed, leaving a product that is mostly hydrogen. The equilibrium conditions are 1 bar and 800 K.

(*a*) Would there be any advantage to carrying out the reaction at pressures above 1 bar?

(*b*) If the equilibrium temperature were raised, would the conversion of CO be increased?

(*c*) For the given equilibrium conditions, determine the molar ratio of steam to "water gas" ($H_2 + CO$) required to produce a *product* gas containing only 2-mol-% CO after cooling to 20°C, where the unreacted H_2O has been virtually all condensed out.

(*d*) Is there any danger that solid carbon will form at the equilibrium conditions by the reaction

$$2CO(g) \rightarrow CO_2(g) + C(s)$$

15.39 One method for the manufacture of "synthesis gas" is the catalytic reforming of methane with steam:

$$CH_4(g) + H_2O(g) \rightarrow CO(g) + 3H_2(g)$$

The only other reaction considered is

$$CO(g) + H_2O(g) \rightarrow CO_2(g) + H_2(g)$$

Assume equilibrium is attained for both reactions at 1 bar and 1,300 K.

(*a*) Would it be better to carry out the reaction at pressures above 1 bar?

(*b*) Would it be better to carry out the reaction at temperatures below 1,300 K?

(*c*) Estimate the molar ratio of hydrogen to carbon monoxide in the synthesis gas if the feed consists of an equimolar mixture of steam and methane.

(*d*) Repeat part (*c*) for a steam to methane mole ratio in the feed of 2.

(*e*) How could the feed composition be altered to yield a lower ratio of hydrogen to carbon monoxide in the synthesis gas than is obtained in part (*c*).

(*f*) Is there any danger that carbon will deposit by the reaction $2CO \rightarrow C + CO_2$ under conditions of part (*c*)? Part (*d*)? If so, how could the feed be altered to prevent carbon deposition?

15.40 Set up the equations required for solution of Example 15.13 by the method of equilibrium constants. Verify that your equations yield the same equilibrium compositions as given in the example.

15.41 Ethylene oxide as a vapor and water as liquid, both at 25°C and 101.33 kPa, react to form an aqueous solution of ethylene glycol (1,2-ethanediol) at the same conditions:

$$\langle(CH_2)_2\rangle O + H_2O \rightarrow CH_2OH.CH_2OH$$

If the initial molar ratio of ethylene oxide to water is 3.0, determine the equilibrium conversion of ethylene oxide to ethylene glycol.

At equilibrium the system consists of liquid and vapor in equilibrium, and the intensive state of the system is fixed by the specification of T and P. Therefore, one must first determine the phase compositions, independent of the ratio of reactants. These results may then be applied in the material-balance equations to find the equilibrium conversion.

Choose as standard states for water and ethylene glycol the pure liquids at 1 bar and for ethylene oxide the pure ideal gas at 1 bar. Assume that the Lewis/Randall rule applies to the water in the liquid phase and that the vapor phase is an ideal gas. The partial pressure of ethylene oxide over the liquid phase is given by

$$p_i/\text{kPa} = 415x_i$$

The vapor pressure of ethylene glycol at 25°C is so low that its concentration in the vapor phase is negligible.

CHAPTER

SIXTEEN

THERMODYNAMIC ANALYSIS OF PROCESSES

The object of this chapter is the evaluation of real processes from the thermodynamic point of view. No new fundamental ideas are needed; the method is based on a combination of the first and second laws. Hence, the chapter affords a review of the principles of thermodynamics.

Although applications of thermodynamics to processes are often based on the assumption of reversibility, real irreversible processes are nevertheless amenable to thermodynamic analysis. The goal of such an analysis is to determine how efficiently energy is used or produced and to show quantitatively the effect of inefficiencies in each step of a process. The cost of energy is of concern in any manufacturing operation, and the first step in any attempt to reduce energy requirements is to determine where and to what extent energy is wasted through process irreversibilities. The treatment here is limited to steady-state flow processes, because of their overwhelming industrial preponderance.

16.1 SECOND-LAW RELATION FOR STEADY-STATE FLOW PROCESSES

The general energy balance for steady-state flow processes is given by Eq. (7.8):

$$\Delta[(H + \tfrac{1}{2}u^2 + zg)\dot{m}]_{\text{fs}} = \dot{Q} - \dot{W}_s \tag{7.8}$$

We can also write a general second-law relation for such processes. The basic requirement is given by Eq. (5.19), $\Delta S_{\text{total}} \geq 0$. Thus, with respect to the control volume for a steady-flow process in surroundings at temperature T_0:

$$\left\{\begin{array}{l}\text{Net rate of entropy}\\ \text{transport out by}\\ \text{flowing streams}\end{array}\right\} + \left\{\begin{array}{l}\text{rate of entropy change}\\ \text{of surroundings}\\ \text{from heat transfer}\end{array}\right\} \geq 0$$

Whence

$$\Delta(S\dot{m})_{\text{fs}} + \frac{\dot{Q}_{\text{surr}}}{T_0} \geq 0$$

Since heat transfer with respect to the surroundings is the negative of heat transfer with respect to the system, $\dot{Q}_{\text{surr}} = -\dot{Q}$. Therefore the *rate of entropy generation* $\dot{S}_{\text{total}}$ is defined as

$$\dot{S}_{\text{total}} \equiv \Delta(S\dot{m})_{\text{fs}} - \frac{\dot{Q}}{T_0} \geq 0 \tag{16.1}$$

For the special case of a single stream flowing through the control volume, this becomes

$$\dot{S}_{\text{total}} \equiv \dot{m}\,\Delta S - \frac{\dot{Q}}{T_0} \geq 0 \tag{16.2}$$

Division by $\dot{m}$ provides an equation based on a unit amount of fluid flowing through the control volume:

$$\Delta S_{\text{total}} \equiv \Delta S - \frac{Q}{T_0} \geq 0 \tag{16.3}$$

The equality in the preceding equations applies to reversible processes.

16.2 CALCULATION OF IDEAL WORK

In any steady-state flow process requiring work, there is an absolute minimum amount which must be expended to accomplish the desired change of state of the fluid flowing through the control volume. In a process producing work, there is an absolute maximum amount which may be accomplished as the result of a given change of state of the fluid flowing through the control volume. In either case, this limiting value is called the *ideal work*, W_{ideal}. It is the work resulting when the change of state of the fluid is accomplished *completely reversibly*. The requirement of complete reversibility for a process implies the following:

1. All changes within the control volume are reversible.
2. Heat transfer to or from the surroundings is also reversible.

A completely reversible process is hypothetical, devised solely for determination of the ideal work associated with a given change of state. Its only connection with an actual process is that it brings about the same change of state as the actual process. Our objective is to compare the actual work of a process with the work of the hypothetical reversible process.

The second requirement listed for complete reversibility means that all heat transfer between the system and the surroundings in the hypothetical reversible process must occur at the temperature of the surroundings, here denoted by T_0. This means that at least the part of the system transmitting heat must be at the surroundings temperature T_0. This may require inclusion within the system of Carnot engines or heat pumps to accomplish reversible transfer of heat between temperature levels in the system and the temperature level of the surroundings. Since the Carnot engines and heat pumps are cyclic they contribute nothing to the change of state of the system. An illustration of a hypothetical reversible process is given in Example 16.1.

It is never *necessary* to describe hypothetical processes devised for the calculation of ideal work. All that is required is the realization that such a process may always be imagined. The equation for ideal work is developed with complete generality in the following paragraphs from the first and second laws and the listed requirements of reversibility.

We presume that the process is completely reversible and that the system exists in surroundings that constitute a heat reservoir at the constant temperature T_0. For any completely reversible process, the entropy generation is zero, and Eq. (16.1) becomes

$$\dot{Q} = T_0\,\Delta(S\dot{m})_{\text{fs}}$$

Substitution of this expression for $\dot{Q}$ in the energy balance of Eq. (7.8) gives

$$\Delta[(H + \tfrac{1}{2}u^2 + zg)\dot{m}]_{\text{fs}} = T_0\,\Delta(S\dot{m})_{\text{fs}} - \dot{W}_s(\text{rev})$$

where $\dot{W}_s(\text{rev})$ indicates that the work is for a reversible process. We call this work the *ideal work*, $\dot{W}_{\text{ideal}}$. Thus

$$\dot{W}_{\text{ideal}} = T_0\,\Delta(S\dot{m})_{\text{fs}} - \Delta[(H + \tfrac{1}{2}u^2 + zg)\dot{m}]_{\text{fs}} \tag{16.4}$$

In most applications to chemical processes, the kinetic- and potential-energy terms are negligible compared with the others; in this event Eq. (16.4) is written

$$\boxed{\dot{W}_{\text{ideal}} = T_0\,\Delta(S\dot{m})_{\text{fs}} - \Delta(H\dot{m})_{\text{fs}}} \tag{16.5}$$

For the special case of a single stream flowing through the control volume, Eq. (16.5) becomes

$$\dot{W}_{\text{ideal}} = \dot{m}(T_0\,\Delta S - \Delta H) \tag{16.6}$$

Division by $\dot{m}$ puts this equation on a unit-mass basis

$$W_{\text{ideal}} = T_0\,\Delta S - \Delta H \tag{16.7}$$

Equations (16.4) through (16.7) give the work of completely reversible processes associated with given property changes in the flowing streams. When the same property change occurs in an actual process, the actual work $\dot{W}_s$ (or

W_s) is given by an energy balance, and we can compare the actual work with the ideal work. When $\dot{W}_{\text{ideal}}$ (or W_{ideal}) is positive, it is the *maximum work obtainable* from a given change in the properties of the flowing streams, and is larger than $\dot{W}_s$. In this case we define a thermodynamic efficiency η_t as the ratio of the actual work to the ideal work:

$$\eta_t(\text{work produced}) = \frac{\dot{W}_s}{\dot{W}_{\text{ideal}}} \tag{16.8}$$

When $\dot{W}_{\text{ideal}}$ (or W_{ideal}) is negative, $|\dot{W}_{\text{ideal}}|$ is the *minimum work required* to bring about a given change in the properties of the flowing streams, and is smaller than $|\dot{W}_s|$. In this case, the thermodynamic efficiency is defined as the ratio of the ideal work to the actual work:

$$\eta_t(\text{work required}) = \frac{\dot{W}_{\text{ideal}}}{\dot{W}_s} \tag{16.9}$$

Example 16.1 What is the maximum work that can be obtained in a steady-state flow process from 1 mol of nitrogen (assumed an ideal gas) at 800 K and 50 bar? Take the temperature and pressure of the surroundings as 300 K and 1.0133 bar.

SOLUTION The maximum possible work is obtained from any completely reversible process that reduces the nitrogen to the temperature and pressure of the surroundings, i.e., to 300 K and 1.0133 bar. The maintenance of a final temperature or pressure below that of the surroundings would require work in an amount at least equal to any gain in work from the process as a result of the lower level. The result is obtained directly from Eq. (16.7), where ΔS and ΔH are the molar entropy and enthalpy changes of the nitrogen as its state is changed from 800 K and 50 bar to 300 K and 1.0133 bar. For an ideal gas, enthalpy is independent of pressure, and its change is given by Eq. (4.8):

$$\Delta H = C_{P\text{mh}}(T_2 - T_1)$$

The mean heat capacity is evaluated by Eq. (4.7) with data from Table 4.1. With $T_{\text{am}} = 550$ K, we obtain

$$C_{P_{\text{mh}}}/R = 3.280 + (0.000593)(550) + \frac{4{,}000}{(800)(300)}$$

$$= 3.623$$

Whence

$$\Delta H = (3.623)(8.314)(300 - 800) = -15{,}060 \text{ J mol}^{-1}$$

The entropy change is found from Eq. (5.18):

$$\Delta S = C_{P\text{ms}} \ln \frac{T_2}{T_1} - R \ln \frac{P_2}{P_1}$$

The mean heat capacity is here evaluated by Eq. (5.17) with $T_{\text{lm}} = 509.77$ K:

$$C_{P_{\text{ms}}}/R = 3.280 + (0.000593)(509.77) + \frac{(550)(509.77)(4{,}000)}{(800)^2(300)^2}$$

$$= 3.602$$

Whence

$$\Delta S = (3.602)(8.314) \ln \frac{300}{800} - 8.314 \ln \frac{1.0133}{50} = 3.042 \text{ J mol}^{-1} \text{ K}^{-1}$$

With these values of ΔH and ΔS, Eq. (16.7) becomes

$$W_{\text{ideal}} = (300)(3.042) - (-15{,}060) = 15{,}973 \text{ J mol}^{-1}$$

The significance of this simple calculation becomes evident when we consider in detail the steps of a specific reversible process designed to bring about the same change of state. Suppose the nitrogen is continuously changed to its final state at 1.0133 bar and $T_2 = T_0 = 300$ K by the following two-step process:

1. Reversible, adiabatic expansion (as in a turbine) from the initial state P_1, T_1, H_1 to 1.0133 bar. Suppose the temperature at the end of this isentropic step is T'.
2. Cooling (or heating, if T' is less than T_2) to the final temperature T_2 at a constant pressure of 1.0133 bar.

For step 1, a steady-state flow process, the energy balance is

$$Q - W_s = \Delta H$$

or, since the process is adiabatic

$$W_s = -\Delta H = -(H' - H_1)$$

where H' is the enthalpy at the intermediate state of T' and 1.0133 bar. For maximum work, step 2 must also be reversible, with heat transferred reversibly to the surroundings at T_0. These requirements are met by use of Carnot engines which receive heat from the nitrogen, produce work W_{Carnot}, and reject heat to the surroundings at T_0. Since the temperature of the heat source, the nitrogen, decreases from T' to T_2, the expression for the work of the Carnot engines is written in differential form:

$$dW_{\text{Carnot}} = \frac{T - T_0}{T}(-dQ)$$

The minus sign preceding dQ is required in order that Q refer to the nitrogen. Integration yields

$$W_{\text{Carnot}} = -Q + T_0 \int_{T'}^{T_2} \frac{dQ}{T}$$

Quantity Q, the heat exchanged with the nitrogen, is equal to the enthalpy change $H_2 - H'$. The integral is the change in entropy of the nitrogen as it is cooled by the Carnot engines. Since step 1 occurs at constant entropy, the integral also represents ΔS for both steps. Hence

$$W_{\text{Carnot}} = -(H_2 - H') + T_0 \, \Delta S$$

The sum of W_s and W_{Carnot} gives the ideal work; thus

$$\begin{aligned} W_{\text{ideal}} &= -(H' - H_1) - (H_2 - H') + T_0 \, \Delta S \\ &= -(H_2 - H_1) + T_0 \, \Delta S \\ &= -\Delta H + T_0 \, \Delta S \end{aligned}$$

which is the same as Eq. (16.7).

This derivation makes clear the difference between W_s, the shaft work of the turbine, and W_{ideal}. The ideal work includes not only the shaft work, but also all work obtainable by the operation of heat engines for the reversible transfer of heat to the surroundings at T_0.

Example 16.2 Rework Example 5.5, making use of the equation for ideal work.

SOLUTION The procedure here is to calculate the maximum possible work W_{ideal} which can be obtained from 1 kg of steam in a flow process as it undergoes a change in state from saturated steam at 100°C to liquid water at 0°C. Now the problem reduces to the question of whether this amount of work is sufficient to operate a Carnot heat pump delivering 2,000 kJ as heat at 200°C and taking heat from the unlimited supply of cooling water at 0°C.

For the steam,

$$\Delta H = 0 - 2{,}676.0 = -2{,}676.0$$

$$\Delta S = 0 - 7.3554 = -7.3554$$

Neglecting kinetic- and potential-energy terms, we have by Eq. (16.7):

$$W_{ideal} = T_0\,\Delta S - \Delta H = (273.15)(-7.3554) - (-2{,}676.0)$$

$$= 666.9 \text{ kJ kg}^{-1}$$

If this amount of work, the maximum obtainable from the steam, is used to derive a Carnot heat pump operating between the temperatures of 0 and 200°C, the heat transferred at the higher temperature is

$$Q = W\frac{T}{T_0 - T} = (666.9)\left(\frac{200 + 273.15}{200 - 0}\right) = 1{,}577.7 \text{ kJ}$$

This is the maximum possible heat release at 200°C; it is less than the claimed value of 2,000 kJ. As in Example 5.5, we conclude that the process described is not possible.

Example 16.3 What is the thermodynamic efficiency of the compression process of Example 7.8 if $T_0 = 300$ K?

SOLUTION Saturated steam at 100 kPa is compressed adiabatically to 300 kPa with a compressor efficiency of 0.75. From the results of Example 7.8, we have:

$$\Delta H = 2{,}959.9 - 2{,}675.4 = 289.5 \text{ kJ kg}^{-1}$$

$$\Delta S = 7.5019 - 7.3598 = 0.1421 \text{ kJ kg}^{-1}\text{ K}^{-1}$$

and

$$W_s = -284.5 \text{ kJ kg}^{-1}$$

Application of Eq. (16.7) gives

$$W_{ideal} = T_0\,\Delta S - \Delta H = (300)(0.1421) - 289.5 = -241.9 \text{ kJ kg}^{-1}$$

Then by Eq. (16.9),

$$\eta_t = \frac{W_{ideal}}{W_s} = \frac{-241.9}{-284.5} = 0.850$$

The *compressor* efficiency η, based on reversible compression to a final state where

$S_2 = S_1$, is different from the *thermodynamic* efficiency η_t, which is based on reversible compression to the *actual* final state where $S_2 > S_1$.

16.3 LOST WORK

The energy that becomes unavailable for work as the result of irreversibilities in a process is called the *lost work*, and is defined as the difference between the ideal work for a process and the actual work of the process. Thus by definition,

$$W_{\text{lost}} \equiv W_{\text{ideal}} - W_s \tag{16.10}$$

In terms of rates this is written

$$\dot{W}_{\text{lost}} \equiv \dot{W}_{\text{ideal}} - \dot{W}_s \tag{16.11}$$

The ideal work rate is given by Eq. (16.4):

$$\dot{W}_{\text{ideal}} = T_0\,\Delta(S\dot{m})_{\text{fs}} - \Delta[(H + \tfrac{1}{2}u^2 + zg)\dot{m}]_{\text{fs}}$$

The actual work rate comes from Eq. (7.8)

$$\dot{W}_s = \dot{Q} - \Delta[(H + \tfrac{1}{2}u^2 + zg)\dot{m}]_{\text{fs}}$$

The difference between these two equations gives

$$\boxed{\dot{W}_{\text{lost}} = T_0\,\Delta(S\dot{m})_{\text{fs}} - \dot{Q}} \tag{16.12}$$

For the special case of a single stream flowing through the control volume,

$$\dot{W}_{\text{lost}} = \dot{m}T_0\,\Delta S - \dot{Q} \tag{16.13}$$

Division of this equation by $\dot{m}$ gives

$$W_{\text{lost}} = T_0\,\Delta S - Q \tag{16.14}$$

where the basis is now a unit amount of fluid flowing through the control volume.

As a result of Eq. (16.1) we can write:

$$T_0\dot{S}_{\text{total}} = T_0\,\Delta(S\dot{m})_{\text{fs}} - \dot{Q}$$

Since the right-hand sides of this equation and of Eq. (16.12) are identical, it follows that

$$\boxed{\dot{W}_{\text{lost}} = T_0\dot{S}_{\text{total}}} \tag{16.15}$$

For flow on the basis of a unit amount of fluid, Eqs. (16.3) and (16.14) give

$$W_{\text{lost}} = T_0\,\Delta S_{\text{total}} \tag{16.16}$$

The second law of thermodynamics, as reflected in Eqs. (16.1) and (16.3), requires

$$\dot{S}_{\text{total}} \geq 0 \qquad \Delta S_{\text{total}} \geq 0$$

Whence

$$\dot{W}_{\text{lost}} \geq 0 \qquad W_{\text{lost}} \geq 0$$

When a process is completely reversible, the equality holds, and the lost work is zero. For irreversible processes the inequality holds, and the lost work, i.e., the energy that becomes unavailable for work, is positive. The engineering significance of this result is clear: The greater the irreversibility of a process, the greater the rate of entropy production and the greater the amount of energy that becomes unavailable for work. Thus every irreversibility carries with it a price.

Example 16.4 What is the lost work associated with the compression process of Example (16.3)?

SOLUTION Since the compression process is adiabatic, Eqs. (16.14) and (16.16) both reduce to

$$W_{\text{lost}} = T_0 \, \Delta S$$

where ΔS is the entropy change of the steam as a result of compression. Taking this value from Example 16.3, we find

$$W_{\text{lost}} = (300)(0.1421) = 42.6 \text{ kJ kg}^{-1}$$

This result is also given by Eq. (16.10), where values are from Examples 16.3:

$$W_{\text{lost}} = W_{\text{ideal}} - W_s = -241.9 - (-284.5) = 42.6 \text{ kJ kg}^{-1}$$

16.4 THERMODYNAMIC ANALYSIS OF STEADY-STATE FLOW PROCESSES

Many processes consist of a number of steps, and lost-work calculations are then made for each step separately. By Eq. (16.15),

$$\dot{W}_{\text{lost}} = T_0 \dot{S}_{\text{total}}$$

Summing over the steps of a process gives

$$\sum \dot{W}_{\text{lost}} = T_0 \sum \dot{S}_{\text{total}}$$

Dividing the latter by the former, we get

$$\frac{\dot{W}_{\text{lost}}}{\sum \dot{W}_{\text{lost}}} = \frac{\dot{S}_{\text{total}}}{\sum \dot{S}_{\text{total}}}$$

Thus an analysis of the lost work, made by calculation of the fraction that each individual lost-work term represents of the total lost work, is the same as an analysis of the rate of entropy generation, made by expressing each individual entropy-generation term as a fraction of the sum of all entropy-generation terms.

An alternative to the lost-work or entropy-generation analysis is a work analysis. For this, we write Eq. (16.11) as

$$\sum \dot{W}_{\text{lost}} = \dot{W}_{\text{ideal}} - \dot{W}_s \tag{16.17}$$

For a work-producing process, all of these work quantities are positive and $\dot{W}_{\text{ideal}} > \dot{W}_s$. We therefore write the preceding equation as

$$\dot{W}_{\text{ideal}} = \dot{W}_s + \sum \dot{W}_{\text{lost}} \tag{16.18}$$

A work analysis then expresses each of the individual work terms on the right as a fraction of $\dot{W}_{\text{ideal}}$.

For a work-requiring process, $\dot{W}_s$ and $\dot{W}_{\text{ideal}}$ are negative, and $|\dot{W}_s| > |\dot{W}_{\text{ideal}}|$. Equation (16.17) is therefore best written:

$$|\dot{W}_s| = |\dot{W}_{\text{ideal}}| + \sum \dot{W}_{\text{lost}} \tag{16.19}$$

A work analysis here expresses each of the individual work terms on the right as a fraction of $|\dot{W}_s|$. A work analysis cannot be carried out in the case where a process is so inefficient that $\dot{W}_{\text{ideal}}$ is positive, indicating that the process should produce work, but $\dot{W}_s$ is negative, indicating that the process in fact requires work. A lost-work or entropy-generation analysis is always possible.

Example 16.5 The operating conditions of a practical steam power plant are described in Example 8.1, parts (*b*) and (*c*). In addition, steam generation is accomplished in a furnace/boiler unit where methane is burned completely to CO_2 and H_2O with 25 percent excess air. The flue gas leaving the furnace has a temperature of 460 K, and $T_0 = 298.15$ K. Make a thermodynamic analysis of the power plant.

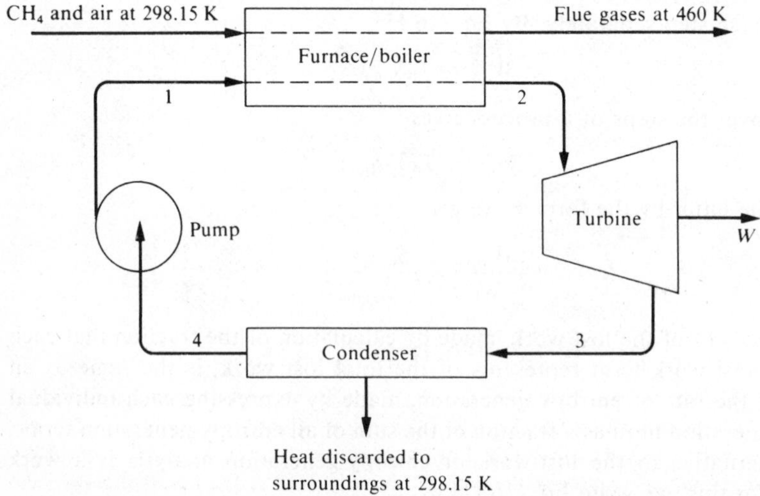

Figure 16.1 Power cycle of Example 16.5.

SOLUTION A flow diagram of the power plant is shown in Fig. 16.1. The conditions and properties for key points in the steam cycle, taken from Example 8.1, are listed in the following table.

Point	State of steam	t/°C	P/kPa	H/kJ kg^{-1}	S/kJ kg^{-1} K^{-1}
1	Subcooled liquid	45.83	8,600	203.4	0.6580
2	Superheated vapor	500	8,600	3,391.6	6.6858
3	Wet vapor, $x = 0.9378$	45.83	10	2,436.0	7.6846
4	Saturated liquid	45.83	10	191.8	0.6493

Since the steam undergoes a cyclic process, the only changes that need be considered for calculation of the ideal work are those of the gases passing through the furnance. The reaction occurring is

$$CH_4 + 2O_2 \rightarrow CO_2 + 2H_2O$$

For this reaction, data from Tables 4.4 and 15.1 give:

$$\Delta H^\circ_{298} = -393{,}509 + (2)(-241{,}818) - (-74{,}520) = -802{,}625 \text{ J}$$

$$\Delta G^\circ_{298} = -394{,}359 + (2)(-228{,}572) - (-50{,}460) = -801{,}043 \text{ J}$$

Moreover,

$$\Delta S^\circ_{298} = \frac{\Delta H^\circ_{298} - \Delta G^\circ_{298}}{298.15} = -5.306 \text{ J K}^{-1}$$

On the basis of 1 mol of methane burned with 25 percent excess air, the air entering the furnace contains:

$$O_2: \; (2)(1.25) = 2.5 \text{ mol}$$

$$N_2: \; (2.5)(79/21) = 9.405 \text{ mol}$$

Total: 11.905 mol air

After complete combustion of the methane, the flue gas contains:

CO_2: 1 mol	$y_{CO_2} = 0.0775$
H_2O: 2 mol	$y_{H_2O} = 0.1550$
O_2: 0.5 mol	$y_{O_2} = 0.0387$
N_2: 9.405 mol	$y_{N_2} = 0.7288$

Total: 12.905 mol flue gas

The change of state that occurs in the furnace leads from methane and air at atmospheric pressure and 298.15 K, the temperature of the surroundings, to flue gas at atmospheric pressure and 460 K. For the purpose of calculating ΔH and ΔS for this change of state, we devise the path shown in Fig. 16.2. The assumption of ideal gases is entirely reasonable here, and on this basis we calculate ΔH and ΔS for each of the four steps shown in Fig. 16.2.

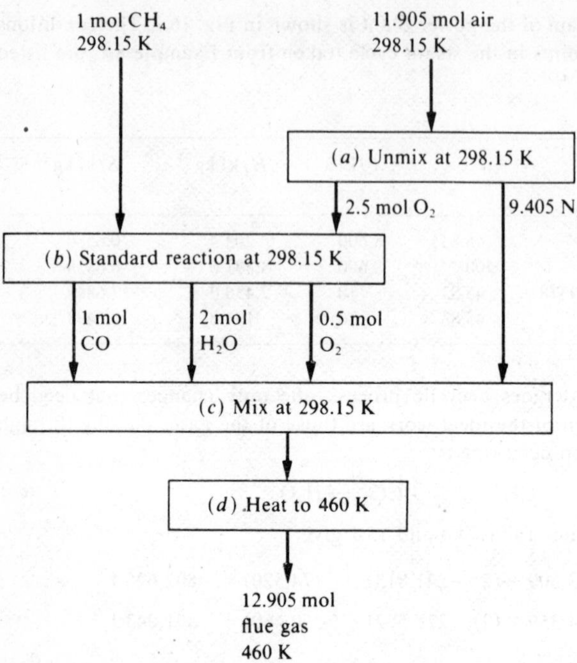

Figure 16.2 Calculation path for combustion process of Example 16.5.

Step a: For unmixing the entering air, we have by Eq. (13.47) and by Eq. (13.45) with a change of sign,

$$\Delta H_a = 0$$

$$\begin{aligned}\Delta S_a &= nR \sum y_i \ln y_i \\ &= (11.905)(8.314)(0.21 \ln 0.21 + 0.79 \ln 0.79) \\ &= -50.870 \text{ J K}^{-1}\end{aligned}$$

Step b: For the standard reaction at 298.15 K,

$$\Delta H_b = \Delta H^\circ_{298} = -802{,}625 \text{ J}$$

$$\Delta S_b = \Delta S^\circ_{298} = -5.306 \text{ J K}^{-1}$$

Step c: For mixing to form the flue gas,

$$\Delta H_c = 0$$

$$\begin{aligned}\Delta S_c &= -nR \sum y_i \ln y_i \\ &= -(12.905)(8.314)(0.0775 \ln 0.0775 + 0.1550 \ln 0.1550 \\ &\quad + 0.0387 \ln 0.0387 + 0.7288 \ln 0.7288) \\ &= 90.510 \text{ J K}^{-1}\end{aligned}$$

Step d: For the heating step, we calculate mean heat capacities between 298.15 and 460 K by Eqs. (4.7) and (5.17) with data from Table 4.1. The results in J mol^{-1} K^{-1} are summarized as follows:

	$C_{P_{mh}}$	$C_{P_{ms}}$
CO_2	41.649	41.377
H_2O	34.153	34.106
N_2	29.381	29.360
O_2	30.473	30.405

We multiply each individual heat capacity by the number of moles of that species in the flue gas and sum over all species. This gives total mean heat capacities for the 12.905 mol of mixture:

$$C^t_{P_{mh}} = 401.520 \qquad \text{and} \qquad C^t_{P_{ms}} = 400.922 \text{ J K}^{-1}$$

Then

$$\Delta H_d = C^t_{P_{mh}}(T_2 - T_1) = (401.520)(460 - 298.15)$$
$$= 64{,}986 \text{ J}$$

$$\Delta S_d = C^t_{P_{ms}} \ln \frac{T_2}{T_1} = 400.922 \ln \frac{460}{298.15} = 173.852 \text{ J K}^{-1}$$

For the total process on the basis of 1 mol CH_4 burned,

$$\Delta H = \sum \Delta H_i = 0 - 802{,}625 + 0 + 64{,}986$$
$$= -737{,}639 \text{ J} \qquad \text{or} \qquad -737.64 \text{ kJ}$$
$$\Delta S = \sum \Delta S_i = -50.870 - 5.306 + 90.510 + 173.852$$
$$= 208.186 \text{ J K}^{-1} \qquad \text{or} \qquad 0.2082 \text{ kJ K}^{-1}$$

The steam rate found in Example 8.1 is

$$\dot{m} = 84.75 \text{ kg s}^{-1}$$

An energy balance around the furnace/boiler unit, where heat is transferred from the combustion gases to the steam, allows calculation of the entering methane rate $\dot{n}_{CH_4}$:

$$(84.75)(3{,}391.6 - 203.4) + \dot{n}_{CH_4}(-737.64) = 0$$

Whence

$$\dot{n}_{CH_4} = 366.30 \text{ mol s}^{-1}$$

The ideal work, given by Eq. (16.6), is

$$\dot{W}_{\text{ideal}} = 366.30[(298.15)(0.2082) - (-737.64)]$$
$$= 292.94 \times 10^3 \text{ kJ s}^{-1}$$

or

$$\dot{W}_{\text{ideal}} = 292.94 \times 10^3 \text{ kW}$$

The rate of entropy generation in each of the four units of the power plant is calculated by Eq. (16.1), and the lost work is then given by Eq. (16.15).

Furnace/boiler: We have assumed no heat transfer from the furnace/boiler to the surroundings; therefore $\dot{Q} = 0$. The term $\Delta(S\dot{m})_{\text{fs}}$ is simply the sum of the entropy changes of the two streams multiplied by their rates:

$$\dot{S}_{\text{total}} = (366.30)(0.2082) + (84.75)(6.6858 - 0.6580)$$

$$= 587.12 \text{ kJ s}^{-1}\text{ K}^{-1} \qquad \text{or} \qquad 587.12 \text{ kW K}^{-1}$$

and

$$\dot{W}_{\text{lost}} = T_0 \dot{S}_{\text{total}} = (298.15)(587.12) = 175.05 \times 10^3 \text{ kW}$$

Turbine: For adiabatic operation, we have:

$$\dot{S}_{\text{total}} = (84.75)(7.6846 - 6.6858) = 84.65 \text{ kW K}^{-1}$$

and

$$\dot{W}_{\text{lost}} = (298.15)(84.65) = 25.24 \times 10^3 \text{ kW}$$

Condenser: The condenser transfers heat from the condensing steam to the surroundings at 298.15 K in an amount determined in Example 8.1:

$$\dot{Q}(\text{condenser}) = -190.2 \times 10^3 \text{ kJ s}^{-1}$$

Thus

$$\dot{S}_{\text{total}} = (84.75)(0.6493 - 7.6846) + \frac{190{,}200}{298.15}$$

$$= 41.69 \text{ kW K}^{-1}$$

and

$$\dot{W}_{\text{lost}} = (298.15)(41.69) = 12.32 \times 10^3 \text{ kW}$$

Pump: Since the pump operates adiabatically,

$$\dot{S}_{\text{total}} = (84.75)(0.6580 - 0.6493) = 0.74 \text{ kW K}^{-1}$$

and

$$\dot{W}_{\text{lost}} = 0.22 \times 10^3 \text{ kW}$$

The entropy-generation analysis is as follows:

	kW K^{-1}	Percent of $\sum \dot{S}_{\text{total}}$
$\dot{S}_{\text{total}}$(furnace/boiler)	587.12	82.2
$\dot{S}_{\text{total}}$(turbine)	84.65	11.9
$\dot{S}_{\text{total}}$(condenser)	41.69	5.8
$\dot{S}_{\text{total}}$(pump)	0.74	0.1
$\sum \dot{S}_{\text{total}}$	714.20	100.0

A work analysis is carried out in accord with Eq. (16.18):

$$\dot{W}_{\text{ideal}} = \dot{W}_s + \sum \dot{W}_{\text{lost}}$$

The results of this analysis are shown in the following table:

	kW	Percent of $\dot{W}_{\text{ideal}}$
$\dot{W}_s$ (from Example 8.1)	80.00×10^3	27.3 $(=\eta_t)$
$\dot{W}_{\text{lost}}$(furnace/boiler)	175.05×10^3	59.8
$\dot{W}_{\text{lost}}$(turbine)	25.24×10^3	8.6
$\dot{W}_{\text{lost}}$(condenser)	12.43×10^3	4.2
$\dot{W}_{\text{lost}}$(pump)	0.22×10^3	0.1
Sum = $\dot{W}_{\text{ideal}}$	292.94×10^3	100.0

The thermodynamic efficiency of the power plant is 27.3 percent, and the major source of inefficiency is the furnace/boiler. The combustion process itself accounts for most of the entropy generation in this unit, and the remainder is the result of heat transfer across finite temperature differences.

Example 16.6 Methane is liquefied in a simple Linde system, as shown in Fig. 16.3. The methane enters the compressor at 1 bar and 300 K, and after compression to 60 bar is cooled back to 300 K. The product is saturated liquid methane at 1 bar. The unliquefied methane, also at 1 bar, is returned through a heat exchanger where it is heated to 295 K by the high-pressure methane. A heat leak into the heat exchanger of 5 kJ is assumed for each kilogram of methane entering the compressor. Heat leaks to other parts of the liquefier are assumed negligible. Make a thermodynamic analysis of the process for a surroundings temperature of $T_0 = 300$ K.

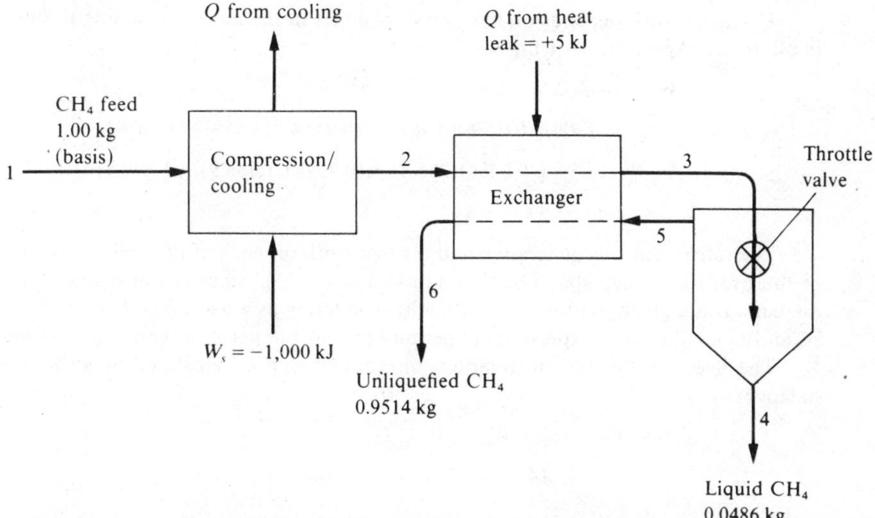

Figure 16.3 Linde liquefaction system of Example 16.6.

SOLUTION Methane compression from 1 to 60 bar is assumed to be carried out in a three-stage machine with inter- and after-cooling to 300 K and a compressor efficiency of 75 percent. The actual work of this compression is estimated as 1,000 kJ per kilogram of methane. The fraction of the methane that is liquefied z is calculated by an energy balance:

$$H_4 z + H_6(1 - z) - H_2 = Q$$

where Q is the heat leak from the surroundings. Solution for z gives

$$z = \frac{H_6 - H_2 - Q}{H_6 - H_4} = \frac{1{,}188.9 - 1{,}140.0 - 5}{1{,}188.9 - 285.4}$$

$$= 0.0486$$

This result may be compared with the value of 0.0541 obtained in Example 9.3 for the same operating conditions, but no heat leak. The properties at the various key points of the process, given in the accompanying table, are either available as data or are calculated by standard methods. Data are from Perry and Green.† The basis of all calculations is 1 kg of methane entering the process.

Point	State of the CH_4	t/K	P/bar	H/kJ kg^{-1}	S/kJ kg^{-1} K^{-1}
1	Superheated vapor	300.0	1	1,199.8	11.629
2	Superheated vapor	300.0	60	1,140.0	9.359
3	Superheated vapor	207.1	60	772.0	7.798
4	Saturated liquid	111.5	1	285.4	4.962
5	Saturated vapor	111.5	1	796.9	9.523
6	Superheated vapor	295.0	1	1,188.9	11.589

The ideal work depends on the overall changes in the methane passing through the liquefier. Application of Eq. (16.5) gives

$$W_{\text{ideal}} = T_0\,\Delta S - \Delta H$$

$$= (300)[(0.0486)(4.962) + (0.9514)(11.589) - 11.629]$$

$$- [(0.0486)(285.4) + (0.9514)(1{,}188.9) - 1{,}199.8]$$

$$= -53.8 \text{ kJ}$$

The rate of entropy generation and the lost work for each of the individual parts of the process are calculated by Eqs. (16.1) and (16.15). Since the flow rate of the methane is not given, we take 1 kg of methane entering as a basis. The rates $\dot{S}$, $\dot{W}_{\text{lost}}$, $\dot{m}$, and $\dot{Q}$ are therefore expressed not per unit of time but per kg of entering methane.

The heat transfer for the compression/cooling step is calculated by an energy balance:

$$\dot{Q} = \Delta H + \dot{W}_s = (H_2 - H_1) + \dot{W}_s$$

$$= (1{,}140.0 - 1{,}199.8) - 1{,}000 = -1{,}059.8 \text{ kJ}$$

† R. H. Perry and D. Green, *Perry's Chemical Engineers' Handbook*, p. 3-203, McGraw-Hill, New York, 1984.

Then

$$\dot{S}_{\text{total}}(\text{compression/cooling}) = S_2 - S_1 - \frac{\dot{Q}}{T_0}$$

Whence

$$\dot{S}_{\text{total}}(\text{compression/cooling}) = 9.359 - 11.629 + \frac{1{,}059.8}{300}$$

$$= 1.2627 \text{ kJ kg}^{-1} \text{ K}^{-1}$$

and

$$\dot{W}_{\text{lost}}(\text{compression/cooling}) = (300)(1.2627) = 378.8 \text{ kJ kg}^{-1}$$

For the exchanger, with $\dot{Q}$ equal to the heat leak,

$$\dot{S}_{\text{total}}(\text{exchanger}) = (S_6 - S_5)(1 - z) + (S_3 - S_2)(1) - \frac{\dot{Q}}{T_0}$$

Whence

$$\dot{S}_{\text{total}}(\text{exchanger}) = (11.589 - 9.523)(0.9514) + (7.798 - 9.359) - \frac{5}{300}$$

$$= 0.3879 \text{ kJ kg}^{-1} \text{ K}^{-1}$$

and

$$\dot{W}_{\text{lost}}(\text{exchanger}) = (300)(0.3879) = 116.4 \text{ kJ kg}^{-1}$$

For the throttle and separator, we assume adiabatic operation; thus

$$\dot{S}_{\text{total}}(\text{throttle}) = S_4 z + S_5(1 - z) - S_3$$

$$= (4.962)(0.0486) + (9.523)(0.9514) - 7.798$$

$$= 1.5033 \text{ kJ kg}^{-1} \text{ K}^{-1}$$

and

$$\dot{W}_{\text{lost}}(\text{throttle}) = (300)(1.5033) = 451.0 \text{ kJ kg}^{-1}$$

Analysis of the process with respect to entropy generation is shown in the following table:

	kJ kg^{-1} K^{-1}	Percent of $\sum \dot{S}_{\text{total}}$
$\dot{S}_{\text{total}}$(compression/cooling)	1.2627	40.0
$\dot{S}_{\text{total}}$(exchanger)	0.3879	12.3
$\dot{S}_{\text{total}}$(throttle)	1.5033	47.7
$\sum \dot{S}_{\text{total}}$	3.1539	100.0

The work analysis, based on Eq. (16.19)

$$|\dot{W}_s| = |\dot{W}_{\text{ideal}}| + \sum \dot{W}_{\text{lost}}$$

is shown in the following table:

	$\mathrm{kW\,kg^{-1}}$	Percent of $\lvert \dot{W}_s \rvert$
$\lvert \dot{W}_{\text{ideal}} \rvert$	53.8	5.4 $(=\eta_t)$
$\dot{W}_{\text{lost}}$(compression/cooling)	378.8	37.9
$\dot{W}_{\text{lost}}$(exchanger)	116.4	11.6
$\dot{W}_{\text{lost}}$(throttle)	451.0	45.1
Sum $= \lvert \dot{W}_s \rvert$	1,000.0	100.0

The largest loss occurs in the throttling step. Elimination of this highly irreversible process in favor of a turbine results in a considerable increase in efficiency.

From the standpoint of energy conservation, the thermodynamic efficiency of a process should be as high as possible, and the entropy generation or lost work as low as possible. The final design of a process depends largely on economic considerations, and the cost of energy is an important factor. The thermodynamic analysis of a specific process shows the locations of the major inefficiencies, and hence the pieces of equipment or steps in the process that could be altered or replaced to advantage. However, this sort of analysis gives no hint as to the nature of the changes that might be made. It merely shows that the present design is wasteful of energy and that there is room for improvement. One function of the chemical engineer is to try to devise a better process and to use ingenuity to keep the capital expenditure low. Each newly devised process may, of course, be analyzed to determine what improvement has been made.

PROBLEMS

16.1 Determine the maximum amount of work that can be obtained in a flow process from 1 kg of steam at 2,000 kPa and 400°C for surrounding conditions of 101.33 kPa and 300 K.

16.2 Water at 300 K and 10,000 kPa flows into a boiler at the rate of $10\ \mathrm{kg\,s^{-1}}$ and is vaporized, producing saturated vapor at 10,000 kPa. What is the maximum fraction of the heat added to the water in the boiler that can be converted into work in a process whose end product is water at its initial conditions if $T_0 = 300$ K? What happens to the rest of the heat? What is the rate of entropy change in the surroundings as a result of the work-producing process? In the system? Total?

16.3 Suppose the heat added to the water in the boiler in Prob. 16.2 comes from a furnace at a temperature of 500°C. What is the total rate of entropy change as a result of the heating process? What is $\dot{W}_{\text{lost}}$?

16.4 What is the ideal-work rate for the expansion process of Example 7.5? What is the thermodynamic efficiency of the process? What is the rate of entropy generation $\dot{S}_{\text{total}}$? What is $\dot{W}_{\text{lost}}$? Take $T_0 = 300$ K.

16.5 What is the ideal work for the compression process of Example 7.9? What is the thermodynamic efficiency of the process? What is ΔS_{total}? What is W_{lost}? Take $T_0 = 293.15$ K.

16.6 What is the ideal work for the pumping process of Example 7.10? What is the thermodynamic efficiency of the process? What is ΔS_{total}? What is W_{lost}? Take $T_0 = 300$ K.

16.7 What is the ideal work for the separation of an equimolar mixture of methane and ethane at 150°C and 5 bar in a steady-flow process into product streams of the pure gases at 35°C and 1 bar if $T_0 = 300$ K?

16.8 What is the work required for the separation of air (21-mol-% oxygen and 79-mol-% nitrogen) at 25°C and 1 bar in a steady-flow process into product streams of pure oxygen and nitrogen, also at 25°C and 1 bar, if the thermodynamic efficiency of the process is 5 percent and if $T_0 = 300$ K?

16.9 An ideal gas at 2,000 kPa is throttled adiabatically to 200 kPa at the rate of 16 mol s^{-1}. Determine $\dot{S}_{\text{total}}$ and $\dot{W}_{\text{lost}}$ if $T_0 = 300$ K.

16.10 A refrigeration system cools a brine solution from 25°C to −15°C at the rate of 20 kg s^{-1}. Heat is discarded to the atmosphere at a temperature of 30°C. What is the power requirement if the thermodynamic efficiency of the system is 0.27? The specific heat of the brine is 3.5 kJ kg^{-1} °C^{-1}.

16.11 An ice plant produces 0.5 kg s^{-1} of flake ice at 0°C from water at 20°C (T_0) in a continuous process. If the latent heat of fusion of water is 333.4 kJ kg^{-1} and if the thermodynamic efficiency of the process is 32 percent, what is the power requirement of the plant?

16.12 Exhaust gas at 375°C and 1 bar from internal-combustion engines flows at the rate of 100 mol s^{-1} into a waste-heat boiler where saturated steam is generated at a pressure of 1,000 kPa. Water enters the boiler at 20°C (T_0), and the exhaust gases leave at 200°C. The heat capacity of the exhaust gases is $C_P/R = 3.34 + 1.12 \times 10^{-3}T$, where T is in kelvins. The steam flows into an adiabatic turbine from which it exhausts at a pressure of 30 kPa. If the turbine efficiency η is 75 percent,
(*a*) What is $\dot{W}_s$, the power output of the turbine?
(*b*) What is the thermodynamic efficiency of the boiler/turbine combination?
(*c*) Determine the $\dot{S}_{\text{total}}$ for the boiler and for the turbine.
(*d*) Express $\dot{W}_{\text{lost}}$(boiler) and $\dot{W}_{\text{lost}}$(turbine) as fractions of $\dot{W}_{\text{ideal}}$, the ideal work of the process.

16.13 Consider the direct transfer of heat from a heat reservoir at T_1 to another heat reservoir at temperature T_2, where $T_1 > T_2 > T_0$. It is not obvious why the lost work of this process should depend on T_0, the temperature of the surroundings, because the surroundings are not involved in the actual heat-transfer process. Through appropriate use of the Carnot-engine formula, show for the transfer of an amount of heat equal to $|Q|$ that

$$W_{\text{lost}} = T_0|Q|\frac{T_1 - T_2}{T_1 T_2} = T_0\,\Delta S_{\text{total}}$$

16.14 An inventor has developed a complicated process for making heat continuously available at an elevated temperature. Saturated steam at 150°C is the only source of energy. Assuming that there is plenty of cooling water available at 300 K, what is the maximum temperature level at which heat in the amount of 1,100 kJ can be made available for each kilogram of steam flowing through the process?

16.15 A plant takes in water at 70(°F), cools it to 32(°F), and freezes it at this temperature, producing 1(lb$_m$)(s)$^{-1}$ of ice. Heat rejection is at 70(°F). The heat of fusion of water is 143.3(Btu)(lb$_m$)$^{-1}$.

(*a*) What is $\dot{W}_{\text{ideal}}$ for the process?

(*b*) What is the power of requirement of a single Carnot heat pump operating between 32 and 70(°F)? What is the thermodynamic efficiency of this process? What is its irreversible feature?

(*c*) What is the power requirement if an ideal ammonia vapor-compression refrigeration cycle is used. *Ideal* here implies isentropic compression, infinite cooling-water rate in the condenser, and minimum heat-transfer driving forces in evaporator and condenser of 0(°F). What is the thermodynamic efficiency of this process? What are its irreversible features?

(*d*) What is the power requirement of a practical ammonia vapor-compression cycle for which the compressor efficiency is 75 percent, the minimum temperature differences in evaporator and condenser are 8(°F), and the temperature rise of the cooling water in the condenser is 20(°F)? Make a thermodynamic analysis of this process.

16.16 Consider a steady-flow process in which the following gas-phase reaction takes place: $CO + \frac{1}{2}O_2 \rightarrow CO_2$. The surroundings are at 300 K.

(*a*) What is W_{ideal} when the reactants enter the process as pure carbon monoxide and as air containing the stoichiometric amount of oxygen, both at 25°C and 1 bar, and the products of complete combustion leave the process at the same conditions?

(*b*) The overall process is exactly the same as in (*a*). However, we now specify that the CO is burned in an adiabatic reactor at 1 bar. What is W_{ideal} for the process of cooling the flue gases to

25°C? What is the irreversible feature of the overall process? What is its thermodynamic efficiency? What has increased in entropy? By how much?

16.17 A chemical plant has saturated steam available at 2,700 kPa, but because of a process change has little use for steam at this pressure. Rather, steam at 1,000 kPa is required. Also available is saturated exhaust steam at 275 kPa. The suggestion is that the 275-kPa steam be compressed to 1,000 kPa, obtaining the necessary work from expansion of the 2,700-kPa steam to 1,000 kPa. The two streams at 1,000 kPa would then be mixed. Determine the rates at which steam at each initial pressure must be supplied to provide enough steam at 1,000 kPa so that upon condensation to saturated liquid heat in the amount of 300 kJ s^{-1} is released,

(*a*) If the process is carried out in a completely reversible manner.

(*b*) If the higher-pressure steam expands in a turbine of 78 percent efficiency and the lower-pressure steam is compressed in a machine of 75 percent efficiency. Make a thermodynamic analysis of this process.

16.18 Make a thermodynamic analysis of the refrigeration cycle described in one of the parts of Prob. 9.3. Assume that the refrigeration effect maintains a heat reservoir at a temperature 9(°F) [5°C] above the evaporation temperature and that T_0 is 9(°F) below the condensation temperature.

16.19 Make a thermodynamic analysis of the refrigeration cycle described in the first paragraph of Prob. 9.5. Assume that the refrigeration effect maintains a heat reservoir at a temperature 9(°F) above the evaporation temperature and that T_0 is 9(°F) below the condensation temperature.

16.20 A colloidal solution enters a single-effect evaporator at 100°C. Water is vaporized from the solution, producing a more concentrated solution and 0.5 kg s^{-1} of saturated-vapor steam at 100°C. This steam is compressed and sent to the heating coils of the evaporator to supply the heat required for its operation. For a minimum heat-transfer driving force across the evaporator coils of 20°C, for a compressor efficiency of 75 percent, and for adiabatic operation, what is the state of the stream leaving the heating coils of the evaporator? For a surroundings temperature of 300 K, make a thermodynamic analysis of the process.

16.21 Refrigeration at a temperature level of 80 K is required for a certain process. A cycle with helium gas as refrigerant operates as follows. Helium at 1 bar is compressed adiabatically to 5 bar, cooled to 25°C by heat transfer to the surroundings, and sent to a countercurrent heat exchanger where it is cooled by returning helium. From there it expands adiabatically to 1 bar in a turbine, which produces work used to help drive the compressor. The helium then enters the refrigerator, where it absorbs enough heat to raise its temperature to 75 K, and finally returns to the compressor by way of the heat exchanger. Assume helium an ideal gas for which $C_P = (5/2)R$. Compressor and turbine efficiencies are 77 percent, and the minimum temperature difference in the exchanger is 6°C. For a refrigeration load of 2 kJ s^{-1}, and assuming no heat leaks from the surroundings, determine the helium circulation rate. For a surroundings temperature of 290 K, make a thermodynamic analysis of the process.

16.22 An elementary nuclear-powered gas-turbine power plant operates as shown in Fig. P16.22. Air entering at point 1 is compressed adiabatically to point 2, heated at constant pressure between points 2 and 3, and expanded adiabatically from point 3 to point 4. Specified conditions are:

Point 1: $t_1 = 20°C$, $P_1 = 1$ bar
Point 2: $P_2 = 4$ bar
Point 3: $t_3 = 540°C$, $P_3 = 4$ bar
Point 4: $P_2 = 1$ bar

The work to drive the compressor W_c comes from the turbine, and the additional work of the turbine W_s is the net work output of the power plant. The compressor and turbine efficiencies are given on the figure. Assume air an ideal gas for which $C_P = (7/2)R$. Including the nuclear reactor as part of the system and treating it as a heat reservoir at 650°C, make a thermodynamic analysis of the process $T_0 = 293.15$ K.

16.23 A design for an ammonia-synthesis plant includes a step that takes ammonia vapor at 200(°F) and 300(psia) and changes its state to saturated liquid ammonia at 14.71(psia). For ammonia in its

Nuclear
reactor
at 650°C
q
2
3
Compressor
η = 0.75
Turbine
η = 0.80
W_c
W_s
1
4

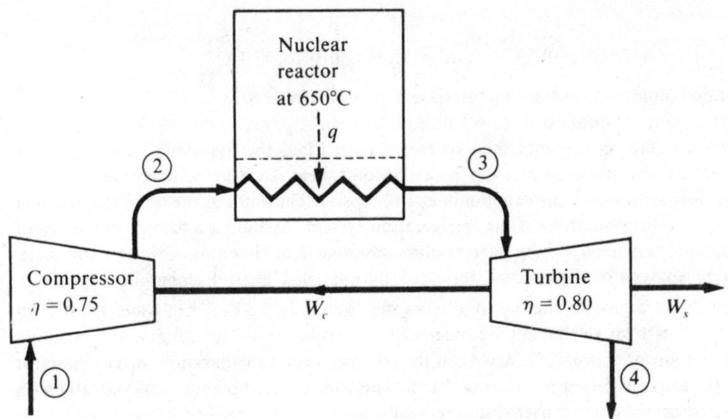

Figure P16.22

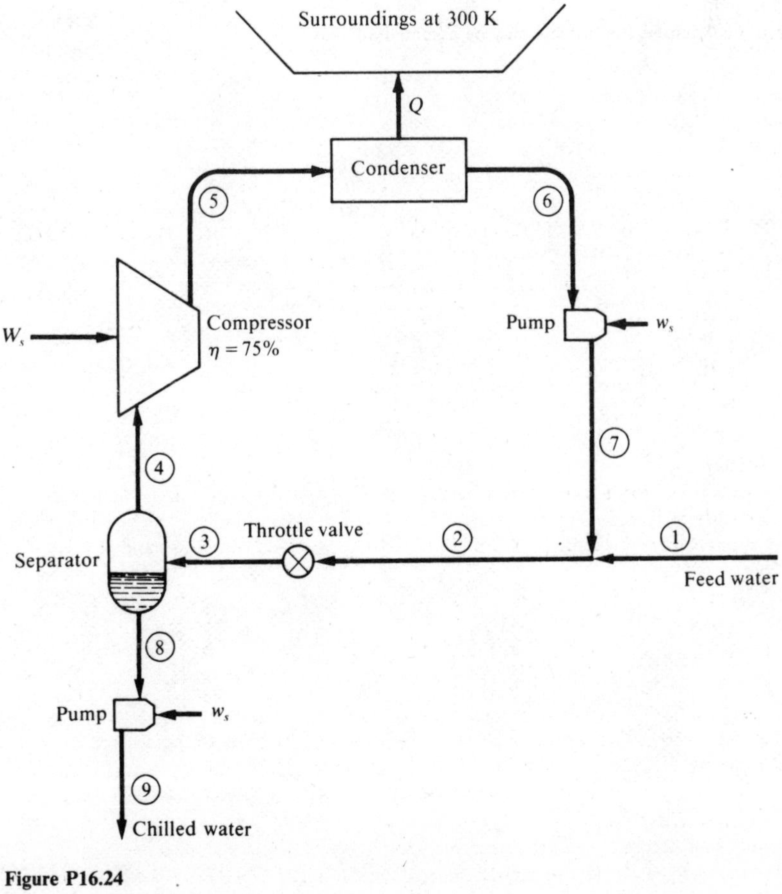

Surroundings at 300 K
Q
Condenser
5
6
Compressor
η = 75%
W_s
Pump
w_s
7
4
Separator
3
Throttle valve
2
1
Feed water
8
Pump
w_s
9
Chilled water

Figure P16.24

initial state:

$$H_1 = 693.5(\text{Btu})(\text{lb}_m)^{-1} \qquad S_1 = 1.2344(\text{Btu})(\text{lb}_m)^{-1}(\text{R})^{-1}$$

Data for saturated ammonia on the same basis are given in Table 9.2.

(*a*) What is W_{ideal} for the process if $T_0 = 530(\text{R})$?

(*b*) The change of state is accomplished in two stages. First, the ammonia vapor expands adiabatically in a turbine to a pressure of 14.71(psia). Second, heat is extracted from the ammonia by refrigeration, yielding saturated liquid ammonia at 14.71(psia). The work of the turbine is applied toward the work required for operation of the refrigeration system. Assuming a turbine efficiency of 0.75 and a thermodynamic efficiency of the refrigeration system of 0.30, determine W_s(net) and make an entropy-generation analysis of the process. Base calculations on $1(\text{lb}_m)$ of ammonia.

16.24 Figure P16.24 shows a process that accomplishes the chilling of 0.5 kg s^{-1} of water from 26 to 4°C. The water acts as its own refrigerant by means of a recycle loop. The compressor (vacuum pump) maintains a suction pressure at point 4 such that the saturation temperature in the separator is 4°C, and discharges at point 5 to a pressure of 6 kPa. The compressor operates adiabatically with an efficiency of 75 percent. The condenser discharges saturated liquid water at 6 kPa.

Make a thermodynamic analysis of the process, considering it to consist of the following parts:

(*a*) Points 6 and 1 to points 4 and 9.

(*b*) Point 4 to point 5.

(*c*) Point 5 to point 6.

Apart from the condenser, the process may be assumed adiabatic.

APPENDIX

A

CONVERSION FACTORS AND VALUES OF THE GAS CONSTANT

Because standard reference books contain data in diverse units, we include Tables A.1 and A.2 to aid the conversion of values from one set of units to another. Those units having no connection with the SI system are enclosed in parentheses. The following definitions are noted:

(ft) ≡ U.S. National Bureau of Standards defined foot ≡ 3.048×10^{-1} m

(in) ≡ U.S. National Bureau of Standards defined inch ≡ 2.54×10^{-2} m

(lb_m) ≡ U.S. National Bureau of Standards defined pound mass (avoirdupois) ≡ 4.5359237×10^{-1} kg

(lb_f) ≡ force to accelerate 1 (lb_m) 32.1740 (ft) s^{-2}

(atm) ≡ standard atmospheric pressure

(psia) ≡ pounds force per square inch absolute pressure

(torr) ≡ pressure exerted by 1 mm mercury at 0°C and standard gravity

(cal) ≡ thermochemical calorie

(Btu) ≡ international steam table British thermal unit

(lb mol) ≡ mass in pounds mass with numerical value equal to the molar mass (molecular weight)

(R) ≡ absolute temperature in Rankines

The conversion factors of Table A.1 are referred to a single basic or derived unit of the SI system. Conversions between other pairs of units for a given quantity are made as in the following example:

$$1 \text{ bar} = 0.986923 \text{ (atm)} = 750.061 \text{ (torr)}$$

Thus

$$1 \text{ (atm)} = \frac{750.061}{0.986923} = 760.00 \text{ (torr)}$$

Table A.1 Conversion factors

Quantity	Conversion
Length	1 m = 100 cm
	= 3.28084(ft)
	= 39.3701(in)
Mass	1 kg = 10^3 g
	= 2.20462(lb_m)
Force	1 N = 1 kg m s^{-2}
	= 10^5(dyne)
	= 0.224809(lb_f)
Pressure	1 bar = 10^5 N m^{-2}
	= 10^5 Pa
	= 10^2 kPa
	= 10^6(dyne)cm^{-2}
	= 0.986923(atm)
	= 14.5038(psia)
	= 750.061(torr)
Volume	1 m^3 = 10^6 cm^3
	= 35.3147$(ft)^3$
Density	1 g cm^{-3} = 10^3 kg m^{-3}
	= 62.4278(lb_m)$(ft)^{-3}$
Energy	1 J = 1 N m
	= 1 m^3 Pa
	= 10^{-5} m^3 bar
	= 10 cm^3 bar
	= 9.86923 cm^3(atm)
	= 10^7(dyne)cm
	= 10^7(erg)
	= 0.239006(cal)
	= $5.12197 \times 10^{-3}$$(ft)^3$(psia)
	= 0.737562(ft)(lb_f)
	= 9.47831×10^{-4}(Btu)
Power	1 kW = 10^3 J s^{-1}
	= 239.006(cal)$(s)^{-1}$
	= 737.562(ft)(lb_f)$(s)^{-1}$
	= 0.94783(Btu)$(s)^{-1}$
	= 1.34102(hp)

Table A.2 Values of the universal gas constant

R = 8.314 J mol^{-1} K^{-1} = 8.314 m^3 Pa mol^{-1} K^{-1}
= 83.14 cm^3 bar mol^{-1} K^{-1} = 8,314 cm^3 kPa mol^{-1} K^{-1} = 82.06 cm^3(atm)mol^{-1} K^{-1}
= 62,356 cm^3(torr)mol^{-1} K^{-1}
= 1.987(cal)mol^{-1} K^{-1} = 1.986(Btu)(lb mol)$^{-1}$(R)$^{-1}$
= 0.7302$(ft)^3$(atm)(lb mol)$^{-1}$(R)$^{-1}$ = 10.73$(ft)^3$(psia)(lb mol)$^{-1}$(R)$^{-1}$
= 1,545(ft)(lb_f)(lb mol)$^{-1}$(R)$^{-1}$

APPENDIX
B
CRITICAL CONSTANTS AND ACENTRIC FACTORS

	T_c/K	P_c/bar	$V_c/10^{-6}\ m^3\ mol^{-1}$	Z_c	ω
Paraffins:					
Methane	190.6	46.0	99.	0.288	0.008
Ethane	305.4	48.8	148.	0.285	0.098
Propane	369.8	42.5	203.	0.281	0.152
n-butane	425.2	38.0	255.	0.274	0.193
Isobutane	408.1	36.5	263.	0.283	0.176
n-Pentane	469.6	33.7	304.	0.262	0.251
Isopentane	460.4	33.8	306.	0.271	0.227
Neopentane	433.8	32.0	303.	0.269	0.197
n-Hexane	507.4	29.7	370.	0.260	0.296
n-Heptane	540.2	27.4	432.	0.263	0.351
n-Octane	568.8	24.8	492.	0.259	0.394
Monoolefins:					
Ethylene	282.4	50.4	129.	0.276	0.085
Propylene	365.0	46.2	181.	0.275	0.148
1-Butene	419.6	40.2	240.	0.277	0.187
1-Pentene	464.7	40.5	300.	0.31	0.245
Miscellaneous organic compounds:					
Acetic acid	594.4	57.9	171.	0.200	0.454
Acetone	508.1	47.0	209.	0.232	0.309
Acetonitrile	547.9	48.3	173.	0.184	0.321
Acetylene	308.3	61.4	113.	0.271	0.184
Benzene	562.1	48.9	259.	0.271	0.212

	T_c/K	P_c/bar	$V_c/10^{-6}$ m^3 mol^{-1}	Z_c	ω
1,3-Butadiene	425.0	43.3	221.	0.270	0.195
Chlorobenzene	632.4	45.2	308.	0.265	0.249
Cyclohexane	553.4	40.7	308.	0.273	0.213
Dichlorodifluoromethane (Freon-12)	385.0	41.2	217.	0.280	0.176
Diethyl ether	466.7	36.4	280.	0.262	0.281
Ethanol	516.2	63.8	167.	0.248	0.635
Ethylene oxide	469.	71.9	140.	0.258	0.200
Methanol	512.6	81.0	118.	0.224	0.559
Methyl chloride	416.3	66.8	139.	0.268	0.156
Methyl ethyl ketone	535.6	41.5	267.	0.249	0.329
Toluene	591.7	41.1	316.	0.264	0.257
Trichlorofluoromethane (Freon-11)	471.2	44.1	248.	0.279	0.188
Trichlorotrifluoroethane (Freon-113)	487.2	34.1	304.	0.256	0.252
Elementary gases:					
Argon	150.8	48.7	74.9	0.291	0.0
Bromine	584.	103.	127.	0.270	0.132
Chlorine	417.	77.	124.	0.275	0.073
Helium 4	5.2	2.27	57.3	0.301	−0.387
Hydrogen	33.2	13.0	65.0	0.305	−0.22
Krypton	209.4	55.0	91.2	0.288	0.0
Neon	44.4	27.6	41.7	0.311	0.0
Nitrogen	126.2	33.9	89.5	0.290	0.040
Oxygen	154.6	50.5	73.4	0.288	0.021
Xenon	289.7	58.4	118.	0.286	0.0
Miscellaneous inorganic compounds:					
Ammonia	405.6	112.8	72.5	0.242	0.250
Carbon dioxide	304.2	73.8	94.0	0.274	0.225
Carbon disulfide	552.	79.	170.	0.293	0.115
Carbon monoxide	132.9	35.0	93.1	0.295	0.049
Carbon tetrachloride	556.4	45.6	276.	0.272	0.194
Chloroform	536.4	55.	239.	0.293	0.216
Hydrazine	653.	147.	96.1	0.260	0.328
Hydrogen chloride	324.6	83.	81.	0.249	0.12
Hydrogen cyanide	456.8	53.9	139.	0.197	0.407
Hydrogen sulfide	373.2	89.4	98.5	0.284	0.100
Nitric oxide (NO)	180.	65.	58.	0.25	0.607
Nitrous oxide (N_2O)	309.6	72.4	97.4	0.274	0.160
Sulfur dioxide	430.8	78.8	122.	0.268	0.251
Sulfur trioxide	491.0	82.	130.	0.26	0.41
Water	647.3	220.5	56.	0.229	0.344

References: A. P. Kudchadker, G. H. Alani, and B. J. Zwolinski, *Chem. Rev.*, **68**: 659 (1968); J. F. Mathews, *Chem. Rev.*, **72**: 71 (1972); R. C. Reid, J. M. Prausnitz, and T. K. Sherwood, "The Properties of Gases and Liquids," 3d ed., McGraw-Hill, New York, 1977; C. A. Passut and R. P. Danner, *Ind. Eng. Chem. Process Des. Develop.*, **12**: 365 (1974).

APPENDIX

C

STEAM TABLES

Table C.1 Properties of Saturated Steam (SI Units)

Table C.2 Properties of Superheated Steam (SI Units)

Table C.3 Properties of Saturated Steam (English Units)

Table C.4 Properties of Superheated Steam (English Units)

All tables are generated by computer from programs based on "The 1976 IFC† Formulation for Industrial Use: A Formulation of the Thermodynamic Properties of Ordinary Water Substance," as published in the "ASME Steam Tables," 4th ed., App. I, pp. 11-29, The Am. Soc. Mech. Engrs., New York, 1979.

† International Formulation Committee.

TABLE C.1. SATURATED STEAM SI UNITS

V = SPECIFIC VOLUME cu cm/g
U = SPECIFIC INTERNAL ENERGY kJ/kg
H = SPECIFIC ENTHALPY kJ/kg
S = SPECIFIC ENTROPY kJ/(kg K)

TEMPERATURE		ABS PRESS	SPECIFIC VOLUME V			INTERNAL ENERGY U			ENTHALPY H			ENTROPY S		
C	K	kPa	SAT LIQUID	EVAP	SAT VAPOR	SAT LIQUID	EVAP	SAT VAPOR	SAT LIQUID	EVAP	SAT VAPOR	SAT LIQUID	EVAP	SAT VAPOR
0	273.15	0.611	1.000	206300	206300	-0.04	2375.7	2375.6	-0.04	2501.7	2501.6	0.0000	9.1578	9.1578
0.01	273.16	0.611	1.000	206200	206200	0.0	2375.6	2375.6	0.00	2501.6	2501.6	0.0	9.1575	9.1575
1	274.15	0.657	1.000	192600	192600	4.17	2372.7	2376.9	4.17	2499.2	2503.4	0.0153	9.1158	9.1311
2	275.15	0.705	1.000	179900	179900	8.39	2369.9	2378.3	8.39	2496.8	2505.2	0.0306	9.0741	9.1047
3	276.15	0.757	1.000	168200	168200	12.60	2367.1	2379.7	12.60	2494.5	2507.1	0.0459	9.0326	9.0785
4	277.15	0.813	1.000	157300	157300	16.80	2364.3	2381.1	16.80	2492.1	2508.9	0.0611	8.9915	9.0526
5	278.15	0.872	1.000	147200	147200	21.01	2361.4	2382.4	21.01	2489.7	2510.7	0.0762	8.9507	9.0269
6	279.15	0.935	1.000	137800	137800	25.21	2358.6	2383.8	25.21	2487.4	2512.6	0.0913	8.9102	9.0014
7	280.15	1.001	1.000	129100	129100	29.41	2355.8	2385.2	29.41	2485.0	2514.4	0.1063	8.8699	8.9762
8	281.15	1.072	1.000	121000	121000	33.60	2353.0	2386.6	33.60	2482.6	2516.2	0.1213	8.8300	8.9513
9	282.15	1.147	1.000	113400	113400	37.80	2350.1	2387.9	37.80	2480.3	2518.1	0.1362	8.7903	8.9265
10	283.15	1.227	1.000	106400	106400	41.99	2347.3	2389.3	41.99	2477.9	2519.9	0.1510	8.7510	8.9020
11	284.15	1.312	1.000	99910	99910	46.18	2344.5	2390.7	46.19	2475.5	2521.7	0.1658	8.7119	8.8776
12	285.15	1.401	1.000	93830	93840	50.38	2341.7	2392.1	50.38	2473.2	2523.6	0.1805	8.6731	8.8536
13	286.15	1.497	1.001	88180	88180	54.56	2338.9	2393.4	54.57	2470.8	2525.4	0.1952	8.6345	8.8297
14	287.15	1.597	1.001	82900	82900	58.75	2336.1	2394.8	58.75	2468.5	2527.2	0.2098	8.5963	8.8060
15	288.15	1.704	1.001	77980	77980	62.94	2333.2	2396.2	62.94	2466.1	2529.1	0.2243	8.5582	8.7826
16	289.15	1.817	1.001	73380	73380	67.12	2330.4	2397.6	67.13	2463.8	2530.9	0.2388	8.5205	8.7593
17	290.15	1.936	1.001	69090	69090	71.31	2327.6	2398.9	71.31	2461.4	2532.7	0.2533	8.4830	8.7363
18	291.15	2.062	1.001	65090	65090	75.49	2324.8	2400.3	75.50	2459.0	2534.5	0.2677	8.4458	8.7135
19	292.15	2.196	1.002	61340	61340	79.68	2322.0	2401.7	79.68	2456.7	2536.4	0.2820	8.4088	8.6908
20	293.15	2.337	1.002	57840	57840	83.86	2319.2	2403.0	83.86	2454.3	2538.2	0.2963	8.3721	8.6684
21	294.15	2.485	1.002	54560	54560	88.04	2316.4	2404.4	88.04	2452.0	2540.0	0.3105	8.3356	8.6462
22	295.15	2.642	1.002	51490	51490	92.22	2313.6	2405.8	92.23	2449.6	2541.8	0.3247	8.2994	8.6241
23	296.15	2.808	1.002	48620	48620	96.40	2310.7	2407.1	96.41	2447.2	2543.6	0.3389	8.2634	8.6023
24	297.15	2.982	1.003	45920	45930	100.6	2307.9	2408.5	100.6	2444.9	2545.5	0.3530	8.2277	8.5806
25	298.15	3.166	1.003	43400	43400	104.8	2305.1	2409.9	104.8	2442.5	2547.3	0.3670	8.1922	8.5592
26	299.15	3.360	1.003	41030	41030	108.9	2302.3	2411.2	108.9	2440.2	2549.1	0.3810	8.1569	8.5379
27	300.15	3.564	1.003	38810	38810	113.1	2299.5	2412.6	113.1	2437.8	2550.9	0.3949	8.1218	8.5168
28	301.15	3.778	1.004	36730	36730	117.3	2296.7	2414.0	117.3	2435.4	2552.7	0.4088	8.0870	8.4959
29	302.15	4.004	1.004	34770	34770	121.5	2293.8	2415.3	121.5	2433.1	2554.5	0.4227	8.0524	8.4751

30	303.15	4.241	1.004	32930	32930	125.7	2291.0	2416.7	125.7	2430.7	2556.4	0.4365	8.0180	8.4546
31	304.15	4.491	1.005	31200	31200	129.8	2288.2	2418.0	129.8	2428.3	2558.2	0.4503	7.9839	8.4342
32	305.15	4.753	1.005	29570	29570	134.0	2285.4	2419.4	134.0	2425.9	2560.0	0.4640	7.9500	8.4140
33	306.15	5.029	1.005	28040	28040	138.2	2282.6	2420.8	138.2	2423.6	2561.8	0.4777	7.9163	8.3939
34	307.15	5.318	1.006	26600	26600	142.4	2279.7	2422.1	142.4	2421.2	2563.6	0.4913	7.8828	8.3740
35	308.15	5.622	1.006	25240	25240	146.6	2276.9	2423.5	146.6	2418.8	2565.4	0.5049	7.8495	8.3543
36	309.15	5.940	1.006	23970	23970	150.7	2274.1	2424.8	150.7	2416.4	2567.2	0.5184	7.8164	8.3348
37	310.15	6.274	1.007	22760	22760	154.9	2271.3	2426.2	154.9	2414.1	2569.0	0.5319	7.7835	8.3154
38	311.15	6.624	1.007	21630	21630	159.1	2268.4	2427.5	159.1	2411.7	2570.8	0.5453	7.7509	8.2962
39	312.15	6.991	1.007	20560	20560	163.3	2265.6	2428.9	163.3	2409.3	2572.6	0.5588	7.7184	8.2772
40	313.15	7.375	1.008	19550	19550	167.4	2262.8	2430.2	167.5	2406.9	2574.4	0.5721	7.6861	8.2583
41	314.15	7.777	1.008	18590	18590	171.6	2259.9	2431.6	171.6	2404.5	2576.2	0.5854	7.6541	8.2395
42	315.15	8.198	1.009	17690	17690	175.8	2257.1	2432.9	175.8	2402.1	2577.9	0.5987	7.6222	8.2209
43	316.15	8.639	1.009	16840	16840	180.0	2254.3	2434.2	180.0	2399.7	2579.7	0.6120	7.5905	8.2025
44	317.15	9.100	1.009	16040	16040	184.2	2251.4	2435.6	184.2	2397.3	2581.5	0.6252	7.5590	8.1842
45	318.15	9.582	1.010	15280	15280	188.3	2248.6	2436.9	188.4	2394.9	2583.3	0.6383	7.5277	8.1661
46	319.15	10.09	1.010	14560	14560	192.5	2245.7	2438.3	192.5	2392.5	2585.1	0.6514	7.4966	8.1481
47	320.15	10.61	1.011	13880	13880	196.7	2242.9	2439.6	196.7	2390.1	2586.9	0.6645	7.4657	8.1302
48	321.15	11.16	1.011	13230	13230	200.9	2240.0	2440.9	200.9	2387.7	2588.6	0.6776	7.4350	8.1125
49	322.15	11.74	1.012	12620	12620	205.1	2237.2	2442.3	205.1	2385.3	2590.4	0.6906	7.4044	8.0950
50	323.15	12.34	1.012	12040	12050	209.2	2234.3	2443.6	209.3	2382.9	2592.2	0.7035	7.3741	8.0776
51	324.15	12.96	1.013	11500	11500	213.4	2231.5	2444.9	213.4	2380.5	2593.9	0.7164	7.3439	8.0603
52	325.15	13.61	1.013	10980	10980	217.6	2228.6	2446.2	217.6	2378.1	2595.7	0.7293	7.3138	8.0432
53	326.15	14.29	1.014	10490	10490	221.8	2225.8	2447.6	221.8	2375.7	2597.5	0.7422	7.2840	8.0262
54	327.15	15.00	1.014	10020	10020	226.0	2222.9	2448.9	226.0	2373.2	2599.2	0.7550	7.2543	8.0093
55	328.15	15.74	1.015	9577.9	9578.9	230.2	2220.0	2450.2	230.2	2370.8	2601.0	0.7677	7.2248	7.9925
56	329.15	16.51	1.015	9157.7	9158.7	234.3	2217.2	2451.5	234.4	2368.4	2602.7	0.7804	7.1955	7.9759
57	330.15	17.31	1.016	8758.7	8759.8	238.5	2214.3	2452.8	238.5	2365.9	2604.5	0.7931	7.1663	7.9595
58	331.15	18.15	1.016	8379.8	8380.8	242.7	2211.4	2454.1	242.7	2363.5	2606.2	0.8058	7.1373	7.9431
59	332.15	19.02	1.017	8019.7	8020.8	246.9	2208.6	2455.4	246.9	2361.1	2608.0	0.8184	7.1085	7.9269
60	333.15	19.92	1.017	7677.5	7678.5	251.1	2205.7	2456.8	251.1	2358.6	2609.7	0.8310	7.0798	7.9108
61	334.15	20.86	1.018	7352.1	7353.2	255.3	2202.8	2458.1	255.3	2356.2	2611.4	0.8435	7.0513	7.8948
62	335.15	21.84	1.018	7042.7	7043.7	259.4	2199.9	2459.4	259.5	2353.7	2613.2	0.8560	7.0230	7.8790
63	336.15	22.86	1.019	6748.2	6749.3	263.6	2197.0	2460.7	263.6	2351.3	2614.9	0.8685	6.9948	7.8633
64	337.15	23.91	1.019	6468.0	6469.0	267.8	2194.1	2462.0	267.8	2348.8	2616.6	0.8809	6.9667	7.8477
65	338.15	25.01	1.020	6201.3	6202.3	272.0	2191.2	2463.2	272.0	2346.3	2618.4	0.8933	6.9388	7.8322
66	339.15	26.15	1.020	5947.2	5948.2	276.2	2188.3	2464.5	276.2	2343.9	2620.1	0.9057	6.9111	7.8168
67	340.15	27.33	1.021	5705.2	5706.2	280.4	2185.4	2465.8	280.4	2341.4	2621.8	0.9180	6.8835	7.8015
68	341.15	28.56	1.022	5474.6	5475.6	284.6	2182.5	2467.1	284.6	2338.9	2623.5	0.9303	6.8561	7.7864
69	342.15	29.84	1.022	5254.8	5255.8	288.8	2179.6	2468.4	288.8	2336.4	2625.2	0.9426	6.8288	7.7714
70	343.15	31.16	1.023	5045.2	5046.3	292.9	2176.7	2469.7	293.0	2334.0	2626.9	0.9548	6.8017	7.7565
71	344.15	32.53	1.023	4845.4	4846.4	297.1	2173.8	2470.9	297.2	2331.5	2628.6	0.9670	6.7747	7.7417
72	345.15	33.96	1.024	4654.7	4655.7	301.3	2170.9	2472.2	301.4	2329.0	2630.3	0.9792	6.7478	7.7270
73	346.15	35.43	1.025	4472.7	4473.7	305.5	2168.0	2473.5	305.5	2326.5	2632.0	0.9913	6.7211	7.7124
74	347.15	36.96	1.025	4299.0	4300.0	309.7	2165.1	2474.8	309.7	2324.0	2633.7	1.0034	6.6945	7.6979

TABLE C.1. SATURATED STEAM SI UNITS (Continued)

TEMPERATURE		ABS PRESS	SPECIFIC VOLUME V			INTERNAL ENERGY U			ENTHALPY H			ENTROPY S		
C	K	kPa	SAT LIQUID	EVAP	SAT VAPOR	SAT LIQUID	EVAP	SAT VAPOR	SAT LIQUID	EVAP	SAT VAPOR	SAT LIQUID	EVAP	SAT VAPOR
75	348.15	38.55	1.026	4133.1	4134.1	313.9	2162.1	2476.0	313.9	2321.5	2635.4	1.0154	6.6681	7.6835
76	349.15	40.19	1.027	3974.6	3975.7	318.1	2159.2	2477.3	318.1	2318.9	2637.1	1.0275	6.6418	7.6693
77	350.15	41.89	1.027	3823.3	3824.3	322.3	2156.3	2478.5	322.3	2316.4	2638.7	1.0395	6.6156	7.6551
78	351.15	43.65	1.028	3678.6	3679.6	326.5	2153.3	2479.8	326.5	2313.9	2640.4	1.0514	6.5896	7.6410
79	352.15	45.47	1.029	3540.3	3541.3	330.7	2150.4	2481.1	330.7	2311.4	2642.1	1.0634	6.5637	7.6271
80	353.15	47.36	1.029	3408.1	3409.1	334.9	2147.4	2482.3	334.9	2308.8	2643.8	1.0753	6.5380	7.6132
81	354.15	49.31	1.030	3281.6	3282.6	339.1	2144.5	2483.5	339.1	2306.3	2645.4	1.0871	6.5123	7.5995
82	355.15	51.33	1.031	3160.6	3161.6	343.3	2141.5	2484.8	343.3	2303.8	2647.1	1.0990	6.4868	7.5858
83	356.15	53.42	1.031	3044.8	3045.8	347.5	2138.6	2486.0	347.5	2301.2	2648.7	1.1108	6.4615	7.5722
84	357.15	55.57	1.032	2933.9	2935.0	351.7	2135.6	2487.3	351.7	2298.6	2650.4	1.1225	6.4362	7.5587
85	358.15	57.80	1.033	2827.8	2828.8	355.9	2132.6	2488.5	355.9	2296.1	2652.0	1.1343	6.4111	7.5454
86	359.15	60.11	1.033	2726.1	2727.2	360.1	2129.7	2489.7	360.1	2293.5	2653.6	1.1460	6.3861	7.5321
87	360.15	62.49	1.034	2628.8	2629.8	364.3	2126.7	2490.9	364.3	2290.9	2655.3	1.1577	6.3612	7.5189
88	361.15	64.95	1.035	2535.4	2536.5	368.5	2123.7	2492.2	368.5	2288.4	2656.9	1.1693	6.3365	7.5058
89	362.15	67.49	1.035	2446.0	2447.0	372.7	2120.7	2493.4	372.7	2285.8	2658.5	1.1809	6.3119	7.4928
90	363.15	70.11	1.036	2360.3	2361.3	376.9	2117.7	2494.6	376.9	2283.2	2660.1	1.1925	6.2873	7.4799
91	364.15	72.81	1.037	2278.0	2279.1	381.1	2114.7	2495.8	381.1	2280.6	2661.7	1.2041	6.2629	7.4670
92	365.15	75.61	1.038	2199.2	2200.2	385.3	2111.7	2497.0	385.4	2278.0	2663.4	1.2156	6.2387	7.4543
93	366.15	78.49	1.038	2123.5	2124.5	389.5	2108.7	2498.2	389.6	2275.4	2665.0	1.2271	6.2145	7.4416
94	367.15	81.46	1.039	2050.9	2051.9	393.7	2105.7	2499.4	393.8	2272.8	2666.6	1.2386	6.1905	7.4291
95	368.15	84.53	1.040	1981.2	1982.2	397.9	2102.7	2500.6	398.0	2270.2	2668.1	1.2501	6.1665	7.4166
96	369.15	87.69	1.041	1914.3	1915.3	402.1	2099.7	2501.8	402.2	2267.5	2669.7	1.2615	6.1427	7.4042
97	370.15	90.94	1.041	1850.0	1851.0	406.3	2096.6	2503.0	406.4	2264.9	2671.3	1.2729	6.1190	7.3919
98	371.15	94.30	1.042	1788.3	1789.3	410.5	2093.6	2504.1	410.6	2262.2	2672.9	1.2842	6.0954	7.3796
99	372.15	97.76	1.043	1729.0	1730.0	414.7	2090.6	2505.3	414.8	2259.6	2674.4	1.2956	6.0719	7.3675
100	373.15	101.33	1.044	1672.0	1673.0	419.0	2087.5	2506.5	419.1	2256.9	2676.0	1.3069	6.0485	7.3554
102	375.15	108.78	1.045	1564.5	1565.5	427.4	2081.4	2508.8	427.5	2251.6	2679.1	1.3294	6.0021	7.3315
104	377.15	116.68	1.047	1465.1	1466.2	435.8	2075.3	2511.1	435.9	2246.3	2682.2	1.3518	5.9560	7.3078
106	379.15	125.04	1.049	1373.1	1374.2	444.3	2069.2	2513.4	444.4	2240.9	2685.3	1.3742	5.9104	7.2845
108	381.15	133.90	1.050	1287.9	1288.9	452.7	2063.0	2515.7	452.9	2235.4	2688.3	1.3964	5.8651	7.2615
110	383.15	143.27	1.052	1208.9	1209.9	461.2	2056.8	2518.0	461.3	2230.0	2691.3	1.4185	5.8203	7.2388
112	385.15	153.16	1.054	1135.6	1136.6	469.6	2050.6	2520.2	469.8	2224.5	2694.3	1.4405	5.7758	7.2164
114	387.15	163.62	1.055	1067.5	1068.5	478.1	2044.3	2522.4	478.3	2219.0	2697.2	1.4624	5.7318	7.1942
116	389.15	174.65	1.057	1004.2	1005.2	486.6	2038.1	2524.6	486.7	2213.4	2700.2	1.4842	5.6881	7.1723
118	391.15	186.28	1.059	945.3	946.3	495.0	2031.8	2526.8	495.2	2207.9	2703.1	1.5060	5.6447	7.1507
120	393.15	198.54	1.061	890.5	891.5	503.5	2025.4	2529.0	503.7	2202.2	2706.0	1.5276	5.6017	7.1293
122	395.15	211.45	1.062	839.4	840.5	512.0	2019.1	2531.1	512.2	2196.6	2708.8	1.5491	5.5590	7.1082
124	397.15	225.04	1.064	791.8	792.8	520.5	2012.7	2533.2	520.7	2190.9	2711.6	1.5706	5.5167	7.0873
126	399.15	239.33	1.066	747.3	748.4	529.0	2006.3	2535.3	529.2	2185.2	2714.4	1.5919	5.4747	7.0666
128	401.15	254.35	1.068	705.8	706.9	537.5	1999.9	2537.4	537.8	2179.4	2717.2	1.6132	5.4330	7.0462

130	403.15	270.13	1.070	667.1	668.1	546.0	1993.4	2539.4	546.3	2173.6	2719.9	1.6344	5.3917	7.0261
132	405.15	286.70	1.072	630.8	631.9	554.5	1986.9	2541.4	554.8	2167.8	2722.6	1.6555	5.3507	7.0061
134	407.15	304.07	1.074	596.9	598.0	563.1	1980.4	2543.4	563.4	2161.9	2725.3	1.6765	5.3099	6.9864
136	409.15	322.29	1.076	565.1	566.2	571.6	1973.8	2545.4	572.0	2155.9	2727.9	1.6974	5.2695	6.9669
138	411.15	341.38	1.078	535.3	536.4	580.2	1967.2	2547.4	580.5	2150.0	2730.5	1.7182	5.2293	6.9475
140	413.15	361.38	1.080	507.4	508.5	588.7	1960.6	2549.3	589.1	2144.0	2733.1	1.7390	5.1894	6.9284
142	415.15	382.31	1.082	481.2	482.3	597.3	1953.9	2551.2	597.7	2137.9	2735.6	1.7597	5.1499	6.9095
144	417.15	404.20	1.084	456.6	457.7	605.9	1947.2	2553.1	606.3	2131.8	2738.1	1.7803	5.1105	6.8908
146	419.15	427.09	1.086	433.5	434.6	614.4	1940.5	2554.9	614.9	2125.7	2740.6	1.8008	5.0715	6.8723
148	421.15	451.01	1.089	411.8	412.9	623.0	1933.7	2556.8	623.5	2119.5	2743.0	1.8213	5.0327	6.8539
150	423.15	476.00	1.091	391.4	392.4	631.6	1926.9	2558.6	632.1	2113.2	2745.4	1.8416	4.9941	6.8358
152	425.15	502.08	1.093	372.1	373.2	640.2	1920.1	2560.3	640.8	2106.9	2747.7	1.8619	4.9558	6.8178
154	427.15	529.29	1.095	354.0	355.1	648.9	1913.2	2562.1	649.4	2100.6	2750.0	1.8822	4.9178	6.8000
156	429.15	557.67	1.098	336.9	338.0	657.5	1906.3	2563.8	658.1	2094.2	2752.3	1.9023	4.8800	6.7823
158	431.15	587.25	1.100	320.8	321.9	666.1	1899.3	2565.5	666.8	2087.7	2754.5	1.9224	4.8424	6.7648
160	433.15	618.06	1.102	305.7	306.8	674.8	1892.3	2567.1	675.5	2081.3	2756.7	1.9425	4.8050	6.7475
162	435.15	650.16	1.105	291.3	292.4	683.5	1885.3	2568.8	684.2	2074.7	2758.9	1.9624	4.7679	6.7303
164	437.15	683.56	1.107	277.8	278.9	692.1	1878.2	2570.4	692.9	2068.1	2761.0	1.9823	4.7309	6.7133
166	439.15	718.31	1.109	265.0	266.1	700.8	1871.1	2571.9	701.6	2061.4	2763.1	2.0022	4.6942	6.6964
168	441.15	754.45	1.112	252.9	254.0	709.5	1863.9	2573.4	710.4	2054.7	2765.1	2.0219	4.6577	6.6796
170	443.15	792.02	1.114	241.4	242.6	718.2	1856.7	2574.9	719.1	2047.9	2767.1	2.0416	4.6214	6.6630
172	445.15	831.06	1.117	230.6	231.7	727.0	1849.5	2576.4	727.9	2041.1	2769.0	2.0613	4.5853	6.6465
174	447.15	871.60	1.120	220.3	221.5	735.7	1842.2	2577.8	736.7	2034.2	2770.9	2.0809	4.5493	6.6302
176	449.15	913.68	1.122	210.6	211.7	744.4	1834.8	2579.3	745.5	2027.3	2772.7	2.1004	4.5136	6.6140
178	451.15	957.36	1.125	201.4	202.5	753.2	1827.4	2580.6	754.3	2020.2	2774.5	2.1199	4.4780	6.5979
180	453.15	1002.7	1.128	192.7	193.8	762.0	1820.0	2581.9	763.1	2013.1	2776.3	2.1393	4.4426	6.5819
182	455.15	1049.6	1.130	184.4	185.5	770.8	1812.5	2583.2	772.0	2006.0	2778.0	2.1587	4.4074	6.5660
184	457.15	1098.3	1.133	176.5	177.6	779.6	1804.9	2584.5	780.8	1998.8	2779.6	2.1780	4.3723	6.5503
186	459.15	1148.8	1.136	169.0	170.2	788.4	1797.3	2585.7	789.7	1991.5	2781.2	2.1972	4.3374	6.5346
188	461.15	1201.0	1.139	161.9	163.1	797.2	1789.7	2586.9	798.6	1984.2	2782.8	2.2164	4.3026	6.5191
190	463.15	1255.1	1.142	155.2	156.3	806.1	1782.0	2588.1	807.5	1976.7	2784.3	2.2356	4.2680	6.5036
192	465.15	1311.1	1.144	148.8	149.9	814.9	1774.2	2589.2	816.5	1969.3	2785.7	2.2547	4.2336	6.4883
194	467.15	1369.0	1.147	142.6	143.8	823.8	1766.4	2590.2	825.4	1961.7	2787.1	2.2738	4.1993	6.4730
196	469.15	1428.9	1.150	136.8	138.0	832.7	1758.6	2591.3	834.4	1954.1	2788.4	2.2928	4.1651	6.4578
198	471.15	1490.9	1.153	131.3	132.4	841.6	1750.6	2592.3	843.4	1946.4	2789.7	2.3117	4.1310	6.4428
200	473.15	1554.9	1.156	126.0	127.2	850.6	1742.6	2593.2	852.4	1938.6	2790.9	2.3307	4.0971	6.4278
202	475.15	1621.0	1.160	121.0	122.1	859.5	1734.6	2594.1	861.4	1930.7	2792.1	2.3495	4.0633	6.4128
204	477.15	1689.3	1.163	116.2	117.3	868.5	1726.5	2595.0	870.5	1922.8	2793.2	2.3684	4.0296	6.3980
206	479.15	1759.8	1.166	111.6	112.8	877.5	1718.3	2595.8	879.5	1914.7	2794.3	2.3872	3.9961	6.3832
208	481.15	1832.6	1.169	107.2	108.4	886.5	1710.1	2596.6	888.6	1906.6	2795.3	2.4059	3.9626	6.3686
210	483.15	1907.7	1.173	103.1	104.2	895.5	1701.8	2597.3	897.7	1898.5	2796.2	2.4247	3.9293	6.3539
212	485.15	1985.2	1.176	99.09	100.26	904.5	1693.5	2598.0	906.9	1890.2	2797.1	2.4434	3.8960	6.3394
214	487.15	2065.1	1.179	95.28	96.46	913.6	1685.1	2598.7	916.0	1881.8	2797.9	2.4620	3.8629	6.3249
216	489.15	2147.5	1.183	91.65	92.83	922.7	1676.6	2599.3	925.2	1873.4	2798.6	2.4806	3.8298	6.3104
218	491.15	2232.4	1.186	88.17	89.36	931.8	1668.0	2599.8	934.4	1864.9	2799.3	2.4992	3.7968	6.2960

TABLE C.1. SATURATED STEAM SI UNITS (Continued)

TEMPERATURE		ABS PRESS	SPECIFIC VOLUME V			INTERNAL ENERGY U			ENTHALPY H			ENTROPY S		
C	K	kPa	SAT LIQUID	EVAP	SAT VAPOR	SAT LIQUID	EVAP	SAT VAPOR	SAT LIQUID	EVAP	SAT VAPOR	SAT LIQUID	EVAP	SAT VAPOR
220	493.15	2319.8	1.190	84.85	86.04	940.9	1659.4	2600.3	943.7	1856.2	2799.9	2.5178	3.7639	6.2817
222	495.15	2409.9	1.194	81.67	82.86	950.1	1650.7	2600.8	952.9	1847.5	2800.5	2.5363	3.7311	6.2674
224	497.15	2502.7	1.197	78.62	79.82	959.2	1642.0	2601.2	962.2	1838.7	2800.9	2.5548	3.6984	6.2532
226	499.15	2598.2	1.201	75.71	76.91	968.4	1633.1	2601.5	971.5	1829.8	2801.4	2.5733	3.6657	6.2390
228	501.15	2696.5	1.205	72.92	74.12	977.6	1624.2	2601.8	980.9	1820.8	2801.7	2.5917	3.6331	6.2249
230	503.15	2797.6	1.209	70.24	71.45	986.9	1615.2	2602.1	990.3	1811.7	2802.0	2.6102	3.6006	6.2107
232	505.15	2901.6	1.213	67.68	68.89	996.2	1606.1	2602.3	999.7	1802.5	2802.2	2.6286	3.5681	6.1967
234	507.15	3008.6	1.217	65.22	66.43	1005.4	1597.0	2602.4	1009.1	1793.2	2802.3	2.6470	3.5356	6.1826
236	509.15	3118.6	1.221	62.86	64.08	1014.8	1587.7	2602.5	1018.6	1783.8	2802.3	2.6653	3.5033	6.1686
238	511.15	3231.7	1.225	60.60	61.82	1024.1	1578.4	2602.5	1028.1	1774.2	2802.3	2.6837	3.4709	6.1546
240	513.15	3347.8	1.229	58.43	59.65	1033.5	1569.0	2602.5	1037.6	1764.6	2802.2	2.7020	3.4386	6.1406
242	515.15	3467.2	1.233	56.34	57.57	1042.9	1559.5	2602.4	1047.2	1754.9	2802.0	2.7203	3.4063	6.1266
244	517.15	3589.8	1.238	54.34	55.58	1052.3	1549.9	2602.2	1056.8	1745.0	2801.8	2.7386	3.3740	6.1127
246	519.15	3715.7	1.242	52.41	53.66	1061.8	1540.2	2602.0	1066.4	1735.0	2801.4	2.7569	3.3418	6.0987
248	521.15	3844.9	1.247	50.56	51.81	1071.3	1530.5	2601.8	1076.1	1724.9	2801.0	2.7752	3.3096	6.0848
250	523.15	3977.6	1.251	48.79	50.04	1080.8	1520.6	2601.4	1085.8	1714.7	2800.4	2.7935	3.2773	6.0708
252	525.15	4113.7	1.256	47.08	48.33	1090.4	1510.6	2601.0	1095.5	1704.3	2799.8	2.8118	3.2451	6.0569
254	527.15	4253.4	1.261	45.43	46.69	1100.0	1500.5	2600.5	1105.3	1693.8	2799.1	2.8300	3.2129	6.0429
256	529.15	4396.7	1.266	43.85	45.11	1109.6	1490.4	2600.0	1115.2	1683.2	2798.3	2.8483	3.1807	6.0290
258	531.15	4543.7	1.271	42.33	43.60	1119.3	1480.1	2599.3	1125.0	1672.4	2797.4	2.8666	3.1484	6.0150
260	533.15	4694.3	1.276	40.86	42.13	1129.0	1469.7	2598.6	1134.9	1661.5	2796.4	2.8848	3.1161	6.0010
262	535.15	4848.8	1.281	39.44	40.73	1138.7	1459.2	2597.8	1144.9	1650.4	2795.3	2.9031	3.0838	5.9869
264	537.15	5007.1	1.286	38.08	39.37	1148.5	1448.5	2597.0	1154.9	1639.2	2794.1	2.9214	3.0515	5.9729
266	539.15	5169.3	1.291	36.77	38.06	1158.3	1437.8	2596.1	1165.0	1627.8	2792.8	2.9397	3.0191	5.9588
268	541.15	5335.5	1.297	35.51	36.80	1168.2	1426.9	2595.0	1175.1	1616.3	2791.4	2.9580	2.9866	5.9446
270	543.15	5505.8	1.303	34.29	35.59	1178.1	1415.9	2593.9	1185.2	1604.6	2789.9	2.9763	2.9541	5.9304
272	545.15	5680.2	1.308	33.11	34.42	1188.0	1404.7	2592.7	1195.4	1592.8	2788.2	2.9947	2.9215	5.9162
274	547.15	5858.7	1.314	31.97	33.29	1198.0	1393.4	2591.4	1205.7	1580.8	2786.5	3.0131	2.8889	5.9019
276	549.15	6041.5	1.320	30.88	32.20	1208.0	1382.0	2590.1	1216.0	1568.5	2784.6	3.0314	2.8561	5.8876
278	551.15	6228.7	1.326	29.82	31.14	1218.1	1370.4	2588.6	1226.4	1556.2	2782.6	3.0499	2.8233	5.8731
280	553.15	6420.2	1.332	28.79	30.13	1228.3	1358.7	2587.0	1236.8	1543.6	2780.4	3.0683	2.7903	5.8586
282	555.15	6616.1	1.339	27.81	29.14	1238.5	1346.8	2585.3	1247.3	1530.8	2778.1	3.0868	2.7573	5.8440
284	557.15	6816.6	1.345	26.85	28.20	1248.7	1334.8	2583.5	1257.9	1517.8	2775.7	3.1053	2.7241	5.8294
286	559.15	7021.8	1.352	25.93	27.28	1259.0	1322.6	2581.6	1268.5	1504.6	2773.2	3.1238	2.6908	5.8146
288	561.15	7231.5	1.359	25.03	26.39	1269.4	1310.2	2579.6	1279.2	1491.2	2770.5	3.1424	2.6573	5.7997
290	563.15	7446.1	1.366	24.17	25.54	1279.8	1297.7	2577.5	1290.0	1477.6	2767.6	3.1611	2.6237	5.7848
292	565.15	7665.4	1.373	23.33	24.71	1290.3	1284.9	2575.3	1300.9	1463.8	2764.6	3.1798	2.5899	5.7697
294	567.15	7889.7	1.381	22.52	23.90	1300.9	1272.0	2572.9	1311.8	1449.7	2761.5	3.1985	2.5560	5.7545
296	569.15	8118.9	1.388	21.74	23.13	1311.5	1258.9	2570.4	1322.8	1435.4	2758.2	3.2173	2.5218	5.7392
298	571.15	8353.2	1.396	20.98	22.38	1322.2	1245.6	2567.8	1333.9	1420.8	2754.7	3.2362	2.4875	5.7237

300	573.15	8592.7	1.404	20.24	21.65	1333.0	1232.0	2565.0	1345.1	1406.0	2751.0	3.2552	2.4529	5.7081
302	575.15	8837.4	1.412	19.53	20.94	1343.8	1218.3	2562.1	1356.3	1390.9	2747.2	3.2742	2.4182	5.6924
304	577.15	9087.3	1.421	18.84	20.26	1354.8	1204.3	2559.1	1367.7	1375.5	2743.2	3.2933	2.3832	5.6765
306	579.15	9342.7	1.430	18.17	19.60	1365.8	1190.1	2555.9	1379.1	1359.8	2739.0	3.3125	2.3479	5.6604
308	581.15	9603.6	1.439	17.52	18.96	1376.9	1175.6	2552.5	1390.7	1343.9	2734.6	3.3318	2.3124	5.6442
310	583.15	9870.0	1.448	16.89	18.33	1388.1	1161.0	2549.1	1402.4	1327.6	2730.0	3.3512	2.2766	5.6278
312	585.15	10142.1	1.458	16.27	17.73	1399.4	1146.0	2545.4	1414.2	1311.0	2725.2	3.3707	2.2404	5.6111
314	587.15	10420.0	1.468	15.68	17.14	1410.8	1130.8	2541.6	1426.1	1294.1	2720.2	3.3903	2.2040	5.5943
316	589.15	10703.	1.478	15.09	16.57	1422.3	1115.2	2537.5	1438.1	1276.8	2714.9	3.4101	2.1672	5.5772
318	591.15	10993.4	1.488	14.53	16.02	1433.9	1099.4	2533.3	1450.3	1259.1	2709.4	3.4300	2.1300	5.5599
320	593.15	11289.1	1.500	13.98	15.48	1445.7	1083.2	2528.9	1462.6	1241.1	2703.7	3.4500	2.0923	5.5423
322	595.15	11591.0	1.511	13.44	14.96	1457.5	1066.7	2524.3	1475.1	1222.6	2697.6	3.4702	2.0542	5.5244
324	597.15	11899.2	1.523	12.92	14.45	1469.5	1049.9	2519.4	1487.7	1203.6	2691.3	3.4906	2.0156	5.5062
326	599.15	12213.7	1.535	12.41	13.95	1481.7	1032.6	2514.3	1500.4	1184.2	2684.6	3.5111	1.9764	5.4876
328	601.15	12534.8	1.548	11.91	13.46	1494.0	1014.8	2508.8	1513.4	1164.2	2677.6	3.5319	1.9367	5.4685
330	603.15	12862.5	1.561	11.43	12.99	1506.4	996.7	2503.1	1526.5	1143.6	2670.2	3.5528	1.8962	5.4490
332	605.15	13197.0	1.575	10.95	12.53	1519.1	978.0	2497.0	1539.9	1122.5	2662.3	3.5740	1.8550	5.4290
334	607.15	13538.3	1.590	10.49	12.08	1531.9	958.7	2490.6	1553.4	1100.7	2654.1	3.5955	1.8129	5.4084
336	609.15	13886.7	1.606	10.03	11.63	1544.9	938.9	2483.7	1567.2	1078.1	2645.3	3.6172	1.7700	5.3872
338	611.15	14242.3	1.622	9.58	11.20	1558.1	918.4	2476.4	1581.2	1054.8	2636.0	3.6392	1.7261	5.3653
340	613.15	14605.2	1.639	9.14	10.78	1571.5	897.2	2468.7	1595.5	1030.7	2626.2	3.6616	1.6811	5.3427
342	615.15	14975.5	1.657	8.71	10.37	1585.2	875.2	2460.5	1610.0	1005.7	2615.7	3.6844	1.6350	5.3194
344	617.15	15353.5	1.676	8.286	9.962	1599.2	852.5	2451.7	1624.9	979.7	2604.7	3.7075	1.5877	5.2952
346	619.15	15739.3	1.696	7.870	9.566	1613.5	828.9	2442.4	1640.2	952.8	2593.0	3.7311	1.5391	5.2702
348	621.15	16133.1	1.718	7.461	9.178	1628.1	804.5	2432.6	1655.8	924.8	2580.7	3.7553	1.4891	5.2444
350	623.15	16535.1	1.741	7.058	8.799	1643.0	779.2	2422.2	1671.8	895.9	2567.7	3.7801	1.4375	5.2177
352	625.15	16945.5	1.766	6.654	8.420	1659.4	751.5	2410.8	1689.3	864.2	2553.5	3.8071	1.3822	5.1893
354	627.15	17364.4	1.794	6.252	8.045	1676.3	722.4	2398.7	1707.5	830.9	2538.4	3.8349	1.3247	5.1596
356	629.15	17792.2	1.824	5.850	7.674	1693.4	692.2	2385.6	1725.9	796.2	2522.1	3.8629	1.2654	5.1283
358	631.15	18229.0	1.858	5.448	7.306	1710.8	660.5	2371.4	1744.7	759.9	2504.6	3.8915	1.2037	5.0953
360	633.15	18675.1	1.896	5.044	6.940	1728.8	627.1	2355.8	1764.2	721.3	2485.4	3.9210	1.1390	5.0600
361	634.15	18901.7	1.917	4.840	6.757	1738.0	609.5	2347.5	1774.2	701.0	2475.2	3.9362	1.1052	5.0414
362	635.15	19130.7	1.939	4.634	6.573	1747.5	591.2	2338.7	1784.6	679.8	2464.4	3.9518	1.0702	5.0220
363	636.15	19362.1	1.963	4.425	6.388	1757.3	572.1	2329.3	1795.3	657.8	2453.0	3.9679	1.0338	5.0017
364	637.15	19596.1	1.988	4.213	6.201	1767.4	552.0	2319.4	1806.4	634.6	2440.9	3.9846	0.9958	4.9804
365	638.15	19832.6	2.016	3.996	6.012	1778.0	530.8	2308.8	1818.0	610.0	2428.0	4.0021	0.9558	4.9579
366	639.15	20071.6	2.046	3.772	5.819	1789.1	508.2	2297.3	1830.2	583.9	2414.1	4.0205	0.9134	4.9339
367	640.15	20313.2	2.080	3.540	5.621	1801.0	483.8	2284.8	1843.2	555.7	2399.0	4.0401	0.8680	4.9081
368	641.15	20557.5	2.118	3.298	5.416	1813.8	457.3	2271.1	1857.3	525.1	2382.4	4.0613	0.8189	4.8801
369	642.15	20804.4	2.162	3.039	5.201	1827.8	427.9	2255.7	1872.8	491.1	2363.9	4.0846	0.7647	4.8492
370	643.15	21054.0	2.214	2.759	4.973	1843.6	394.5	2238.1	1890.2	452.6	2342.8	4.1108	0.7036	4.8144
371	644.15	21306.4	2.278	2.446	4.723	1862.0	355.3	2217.3	1910.5	407.4	2317.9	4.1414	0.6324	4.7738
372	645.15	21561.6	2.364	2.075	4.439	1884.6	306.6	2191.2	1935.6	351.4	2287.0	4.1794	0.5446	4.7240
373	646.15	21819.7	2.496	1.588	4.084	1916.0	238.9	2154.9	1970.5	273.5	2244.0	4.2325	0.4233	4.6559
374	647.15	22080.5	2.843	0.623	3.466	1983.9	95.7	2079.7	2046.7	109.5	2156.2	4.3493	0.1692	4.5185
374.15	647.30	22120.0	3.170	0.000	3.170	2037.3	0.0	2037.3	2107.4	0.0	2107.4	4.4429	0.0000	4.4429

TABLE C.2. SUPERHEATED STEAM SI UNITS

ABS PRESS KPA (SAT TEMP) DEG C		SAT WATER	SAT STEAM	TEMPERATURE, DEG C (TEMPERATURE, K) 75 (348.15)	100 (373.15)	125 (398.15)	150 (423.15)	175 (448.15)	200 (473.15)	225 (498.15)	250 (523.15)
1 (6.98)	V	1.000	129200	160640	172180	183720	195270	206810	218350	229890	241430
	U	29.334	2385.2	2480.8	2516.4	2552.3	2588.5	2624.9	2661.7	2698.8	2736.3
	H	29.335	2514.4	2641.5	2688.6	2736.0	2783.7	2831.7	2880.1	2928.7	2977.7
	S	0.1060	8.9767	9.3828	9.5136	9.6365	9.7527	9.8629	9.9679	10.0681	10.1641
10 (45.83)	V	1.010	14670	16030	17190	18350	19510	20660	21820	22980	24130
	U	191.822	2438.0	2479.7	2515.6	2551.6	2588.0	2624.5	2661.4	2698.6	2736.1
	H	191.832	2584.8	2640.0	2687.5	2735.2	2783.1	2831.2	2879.6	2928.4	2977.4
	S	0.6493	8.1511	8.3168	8.4486	8.5722	8.6888	8.7994	8.9045	9.0049	9.1010
20 (60.09)	V	1.017	7649.8	8000.0	8584.7	9167.1	9748.0	10320	10900	11480	12060
	U	251.432	2456.9	2478.4	2514.6	2550.9	2587.4	2624.1	2661.0	2698.3	2735.8
	H	251.453	2609.9	2638.4	2686.3	2734.2	2782.3	2830.6	2879.2	2928.0	2977.1
	S	0.8321	7.9094	7.9933	8.1261	8.2504	8.3676	8.4785	8.5839	8.6844	8.7806
30 (69.12)	V	1.022	5229.3	5322.0	5714.4	6104.6	6493.2	6880.8	7267.5	7653.8	8039.7
	U	289.271	2468.6	2477.1	2513.6	2550.2	2586.8	2623.6	2660.7	2698.0	2735.6
	H	289.302	2625.4	2636.8	2685.1	2733.3	2781.6	2830.0	2878.7	2927.6	2976.8
	S	0.9441	7.7695	7.8024	7.9363	8.0614	8.1791	8.2903	8.3960	8.4967	8.5930
40 (75.89)	V	1.027	3993.4	....	4279.2	4573.3	4865.8	5157.2	5447.8	5738.0	6027.7
	U	317.609	2477.1	....	2512.6	2549.4	2586.2	2623.2	2660.3	2697.7	2735.4
	H	317.650	2636.9	....	2683.8	2732.3	2780.9	2829.5	2878.2	2927.2	2976.5
	S	1.0261	7.6709	....	7.8009	7.9268	8.0450	8.1566	8.2624	8.3633	8.4598
50 (81.35)	V	1.030	3240.2	....	3418.1	3654.5	3889.3	4123.0	4356.0	4588.5	4820.5
	U	340.513	2484.0	....	2511.7	2548.6	2585.6	2622.7	2659.9	2697.4	2735.1
	H	340.564	2646.0	....	2682.6	2731.4	2780.1	2828.9	2877.7	2926.8	2976.1
	S	1.0912	7.5947	....	7.6953	7.8219	7.9406	8.0526	8.1587	8.2598	8.3564
75 (91.79)	V	1.037	2216.9	....	2269.8	2429.4	2587.3	2744.2	2900.2	3055.8	3210.9
	U	384.374	2496.7	....	2509.2	2546.7	2584.2	2621.6	2659.0	2696.7	2734.5
	H	384.451	2663.0	....	2679.4	2728.9	2778.2	2827.4	2876.6	2925.8	2975.3
	S	1.2131	7.4570	....	7.5014	7.6300	7.7500	7.8629	7.9697	8.0712	8.1681
100 (99.63)	V	1.043	1693.7	....	1695.5	1816.7	1936.3	2054.7	2172.3	2289.4	2406.1
	U	417.406	2506.1	....	2506.6	2544.8	2582.7	2620.4	2658.1	2695.9	2733.9
	H	417.511	2675.4	....	2676.2	2726.5	2776.3	2825.9	2875.4	2924.9	2974.5
	S	1.3027	7.3598	....	7.3618	7.4923	7.6137	7.7275	7.8349	7.9369	8.0342

101.325 (100.00)	V	1.044	1673.0	....	1673.0	1792.7	1910.7	2027.7	2143.8	2259.3	2374.5
	U	418.959	2506.5	....	2506.5	2544.7	2582.6	2620.4	2658.1	2695.9	2733.9
	H	419.064	2676.0	....	2676.0	2726.4	2776.2	2825.8	2875.3	2924.8	2974.5
	S	1.3069	7.3554	....	7.3554	7.4860	7.6075	7.7213	7.8288	7.9308	8.0280
125 (105.99)	V	1.049	1374.6	....	....	1449.1	1545.6	1641.0	1735.6	1829.6	1923.2
	U	444.224	2513.4	....	....	2542.9	2581.2	2619.3	2657.2	2695.2	2733.3
	H	444.356	2685.2	....	....	2724.0	2774.4	2824.4	2874.2	2923.9	2973.7
	S	1.3740	7.2847	....	....	7.3844	7.5072	7.6219	7.7300	7.8324	7.9300
150 (111.37)	V	1.053	1159.0	....	....	1204.0	1285.2	1365.2	1444.4	1523.0	1601.3
	U	466.968	2519.5	....	....	2540.9	2579.7	2618.1	2656.3	2694.4	2732.7
	H	467.126	2693.4	....	....	2721.5	2772.5	2822.9	2872.9	2922.9	2972.9
	S	1.4336	7.2234	....	....	7.2953	7.4194	7.5352	7.6439	7.7468	7.8447
175 (116.06)	V	1.057	1003.34	....	....	1028.8	1099.1	1168.2	1236.4	1304.1	1371.3
	U	486.815	2524.7	....	....	2538.9	2578.2	2616.9	2655.3	2693.7	2732.1
	H	487.000	2700.3	....	....	2719.0	2770.5	2821.3	2871.7	2921.9	2972.0
	S	1.4849	7.1716	....	....	7.2191	7.3447	7.4614	7.5708	7.6741	7.7724
200 (120.23)	V	1.061	885.44	....	....	897.47	959.54	1020.4	1080.4	1139.8	1198.9
	U	504.489	2529.2	....	....	2536.9	2576.6	2615.7	2654.4	2692.9	2731.4
	H	504.701	2706.3	....	....	2716.4	2768.5	2819.8	2870.5	2920.9	2971.2
	S	1.5301	7.1268	....	....	7.1523	7.2794	7.3971	7.5072	7.6110	7.7096
225 (123.99)	V	1.064	792.97	....	....	795.25	850.97	905.44	959.06	1012.1	1064.7
	U	520.465	2533.2	....	....	2534.8	2575.1	2614.5	2653.5	2692.2	2730.8
	H	520.705	2711.6	....	....	2713.8	2766.5	2818.2	2869.3	2919.9	2970.4
	S	1.5705	7.0873	....	....	7.0928	7.2213	7.3400	7.4508	7.5551	7.6540
250 (127.43)	V	1.068	718.44	....	....	....	764.09	813.47	861.98	909.91	957.41
	U	535.077	2536.8	....	....	....	2573.5	2613.3	2652.5	2691.4	2730.2
	H	535.343	2716.4	....	....	....	2764.5	2816.7	2868.0	2918.9	2969.6
	S	1.6071	7.0520	....	....	....	7.1689	7.2886	7.4001	7.5050	7.6042
275 (130.60)	V	1.071	657.04	....	....	....	693.00	738.21	782.55	826.29	869.61
	U	548.564	2540.0	....	....	....	2571.9	2612.1	2651.6	2690.7	2729.6
	H	548.858	2720.7	....	....	....	2762.5	2815.1	2866.8	2917.9	2968.7
	S	1.6407	7.0201	....	....	....	7.1211	7.2419	7.3541	7.4594	7.5590
300 (133.54)	V	1.073	605.56	....	....	....	633.74	675.49	716.35	756.60	796.44
	U	561.107	2543.0	....	....	....	2570.3	2610.8	2650.6	2689.9	2729.0
	H	561.429	2724.7	....	....	....	2760.4	2813.5	2865.5	2916.9	2967.9
	S	1.6716	6.9909	....	....	....	7.0771	7.1990	7.3119	7.4177	7.5176

TABLE C.2. SUPERHEATED STEAM SI UNITS (Continued)

ABS PRESS KPA (SAT TEMP) DEG C		SAT WATER	SAT STEAM	TEMPERATURE, DEG C (TEMPERATURE, K) 300 (573.15)	350 (623.15)	400 (673.15)	450 (723.15)	500 (773.15)	550 (823.15)	600 (873.15)	650 (923.15)
1	V	1.000	129200	264500	287580	310660	333730	356810	379880	402960	426040
(6.98)	U	29.334	2385.2	2812.3	2889.9	2969.1	3049.9	3132.4	3216.7	3302.6	3390.3
	H	29.335	2514.4	3076.8	3177.5	3279.7	3383.6	3489.2	3596.5	3705.6	3816.4
	S	0.1060	8.9767	10.3450	10.5133	10.6711	10.8200	10.9612	11.0957	11.2243	11.3476
10	V	1.010	14670	26440	28750	31060	33370	35670	37980	40290	42600
(45.83)	U	191.822	2438.0	2812.2	2889.8	2969.0	3049.8	3132.3	3216.6	3302.6	3390.3
	H	191.832	2584.8	3076.6	3177.3	3279.6	3383.5	3489.1	3596.5	3705.5	3816.3
	S	0.6493	8.1511	9.2820	9.4504	9.6083	9.7572	9.8984	10.0329	10.1616	10.2849
20	V	1.017	7649.8	13210	14370	15520	16680	17830	18990	20140	21300
(60.09)	U	251.432	2456.9	2812.0	2889.6	2968.9	3049.7	3132.3	3216.5	3302.5	3390.2
	H	251.453	2609.9	3076.4	3177.1	3279.4	3383.4	3489.0	3596.4	3705.4	3816.2
	S	0.8321	7.9094	8.9618	9.1303	9.2882	9.4372	9.5784	9.7130	9.8416	9.9650
30	V	1.022	5229.3	8810.8	9581.2	10350	11120	11890	12660	13430	14190
(69.12)	U	289.271	2468.6	2811.8	2889.5	2968.7	3049.6	3132.2	3216.5	3302.5	3390.2
	H	289.302	2625.4	3076.1	3176.9	3279.3	3383.3	3488.9	3596.3	3705.4	3816.2
	S	0.9441	7.7695	8.7744	8.9430	9.1010	9.2499	9.3912	9.5257	9.6544	9.7778
40	V	1.027	3993.4	6606.5	7184.6	7762.5	8340.1	8917.6	9494.9	10070	10640
(75.89)	U	317.609	2477.1	2811.6	2889.4	2968.6	3049.5	3132.1	3216.4	3302.4	3390.1
	H	317.650	2636.9	3075.9	3176.8	3279.1	3383.1	3488.8	3596.2	3705.3	3816.1
	S	1.0261	7.6709	8.6413	8.8100	8.9680	9.1170	9.2583	9.3929	9.5216	9.6450
50	V	1.030	3240.2	5283.9	5746.7	6209.1	6671.4	7133.5	7595.5	8057.4	8519.2
(81.35)	U	340.513	2484.0	2811.5	2889.2	2968.5	3049.4	3132.0	3216.3	3302.3	3390.1
	H	340.564	2646.0	3075.7	3176.6	3279.0	3383.0	3488.7	3596.1	3705.2	3816.0
	S	1.0912	7.5947	8.5380	8.7068	8.8649	9.0139	9.1552	9.2898	9.4185	9.5419
75	V	1.037	2216.9	3520.5	3829.4	4138.0	4446.4	4754.7	5062.8	5370.9	5678.9
(91.79)	U	384.374	2496.7	2811.0	2888.9	2968.2	3049.2	3131.8	3216.1	3302.2	3389.9
	H	384.451	2663.0	3075.1	3176.1	3278.6	3382.7	3488.4	3595.8	3705.0	3815.9
	S	1.2131	7.4570	8.3502	8.5191	8.6773	8.8265	8.9678	9.1025	9.2312	9.3546
100	V	1.043	1693.7	2638.7	2870.8	3102.5	3334.0	3565.3	3796.5	4027.7	4258.8
(99.63)	U	417.406	2506.1	2810.6	2888.6	2968.0	3049.0	3131.6	3216.0	3302.0	3389.8
	H	417.511	2675.4	3074.5	3175.6	3278.2	3382.4	3488.1	3595.6	3704.8	3815.7
	S	1.3027	7.3598	8.2166	8.3858	8.5442	8.6934	8.8348	8.9695	9.0982	9.2217

101.325 (100.00)	V	1.044	1673.0	2604.2	2833.2	3061.9	3290.3	3518.7	3746.9	3975.0	4203.1
	U	418.959	2506.5	2810.6	2888.5	2968.0	3048.9	3131.6	3215.9	3302.0	3389.8
	H	419.064	2676.0	3074.4	3175.6	3278.2	3382.3	3488.1	3595.6	3704.8	3815.7
	S	1.3069	7.3554	8.2105	8.3797	8.5381	8.6873	8.8287	8.9634	9.0922	9.2156
125 (105.99)	V	1.049	1374.6	2109.7	2295.6	2481.2	2666.5	2851.7	3036.8	3221.8	3406.7
	U	444.224	2513.4	2810.2	2888.2	2967.7	3048.7	3131.4	3215.8	3301.9	3389.7
	H	444.356	2685.2	3073.9	3175.2	3277.8	3382.0	3487.9	3595.4	3704.6	3815.5
	S	1.3740	7.2847	8.1129	8.2823	8.4408	8.5901	8.7316	8.8663	8.9951	9.1186
150 (111.37)	V	1.053	1159.0	1757.0	1912.2	2066.9	2221.5	2375.9	2530.2	2684.5	2838.6
	U	466.968	2519.5	2809.7	2887.9	2967.4	3048.5	3131.2	3215.6	3301.7	3389.5
	H	467.126	2693.4	3073.3	3174.7	3277.5	3381.7	3487.6	3595.1	3704.4	3815.3
	S	1.4336	7.2234	8.0280	8.1976	8.3562	8.5056	8.6472	8.7819	8.9108	9.0343
175 (116.06)	V	1.057	1003.34	1505.1	1638.3	1771.1	1903.7	2036.1	2168.4	2300.7	2432.9
	U	486.815	2524.7	2809.3	2887.5	2967.1	3048.3	3131.0	3215.4	3301.6	3389.4
	H	487.000	2700.3	3072.7	3174.2	3277.1	3381.4	3487.3	3594.9	3704.2	3815.1
	S	1.4849	7.1716	7.9561	8.1259	8.2847	8.4341	8.5758	8.7106	8.8394	8.9630
200 (120.23)	V	1.061	885.44	1316.2	1432.8	1549.2	1665.3	1781.2	1897.1	2012.9	2128.6
	U	504.489	2529.2	2808.8	2887.2	2966.9	3048.0	3130.8	3215.3	3301.4	3389.2
	H	504.701	2706.3	3072.1	3173.8	3276.7	3381.1	3487.0	3594.7	3704.0	3815.0
	S	1.5301	7.1268	7.8937	8.0638	8.2226	8.3722	8.5139	8.6487	8.7776	8.9012
225 (123.99)	V	1.064	792.97	1169.2	1273.1	1376.6	1479.9	1583.0	1686.0	1789.0	1891.9
	U	520.465	2533.2	2808.4	2886.9	2966.6	3047.8	3130.6	3215.1	3301.2	3389.1
	H	520.705	2711.6	3071.5	3173.3	3276.3	3380.8	3486.8	3594.4	3703.8	3814.8
	S	1.5705	7.0873	7.8385	8.0088	8.1679	8.3175	8.4593	8.5942	8.7231	8.8467
250 (127.43)	V	1.068	718.44	1051.6	1145.2	1238.5	1331.5	1424.4	1517.2	1609.9	1702.5
	U	535.077	2536.8	2808.0	2886.5	2966.3	3047.6	3130.4	3214.9	3301.1	3389.0
	H	535.343	2716.4	3070.9	3172.8	3275.9	3380.4	3486.5	3594.2	3703.6	3814.6
	S	1.6071	7.0520	7.7891	7.9597	8.1188	8.2686	8.4104	8.5453	8.6743	8.7980
275 (130.60)	V	1.071	657.04	955.45	1040.7	1125.5	1210.2	1294.7	1379.0	1463.3	1547.6
	U	548.564	2540.0	2807.5	2886.2	2966.0	3047.3	3130.2	3214.7	3300.9	3388.8
	H	548.858	2720.7	3070.3	3172.4	3275.5	3380.1	3486.2	3594.0	3703.4	3814.4
	S	1.6407	7.0201	7.7444	7.9151	8.0744	8.2243	8.3661	8.5011	8.6301	8.7538
300 (133.54)	V	1.073	605.56	875.29	953.52	1031.4	1109.0	1186.5	1263.9	1341.2	1418.5
	U	561.107	2543.0	2807.1	2885.8	2965.8	3047.1	3130.0	3214.5	3300.8	3388.7
	H	561.429	2724.7	3069.7	3171.9	3275.2	3379.8	3486.0	3593.7	3703.2	3814.2
	S	1.6716	6.9909	7.7034	7.8744	8.0338	8.1838	8.3257	8.4608	8.5898	8.7135

TABLE C.2. SUPERHEATED STEAM SI UNITS (Continued)

ABS PRESS KPA (SAT TEMP) DEG C		SAT WATER	SAT STEAM	TEMPERATURE, DEG C (TEMPERATURE, K) 150 (423.15)	175 (448.15)	200 (473.15)	220 (493.15)	240 (513.15)	260 (533.15)	280 (553.15)	300 (573.15)
325 (136.29)	V	1.076	561.75	583.58	622.41	660.33	690.22	719.81	749.18	778.39	807.47
	U	572.847	2545.7	2568.7	2609.6	2649.6	2681.2	2712.7	2744.0	2775.3	2806.6
	H	573.197	2728.3	2758.4	2811.9	2864.2	2905.6	2946.6	2987.5	3028.2	3069.0
	S	1.7004	6.9640	7.0363	7.1592	7.2729	7.3585	7.4400	7.5181	7.5933	7.6657
350 (138.87)	V	1.079	524.00	540.58	576.90	612.31	640.18	667.75	695.09	722.27	749.33
	U	583.892	2548.2	2567.1	2608.3	2648.6	2680.4	2712.0	2743.4	2774.8	2806.2
	H	584.270	2731.6	2756.3	2810.3	2863.0	2904.5	2945.7	2986.7	3027.6	3068.4
	S	1.7273	6.9392	6.9982	7.1222	7.2366	7.3226	7.4045	7.4828	7.5581	7.6307
375 (141.31)	V	1.081	491.13	503.29	537.46	570.69	596.81	622.62	648.22	673.64	698.94
	U	594.332	2550.6	2565.4	2607.1	2647.7	2679.6	2711.3	2742.8	2774.3	2805.7
	H	594.737	2734.7	2754.1	2808.6	2861.7	2903.4	2944.8	2985.9	3026.9	3067.8
	S	1.7526	6.9160	6.9624	7.0875	7.2027	7.2891	7.3713	7.4499	7.5254	7.5981
400 (143.62)	V	1.084	462.22	470.66	502.93	534.26	558.85	583.14	607.20	631.09	654.85
	U	604.237	2552.7	2563.7	2605.8	2646.7	2678.8	2710.6	2742.2	2773.7	2805.3
	H	604.670	2737.6	2752.0	2807.0	2860.4	2902.3	2943.9	2985.1	3026.2	3067.2
	S	1.7764	6.8943	6.9285	7.0548	7.1708	7.2576	7.3402	7.4190	7.4947	7.5675
425 (145.82)	V	1.086	436.61	441.85	472.47	502.12	525.36	548.30	571.01	593.54	615.95
	U	613.667	2554.8	2562.0	2604.5	2645.7	2678.0	2709.9	2741.6	2773.2	2804.8
	H	614.128	2740.3	2749.8	2805.3	2859.1	2901.2	2942.9	2984.3	3025.5	3066.6
	S	1.7990	6.8739	6.8965	7.0239	7.1407	7.2280	7.3108	7.3899	7.4657	7.5388
450 (147.92)	V	1.088	413.75	416.24	445.38	473.55	495.59	517.33	538.83	560.17	581.37
	U	622.672	2556.7	2560.3	2603.2	2644.7	2677.1	2709.2	2741.0	2772.7	2804.4
	H	623.162	2742.9	2747.7	2803.7	2857.8	2900.2	2942.0	2983.5	3024.8	3066.0
	S	1.8204	6.8547	6.8660	6.9946	7.1121	7.1999	7.2831	7.3624	7.4384	7.5116
475 (149.92)	V	1.091	393.22	393.31	421.14	447.97	468.95	489.62	510.05	530.30	550.43
	U	631.294	2558.5	2558.6	2601.9	2643.7	2676.3	2708.5	2740.4	2772.2	2803.9
	H	631.812	2745.3	2745.5	2802.0	2856.5	2899.1	2941.1	2982.7	3024.1	3065.4
	S	1.8408	6.8365	6.8369	6.9667	7.0850	7.1732	7.2567	7.3363	7.4125	7.4858
500 (151.84)	V	1.093	374.68	....	399.31	424.96	444.97	464.67	484.14	503.43	522.58
	U	639.569	2560.2	....	2600.6	2642.7	2675.5	2707.8	2739.8	2771.7	2803.5
	H	640.116	2747.5	....	2800.3	2855.1	2898.0	2940.1	2981.9	3023.4	3064.8
	S	1.8604	6.8192	....	6.9400	7.0592	7.1478	7.2317	7.3115	7.3879	7.4614

525	V	1.095	357.84	....	379.56	404.13	423.28	442.11	460.70	479.11	497.38
(153.69)	U	647.528	2561.8	....	2599.3	2641.6	2674.6	2707.1	2739.2	2771.2	2803.0
	H	648.103	2749.7	....	2798.6	2853.8	2896.8	2939.2	2981.1	3022.7	3064.1
	S	1.8790	6.8027	....	6.9145	7.0345	7.1236	7.2078	7.2879	7.3645	7.4381
550	V	1.097	342.48	....	361.60	385.19	403.55	421.59	439.38	457.00	474.48
(155.47)	U	655.199	2563.3	....	2598.0	2640.6	2673.8	2706.4	2738.6	2770.6	2802.6
	H	655.802	2751.7	....	2796.8	2852.5	2895.7	2938.3	2980.3	3022.0	3063.5
	S	1.8970	6.7870	....	6.8900	7.0108	7.1004	7.1849	7.2653	7.3421	7.4158
575	V	1.099	328.41	....	345.20	367.90	385.54	402.85	419.92	436.81	453.56
(157.18)	U	662.603	2564.8	....	2596.6	2639.6	2672.9	2705.7	2738.0	2770.1	2802.1
	H	663.235	2753.6	....	2795.1	2851.1	2894.6	2937.3	2979.5	3021.3	3062.9
	S	1.9142	6.7720	....	6.8664	6.9880	7.0781	7.1630	7.2436	7.3206	7.3945
600	V	1.101	315.47	....	330.16	352.04	369.03	385.68	402.08	418.31	434.39
(158.84)	U	669.762	2566.2	....	2595.3	2638.5	2672.1	2705.0	2737.4	2769.6	2801.6
	H	670.423	2755.5	....	2793.3	2849.7	2893.5	2936.4	2978.7	3020.6	3062.3
	S	1.9308	6.7575	....	6.8437	6.9662	7.0567	7.1419	7.2228	7.3000	7.3740
625	V	1.103	303.54	....	316.31	337.45	353.83	369.87	385.67	401.28	416.75
(160.44)	U	676.695	2567.5	....	2593.9	2637.5	2671.2	2704.2	2736.8	2769.1	2801.2
	H	677.384	2757.2	....	2791.6	2848.4	2892.3	2935.4	2977.8	3019.9	3061.7
	S	1.9469	6.7437	....	6.8217	6.9451	7.0361	7.1217	7.2028	7.2802	7.3544
650	V	1.105	292.49	....	303.53	323.98	339.80	355.29	370.52	385.56	400.47
(161.99)	U	683.417	2568.7	....	2592.5	2636.4	2670.3	2703.5	2736.2	2768.5	2800.7
	H	684.135	2758.9	....	2789.8	2847.0	2891.2	2934.4	2977.0	3019.2	3061.0
	S	1.9623	6.7304	....	6.8004	6.9247	7.0162	7.1021	7.1835	7.2611	7.3355
675	V	1.106	282.23	....	291.69	311.51	326.81	341.78	356.49	371.01	385.39
(163.49)	U	689.943	2570.0	....	2591.1	2635.4	2669.5	2702.8	2735.6	2768.0	2800.3
	H	690.689	2760.5	....	2788.0	2845.6	2890.1	2933.5	2976.2	3018.5	3060.4
	S	1.9773	6.7176	....	6.7798	6.9050	6.9970	7.0833	7.1650	7.2428	7.3173
700	V	1.108	272.68	....	280.69	299.92	314.75	329.23	343.46	357.50	371.39
(164.96)	U	696.285	2571.1	....	2589.7	2634.3	2668.6	2702.1	2735.0	2767.5	2799.8
	H	697.061	2762.0	....	2786.2	2844.2	2888.9	2932.5	2975.4	3017.7	3059.8
	S	1.9918	6.7052	....	6.7598	6.8859	6.9784	7.0651	7.1470	7.2250	7.2997
725	V	1.110	263.77	....	270.45	289.13	303.51	317.55	331.33	344.92	358.36
(166.38)	U	702.457	2572.2	....	2588.3	2633.2	2667.7	2701.3	2734.3	2767.0	2799.3
	H	703.261	2763.4	....	2784.4	2842.8	2887.7	2931.5	2974.6	3017.0	3059.1
	S	2.0059	6.6932	....	6.7404	6.8673	6.9604	7.0474	7.1296	7.2078	7.2827

TABLE C.2. SUPERHEATED STEAM SI UNITS (Continued)

ABS PRESS KPA (SAT TEMP) DEG C		SAT WATER	SAT STEAM	TEMPERATURE, DEG C (TEMPERATURE, K) 325 (598.15)	350 (623.15)	400 (673.15)	450 (723.15)	500 (773.15)	550 (823.15)	600 (873.15)	650 (923.15)
325 (136.29)	V	1.076	561.75	843.68	879.78	951.73	1023.5	1095.0	1166.5	1237.9	1309.2
	U	572.847	2545.7	2845.9	2885.5	2965.5	3046.9	3129.8	3214.4	3300.6	3388.6
	H	573.197	2728.3	3120.1	3171.4	3274.8	3379.5	3485.7	3593.5	3702.9	3814.1
	S	1.7004	6.9640	7.7530	7.8369	7.9965	8.1465	8.2885	8.4236	8.5527	8.6764
350 (138.87)	V	1.079	524.00	783.01	816.57	883.45	950.11	1016.6	1083.0	1149.3	1215.6
	U	583.892	2548.2	2845.6	2885.1	2965.2	3046.6	3129.6	3214.2	3300.5	3388.4
	H	584.270	2731.6	3119.6	3170.9	3274.4	3379.2	3485.4	3593.3	3702.7	3813.9
	S	1.7273	6.9392	7.7181	7.8022	7.9619	8.1120	8.2540	8.3892	8.5183	8.6421
375 (141.31)	V	1.081	491.13	730.42	761.79	824.28	886.54	948.66	1010.7	1072.6	1134.5
	U	594.332	2550.6	2845.2	2884.8	2964.9	3046.4	3129.4	3214.0	3300.3	3388.3
	H	594.737	2734.7	3119.1	3170.5	3274.0	3378.8	3485.1	3593.0	3702.6	3813.7
	S	1.7526	6.9160	7.6856	7.7698	7.9296	8.0798	8.2219	8.3571	8.4863	8.6101
400 (143.62)	V	1.084	462.22	684.41	713.85	772.50	830.92	889.19	947.35	1005.4	1063.4
	U	604.237	2552.7	2844.8	2884.5	2964.6	3046.2	3129.2	3213.8	3300.2	3388.2
	H	604.670	2737.6	3118.5	3170.0	3273.6	3378.5	3484.9	3592.8	3702.3	3813.5
	S	1.7764	6.8943	7.6552	7.7395	7.8994	8.0497	8.1919	8.3271	8.4563	8.5802
425 (145.82)	V	1.086	436.61	643.81	671.56	726.81	781.84	836.72	891.49	946.17	1000.8
	U	613.667	2554.8	2844.4	2884.1	2964.4	3045.9	3129.0	3213.7	3300.0	3388.0
	H	614.128	2740.3	3118.0	3169.5	3273.3	3378.2	3484.6	3592.5	3702.1	3813.4
	S	1.7990	6.8739	7.6265	7.7109	7.8710	8.0214	8.1636	8.2989	8.4282	8.5520
450 (147.92)	V	1.088	413.75	607.73	633.97	686.20	738.21	790.07	841.83	893.50	945.10
	U	622.672	2556.7	2844.0	2883.8	2964.1	3045.7	3128.8	3213.5	3299.8	3387.9
	H	623.162	2742.9	3117.5	3169.1	3272.9	3377.9	3484.3	3592.3	3701.9	3813.2
	S	1.8204	6.8547	7.5995	7.6840	7.8442	7.9947	8.1370	8.2723	8.4016	8.5255
475 (149.92)	V	1.091	393.22	575.44	600.33	649.87	699.18	748.34	797.40	846.37	895.27
	U	631.294	2558.5	2843.6	2883.4	2963.8	3045.4	3128.6	3213.3	3299.7	3387.7
	H	631.812	2745.3	3116.9	3168.6	3272.5	3377.6	3484.0	3592.1	3701.7	3813.0
	S	1.8408	6.8365	7.5739	7.6585	7.8189	7.9694	8.1118	8.2472	8.3765	8.5004
500 (151.84)	V	1.093	374.68	546.38	570.05	617.16	664.05	710.78	757.41	803.95	850.42
	U	639.569	2560.2	2843.2	2883.1	2963.5	3045.2	3128.4	3213.1	3299.5	3387.6
	H	640.116	2747.5	3116.4	3168.1	3272.1	3377.2	3483.8	3591.8	3701.5	3812.8
	S	1.8604	6.8192	7.5496	7.6343	7.7948	7.9454	8.0879	8.2233	8.3526	8.4766

525 (153.69)	V	1.095	357.84	520.08	542.66	587.58	632.26	676.80	721.23	765.57	809.85
	U	647.528	2561.8	2842.8	2882.7	2963.2	3045.0	3128.2	3213.0	3299.4	3387.5
	H	648.103	2749.7	3115.9	3167.6	3271.7	3376.9	3483.5	3591.6	3701.3	3812.6
	S	1.8790	6.8027	7.5264	7.6112	7.7719	7.9226	8.0651	8.2006	8.3299	8.4539
550 (155.47)	V	1.097	342.48	496.18	517.76	560.68	603.37	645.91	688.34	730.68	772.96
	U	655.199	2563.3	2842.4	2882.4	2963.0	3044.7	3128.0	3212.8	3299.2	3387.3
	H	655.802	2751.7	3115.3	3167.2	3271.3	3376.6	3483.2	3591.4	3701.1	3812.5
	S	1.8970	6.7870	7.5043	7.5892	7.7500	7.9008	8.0433	8.1789	8.3083	8.4323
575 (157.18)	V	1.099	328.41	474.36	495.03	536.12	576.98	617.70	658.30	698.83	739.28
	U	662.603	2564.8	2842.0	2882.1	2962.7	3044.5	3127.8	3212.6	3299.1	3387.2
	H	663.235	2753.6	3114.8	3166.7	3271.0	3376.3	3482.9	3591.1	3700.9	3812.3
	S	1.9142	6.7720	7.4831	7.5681	7.7290	7.8799	8.0226	8.1581	8.2876	8.4116
600 (158.84)	V	1.101	315.47	454.35	474.19	513.61	552.80	591.84	630.78	669.63	708.41
	U	669.762	2566.2	2841.6	2881.7	2962.4	3044.3	3127.6	3212.4	3298.9	3387.1
	H	670.423	2755.5	3114.3	3166.2	3270.6	3376.0	3482.7	3590.9	3700.7	3812.1
	S	1.9308	6.7575	7.4628	7.5479	7.7090	7.8600	8.0027	8.1383	8.2678	8.3919
625 (160.44)	V	1.103	303.54	435.94	455.01	492.89	530.55	568.05	605.45	642.76	680.01
	U	676.695	2567.5	2841.2	2881.4	2962.1	3044.0	3127.4	3212.2	3298.8	3386.9
	H	677.384	2757.2	3113.7	3165.7	3270.2	3375.6	3482.4	3590.7	3700.5	3811.9
	S	1.9469	6.7437	7.4433	7.5285	7.6897	7.8408	7.9836	8.1192	8.2488	8.3729
650 (161.99)	V	1.105	292.49	418.95	437.31	473.78	510.01	546.10	582.07	617.96	653.79
	U	683.417	2568.7	2840.9	2881.0	2961.8	3043.8	3127.2	3212.1	3298.6	3386.8
	H	684.135	2758.9	3113.2	3165.3	3269.8	3375.3	3482.1	3590.4	3700.3	3811.8
	S	1.9623	6.7304	7.4245	7.5099	7.6712	7.8224	7.9652	8.1009	8.2305	8.3546
675 (163.49)	V	1.106	282.23	403.22	420.92	456.07	491.00	525.77	560.43	595.00	629.51
	U	689.943	2570.0	2840.5	2880.7	2961.6	3043.6	3127.0	3211.9	3298.5	3386.7
	H	690.689	2760.5	3112.6	3164.8	3269.4	3375.0	3481.8	3590.2	3700.1	3811.6
	S	1.9773	6.7176	7.4064	7.4919	7.6534	7.8046	7.9475	8.0833	8.2129	8.3371
700 (164.96)	V	1.108	272.68	388.61	405.71	439.64	473.34	506.89	540.33	573.68	606.97
	U	696.285	2571.1	2840.1	2880.3	2961.3	3043.3	3126.8	3211.7	3298.3	3386.5
	H	697.061	2762.0	3112.1	3164.3	3269.0	3374.7	3481.6	3589.9	3699.9	3811.4
	S	1.9918	6.7052	7.3890	7.4745	7.6362	7.7875	7.9305	8.0663	8.1959	8.3201
725 (166.38)	V	1.110	263.77	375.01	391.54	424.33	456.90	489.31	521.61	553.83	585.99
	U	702.457	2572.2	2839.7	2880.0	2961.0	3043.1	3126.6	3211.5	3298.1	3386.4
	H	703.261	2763.4	3111.5	3163.8	3268.7	3374.3	3481.3	3589.7	3699.7	3811.2
	S	2.0059	6.6932	7.3721	7.4578	7.6196	7.7710	7.9140	8.0499	8.1796	8.3038

TABLE C.2. SUPERHEATED STEAM SI UNITS (Continued)

ABS PRESS KPA (SAT TEMP) DEG C		SAT WATER	SAT STEAM	TEMPERATURE, DEG C (TEMPERATURE, K) 175 (448.15)	200 (473.15)	220 (493.15)	240 (513.15)	260 (533.15)	280 (553.15)	300 (573.15)	325 (598.15)
750 (167.76)	V	1.112	255.43	260.88	279.05	293.03	306.65	320.01	333.17	346.19	362.32
	U	708.467	2573.3	2586.9	2632.1	2666.8	2700.6	2733.7	2766.4	2798.9	2839.3
	H	709.301	2764.8	2782.5	2841.4	2886.6	2930.6	2973.7	3016.3	3058.5	3111.0
	S	2.0195	6.6817	6.7215	6.8494	6.9429	7.0303	7.1128	7.1912	7.2662	7.3558
775 (169.10)	V	1.113	247.61	251.93	269.63	283.22	296.45	309.41	322.19	334.81	350.44
	U	714.326	2574.3	2585.4	2631.0	2665.9	2699.8	2733.1	2765.9	2798.4	2838.9
	H	715.189	2766.2	2780.7	2840.0	2885.4	2929.6	2972.9	3015.6	3057.9	3110.5
	S	2.0328	6.6705	6.7031	6.8319	6.9259	7.0137	7.0965	7.1751	7.2502	7.3400
800 (170.41)	V	1.115	240.26	243.53	260.79	274.02	286.88	299.48	311.89	324.14	339.31
	U	720.043	2575.3	2584.0	2629.9	2665.0	2699.1	2732.5	2765.4	2797.9	2838.5
	H	720.935	2767.5	2778.8	2838.6	2884.2	2928.6	2972.1	3014.9	3057.3	3109.9
	S	2.0457	6.6596	6.6851	6.8148	6.9094	6.9976	7.0807	7.1595	7.2348	7.3247
825 (171.69)	V	1.117	233.34	235.64	252.48	265.37	277.90	290.15	302.21	314.12	328.85
	U	725.625	2576.2	2582.5	2628.8	2664.1	2698.4	2731.8	2764.8	2797.5	2838.1
	H	726.547	2768.7	2776.9	2837.1	2883.1	2927.6	2971.2	3014.1	3056.6	3109.4
	S	2.0583	6.6491	6.6675	6.7982	6.8933	6.9819	7.0653	7.1443	7.2197	7.3098
850 (172.94)	V	1.118	226.81	228.21	244.66	257.24	269.44	281.37	293.10	304.68	319.00
	U	731.080	2577.1	2581.1	2627.7	2663.2	2697.6	2731.2	2764.3	2797.0	2837.7
	H	732.031	2769.9	2775.1	2835.7	2881.9	2926.6	2970.4	3013.4	3056.0	3108.8
	S	2.0705	6.6388	6.6504	6.7820	6.8777	6.9666	7.0503	7.1295	7.2051	7.2954
875 (174.16)	V	1.120	220.65	221.20	237.29	249.56	261.46	273.09	284.51	295.79	309.72
	U	736.415	2578.0	2579.6	2626.6	2662.3	2696.8	2730.6	2763.7	2796.5	2837.3
	H	737.394	2771.0	2773.1	2834.2	2880.7	2925.6	2969.5	3012.7	3055.3	3108.3
	S	2.0825	6.6289	6.6336	6.7662	6.8624	6.9518	7.0357	7.1152	7.1909	7.2813
900 (175.36)	V	1.121	214.81	....	230.32	242.31	253.93	265.27	276.40	287.39	300.96
	U	741.635	2578.8	....	2625.5	2661.4	2696.1	2729.9	2763.2	2796.1	2836.9
	H	742.644	2772.1	....	2832.7	2879.5	2924.6	2968.7	3012.0	3054.7	3107.7
	S	2.0941	6.6192	....	6.7508	6.8475	6.9373	7.0215	7.1012	7.1771	7.2676
925 (176.53)	V	1.123	209.28	....	223.73	235.46	246.80	257.87	268.73	279.44	292.66
	U	746.746	2579.6	....	2624.3	2660.5	2695.3	2729.3	2762.6	2795.6	2836.5
	H	747.784	2773.2	....	2831.3	2878.3	2923.6	2967.8	3011.2	3054.1	3107.2
	S	2.1055	6.6097	....	6.7357	6.8329	6.9231	7.0076	7.0875	7.1636	7.2543

P (T sat)		Sat. liq.	Sat. vap.								
950	V	1.124	204.03	....	217.48	228.96	240.05	250.86	261.46	271.91	284.81
	U	751.754	2580.4	....	2623.2	2659.5	2694.6	2728.7	2762.1	2795.1	2836.0
(177.67)	H	752.822	2774.2	....	2829.8	2877.0	2922.6	2967.0	3010.5	3053.4	3106.6
	S	2.1166	6.6005	....	6.7209	6.8187	6.9093	6.9941	7.0742	7.1505	7.2413
975	V	1.126	199.04	....	211.55	222.79	233.64	244.20	254.56	264.76	277.35
	U	756.663	2581.1	....	2622.0	2658.6	2693.8	2728.0	2761.5	2794.6	2835.6
(178.79)	H	757.761	2775.2	....	2828.3	2875.8	2921.6	2966.1	3009.7	3052.8	3106.1
	S	2.1275	6.5916	....	6.7064	6.8048	6.8958	6.9809	7.0612	7.1377	7.2286
1000	V	1.127	194.29	....	205.92	216.93	227.55	237.89	248.01	257.98	270.27
	U	761.478	2581.9	....	2620.9	2657.7	2693.0	2727.4	2761.0	2794.2	2835.2
(179.88)	H	762.605	2776.2	....	2826.8	2874.6	2920.6	2965.2	3009.0	3052.1	3105.5
	S	2.1382	6.5828	....	6.6922	6.7911	6.8825	6.9680	7.0485	7.1251	7.2163
1050	V	1.130	185.45	....	195.45	206.04	216.24	226.15	235.84	245.37	257.12
	U	770.843	2583.3	....	2618.5	2655.8	2691.5	2726.1	2759.9	2793.2	2834.4
(182.02)	H	772.029	2778.0	....	2823.8	2872.1	2918.5	2963.5	3007.5	3050.8	3104.4
	S	2.1588	6.5659	....	6.6645	6.7647	6.8569	6.9430	7.0240	7.1009	7.1924
1100	V	1.133	177.38	....	185.92	196.14	205.96	215.47	224.77	233.91	245.16
	U	779.878	2584.5	....	2616.2	2653.9	2689.9	2724.7	2758.8	2792.2	2833.6
(184.07)	H	781.124	2779.7	....	2820.7	2869.6	2916.4	2961.8	3006.0	3049.6	3103.3
	S	2.1786	6.5497	....	6.6379	6.7392	6.8323	6.9190	7.0005	7.0778	7.1695
1150	V	1.136	169.99	....	177.22	187.10	196.56	205.73	214.67	223.44	234.25
	U	788.611	2585.8	....	2613.8	2651.9	2688.3	2723.4	2757.7	2791.3	2832.8
(186.05)	H	789.917	2781.3	....	2817.6	2867.1	2914.4	2960.0	3004.5	3048.2	3102.2
	S	2.1977	6.5342	....	6.6122	6.7147	6.8086	6.8959	6.9779	7.0556	7.1476
1200	V	1.139	163.20	....	169.23	178.80	187.95	196.79	205.40	213.85	224.24
	U	797.064	2586.9	....	2611.3	2650.0	2686.7	2722.1	2756.5	2790.3	2832.0
(187.96)	H	798.430	2782.7	....	2814.4	2864.5	2912.2	2958.2	3003.0	3046.9	3101.0
	S	2.2161	6.5194	....	6.5872	6.6909	6.7858	6.8738	6.9562	7.0342	7.1266
1250	V	1.141	156.93	....	161.88	171.17	180.02	188.56	196.88	205.02	215.03
	U	805.259	2588.0	....	2608.9	2648.0	2685.1	2720.8	2755.4	2789.3	2831.1
(189.81)	H	806.685	2784.1	....	2811.2	2861.9	2910.1	2956.5	3001.5	3045.6	3099.9
	S	2.2338	6.5050	....	6.5630	6.6680	6.7637	6.8523	6.9353	7.0136	7.1064
1300	V	1.144	151.13	....	155.09	164.11	172.70	180.97	189.01	196.87	206.53
	U	813.213	2589.0	....	2606.4	2646.0	2683.5	2719.4	2754.3	2788.4	2830.3
(191.61)	H	814.700	2785.4	....	2808.0	2859.3	2908.0	2954.7	3000.0	3044.3	3098.8
	S	2.2510	6.4913	....	6.5394	6.6457	6.7424	6.8316	6.9151	6.9938	7.0869

TABLE C.2. SUPERHEATED STEAM SI UNITS (Continued)

ABS PRESS KPA (SAT TEMP) DEG C		SAT WATER	SAT STEAM	350 (623.15)	375 (648.15)	400 (673.15)	450 (723.15)	500 (773.15)	550 (823.15)	600 (873.15)	650 (923.15)
				TEMPERATURE, DEG C (TEMPERATURE, K)							
750 (167.76)	V	1.112	255.43	378.31	394.22	410.05	441.55	472.90	504.15	535.30	566.40
	U	708.467	2573.3	2879.6	2920.1	2960.7	3042.9	3126.3	3211.4	3298.0	3386.2
	H	709.301	2764.8	3163.4	3215.7	3268.3	3374.0	3481.0	3589.5	3699.5	3811.0
	S	2.0195	6.6817	7.4416	7.5240	7.6035	7.7550	7.8981	8.0340	8.1637	8.2880
775 (169.10)	V	1.113	247.61	365.94	381.35	396.69	427.20	457.56	487.81	517.97	548.07
	U	714.326	2574.3	2879.3	2919.8	2960.4	3042.6	3126.1	3211.2	3297.8	3386.1
	H	715.189	2766.2	3162.9	3215.3	3267.9	3373.7	3480.8	3589.2	3699.3	3810.9
	S	2.0328	6.6705	7.4259	7.5084	7.5880	7.7396	7.8827	8.0187	8.1484	8.2727
800 (170.41)	V	1.115	240.26	354.34	369.29	384.16	413.74	443.17	472.49	501.72	530.89
	U	720.043	2575.3	2878.9	2919.5	2960.2	3042.4	3125.9	3211.0	3297.7	3386.0
	H	720.935	2767.5	3162.4	3214.9	3267.5	3373.4	3480.5	3589.0	3699.1	3810.7
	S	2.0457	6.6596	7.4107	7.4932	7.5729	7.7246	7.8678	8.0038	8.1336	8.2579
825 (171.69)	V	1.117	233.34	343.45	357.96	372.39	401.10	429.65	458.10	486.46	514.76
	U	725.625	2576.2	2878.6	2919.1	2959.9	3042.2	3125.7	3210.8	3297.5	3385.8
	H	726.547	2768.7	3161.9	3214.5	3267.1	3373.1	3480.2	3588.8	3698.8	3810.5
	S	2.0583	6.6491	7.3959	7.4786	7.5583	7.7101	7.8533	7.9894	8.1192	8.2436
850 (172.94)	V	1.118	226.81	333.20	347.29	361.31	389.20	416.93	444.56	472.09	499.57
	U	731.080	2577.1	2878.2	2918.8	2959.6	3041.9	3125.5	3210.7	3297.4	3385.7
	H	732.031	2769.9	3161.4	3214.0	3266.7	3372.7	3479.9	3588.5	3698.6	3810.3
	S	2.0705	6.6388	7.3815	7.4643	7.5441	7.6960	7.8393	7.9754	8.1053	8.2296
875 (174.16)	V	1.120	220.65	323.53	337.24	350.87	377.98	404.94	431.79	458.55	485.25
	U	736.415	2578.0	2877.9	2918.5	2959.3	3041.7	3125.3	3210.5	3297.2	3385.6
	H	737.394	2771.0	3161.0	3213.6	3266.3	3372.4	3479.7	3588.3	3698.4	3810.2
	S	2.0825	6.6289	7.3676	7.4504	7.5303	7.6823	7.8257	7.9618	8.0917	8.2161
900 (175.36)	V	1.121	214.81	314.40	327.74	341.01	367.39	393.61	419.73	445.76	471.72
	U	741.635	2578.8	2877.5	2918.2	2959.0	3041.4	3125.1	3210.3	3297.1	3385.4
	H	742.644	2772.1	3160.5	3213.2	3266.0	3372.1	3479.4	3588.1	3698.2	3810.0
	S	2.0941	6.6192	7.3540	7.4370	7.5169	7.6689	7.8124	7.9486	8.0785	8.2030
925 (176.53)	V	1.123	209.28	305.76	318.75	331.68	357.36	382.90	408.32	433.66	458.93
	U	746.746	2579.6	2877.2	2917.9	2958.8	3041.2	3124.9	3210.1	3296.9	3385.3
	H	747.784	2773.2	3160.0	3212.7	3265.6	3371.8	3479.1	3587.8	3698.0	3809.8
	S	2.1055	6.6097	7.3408	7.4238	7.5038	7.6560	7.7995	7.9357	8.0657	8.1902

950 (177.67)	V	1.124	204.03	297.57	310.24	322.84	347.87	372.74	397.51	422.19	446.81
	U	751.754	2580.4	2876.8	2917.6	2958.5	3041.0	3124.7	3209.9	3296.7	3385.1
	H	752.822	2774.2	3159.5	3212.3	3265.2	3371.5	3478.8	3587.6	3697.8	3809.6
	S	2.1166	6.6005	7.3279	7.4110	7.4911	7.6433	7.7869	7.9232	8.0532	8.1777
975 (178.79)	V	1.126	199.04	289.81	302.17	314.45	338.86	363.11	387.26	411.32	435.31
	U	756.663	2581.1	2876.5	2917.3	2958.2	3040.7	3124.5	3209.8	3296.6	3385.0
	H	757.761	2775.2	3159.0	3211.9	3264.8	3371.1	3478.6	3587.3	3697.6	3809.4
	S	2.1275	6.5916	7.3154	7.3986	7.4787	7.6310	7.7747	7.9110	8.0410	8.1656
1000 (179.88)	V	1.127	194.29	282.43	294.50	306.49	330.30	353.96	377.52	400.98	424.38
	U	761.478	2581.9	2876.1	2917.0	2957.9	3040.5	3124.3	3209.6	3296.4	3384.9
	H	762.605	2776.2	3158.5	3211.5	3264.4	3370.8	3478.3	3587.1	3697.4	3809.3
	S	2.1382	6.5828	7.3031	7.3864	7.4665	7.6190	7.7627	7.8991	8.0292	8.1537
1050 (182.02)	V	1.130	185.45	268.74	280.25	291.69	314.41	336.97	359.43	381.79	404.10
	U	770.843	2583.3	2875.4	2916.3	2957.4	3040.0	3123.9	3209.2	3296.1	3384.6
	H	772.029	2778.0	3157.6	3210.6	3263.6	3370.2	3477.7	3586.6	3697.0	3808.9
	S	2.1588	6.5659	7.2795	7.3629	7.4432	7.5958	7.7397	7.8762	8.0063	8.1309
1100 (184.07)	V	1.133	177.38	256.28	267.30	278.24	299.96	321.53	342.98	364.35	385.65
	U	779.878	2584.5	2874.7	2915.7	2956.8	3039.6	3123.5	3208.9	3295.8	3384.3
	H	781.124	2779.7	3156.6	3209.7	3262.9	3369.5	3477.2	3586.2	3696.6	3808.5
	S	2.1786	6.5497	7.2569	7.3405	7.4209	7.5737	7.7177	7.8543	7.9845	8.1092
1150 (186.05)	V	1.136	169.99	244.91	255.47	265.96	286.77	307.42	327.97	348.42	368.81
	U	788.611	2585.8	2874.0	2915.1	2956.2	3039.1	3123.1	3208.5	3295.5	3384.1
	H	789.917	2781.3	3155.6	3208.9	3262.1	3368.9	3476.6	3585.7	3696.2	3808.2
	S	2.1977	6.5342	7.2352	7.3190	7.3995	7.5525	7.6966	7.8333	7.9636	8.0883
1200 (187.96)	V	1.139	163.20	234.49	244.63	254.70	274.68	294.50	314.20	333.82	353.38
	U	797.064	2586.9	2873.3	2914.4	2955.7	3038.6	3122.7	3208.2	3295.2	3383.8
	H	798.430	2782.7	3154.6	3208.0	3261.3	3368.2	3476.1	3585.2	3695.8	3807.8
	S	2.2161	6.5194	7.2144	7.2983	7.3790	7.5323	7.6765	7.8132	7.9436	8.0684
1250 (189.81)	V	1.141	156.93	224.90	234.66	244.35	263.55	282.60	301.54	320.39	339.18
	U	805.259	2588.0	2872.5	2913.8	2955.1	3038.1	3122.3	3207.8	3294.9	3383.5
	H	806.685	2784.1	3153.7	3207.1	3260.5	3367.6	3475.5	3584.7	3695.4	3807.5
	S	2.2338	6.5050	7.1944	7.2785	7.3593	7.5128	7.6571	7.7940	7.9244	8.0493
1300 (191.61)	V	1.144	151.13	216.05	225.46	234.79	253.28	271.62	289.85	307.99	326.07
	U	813.213	2589.0	2871.8	2913.2	2954.5	3037.7	3121.9	3207.5	3294.6	3383.2
	H	814.700	2785.4	3152.7	3206.3	3259.7	3366.9	3475.0	3584.3	3695.0	3807.1
	S	2.2510	6.4913	7.1751	7.2594	7.3404	7.4940	7.6385	7.7754	7.9060	8.0309

TABLE C.2. SUPERHEATED STEAM SI UNITS (Continued)

ABS PRESS KPA (SAT TEMP) DEG C		SAT WATER	SAT STEAM	TEMPERATURE, DEG C (TEMPERATURE, K) 200 (473.15)	225 (498.15)	250 (523.15)	275 (548.15)	300 (573.15)	325 (598.15)	350 (623.15)	375 (648.15)
	V	1.146	145.74	148.79	159.70	169.96	179.79	189.33	198.66	207.85	216.93
1350	U	820.944	2589.9	2603.9	2653.6	2700.1	2744.4	2787.4	2829.5	2871.1	2912.5
(193.35)	H	822.491	2786.6	2804.7	2869.2	2929.5	2987.1	3043.0	3097.7	3151.7	3205.4
	S	2.2676	6.4780	6.5165	6.6493	6.7675	6.8750	6.9746	7.0681	7.1566	7.2410
	V	1.149	140.72	142.94	153.57	163.55	173.08	182.32	191.35	200.24	209.02
1400	U	828.465	2590.8	2601.3	2651.7	2698.6	2743.2	2786.4	2828.6	2870.4	2911.9
(195.04)	H	830.074	2787.8	2801.4	2866.7	2927.6	2985.5	3041.6	3096.5	3150.7	3204.5
	S	2.2837	6.4651	6.4941	6.6285	6.7477	6.8560	6.9561	7.0499	7.1386	7.2233
	V	1.151	136.04	137.48	147.86	157.57	166.83	175.79	184.54	193.15	201.65
1450	U	835.791	2591.6	2598.7	2649.7	2697.1	2742.0	2785.4	2827.8	2869.7	2911.3
(196.69)	H	837.460	2788.9	2798.1	2864.1	2925.5	2983.9	3040.3	3095.4	3149.7	3203.6
	S	2.2993	6.4526	6.4722	6.6082	6.7286	6.8376	6.9381	7.0322	7.1212	7.2061
	V	1.154	131.66	132.38	142.53	151.99	161.00	169.70	178.19	186.53	194.77
1500	U	842.933	2592.4	2596.1	2647.7	2695.5	2740.8	2784.4	2826.9	2868.9	2910.6
(198.29)	H	844.663	2789.9	2794.7	2861.5	2923.5	2982.3	3038.9	3094.2	3148.7	3202.8
	S	2.3145	6.4406	6.4508	6.5885	6.7099	6.8196	6.9207	7.0152	7.1044	7.1894
	V	1.156	127.55	127.61	137.54	146.77	155.54	164.00	172.25	180.34	188.33
1550	U	849.901	2593.2	2593.5	2645.8	2694.0	2739.5	2783.4	2826.1	2868.2	2910.0
(199.85)	H	851.694	2790.8	2791.3	2858.9	2921.5	2980.6	3037.6	3093.1	3147.7	3201.9
	S	2.3292	6.4289	6.4298	6.5692	6.6917	6.8022	6.9038	6.9986	7.0881	7.1733
	V	1.159	123.69	....	132.85	141.87	150.42	158.66	166.68	174.54	182.30
1600	U	856.707	2593.8	....	2643.7	2692.4	2738.3	2782.4	2825.2	2867.5	2909.3
(201.37)	H	858.561	2791.7	....	2856.3	2919.4	2979.0	3036.2	3091.9	3146.7	3201.0
	S	2.3436	6.4175	....	6.5503	6.6740	6.7852	6.8873	6.9825	7.0723	7.1577
	V	1.161	120.05	....	128.45	137.27	145.61	153.64	161.44	169.09	176.63
1650	U	863.359	2594.5	....	2641.7	2690.9	2737.1	2781.3	2824.4	2866.7	2908.7
(202.86)	H	865.275	2792.6	....	2853.6	2917.4	2977.3	3034.8	3090.8	3145.7	3200.1
	S	2.3576	6.4065	....	6.5319	6.6567	6.7687	6.8713	6.9669	7.0569	7.1425
	V	1.163	116.62	....	124.31	132.94	141.09	148.91	156.51	163.96	171.30
1700	U	869.866	2595.1	....	2639.6	2689.3	2735.8	2780.3	2823.5	2866.0	2908.0
(204.31)	H	871.843	2793.4	....	2851.0	2915.3	2975.6	3033.5	3089.6	3144.7	3199.2
	S	2.3713	6.3957	....	6.5138	6.6398	6.7526	6.8557	6.9516	7.0419	7.1277

1750 (205.72)	V	1.166	113.38	....	120.39	128.85	136.82	144.45	151.87	159.12	166.27
	U	876.234	2595.7	....	2637.6	2687.7	2734.5	2779.3	2822.7	2865.3	2907.4
	H	878.274	2794.1	....	2848.2	2913.2	2974.0	3032.1	3088.4	3143.7	3198.4
	S	2.3846	6.3853	....	6.4961	6.6233	6.7368	6.8405	6.9368	7.0273	7.1133
1800 (207.11)	V	1.168	110.32	....	116.69	124.99	132.78	140.24	147.48	154.55	161.51
	U	882.472	2596.3	....	2635.5	2686.1	2733.3	2778.2	2821.8	2864.5	2906.7
	H	884.574	2794.8	....	2845.5	2911.0	2972.3	3030.7	3087.3	3142.7	3197.5
	S	2.3976	6.3751	....	6.4787	6.6071	6.7214	6.8257	6.9223	7.0131	7.0993
1850 (208.47)	V	1.170	107.41	....	113.19	121.33	128.96	136.26	143.33	150.23	157.02
	U	888.585	2596.8	....	2633.3	2684.4	2732.0	2777.2	2820.9	2863.8	2906.1
	H	890.750	2795.5	....	2842.8	2908.9	2970.6	3029.3	3086.1	3141.7	3196.6
	S	2.4103	6.3651	....	6.4616	6.5912	6.7064	6.8112	6.9082	6.9993	7.0856
1900 (209.80)	V	1.172	104.65	....	109.87	117.87	125.35	132.49	139.39	146.14	152.76
	U	894.580	2597.3	....	2631.2	2682.8	2730.7	2776.2	2820.1	2863.0	2905.4
	H	896.807	2796.1	....	2840.0	2906.7	2968.8	3027.9	3084.9	3140.7	3195.7
	S	2.4228	6.3554	....	6.4448	6.5757	6.6917	6.7970	6.8944	6.9857	7.0723
1950 (211.10)	V	1.174	102.031	....	106.72	114.58	121.91	128.90	135.66	142.25	148.72
	U	900.461	2597.7	....	2629.0	2681.1	2729.4	2775.1	2819.2	2862.3	2904.8
	H	902.752	2796.7	....	2837.1	2904.6	2967.1	3026.5	3083.7	3139.7	3194.8
	S	2.4349	6.3459	....	6.4283	6.5604	6.6772	6.7831	6.8809	6.9725	7.0593
2000 (212.37)	V	1.177	99.536	....	103.72	111.45	118.65	125.50	132.11	138.56	144.89
	U	906.236	2598.2	....	2626.9	2679.5	2728.1	2774.0	2818.3	2861.5	2904.1
	H	908.589	2797.2	....	2834.3	2902.4	2965.4	3025.0	3082.5	3138.6	3193.9
	S	2.4469	6.3366	....	6.4120	6.5454	6.6631	6.7696	6.8677	6.9596	7.0466
2100 (214.85)	V	1.181	94.890	....	98.147	105.64	112.59	119.18	125.53	131.70	137.76
	U	917.479	2598.9	....	2622.4	2676.1	2725.4	2771.9	2816.5	2860.0	2902.8
	H	919.959	2798.2	....	2828.5	2897.9	2961.9	3022.2	3080.1	3136.6	3192.1
	S	2.4700	6.3187	....	6.3802	6.5162	6.6356	6.7432	6.8422	6.9347	7.0220
2200 (217.24)	V	1.185	90.652	....	93.067	100.35	107.07	113.43	119.53	125.47	131.28
	U	928.346	2599.6	....	2617.9	2672.7	2722.7	2769.7	2814.7	2858.5	2901.5
	H	930.953	2799.1	....	2822.7	2893.4	2958.3	3019.3	3077.7	3134.5	3190.3
	S	2.4922	6.3015	....	6.3492	6.4879	6.6091	6.7179	6.8177	6.9107	6.9985
2300 (219.55)	V	1.189	86.769	....	88.420	95.513	102.03	108.18	114.06	119.77	125.36
	U	938.866	2600.2	....	2613.3	2669.2	2720.0	2767.6	2812.9	2857.0	2900.2
	H	941.601	2799.8	....	2816.7	2888.9	2954.7	3016.4	3075.3	3132.4	3188.5
	S	2.5136	6.2849	....	6.3190	6.4605	6.5835	6.6935	6.7941	6.8877	6.9759

TABLE C.2. SUPERHEATED STEAM SI UNITS (Continued)

ABS PRESS KPA (SAT TEMP) DEG C		SAT WATER	SAT STEAM	TEMPERATURE, DEG C (TEMPERATURE, K) 400 (673.15)	425 (698.15)	450 (723.15)	475 (748.15)	500 (773.15)	550 (823.15)	600 (873.15)	650 (923.15)
1350 (193.35)	V	1.146	145.74	225.94	234.88	243.78	252.63	261.46	279.03	296.51	313.93
	U	820.944	2589.9	2953.9	2995.5	3037.2	3079.2	3121.5	3207.1	3294.3	3383.0
	H	822.491	2786.6	3259.0	3312.6	3366.3	3420.2	3474.4	3583.8	3694.5	3806.8
	S	2.2676	6.4780	7.3221	7.4003	7.4759	7.5493	7.6205	7.7576	7.8882	8.0132
1400 (195.04)	V	1.149	140.72	217.72	226.35	234.95	243.50	252.02	268.98	285.85	302.66
	U	828.465	2590.8	2953.4	2994.9	3036.7	3078.7	3121.1	3206.8	3293.9	3382.7
	H	830.074	2787.8	3258.2	3311.8	3365.6	3419.6	3473.9	3583.3	3694.1	3806.4
	S	2.2837	6.4651	7.3045	7.3828	7.4585	7.5319	7.6032	7.7404	7.8710	7.9961
1450 (196.69)	V	1.151	136.04	210.06	218.42	226.72	234.99	243.23	259.62	275.93	292.16
	U	835.791	2591.6	2952.8	2994.4	3036.2	3078.3	3120.7	3206.4	3293.6	3382.4
	H	837.460	2788.9	3257.4	3311.1	3365.0	3419.0	3473.3	3582.9	3693.7	3806.1
	S	2.2993	6.4526	7.2874	7.3658	7.4416	7.5151	7.5865	7.7237	7.8545	7.9796
1500 (198.29)	V	1.154	131.66	202.92	211.01	219.05	227.06	235.03	250.89	266.66	282.37
	U	842.933	2592.4	2952.2	2993.9	3035.8	3077.9	3120.3	3206.0	3293.3	3382.1
	H	844.663	2789.9	3256.6	3310.4	3364.3	3418.4	3472.8	3582.4	3693.3	3805.7
	S	2.3145	6.4406	7.2709	7.3494	7.4253	7.4989	7.5703	7.7077	7.8385	7.9636
1550 (199.85)	V	1.156	127.55	196.24	204.08	211.87	219.63	227.35	242.72	258.00	273.21
	U	849.901	2593.2	2951.7	2993.4	3035.3	3077.4	3119.8	3205.7	3293.0	3381.9
	H	851.694	2790.8	3255.8	3309.7	3363.7	3417.8	3472.2	3581.9	3692.9	3805.3
	S	2.3292	6.4289	7.2550	7.3336	7.4095	7.4832	7.5547	7.6921	7.8230	7.9482
1600 (201.37)	V	1.159	123.69	189.97	197.58	205.15	212.67	220.16	235.06	249.87	264.62
	U	856.707	2593.8	2951.1	2992.9	3034.8	3077.0	3119.4	3205.3	3292.7	3381.6
	H	858.561	2791.7	3255.0	3309.0	3363.0	3417.2	3471.7	3581.4	3692.5	3805.0
	S	2.3436	6.4175	7.2394	7.3182	7.3942	7.4679	7.5395	7.6770	7.8080	7.9333
1650 (202.86)	V	1.161	120.05	184.09	191.48	198.82	206.13	213.40	227.86	242.24	256.55
	U	863.359	2594.5	2950.5	2992.3	3034.3	3076.5	3119.0	3205.0	3292.4	3381.3
	H	865.275	2792.6	3254.2	3308.3	3362.4	3416.7	3471.1	3581.0	3692.1	3804.6
	S	2.3576	6.4065	7.2244	7.3032	7.3794	7.4531	7.5248	7.6624	7.7934	7.9188
1700 (204.31)	V	1.163	116.62	178.55	185.74	192.87	199.97	207.04	221.09	235.06	248.96
	U	869.866	2595.1	2949.9	2991.8	3033.9	3076.1	3118.6	3204.6	3292.1	3381.0
	H	871.843	2793.4	3253.5	3307.6	3361.7	3416.1	3470.6	3580.5	3691.7	3804.3
	S	2.3713	6.3957	7.2098	7.2887	7.3649	7.4388	7.5105	7.6482	7.7793	7.9047

1750 (205.72)	V	1.166	113.38	173.32	180.32	187.26	194.17	201.04	214.71	228.28	241.80
	U	876.234	2595.7	2949.3	2991.3	3033.4	3075.7	3118.2	3204.3	3291.8	3380.8
	H	878.274	2794.1	3252.7	3306.9	3361.1	3415.5	3470.0	3580.0	3691.3	3803.9
	S	2.3846	6.3853	7.1955	7.2746	7.3509	7.4248	7.4965	7.6344	7.7656	7.8910
1800 (207.11)	V	1.168	110.32	168.39	175.20	181.97	188.69	195.38	208.68	221.89	235.03
	U	882.472	2596.3	2948.8	2990.8	3032.9	3075.2	3117.8	3203.9	3291.5	3380.5
	H	884.574	2794.8	3251.9	3306.1	3360.4	3414.9	3469.5	3579.5	3690.9	3803.6
	S	2.3976	6.3751	7.1816	7.2608	7.3372	7.4112	7.4830	7.6209	7.7522	7.8777
1850 (208.47)	V	1.170	107.41	163.73	170.37	176.96	183.50	190.02	202.97	215.84	228.64
	U	888.585	2596.8	2948.2	2990.3	3032.4	3074.8	3117.4	3203.6	3291.1	3380.2
	H	890.750	2795.5	3251.1	3305.4	3359.8	3414.3	3468.9	3579.1	3690.4	3803.2
	S	2.4103	6.3651	7.1681	7.2474	7.3239	7.3980	7.4698	7.6079	7.7392	7.8648
1900 (209.80)	V	1.172	104.65	159.30	165.78	172.21	178.59	184.94	197.57	210.11	222.58
	U	894.580	2597.3	2947.6	2989.7	3031.9	3074.3	3117.0	3203.2	3290.8	3380.0
	H	896.807	2796.1	3250.3	3304.7	3359.1	3413.7	3468.4	3578.6	3690.0	3802.8
	S	2.4228	6.3554	7.1550	7.2344	7.3109	7.3851	7.4570	7.5951	7.7265	7.8522
1950 (211.10)	V	1.174	102.031	155.11	161.43	167.70	173.93	180.13	192.44	204.67	216.83
	U	900.461	2597.7	2947.0	2989.2	3031.5	3073.9	3116.6	3202.9	3290.5	3379.7
	H	902.752	2796.7	3249.5	3304.0	3358.5	3413.1	3467.8	3578.1	3689.6	3802.5
	S	2.4349	6.3459	7.1421	7.2216	7.2983	7.3725	7.4445	7.5827	7.7142	7.8399
2000 (212.37)	V	1.177	99.536	151.13	157.30	163.42	169.51	175.55	187.57	199.50	211.36
	U	906.236	2598.2	2946.4	2988.7	3031.0	3073.5	3116.2	3202.5	3290.2	3379.4
	H	908.589	2797.2	3248.7	3303.3	3357.8	3412.5	3467.3	3577.6	3689.2	3802.1
	S	2.4469	6.3366	7.1296	7.2092	7.2859	7.3602	7.4323	7.5706	7.7022	7.8279
2100 (214.85)	V	1.181	94.890	143.73	149.63	155.48	161.28	167.06	178.53	189.91	201.22
	U	917.479	2598.9	2945.3	2987.6	3030.0	3072.6	3115.3	3201.8	3289.6	3378.9
	H	919.959	2798.2	3247.1	3301.8	3356.5	3411.3	3466.2	3576.7	3688.4	3801.4
	S	2.4700	6.3187	7.1053	7.1851	7.2621	7.3365	7.4087	7.5472	7.6789	7.8048
2200 (217.24)	V	1.185	90.652	137.00	142.65	148.25	153.81	159.34	170.30	181.19	192.00
	U	928.346	2599.6	2944.1	2986.6	3029.1	3071.7	3114.5	3201.1	3289.0	3378.3
	H	930.953	2799.1	3245.5	3300.4	3355.2	3410.1	3465.1	3575.7	3687.6	3800.7
	S	2.4922	6.3015	7.0821	7.1621	7.2393	7.3139	7.3862	7.5249	7.6568	7.7827
2300 (219.55)	V	1.189	86.769	130.85	136.28	141.65	146.99	152.28	162.80	173.22	183.58
	U	938.866	2600.2	2942.9	2985.5	3028.1	3070.8	3113.7	3200.4	3288.3	3377.8
	H	941.601	2799.8	3243.9	3299.0	3353.9	3408.9	3464.0	3574.8	3686.7	3800.0
	S	2.5136	6.2849	7.0598	7.1401	7.2174	7.2922	7.3646	7.5035	7.6355	7.7616

TABLE C.2. SUPERHEATED STEAM SI UNITS (Continued)

ABS PRESS KPA (SAT TEMP) DEG C		SAT WATER	SAT STEAM	225 (498.15)	250 (523.15)	275 (548.15)	300 (573.15)	325 (598.15)	350 (623.15)	375 (648.15)	400 (673.15)
				TEMPERATURE, DEG C (TEMPERATURE, K)							
2400	V	1.193	83.199	84.149	91.075	97.411	103.36	109.05	114.55	119.93	125.22
(221.78)	U	949.066	2600.7	2608.6	2665.6	2717.3	2765.4	2811.1	2855.4	2898.8	2941.7
	H	951.929	2800.4	2810.6	2884.2	2951.1	3013.4	3072.8	3130.4	3186.7	3242.3
	S	2.5343	6.2690	6.2894	6.4338	6.5586	6.6699	6.7714	6.8656	6.9542	7.0384
2500	V	1.197	79.905	80.210	86.985	93.154	98.925	104.43	109.75	114.94	120.04
(223.94)	U	958.969	2601.2	2603.8	2662.0	2714.5	2763.1	2809.3	2853.9	2897.5	2940.6
	H	961.962	2800.9	2804.3	2879.5	2947.4	3010.4	3070.4	3128.2	3184.8	3240.7
	S	2.5543	6.2536	6.2604	6.4077	6.5345	6.6470	6.7494	6.8442	6.9333	7.0178
2600	V	1.201	76.856	....	83.205	89.220	94.830	100.17	105.32	110.33	115.26
(226.04)	U	968.597	2601.5	....	2658.4	2711.7	2760.9	2807.4	2852.3	2896.1	2939.4
	H	971.720	2801.4	....	2874.7	2943.6	3007.4	3067.9	3126.1	3183.0	3239.0
	S	2.5736	6.2387	....	6.3823	6.5110	6.6249	6.7281	6.8236	6.9131	6.9979
2700	V	1.205	74.025	....	79.698	85.575	91.036	96.218	101.21	106.07	110.83
(228.07)	U	977.968	2601.8	....	2654.7	2708.8	2758.6	2805.6	2850.7	2894.8	2938.2
	H	981.222	2801.7	....	2869.9	2939.8	3004.4	3065.4	3124.0	3181.2	3237.4
	S	2.5924	6.2244	....	6.3575	6.4882	6.6034	6.7075	6.8036	6.8935	6.9787
2800	V	1.209	71.389	....	76.437	82.187	87.510	92.550	97.395	102.10	106.71
(230.05)	U	987.100	2602.1	....	2650.9	2705.9	2756.3	2803.7	2849.2	2893.4	2937.0
	H	990.485	2802.0	....	2864.9	2936.0	3001.3	3062.8	3121.9	3179.3	3235.8
	S	2.6106	6.2104	....	6.3331	6.4659	6.5824	6.6875	6.7842	6.8746	6.9601
2900	V	1.213	68.928	....	73.395	79.029	84.226	89.133	93.843	98.414	102.88
(231.97)	U	996.008	2602.3	....	2647.1	2702.9	2754.0	2801.8	2847.6	2892.0	2935.8
	H	999.524	2802.2	....	2859.9	2932.1	2998.2	3060.3	3119.7	3177.4	3234.1
	S	2.6283	6.1969	....	6.3092	6.4441	6.5621	6.6681	6.7654	6.8563	6.9421
3000	V	1.216	66.626	....	70.551	76.078	81.159	85.943	90.526	94.969	99.310
(233.84)	U	1004.7	2602.4	....	2643.2	2700.0	2751.6	2799.9	2846.0	2890.7	2934.6
	H	1008.4	2802.3	....	2854.8	2928.2	2995.1	3057.7	3117.5	3175.6	3232.5
	S	2.6455	6.1837	....	6.2857	6.4228	6.5422	6.6491	6.7471	6.8385	6.9246
3100	V	1.220	64.467	....	67.885	73.315	78.287	82.958	87.423	91.745	95.965
(235.67)	U	1013.2	2602.5	....	2639.2	2697.0	2749.2	2797.9	2844.3	2889.3	2933.4
	H	1017.0	2802.3	....	2849.6	2924.2	2991.9	3055.1	3115.4	3173.7	3230.8
	S	2.6623	6.1709	....	6.2626	6.4019	6.5227	6.6307	6.7294	6.8212	6.9077

3200	V	1.224	62.439	....	65.380	70.721	75.593	80.158	84.513	88.723	92.829
	U	1021.5	2602.5	....	2635.2	2693.9	2746.8	2796.0	2842.7	2887.9	2932.1
(237.45)	H	1025.4	2802.3	....	2844.4	2920.2	2988.7	3052.5	3113.2	3171.8	3229.2
	S	2.6786	6.1585	....	6.2398	6.3815	6.5037	6.6127	6.7120	6.8043	6.8912
3300	V	1.227	60.529	....	63.021	68.282	73.061	77.526	81.778	85.883	89.883
	U	1029.7	2602.5	....	2631.1	2690.8	2744.4	2794.0	2841.1	2886.5	2930.9
(239.18)	H	1033.7	2802.3	....	2839.0	2916.1	2985.5	3049.9	3110.9	3169.9	3227.5
	S	2.6945	6.1463	....	6.2173	6.3614	6.4851	6.5951	6.6952	6.7879	6.8752
3400	V	1.231	58.728	....	60.796	65.982	70.675	75.048	79.204	83.210	87.110
	U	1037.6	2602.5	....	2626.9	2687.7	2741.9	2792.0	2839.4	2885.1	2929.7
(240.88)	H	1041.8	2802.1	....	2833.6	2912.0	2982.2	3047.2	3108.7	3168.0	3225.9
	S	2.7101	6.1344	....	6.1951	6.3416	6.4669	6.5779	6.6787	6.7719	6.8595
3500	V	1.235	57.025	....	58.693	63.812	68.424	72.710	76.776	80.689	84.494
	U	1045.4	2602.4	....	2622.7	2684.5	2739.5	2790.0	2837.8	2883.7	2928.4
(242.54)	H	1049.8	2802.0	....	2828.1	2907.8	2979.0	3044.5	3106.5	3166.1	3224.2
	S	2.7253	6.1228	....	6.1732	6.3221	6.4491	6.5611	6.6626	6.7563	6.8443
3600	V	1.238	55.415	....	56.702	61.759	66.297	70.501	74.482	78.308	82.024
	U	1053.1	2602.2	....	2618.4	2681.3	2737.0	2788.0	2836.1	2882.3	2927.2
(244.16)	H	1057.6	2801.7	....	2822.5	2903.6	2975.6	3041.8	3104.2	3164.2	3222.5
	S	2.7401	6.1115	....	6.1514	6.3030	6.4315	6.5446	6.6468	6.7411	6.8294
3700	V	1.242	53.888	....	54.812	59.814	64.282	68.410	72.311	76.055	79.687
	U	1060.6	2602.1	....	2614.0	2678.0	2734.4	2786.0	2834.4	2880.8	2926.0
(245.75)	H	1065.2	2801.4	....	2816.8	2899.3	2972.3	3039.1	3102.0	3162.2	3220.8
	S	2.7547	6.1004	....	6.1299	6.2841	6.4143	6.5284	6.6314	6.7262	6.8149
3800	V	1.245	52.438	....	53.017	57.968	62.372	66.429	70.254	73.920	77.473
	U	1068.0	2601.9	....	2609.5	2674.7	2731.9	2783.9	2832.7	2879.4	2924.7
(247.31)	H	1072.7	2801.1	....	2811.0	2895.0	2968.9	3036.4	3099.7	3160.3	3219.1
	S	2.7689	6.0896	....	6.1085	6.2654	6.3973	6.5126	6.6163	6.7117	6.8007
3900	V	1.249	51.061	....	51.308	56.215	60.558	64.547	68.302	71.894	75.372
	U	1075.3	2601.6	....	2605.0	2671.4	2729.3	2781.9	2831.0	2877.9	2923.5
(248.84)	H	1080.1	2800.8	....	2805.1	2890.6	2965.5	3033.6	3097.4	3158.3	3217.4
	S	2.7828	6.0789	....	6.0872	6.2470	6.3806	6.4970	6.6015	6.6974	6.7868
4000	V	1.252	49.749	....	....	54.546	58.833	62.759	66.446	69.969	73.376
	U	1082.4	2601.3	....	....	2668.0	2726.7	2779.8	2829.3	2876.5	2922.2
(250.33)	H	1087.4	2800.3	....	....	2886.1	2962.0	3030.8	3095.1	3156.4	3215.7
	S	2.7965	6.0685	....	....	6.2288	6.3642	6.4817	6.5870	6.6834	6.7733

TABLE C.2. SUPERHEATED STEAM SI UNITS (Continued)

ABS PRESS KPA (SAT TEMP) DEG C		SAT WATER	SAT STEAM	TEMPERATURE, DEG C (TEMPERATURE, K) 425 (698.15)	450 (723.15)	475 (748.15)	500 (773.15)	525 (798.15)	550 (823.15)	600 (873.15)	650 (923.15)
	V	1.193	83.199	130.44	135.61	140.73	145.82	150.88	155.91	165.92	175.86
2400	U	949.066	2600.7	2984.5	3027.1	3069.9	3112.9	3156.1	3199.6	3287.7	3377.2
(221.78)	H	951.929	2800.4	3297.5	3352.6	3407.7	3462.9	3518.2	3573.8	3685.9	3799.3
	S	2.5343	6.2690	7.1189	7.1964	7.2713	7.3439	7.4144	7.4830	7.6152	7.7414
	V	1.197	79.905	125.07	130.04	134.97	139.87	144.74	149.58	159.21	168.76
2500	U	958.969	2601.2	2983.4	3026.2	3069.0	3112.1	3155.4	3198.9	3287.1	3376.7
(223.94)	H	961.962	2800.9	3296.1	3351.3	3406.5	3461.7	3517.2	3572.9	3685.1	3798.6
	S	2.5543	6.2536	7.0986	7.1763	7.2513	7.3240	7.3946	7.4633	7.5956	7.7220
	V	1.201	76.856	120.11	124.91	129.66	134.38	139.07	143.74	153.01	162.21
2600	U	968.597	2601.5	2982.3	3025.2	3068.1	3111.2	3154.6	3198.2	3286.5	3376.1
(226.04)	H	971.720	2801.4	3294.6	3349.9	3405.3	3460.6	3516.2	3571.9	3684.3	3797.9
	S	2.5736	6.2387	7.0789	7.1568	7.2320	7.3048	7.3755	7.4443	7.5768	7.7033
	V	1.205	74.025	115.52	120.15	124.74	129.30	133.82	138.33	147.27	156.14
2700	U	977.968	2601.8	2981.2	3024.2	3067.2	3110.4	3153.8	3197.5	3285.8	3375.6
(228.07)	H	981.222	2801.7	3293.1	3348.6	3404.0	3459.5	3515.2	3571.0	3683.5	3797.1
	S	2.5924	6.2244	7.0600	7.1381	7.2134	7.2863	7.3571	7.4260	7.5587	7.6853
	V	1.209	71.389	111.25	115.74	120.17	124.58	128.95	133.30	141.94	150.50
2800	U	987.100	2602.1	2980.2	3023.2	3066.3	3109.6	3153.1	3196.8	3285.2	3375.0
(230.05)	H	990.485	2802.0	3291.7	3347.3	3402.8	3458.4	3514.1	3570.0	3682.6	3796.4
	S	2.6106	6.2104	7.0416	7.1199	7.1954	7.2685	7.3394	7.4084	7.5412	7.6679
	V	1.213	68.928	107.28	111.62	115.92	120.18	124.42	128.62	136.97	145.26
2900	U	996.008	2602.3	2979.1	3022.3	3065.5	3108.8	3152.3	3196.1	3284.6	3374.5
(231.97)	H	999.524	2802.2	3290.2	3346.0	3401.6	3457.3	3513.1	3569.1	3681.8	3795.7
	S	2.6283	6.1969	7.0239	7.1024	7.1780	7.2512	7.3222	7.3913	7.5243	7.6511
	V	1.216	66.626	103.58	107.79	111.95	116.08	120.18	124.26	132.34	140.36
3000	U	1004.7	2602.4	2978.0	3021.3	3064.6	3107.9	3151.5	3195.4	3284.0	3373.9
(233.84)	H	1008.4	2802.3	3288.7	3344.6	3400.4	3456.2	3512.1	3568.1	3681.0	3795.0
	S	2.6455	6.1837	7.0067	7.0854	7.1612	7.2345	7.3056	7.3748	7.5079	7.6349
	V	1.220	64.467	100.11	104.20	108.24	112.24	116.22	120.17	128.01	135.78
3100	U	1013.2	2602.5	2976.9	3020.3	3063.7	3107.1	3150.8	3194.7	3283.3	3373.4
(235.67)	H	1017.0	2802.3	3287.3	3343.3	3399.2	3455.1	3511.0	3567.2	3680.2	3794.3
	S	2.6623	6.1709	6.9900	7.0689	7.1448	7.2183	7.2895	7.3588	7.4920	7.6191

3200 (237.45)	V	1.224	62.439	96.859	100.83	104.76	108.65	112.51	116.34	123.95	131.48
	U	1021.5	2602.5	2975.9	3019.3	3062.8	3106.3	3150.0	3193.9	3282.7	3372.8
	H	1025.4	2802.3	3285.8	3342.0	3398.0	3454.0	3510.0	3566.2	3679.3	3793.6
	S	2.6786	6.1585	6.9738	7.0528	7.1290	7.2026	7.2739	7.3433	7.4767	7.6039
3300 (239.18)	V	1.227	60.529	93.805	97.668	101.49	105.27	109.02	112.74	120.13	127.45
	U	1029.7	2602.5	2974.8	3018.3	3061.9	3105.5	3149.2	3193.2	3282.1	3372.3
	H	1033.7	2802.3	3284.3	3340.6	3396.8	3452.8	3509.0	3565.3	3678.5	3792.9
	S	2.6945	6.1463	6.9580	7.0373	7.1136	7.1873	7.2588	7.3282	7.4618	7.5891
3400 (240.88)	V	1.231	58.728	90.930	94.692	98.408	102.09	105.74	109.36	116.54	123.65
	U	1037.6	2602.5	2973.7	3017.4	3061.0	3104.6	3148.4	3192.5	3281.5	3371.7
	H	1041.8	2802.1	3282.8	3339.3	3395.5	3451.7	3507.9	3564.3	3677.7	3792.1
	S	2.7101	6.1344	6.9426	7.0221	7.0986	7.1724	7.2440	7.3136	7.4473	7.5747
3500 (242.54)	V	1.235	57.025	88.220	91.886	95.505	99.088	102.64	106.17	113.15	120.07
	U	1045.4	2602.4	2972.6	3016.4	3060.1	3103.8	3147.7	3191.8	3280.8	3371.2
	H	1049.8	2802.0	3281.3	3338.0	3394.3	3450.6	3506.9	3563.4	3676.9	3791.4
	S	2.7253	6.1228	6.9277	7.0074	7.0840	7.1580	7.2297	7.2993	7.4332	7.5607
3600 (244.16)	V	1.238	55.415	85.660	89.236	92.764	96.255	99.716	103.15	109.96	116.69
	U	1053.1	2602.2	2971.5	3015.4	3059.2	3103.0	3146.9	3191.1	3280.2	3370.6
	H	1057.6	2801.7	3279.8	3336.6	3393.1	3449.5	3505.9	3562.4	3676.1	3790.7
	S	2.7401	6.1115	6.9131	6.9930	7.0698	7.1439	7.2157	7.2854	7.4195	7.5471
3700 (245.75)	V	1.242	53.888	83.238	86.728	90.171	93.576	96.950	100.30	106.93	113.49
	U	1060.6	2602.1	2970.4	3014.4	3058.2	3102.1	3146.1	3190.4	3279.6	3370.1
	H	1065.2	2801.4	3278.4	3335.3	3391.9	3448.4	3504.9	3561.5	3675.2	3790.0
	S	2.7547	6.1004	6.8989	6.9790	7.0559	7.1302	7.2021	7.2719	7.4061	7.5339
3800 (247.31)	V	1.245	52.438	80.944	84.353	87.714	91.038	94.330	97.596	104.06	110.46
	U	1068.0	2601.9	2969.3	3013.4	3057.3	3101.3	3145.4	3189.6	3279.0	3369.5
	H	1072.7	2801.1	3276.8	3333.9	3390.7	3447.2	3503.8	3560.5	3674.4	3789.3
	S	2.7689	6.0896	6.8849	6.9653	7.0424	7.1168	7.1888	7.2587	7.3931	7.5210
3900 (248.84)	V	1.249	51.061	78.767	82.099	85.383	88.629	91.844	95.033	101.35	107.59
	U	1075.3	2601.6	2968.2	3012.4	3056.4	3100.5	3144.6	3188.9	3278.3	3369.0
	H	1080.1	2800.8	3275.3	3332.6	3389.4	3446.1	3502.8	3559.5	3673.6	3788.6
	S	2.7828	6.0789	6.8713	6.9519	7.0292	7.1037	7.1759	7.2459	7.3804	7.5084
4000 (250.33)	V	1.252	49.749	76.698	79.958	83.169	86.341	89.483	92.598	98.763	104.86
	U	1082.4	2601.3	2967.0	3011.4	3055.5	3099.6	3143.8	3188.2	3277.7	3368.4
	H	1087.4	2800.3	3273.8	3331.2	3388.2	3445.0	3501.7	3558.6	3672.8	3787.9
	S	2.7965	6.0685	6.8581	6.9388	7.0163	7.0909	7.1632	7.2333	7.3680	7.4961

TABLE C.2. SUPERHEATED STEAM SI UNITS (Continued)

ABS PRESS KPA (SAT TEMP) DEG C		SAT WATER	SAT STEAM	TEMPERATURE, DEG C (TEMPERATURE, K) 260 (533.15)	275 (548.15)	300 (573.15)	325 (598.15)	350 (623.15)	375 (648.15)	400 (673.15)	425 (698.15)
4100 (251.80)	V	1.256	48.500	50.150	52.955	57.191	61.057	64.680	68.137	71.476	74.730
	U	1089.4	2601.0	2624.6	2664.5	2724.0	2777.7	2827.6	2875.0	2920.9	2965.9
	H	1094.6	2799.9	2830.3	2881.6	2958.5	3028.0	3092.8	3154.4	3214.0	3272.3
	S	2.8099	6.0583	6.1157	6.2107	6.3480	6.4667	6.5727	6.6697	6.7600	6.8450
4200 (253.24)	V	1.259	47.307	48.654	51.438	55.625	59.435	62.998	66.392	69.667	72.856
	U	1096.3	2600.7	2620.4	2661.0	2721.4	2775.6	2825.8	2873.6	2919.7	2964.8
	H	1101.6	2799.4	2824.8	2877.1	2955.0	3025.2	3090.4	3152.4	3212.3	3270.8
	S	2.8231	6.0482	6.0962	6.1929	6.3320	6.4519	6.5587	6.6563	6.7469	6.8323
4300 (254.66)	V	1.262	46.168	47.223	49.988	54.130	57.887	61.393	64.728	67.942	71.069
	U	1103.1	2600.3	2616.2	2657.5	2718.7	2773.4	2824.1	2872.1	2918.4	2963.7
	H	1108.5	2798.9	2819.2	2872.4	2951.4	3022.3	3088.1	3150.4	3210.5	3269.3
	S	2.8360	6.0383	6.0768	6.1752	6.3162	6.4373	6.5450	6.6431	6.7341	6.8198
4400 (256.05)	V	1.266	45.079	45.853	48.601	52.702	56.409	59.861	63.139	66.295	69.363
	U	1109.8	2599.9	2611.8	2653.9	2716.0	2771.3	2822.3	2870.6	2917.1	2962.5
	H	1115.4	2798.3	2813.6	2867.8	2947.8	3019.5	3085.7	3148.4	3208.8	3267.7
	S	2.8487	6.0286	6.0575	6.1577	6.3006	6.4230	6.5315	6.6301	6.7216	6.8076
4500 (257.41)	V	1.269	44.037	44.540	47.273	51.336	54.996	58.396	61.620	64.721	67.732
	U	1116.4	2599.5	2607.4	2650.3	2713.2	2769.1	2820.5	2869.1	2915.8	2961.4
	H	1122.1	2797.7	2807.9	2863.0	2944.2	3016.6	3083.3	3146.4	3207.1	3266.2
	S	2.8612	6.0191	6.0382	6.1403	6.2852	6.4088	6.5182	6.6174	6.7093	6.7955
4600 (258.75)	V	1.272	43.038	43.278	46.000	50.027	53.643	56.994	60.167	63.215	66.172
	U	1122.9	2599.1	2602.9	2646.6	2710.4	2766.9	2818.7	2867.6	2914.5	2960.3
	H	1128.8	2797.0	2802.0	2858.2	2940.5	3013.7	3080.9	3144.4	3205.3	3264.7
	S	2.8735	6.0097	6.0190	6.1230	6.2700	6.3949	6.5050	6.6049	6.6972	6.7838
4700 (260.07)	V	1.276	42.081	....	44.778	48.772	52.346	55.651	58.775	61.773	64.679
	U	1129.3	2598.6	....	2642.9	2707.6	2764.7	2816.9	2866.1	2913.2	2959.1
	H	1135.3	2796.4	....	2853.3	2936.8	3010.7	3078.5	3142.3	3203.6	3263.1
	S	2.8855	6.0004	....	6.1058	6.2549	6.3811	6.4921	6.5926	6.6853	6.7722
4800 (261.37)	V	1.279	41.161	....	43.604	47.569	51.103	54.364	57.441	60.390	63.247
	U	1135.6	2598.1	....	2639.1	2704.8	2762.5	2815.1	2864.6	2911.9	2958.0
	H	1141.8	2795.7	....	2848.4	2933.1	3007.8	3076.1	3140.3	3201.8	3261.6
	S	2.8974	5.9913	....	6.0887	6.2399	6.3675	6.4794	6.5805	6.6736	6.7608

4900 (262.65)	V	1.282	40.278	....	42.475	46.412	49.909	53.128	56.161	59.064	61.874
	U	1141.9	2597.6	....	2635.2	2701.9	2760.2	2813.3	2863.0	2910.6	2956.9
	H	1148.2	2794.9	....	2843.3	2929.3	3004.8	3073.6	3138.2	3200.0	3260.0
	S	2.9091	5.9823	....	6.0717	6.2252	6.3541	6.4669	6.5685	6.6621	6.7496
5000 (263.91)	V	1.286	39.429	....	41.388	45.301	48.762	51.941	54.932	57.791	60.555
	U	1148.0	2597.0	....	2631.3	2699.0	2758.0	2811.5	2861.5	2909.3	2955.7
	H	1154.5	2794.2	....	2838.2	2925.5	3001.8	3071.2	3136.2	3198.3	3258.5
	S	2.9206	5.9735	....	6.0547	6.2105	6.3408	6.4545	6.5568	6.6508	6.7386
5100 (265.15)	V	1.289	38.611	....	40.340	44.231	47.660	50.801	53.750	56.567	59.288
	U	1154.1	2596.5	....	2627.3	2696.1	2755.7	2809.6	2860.0	2908.0	2954.5
	H	1160.7	2793.4	....	2833.1	2921.7	2998.7	3068.7	3134.1	3196.5	3256.9
	S	2.9319	5.9648	....	6.0378	6.1960	6.3277	6.4423	6.5452	6.6396	6.7278
5200 (266.37)	V	1.292	37.824	....	39.330	43.201	46.599	49.703	52.614	55.390	58.070
	U	1160.1	2595.9	....	2623.3	2693.1	2753.4	2807.8	2858.4	2906.7	2953.4
	H	1166.8	2792.6	....	2827.8	2917.8	2995.7	3066.2	3132.0	3194.7	3255.4
	S	2.9431	5.9561	....	6.0210	6.1815	6.3147	6.4302	6.5338	6.6287	6.7172
5300 (267.58)	V	1.296	37.066	....	38.354	42.209	45.577	48.647	51.520	54.257	56.897
	U	1166.1	2595.3	....	2619.2	2690.1	2751.0	2805.9	2856.9	2905.3	2952.2
	H	1172.9	2791.7	....	2822.5	2913.8	2992.6	3063.7	3129.9	3192.9	3253.8
	S	2.9541	5.9476	....	6.0041	6.1672	6.3018	6.4183	6.5225	6.6179	6.7067
5400 (268.76)	V	1.299	36.334	....	37.411	41.251	44.591	47.628	50.466	53.166	55.768
	U	1171.9	2594.6	....	2615.0	2687.1	2748.7	2804.0	2855.3	2904.0	2951.1
	H	1178.9	2790.8	....	2817.0	2909.8	2989.5	3061.2	3127.8	3191.1	3252.2
	S	2.9650	5.9392	....	5.9873	6.1530	6.2891	6.4066	6.5114	6.6072	6.6963
5500 (269.93)	V	1.302	35.628	....	36.499	40.327	43.641	46.647	49.450	52.115	54.679
	U	1177.7	2594.0	....	2610.8	2684.0	2746.3	2802.1	2853.7	2902.7	2949.9
	H	1184.9	2789.9	....	2811.5	2905.8	2986.4	3058.7	3125.7	3189.3	3250.6
	S	2.9757	5.9309	....	5.9705	6.1388	6.2765	6.3949	6.5004	6.5967	6.6862
5600 (271.09)	V	1.306	34.946	....	35.617	39.434	42.724	45.700	48.470	51.100	53.630
	U	1183.5	2593.3	....	2606.5	2680.9	2744.0	2800.2	2852.1	2901.3	2948.7
	H	1190.8	2789.0	....	2805.9	2901.7	2983.2	3056.1	3123.6	3187.5	3249.0
	S	2.9863	5.9227	....	5.9537	6.1248	6.2640	6.3834	6.4896	6.5863	6.6761
5700 (272.22)	V	1.309	34.288	....	34.761	38.571	41.838	44.785	47.525	50.121	52.617
	U	1189.1	2592.6	....	2602.1	2677.8	2741.6	2798.3	2850.5	2899.9	2947.5
	H	1196.6	2788.0	....	2800.2	2897.6	2980.0	3053.5	3121.4	3185.6	3247.5
	S	2.9968	5.9146	....	5.9369	6.1108	6.2516	6.3720	6.4789	6.5761	6.6663

TABLE C.2. SUPERHEATED STEAM SI UNITS (Continued)

ABS PRESS KPA (SAT TEMP) DEG C		SAT WATER	SAT STEAM	TEMPERATURE, DEG C (TEMPERATURE, K) 450 (723.15)	475 (748.15)	500 (773.15)	525 (798.15)	550 (823.15)	575 (848.15)	600 (873.15)	650 (923.15)
4100 (251.80)	V	1.256	48.500	77.921	81.062	84.165	87.236	90.281	93.303	96.306	102.26
	U	1089.4	2601.0	3010.4	3054.6	3098.8	3143.0	3187.5	3232.1	3277.1	3367.9
	H	1094.6	2799.9	3329.9	3387.0	3443.9	3500.7	3557.6	3614.7	3671.9	3787.1
	S	2.8099	6.0583	6.9260	7.0037	7.0785	7.1508	7.2210	7.2893	7.3558	7.4842
4200 (253.24)	V	1.259	47.307	75.981	79.056	82.092	85.097	88.075	91.030	93.966	99.787
	U	1096.3	2600.7	3009.4	3053.7	3097.9	3142.3	3186.8	3231.5	3276.5	3367.3
	H	1101.6	2799.4	3328.5	3385.7	3442.7	3499.7	3556.7	3613.8	3671.1	3786.4
	S	2.8231	6.0482	6.9135	6.9913	7.0662	7.1387	7.2090	7.2774	7.3440	7.4724
4300 (254.66)	V	1.262	46.168	74.131	77.143	80.116	83.057	85.971	88.863	91.735	97.428
	U	1103.1	2600.3	3008.4	3052.8	3097.1	3141.5	3186.0	3230.8	3275.8	3366.8
	H	1108.5	2798.9	3327.1	3384.5	3441.6	3498.6	3555.7	3612.9	3670.3	3785.7
	S	2.8360	6.0383	6.9012	6.9792	7.0543	7.1269	7.1973	7.2658	7.3324	7.4610
4400 (256.05)	V	1.266	45.079	72.365	75.317	78.229	81.110	83.963	86.794	89.605	95.177
	U	1109.8	2599.9	3007.4	3051.9	3096.3	3140.7	3185.3	3230.1	3275.2	3366.2
	H	1115.4	2798.3	3325.8	3383.3	3440.5	3497.6	3554.7	3612.0	3669.5	3785.0
	S	2.8487	6.0286	6.8892	6.9674	7.0426	7.1153	7.1858	7.2544	7.3211	7.4498
4500 (257.41)	V	1.269	44.037	70.677	73.572	76.427	79.249	82.044	84.817	87.570	93.025
	U	1116.4	2599.5	3006.3	3050.9	3095.4	3139.9	3184.6	3229.5	3274.6	3365.7
	H	1122.1	2797.7	3324.4	3382.0	3439.3	3496.6	3553.8	3611.1	3668.6	3784.3
	S	2.8612	6.0191	6.8774	6.9558	7.0311	7.1040	7.1746	7.2432	7.3100	7.4388
4600 (258.75)	V	1.272	43.038	69.063	71.903	74.702	77.469	80.209	82.926	85.623	90.967
	U	1122.9	2599.1	3005.3	3050.0	3094.6	3139.2	3183.9	3228.8	3273.9	3365.1
	H	1128.8	2797.0	3323.0	3380.8	3438.2	3495.5	3552.8	3610.2	3667.8	3783.6
	S	2.8735	6.0097	6.8659	6.9444	7.0199	7.0928	7.1636	7.2323	7.2991	7.4281
4700 (260.07)	V	1.276	42.081	67.517	70.304	73.051	75.765	78.452	81.116	83.760	88.997
	U	1129.3	2598.6	3004.3	3049.1	3093.7	3138.4	3183.1	3228.1	3273.3	3364.6
	H	1135.3	2796.4	3321.6	3379.5	3437.1	3494.5	3551.9	3609.3	3667.0	3782.9
	S	2.8855	6.0004	6.8545	6.9332	7.0089	7.0819	7.1527	7.2215	7.2885	7.4176
4800 (261.37)	V	1.279	41.161	66.036	68.773	71.469	74.132	76.768	79.381	81.973	87.109
	U	1135.6	2598.1	3003.3	3048.2	3092.9	3137.6	3182.4	3227.4	3272.7	3364.0
	H	1141.8	2795.7	3320.3	3378.3	3435.9	3493.4	3550.9	3608.5	3666.2	3782.1
	S	2.8974	5.9913	6.8434	6.9223	6.9981	7.0712	7.1422	7.2110	7.2781	7.4072

4900	V	1.282	40.278	64.615	67.303	69.951	72.565	75.152	77.716	80.260	85.298
(262.65)	U	1141.9	2597.6	3002.3	3047.2	3092.0	3136.8	3181.7	3226.8	3272.0	3363.5
	H	1148.2	2794.9	3318.9	3377.0	3434.8	3492.4	3549.9	3607.6	3665.3	3781.4
	S	2.9091	5.9823	6.8324	6.9115	6.9874	7.0607	7.1318	7.2007	7.2678	7.3971
5000	V	1.286	39.429	63.250	65.893	68.494	71.061	73.602	76.119	78.616	83.559
(263.91)	U	1148.0	2597.0	3001.2	3046.3	3091.2	3136.0	3181.0	3226.1	3271.4	3362.9
	H	1154.5	2794.2	3317.5	3375.8	3433.7	3491.3	3549.0	3606.7	3664.5	3780.7
	S	2.9206	5.9735	6.8217	6.9009	6.9770	7.0504	7.1215	7.1906	7.2578	7.3872
5100	V	1.289	38.611	61.940	64.537	67.094	69.616	72.112	74.584	77.035	81.888
(265.15)	U	1154.1	2596.5	3000.2	3045.4	3090.3	3135.3	3180.2	3225.4	3270.8	3362.4
	H	1160.7	2793.4	3316.1	3374.5	3432.5	3490.3	3548.0	3605.8	3663.7	3780.0
	S	2.9319	5.9648	6.8111	6.8905	6.9668	7.0403	7.1115	7.1807	7.2479	7.3775
5200	V	1.292	37.824	60.679	63.234	65.747	68.227	70.679	73.108	75.516	80.282
(266.37)	U	1160.1	2595.9	2999.2	3044.5	3089.5	3134.5	3179.5	3224.7	3270.2	3361.8
	H	1166.8	2792.6	3314.7	3373.3	3431.4	3489.3	3547.1	3604.9	3662.8	3779.3
	S	2.9431	5.9561	6.8007	6.8803	6.9567	7.0304	7.1017	7.1709	7.2382	7.3679
5300	V	1.296	37.066	59.466	61.980	64.452	66.890	69.300	71.687	74.054	78.736
(267.58)	U	1166.1	2595.3	2998.2	3043.5	3088.6	3133.7	3178.8	3224.1	3269.5	3361.3
	H	1172.9	2791.7	3313.3	3372.0	3430.2	3488.2	3546.1	3604.0	3662.0	3778.6
	S	2.9541	5.9476	6.7905	6.8703	6.9468	7.0206	7.0920	7.1613	7.2287	7.3585
5400	V	1.299	36.334	58.297	60.772	63.204	65.603	67.973	70.320	72.646	77.248
(268.76)	U	1171.9	2594.6	2997.1	3042.6	3087.8	3132.9	3178.1	3223.4	3268.9	3360.7
	H	1178.9	2790.8	3311.9	3370.8	3429.1	3487.2	3545.1	3603.1	3661.2	3777.8
	S	2.9650	5.9392	6.7804	6.8604	6.9371	7.0110	7.0825	7.1519	7.2194	7.3493
5500	V	1.302	35.628	57.171	59.608	62.002	64.362	66.694	69.002	71.289	75.814
(269.93)	U	1177.7	2594.0	2996.1	3041.7	3086.9	3132.1	3177.3	3222.7	3268.3	3360.2
	H	1184.9	2789.9	3310.5	3369.5	3427.9	3486.1	3544.2	3602.2	3660.4	3777.1
	S	2.9757	5.9309	6.7705	6.8507	6.9275	7.0015	7.0731	7.1426	7.2102	7.3402
5600	V	1.306	34.946	56.085	58.486	60.843	63.165	65.460	67.731	69.981	74.431
(271.09)	U	1183.5	2593.3	2995.0	3040.7	3086.1	3131.3	3176.6	3222.0	3267.6	3359.6
	H	1190.8	2789.0	3309.1	3368.2	3426.8	3485.1	3543.2	3601.3	3659.5	3776.4
	S	2.9863	5.9227	6.7607	6.8411	6.9181	6.9922	7.0639	7.1335	7.2011	7.3313
5700	V	1.309	34.288	55.038	57.403	59.724	62.011	64.270	66.504	68.719	73.096
(272.22)	U	1189.1	2592.6	2994.0	3039.8	3085.2	3130.5	3175.9	3221.3	3267.0	3359.1
	H	1196.6	2788.0	3307.7	3367.0	3425.6	3484.0	3542.2	3600.4	3658.7	3775.7
	S	2.9968	5.9146	6.7511	6.8316	6.9088	6.9831	7.0549	7.1245	7.1923	7.3226

TABLE C.2. SUPERHEATED STEAM SI UNITS (Continued)

ABS PRESS KPA (SAT TEMP) DEG C		SAT WATER	SAT STEAM	TEMPERATURE, DEG C (TEMPERATURE, K) 280 (553.15)	290 (563.15)	300 (573.15)	325 (598.15)	350 (623.15)	375 (648.15)	400 (673.15)	425 (698.15)
5800 (273.35)	V	1.312	33.651	34.756	36.301	37.736	40.982	43.902	46.611	49.176	51.638
	U	1194.7	2591.9	2614.4	2645.7	2674.6	2739.1	2796.3	2848.9	2898.6	2946.4
	H	1202.3	2787.0	2816.0	2856.3	2893.5	2976.8	3051.0	3119.3	3183.8	3245.9
	S	3.0071	5.9066	5.9592	6.0314	6.0969	6.2393	6.3608	6.4683	6.5660	6.6565
5900 (274.46)	V	1.315	33.034	33.953	35.497	36.928	40.154	43.048	45.728	48.262	50.693
	U	1200.3	2591.1	2610.2	2642.1	2671.4	2736.7	2794.4	2847.3	2897.2	2945.2
	H	1208.0	2786.0	2810.5	2851.5	2889.3	2973.6	3048.4	3117.1	3182.0	3244.3
	S	3.0172	5.8986	5.9431	6.0166	6.0830	6.2272	6.3496	6.4578	6.5560	6.6469
6000 (275.55)	V	1.319	32.438	33.173	34.718	36.145	39.353	42.222	44.874	47.379	49.779
	U	1205.8	2590.4	2605.9	2638.4	2668.1	2734.2	2792.4	2845.7	2895.8	2944.0
	H	1213.7	2785.0	2804.9	2846.7	2885.0	2970.4	3045.8	3115.0	3180.1	3242.6
	S	3.0273	5.8908	5.9270	6.0017	6.0692	6.2151	6.3386	6.4475	6.5462	6.6374
6100 (276.63)	V	1.322	31.860	32.415	33.962	35.386	38.577	41.422	44.048	46.524	48.895
	U	1211.2	2589.6	2601.5	2634.6	2664.8	2731.7	2790.4	2844.1	2894.5	2942.8
	H	1219.3	2783.9	2799.3	2841.8	2880.7	2967.1	3043.1	3112.8	3178.3	3241.0
	S	3.0372	5.8830	5.9108	5.9869	6.0555	6.2031	6.3277	6.4373	6.5364	6.6280
6200 (277.70)	V	1.325	31.300	31.679	33.227	34.650	37.825	40.648	43.248	45.697	48.039
	U	1216.6	2588.8	2597.1	2630.8	2661.5	2729.2	2788.5	2842.4	2893.1	2941.6
	H	1224.8	2782.9	2793.5	2836.8	2876.3	2963.8	3040.5	3110.6	3176.4	3239.4
	S	3.0471	5.8753	5.8946	5.9721	6.0418	6.1911	6.3168	6.4272	6.5268	6.6188
6300 (278.75)	V	1.328	30.757	30.962	32.514	33.935	37.0[illegible]	39.898	42.473	44.895	47.210
	U	1221.9	2588.0	2592.6	2626.9	2658.1	2726.7	2786.5	2840.8	2891.7	2940.4
	H	1230.3	2781.8	2787.6	2831.7	2871.9	2960.4	3037.8	3108.4	3174.5	3237.8
	S	3.0568	5.8677	5.8783	5.9573	6.0281	6.1793	6.3061	6.4172	6.5173	6.6096
6400 (279.79)	V	1.332	30.230	30.265	31.821	33.241	36.390	39.170	41.722	44.119	46.407
	U	1227.2	2587.2	2587.9	2623.0	2654.7	2724.2	2784.4	2839.1	2890.3	2939.2
	H	1235.7	2780.6	2781.6	2826.6	2867.5	2957.1	3035.1	3106.2	3172.7	3236.2
	S	3.0664	5.8601	5.8619	5.9425	6.0144	6.1675	6.2955	6.4072	6.5079	6.6006
6500 (280.82)	V	1.335	29.719	....	31.146	32.567	35.704	38.465	40.994	43.366	45.629
	U	1232.5	2586.3	....	2619.0	2651.2	2721.6	2782.4	2837.5	2888.9	2938.0
	H	1241.1	2779.5	....	2821.4	2862.9	2953.7	3032.4	3103.9	3170.8	3234.5
	S	3.0759	5.8527	....	5.9277	6.0008	6.1558	6.2849	6.3974	6.4986	6.5917

	V	1.338	29.223	....	30.490	31.911	35.038	37.781	40.287	42.636	44.874
6600	U	1237.6	2585.5	....	2614.9	2647.7	2719.0	2780.4	2835.8	2887.5	2936.7
(281.84)	H	1246.5	2778.3	....	2816.1	2858.4	2950.2	3029.7	3101.7	3168.9	3232.9
	S	3.0853	5.8452	....	5.9129	5.9872	6.1442	6.2744	6.3877	6.4894	6.5828
	V	1.342	28.741	....	29.850	31.273	34.391	37.116	39.601	41.927	44.141
6700	U	1242.8	2584.6	....	2610.8	2644.2	2716.4	2778.3	2834.1	2886.1	2935.5
(282.84)	H	1251.8	2777.1	....	2810.8	2853.7	2946.8	3027.0	3099.5	3167.0	3231.3
	S	3.0946	5.8379	....	5.8980	5.9736	6.1326	6.2640	6.3781	6.4803	6.5741
	V	1.345	28.272	....	29.226	30.652	33.762	36.470	38.935	41.239	43.430
6800	U	1247.9	2583.7	....	2606.6	2640.6	2713.7	2776.2	2832.4	2884.7	2934.3
(283.84)	H	1257.0	2775.9	....	2805.3	2849.0	2943.3	3024.2	3097.2	3165.1	3229.6
	S	3.1038	5.8306	....	5.8830	5.9599	6.1211	6.2537	6.3686	6.4713	6.5655
	V	1.351	27.373	....	28.024	29.457	32.556	35.233	37.660	39.922	42.068
7000	U	1258.0	2581.8	....	2597.9	2633.2	2708.4	2772.1	2829.0	2881.8	2931.8
(285.79)	H	1267.4	2773.5	....	2794.1	2839.4	2936.3	3018.7	3092.7	3161.2	3226.3
	S	3.1219	5.8162	....	5.8530	5.9327	6.0982	6.2333	6.3497	6.4536	6.5485
	V	1.358	26.522	....	26.878	28.321	31.413	34.063	36.454	38.676	40.781
7200	U	1267.9	2579.9	....	2589.0	2625.6	2702.9	2767.8	2825.6	2878.9	2929.4
(287.70)	H	1277.6	2770.9	....	2782.5	2829.5	2929.1	3013.1	3088.1	3157.4	3223.0
	S	3.1397	5.8020	....	5.8226	5.9054	6.0755	6.2132	6.3312	6.4362	6.5319
	V	1.364	25.715	....	25.781	27.238	30.328	32.954	35.312	37.497	39.564
7400	U	1277.6	2578.0	....	2579.7	2617.8	2697.3	2763.5	2822.1	2876.0	2926.9
(289.57)	H	1287.7	2768.3	....	2770.5	2819.3	2921.8	3007.4	3083.4	3153.5	3219.6
	S	3.1571	5.7880	....	5.7919	5.8779	6.0530	6.1933	6.3130	6.4190	6.5156
	V	1.371	24.949	....	....	26.204	29.297	31.901	34.229	36.380	38.409
7600	U	1287.2	2575.9	....	....	2609.7	2691.7	2759.2	2818.6	2873.1	2924.3
(291.41)	H	1297.6	2765.5	....	....	2808.8	2914.3	3001.6	3078.7	3149.6	3216.3
	S	3.1742	5.7742	....	....	5.8503	6.0306	6.1737	6.2950	6.4022	6.4996
	V	1.378	24.220	....	....	25.214	28.315	30.900	33.200	35.319	37.314
7800	U	1296.7	2573.8	....	....	2601.3	2685.9	2754.8	2815.1	2870.1	2921.8
(293.21)	H	1307.4	2762.8	....	....	2798.0	2906.7	2995.8	3074.0	3145.6	3212.9
	S	3.1911	5.7605	....	....	5.8224	6.0082	6.1542	6.2773	6.3857	6.4839
	V	1.384	23.525	....	....	24.264	27.378	29.948	32.222	34.310	36.273
8000	U	1306.0	2571.7	....	....	2592.7	2679.9	2750.3	2811.5	2867.1	2919.3
(294.97)	H	1317.1	2759.9	....	....	2786.8	2899.0	2989.9	3069.2	3141.6	3209.5
	S	3.2076	5.7471	....	....	5.7942	5.9860	6.1349	6.2599	6.3694	6.4684

TABLE C.2. SUPERHEATED STEAM SI UNITS (Continued)

ABS PRESS KPA (SAT TEMP) DEG C		SAT WATER	SAT STEAM	TEMPERATURE, DEG C (TEMPERATURE, K) 450 (723.15)	475 (748.15)	500 (773.15)	525 (798.15)	550 (823.15)	575 (848.15)	600 (873.15)	650 (923.15)
5800 (273.35)	V	1.312	33.651	54.026	56.357	58.644	60.896	63.120	65.320	67.500	71.807
	U	1194.7	2591.9	2992.9	3038.8	3084.4	3129.8	3175.2	3220.7	3266.4	3358.5
	H	1202.3	2787.0	3306.3	3365.7	3424.5	3483.0	3541.2	3599.5	3657.9	3775.0
	S	3.0071	5.9066	6.7416	6.8223	6.8996	6.9740	7.0460	7.1157	7.1835	7.3139
5900 (274.46)	V	1.315	33.034	53.048	55.346	57.600	59.819	62.010	64.176	66.322	70.563
	U	1200.3	2591.1	2991.9	3037.9	3083.5	3129.0	3174.4	3220.0	3265.7	3357.9
	H	1208.0	2786.0	3304.9	3364.4	3423.3	3481.9	3540.3	3598.6	3657.0	3774.3
	S	3.0172	5.8986	6.7322	6.8132	6.8906	6.9652	7.0372	7.1070	7.1749	7.3054
6000 (275.55)	V	1.319	32.438	52.103	54.369	56.592	58.778	60.937	63.071	65.184	69.359
	U	1205.8	2590.4	2990.8	3036.9	3082.6	3128.2	3173.7	3219.3	3265.1	3357.4
	H	1213.7	2785.0	3303.5	3363.2	3422.2	3480.8	3539.3	3597.7	3656.2	3773.5
	S	3.0273	5.8908	6.7230	6.8041	6.8818	6.9564	7.0285	7.0985	7.1664	7.2971
6100 (276.63)	V	1.322	31.860	51.189	53.424	55.616	57.771	59.898	62.001	64.083	68.196
	U	1211.2	2589.6	2989.8	3036.0	3081.8	3127.4	3173.0	3218.6	3264.5	3356.8
	H	1219.3	2783.9	3302.0	3361.9	3421.0	3479.8	3538.3	3596.8	3655.4	3772.8
	S	3.0372	5.8830	6.7139	6.7952	6.8730	6.9478	7.0200	7.0900	7.1581	7.2889
6200 (277.70)	V	1.325	31.300	50.304	52.510	54.671	56.797	58.894	60.966	63.018	67.069
	U	1216.6	2588.8	2988.7	3035.0	3080.9	3126.6	3172.2	3218.0	3263.8	3356.3
	H	1224.8	2782.9	3300.6	3360.6	3419.9	3478.7	3537.4	3595.9	3654.5	3772.1
	S	3.0471	5.8753	6.7049	6.7864	6.8644	6.9393	7.0116	7.0817	7.1498	7.2808
6300 (278.75)	V	1.328	30.757	49.447	51.624	53.757	55.853	57.921	59.964	61.986	65.979
	U	1221.9	2588.0	2987.7	3034.1	3080.1	3125.8	3171.5	3217.3	3263.2	3355.7
	H	1230.3	2781.8	3299.2	3359.3	3418.7	3477.7	3536.4	3595.0	3653.7	3771.4
	S	3.0568	5.8677	6.6960	6.7778	6.8559	6.9309	7.0034	7.0735	7.1417	7.2728
6400 (279.79)	V	1.332	30.230	48.617	50.767	52.871	54.939	56.978	58.993	60.987	64.922
	U	1227.2	2587.2	2986.6	3033.1	3079.2	3125.0	3170.8	3216.6	3262.6	3355.2
	H	1235.7	2780.6	3297.7	3358.0	3417.6	3476.6	3535.4	3594.1	3652.9	3770.7
	S	3.0664	5.8601	6.6872	6.7692	6.8475	6.9226	6.9952	7.0655	7.1337	7.2649
6500 (280.82)	V	1.335	29.719	47.812	49.935	52.012	54.053	56.065	58.052	60.018	63.898
	U	1232.5	2586.3	2985.5	3032.2	3078.3	3124.2	3170.0	3215.9	3261.9	3354.6
	H	1241.1	2779.5	3296.3	3356.8	3416.4	3475.6	3534.4	3593.2	3652.1	3770.0
	S	3.0759	5.8527	6.6786	6.7608	6.8392	6.9145	6.9871	7.0575	7.1258	7.2572

6600 (281.84)	V	1.338	29.223	47.031	49.129	51.180	53.194	55.179	57.139	59.079	62.905
	U	1237.6	2585.5	2984.5	3031.2	3077.4	3123.4	3169.3	3215.2	3261.3	3354.1
	H	1246.5	2778.3	3294.9	3355.5	3415.2	3474.5	3533.5	3592.3	3651.2	3769.2
	S	3.0853	5.8452	6.6700	6.7524	6.8310	6.9064	6.9792	7.0497	7.1181	7.2495
6700 (282.84)	V	1.342	28.741	46.274	48.346	50.372	52.361	54.320	56.254	58.168	61.942
	U	1242.8	2584.6	2983.4	3030.3	3076.6	3122.6	3168.6	3214.5	3260.7	3353.5
	H	1251.8	2777.1	3293.4	3354.2	3414.1	3473.4	3532.5	3591.4	3650.4	3768.5
	S	3.0946	5.8379	6.6616	6.7442	6.8229	6.8985	6.9714	7.0419	7.1104	7.2420
6800 (283.84)	V	1.345	28.272	45.539	47.587	49.588	51.552	53.486	55.395	57.283	61.007
	U	1247.9	2583.7	2982.3	3029.3	3075.7	3121.8	3167.8	3213.9	3260.0	3353.0
	H	1257.0	2775.9	3292.0	3352.9	3412.9	3472.4	3531.5	3590.5	3649.6	3767.8
	S	3.1038	5.8306	6.6532	6.7361	6.8150	6.8907	6.9636	7.0343	7.1028	7.2345
7000 (285.79)	V	1.351	27.373	44.131	46.133	48.086	50.003	51.889	53.750	55.590	59.217
	U	1258.0	2581.8	2980.1	3027.4	3074.0	3120.2	3166.3	3212.5	3258.8	3351.9
	H	1267.4	2773.5	3289.1	3350.3	3410.6	3470.2	3529.6	3588.7	3647.9	3766.4
	S	3.1219	5.8162	6.6368	6.7201	6.7993	6.8753	6.9485	7.0193	7.0880	7.2200
7200 (287.70)	V	1.358	26.522	42.802	44.759	46.668	48.540	50.381	52.197	53.991	57.527
	U	1267.9	2579.9	2978.0	3025.4	3072.2	3118.6	3164.9	3211.1	3257.5	3350.7
	H	1277.6	2770.9	3286.1	3347.7	3408.2	3468.1	3527.6	3586.9	3646.2	3764.9
	S	3.1397	5.8020	6.6208	6.7044	6.7840	6.8602	6.9337	7.0047	7.0735	7.2058
7400 (289.57)	V	1.364	25.715	41.544	43.460	45.327	47.156	48.954	50.727	52.478	55.928
	U	1277.6	2578.0	2975.8	3023.5	3070.4	3117.0	3163.4	3209.8	3256.2	3349.6
	H	1287.7	2768.3	3283.2	3345.1	3405.9	3466.0	3525.7	3585.1	3644.5	3763.5
	S	3.1571	5.7880	6.6050	6.6892	6.7691	6.8456	6.9192	6.9904	7.0594	7.1919
7600 (291.41)	V	1.371	24.949	40.351	42.228	44.056	45.845	47.603	49.335	51.045	54.413
	U	1287.2	2575.9	2973.6	3021.5	3068.7	3115.4	3161.9	3208.4	3254.9	3348.5
	H	1297.6	2765.5	3280.3	3342.5	3403.5	3463.8	3523.7	3583.3	3642.9	3762.1
	S	3.1742	5.7742	6.5896	6.6742	6.7545	6.8312	6.9051	6.9765	7.0457	7.1784
7800 (293.21)	V	1.378	24.220	39.220	41.060	42.850	44.601	46.320	48.014	49.686	52.976
	U	1296.7	2573.8	2971.4	3019.6	3066.9	3113.8	3160.4	3207.0	3253.7	3347.4
	H	1307.4	2762.8	3277.3	3339.8	3401.1	3461.7	3521.7	3581.5	3641.2	3760.6
	S	3.1911	5.7605	6.5745	6.6596	6.7402	6.8172	6.8913	6.9629	7.0322	7.1652
8000 (294.97)	V	1.384	23.525	38.145	39.950	41.704	43.419	45.102	46.759	48.394	51.611
	U	1306.0	2571.7	2969.2	3017.6	3065.1	3112.2	3158.9	3205.6	3252.4	3346.3
	H	1317.1	2759.9	3274.3	3337.2	3398.8	3459.5	3519.7	3579.7	3639.5	3759.2
	S	3.2076	5.7471	6.5597	6.6452	6.7262	6.8035	6.8778	6.9496	7.0191	7.1523

TABLE C.2. SUPERHEATED STEAM SI UNITS (Continued)

ABS PRESS KPA (SAT TEMP) DEG C		SAT WATER	SAT STEAM	TEMPERATURE, DEG C (TEMPERATURE, K) 300 (573.15)	320 (593.15)	340 (613.15)	360 (633.15)	380 (653.15)	400 (673.15)	425 (698.15)	450 (723.15)
8200 (296.70)	V	1.391	22.863	23.350	25.916	28.064	29.968	31.715	33.350	35.282	37.121
	U	1315.2	2569.5	2583.7	2657.7	2718.5	2771.5	2819.5	2864.1	2916.7	2966.9
	H	1326.6	2757.0	2775.2	2870.2	2948.6	3017.2	3079.5	3137.6	3206.0	3271.3
	S	3.2239	5.7338	5.7656	5.9288	6.0588	6.1689	6.2659	6.3534	6.4532	6.5452
8400 (298.39)	V	1.398	22.231	22.469	25.058	27.203	29.094	30.821	32.435	34.337	36.147
	U	1324.3	2567.2	2574.4	2651.1	2713.4	2767.3	2816.0	2861.1	2914.1	2964.7
	H	1336.1	2754.0	2763.1	2861.6	2941.9	3011.7	3074.8	3133.5	3202.6	3268.3
	S	3.2399	5.7207	5.7366	5.9056	6.0388	6.1509	6.2491	6.3376	6.4383	6.5309
8600 (300.06)	V	1.404	21.627	....	24.236	26.380	28.258	29.968	31.561	33.437	35.217
	U	1333.3	2564.9	....	2644.3	2708.1	2763.1	2812.4	2858.0	2911.5	2962.4
	H	1345.4	2750.9	....	2852.7	2935.0	3006.1	3070.1	3129.4	3199.1	3265.3
	S	3.2557	5.7076	....	5.8823	6.0189	6.1330	6.2326	6.3220	6.4236	6.5168
8800 (301.70)	V	1.411	21.049	....	23.446	25.592	27.459	29.153	30.727	32.576	34.329
	U	1342.2	2562.6	....	2637.3	2702.8	2758.8	2808.8	2854.9	2908.9	2960.1
	H	1354.6	2747.8	....	2843.6	2928.0	3000.4	3065.3	3125.3	3195.6	3262.2
	S	3.2713	5.6948	....	5.8590	5.9990	6.1152	6.2162	6.3067	6.4092	6.5030
9000 (303.31)	V	1.418	20.495	....	22.685	24.836	26.694	28.372	29.929	31.754	33.480
	U	1351.0	2560.1	....	2630.1	2697.4	2754.4	2805.2	2851.8	2906.3	2957.8
	H	1363.7	2744.6	....	2834.3	2920.9	2994.7	3060.5	3121.2	3192.0	3259.2
	S	3.2867	5.6820	....	5.8355	5.9792	6.0976	6.2000	6.2915	6.3949	6.4894
9200 (304.89)	V	1.425	19.964	....	21.952	24.110	25.961	27.625	29.165	30.966	32.668
	U	1359.7	2557.7	....	2622.7	2691.9	2750.0	2801.5	2848.7	2903.6	2955.5
	H	1372.8	2741.3	....	2824.7	2913.7	2988.9	3055.7	3117.0	3188.5	3256.1
	S	3.3018	5.6694	....	5.8118	5.9594	6.0801	6.1840	6.2765	6.3808	6.4760
9400 (306.44)	V	1.432	19.455	....	21.245	23.412	25.257	26.909	28.433	30.212	31.891
	U	1368.2	2555.2	....	2615.1	2686.3	2745.6	2797.8	2845.5	2900.9	2953.2
	H	1381.7	2738.0	....	2814.8	2906.3	2983.0	3050.7	3112.8	3184.9	3253.0
	S	3.3168	5.6568	....	5.7879	5.9397	6.0627	6.1681	6.2617	6.3669	6.4628
9600 (307.97)	V	1.439	18.965	....	20.561	22.740	24.581	26.221	27.731	29.489	31.145
	U	1376.7	2552.6	....	2607.3	2680.5	2741.0	2794.1	2842.3	2898.2	2950.9
	H	1390.6	2734.7	....	2804.7	2898.8	2977.0	3045.8	3108.5	3181.3	3249.9
	S	3.3315	5.6444	....	5.7637	5.9199	6.0454	6.1524	6.2470	6.3532	6.4498

9800 (309.48)	V	1.446	18.494	....	19.899	22.093	23.931	25.561	27.056	28.795	30.429
	U	1385.2	2550.0	....	2599.2	2674.7	2736.4	2790.3	2839.1	2895.5	2948.6
	H	1399.3	2731.2	....	2794.3	2891.2	2971.0	3040.8	3104.2	3177.7	3246.8
	S	3.3461	5.6321	....	5.7393	5.9001	6.0282	6.1368	6.2325	6.3397	6.4369
10000 (310.96)	V	1.453	18.041	....	19.256	21.468	23.305	24.926	26.408	28.128	29.742
	U	1393.5	2547.3	....	2590.9	2668.7	2731.8	2786.4	2835.8	2892.8	2946.2
	H	1408.0	2727.7	....	2783.5	2883.4	2964.8	3035.7	3099.9	3174.1	3243.6
	S	3.3605	5.6198	....	5.7145	5.8803	6.0110	6.1213	6.2182	6.3264	6.4243
10200 (312.42)	V	1.460	17.605	....	18.632	20.865	22.702	24.315	25.785	27.487	29.081
	U	1401.8	2544.6	....	2582.3	2662.6	2727.0	2782.6	2832.6	2890.0	2943.9
	H	1416.7	2724.2	....	2772.3	2875.4	2958.6	3030.6	3095.6	3170.4	3240.5
	S	3.3748	5.6076	....	5.6894	5.8604	5.9940	6.1059	6.2040	6.3131	6.4118
10400 (313.86)	V	1.467	17.184	....	18.024	20.282	22.121	23.726	25.185	26.870	28.446
	U	1410.0	2541.8	....	2573.4	2656.3	2722.2	2778.7	2829.3	2887.3	2941.5
	H	1425.2	2720.6	....	2760.8	2867.2	2952.3	3025.4	3091.2	3166.7	3237.3
	S	3.3889	5.5955	....	5.6638	5.8404	5.9769	6.0907	6.1899	6.3001	6.3994
10600 (315.27)	V	1.474	16.778	....	17.432	19.717	21.560	23.159	24.607	26.276	27.834
	U	1418.1	2539.0	....	2564.1	2649.9	2717.4	2774.7	2825.9	2884.5	2939.1
	H	1433.7	2716.9	....	2748.9	2858.9	2945.9	3020.2	3086.8	3163.0	3234.1
	S	3.4029	5.5835	....	5.6376	5.8203	5.9599	6.0755	6.1759	6.2872	6.3872
10800 (316.67)	V	1.481	16.385	....	16.852	19.170	21.018	22.612	24.050	25.703	27.245
	U	1426.2	2536.2	....	2554.5	2643.4	2712.4	2770.7	2822.6	2881.7	2936.7
	H	1442.2	2713.1	....	2736.5	2850.4	2939.4	3014.9	3082.3	3159.3	3230.9
	S	3.4167	5.5715	....	5.6109	5.8000	5.9429	6.0604	6.1621	6.2744	6.3752
11000 (318.05)	V	1.489	16.006	....	16.285	18.639	20.494	22.083	23.512	25.151	26.676
	U	1434.2	2533.2	....	2544.4	2636.7	2707.4	2766.7	2819.2	2878.9	2934.3
	H	1450.6	2709.3	....	2723.5	2841.7	2932.8	3009.6	3077.8	3155.5	3227.7
	S	3.4304	5.5595	....	5.5835	5.7797	5.9259	6.0454	6.1483	6.2617	6.3633
11200 (319.40)	V	1.496	15.639	....	15.726	18.124	19.987	21.573	22.993	24.619	26.128
	U	1442.1	2530.3	....	2533.8	2629.8	2702.2	2762.6	2815.8	2876.0	2931.8
	H	1458.9	2705.4	....	2710.0	2832.8	2926.1	3004.2	3073.3	3151.7	3224.5
	S	3.4440	5.5476	....	5.5553	5.7591	5.9090	6.0305	6.1347	6.2491	6.3515
11400 (320.74)	V	1.504	15.284	....	....	17.622	19.495	21.079	22.492	24.104	25.599
	U	1450.0	2527.2	....	....	2622.7	2697.0	2758.4	2812.3	2873.1	2929.4
	H	1467.2	2701.5	....	....	2823.6	2919.3	2998.7	3068.7	3147.9	3221.2
	S	3.4575	5.5357	....	....	5.7383	5.8920	6.0156	6.1211	6.2367	6.3399

TABLE C.2. SUPERHEATED STEAM SI UNITS (Continued)

ABS PRESS KPA (SAT TEMP) DEG C		SAT WATER	SAT STEAM	TEMPERATURE, DEG C (TEMPERATURE, K)							
				475 (748.15)	500 (773.15)	525 (798.15)	550 (823.15)	575 (848.15)	600 (873.15)	625 (898.15)	650 (923.15)
8200 (296.70)	V	1.391	22.863	38.893	40.614	42.295	43.943	45.566	47.166	48.747	50.313
	U	1315.2	2569.5	3015.6	3063.3	3110.5	3157.4	3204.3	3251.1	3298.1	3345.2
	H	1326.6	2757.0	3334.5	3396.4	3457.3	3517.8	3577.9	3637.9	3697.8	3757.7
	S	3.2239	5.7338	6.6311	6.7124	6.7900	6.8646	6.9365	7.0062	7.0739	7.1397
8400 (298.39)	V	1.398	22.231	37.887	39.576	41.224	42.839	44.429	45.996	47.544	49.076
	U	1324.3	2567.2	3013.6	3061.6	3108.9	3155.9	3202.9	3249.8	3296.9	3344.1
	H	1336.1	2754.0	3331.9	3394.0	3455.2	3515.8	3576.1	3636.2	3696.2	3756.3
	S	3.2399	5.7207	6.6173	6.6990	6.7769	6.8516	6.9238	6.9936	7.0614	7.1274
8600 (300.06)	V	1.404	21.627	36.928	38.586	40.202	41.787	43.345	44.880	46.397	47.897
	U	1333.3	2564.9	3011.6	3059.8	3107.3	3154.4	3201.5	3248.5	3295.7	3342.9
	H.	1345.4	2750.9	3329.2	3391.6	3453.0	3513.8	3574.3	3634.5	3694.7	3754.9
	S	3.2557	5.7076	6.6037	6.6858	6.7639	6.8390	6.9113	6.9813	7.0492	7.1153
8800 (301.70)	V	1.411	21.049	36.011	37.640	39.228	40.782	42.310	43.815	45.301	46.771
	U	1342.2	2562.6	3009.6	3058.0	3105.6	3152.9	3200.1	3247.2	3294.5	3341.8
	H	1354.6	2747.8	3326.5	3389.2	3450.8	3511.8	3572.4	3632.8	3693.1	3753.4
	S	3.2713	5.6948	6.5904	6.6728	6.7513	6.8265	6.8990	6.9692	7.0373	7.1035
9000 (303.31)	V	1.418	20.495	35.136	36.737	38.296	39.822	41.321	42.798	44.255	45.695
	U	1351.0	2560.1	3007.6	3056.1	3104.0	3151.4	3198.7	3246.0	3293.3	3340.7
	H	1363.7	2744.6	3323.8	3386.8	3448.7	3509.8	3570.6	3631.1	3691.6	3752.0
	S	3.2867	5.6820	6.5773	6.6600	6.7388	6.8143	6.8870	6.9574	7.0256	7.0919
9200 (304.89)	V	1.425	19.964	34.298	35.872	37.405	38.904	40.375	41.824	43.254	44.667
	U	1359.7	2557.7	3005.6	3054.3	3102.3	3149.9	3197.3	3244.7	3292.1	3339.6
	H	1372.8	2741.3	3321.1	3384.4	3446.5	3507.8	3568.8	3629.5	3690.0	3750.5
	S	3.3018	5.6694	6.5644	6.6475	6.7266	6.8023	6.8752	6.9457	7.0141	7.0806
9400 (306.44)	V	1.432	19.455	33.495	35.045	36.552	38.024	39.470	40.892	42.295	43.682
	U	1368.2	2555.2	3003.5	3052.5	3100.7	3148.4	3195.9	3243.4	3290.9	3338.5
	H	1381.7	2738.0	3318.4	3381.9	3444.3	3505.9	3566.9	3627.8	3688.4	3749.1
	S	3.3168	5.6568	6.5517	6.6352	6.7146	6.7906	6.8637	6.9343	7.0029	7.0695
9600 (307.97)	V	1.439	18.965	32.726	34.252	35.734	37.182	38.602	39.999	41.377	42.738
	U	1376.7	2552.6	3001.5	3050.7	3099.0	3146.9	3194.5	3242.1	3289.7	3337.4
	H	1390.6	2734.7	3315.6	3379.5	3442.1	3503.9	3565.1	3626.1	3686.9	3747.6
	S	3.3315	5.6444	6.5392	6.6231	6.7028	6.7790	6.8523	6.9231	6.9918	7.0585

	V	1.446	18.494	31.988	33.491	34.949	36.373	37.769	39.142	40.496	41.832
9800	U	1385.2	2550.0	2999.4	3048.8	3097.4	3145.4	3193.1	3240.8	3288.5	3336.2
(309.48)	H	1399.3	2731.2	3312.9	3377.0	3439.9	3501.9	3563.3	3624.4	3685.3	3746.2
	S	3.3461	5.6321	6.5268	6.6112	6.6912	6.7676	6.8411	6.9121	6.9810	7.0478
	V	1.453	18.041	31.280	32.760	34.196	35.597	36.970	38.320	39.650	40.963
10000	U	1393.5	2547.3	2997.4	3047.0	3095.7	3143.9	3191.7	3239.5	3287.3	3335.1
(310.96)	H	1408.0	2727.7	3310.1	3374.6	3437.7	3499.8	3561.4	3622.7	3683.8	3744.7
	S	3.3605	5.6198	6.5147	6.5994	6.6797	6.7564	6.8302	6.9013	6.9703	7.0373
	V	1.460	17.605	30.599	32.058	33.472	34.851	36.202	37.530	38.837	40.128
10200	U	1401.8	2544.6	2995.3	3045.2	3094.0	3142.3	3190.3	3238.2	3286.1	3334.0
(312.42)	H	1416.7	2724.2	3307.4	3372.1	3435.5	3497.8	3559.6	3621.0	3682.2	3743.3
	S	3.3748	5.6076	6.5027	6.5879	6.6685	6.7454	6.8194	6.8907	6.9598	7.0269
	V	1.467	17.184	29.943	31.382	32.776	34.134	35.464	36.770	38.056	39.325
10400	U	1410.0	2541.8	2993.2	3043.3	3092.4	3140.8	3188.9	3236.9	3284.8	3332.9
(313.86)	H	1425.2	2720.6	3304.6	3369.7	3433.2	3495.8	3557.8	3619.3	3680.6	3741.8
	S	3.3889	5.5955	6.4909	6.5765	6.6574	6.7346	6.8087	6.8803	6.9495	7.0167
	V	1.474	16.778	29.313	30.732	32.106	33.444	34.753	36.039	37.304	38.552
10600	U	1418.1	2539.0	2991.1	3041.4	3090.7	3139.3	3187.5	3235.6	3283.6	3331.7
(315.27)	H	1433.7	2716.9	3301.8	3367.2	3431.0	3493.8	3555.9	3617.6	3679.1	3740.4
	S	3.4029	5.5835	6.4793	6.5652	6.6465	6.7239	6.7983	6.8700	6.9394	7.0067
	V	1.481	16.385	28.706	30.106	31.461	32.779	34.069	35.335	36.580	37.808
10800	U	1426.2	2536.2	2989.0	3039.6	3089.0	3137.8	3186.1	3234.3	3282.4	3330.6
(316.67)	H	1442.2	2713.1	3299.0	3364.7	3428.8	3491.8	3554.1	3615.9	3677.5	3738.9
	S	3.4167	5.5715	6.4678	6.5542	6.6357	6.7134	6.7880	6.8599	6.9294	6.9969
	V	1.489	16.006	28.120	29.503	30.839	32.139	33.410	34.656	35.882	37.091
11000	U	1434.2	2533.2	2986.9	3037.7	3087.3	3136.2	3184.7	3233.0	3281.2	3329.5
(318.05)	H	1450.6	2709.3	3296.2	3362.2	3426.5	3489.7	3552.2	3614.2	3675.9	3737.5
	S	3.4304	5.5595	6.4564	6.5432	6.6251	6.7031	6.7779	6.8499	6.9196	6.9872
	V	1.496	15.639	27.555	28.921	30.240	31.521	32.774	34.002	35.210	36.400
11200	U	1442.1	2530.3	2984.8	3035.8	3085.6	3134.7	3183.3	3231.7	3280.0	3328.4
(319.40)	H	1458.9	2705.4	3293.4	3359.7	3424.3	3487.7	3550.4	3612.5	3674.4	3736.0
	S	3.4440	5.5476	6.4452	6.5324	6.6147	6.6929	6.7679	6.8401	6.9099	6.9777
	V	1.504	15.284	27.010	28.359	29.661	30.925	32.160	33.370	34.560	35.733
11400	U	1450.0	2527.2	2982.6	3033.9	3083.9	3133.1	3181.9	3230.4	3278.8	3327.2
(320.74)	H	1467.2	2701.5	3290.5	3357.2	3422.1	3485.7	3548.5	3610.8	3672.8	3734.6
	S	3.4575	5.5357	6.4341	6.5218	6.6043	6.6828	6.7580	6.8304	6.9004	6.9683

TABLE C.2. SUPERHEATED STEAM SI UNITS (Continued)

ABS PRESS KPA (SAT TEMP) DEG C		SAT WATER	SAT STEAM	TEMPERATURE, DEG C (TEMPERATURE, K) 340 (613.15)	360 (633.15)	380 (653.15)	400 (673.15)	420 (693.15)	440 (713.15)	460 (733.15)	480 (753.15)
11600 (322.06)	V	1.511	14.940	17.134	19.019	20.601	22.007	23.298	24.507	25.655	26.755
	U	1457.9	2524.1	2615.5	2691.7	2754.2	2808.8	2858.4	2904.7	2948.7	2990.9
	H	1475.4	2697.4	2814.2	2912.3	2993.2	3064.1	3128.7	3189.0	3246.3	3301.3
	S	3.4708	5.5239	5.7174	5.8749	6.0007	6.1077	6.2022	6.2880	6.3672	6.4413
11800 (323.36)	V	1.519	14.607	16.658	18.557	20.139	21.538	22.820	24.019	25.155	26.243
	U	1465.7	2521.0	2608.0	2686.3	2750.0	2805.3	2855.4	2902.1	2946.3	2988.8
	H	1483.6	2693.3	2804.6	2905.3	2987.6	3059.5	3124.7	3185.5	3243.1	3298.5
	S	3.4840	5.5121	5.6962	5.8579	5.9860	6.0943	6.1898	6.2763	6.3561	6.4305
12000 (324.65)	V	1.527	14.283	16.193	18.108	19.691	21.084	22.357	23.546	24.672	25.748
	U	1473.4	2517.8	2600.4	2680.8	2745.7	2801.8	2852.4	2899.4	2944.0	2986.7
	H	1491.8	2689.2	2794.7	2898.1	2982.0	3054.8	3120.7	3182.0	3240.0	3295.7
	S	3.4972	5.5002	5.6747	5.8408	5.9712	6.0810	6.1775	6.2647	6.3450	6.4199
12200 (325.91)	V	1.535	13.969	15.740	17.672	19.256	20.645	21.910	23.089	24.204	25.269
	U	1481.2	2514.5	2592.5	2675.2	2741.4	2798.2	2849.3	2896.8	2941.6	2984.6
	H	1499.9	2684.9	2784.5	2890.8	2976.3	3050.0	3116.6	3178.5	3236.9	3292.9
	S	3.5102	5.4884	5.6528	5.8236	5.9565	6.0678	6.1653	6.2532	6.3341	6.4094
12400 (327.17)	V	1.543	13.664	15.296	17.248	18.835	20.219	21.476	22.646	23.751	24.806
	U	1488.8	2511.1	2584.4	2669.5	2737.0	2794.6	2846.3	2894.1	2939.2	2982.5
	H	1508.0	2680.6	2774.0	2883.4	2970.5	3045.3	3112.6	3174.9	3233.8	3290.1
	S	3.5232	5.4765	5.6307	5.8063	5.9418	6.0546	6.1532	6.2418	6.3232	6.3990
12600 (328.40)	V	1.551	13.367	14.861	16.836	18.426	19.805	21.055	22.217	23.312	24.357
	U	1496.5	2507.7	2576.0	2663.7	2732.5	2790.9	2843.2	2891.4	2936.9	2980.4
	H	1516.0	2676.1	2763.2	2875.8	2964.6	3040.5	3108.5	3171.3	3230.6	3287.3
	S	3.5361	5.4646	5.6082	5.7889	5.9272	6.0416	6.1411	6.2305	6.3125	6.3888
12800 (329.62)	V	1.559	13.078	14.434	16.433	18.028	19.404	20.648	21.801	22.887	23.922
	U	1504.1	2504.2	2567.3	2657.7	2728.0	2787.2	2840.1	2888.7	2934.5	2978.2
	H	1524.0	2671.6	2752.1	2868.1	2958.7	3035.6	3104.3	3167.7	3227.4	3284.4
	S	3.5488	5.4527	5.5852	5.7715	5.9125	6.0285	6.1291	6.2193	6.3019	6.3786
13000 (330.83)	V	1.567	12.797	14.015	16.041	17.642	19.015	20.252	21.397	22.474	23.500
	U	1511.6	2500.6	2558.4	2651.6	2723.4	2783.5	2836.9	2886.0	2932.1	2976.1
	H	1532.0	2667.0	2740.6	2860.2	2952.7	3030.7	3100.2	3164.1	3224.2	3281.6
	S	3.5616	5.4408	5.5618	5.7539	5.8979	6.0155	6.1173	6.2082	6.2913	6.3685

13200 (332.02)	V	1.576	12.523	13.602	15.659	17.266	18.637	19.868	21.006	22.074	23.091
	U	1519.2	2497.0	2549.1	2645.4	2718.7	2779.8	2833.7	2883.2	2929.6	2973.9
	H	1540.0	2662.3	2728.7	2852.1	2946.6	3025.8	3096.0	3160.5	3221.0	3278.7
	S	3.5742	5.4288	5.5378	5.7362	5.8832	6.0026	6.1054	6.1972	6.2809	6.3585
13400 (333.19)	V	1.584	12.256	13.195	15.285	16.900	18.270	19.495	20.625	21.686	22.694
	U	1526.7	2493.2	2539.5	2639.1	2714.0	2776.0	2830.6	2880.5	2927.2	2971.7
	H	1547.9	2657.4	2716.3	2843.9	2940.5	3020.8	3091.8	3156.8	3217.8	3275.8
	S	3.5868	5.4168	5.5133	5.7183	5.8685	5.9897	6.0937	6.1862	6.2705	6.3486
13600 (334.36)	V	1.593	11.996	12.793	14.920	16.544	17.912	19.133	20.256	21.309	22.308
	U	1534.2	2489.4	2529.4	2632.6	2709.2	2772.2	2827.3	2877.7	2924.8	2969.5
	H	1555.8	2652.5	2703.4	2835.5	2934.2	3015.8	3087.6	3153.2	3214.6	3272.9
	S	3.5993	5.4047	5.4880	5.7002	5.8538	5.9769	6.0820	6.1753	6.2603	6.3388
13800 (335.51)	V	1.602	11.743	12.394	14.563	16.196	17.565	18.781	19.897	20.942	21.933
	U	1541.6	2485.5	2518.9	2625.9	2704.3	2768.3	2824.1	2874.9	2922.3	2967.4
	H	1563.7	2647.5	2689.9	2826.9	2927.9	3010.7	3083.3	3149.5	3211.3	3270.0
	S	3.6118	5.3925	5.4619	5.6820	5.8391	5.9641	6.0704	6.1645	6.2501	6.3291
14000 (336.64)	V	1.611	11.495	11.997	14.213	15.858	17.227	18.438	19.549	20.586	21.569
	U	1549.1	2481.4	2507.7	2619.1	2699.4	2764.4	2820.8	2872.1	2919.8	2965.2
	H	1571.6	2642.4	2675.7	2818.1	2921.4	3005.6	3079.0	3145.8	3208.1	3267.1
	S	3.6242	5.3803	5.4348	5.6636	5.8243	5.9513	6.0588	6.1538	6.2399	6.3194
14200 (337.76)	V	1.620	11.253	11.600	13.871	15.528	16.897	18.105	19.209	20.240	21.215
	U	1556.5	2477.3	2495.9	2612.1	2694.4	2760.4	2817.6	2869.3	2917.4	2962.9
	H	1579.5	2637.1	2660.6	2809.1	2914.9	3000.4	3074.6	3142.0	3204.8	3264.2
	S	3.6366	5.3679	5.4064	5.6449	5.8095	5.9385	6.0473	6.1431	6.2299	6.3099
14400 (338.87)	V	1.629	11.017	11.199	13.535	15.206	16.576	17.781	18.879	19.903	20.871
	U	1563.9	2473.1	2483.2	2605.0	2689.3	2756.4	2814.2	2866.4	2914.9	2960.7
	H	1587.4	2631.8	2644.4	2799.9	2908.3	2995.1	3070.3	3138.3	3201.5	3261.3
	S	3.6490	5.3555	5.3762	5.6260	5.7947	5.9258	6.0358	6.1325	6.2199	6.3004
14600 (339.97)	V	1.638	10.786	10.791	13.205	14.891	16.264	17.465	18.558	19.576	20.536
	U	1571.3	2468.8	2469.1	2597.6	2684.1	2752.4	2810.9	2863.6	2912.4	2958.5
	H	1595.3	2626.3	2626.6	2790.4	2901.5	2989.9	3065.9	3134.5	3198.2	3258.3
	S	3.6613	5.3431	5.3436	5.6069	5.7797	5.9130	6.0244	6.1220	6.2100	6.2910
14800 (341.06)	V	1.648	10.561	....	12.881	14.583	15.959	17.157	18.245	19.256	20.210
	U	1578.7	2464.4	....	2590.1	2678.9	2748.3	2807.5	2860.7	2909.8	2956.3
	H	1603.1	2620.7	....	2780.7	2894.7	2984.5	3061.5	3130.7	3194.8	3255.4
	S	3.6736	5.3305	....	5.5874	5.7648	5.9003	6.0130	6.1115	6.2002	6.2817

TABLE C.2. SUPERHEATED STEAM SI UNITS (Continued)

ABS PRESS KPA (SAT TEMP) DEG C		SAT WATER	SAT STEAM	TEMPERATURE, DEG C (TEMPERATURE, K) 500 (773.15)	520 (793.15)	540 (813.15)	560 (833.15)	580 (853.15)	600 (873.15)	625 (898.15)	650 (923.15)
	V	1.511	14.940	27.817	28.848	29.855	30.840	31.808	32.761	33.934	35.089
11600	U	1457.9	2524.1	3032.0	3072.3	3111.9	3151.2	3190.2	3229.1	3277.6	3326.1
(322.06)	H	1475.4	2697.4	3354.7	3406.9	3458.2	3508.9	3559.2	3609.1	3671.2	3733.1
	S	3.4708	5.5239	6.5113	6.5779	6.6419	6.7034	6.7630	6.8209	6.8910	6.9590
	V	1.519	14.607	27.293	28.312	29.305	30.278	31.232	32.172	33.328	34.467
11800	U	1465.7	2521.0	3030.1	3070.5	3110.3	3149.7	3188.8	3227.8	3276.4	3325.0
(323.36)	H	1483.6	2693.3	3352.2	3404.6	3456.1	3507.0	3557.3	3607.4	3669.6	3731.7
	S	3.4840	5.5121	6.5009	6.5678	6.6320	6.6938	6.7535	6.8115	6.8818	6.9499
	V	1.527	14.283	26.786	27.793	28.774	29.734	30.676	31.603	32.743	33.865
12000	U	1473.4	2517.8	3028.2	3068.8	3108.7	3148.2	3187.4	3226.4	3275.2	3323.8
(324.65)	H	1491.8	2689.2	3349.6	3402.3	3454.0	3505.0	3555.5	3605.7	3668.1	3730.2
	S	3.4972	5.5002	6.4906	6.5578	6.6222	6.6842	6.7441	6.8022	6.8727	6.9409
	V	1.535	13.969	26.296	27.291	28.260	29.208	30.138	31.052	32.177	33.283
12200	U	1481.2	2514.5	3026.3	3067.0	3107.1	3146.7	3186.0	3225.1	3273.9	3322.7
(325.91)	H	1499.9	2684.9	3347.1	3400.0	3451.8	3503.0	3553.7	3604.0	3666.5	3728.8
	S	3.5102	5.4884	6.4805	6.5480	6.6126	6.6747	6.7348	6.7931	6.8637	6.9321
	V	1.543	13.664	25.821	26.805	27.763	28.699	29.617	30.519	31.629	32.720
12400	U	1488.8	2511.1	3024.4	3065.2	3105.4	3145.2	3184.6	3223.8	3272.7	3321.6
(327.17)	H	1508.0	2680.6	3344.5	3397.6	3449.7	3501.0	3551.8	3602.3	3664.9	3727.3
	S	3.5232	5.4765	6.4704	6.5382	6.6030	6.6654	6.7257	6.7841	6.8548	6.9234
	V	1.551	13.367	25.362	26.334	27.281	28.206	29.112	30.003	31.098	32.175
12600	U	1496.5	2507.7	3022.4	3063.5	3103.8	3143.7	3183.2	3222.5	3271.5	3320.4
(328.40)	H	1516.0	2676.1	3342.0	3395.3	3447.6	3499.1	3550.0	3600.5	3663.3	3725.8
	S	3.5361	5.4646	6.4605	6.5286	6.5936	6.6562	6.7166	6.7752	6.8461	6.9148
	V	1.559	13.078	24.916	25.878	26.814	27.728	28.623	29.503	30.584	31.647
12800	U	1504.1	2504.2	3020.5	3061.7	3102.2	3142.2	3181.8	3221.2	3270.3	3319.3
(329.62)	H	1524.0	2671.6	3339.4	3393.0	3445.4	3497.1	3548.2	3598.8	3661.7	3724.4
	S	3.5488	5.4527	6.4507	6.5190	6.5843	6.6471	6.7077	6.7664	6.8375	6.9063
	V	1.567	12.797	24.485	25.437	26.362	27.265	28.150	29.019	30.086	31.135
13000	U	1511.6	2500.6	3018.5	3059.9	3100.6	3140.6	3180.4	3219.9	3269.0	3318.2
(330.83)	H	1532.0	2667.0	3336.8	3390.6	3443.3	3495.1	3546.3	3597.1	3660.2	3722.9
	S	3.5616	5.4408	6.4409	6.5096	6.5752	6.6381	6.6989	6.7577	6.8289	6.8979

P (Tsat)											
13200 (332.02)	V	1.576	12.523	24.066	25.008	25.923	26.816	27.690	28.549	29.603	30.639
	U	1519.2	2497.0	3016.6	3058.1	3098.9	3139.1	3178.9	3218.5	3267.8	3317.0
	H	1540.0	2662.3	3334.3	3388.3	3441.1	3493.1	3544.5	3595.4	3658.6	3721.5
	S	3.5742	5.4288	6.4313	6.5003	6.5661	6.6292	6.6902	6.7492	6.8205	6.8896
13400 (333.19)	V	1.584	12.256	23.659	24.592	25.497	26.380	27.245	28.093	29.134	30.157
	U	1526.7	2493.2	3014.6	3056.4	3097.3	3137.6	3177.5	3217.2	3266.6	3315.9
	H	1547.9	2657.4	3331.7	3385.9	3438.9	3491.1	3542.6	3593.6	3657.0	3720.0
	S	3.5868	5.4168	6.4218	6.4910	6.5571	6.6205	6.6816	6.7407	6.8122	6.8814
13600 (334.36)	V	1.593	11.996	23.265	24.188	25.084	25.958	26.812	27.651	28.679	29.690
	U	1534.2	2489.4	3012.7	3054.6	3095.6	3136.1	3176.1	3215.9	3265.4	3314.8
	H	1555.8	2652.5	3329.1	3383.5	3436.8	3489.1	3540.7	3591.9	3655.4	3718.5
	S	3.5993	5.4047	6.4124	6.4819	6.5482	6.6118	6.6731	6.7323	6.8040	6.8734
13800 (335.51)	V	1.602	11.743	22.882	23.796	24.683	25.547	26.392	27.221	28.238	29.236
	U	1541.6	2485.5	3010.7	3052.8	3094.0	3134.5	3174.7	3214.5	3264.1	3313.6
	H	1563.7	2647.5	3326.4	3381.2	3434.6	3487.1	3538.9	3590.2	3653.8	3717.1
	S	3.6118	5.3925	6.4030	6.4729	6.5394	6.6032	6.6646	6.7241	6.7959	6.8654
14000 (336.64)	V	1.611	11.495	22.509	23.415	24.293	25.148	25.984	26.804	27.809	28.795
	U	1549.1	2481.4	3008.7	3051.0	3092.3	3133.0	3173.3	3213.2	3262.9	3312.5
	H	1571.6	2642.4	3323.8	3378.8	3432.4	3485.1	3537.0	3588.5	3652.2	3715.6
	S	3.6242	5.3803	6.3937	6.4639	6.5307	6.5947	6.6563	6.7159	6.7879	6.8575
14200 (337.76)	V	1.620	11.253	22.147	23.045	23.914	24.760	25.587	26.398	27.392	28.367
	U	1556.5	2477.3	3006.7	3049.1	3090.6	3131.5	3171.8	3211.9	3261.7	3311.3
	H	1579.5	2637.1	3321.2	3376.4	3430.2	3483.1	3535.2	3586.7	3650.6	3714.1
	S	3.6366	5.3679	6.3846	6.4550	6.5221	6.5863	6.6481	6.7078	6.7800	6.8497
14400 (338.87)	V	1.629	11.017	21.796	22.685	23.546	24.384	25.202	26.004	26.986	27.951
	U	1563.9	2473.1	3004.7	3047.3	3089.0	3129.9	3170.4	3210.5	3260.4	3310.2
	H	1587.4	2631.8	3318.6	3374.0	3428.0	3481.1	3533.3	3585.0	3649.0	3712.7
	S	3.6490	5.3555	6.3755	6.4463	6.5136	6.5780	6.6400	6.6998	6.7722	6.8420
14600 (339.97)	V	1.638	10.786	21.453	22.335	23.187	24.017	24.827	25.620	26.592	27.545
	U	1571.3	2468.8	3002.7	3045.5	3087.3	3128.4	3169.0	3209.2	3259.2	3309.0
	H	1595.3	2626.3	3315.9	3371.6	3425.8	3479.0	3531.4	3583.3	3647.5	3711.2
	S	3.6613	5.3431	6.3665	6.4376	6.5051	6.5697	6.6319	6.6919	6.7644	6.8344
14800 (341.06)	V	1.648	10.561	21.120	21.994	22.839	23.660	24.462	25.247	26.209	27.151
	U	1578.7	2464.4	3000.7	3043.7	3085.6	3126.8	3167.5	3207.9	3258.0	3307.9
	H	1603.1	2620.7	3313.3	3369.2	3423.6	3477.0	3529.6	3581.5	3645.9	3709.7
	S	3.6736	5.3305	6.3575	6.4289	6.4968	6.5616	6.6239	6.6841	6.7568	6.8269

TABLE C.2. SUPERHEATED STEAM SI UNITS (Continued)

ABS PRESS KPA (SAT TEMP) DEG C		SAT WATER	SAT STEAM	TEMPERATURE, DEG C (TEMPERATURE, K) 360 (633.15)	370 (643.15)	380 (653.15)	400 (673.15)	420 (693.15)	440 (713.15)	460 (733.15)	480 (753.15)
15000	V	1.658	10.340	12.562	13.481	14.282	15.661	16.857	17.940	18.946	19.893
(342.13)	U	1586.1	2459.9	2582.3	2631.4	2673.5	2744.2	2804.2	2857.8	2907.3	2954.0
	H	1611.0	2615.0	2770.8	2833.6	2887.7	2979.1	3057.0	3126.9	3191.5	3252.4
	S	3.6859	5.3178	5.5677	5.6662	5.7497	5.8876	6.0016	6.1010	6.1904	6.2724
15200	V	1.668	10.1246	12.248	13.181	13.988	15.371	16.565	17.643	18.643	19.584
(343.19)	U	1593.5	2455.3	2574.4	2624.9	2668.0	2740.0	2800.7	2854.9	2904.8	2951.7
	H	1618.9	2609.2	2760.5	2825.3	2880.7	2973.7	3052.5	3123.0	3188.1	3249.4
	S	3.6981	5.3051	5.5476	5.6491	5.7345	5.8749	5.9903	6.0907	6.1807	6.2632
15400	V	1.678	9.9136	11.939	12.885	13.700	15.087	16.279	17.354	18.348	19.283
(344.24)	U	1600.9	2450.6	2566.2	2618.3	2662.5	2735.8	2797.3	2851.9	2902.2	2949.5
	H	1626.8	2603.3	2750.0	2816.7	2873.5	2968.2	3048.0	3119.2	3184.8	3246.4
	S	3.7103	5.2922	5.5272	5.6317	5.7193	5.8622	5.9791	6.0803	6.1711	6.2541
15600	V	1.689	9.7072	11.635	12.595	13.418	14.810	16.001	17.071	18.060	18.989
(345.28)	U	1608.3	2445.8	2557.7	2611.5	2656.8	2731.6	2793.8	2849.0	2899.6	2947.2
	H	1634.7	2597.3	2739.2	2808.0	2866.2	2962.6	3043.4	3115.3	3181.4	3243.4
	S	3.7226	5.2793	5.5063	5.6142	5.7040	5.8495	5.9678	6.0701	6.1615	6.2450
15800	V	1.699	9.5052	11.334	12.311	13.142	14.539	15.729	16.796	17.780	18.703
(346.31)	U	1615.7	2440.9	2549.0	2604.6	2651.1	2727.2	2790.3	2846.0	2897.0	2944.9
	H	1642.6	2591.1	2728.0	2799.1	2858.7	2957.0	3038.8	3111.4	3178.0	3240.4
	S	3.7348	5.2663	5.4851	5.5964	5.6885	5.8367	5.9566	6.0598	6.1519	6.2360
16000	V	1.710	9.3075	11.036	12.030	12.871	14.275	15.464	16.527	17.506	18.423
(347.33)	U	1623.2	2436.0	2539.9	2597.4	2645.2	2722.9	2786.8	2843.0	2894.5	2942.6
	H	1650.5	2584.9	2716.5	2789.9	2851.1	2951.3	3034.2	3107.5	3174.5	3237.4
	S	3.7471	5.2531	5.4634	5.5784	5.6729	5.8240	5.9455	6.0497	6.1425	6.2270
16200	V	1.722	9.1141	10.742	11.755	12.605	14.016	15.205	16.265	17.239	18.151
(348.34)	U	1630.6	2430.9	2530.6	2590.1	2639.2	2718.4	2783.2	2840.0	2891.8	2940.3
	H	1658.5	2578.5	2704.6	2780.6	2843.4	2945.5	3029.5	3103.5	3171.1	3234.4
	S	3.7594	5.2399	5.4411	5.5602	5.6572	5.8112	5.9343	6.0395	6.1330	6.2181
16400	V	1.733	8.9247	10.450	11.483	12.344	13.763	14.952	16.008	16.978	17.885
(349.33)	U	1638.1	2425.7	2520.9	2582.6	2633.1	2714.0	2779.6	2837.0	2889.2	2938.0
	H	1666.5	2572.1	2692.3	2771.0	2835.5	2939.7	3024.8	3099.5	3167.7	3231.3
	S	3.7717	5.2267	5.4183	5.5417	5.6413	5.7984	5.9232	6.0294	6.1237	6.2093

16600 (350.32)	V	1.745	8.7385	10.160	11.216	12.087	13.515	14.704	15.758	16.724	17.625
	U	1645.5	2420.4	2510.9	2574.9	2626.9	2709.4	2776.0	2833.9	2886.6	2935.7
	H	1674.5	2565.5	2679.5	2761.1	2827.5	2933.8	3020.1	3095.5	3164.2	3228.2
	S	3.7842	5.2132	5.3949	5.5228	5.6253	5.7856	5.9120	6.0194	6.1143	6.2005
16800 (351.30)	V	1.757	8.5535	9.8712	10.952	11.835	13.272	14.462	15.514	16.475	17.372
	U	1653.5	2414.9	2500.4	2567.0	2620.5	2704.8	2772.4	2830.9	2883.9	2933.3
	H	1683.0	2558.6	2666.2	2751.0	2819.4	2927.8	3015.3	3091.5	3160.7	3225.2
	S	3.7974	5.1994	5.3708	5.5037	5.6091	5.7728	5.9009	6.0093	6.1050	6.1918
17000 (352.26)	V	1.770	8.3710	9.5837	10.691	11.588	13.034	14.225	15.274	16.232	17.124
	U	1661.6	2409.3	2489.5	2558.9	2614.0	2700.2	2768.7	2827.8	2881.3	2931.0
	H	1691.7	2551.6	2652.4	2740.7	2811.0	2921.7	3010.5	3087.5	3157.2	3222.1
	S	3.8107	5.1855	5.3458	5.4842	5.5928	5.7599	5.8899	5.9993	6.0958	6.1831
17200 (353.22)	V	1.783	8.1912	9.2964	10.434	11.344	12.801	13.994	15.041	15.994	16.882
	U	1669.7	2403.5	2478.0	2550.6	2607.4	2695.4	2764.9	2824.7	2878.6	2928.6
	H	1700.4	2544.4	2637.9	2730.0	2802.5	2915.6	3005.6	3083.4	3153.7	3219.0
	S	3.8240	5.1713	5.3200	5.4644	5.5762	5.7469	5.8788	5.9894	6.0866	6.1745
17400 (354.17)	V	1.796	8.0140	9.0084	10.179	11.104	12.573	13.767	14.812	15.762	16.645
	U	1677.7	2397.6	2466.0	2542.0	2600.6	2690.7	2761.2	2821.6	2875.9	2926.3
	H	1709.0	2537.1	2622.7	2719.1	2793.8	2909.4	3000.7	3079.3	3150.2	3215.9
	S	3.8372	5.1570	5.2931	5.4442	5.5595	5.7340	5.8677	5.9794	6.0775	6.1659
17600 (355.11)	V	1.810	7.8394	8.7187	9.9272	10.868	12.348	13.545	14.588	15.535	16.414
	U	1685.8	2391.6	2453.2	2533.1	2593.7	2685.8	2757.4	2818.4	2873.2	2923.9
	H	1717.6	2529.5	2606.6	2707.8	2784.9	2903.1	2995.8	3075.2	3146.6	3212.8
	S	3.8504	5.1425	5.2649	5.4235	5.5425	5.7209	5.8566	5.9695	6.0684	6.1574
17800 (356.04)	V	1.825	7.6674	1.9676	9.6775	10.635	12.129	13.328	14.369	15.313	16.188
	U	1693.7	2385.3	1743.6	2523.9	2586.5	2680.9	2753.6	2815.2	2870.5	2921.5
	H	1726.2	2521.8	1778.6	2696.2	2775.8	2896.8	2990.8	3071.0	3143.0	3209.6
	S	3.8635	5.1278	3.9466	5.4024	5.5253	5.7078	5.8456	5.9597	6.0593	6.1489
18000 (356.96)	V	1.840	7.4977	1.9465	9.4298	10.405	11.913	13.115	14.155	15.096	15.966
	U	1701.7	2378.9	1737.4	2514.5	2579.3	2675.9	2749.7	2812.1	2867.7	2919.1
	H	1734.8	2513.9	1772.4	2684.2	2766.6	2890.3	2985.8	3066.9	3139.4	3206.5
	S	3.8765	5.1128	3.9362	5.3808	5.5079	5.6947	5.8345	5.9498	6.0502	6.1405
18200 (357.87)	V	1.856	7.3302	1.9285	9.1838	10.178	11.701	12.906	13.945	14.883	15.750
	U	1709.7	2372.3	1732.0	2504.7	2571.8	2670.8	2745.8	2808.8	2865.0	2916.7
	H	1743.4	2505.8	1767.1	2671.9	2757.1	2883.8	2980.7	3062.7	3135.8	3203.3
	S	3.8896	5.0975	3.9272	5.3587	5.4902	5.6815	5.8234	5.9400	6.0412	6.1321

TABLE C.2. SUPERHEATED STEAM SI UNITS (Continued)

ABS PRESS KPA (SAT TEMP) DEG C		SAT WATER	SAT STEAM	TEMPERATURE, DEG C (TEMPERATURE, K) 500 (773.15)	520 (793.15)	540 (813.15)	560 (833.15)	580 (853.15)	600 (873.15)	625 (898.15)	650 (923.15)
15000 (342.13)	V	1.658	10.340	20.795	21.662	22.499	23.313	24.107	24.884	25.835	26.768
	U	1586.1	2459.9	2998.7	3041.8	3084.0	3125.3	3166.1	3206.5	3256.7	3306.7
	H	1611.0	2615.0	3310.6	3366.8	3421.4	3475.0	3527.7	3579.8	3644.3	3708.3
	S	3.6859	5.3178	6.3487	6.4204	6.4885	6.5535	6.6160	6.6764	6.7492	6.8195
15200 (343.19)	V	1.668	10.1246	20.479	21.339	22.169	22.975	23.761	24.530	25.472	26.394
	U	1593.5	2455.3	2996.6	3040.0	3082.3	3123.7	3164.6	3205.2	3255.5	3305.6
	H	1618.9	2609.2	3307.9	3364.4	3419.2	3473.0	3525.8	3578.0	3642.7	3706.8
	S	3.6981	5.3051	6.3399	6.4119	6.4803	6.5455	6.6082	6.6687	6.7417	6.8121
15400 (344.24)	V	1.678	9.9136	20.171	21.024	21.847	22.645	23.424	24.185	25.117	26.030
	U	1600.9	2450.6	2994.6	3038.2	3080.6	3122.2	3163.2	3203.8	3254.3	3304.5
	H	1626.8	2603.3	3305.2	3361.9	3417.0	3470.9	3523.9	3576.3	3641.1	3705.3
	S	3.7103	5.2922	6.3311	6.4035	6.4721	6.5376	6.6005	6.6612	6.7343	6.8049
15600 (345.28)	V	1.689	9.7072	19.871	20.717	21.533	22.324	23.095	23.850	24.772	25.676
	U	1608.3	2445.8	2992.6	3036.3	3078.9	3120.6	3161.8	3202.5	3253.0	3303.3
	H	1634.7	2597.3	3302.6	3359.5	3414.8	3468.9	3522.0	3574.5	3639.5	3703.8
	S	3.7226	5.2793	6.3225	6.3952	6.4641	6.5298	6.5928	6.6536	6.7270	6.7977
15800 (346.31)	V	1.699	9.5052	19.579	20.418	21.227	22.011	22.775	23.523	24.436	25.330
	U	1615.7	2440.9	2990.5	3034.5	3077.2	3119.1	3160.3	3201.1	3251.8	3302.1
	H	1642.6	2591.1	3299.9	3357.1	3412.6	3466.8	3520.2	3572.8	3637.9	3702.4
	S	3.7348	5.2663	6.3139	6.3869	6.4561	6.5220	6.5852	6.6462	6.7197	6.7905
16000 (347.33)	V	1.710	9.3075	19.293	20.126	20.928	21.706	22.463	23.204	24.108	24.994
	U	1623.2	2436.0	2988.4	3032.6	3075.5	3117.5	3158.9	3199.8	3250.5	3301.0
	H	1650.5	2584.9	3297.1	3354.6	3410.3	3464.8	3518.3	3571.0	3636.3	3700.9
	S	3.747,1	5.2531	6.3054	6.3787	6.4481	6.5143	6.5777	6.6389	6.7125	6.7835
16200 (348.34)	V	1.722	9.1141	19.015	19.841	20.637	21.409	22.159	22.892	23.788	24.665
	U	1630.6	2430.9	2986.4	3030.7	3073.8	3115.9	3157.4	3198.4	3249.3	3299.8
	H	1658.5	2578.5	3294.4	3352.1	3408.1	3462.7	3516.4	3569.3	3634.6	3699.4
	S	3.7594	5.2399	6.2969	6.3706	6.4403	6.5067	6.5703	6.6316	6.7054	6.7765
16400 (349.33)	V	1.733	8.9247	18.743	19.564	20.354	21.118	21.862	22.589	23.476	24.344
	U	1638.1	2425.7	2984.3	3028.8	3072.1	3114.3	3155.9	3197.1	3248.0	3298.7
	H	1666.5	2572.1	3291.7	3349.7	3405.9	3460.7	3514.5	3567.5	3633.0	3697.9
	S	3.7717	5.2267	6.2885	6.3625	6.4325	6.4991	6.5629	6.6243	6.6983	6.7696

16600 (350.32)	V	1.745	8.7385	18.478	19.293	20.076	20.835	21.572	22.292	23.172	24.032
	U	1645.5	2420.4	2982.2	3027.0	3070.4	3112.8	3154.5	3195.7	3246.8	3297.5
	H	1674.5	2565.5	3289.0	3347.2	3403.6	3458.6	3512.6	3565.8	3631.4	3696.5
	S	3.7842	5.2132	6.2801	6.3545	6.4247	6.4916	6.5556	6.6172	6.6913	6.7628
16800 (351.30)	V	1.757	8.5535	18.219	19.028	19.806	20.558	21.289	22.003	22.875	23.726
	U	1653.5	2414.9	2980.1	3025.1	3068.6	3111.2	3153.0	3194.3	3245.5	3296.4
	H	1683.0	2558.6	3286.2	3344.7	3401.4	3456.6	3510.7	3564.0	3629.8	3695.0
	S	3.7974	5.1994	6.2718	6.3466	6.4171	6.4841	6.5483	6.6101	6.6844	6.7560
17000 (352.26)	V	1.770	8.3710	17.966	18.770	19.542	20.288	21.013	21.721	22.584	23.428
	U	1661.6	2409.3	2978.0	3023.2	3066.9	3109.6	3151.6	3193.0	3244.3	3295.2
	H	1691.7	2551.6	3283.5	3342.3	3399.1	3454.5	3508.8	3562.2	3628.2	3693.5
	S	3.8107	5.1855	6.2636	6.3387	6.4095	6.4768	6.5411	6.6031	6.6776	6.7493
17200 (353.22)	V	1.783	8.1912	17.719	18.517	19.283	20.024	20.743	21.445	22.301	23.137
	U	1669.7	2403.5	2975.9	3021.3	3065.2	3108.0	3150.1	3191.6	3243.0	3294.1
	H	1700.4	2544.4	3280.7	3339.8	3396.9	3452.4	3506.9	3560.5	3626.6	3692.0
	S	3.8240	5.1713	6.2554	6.3308	6.4019	6.4694	6.5340	6.5961	6.6708	6.7426
17400 (354.17)	V	1.796	8.0140	17.478	18.270	19.031	19.766	20.480	21.175	22.024	22.852
	U	1677.7	2397.6	2973.8	3019.4	3063.4	3106.4	3148.6	3190.3	3241.8	3292.9
	H	1709.0	2537.1	3277.9	3337.3	3394.6	3450.4	3505.0	3558.7	3625.0	3690.5
	S	3.8372	5.1570	6.2473	6.3231	6.3944	6.4622	6.5269	6.5892	6.6640	6.7360
17600 (355.11)	V	1.810	7.8394	17.241	18.029	18.785	19.514	20.222	20.912	21.753	22.574
	U	1685.8	2391.6	2971.7	3017.5	3061.7	3104.8	3147.1	3188.9	3240.5	3291.7
	H	1717.6	2529.5	3275.2	3334.8	3392.3	3448.3	3503.1	3556.9	3623.4	3689.0
	S	3.8504	5.1425	6.2392	6.3153	6.3870	6.4550	6.5199	6.5824	6.6574	6.7295
17800 (356.04)	V	1.825	7.6674	17.011	17.793	18.544	19.268	19.970	20.654	21.488	22.303
	U	1693.7	2385.3	2969.6	3015.6	3060.0	3103.2	3145.7	3187.5	3239.2	3290.6
	H	1726.2	2521.8	3272.4	3332.3	3390.0	3446.2	3501.1	3555.2	3621.7	3687.6
	S	3.8635	5.1278	6.2311	6.3076	6.3796	6.4478	6.5130	6.5756	6.6507	6.7230
18000 (356.96)	V	1.840	7.4977	16.785	17.563	18.308	19.027	19.724	20.403	21.230	22.037
	U	1701.7	2378.9	2967.4	3013.6	3058.2	3101.6	3144.2	3186.1	3238.0	3289.4
	H	1734.8	2513.9	3269.6	3329.8	3387.8	3444.1	3499.2	3553.4	3620.1	3686.1
	S	3.8765	5.1128	6.2232	6.3000	6.3722	6.4407	6.5061	6.5688	6.6442	6.7166
18200 (357.87)	V	1.856	7.3302	16.564	17.337	18.077	18.791	19.483	20.156	20.977	21.777
	U	1709.7	2372.3	2965.3	3011.7	3056.5	3100.0	3142.7	3184.8	3236.7	3288.2
	H	1743.4	2505.8	3266.8	3327.2	3385.5	3442.0	3497.3	3551.6	3618.5	3684.6
	S	3.8896	5.0975	6.2152	6.2924	6.3650	6.4337	6.4992	6.5622	6.6377	6.7103

TABLE C.2. SUPERHEATED STEAM SI UNITS (Continued)

ABS PRESS KPA (SAT TEMP) DEG C		SAT WATER	SAT STEAM	TEMPERATURE, DEG C (TEMPERATURE, K)							
				380 (653.15)	390 (663.15)	400 (673.15)	410 (683.15)	420 (693.15)	440 (713.15)	460 (733.15)	480 (753.15)
	V	1.872	7.1647	9.9544	10.782	11.493	12.125	12.701	13.740	14.675	15.538
18400	U	1717.7	2365.6	2564.2	2619.5	2665.7	2705.9	2741.9	2805.6	2862.2	2914.3
(358.77)	H	1752.1	2497.4	2747.4	2817.8	2877.2	2929.0	2975.6	3058.4	3132.2	3200.2
	S	3.9028	5.0820	5.4722	5.5794	5.6682	5.7446	5.8124	5.9302	6.0323	6.1237
	V	1.889	7.0009	9.7333	10.572	11.288	11.923	12.501	13.539	14.471	15.330
18600	U	1725.7	2358.5	2556.4	2613.2	2660.5	2701.4	2737.9	2802.4	2859.4	2911.8
(359.67)	H	1760.9	2488.8	2737.4	2809.8	2870.4	2923.2	2970.4	3054.2	3128.6	3197.0
	S	3.9160	5.0661	5.4540	5.5641	5.6548	5.7326	5.8013	5.9204	6.0233	6.1154
	V	1.907	6.8386	9.5146	10.365	11.087	11.725	12.304	13.341	14.271	15.127
18800	U	1733.8	2351.3	2548.3	2606.8	2655.2	2696.8	2733.9	2799.1	2856.6	2909.4
(360.55)	H	1769.7	2479.8	2727.2	2801.7	2863.6	2917.3	2965.2	3049.9	3124.9	3193.8
	S	3.9294	5.0498	5.4354	5.5486	5.6413	5.7205	5.7902	5.9106	6.0144	6.1071
	V	1.926	6.6775	9.2983	10.160	10.889	11.531	12.111	13.148	14.075	14.928
19000	U	1742.1	2343.8	2540.1	2600.3	2649.8	2692.2	2729.9	2795.8	2853.8	2906.9
(361.43)	H	1778.7	2470.6	2716.8	2793.4	2856.7	2911.3	2960.0	3045.6	3121.3	3190.6
	S	3.9429	5.0332	5.4166	5.5330	5.6278	5.7083	5.7791	5.9009	6.0056	6.0988
	V	1.946	6.5173	9.0841	9.9593	10.695	11.340	11.922	12.959	13.884	14.733
19200	U	1750.4	2335.9	2531.6	2593.7	2644.3	2687.6	2725.8	2792.5	2851.0	2904.5
(362.30)	H	1787.8	2461.1	2706.0	2784.9	2849.7	2905.3	2954.7	3041.3	3117.6	3187.4
	S	3.9566	5.0160	5.3973	5.5172	5.6141	5.6962	5.7679	5.8912	5.9967	6.0906
	V	1.967	6.3575	8.8720	9.7608	10.504	11.152	11.736	12.773	13.696	14.542
19400	U	1758.9	2327.8	2522.9	2586.9	2638.8	2682.8	2721.6	2789.1	2848.2	2902.0
(363.16)	H	1797.0	2451.1	2695.0	2776.3	2842.5	2899.2	2949.3	3036.9	3113.9	3184.1
	S	3.9706	4.9983	5.3777	5.5012	5.6004	5.6840	5.7568	5.8814	5.9879	6.0825
	V	1.989	6.1978	8.6617	9.5649	10.315	10.968	11.553	12.590	13.511	14.355
19600	U	1767.6	2319.2	2514.0	2580.0	2633.1	2678.0	2717.5	2785.8	2845.3	2899.5
(364.02)	H	1806.5	2440.7	2683.7	2767.5	2835.3	2893.0	2943.9	3032.5	3110.2	3180.9
	S	3.9849	4.9801	5.3577	5.4850	5.5866	5.6717	5.7456	5.8717	5.9791	6.0743
	V	2.012	6.0377	8.4530	9.3715	10.130	10.786	11.374	12.412	13.331	14.171
19800	U	1776.5	2310.3	2504.7	2572.9	2627.4	2673.2	2713.2	2782.4	2842.5	2897.0
(364.86)	H	1816.3	2429.8	2672.1	2758.5	2828.0	2886.8	2938.4	3028.1	3106.4	3177.6
	S	3.9996	4.9610	5.3373	5.4686	5.5726	5.6594	5.7345	5.8620	5.9704	6.0662

P (T)											
20000 (365.70)	V	2.037	5.8766	8.2458	9.1805	9.9470	10.608	11.197	12.236	13.154	13.991
	U	1785.7	2300.8	2495.2	2565.7	2621.6	2668.3	2709.0	2778.9	2839.6	2894.5
	H	1826.5	2418.4	2660.2	2749.3	2820.5	2880.4	2932.9	3023.7	3102.7	3174.4
	S	4.0149	4.9412	5.3165	5.4520	5.5585	5.6470	5.7232	5.8523	5.9616	6.0581
20200 (366.53)	V	2.064	5.7138	8.0399	8.9918	9.7669	10.432	11.024	12.064	12.980	13.815
	U	1795.3	2290.8	2485.4	2558.3	2615.6	2663.3	2704.7	2775.5	2836.7	2892.0
	H	1837.0	2406.2	2647.9	2740.0	2812.9	2874.0	2927.4	3019.2	3098.9	3171.1
	S	4.0308	4.9204	5.2951	5.4352	5.5444	5.6345	5.7120	5.8427	5.9529	6.0501
20400 (367.36)	V	2.093	5.5484	7.8352	8.8052	9.5894	10.260	10.854	11.895	12.809	13.642
	U	1805.4	2280.1	2475.3	2550.8	2609.6	2658.3	2700.3	2772.0	2833.8	2889.5
	H	1848.1	2393.3	2635.2	2730.4	2805.2	2867.6	2921.7	3014.7	3095.1	3167.8
	S	4.0474	4.8984	5.2733	5.4181	5.5300	5.6220	5.7007	5.8330	5.9442	6.0421
20600 (368.17)	V	2.125	5.3793	7.6269	8.6206	9.4144	10.090	10.687	11.729	12.642	13.472
	U	1816.1	2268.5	2464.7	2543.1	2603.4	2653.1	2695.9	2768.5	2830.8	2887.0
	H	1859.9	2379.3	2621.9	2720.7	2797.4	2861.0	2916.0	3010.1	3091.3	3164.5
	S	4.0651	4.8750	5.2505	5.4007	5.5156	5.6094	5.6894	5.8233	5.9355	6.0341
20800 (368.98)	V	2.161	5.2050	7.4179	8.4380	9.2417	9.9225	10.522	11.566	12.478	13.305
	U	1827.6	2256.0	2453.4	2535.2	2597.2	2648.0	2691.4	2765.0	2827.9	2884.4
	H	1872.5	2364.2	2607.7	2710.7	2789.4	2854.3	2910.3	3005.6	3087.4	3161.2
	S	4.0841	4.8498	5.2265	5.3831	5.5010	5.5967	5.6781	5.8136	5.9269	6.0261
21000 (369.78)	V	2.202	5.0234	7.2076	8.2572	9.0714	9.7577	10.360	11.405	12.316	13.142
	U	1840.0	2242.1	2441.5	2527.1	2590.8	2642.7	2686.9	2761.5	2824.9	2881.9
	H	1886.3	2347.6	2592.8	2700.5	2781.3	2847.6	2904.5	3001.0	3083.6	3157.8
	S	4.1048	4.8223	5.2015	5.3652	5.4863	5.5840	5.6667	5.8040	5.9182	6.0182
21200 (370.58)	V	2.249	4.8314	6.9956	8.0781	8.9032	9.5953	10.201	11.248	12.158	12.981
	U	1853.9	2226.5	2428.9	2518.8	2584.4	2637.4	2682.4	2757.9	2822.0	2879.3
	H	1901.5	2328.9	2577.2	2690.1	2773.1	2840.8	2898.6	2996.4	3079.7	3154.5
	S	4.1279	4.7917	5.1754	5.3471	5.4714	5.5712	5.6553	5.7943	5.9096	6.0103
21400 (371.37)	V	2.306	4.6239	6.7811	7.9007	8.7372	9.4354	10.044	11.093	12.003	12.824
	U	1869.7	2208.4	2415.6	2510.4	2577.8	2631.9	2677.8	2754.3	2819.0	2876.7
	H	1919.0	2307.4	2560.7	2679.4	2764.7	2833.9	2892.7	2991.7	3075.8	3151.1
	S	4.1543	4.7569	5.1481	5.3286	5.4563	5.5583	5.6438	5.7846	5.9010	6.0024
21600 (372.15)	V	2.379	4.3918	6.5629	7.7248	8.5732	9.2777	9.8898	10.941	11.850	12.669
	U	1888.6	2186.7	2401.5	2501.7	2571.1	2626.5	2673.1	2750.7	2815.9	2874.1
	H	1940.0	2281.5	2543.2	2668.5	2756.2	2826.9	2886.7	2987.0	3071.9	3147.8
	S	4.1861	4.7154	5.1192	5.3098	5.4411	5.5453	5.6323	5.7750	5.8924	5.9945

TABLE C.2. SUPERHEATED STEAM SI UNITS (Continued)

ABS PRESS KPA (SAT TEMP) DEG C		SAT WATER	SAT STEAM	TEMPERATURE, DEG C (TEMPERATURE, K) 500 (773.15)	520 (793.15)	540 (813.15)	560 (833.15)	580 (853.15)	600 (873.15)	625 (898.15)	650 (923.15)
18400 (358.77)	V	1.872	7.1647	16.348	17.116	17.852	18.560	19.247	19.915	20.729	21.523
	U	1717.7	2365.6	2963.2	3009.8	3054.7	3098.4	3141.2	3183.4	3235.5	3287.1
	H	1752.1	2497.4	3263.9	3324.7	3383.2	3439.9	3495.4	3549.8	3616.9	3683.1
	S	3.9028	5.0820	6.2073	6.2849	6.3577	6.4267	6.4924	6.5555	6.6312	6.7040
18600 (359.67)	V	1.889	7.0009	16.136	16.900	17.631	18.335	19.017	19.680	20.487	21.274
	U	1725.7	2358.5	2961.0	3007.8	3052.9	3096.8	3139.7	3182.0	3234.2	3285.9
	H	1760.9	2488.8	3261.1	3322.2	3380.9	3437.8	3493.4	3548.1	3615.3	3681.6
	S	3.9160	5.0661	6.1995	6.2774	6.3505	6.4197	6.4857	6.5490	6.6248	6.6977
18800 (360.55)	V	1.907	6.8386	15.929	16.689	17.415	18.114	18.791	19.449	20.250	21.030
	U	1733.8	2351.3	2958.8	3005.9	3051.2	3095.2	3138.3	3180.6	3232.9	3284.7
	H	1769.7	2479.8	3258.3	3319.6	3378.6	3435.7	3491.5	3546.3	3613.6	3680.1
	S	3.9294	5.0498	6.1916	6.2700	6.3434	6.4128	6.4790	6.5424	6.6185	6.6915
19000 (361.43)	V	1.926	6.6775	15.726	16.481	17.203	17.898	18.570	19.223	20.018	20.792
	U	1742.1	2343.8	2956.7	3003.9	3049.4	3093.6	3136.8	3179.2	3231.7	3283.6
	H	1778.7	2470.6	3255.4	3317.1	3376.3	3433.6	3489.6	3544.5	3612.0	3678.6
	S	3.9429	5.0332	6.1839	6.2626	6.3363	6.4060	6.4723	6.5360	6.6122	6.6854
19200 (362.30)	V	1.946	6.5173	15.527	16.279	16.996	17.686	18.353	19.002	19.791	20.559
	U	1750.4	2335.9	2954.5	3002.0	3047.6	3091.9	3135.3	3177.9	3230.4	3282.4
	H	1787.8	2461.1	3252.6	3314.5	3374.0	3431.5	3487.6	3542.7	3610.4	3677.1
	S	3.9566	5.0160	6.1761	6.2552	6.3292	6.3992	6.4657	6.5295	6.6059	6.6793
19400 (363.16)	V	1.967	6.3575	15.333	16.080	16.793	17.479	18.141	18.785	19.568	20.330
	U	1758.9	2327.8	2952.3	3000.0	3045.9	3090.3	3133.8	3176.5	3229.1	3281.2
	H	1797.0	2451.1	3249.7	3312.0	3371.6	3429.4	3485.7	3540.9	3608.7	3675.6
	S	3.9706	4.9983	6.1684	6.2479	6.3222	6.3924	6.4592	6.5231	6.5997	6.6732
19600 (364.02)	V	1.989	6.1978	15.142	15.885	16.594	17.275	17.934	18.573	19.350	20.106
	U	1767.6	2319.2	2950.1	2998.0	3044.1	3088.7	3132.3	3175.1	3227.8	3280.1
	H	1806.5	2440.7	3246.9	3309.4	3369.3	3427.3	3483.8	3539.1	3607.1	3674.1
	S	3.9849	4.9801	6.1608	6.2406	6.3153	6.3857	6.4527	6.5168	6.5936	6.6672
19800 (364.86)	V	2.012	6.0377	14.955	15.694	16.399	17.076	17.730	18.365	19.136	19.887
	U	1776.5	2310.3	2947.9	2996.1	3042.3	3087.1	3130.8	3173.7	3226.6	3278.9
	H	1816.3	2429.8	3244.0	3306.8	3367.0	3425.2	3481.8	3537.3	3605.5	3672.6
	S	3.9996	4.9610	6.1532	6.2334	6.3084	6.3790	6.4462	6.5105	6.5875	6.6613

P (T)											
20000 (365.70)	V	2.037	5.8766	14.771	15.507	16.208	16.881	17.531	18.161	18.927	19.672
	U	1785.7	2300.8	2945.7	2994.1	3040.5	3085.4	3129.3	3172.3	3225.3	3277.7
	H	1826.5	2418.4	3241.1	3304.2	3364.7	3423.0	3479.9	3535.5	3603.8	3671.1
	S	4.0149	4.9412	6.1456	6.2262	6.3015	6.3724	6.4398	6.5043	6.5814	6.6554
20200 (366.53)	V	2.064	5.7138	14.592	15.324	16.021	16.690	17.335	17.962	18.722	19.461
	U	1795.3	2290.8	2943.4	2992.1	3038.7	3083.8	3127.7	3170.9	3224.0	3276.5
	H	1837.0	2406.2	3238.2	3301.6	3362.3	3420.9	3477.9	3533.7	3602.2	3669.6
	S	4.0308	4.9204	6.1381	6.2191	6.2946	6.3658	6.4334	6.4981	6.5754	6.6495
20400 (367.36)	V	2.093	5.5484	14.415	15.144	15.838	16.502	17.144	17.766	18.521	19.254
	U	1805.4	2280.1	2941.2	2990.1	3036.9	3082.1	3126.2	3169.5	3222.7	3275.4
	H	1848.1	2393.3	3235.3	3299.0	3360.0	3418.8	3476.0	3531.9	3600.6	3668.1
	S	4.0474	4.8984	6.1305	6.2120	6.2878	6.3593	6.4271	6.4919	6.5694	6.6437
20600 (368.17)	V	2.125	5.3793	14.243	14.968	15.658	16.318	16.956	17.574	18.323	19.052
	U	1816.1	2268.5	2939.0	2988.1	3035.1	3080.5	3124.7	3168.1	3221.5	3274.2
	H	1859.9	2379.3	3232.4	3296.4	3357.6	3416.6	3474.0	3530.1	3598.9	3666.6
	S	4.0651	4.8750	6.1231	6.2049	6.2811	6.3528	6.4208	6.4858	6.5635	6.6379
20800 (368.98)	V	2.161	5.2050	14.073	14.795	15.481	16.138	16.771	17.385	18.130	18.853
	U	1827.6	2256.0	2936.7	2986.1	3033.3	3078.8	3123.2	3166.7	3220.2	3273.0
	H	1872.5	2364.2	3229.5	3293.8	3355.3	3414.5	3472.1	3528.3	3597.3	3665.1
	S	4.0841	4.8498	6.1156	6.1978	6.2744	6.3463	6.4146	6.4798	6.5576	6.6322
21000 (369.78)	V	2.202	5.0234	13.907	14.625	15.308	15.961	16.591	17.201	17.940	18.658
	U	1840.0	2242.1	2934.5	2984.1	3031.5	3077.2	3121.7	3165.3	3218.9	3271.8
	H	1886.3	2347.6	3226.5	3291.2	3352.9	3412.4	3470.1	3526.5	3595.6	3663.6
	S	4.1048	4.8223	6.1082	6.1908	6.2677	6.3399	6.4084	6.4737	6.5518	6.6265
21200 (370.58)	V	2.249	4.8314	13.743	14.459	15.138	15.787	16.413	17.019	17.754	18.467
	U	1853.9	2226.5	2932.2	2982.1	3029.6	3075.5	3120.2	3163.9	3217.6	3270.6
	H	1901.5	2328.9	3223.6	3288.6	3350.6	3410.2	3468.1	3524.7	3594.0	3662.1
	S	4.1279	4.7917	6.1008	6.1838	6.2610	6.3335	6.4022	6.4677	6.5460	6.6208
21400 (371.37)	V	2.306	4.6239	13.583	14.295	14.971	15.617	16.239	16.841	17.571	18.279
	U	1869.7	2208.4	2930.0	2980.0	3027.8	3073.9	3118.6	3162.5	3216.3	3269.5
	H	1919.0	2307.4	3220.6	3286.0	3348.2	3408.1	3466.2	3522.9	3592.3	3660.6
	S	4.1543	4.7569	6.0935	6.1769	6.2544	6.3271	6.3961	6.4618	6.5402	6.6152
21600 (372.15)	V	2.379	4.3918	13.425	14.135	14.807	15.450	16.068	16.667	17.392	18.095
	U	1888.6	2186.7	2927.7	2978.0	3026.0	3072.2	3117.1	3161.1	3215.0	3268.3
	H	1940.0	2281.5	3217.7	3283.3	3345.8	3405.9	3464.2	3521.1	3590.7	3659.1
	S	4.1861	4.7154	6.0862	6.1700	6.2478	6.3208	6.3900	6.4559	6.5345	6.6096

TABLE C.2. SUPERHEATED STEAM SI UNITS (Continued)

ABS PRESS KPA (SAT TEMP) DEG C		SAT WATER	SAT STEAM	TEMPERATURE, DEG C (TEMPERATURE, K) 380 (653.15)	390 (663.15)	400 (673.15)	410 (683.15)	420 (693.15)	440 (713.15)	460 (733.15)	480 (753.15)
21800 (372.92)	V	2.483	4.1151	6.3399	7.5503	8.4112	9.1222	9.7379	10.792	11.700	12.517
	U	1913.1	2158.3	2386.3	2492.8	2564.2	2620.9	2668.4	2747.0	2812.9	2871.5
	H	1967.2	2248.0	2524.5	2657.4	2747.6	2829.8	2880.7	2982.3	3068.0	3144.4
	S	4.2276	4.6622	5.0886	5.2906	5.4257	5.5322	5.6207	5.7653	5.8839	5.9867
22000 (373.69)	V	2.671	3.7278	6.1105	7.3771	8.2510	8.9689	9.5883	10.645	11.552	12.368
	U	1952.4	2113.6	2370.0	2483.6	2557.2	2615.2	2663.6	2743.3	2809.9	2868.9
	H	2011.1	2195.6	2504.4	2645.9	2738.8	2812.6	2874.6	2977.5	3064.0	3141.0
	S	4.2947	4.5799	5.0559	5.2711	5.4102	5.5190	5.6091	5.7556	5.8753	5.9789
22120 (374.15)	V	3.170	3.1700	5.9689	7.2738	8.1558	8.8779	9.4995	10.558	11.465	12.279
	U	2037.3	2037.3	2359.4	2478.0	2553.0	2611.8	2660.7	2741.1	2808.0	2867.3
	H	2107.4	2107.4	2491.5	2638.9	2733.4	2808.2	2870.9	2974.6	3061.6	3139.0
	S	4.4429	4.4429	5.0350	5.2593	5.4007	5.5110	5.6021	5.7498	5.8702	5.9742

TABLE C.2. SUPERHEATED STEAM SI UNITS (Continued)

ABS PRESS KPA (SAT TEMP) DEG C		SAT WATER	SAT STEAM	TEMPERATURE, DEG C (TEMPERATURE, K) 500 (773.15)	520 (793.15)	540 (813.15)	560 (833.15)	580 (853.15)	600 (873.15)	625 (898.15)	650 (923.15)
	V	2.483	4.1151	13.271	13.977	14.646	15.285	15.900	16.495	17.216	17.915
21800	U	1913.1	2158.3	2925.4	2976.0	3024.1	3070.5	3115.6	3159.6	3213.7	3267.1
(372.92)	H	1967.2	2248.0	3214.7	3280.7	3343.4	3403.7	3462.2	3519.2	3589.0	3657.6
	S	4.2276	4.6622	6.0789	6.1631	6.2413	6.3146	6.3839	6.4500	6.5288	6.6041
	V	2.671	3.7278	13.119	13.822	14.488	15.124	15.736	16.327	17.043	17.737
22000	U	1952.4	2113.6	2923.1	2973.9	3022.3	3068.9	3114.0	3158.2	3212.4	3265.9
(373.69)	H	2011.1	2195.6	3211.7	3278.0	3341.0	3401.6	3460.2	3517.4	3587.4	3656.1
	S	4.2947	4.5799	6.0716	6.1563	6.2347	6.3083	6.3779	6.4441	6.5231	6.5986
	V	3.170	3.1700	13.029	13.731	14.395	15.029	15.638	16.228	16.941	17.632
22120	U	2037.3	2037.3	2921.7	2972.7	3021.2	3067.8	3113.1	3157.4	3211.7	3265.2
(374.15)	H	2107.4	2107.4	3209.9	3276.4	3339.6	3400.3	3459.0	3516.3	3586.4	3655.2
	S	4.4429	4.4429	6.0672	6.1522	6.2309	6.3046	6.3743	6.4406	6.5198	6.5953

TABLE C.3. SATURATED STEAM ENGLISH UNITS

V = SPECIFIC VOLUME (CU FT)/(LBM)
U = SPECIFIC INTERNAL ENERGY (BTU)/(LBM)
H = SPECIFIC ENTHALPY (BTU)/(LBM)
S = SPECIFIC ENTROPY (BTU)/(LBM)(R)

TEMP	ABS PRESS	SPECIFIC VOLUME V			INTERNAL ENERGY U			ENTHALPY H			ENTROPY S		
T (DEG F)	P (PSIA)	SAT LIQUID	EVAP	SAT VAPOR	SAT LIQUID	EVAP	SAT VAPOR	SAT LIQUID	EVAP	SAT VAPOR	SAT LIQUID	EVAP	SAT VAPOR
32	0.0886	0.01602	3304.6	3304.6	-0.02	1021.3	1021.3	-0.02	1075.5	1075.5	0.0	2.1873	2.1873
34	0.0960	0.01602	3061.9	3061.9	2.00	1020.0	1022.0	2.00	1074.4	1076.4	0.0041	2.1762	2.1802
36	0.1040	0.01602	2839.0	2839.0	4.01	1018.6	1022.6	4.01	1073.2	1077.2	0.0081	2.1651	2.1732
38	0.1125	0.01602	2634.1	2634.2	6.02	1017.3	1023.3	6.02	1072.1	1078.1	0.0122	2.1541	2.1663
40	0.1216	0.01602	2445.8	2445.8	8.03	1015.9	1023.9	8.03	1071.0	1079.0	0.0162	2.1432	2.1594
42	0.1314	0.01602	2272.4	2272.4	10.03	1014.6	1024.6	10.03	1069.8	1079.9	0.0202	2.1325	2.1527
44	0.1419	0.01602	2112.8	2112.8	12.04	1013.2	1025.2	12.04	1068.7	1080.7	0.0242	2.1217	2.1459
46	0.1531	0.01602	1965.7	1965.7	14.05	1011.9	1025.9	14.05	1067.6	1081.6	0.0282	2.1111	2.1393
48	0.1651	0.01602	1830.0	1830.0	16.05	1010.5	1026.6	16.05	1066.4	1082.5	0.0321	2.1006	2.1327
50	0.1780	0.01602	1704.8	1704.8	18.05	1009.2	1027.2	18.05	1065.3	1083.4	0.0361	2.0901	2.1262
52	0.1916	0.01602	1589.2	1589.2	20.06	1007.8	1027.9	20.06	1064.2	1084.2	0.0400	2.0798	2.1197
54	0.2063	0.01603	1482.4	1482.4	22.06	1006.5	1028.5	22.06	1063.1	1085.1	0.0439	2.0695	2.1134
56	0.2218	0.01603	1383.6	1383.6	24.06	1005.1	1029.2	24.06	1061.9	1086.0	0.0478	2.0593	2.1070
58	0.2384	0.01603	1292.2	1292.2	26.06	1003.8	1029.3	26.06	1060.8	1086.9	0.0516	2.0491	2.1008
60	0.2561	0.01603	1207.6	1207.6	28.06	1002.4	1030.5	28.06	1059.7	1087.7	0.0555	2.0391	2.0946
62	0.2749	0.01604	1129.2	1129.2	30.06	1001.1	1031.2	30.06	1058.5	1088.6	0.0593	2.0291	2.0885
64	0.2950	0.01604	1056.5	1056.5	32.06	999.8	1031.8	32.06	1057.4	1089.5	0.0632	2.0192	2.0824
66	0.3163	0.01604	989.0	989.1	34.06	998.4	1032.5	34.06	1056.3	1090.4	0.0670	2.0094	2.0764
68	0.3389	0.01605	926.5	926.5	36.05	997.1	1033.1	36.05	1055.2	1091.2	0.0708	1.9996	2.0704
70	0.3629	0.01605	868.3	868.4	38.05	995.7	1033.8	38.05	1054.0	1092.1	0.0745	1.9900	2.0645
72	0.3884	0.01605	814.3	814.3	40.05	994.4	1034.4	40.05	1052.9	1093.0	0.0783	1.9804	2.0587
74	0.4155	0.01606	764.1	764.1	42.05	993.0	1035.1	42.05	1051.8	1093.8	0.0821	1.9708	2.0529
76	0.4442	0.01606	717.4	717.4	44.04	991.7	1035.7	44.04	1050.7	1094.7	0.0858	1.9614	2.0472
78	0.4746	0.01607	673.8	673.9	46.04	990.3	1036.4	46.04	1049.5	1095.6	0.0895	1.9520	2.0415
80	0.5068	0.01607	633.3	633.3	48.03	989.0	1037.0	48.04	1048.4	1096.4	0.0932	1.9426	2.0359

82	0.5409	0.01608	595.5	595.6	50.03	987.7	1037.7	50.03	1047.3	1097.3	0.0969	1.9334	2.0303
84	0.5770	0.01608	560.3	560.3	52.03	986.3	1038.3	52.03	1046.1	1098.2	0.1006	1.9242	2.0248
86	0.6152	0.01609	527.5	527.5	54.02	985.0	1039.0	54.03	1045.0	1099.0	0.1043	1.9151	2.0193
88	0.6555	0.01609	496.8	496.8	56.02	983.6	1039.6	56.02	1043.9	1099.9	0.1079	1.9060	2.0139
90	0.6981	0.01610	468.1	468.1	58.02	982.3	1040.3	58.02	1042.7	1100.8	0.1115	1.8970	2.0086
92	0.7431	0.01610	441.3	441.3	60.01	980.9	1040.9	60.01	1041.6	1101.6	0.1152	1.8881	2.0033
94	0.7906	0.01611	416.3	416.3	62.01	979.6	1041.6	62.01	1040.5	1102.5	0.1188	1.8792	1.9980
96	0.8407	0.01612	392.8	392.9	64.00	978.2	1042.2	64.01	1039.3	1103.3	0.1224	1.8704	1.9928
98	0.8936	0.01612	370.9	370.9	66.00	976.9	1042.9	66.00	1038.2	1104.2	0.1260	1.8617	1.9876
100	0.9492	0.01613	350.4	350.4	68.00	975.5	1043.5	68.00	1037.1	1105.1	0.1295	1.8530	1.9825
102	1.0079	0.01614	331.1	331.1	69.99	974.2	1044.2	70.00	1035.9	1105.9	0.1331	1.8444	1.9775
104	1.0697	0.01614	313.1	313.1	71.99	972.8	1044.8	71.99	1034.8	1106.8	0.1366	1.8358	1.9725
106	1.1347	0.01615	296.2	296.2	73.98	971.5	1045.4	73.99	1033.6	1107.6	0.1402	1.8273	1.9675
108	1.2030	0.01616	280.3	280.3	75.98	970.1	1046.1	75.98	1032.5	1108.5	0.1437	1.8188	1.9626
110	1.275	0.01617	265.4	265.4	77.98	968.8	1046.7	77.98	1031.4	1109.3	0.1472	1.8105	1.9577
112	1.351	0.01617	251.4	251.4	79.97	967.4	1047.4	79.98	1030.2	1110.2	0.1507	1.8021	1.9528
114	1.430	0.01618	238.2	238.2	81.97	966.0	1048.0	81.97	1029.1	1111.0	0.1542	1.7938	1.9480
116	1.513	0.01619	225.8	225.9	83.97	964.7	1048.6	83.97	1027.9	1111.9	0.1577	1.7856	1.9433
118	1.601	0.01620	214.2	214.2	85.96	963.3	1049.3	85.97	1026.8	1112.7	0.1611	1.7774	1.9386
120	1.693	0.01620	203.25	203.26	87.96	962.0	1049.9	87.97	1025.6	1113.6	0.1646	1.7693	1.9339
122	1.789	0.01621	192.94	192.95	89.96	960.6	1050.6	89.96	1024.5	1114.4	0.1680	1.7613	1.9293
124	1.890	0.01622	183.23	183.24	91.96	959.2	1051.2	91.96	1023.3	1115.3	0.1715	1.7533	1.9247
126	1.996	0.01623	174.08	174.09	93.95	957.9	1051.8	93.96	1022.2	1116.1	0.1749	1.7453	1.9202
128	2.107	0.01624	165.45	165.47	95.95	956.5	1052.4	95.96	1021.0	1117.0	0.1783	1.7374	1.9157
130	2.223	0.01625	157.32	157.33	97.95	955.1	1053.1	97.96	1019.8	1117.8	0.1817	1.7295	1.9112
132	2.345	0.01626	149.64	149.66	99.95	953.8	1053.7	99.95	1018.7	1118.6	0.1851	1.7217	1.9068
134	2.472	0.01626	142.40	142.41	101.94	952.4	1054.3	101.95	1017.5	1119.5	0.1884	1.7140	1.9024
136	2.605	0.01627	135.55	135.57	103.94	951.0	1055.0	103.95	1016.4	1120.3	0.1918	1.7063	1.8980
138	2.744	0.01628	129.09	129.11	105.94	949.6	1055.6	105.95	1015.2	1121.1	0.1951	1.6986	1.8937
140	2.889	0.01629	122.98	123.00	107.94	948.3	1056.2	107.95	1014.0	1122.0	0.1985	1.6910	1.8895
142	3.041	0.01630	117.21	117.22	109.94	946.9	1056.8	109.95	1012.9	1122.8	0.2018	1.6834	1.8852
144	3.200	0.01631	111.74	111.76	111.94	945.5	1057.5	111.95	1011.7	1123.6	0.2051	1.6759	1.8810
146	3.365	0.01632	106.58	106.59	113.94	944.1	1058.1	113.95	1010.5	1124.5	0.2084	1.6684	1.8769
148	3.538	0.01633	101.68	101.70	115.94	942.8	1058.7	115.95	1009.3	1125.3	0.2117	1.6610	1.8727
150	3.718	0.01634	97.05	97.07	117.94	941.4	1059.3	117.95	1008.2	1126.1	0.2150	1.6536	1.8686
152	3.906	0.01635	92.66	92.68	119.94	940.0	1059.9	119.95	1007.0	1126.9	0.2183	1.6463	1.8646
154	4.102	0.01636	88.50	88.52	121.94	938.6	1060.5	121.95	1005.8	1127.7	0.2216	1.6390	1.8606
156	4.307	0.01637	84.56	84.57	123.94	937.2	1061.2	123.95	1004.6	1128.6	0.2248	1.6318	1.8566
158	4.520	0.01638	80.82	80.83	125.94	935.8	1061.8	125.96	1003.4	1129.4	0.2281	1.6245	1.8526
160	4.741	0.01640	77.27	77.29	127.94	934.4	1062.4	127.96	1002.2	1130.2	0.2313	1.6174	1.8487

TABLE C.3. SATURATED STEAM ENGLISH UNITS (Continued)

TEMP	ABS PRESS	SPECIFIC VOLUME V			INTERNAL ENERGY U			ENTHALPY H			ENTROPY S		
T (DEG F)	P (PSIA)	SAT LIQUID	EVAP	SAT VAPOR	SAT LIQUID	EVAP	SAT VAPOR	SAT LIQUID	EVAP	SAT VAPOR	SAT LIQUID	EVAP	SAT VAPOR
162	4.972	0.01641	73.90	73.92	129.95	933.0	1063.0	129.96	1001.0	1131.0	0.2345	1.6103	1.8448
164	5.212	0.01642	70.70	70.72	131.95	931.6	1063.6	131.96	999.8	1131.8	0.2377	1.6032	1.8409
166	5.462	0.01643	67.67	67.68	133.95	930.2	1064.2	133.97	998.6	1132.6	0.2409	1.5961	1.8371
168	5.722	0.01644	64.78	64.80	135.95	928.8	1064.8	135.97	997.4	1133.4	0.2441	1.5892	1.8333
170	5.993	0.01645	62.04	62.06	137.96	927.4	1065.4	137.97	996.2	1134.2	0.2473	1.5822	1.8295
172	6.274	0.01646	59.43	59.45	139.96	926.0	1066.0	139.98	995.0	1135.0	0.2505	1.5753	1.8258
174	6.566	0.01647	56.95	56.97	141.96	924.6	1066.6	141.98	993.8	1135.8	0.2537	1.5684	1.8221
176	6.869	0.01649	54.59	54.61	143.97	923.2	1067.2	143.99	992.6	1136.6	0.2568	1.5616	1.8184
178	7.184	0.01650	52.35	52.36	145.97	921.8	1067.8	145.99	991.4	1137.4	0.2600	1.5548	1.8147
180	7.511	0.01651	50.21	50.22	147.98	920.4	1068.4	148.00	990.2	1138.2	0.2631	1.5480	1.8111
182	7.850	0.01652	48.17	48.19	149.98	919.0	1069.0	150.01	989.0	1139.0	0.2662	1.5413	1.8075
184	8.203	0.01653	46.23	46.25	151.99	917.6	1069.6	152.01	987.8	1139.8	0.2694	1.5346	1.8040
186	8.568	0.01655	44.38	44.40	153.99	916.2	1070.2	154.02	986.5	1140.5	0.2725	1.5279	1.8004
188	8.947	0.01656	42.62	42.64	156.00	914.7	1070.7	156.03	985.3	1141.3	0.2756	1.5213	1.7969
190	9.340	0.01657	40.94	40.96	158.01	913.3	1071.3	158.04	984.1	1142.1	0.2787	1.5148	1.7934
192	9.747	0.01658	39.34	39.35	160.02	911.9	1071.9	160.05	982.8	1142.9	0.2818	1.5082	1.7900
194	10.168	0.01660	37.81	37.82	162.02	910.5	1072.5	162.05	981.6	1143.7	0.2848	1.5017	1.7865
196	10.605	0.01661	36.35	36.36	164.03	909.0	1073.1	164.06	980.4	1144.4	0.2879	1.4952	1.7831
198	11.058	0.01662	34.95	34.97	166.04	907.6	1073.6	166.08	979.1	1145.2	0.2910	1.4888	1.7798
200	11.526	0.01664	33.62	33.64	168.05	906.2	1074.2	168.09	977.9	1146.0	0.2940	1.4824	1.7764
202	12.011	0.01665	32.35	32.37	170.06	904.7	1074.8	170.10	976.6	1146.7	0.2971	1.4760	1.7731
204	12.512	0.01666	31.13	31.15	172.07	903.3	1075.3	172.11	975.4	1147.5	0.3001	1.4697	1.7698
206	13.031	0.01668	29.97	29.99	174.08	901.8	1075.9	174.12	974.1	1148.2	0.3031	1.4634	1.7665
208	13.568	0.01669	28.86	28.88	176.09	900.4	1076.5	176.14	972.8	1149.0	0.3061	1.4571	1.7632
210	14.123	0.01670	27.80	27.82	178.11	898.9	1077.0	178.15	971.6	1149.7	0.3091	1.4509	1.7600
212	14.696	0.01672	26.78	26.80	180.12	897.5	1077.6	180.17	970.3	1150.5	0.3121	1.4447	1.7568
215	15.592	0.01674	25.34	25.36	183.14	895.3	1078.4	183.19	968.4	1151.6	0.3166	1.4354	1.7520
220	17.186	0.01678	23.13	23.15	188.18	891.6	1079.8	188.23	965.2	1153.4	0.3241	1.4201	1.7442
225	18.912	0.01681	21.15	21.17	193.22	888.0	1081.2	193.28	962.0	1155.3	0.3315	1.4051	1.7365
230	20.78	0.01685	19.364	19.381	198.27	884.3	1082.5	198.33	958.7	1157.1	0.3388	1.3902	1.7290
235	22.79	0.01689	17.756	17.773	203.32	880.5	1083.9	203.39	955.4	1158.8	0.3461	1.3754	1.7215
240	24.97	0.01693	16.304	16.321	208.37	876.8	1085.2	208.45	952.1	1160.6	0.3533	1.3609	1.7142
245	27.31	0.01697	14.991	15.008	213.43	873.1	1086.5	213.52	948.8	1162.3	0.3606	1.3465	1.7070
250	29.82	0.01701	13.802	13.819	218.50	869.3	1087.8	218.59	945.4	1164.0	0.3677	1.3323	1.7000
255	32.53	0.01705	12.724	12.741	223.57	865.5	1089.0	223.67	942.1	1165.7	0.3748	1.3182	1.6930

260	35.43	0.01709	11.745	11.762	228.64	861.6	1090.3	228.76	938.6	1167.4	0.3819	1.3043	1.6862
265	38.53	0.01713	10.854	10.871	233.73	857.8	1091.5	233.85	935.2	1169.0	0.3890	1.2905	1.6795
270	41.86	0.01717	10.042	10.060	238.82	853.9	1092.7	238.95	931.7	1170.6	0.3960	1.2769	1.6729
275	45.41	0.01722	9.302	9.320	243.91	850.0	1093.9	244.06	928.2	1172.2	0.4029	1.2634	1.6663
280	49.20	0.01726	8.627	8.644	249.01	846.1	1095.1	249.17	924.6	1173.8	0.4098	1.2501	1.6599
285	53.24	0.01731	8.009	8.026	254.12	842.1	1096.2	254.29	921.0	1175.3	0.4167	1.2368	1.6536
290	57.55	0.01736	7.443	7.460	259.24	838.1	1097.4	259.43	917.4	1176.8	0.4236	1.2238	1.6473
295	62.13	0.01740	6.924	6.942	264.37	834.1	1098.5	264.57	913.7	1178.3	0.4304	1.2108	1.6412
300	67.01	0.01745	6.448	6.466	269.50	830.1	1099.6	269.71	910.0	1179.7	0.4372	1.1979	1.6351
305	72.18	0.01750	6.011	6.028	274.64	826.0	1100.6	274.87	906.3	1181.1	0.4439	1.1852	1.6291
310	77.67	0.01755	5.608	5.626	279.79	821.9	1101.7	280.04	902.5	1182.5	0.4506	1.1726	1.6232
315	83.48	0.01760	5.238	5.255	284.94	817.7	1102.7	285.21	898.7	1183.9	0.4573	1.1601	1.6174
320	89.64	0.01766	4.896	4.914	290.11	813.6	1103.7	290.40	894.8	1185.2	0.4640	1.1477	1.6116
325	96.16	0.01771	4.581	4.598	295.28	809.4	1104.6	295.60	890.9	1186.5	0.4706	1.1354	1.6059
330	103.05	0.01776	4.289	4.307	300.47	805.1	1105.6	300.81	886.9	1187.7	0.4772	1.1231	1.6003
335	110.32	0.01782	4.020	4.037	305.66	800.8	1106.5	306.03	882.9	1188.9	0.4837	1.1110	1.5947
340	117.99	0.01787	3.770	3.788	310.87	796.5	1107.4	311.26	878.8	1190.1	0.4902	1.0990	1.5892
345	126.08	0.01793	3.539	3.556	316.08	792.2	1108.2	316.50	874.7	1191.2	0.4967	1.0871	1.5838
350	134.60	0.01799	3.324	3.342	321.31	787.8	1109.1	321.76	870.6	1192.3	0.5032	1.0752	1.5784
355	143.57	0.01805	3.124	3.143	326.55	783.3	1109.9	327.03	866.3	1193.4	0.5097	1.0634	1.5731
360	153.01	0.01811	2.939	2.957	331.79	778.9	1110.7	332.31	862.1	1194.4	0.5161	1.0517	1.5678
365	162.93	0.01817	2.767	2.785	337.05	774.3	1111.4	337.60	857.8	1195.4	0.5225	1.0401	1.5626
370	173.34	0.01823	2.606	2.624	342.33	769.8	1112.1	342.91	853.4	1196.3	0.5289	1.0286	1.5575
375	184.27	0.01830	2.457	2.475	347.61	765.2	1112.8	348.24	849.0	1197.2	0.5352	1.0171	1.5523
380	195.73	0.01836	2.317	2.335	352.91	760.5	1113.5	353.58	844.5	1198.0	0.5416	1.0057	1.5473
385	207.74	0.01843	2.187	2.205	358.22	755.9	1114.1	358.93	839.9	1198.8	0.5479	0.9944	1.5422
390	220.32	0.01850	2.065	2.083	363.55	751.1	1114.7	364.30	835.3	1199.6	0.5542	0.9831	1.5372
395	233.49	0.01857	1.9510	1.9695	368.89	746.3	1115.2	369.69	830.6	1200.3	0.5604	0.9718	1.5323
400	247.26	0.01864	1.8444	1.8630	374.24	741.5	1115.7	375.09	825.9	1201.0	0.5667	0.9607	1.5274
405	261.65	0.01871	1.7445	1.7633	379.61	736.6	1116.2	380.52	821.1	1201.6	0.5729	0.9496	1.5225
410	276.69	0.01878	1.6510	1.6697	384.99	731.7	1116.7	385.96	816.2	1202.1	0.5791	0.9385	1.5176
415	292.40	0.01886	1.5632	1.5820	390.40	726.7	1117.1	391.42	811.2	1202.7	0.5853	0.9275	1.5128
420	308.78	0.01894	1.4808	1.4997	395.81	721.6	1117.4	396.90	806.2	1203.1	0.5915	0.9165	1.5080
425	325.87	0.01901	1.4033	1.4224	401.25	716.5	1117.8	402.40	801.1	1203.5	0.5977	0.9055	1.5032
430	343.67	0.01909	1.3306	1.3496	406.70	711.3	1118.0	407.92	796.0	1203.9	0.6038	0.8946	1.4985
435	362.23	0.01918	1.2621	1.2812	412.18	706.1	1118.3	413.46	790.7	1204.2	0.6100	0.8838	1.4937
440	381.54	0.01926	1.1976	1.2169	417.67	700.8	1118.5	419.03	785.4	1204.4	0.6161	0.8729	1.4890
445	401.64	0.01934	1.1369	1.1562	423.18	695.5	1118.7	424.62	780.0	1204.6	0.6222	0.8621	1.4843
450	422.55	0.01943	1.0796	1.0991	428.71	690.1	1118.8	430.23	774.5	1204.7	0.6283	0.8514	1.4797
455	444.28	0.0195	1.0256	1.0451	434.27	684.6	1118.9	435.87	768.9	1204.8	0.6344	0.8406	1.4750

TABLE C.3. SATURATED STEAM ENGLISH UNITS (Continued)

TEMP	ABS PRESS	SPECIFIC VOLUME V			INTERNAL ENERGY U			ENTHALPY H			ENTROPY S		
T (DEG F)	P (PSIA)	SAT LIQUID	EVAP	SAT VAPOR	SAT LIQUID	EVAP	SAT VAPOR	SAT LIQUID	EVAP	SAT VAPOR	SAT LIQUID	EVAP	SAT VAPOR
460	466.87	0.0196	0.9746	0.9942	439.84	679.0	1118.9	441.54	763.2	1204.8	0.6405	0.8299	1.4704
465	490.32	0.0197	0.9265	0.9462	445.44	673.4	1118.9	447.23	757.5	1204.7	0.6466	0.8192	1.4657
470	514.67	0.0198	0.8810	0.9008	451.06	667.7	1118.8	452.95	751.6	1204.6	0.6527	0.8084	1.4611
475	539.94	0.0199	0.8379	0.8578	456.71	662.0	1118.7	458.70	745.7	1204.4	0.6587	0.7977	1.4565
480	566.15	0.0200	0.7972	0.8172	462.39	656.1	1118.5	464.48	739.6	1204.1	0.6648	0.7871	1.4518
485	593.32	0.0201	0.7586	0.7787	468.09	650.2	1118.3	470.29	733.5	1203.8	0.6708	0.7764	1.4472
490	621.48	0.0202	0.7220	0.7422	473.82	644.2	1118.0	476.14	727.2	1203.3	0.6769	0.7657	1.4426
495	650.65	0.0203	0.6874	0.7077	479.57	638.0	1117.6	482.02	720.8	1202.8	0.6830	0.7550	1.4380
500	680.86	0.0204	0.6545	0.6749	485.36	631.8	1117.2	487.94	714.3	1202.2	0.6890	0.7443	1.4333
505	712.12	0.0205	0.6233	0.6438	491.2	625.6	1116.7	493.9	707.7	1201.6	0.6951	0.7336	1.4286
510	744.47	0.0207	0.5936	0.6143	497.0	619.2	1116.2	499.9	700.9	1200.8	0.7012	0.7228	1.4240
515	777.93	0.0208	0.5654	0.5862	502.9	612.7	1115.6	505.9	694.1	1200.0	0.7072	0.7120	1.4193
520	812.53	0.0209	0.5386	0.5596	508.8	606.1	1114.9	512.0	687.0	1199.0	0.7133	0.7013	1.4146
525	848.28	0.0210	0.5131	0.5342	514.8	599.3	1114.2	518.1	679.9	1198.0	0.7194	0.6904	1.4098
530	885.23	0.0212	0.4889	0.5100	520.8	592.5	1113.3	524.3	672.6	1196.9	0.7255	0.6796	1.4051
535	923.39	0.0213	0.4657	0.4870	526.9	585.6	1112.4	530.5	665.1	1195.6	0.7316	0.6686	1.4003
540	962.79	0.0215	0.4437	0.4651	532.9	578.5	1111.4	536.8	657.5	1194.3	0.7378	0.6577	1.3954
545	1003.5	0.0216	0.4226	0.4442	539.1	571.2	1110.3	543.1	649.7	1192.8	0.7439	0.6467	1.3906
550	1045.4	0.0218	0.4026	0.4243	545.3	563.9	1109.1	549.5	641.8	1191.2	0.7501	0.6356	1.3856
555	1088.7	0.0219	0.3834	0.4053	551.5	556.4	1107.9	555.9	633.6	1189.5	0.7562	0.6244	1.3807
560	1133.4	0.0221	0.3651	0.3871	557.8	548.7	1106.5	562.4	625.3	1187.7	0.7625	0.6132	1.3757
565	1179.4	0.0222	0.3475	0.3698	564.1	540.9	1105.0	569.0	616.8	1185.7	0.7687	0.6019	1.3706
570	1226.9	0.0224	0.3308	0.3532	570.5	532.9	1103.4	575.6	608.0	1183.6	0.7750	0.5905	1.3654
575	1275.8	0.0226	0.3147	0.3373	577.0	524.8	1101.7	582.3	599.1	1181.4	0.7813	0.5790	1.3602
580	1326.2	0.0228	0.2994	0.3222	583.5	516.4	1099.9	589.1	589.9	1179.0	0.7876	0.5673	1.3550
585	1378.1	0.0230	0.2846	0.3076	590.1	507.9	1098.0	596.0	580.4	1176.4	0.7940	0.5556	1.3496
590	1431.5	0.0232	0.2705	0.2937	596.8	499.1	1095.9	602.9	570.8	1173.7	0.8004	0.5437	1.3442
595	1486.6	0.0234	0.2569	0.2803	603.5	490.2	1093.7	610.0	560.8	1170.8	0.8069	0.5317	1.3386
600	1543.2	0.0236	0.2438	0.2675	610.4	481.0	1091.3	617.1	550.6	1167.7	0.8134	0.5196	1.3330
605	1601.5	0.0239	0.2313	0.2551	617.3	471.5	1088.8	624.4	540.0	1164.4	0.8200	0.5072	1.3273
610	1661.6	0.0241	0.2191	0.2433	624.4	461.8	1086.1	631.8	529.2	1160.9	0.8267	0.4947	1.3214
615	1723.3	0.0244	0.2075	0.2318	631.5	451.8	1083.3	639.3	517.9	1157.2	0.8334	0.4819	1.3154
620	1786.9	0.0247	0.1961	0.2208	638.8	441.4	1080.2	646.9	506.3	1153.2	0.8403	0.4689	1.3092
625	1852.2	0.0250	0.1852	0.2102	646.2	430.7	1076.8	654.7	494.2	1148.9	0.8472	0.4556	1.3028
630	1919.5	0.0253	0.1746	0.1999	653.7	419.5	1073.2	662.7	481.6	1144.2	0.8542	0.4419	1.2962

635	1988.7	0.0256	0.1643	0.1899	661.4	407.9	1069.3	670.8	468.4	1139.2	0.8614	0.4279	1.2893
640	2059.9	0.0259	0.1543	0.1802	669.2	395.8	1065.0	679.1	454.6	1133.7	0.8686	0.4134	1.2821
645	2133.1	0.0263	0.1445	0.1708	677.3	383.1	1060.4	687.7	440.2	1127.8	0.8761	0.3985	1.2746
650	2208.4	0.0267	0.1350	0.1617	685.5	369.8	1055.3	696.4	425.0	1121.4	0.8837	0.3830	1.2667
655	2285.9	0.0272	0.1257	0.1529	694.0	355.8	1049.8	705.5	409.0	1114.5	0.8915	0.3670	1.2584
660	2365.7	0.0277	0.1166	0.1443	702.8	341.0	1043.9	714.9	392.1	1107.0	0.8995	0.3502	1.2498
662	2398.2	0.0279	0.1131	0.1409	706.4	335.0	1041.4	718.7	385.2	1103.9	0.9029	0.3433	1.2462
664	2431.1	0.0281	0.1095	0.1376	710.2	328.5	1038.7	722.9	377.7	1100.6	0.9064	0.3361	1.2425
666	2464.4	0.0283	0.1059	0.1342	714.2	321.7	1035.9	727.1	370.0	1097.1	0.9100	0.3286	1.2387
668	2498.1	0.0286	0.1023	0.1309	718.3	314.8	1033.0	731.5	362.1	1093.5	0.9137	0.3210	1.2347
670	2532.2	0.0288	0.0987	0.1275	722.3	307.7	1030.0	735.8	354.0	1089.8	0.9174	0.3133	1.2307
672	2566.6	0.0291	0.0951	0.1242	726.4	300.5	1026.9	740.2	345.7	1085.9	0.9211	0.3054	1.2266
674	2601.5	0.0294	0.0916	0.1210	730.5	293.1	1023.6	744.7	337.2	1081.9	0.9249	0.2974	1.2223
676	2636.8	0.0297	0.0880	0.1177	734.7	285.5	1020.2	749.2	328.5	1077.6	0.9287	0.2892	1.2179
678	2672.5	0.0300	0.0844	0.1144	738.9	277.7	1016.6	753.8	319.4	1073.2	0.9326	0.2807	1.2133
680	2708.6	0.0304	0.0808	0.1112	743.2	269.6	1012.8	758.5	310.1	1068.5	0.9365	0.2720	1.2086
682	2745.1	0.0307	0.0772	0.1079	747.7	261.2	1008.8	763.3	300.4	1063.6	0.9406	0.2631	1.2036
684	2782.1	0.0311	0.0735	0.1046	752.2	252.4	1004.6	768.2	290.2	1058.4	0.9447	0.2537	1.1984
686	2819.5	0.0316	0.0698	0.1013	756.9	243.1	1000.0	773.4	279.5	1052.9	0.9490	0.2439	1.1930
688	2857.4	0.0320	0.0659	0.0980	761.8	233.3	995.2	778.8	268.2	1047.0	0.9535	0.2337	1.1872
690	2895.7	0.0326	0.0620	0.0946	767.0	222.9	989.9	784.5	256.1	1040.6	0.9583	0.2227	1.1810
692	2934.5	0.0331	0.0580	0.0911	772.5	211.6	984.1	790.5	243.1	1033.6	0.9634	0.2110	1.1744
694	2973.7	0.0338	0.0537	0.0875	778.5	199.2	977.7	797.1	228.8	1025.9	0.9689	0.1983	1.1671
696	3013.4	0.0345	0.0492	0.0837	785.1	185.4	970.5	804.4	212.8	1017.2	0.9749	0.1841	1.1591
698	3053.6	0.0355	0.0442	0.0797	792.6	169.6	962.2	812.6	194.6	1007.2	0.9818	0.1681	1.1499
700	3094.3	0.0366	0.0386	0.0752	801.5	150.7	952.1	822.4	172.7	995.2	0.9901	0.1490	1.1390
702	3135.5	0.0382	0.0317	0.0700	812.8	126.3	939.1	835.0	144.7	979.7	1.0006	0.1246	1.1252
704	3177.2	0.0411	0.0219	0.0630	830.1	89.1	919.2	854.2	102.0	956.2	1.0169	0.0876	1.1046
705.47	3208.2	0.0508	0.0000	0.0508	875.9	-0.0	875.9	906.0	-0.0	906.0	1.0612	0.0000	1.0612

TABLE C.4. SUPERHEATED STEAM ENGLISH UNITS

ABS PRESS PSIA (SAT TEMP)		SAT WATER	SAT STEAM	200	250	300	350	400	450	500
				TEMPERATURE, DEG F						
1 (101.74)	V	0.0161	333.60	392.5	422.4	452.3	482.1	511.9	541.7	571.5
	U	69.73	1044.1	1077.5	1094.7	1112.0	1129.5	1147.1	1164.9	1182.8
	H	69.73	1105.8	1150.2	1172.9	1195.7	1218.7	1241.8	1265.1	1288.6
	S	0.1326	1.9781	2.0509	2.0841	2.1152	2.1445	2.1722	2.1985	2.2237
5 (162.24)	V	0.0164	73.532	78.14	84.21	90.24	96.25	102.2	108.2	114.2
	U	130.18	1063.1	1076.3	1093.8	1111.3	1128.9	1146.7	1164.5	1182.6
	H	130.20	1131.1	1148.6	1171.7	1194.8	1218.0	1241.3	1264.7	1288.2
	S	0.2349	1.8443	1.8716	1.9054	1.9369	1.9664	1.9943	2.0208	2.0460
10 (193.21)	V	0.0166	38.420	38.84	41.93	44.98	48.02	51.03	54.04	57.04
	U	161.23	1072.3	1074.7	1092.6	1110.4	1128.3	1146.1	1164.1	1182.2
	H	161.26	1143.3	1146.6	1170.2	1193.7	1217.1	1240.6	1264.1	1287.8
	S	0.2836	1.7879	1.7928	1.8273	1.8593	1.8892	1.9173	1.9439	1.9692
14.696 (212.00)	V	0.0167	26.799	....	28.42	30.52	32.60	34.67	36.72	38.77
	U	180.12	1077.6	....	1091.5	1109.6	1127.6	1145.7	1163.7	1181.9
	H	180.17	1150.5	....	1168.8	1192.6	1216.3	1239.9	1263.6	1287.4
	S	0.3121	1.7568	....	1.7833	1.8158	1.8460	1.8743	1.9010	1.9265
15 (213.03)	V	0.0167	26.290	....	27.84	29.90	31.94	33.96	35.98	37.98
	U	181.16	1077.9	....	1091.4	1109.5	1127.6	1145.6	1163.7	1181.9
	H	181.21	1150.9	....	1168.7	1192.5	1216.2	1239.9	1263.6	1287.3
	S	0.3137	1.7552	....	1.7809	1.8134	1.8436	1.8720	1.8988	1.9242
20 (227.96)	V	0.0168	20.087	....	20.79	22.36	23.90	25.43	26.95	28.46
	U	196.21	1082.0	....	1090.2	1108.6	1126.9	1145.1	1163.3	1181.6
	H	196.27	1156.3	....	1167.1	1191.4	1215.4	1239.2	1263.0	1286.9
	S	0.3358	1.7320	....	1.7475	1.7805	1.8111	1.8397	1.8666	1.8921
25 (240.07)	V	0.0169	16.301	....	16.56	17.83	19.08	20.31	21.53	22.74
	U	208.44	1085.2	....	1089.0	1107.7	1126.2	1144.6	1162.9	1181.2
	H	208.52	1160.6	....	1165.6	1190.2	1214.5	1238.5	1262.5	1286.4
	S	0.3535	1.7141	....	1.7212	1.7547	1.7856	1.8145	1.8415	1.8672
30 (250.34)	V	0.0170	13.744	....	....	14.81	15.86	16.89	17.91	18.93
	U	218.84	1087.9	....	....	1106.8	1125.5	1144.0	1162.5	1180.9
	H	218.93	1164.1	....	....	1189.0	1213.6	1237.8	1261.9	1286.0
	S	0.3682	1.6995	....	....	1.7334	1.7647	1.7937	1.8210	1.8467

35 (259.29)	V	0.0171	11.896	…	…	12.65	13.56	14.45	15.33	16.21
	U	227.92	1090.1	…	…	1105.9	1124.8	1143.5	1162.0	1180.5
	H	228.03	1167.1	…	…	1187.8	1212.7	1237.1	1261.3	1285.5
	S	0.3809	1.6872	…	…	1.7152	1.7468	1.7761	1.8035	1.8294
40 (267.25)	V	0.0172	10.497	…	…	11.04	11.84	12.62	13.40	14.16
	U	236.02	1092.1	…	…	1104.9	1124.1	1142.9	1161.6	1180.2
	H	236.14	1169.8	…	…	1186.6	1211.7	1236.4	1260.8	1285.0
	S	0.3921	1.6765	…	…	1.6992	1.7312	1.7608	1.7883	1.8143
45 (274.44)	V	0.0172	9.399	…	…	9.777	10.50	11.20	11.89	12.58
	U	243.34	1093.8	…	…	1104.0	1123.4	1142.4	1161.2	1179.8
	H	243.49	1172.0	…	…	1185.4	1210.8	1235.7	1260.2	1284.6
	S	0.4021	1.6671	…	…	1.6849	1.7173	1.7471	1.7749	1.8009
50 (281.01)	V	0.0173	8.514	…	…	8.769	9.424	10.06	10.69	11.31
	U	250.05	1095.3	…	…	1103.0	1122.7	1141.8	1160.7	1179.5
	H	250.21	1174.1	…	…	1184.1	1209.9	1234.9	1259.6	1284.1
	S	0.4112	1.6586	…	…	1.6720	1.7048	1.7349	1.7628	1.7890
55 (287.08)	V	0.0173	7.785	…	…	7.945	8.546	9.130	9.702	10.27
	U	256.25	1096.7	…	…	1102.0	1121.9	1141.3	1160.3	1179.1
	H	256.43	1175.9	…	…	1182.8	1208.9	1234.2	1259.1	1283.6
	S	0.4196	1.6510	…	…	1.6601	1.6934	1.7237	1.7518	1.7781
60 (292.71)	V	0.0174	7.174	…	…	7.257	7.815	8.354	8.881	9.400
	U	262.02	1098.0	…	…	1101.0	1121.2	1140.7	1159.9	1178.8
	H	262.21	1177.6	…	…	1181.6	1208.0	1233.5	1258.5	1283.2
	S	0.4273	1.6440	…	…	1.6492	1.6829	1.7134	1.7417	1.7681
65 (297.98)	V	0.0174	6.653	…	…	6.675	7.195	7.697	8.186	8.667
	U	267.42	1099.1	…	…	1100.0	1120.4	1140.2	1159.4	1178.4
	H	267.63	1179.1	…	…	1180.3	1207.0	1232.7	1257.9	1282.7
	S	0.4344	1.6375	…	…	1.6390	1.6731	1.7040	1.7324	1.7589
70 (302.93)	V	0.0175	6.205	…	…	…	6.664	7.133	7.590	8.039
	U	272.51	1100.2	…	…	…	1119.7	1139.6	1159.0	1178.1
	H	272.74	1180.6	…	…	…	1206.0	1232.0	1257.3	1282.2
	S	0.4411	1.6316	…	…	…	1.6640	1.6951	1.7237	1.7504
75 (307.61)	V	0.0175	5.814	…	…	…	6.204	6.645	7.074	7.494
	U	277.32	1101.2	…	…	…	1118.9	1139.0	1158.5	1177.7
	H	277.56	1181.9	…	…	…	1205.0	1231.2	1256.7	1281.7
	S	0.4474	1.6260	…	…	…	1.6554	1.6868	1.7156	1.7424

TABLE C.4. SUPERHEATED STEAM ENGLISH UNITS (Continued)

ABS PRESS PSIA (SAT TEMP)		SAT WATER	SAT STEAM	TEMPERATURE, DEG F 600	700	800	900	1000	1100	1200
1	V	0.0161	333.60	631.1	690.7	750.3	809.9	869.5	929.0	988.6
(101.74)	U	69.73	1044.1	1219.3	1256.7	1294.9	1334.0	1374.0	1414.9	1456.7
	H	69.73	1105.8	1336.1	1384.5	1433.7	1483.8	1534.9	1586.8	1639.7
	S	0.1326	1.9781	2.2708	2.3144	2.3551	2.3934	2.4296	2.4640	2.4969
5	V	0.0164	73.532	126.1	138.1	150.0	161.9	173.9	185.8	197.7
(162.24)	U	130.18	1063.1	1219.2	1256.5	1294.8	1333.9	1373.9	1414.8	1456.7
	H	130.20	1131.1	1335.9	1384.3	1433.6	1483.7	1534.7	1586.7	1639.6
	S	0.2349	1.8443	2.0932	2.1369	2.1776	2.2159	2.2521	2.2866	2.3194
10	V	0.0166	38.420	63.03	69.00	74.98	80.94	86.91	92.87	98.84
(193.21)	U	161.23	1072.3	1218.9	1256.4	1294.6	1333.7	1373.8	1414.7	1456.6
	H	161.26	1143.3	1335.5	1384.0	1433.4	1483.5	1534.6	1586.6	1639.5
	S	0.2836	1.7879	2.0166	2.0603	2.1011	2.1394	2.1757	2.2101	2.2430
14.696	V	0.0167	26.799	42.86	46.93	51.00	55.06	59.13	63.19	67.25
(212.00)	U	180.12	1077.6	1218.7	1256.2	1294.5	1333.6	1373.7	1414.6	1456.5
	H	180.17	1150.5	1335.2	1383.8	1433.2	1483.4	1534.5	1586.5	1639.4
	S	0.3121	1.7568	1.9739	2.0177	2.0585	2.0969	2.1331	2.1676	2.2005
15	V	0.0167	26.290	41.99	45.98	49.96	53.95	57.93	61.90	65.88
(213.03)	U	181.16	1077.9	1218.7	1256.2	1294.5	1333.6	1373.7	1414.6	1456.5
	H	181.21	1150.9	1335.2	1383.8	1433.2	1483.4	1534.5	1586.5	1639.4
	S	0.3137	1.7552	1.9717	2.0155	2.0563	2.0946	2.1309	2.1653	2.1982
20	V	0.0168	20.087	31.47	34.46	37.46	40.45	43.43	46.42	49.40
(227.96)	U	196.21	1082.0	1218.4	1256.0	1294.3	1333.5	1373.6	1414.5	1456.4
	H	196.27	1156.3	1334.9	1383.5	1432.9	1483.2	1534.3	1586.3	1639.3
	S	0.3358	1.7320	1.9397	1.9836	2.0244	2.0628	2.0991	2.1336	2.1665
25	V	0.0169	16.301	25.15	27.56	29.95	32.35	34.74	37.13	39.52
(240.07)	U	208.44	1085.2	1218.2	1255.8	1294.2	1333.4	1373.5	1414.4	1456.3
	H	208.52	1160.6	1334.6	1383.3	1432.7	1483.0	1534.2	1586.2	1639.2
	S	0.3535	1.7141	1.9149	1.9588	1.9997	2.0381	2.0744	2.1089	2.1418
30	V	0.0170	13.744	20.95	22.95	24.95	26.95	28.94	30.94	32.93
(250.34)	U	218.84	1087.9	1218.0	1255.6	1294.0	1333.2	1373.3	1414.3	1456.3
	H	218.93	1164.1	1334.2	1383.0	1432.5	1482.8	1534.0	1586.1	1639.0
	S	0.3682	1.6995	1.8946	1.9386	1.9795	2.0179	2.0543	2.0888	2.1217

35 (259.29)	V	0.0171	11.896	17.94	19.66	21.38	23.09	24.80	26.51	28.22
	U	227.92	1090.1	1217.7	1255.4	1293.9	1333.1	1373.2	1414.3	1456.2
	H	228.03	1167.1	1333.9	1382.8	1432.3	1482.7	1533.9	1586.0	1638.9
	S	0.3809	1.6872	1.8774	1.9214	1.9624	2.0009	2.0372	2.0717	2.1046
40 (267.25)	V	0.0172	10.497	15.68	17.19	18.70	20.20	21.70	23.19	24.69
	U	236.02	1092.1	1217.5	1255.3	1293.7	1333.0	1373.1	1414.2	1456.1
	H	236.14	1169.8	1333.6	1382.5	1432.1	1482.5	1533.7	1585.8	1638.8
	S	0.3921	1.6765	1.8624	1.9065	1.9476	1.9860	2.0224	2.0569	2.0899
45 (274.44)	V	0.0172	9.399	13.93	15.28	16.61	17.95	19.28	20.61	21.94
	U	243.34	1093.8	1217.2	1255.1	1293.6	1332.9	1373.0	1414.1	1456.0
	H	243.49	1172.0	1333.3	1382.3	1431.9	1482.3	1533.6	1585.7	1638.7
	S	0.4021	1.6671	1.8492	1.8934	1.9345	1.9730	2.0093	2.0439	2.0768
50 (281.01)	V	0.0173	8.514	12.53	13.74	14.95	16.15	17.35	18.55	19.75
	U	250.05	1095.3	1217.0	1254.9	1293.4	1332.7	1372.9	1414.0	1455.9
	H	250.21	1174.1	1332.9	1382.0	1431.7	1482.2	1533.4	1585.6	1638.6
	S	0.4112	1.6586	1.8374	1.8816	1.9227	1.9613	1.9977	2.0322	2.0652
55 (287.08)	V	0.0173	7.785	11.38	12.48	13.58	14.68	15.77	16.86	17.95
	U	256.25	1096.7	1216.8	1254.7	1293.3	1332.6	1372.8	1413.9	1455.8
	H	256.43	1175.9	1332.6	1381.8	1431.5	1482.0	1533.3	1585.5	1638.5
	S	0.4196	1.6510	1.8266	1.8710	1.9121	1.9507	1.9871	2.0216	2.0546
60 (292.71)	V	0.0174	7.174	10.42	11.44	12.45	13.45	14.45	15.45	16.45
	U	262.02	1098.0	1216.5	1254.5	1293.1	1332.5	1372.7	1413.8	1455.8
	H	262.21	1177.6	1332.3	1381.5	1431.3	1481.8	1533.2	1585.3	1638.4
	S	0.4273	1.6440	1.8168	1.8612	1.9024	1.9410	1.9774	2.0120	2.0450
65 (297.98)	V	0.0174	6.653	9.615	10.55	11.48	12.41	13.34	14.26	15.18
	U	267.42	1099.1	1216.3	1254.3	1293.0	1332.4	1372.6	1413.7	1455.7
	H	267.63	1179.1	1331.9	1381.3	1431.1	1481.6	1533.0	1585.2	1638.3
	S	0.4344	1.6375	1.8077	1.8522	1.8935	1.9321	1.9685	2.0031	2.0361
70 (302.93)	V	0.0175	6.205	8.922	9.793	10.66	11.52	12.38	13.24	14.10
	U	272.51	1100.2	1216.0	1254.1	1292.8	1332.2	1372.5	1413.6	1455.6
	H	272.74	1180.6	1331.6	1381.0	1430.9	1481.5	1532.9	1585.1	1638.2
	S	0.4411	1.6316	1.7993	1.8439	1.8852	1.9238	1.9603	1.9949	2.0279
75 (307.61)	V	0.0175	5.814	8.320	9.135	9.945	10.75	11.55	12.35	13.15
	U	277.32	1101.2	1215.8	1254.0	1292.7	1332.1	1372.4	1413.5	1455.5
	H	277.56	1181.9	1331.3	1380.7	1430.7	1481.3	1532.7	1585.0	1638.1
	S	0.4474	1.6260	1.7915	1.8361	1.8774	1.9161	1.9526	1.9872	2.0202

TABLE C.4. SUPERHEATED STEAM ENGLISH UNITS (Continued)

ABS PRESS PSIA (SAT TEMP)		SAT WATER	SAT STEAM	340	360	380	400	420	450	500
				TEMPERATURE, DEG F						
80 (312.04)	V	0.0176	5.471	5.715	5.885	6.053	6.218	6.381	6.622	7.018
	U	281.89	1102.1	1114.0	1122.3	1130.4	1138.4	1146.3	1158.1	1177.4
	H	282.15	1183.1	1198.6	1209.4	1220.0	1230.5	1240.8	1256.1	1281.3
	S	0.4534	1.6208	1.6405	1.6539	1.6667	1.6790	1.6909	1.7080	1.7349
85 (316.26)	V	0.0176	5.167	5.364	5.525	5.684	5.840	5.995	6.223	6.597
	U	286.24	1102.9	1113.1	1121.5	1129.7	1137.8	1145.8	1157.6	1177.0
	H	286.52	1184.2	1197.5	1208.4	1219.1	1229.7	1240.1	1255.5	1280.8
	S	0.4590	1.6159	1.6328	1.6463	1.6592	1.6716	1.6836	1.7008	1.7279
90 (320.28)	V	0.0177	4.895	5.051	5.205	5.356	5.505	5.652	5.869	6.223
	U	290.40	1103.7	1112.3	1120.8	1129.1	1137.2	1145.3	1157.2	1176.7
	H	290.69	1185.3	1196.4	1207.5	1218.3	1228.9	1239.4	1254.9	1280.3
	S	0.4643	1.6113	1.6254	1.6391	1.6521	1.6646	1.6767	1.6940	1.7212
95 (324.13)	V	0.0177	4.651	4.771	4.919	5.063	5.205	5.345	5.551	5.889
	U	294.38	1104.5	1111.4	1120.0	1128.4	1136.6	1144.7	1156.7	1176.3
	H	294.70	1186.2	1195.3	1206.5	1217.4	1228.1	1238.7	1254.3	1279.8
	S	0.4694	1.6069	1.6184	1.6322	1.6453	1.6580	1.6701	1.6876	1.7149
100 (327.82)	V	0.0177	4.431	4.519	4.660	4.799	4.935	5.068	5.266	5.588
	U	298.21	1105.2	1110.6	1119.2	1127.7	1136.0	1144.2	1156.3	1175.9
	H	298.54	1187.2	1194.2	1205.5	1216.5	1227.4	1238.0	1253.7	1279.3
	S	0.4743	1.6027	1.6116	1.6255	1.6389	1.6516	1.6638	1.6814	1.7088
105 (331.37)	V	0.0178	4.231	4.291	4.427	4.560	4.690	4.818	5.007	5.315
	U	301.89	1105.8	1109.7	1118.5	1127.0	1135.4	1143.7	1155.8	1175.6
	H	302.24	1188.0	1193.1	1204.5	1215.6	1226.6	1237.3	1253.1	1278.8
	S	0.4790	1.5988	1.6051	1.6192	1.6326	1.6455	1.6578	1.6755	1.7031
110 (334.79)	V	0.0178	4.048	4.083	4.214	4.343	4.468	4.591	4.772	5.068
	U	305.44	1106.5	1108.8	1117.7	1126.4	1134.8	1143.1	1155.3	1175.2
	H	305.80	1188.9	1191.9	1203.5	1214.7	1225.8	1236.6	1252.5	1278.3
	S	0.4834	1.5950	1.5988	1.6131	1.6267	1.6396	1.6521	1.6698	1.6975
115 (338.08)	V	0.0179	3.881	3.894	4.020	4.144	4.265	4.383	4.558	4.841
	U	308.87	1107.0	1107.9	1116.9	1125.7	1134.2	1142.6	1154.8	1174.8
	H	309.25	1189.6	1190.8	1202.5	1213.8	1225.0	1235.8	1251.8	1277.9
	S	0.4877	1.5913	1.5928	1.6072	1.6209	1.6340	1.6465	1.6644	1.6922

120 (341.27)	V	0.0179	3.728	....	3.842	3.962	4.079	4.193	4.361	4.634
	U	312.19	1107.6	....	1116.1	1124.9	1133.6	1142.0	1154.4	1174.5
	H	312.58	1190.4	....	1201.4	1212.9	1224.1	1235.1	1251.2	1277.4
	S	0.4919	1.5879	....	1.6015	1.6154	1.6286	1.6412	1.6592	1.6872
125 (344.35)	V	0.0179	3.586	....	3.679	3.794	3.907	4.018	4.180	4.443
	U	315.40	1108.1	....	1115.3	1124.2	1132.9	1141.4	1153.9	1174.1
	H	315.82	1191.1	....	1200.4	1212.0	1223.3	1234.4	1250.6	1276.9
	S	0.4959	1.5845	....	1.5960	1.6100	1.6233	1.6360	1.6541	1.6823
130 (347.33)	V	0.0180	3.454	....	3.527	3.639	3.749	3.856	4.013	4.267
	U	318.52	1108.6	....	1114.5	1123.5	1132.3	1140.9	1153.4	1173.7
	H	318.95	1191.7	....	1199.4	1211.1	1222.5	1233.6	1249.9	1276.4
	S	0.4998	1.5813	....	1.5907	1.6048	1.6182	1.6310	1.6493	1.6775
135 (350.23)	V	0.0180	3.332	....	3.387	3.496	3.602	3.706	3.858	4.104
	U	321.55	1109.1	....	1113.7	1122.8	1131.7	1140.3	1152.9	1173.3
	H	322.00	1192.4	....	1198.3	1210.1	1221.6	1232.9	1249.3	1275.8
	S	0.5035	1.5782	....	1.5855	1.5997	1.6133	1.6262	1.6446	1.6730
140 (353.04)	V	0.0180	3.219	....	3.257	3.363	3.466	3.567	3.714	3.953
	U	324.49	1109.6	....	1112.9	1122.1	1131.0	1139.7	1152.4	1172.9
	H	324.96	1193.0	....	1197.2	1209.2	1220.8	1232.1	1248.7	1275.3
	S	0.5071	1.5752	....	1.5804	1.5948	1.6085	1.6215	1.6400	1.6686
145 (355.77)	V	0.0181	3.113	....	3.135	3.239	3.339	3.437	3.580	3.812
	U	327.36	1110.0	....	1112.0	1121.3	1130.4	1139.1	1151.9	1172.6
	H	327.84	1193.5	....	1196.1	1208.2	1220.0	1231.4	1248.0	1274.8
	S	0.5107	1.5723	....	1.5755	1.5901	1.6039	1.6170	1.6356	1.6643
150 (358.43)	V	0.0181	3.014	....	3.022	3.123	3.221	3.316	3.455	3.680
	U	330.15	1110.4	....	1111.2	1120.6	1129.7	1138.6	1151.4	1172.2
	H	330.65	1194.1	....	1195.1	1207.3	1219.1	1230.6	1247.4	1274.3
	S	0.5141	1.5695	....	1.5707	1.5854	1.5993	1.6126	1.6313	1.6602
155 (361.02)	V	0.0181	2.921	....	....	3.014	3.110	3.203	3.339	3.557
	U	332.87	1110.8	....	....	1119.8	1129.0	1138.0	1150.9	1171.8
	H	333.39	1194.6	....	....	1206.3	1218.2	1229.8	1246.7	1273.8
	S	0.5174	1.5668	....	....	1.5809	1.5949	1.6083	1.6271	1.6561
160 (363.55)	V	0.0182	2.834	....	....	2.913	3.006	3.097	3.229	3.441
	U	335.53	1111.2	....	....	1119.1	1128.4	1137.4	1150.4	1171.4
	H	336.07	1195.1	....	....	1205.3	1217.4	1229.1	1246.0	1273.3
	S	0.5206	1.5641	....	....	1.5764	1.5906	1.6041	1.6231	1.6522

TABLE C.4. SUPERHEATED STEAM ENGLISH UNITS (Continued)

ABS PRESS PSIA (SAT TEMP)		SAT WATER	SAT STEAM	TEMPERATURE, DEG F 600	700	800	900	1000	1100	1200
80 (312.04)	V	0.0176	5.471	7.794	8.560	9.319	10.08	10.83	11.58	12.33
	U	281.89	1102.1	1215.5	1253.8	1292.5	1332.0	1372.3	1413.4	1455.4
	H	282.15	1183.1	1330.9	1380.5	1430.5	1481.1	1532.6	1584.9	1638.0
	S	0.4534	1.6208	1.7842	1.8289	1.8702	1.9089	1.9454	1.9800	2.0131
85 (316.26)	V	0.0176	5.167	7.330	8.052	8.768	9.480	10.19	10.90	11.60
	U	286.24	1102.9	1215.3	1253.6	1292.4	1331.9	1372.2	1413.3	1455.4
	H	286.52	1184.2	1330.6	1380.2	1430.3	1481.0	1532.4	1584.7	1637.9
	S	0.4590	1.6159	1.7772	1.8220	1.8634	1.9021	1.9386	1.9733	2.0063
90 (320.28)	V	0.0177	4.895	6.917	7.600	8.277	8.950	9.621	10.29	10.96
	U	290.40	1103.7	1215.0	1253.4	1292.2	1331.7	1372.0	1413.2	1455.3
	H	290.69	1185.3	1330.2	1380.0	1430.1	1480.8	1532.3	1584.6	1637.8
	S	0.4643	1.6113	1.7707	1.8156	1.8570	1.8957	1.9323	1.9669	2.0000
95 (324.13)	V	0.0177	4.651	6.548	7.196	7.838	8.477	9.113	9.747	10.38
	U	294.38	1104.5	1214.8	1253.2	1292.1	1331.6	1371.9	1413.1	1455.2
	H	294.70	1186.2	1329.9	1379.7	1429.9	1480.6	1532.1	1584.5	1637.7
	S	0.4694	1.6069	1.7645	1.8094	1.8509	1.8897	1.9262	1.9609	1.9940
100 (327.82)	V	0.0177	4.431	6.216	6.833	7.443	8.050	8.655	9.258	9.860
	U	298.21	1105.2	1214.5	1253.0	1291.9	1331.5	1371.8	1413.0	1455.1
	H	298.54	1187.2	1329.6	1379.5	1429.7	1480.4	1532.0	1584.4	1637.6
	S	0.4743	1.6027	1.7586	1.8036	1.8451	1.8839	1.9205	1.9552	1.9883
105 (331.37)	V	0.0178	4.231	5.915	6.504	7.086	7.665	8.241	8.816	9.389
	U	301.89	1105.8	1214.3	1252.8	1291.8	1331.3	1371.7	1412.9	1455.0
	H	302.24	1188.0	1329.2	1379.2	1429.4	1480.3	1531.8	1584.2	1637.5
	S	0.4790	1.5988	1.7530	1.7981	1.8396	1.8785	1.9151	1.9498	1.9828
110 (334.79)	V	0.0178	4.048	5.642	6.205	6.761	7.314	7.865	8.413	8.961
	U	305.44	1106.5	1214.0	1252.7	1291.6	1331.2	1371.6	1412.8	1455.0
	H	305.80	1188.9	1328.9	1378.9	1429.2	1480.1	1531.7	1584.1	1637.4
	S	0.4834	1.5950	1.7476	1.7928	1.8344	1.8732	1.9099	1.9446	1.9777
115 (338.08)	V	0.0179	3.881	5.392	5.932	6.465	6.994	7.521	8.046	8.570
	U	308.87	1107.0	1213.8	1252.5	1291.5	1331.1	1371.5	1412.8	1454.9
	H	309.25	1189.6	1328.6	1378.7	1429.0	1479.9	1531.6	1584.0	1637.2
	S	0.4877	1.5913	1.7425	1.7877	1.8294	1.8682	1.9049	1.9396	1.9727

120 (341.27)	V	0.0179	3.728	5.164	5.681	6.193	6.701	7.206	7.710	8.212
	U	312.19	1107.6	1213.5	1252.3	1291.3	1331.0	1371.4	1412.7	1454.8
	H	312.58	1190.4	1328.2	1378.4	1428.8	1479.8	1531.4	1583.9	1637.1
	S	0.4919	1.5879	1.7376	1.7829	1.8246	1.8635	1.9001	1.9349	1.9680
125 (344.35)	V	0.0179	3.586	4.953	5.451	5.943	6.431	6.916	7.400	7.882
	U	315.40	1108.1	1213.3	1252.1	1291.2	1330.8	1371.3	1412.6	1454.7
	H	315.82	1191.1	1327.9	1378.2	1428.6	1479.6	1531.3	1583.7	1637.0
	S	0.4959	1.5845	1.7328	1.7782	1.8199	1.8589	1.8955	1.9303	1.9634
130 (347.33)	V	0.0180	3.454	4.759	5.238	5.712	6.181	6.649	7.114	7.578
	U	318.52	1108.6	1213.0	1251.9	1291.0	1330.7	1371.2	1412.5	1454.6
	H	318.95	1191.7	1327.5	1377.9	1428.4	1479.4	1531.1	1583.6	1636.9
	S	0.4998	1.5813	1.7283	1.7737	1.8155	1.8545	1.8911	1.9259	1.9591
135 (350.23)	V	0.0180	3.332	4.579	5.042	5.498	5.951	6.401	6.849	7.296
	U	321.55	1109.1	1212.8	1251.7	1290.9	1330.6	1371.1	1412.4	1454.5
	H	322.00	1192.4	1327.2	1377.7	1428.2	1479.2	1531.0	1583.5	1636.8
	S	0.5035	1.5782	1.7239	1.7694	1.8112	1.8502	1.8869	1.9217	1.9548
140 (353.04)	V	0.0180	3.219	4.412	4.859	5.299	5.736	6.171	6.604	7.035
	U	324.49	1109.6	1212.5	1251.5	1290.7	1330.5	1371.0	1412.3	1454.5
	H	324.96	1193.0	1326.8	1377.4	1428.0	1479.1	1530.8	1583.4	1636.7
	S	0.5071	1.5752	1.7196	1.7652	1.8071	1.8461	1.8828	1.9176	1.9508
145 (355.77)	V	0.0181	3.113	4.256	4.689	5.115	5.537	5.957	6.375	6.791
	U	327.36	1110.0	1212.3	1251.3	1290.6	1330.3	1370.9	1412.2	1454.4
	H	327.84	1193.5	1326.5	1377.1	1427.8	1478.9	1530.7	1583.2	1636.6
	S	0.5107	1.5723	1.7155	1.7612	1.8031	1.8421	1.8789	1.9137	1.9469
150 (358.43)	V	0.0181	3.014	4.111	4.530	4.942	5.351	5.757	6.161	6.564
	U	330.15	1110.4	1212.0	1251.1	1290.4	1330.2	1370.7	1412.1	1454.3
	H	330.65	1194.1	1326.1	1376.9	1427.6	1478.7	1530.5	1583.1	1636.5
	S	0.5141	1.5695	1.7115	1.7573	1.7992	1.8383	1.8751	1.9099	1.9431
155 (361.02)	V	0.0181	2.921	3.975	4.381	4.781	5.177	5.570	5.961	6.352
	U	332.87	1110.8	1211.8	1251.0	1290.3	1330.1	1370.6	1412.0	1454.2
	H	333.39	1194.6	1325.8	1376.6	1427.4	1478.6	1530.4	1583.0	1636.4
	S	0.5174	1.5668	1.7077	1.7535	1.7955	1.8346	1.8714	1.9062	1.9394
160 (363.55)	V	0.0182	2.834	3.848	4.242	4.629	5.013	5.395	5.774	6.152
	U	335.53	1111.2	1211.5	1250.8	1290.1	1330.0	1370.5	1411.9	1454.1
	H	336.07	1195.1	1325.4	1376.4	1427.2	1478.4	1530.3	1582.9	1636.3
	S	0.5206	1.5641	1.7039	1.7499	1.7919	1.8310	1.8678	1.9027	1.9359

TABLE C.4. SUPERHEATED STEAM ENGLISH UNITS (Continued)

ABS PRESS PSIA (SAT TEMP)		SAT WATER	SAT STEAM	TEMPERATURE, DEG F 400	420	440	460	480	500	550
165	V	0.0182	2.751	2.908	2.997	3.083	3.168	3.251	3.333	3.533
(366.02)	U	338.12	1111.6	1127.7	1136.8	1145.6	1154.2	1162.7	1171.0	1191.3
	H	338.68	1195.6	1216.5	1228.3	1239.7	1251.0	1261.9	1272.8	1299.2
	S	0.5238	1.5616	1.5864	1.6000	1.6129	1.6252	1.6370	1.6484	1.6753
170	V	0.0182	2.674	2.816	2.903	2.987	3.070	3.151	3.231	3.425
(368.42)	U	340.66	1111.9	1127.0	1136.2	1145.1	1153.7	1162.3	1170.6	1191.0
	H	341.24	1196.0	1215.6	1227.5	1239.0	1250.3	1261.4	1272.2	1298.8
	S	0.5269	1.5591	1.5823	1.5960	1.6090	1.6214	1.6333	1.6447	1.6717
175	V	0.0182	2.601	2.729	2.814	2.897	2.977	3.056	3.134	3.324
(370.77)	U	343.15	1112.2	1126.3	1135.6	1144.5	1153.3	1161.8	1170.2	1190.7
	H	343.74	1196.4	1214.7	1226.7	1238.3	1249.7	1260.8	1271.7	1298.4
	S	0.5299	1.5567	1.5783	1.5921	1.6051	1.6176	1.6296	1.6411	1.6682
180	V	0.0183	2.531	2.647	2.730	2.811	2.890	2.967	3.043	3.229
(373.08)	U	345.58	1112.5	1125.6	1134.9	1144.0	1152.8	1161.4	1169.8	1190.4
	H	346.19	1196.9	1213.8	1225.9	1237.6	1249.0	1260.2	1271.2	1297.9
	S	0.5328	1.5543	1.5743	1.5882	1.6014	1.6140	1.6260	1.6376	1.6647
184	V	0.0183	2.465	2.570	2.651	2.730	2.807	2.883	2.957	3.138
(375.33)	U	347.96	1112.8	1124.9	1134.3	1143.4	1152.3	1160.9	1169.4	1190.1
	H	348.58	1197.2	1212.9	1225.1	1236.9	1248.4	1259.6	1270.7	1297.5
	S	0.5356	1.5520	1.5705	1.5845	1.5978	1.6104	1.6225	1.6341	1.6614
190	V	0.0183	2.403	2.496	2.576	2.654	2.729	2.803	2.876	3.052
(377.53)	U	350.29	1113.1	1124.2	1133.7	1142.9	1151.8	1160.5	1169.0	1189.8
	H	350.94	1197.6	1212.0	1224.3	1236.2	1247.7	1259.0	1270.1	1297.1
	S	0.5384	1.5498	1.5667	1.5808	1.5942	1.6069	1.6191	1.6307	1.6581
195	V	0.0184	2.344	2.426	2.505	2.581	2.655	2.727	2.798	2.971
(379.69)	U	352.58	1113.4	1123.5	1133.1	1142.3	1151.3	1160.0	1168.6	1189.4
	H	353.24	1198.0	1211.1	1223.4	1235.4	1247.1	1258.4	1269.6	1296.6
	S	0.5412	1.5476	1.5630	1.5772	1.5907	1.6035	1.6157	1.6274	1.6549
200	V	0.0184	2.287	2.360	2.437	2.511	2.584	2.655	2.725	2.894
(381.80)	U	354.82	1113.7	1122.8	1132.4	1141.7	1150.8	1159.6	1168.2	1189.1
	H	355.51	1198.3	1210.1	1222.6	1234.7	1246.4	1257.9	1269.0	1296.2
	S	0.5438	1.5454	1.5593	1.5737	1.5872	1.6001	1.6124	1.6242	1.6518

	V	0.0184	2.233	2.297	2.372	2.446	2.517	2.587	2.655	2.820
205	U	357.03	1113.9	1122.1	1131.8	1141.2	1150.3	1159.1	1167.8	1188.8
(383.88)	H	357.73	1198.7	1209.2	1221.8	1234.0	1245.8	1257.3	1268.5	1295.8
	S	0.5465	1.5434	1.5557	1.5702	1.5839	1.5969	1.6092	1.6211	1.6488
	V	0.0184	2.182	2.236	2.311	2.383	2.453	2.521	2.588	2.750
210	U	359.20	1114.2	1121.3	1131.2	1140.6	1149.8	1158.7	1167.4	1188.5
(385.92)	H	359.91	1199.0	1208.2	1221.0	1233.2	1245.1	1256.7	1268.0	1295.3
	S	0.5490	1.5413	1.5522	1.5668	1.5806	1.5936	1.6061	1.6180	1.6458
	V	0.0185	2.133	2.179	2.252	2.323	2.392	2.459	2.524	2.684
215	U	361.32	1114.4	1120.6	1130.5	1140.0	1149.3	1158.2	1167.0	1188.1
(387.91)	H	362.06	1199.3	1207.3	1220.1	1232.5	1244.4	1256.0	1267.4	1294.9
	S	0.5515	1.5393	1.5487	1.5634	1.5773	1.5905	1.6030	1.6149	1.6429
	V	0.0185	2.086	2.124	2.196	2.266	2.333	2.399	2.464	2.620
220	U	363.41	1114.6	1119.9	1129.9	1139.5	1148.7	1157.8	1166.6	1187.8
(389.88)	H	364.17	1199.6	1206.3	1219.3	1231.7	1243.7	1255.4	1266.9	1294.5
	S	0.5540	1.5374	1.5453	1.5601	1.5741	1.5873	1.5999	1.6120	1.6400
	V	0.0185	2.041	2.071	2.143	2.211	2.278	2.342	2.406	2.559
225	U	365.47	1114.9	1119.1	1129.2	1138.9	1148.2	1157.3	1166.1	1187.5
(391.80)	H	366.24	1199.9	1205.4	1218.4	1230.9	1243.1	1254.8	1266.3	1294.0
	S	0.5564	1.5354	1.5419	1.5569	1.5710	1.5843	1.5969	1.6090	1.6372
	V	0.0185	1.9985	2.021	2.091	2.159	2.224	2.288	2.350	2.501
230	U	367.49	1115.1	1118.4	1128.5	1138.3	1147.7	1156.8	1165.7	1187.2
(393.70)	H	368.28	1200.1	1204.4	1217.5	1230.2	1242.4	1254.2	1265.7	1293.6
	S	0.5588	1.5336	1.5385	1.5537	1.5679	1.5813	1.5940	1.6062	1.6344
	V	0.0186	1.9573	1.973	2.042	2.109	2.173	2.236	2.297	2.445
235	U	369.48	1115.3	1117.6	1127.9	1137.7	1147.2	1156.4	1165.3	1186.8
(395.56)	H	370.29	1200.4	1203.4	1216.7	1229.4	1241.7	1253.6	1265.2	1293.1
	S	0.5611	1.5317	1.5353	1.5505	1.5648	1.5783	1.5911	1.6033	1.6317
	V	0.0186	1.9177	1.927	1.995	2.061	2.124	2.186	2.246	2.391
240	U	371.45	1115.5	1116.8	1127.2	1137.1	1146.6	1155.9	1164.9	1186.5
(397.39)	H	372.27	1200.6	1202.4	1215.8	1228.6	1241.0	1253.0	1264.6	1292.7
	S	0.5634	1.5299	1.5320	1.5474	1.5618	1.5754	1.5883	1.6006	1.6291
	V	0.0186	1.8797	1.882	1.950	2.015	2.077	2.138	2.197	2.340
245	U	373.38	1115.6	1116.1	1126.5	1136.5	1146.1	1155.4	1164.4	1186.2
(399.19)	H	374.22	1200.9	1201.4	1214.9	1227.8	1240.3	1252.3	1264.1	1292.3
	S	0.5657	1.5281	1.5288	1.5443	1.5588	1.5725	1.5855	1.5978	1.6265

TABLE C.4. SUPERHEATED STEAM ENGLISH UNITS (Continued)

ABS PRESS PSIA (SAT TEMP)		SAT WATER	SAT STEAM	600	700	800	900	1000	1100	1200
				TEMPERATURE, DEG F						
165 (366.02)	V	0.0182	2.751	3.728	4.111	4.487	4.860	5.230	5.598	5.965
	U	338.12	1111.6	1211.3	1250.6	1289.9	1329.8	1370.4	1411.8	1454.1
	H	338.68	1195.6	1325.1	1376.1	1427.0	1478.2	1530.1	1582.7	1636.2
	S	0.5238	1.5616	1.7003	1.7463	1.7884	1.8275	1.8643	1.8992	1.9324
170 (368.42)	V	0.0182	2.674	3.616	3.988	4.354	4.715	5.075	5.432	5.789
	U	340.66	1111.9	1211.0	1250.4	1289.8	1329.7	1370.3	1411.7	1454.0
	H	341.24	1196.0	1324.7	1375.8	1426.8	1478.0	1530.0	1582.6	1636.1
	S	0.5269	1.5591	1.6968	1.7428	1.7850	1.8241	1.8610	1.8959	1.9291
175 (370.77)	V	0.0182	2.601	3.510	3.872	4.227	4.579	4.929	5.276	5.623
	U	343.15	1112.2	1210.7	1250.2	1289.6	1329.6	1370.2	1411.6	1453.9
	H	343.74	1196.4	1324.4	1375.6	1426.5	1477.9	1529.8	1582.5	1636.0
	S	0.5299	1.5567	1.6933	1.7395	1.7816	1.8208	1.8577	1.8926	1.9258
180 (373.08)	V	0.0183	2.531	3.409	3.762	4.108	4.451	4.791	5.129	5.466
	U	345.58	1112.5	1210.5	1250.0	1289.5	1329.4	1370.1	1411.5	1453.8
	H	346.19	1196.9	1324.0	1375.3	1426.3	1477.7	1529.7	1582.4	1635.9
	S	0.5328	1.5543	1.6900	1.7362	1.7784	1.8176	1.8545	1.8894	1.9227
185 (375.33)	V	0.0183	2.465	3.314	3.658	3.996	4.329	4.660	4.989	5.317
	U	347.96	1112.8	1210.2	1249.8	1289.3	1329.3	1370.0	1411.4	1453.7
	H	348.58	1197.2	1323.7	1375.1	1426.1	1477.5	1529.5	1582.3	1635.8
	S	0.5356	1.5520	1.6867	1.7330	1.7753	1.8145	1.8514	1.8864	1.9196
190 (377.53)	V	0.0183	2.403	3.225	3.560	3.889	4.214	4.536	4.857	5.177
	U	350.29	1113.1	1209.9	1249.6	1289.2	1329.2	1369.9	1411.3	1453.7
	H	350.94	1197.6	1323.3	1374.8	1425.9	1477.4	1529.4	1582.1	1635.7
	S	0.5384	1.5498	1.6835	1.7299	1.7722	1.8115	1.8484	1.8834	1.9166
195 (379.69)	V	0.0184	2.344	3.139	3.467	3.788	4.105	4.419	4.732	5.043
	U	352.58	1113.4	1209.7	1249.4	1289.0	1329.1	1369.8	1411.3	1453.6
	H	353.24	1198.0	1323.0	1374.5	1425.7	1477.2	1529.2	1582.0	1635.6
	S	0.5412	1.5476	1.6804	1.7269	1.7692	1.8085	1.8455	1.8804	1.9137
200 (381.80)	V	0.0184	2.287	3.058	3.378	3.691	4.001	4.308	4.613	4.916
	U	354.82	1113.7	1209.4	1249.2	1288.9	1328.9	1369.7	1411.2	1453.5
	H	355.51	1198.3	1322.6	1374.3	1425.5	1477.0	1529.1	1581.9	1635.4
	S	0.5438	1.5454	1.6773	1.7239	1.7663	1.8057	1.8426	1.8776	1.9109

205 (383.88)	V	0.0184	2.233	2.981	3.294	3.600	3.902	4.202	4.499	4.796
	U	357.03	1113.9	1209.2	1249.0	1288.7	1328.8	1369.6	1411.1	1453.4
	H	357.73	1198.7	1322.3	1374.0	1425.3	1476.8	1528.9	1581.8	1635.3
	S	0.5465	1.5434	1.6744	1.7210	1.7635	1.8028	1.8398	1.8748	1.9081
210 (385.92)	V	0.0184	2.182	2.908	3.214	3.513	3.808	4.101	4.392	4.681
	U	359.20	1114.2	1208.9	1248.8	1288.6	1328.7	1369.4	1411.0	1453.3
	H	359.91	1199.0	1321.9	1373.7	1425.1	1476.7	1528.8	1581.6	1635.2
	S	0.5490	1.5413	1.6715	1.7182	1.7607	1.8001	1.8371	1.8721	1.9054
215 (387.91)	V	0.0185	2.133	2.838	3.137	3.430	3.718	4.004	4.289	4.572
	U	361.32	1114.4	1208.6	1248.7	1288.4	1328.6	1369.3	1410.9	1453.2
	H	362.06	1199.3	1321.5	1373.5	1424.9	1476.5	1528.7	1581.5	1635.1
	S	0.5515	1.5393	1.6686	1.7155	1.7580	1.7974	1.8344	1.8694	1.9028
220 (389.88)	V	0.0185	2.086	2.771	3.064	3.350	3.633	3.912	4.190	4.467
	U	363.41	1114.6	1208.4	1248.5	1288.3	1328.4	1369.2	1410.8	1453.2
	H	364.17	1199.6	1321.2	1373.2	1424.7	1476.3	1528.5	1581.4	1635.0
	S	0.5540	1.5374	1.6658	1.7128	1.7553	1.7948	1.8318	1.8668	1.9002
225 (391.80)	V	0.0185	2.041	2.707	2.994	3.275	3.551	3.825	4.097	4.367
	U	365.47	1114.9	1208.1	1248.3	1288.1	1328.3	1369.1	1410.7	1453.1
	H	366.24	1199.9	1320.8	1372.9	1424.5	1476.1	1528.4	1581.3	1634.9
	S	0.5564	1.5354	1.6631	1.7101	1.7527	1.7922	1.8293	1.8643	1.8977
230 (393.70)	V	0.0185	1.9984	2.646	2.928	3.202	3.473	3.741	4.007	4.272
	U	367.49	1115.1	1207.8	1248.1	1288.0	1328.2	1369.0	1410.6	1453.0
	H	368.28	1200.1	1320.4	1372.7	1424.2	1476.0	1528.2	1581.1	1634.8
	S	0.5588	1.5336	1.6604	1.7075	1.7502	1.7897	1.8268	1.8618	1.8952
235 (395.56)	V	0.0186	1.9573	2.588	2.864	3.133	3.398	3.660	3.921	4.180
	U	369.48	1115.3	1207.6	1247.9	1287.8	1328.0	1368.9	1410.5	1452.9
	H	370.29	1200.4	1320.1	1372.4	1424.0	1475.8	1528.1	1581.0	1634.7
	S	0.5611	1.5317	1.6578	1.7050	1.7477	1.7872	1.8243	1.8594	1.8928
240 (397.39)	V	0.0186	1.9177	2.532	2.802	3.066	3.326	3.583	3.839	4.093
	U	371.45	1115.5	1207.3	1247.7	1287.7	1327.9	1368.8	1410.4	1452.8
	H	372.27	1200.6	1319.7	1372.1	1423.8	1475.6	1527.9	1580.9	1634.6
	S	0.5634	1.5299	1.6552	1.7025	1.7452	1.7848	1.8219	1.8570	1.8904
245 (399.19)	V	0.0186	1.8797	2.478	2.744	3.002	3.257	3.509	3.760	4.009
	U	373.38	1115.6	1207.0	1247.5	1287.5	1327.8	1368.7	1410.3	1452.8
	H	374.22	1200.9	1319.4	1371.9	1423.6	1475.5	1527.8	1580.8	1634.5
	S	0.5657	1.5281	1.6527	1.7000	1.7428	1.7824	1.8196	1.8547	1.8881

TABLE C.4. SUPERHEATED STEAM ENGLISH UNITS (Continued)

ABS PRESS PSIA (SAT TEMP)		SAT WATER	SAT STEAM	420	440	460	480	500	520	550
				TEMPERATURE, DEG F						
250 (400.97)	V	0.0187	1.8432	1.907	1.970	2.032	2.092	2.150	2.207	2.291
	U	375.28	1115.8	1125.8	1135.9	1145.6	1154.9	1164.0	1172.9	1185.8
	H	376.14	1201.1	1214.0	1227.1	1239.6	1251.7	1263.5	1275.0	1291.8
	S	0.5679	1.5264	1.5413	1.5559	1.5697	1.5827	1.5951	1.6070	1.6239
255 (402.72)	V	0.0187	1.8080	1.865	1.928	1.989	2.048	2.105	2.161	2.244
	U	377.15	1116.0	1125.1	1135.3	1145.0	1154.5	1163.6	1172.5	1185.5
	H	378.04	1201.3	1213.1	1226.3	1238.9	1251.1	1262.9	1274.5	1291.4
	S	0.5701	1.5247	1.5383	1.5530	1.5669	1.5800	1.5925	1.6044	1.6214
260 (404.44)	V	0.0187	1.7742	1.825	1.887	1.947	2.005	2.062	2.117	2.198
	U	379.00	1116.2	1124.5	1134.7	1144.5	1154.0	1163.1	1172.1	1185.1
	H	379.90	1201.5	1212.2	1225.5	1238.2	1250.4	1262.4	1274.0	1290.9
	S	0.5722	1.5230	1.5353	1.5502	1.5642	1.5774	1.5899	1.6019	1.6189
265 (406.13)	V	0.0187	1.7416	1.786	1.848	1.907	1.964	2.020	2.075	2.154
	U	380.83	1116.3	1123.8	1134.1	1144.0	1153.5	1162.7	1171.7	1184.8
	H	381.74	1201.7	1211.3	1224.7	1237.5	1249.8	1261.8	1273.4	1290.4
	S	0.5743	1.5214	1.5324	1.5474	1.5614	1.5747	1.5873	1.5993	1.6165
270 (407.80)	V	0.0188	1.7101	1.749	1.810	1.868	1.925	1.980	2.034	2.112
	U	382.62	1116.5	1123.1	1133.5	1143.4	1153.0	1162.3	1171.3	1184.5
	H	383.56	1201.9	1210.4	1223.9	1236.7	1249.2	1261.2	1272.9	1290.0
	S	0.5764	1.5197	1.5295	1.5446	1.5588	1.5721	1.5848	1.5969	1.6140
275 (409.45)	V	0.0188	1.6798	1.713	1.773	1.831	1.887	1.941	1.994	2.071
	U	384.40	1116.6	1122.3	1132.8	1142.9	1152.5	1161.8	1170.9	1184.1
	H	385.35	1202.1	1209.5	1223.1	1236.0	1248.5	1260.6	1272.4	1289.5
	S	0.5784	1.5181	1.5266	1.5419	1.5561	1.5696	1.5823	1.5944	1.6117
280 (411.07)	V	0.0188	1.6505	1.678	1.738	1.795	1.850	1.904	1.956	2.032
	U	386.15	1116.7	1121.6	1132.2	1142.3	1152.0	1161.4	1170.5	1183.8
	H	387.12	1202.3	1208.6	1222.2	1235.3	1247.9	1260.0	1271.9	1289.1
	S	0.5805	1.5166	1.5238	1.5391	1.5535	1.5670	1.5798	1.5920	1.6093
285 (412.67)	V	0.0188	1.6222	1.645	1.704	1.760	1.815	1.868	1.919	1.994
	U	387.88	1116.9	1120.9	1131.6	1141.7	1151.5	1160.9	1170.1	1183.4
	H	388.87	1202.4	1207.6	1221.4	1234.6	1247.2	1259.4	1271.3	1288.6
	S	0.5824	1.5150	1.5210	1.5365	1.5509	1.5645	1.5774	1.5897	1.6070

290 (414.25)	V	0.0188	1.5948	1.612	1.671	1.727	1.780	1.833	1.884	1.958
	U	389.59	1117.0	1120.2	1130.9	1141.2	1151.0	1160.5	1169.7	1183.1
	H	390.60	1202.6	1206.7	1220.6	1233.8	1246.6	1258.9	1270.8	1288.1
	S	0.5844	1.5135	1.5182	1.5338	1.5484	1.5621	1.5750	1.5873	1.6048
295 (415.81)	V	0.0189	1.5684	1.581	1.639	1.694	1.747	1.799	1.849	1.922
	U	391.27	1117.1	1119.5	1130.3	1140.6	1150.5	1160.0	1169.3	1182.7
	H	392.30	1202.7	1205.8	1219.7	1233.1	1245.9	1258.3	1270.2	1287.7
	S	0.5863	1.5120	1.5155	1.5312	1.5458	1.5596	1.5726	1.5850	1.6025
300 (417.35)	V	0.0189	1.5427	1.551	1.608	1.663	1.715	1.766	1.816	1.888
	U	392.94	1117.2	1118.7	1129.6	1140.0	1150.0	1159.6	1168.9	1182.4
	H	393.99	1202.9	1204.8	1218.9	1232.3	1245.2	1257.7	1269.7	1287.2
	S	0.5882	1.5105	1.5127	1.5286	1.5433	1.5572	1.5703	1.5827	1.6003
310 (420.36)	V	0.0189	1.4939	....	1.549	1.603	1.655	1.704	1.753	1.823
	U	396.21	1117.5	....	1128.3	1138.9	1149.0	1158.7	1168.1	1181.7
	H	397.30	1203.2	....	1217.2	1230.8	1243.9	1256.5	1268.6	1286.3
	S	0.5920	1.5076	....	1.5234	1.5384	1.5525	1.5657	1.5782	1.5960
320 (423.31)	V	0.0190	1.4480	....	1.494	1.547	1.597	1.646	1.694	1.762
	U	399.41	1117.7	....	1127.0	1137.7	1147.9	1157.8	1167.2	1181.0
	H	400.53	1203.4	....	1215.5	1229.3	1242.5	1255.2	1267.5	1285.3
	S	0.5956	1.5048	....	1.5184	1.5336	1.5478	1.5612	1.5739	1.5918
330 (426.18)	V	0.0190	1.4048	....	1.442	1.494	1.544	1.591	1.638	1.705
	U	402.53	1117.8	....	1125.7	1136.6	1146.9	1156.8	1166.4	1180.2
	H	403.70	1203.6	....	1213.8	1227.8	1241.2	1254.0	1266.4	1284.4
	S	0.5991	1.5021	....	1.5134	1.5289	1.5433	1.5568	1.5696	1.5876
340 (428.98)	V	0.0191	1.3640	....	1.393	1.444	1.493	1.540	1.585	1.651
	U	405.60	1118.0	....	1124.3	1135.4	1145.8	1155.9	1165.6	1179.5
	H	406.80	1203.8	....	1212.0	1226.2	1239.8	1252.8	1265.3	1283.4
	S	0.6026	1.4994	....	1.5086	1.5242	1.5388	1.5525	1.5654	1.5836
350 (431.73)	V	0.0191	1.3255	....	1.347	1.397	1.445	1.491	1.536	1.600
	U	408.59	1118.1	....	1123.0	1134.2	1144.8	1154.9	1164.7	1178.8
	H	409.83	1204.0	....	1210.2	1224.7	1238.4	1251.5	1264.2	1282.4
	S	0.6059	1.4968	....	1.5038	1.5197	1.5344	1.5483	1.5613	1.5797
360 (434.41)	V	0.0192	1.2891	....	1.303	1.353	1.400	1.445	1.489	1.552
	U	411.53	1118.3	....	1121.6	1132.9	1143.7	1154.0	1163.9	1178.1
	H	412.81	1204.1	....	1208.4	1223.1	1237.0	1250.3	1263.1	1281.5
	S	0.6092	1.4943	....	1.4990	1.5152	1.5301	1.5441	1.5573	1.5758

TABLE C.4. SUPERHEATED STEAM ENGLISH UNITS (Continued)

ABS PRESS PSIA (SAT TEMP)		SAT WATER	SAT STEAM	TEMPERATURE, DEG F 600	700	800	900	1000	1100	1200
	V	0.0187	1.8432	2.426	2.687	2.941	3.191	3.438	3.684	3.928
250	U	375.28	1115.8	1206.7	1247.3	1287.3	1327.7	1368.6	1410.2	1452.7
(400.97)	H	376.14	1201.1	1319.0	1371.6	1423.4	1475.3	1527.6	1580.6	1634.4
	S	0.5679	1.5264	1.6502	1.6976	1.7405	1.7801	1.8173	1.8524	1.8858
	V	0.0187	1.8080	2.377	2.633	2.882	3.127	3.370	3.611	3.850
255	U	377.15	1116.0	1206.5	1247.1	1287.2	1327.5	1368.5	1410.1	1452.6
(402.72)	H	378.04	1201.3	1318.6	1371.3	1423.2	1475.1	1527.5	1580.5	1634.3
	S	0.5701	1.5247	1.6477	1.6953	1.7382	1.7778	1.8150	1.8502	1.8836
	V	0.0187	1.7742	2.329	2.581	2.826	3.066	3.304	3.541	3.776
260	U	379.00	1116.2	1206.2	1246.9	1287.0	1327.4	1368.4	1410.0	1452.5
(404.44)	H	379.90	1201.5	1318.2	1371.1	1423.0	1474.9	1527.3	1580.4	1634.2
	S	0.5722	1.5230	1.6453	1.6930	1.7359	1.7756	1.8128	1.8480	1.8814
	V	0.0187	1.7416	2.283	2.531	2.771	3.007	3.241	3.473	3.704
265	U	380.83	1116.3	1205.9	1246.7	1286.9	1327.3	1368.2	1409.9	1452.4
(406.13)	H	381.74	1201.7	1317.9	1370.8	1422.8	1474.8	1527.2	1580.3	1634.1
	S	0.5743	1.5214	1.6430	1.6907	1.7337	1.7734	1.8106	1.8458	1.8792
	V	0.0188	1.7101	2.239	2.482	2.719	2.951	3.181	3.408	3.635
270	U	382.62	1116.5	1205.6	1246.5	1286.7	1327.2	1368.1	1409.8	1452.3
(407.80)	H	383.56	1201.9	1317.5	1370.5	1422.6	1474.6	1527.0	1580.1	1634.0
	S	0.5764	1.5197	1.6406	1.6885	1.7315	1.7713	1.8085	1.8437	1.8771
	V	0.0188	1.6798	2.196	2.436	2.668	2.896	3.122	3.346	3.568
275	U	384.40	1116.6	1205.4	1246.3	1286.6	1327.0	1368.0	1409.8	1452.3
(409.45)	H	385.35	1202.1	1317.1	1370.3	1422.4	1474.4	1526.9	1580.0	1633.9
	S	0.5784	1.5181	1.6384	1.6863	1.7294	1.7691	1.8064	1.8416	1.8750
	V	0.0188	1.6505	2.155	2.391	2.619	2.844	3.066	3.286	3.504
280	U	386.15	1116.7	1205.1	1246.1	1286.4	1326.9	1367.9	1409.7	1452.2
(411.07)	H	387.12	1202.3	1316.8	1370.0	1422.1	1474.2	1526.8	1579.9	1633.8
	S	0.5805	1.5166	1.6361	1.6841	1.7273	1.7671	1.8043	1.8395	1.8730
	V	0.0188	1.6222	2.115	2.348	2.572	2.793	3.011	3.227	3.442
285	U	387.88	1116.9	1204.8	1245.9	1286.3	1326.8	1367.8	1409.6	1452.1
(412.67)	H	388.87	1202.4	1316.4	1369.7	1421.9	1474.1	1526.6	1579.8	1633.6
	S	0.5824	1.5150	1.6339	1.6820	1.7252	1.7650	1.8023	1.8375	1.8710

290 (414.25)	V	0.0188	1.5948	2.077	2.306	2.527	2.744	2.958	3.171	3.382
	U	389.59	1117.0	1204.5	1245.7	1286.1	1326.6	1367.7	1409.5	1452.0
	H	390.60	1202.6	1316.0	1369.5	1421.7	1473.9	1526.5	1579.6	1633.5
	S	0.5844	1.5135	1.6317	1.6799	1.7232	1.7630	1.8003	1.8356	1.8690
295 (415.81)	V	0.0189	1.5684	2.040	2.265	2.483	2.697	2.908	3.117	3.325
	U	391.27	1117.1	1204.3	1245.5	1286.0	1326.5	1367.6	1409.4	1451.9
	H	392.30	1202.7	1315.6	1369.2	1421.5	1473.7	1526.3	1579.5	1633.4
	S	0.5863	1.5120	1.6295	1.6779	1.7211	1.7610	1.7984	1.8336	1.8671
300 (417.35)	V	0.0189	1.5427	2.004	2.226	2.441	2.651	2.859	3.064	3.269
	U	392.94	1117.2	1204.0	1245.3	1285.8	1326.4	1367.5	1409.3	1451.9
	H	393.99	1202.9	1315.2	1368.9	1421.3	1473.6	1526.2	1579.4	1633.3
	S	0.5882	1.5105	1.6274	1.6758	1.7192	1.7591	1.7964	1.8317	1.8652
310 (420.36)	V	0.0189	1.4939	1.936	2.152	2.360	2.564	2.765	2.964	3.162
	U	396.21	1117.5	1203.4	1244.9	1285.5	1326.1	1367.3	1409.1	1451.7
	H	397.30	1203.2	1314.5	1368.4	1420.9	1473.2	1525.9	1579.2	1633.1
	S	0.5920	1.5076	1.6233	1.6719	1.7153	1.7553	1.7927	1.8280	1.8615
320 (423.31)	V	0.0190	1.4480	1.873	2.082	2.284	2.482	2.677	2.871	3.063
	U	399.41	1117.7	1202.8	1244.5	1285.2	1325.9	1367.0	1408.9	1451.5
	H	400.53	1203.4	1313.7	1367.8	1420.4	1472.9	1525.6	1578.9	1632.9
	S	0.5956	1.5048	1.6192	1.6680	1.7116	1.7516	1.7890	1.8243	1.8579
330 (426.18)	V	0.0190	1.4048	1.813	2.017	2.213	2.405	2.595	2.783	2.969
	U	402.53	1117.8	1202.3	1244.1	1284.9	1325.6	1366.8	1408.7	1451.4
	H	403.70	1203.6	1313.0	1367.3	1420.0	1472.5	1525.3	1578.7	1632.7
	S	0.5991	1.5021	1.6153	1.6643	1.7079	1.7480	1.7855	1.8208	1.8544
340 (428.98)	V	0.0191	1.3640	1.756	1.955	2.146	2.333	2.518	2.700	2.881
	U	405.60	1118.0	1201.7	1243.7	1284.6	1325.4	1366.6	1408.5	1451.2
	H	406.80	1203.8	1312.2	1366.7	1419.6	1472.2	1525.0	1578.4	1632.5
	S	0.6026	1.4994	1.6114	1.6606	1.7044	1.7445	1.7820	1.8174	1.8510
350 (431.73)	V	0.0191	1.3255	1.703	1.897	2.083	2.265	2.444	2.622	2.798
	U	408.59	1118.1	1201.1	1243.3	1284.2	1325.1	1366.4	1408.3	1451.0
	H	409.83	1204.0	1311.4	1366.2	1419.2	1471.8	1524.7	1578.2	1632.3
	S	0.6059	1.4968	1.6077	1.6571	1.7009	1.7411	1.7787	1.8141	1.8477
360 (434.41)	V	0.0192	1.2891	1.652	1.842	2.024	2.201	2.375	2.548	2.720
	U	411.53	1118.3	1200.5	1242.9	1283.9	1324.8	1366.2	1408.2	1450.9
	H	412.81	1204.1	1310.6	1365.6	1418.7	1471.5	1524.4	1577.9	1632.1
	S	0.6092	1.4943	1.6040	1.6536	1.6976	1.7379	1.7754	1.8109	1.8445

TABLE C.4. SUPERHEATED STEAM ENGLISH UNITS (Continued)

ABS PRESS PSIA (SAT TEMP)		SAT WATER	SAT STEAM	TEMPERATURE, DEG F 460	480	500	520	540	560	580
370 (437.04)	V	0.0192	1.2546	1.311	1.357	1.402	1.445	1.486	1.527	1.566
	U	414.41	1118.4	1131.7	1142.6	1153.0	1163.0	1172.6	1182.0	1191.0
	H	415.73	1204.3	1221.4	1235.5	1249.0	1261.9	1274.4	1286.5	1298.3
	S	0.6125	1.4918	1.5107	1.5259	1.5401	1.5534	1.5660	1.5780	1.5894
380 (439.61)	V	0.0193	1.2218	1.271	1.317	1.361	1.403	1.444	1.483	1.522
	U	417.24	1118.5	1130.4	1141.5	1152.0	1162.1	1171.8	1181.2	1190.4
	H	418.59	1204.4	1219.8	1234.1	1247.7	1260.8	1273.3	1285.5	1297.4
	S	0.6156	1.4894	1.5063	1.5217	1.5360	1.5495	1.5622	1.5743	1.5858
390 (442.13)	V	0.0193	1.1906	1.233	1.278	1.321	1.363	1.403	1.442	1.480
	U	420.01	1118.6	1129.2	1140.4	1151.0	1161.2	1171.0	1180.5	1189.7
	H	421.40	1204.5	1218.2	1232.6	1246.4	1259.6	1272.3	1284.6	1296.5
	S	0.6187	1.4870	1.5020	1.5176	1.5321	1.5457	1.5585	1.5707	1.5823
400 (444.60)	V	0.0193	1.1610	1.197	1.242	1.284	1.325	1.364	1.403	1.440
	U	422.74	1118.7	1127.9	1139.3	1150.0	1160.3	1170.2	1179.8	1189.1
	H	424.17	1204.6	1216.5	1231.2	1245.1	1258.4	1271.2	1283.6	1295.7
	S	0.6217	1.4847	1.4978	1.5136	1.5282	1.5420	1.5549	1.5672	1.5789
410 (447.02)	V	0.0194	1.1327	1.163	1.207	1.249	1.289	1.328	1.365	1.402
	U	425.41	1118.7	1126.6	1138.1	1149.0	1159.4	1169.4	1179.1	1188.4
	H	426.88	1204.7	1214.8	1229.7	1243.8	1257.2	1270.2	1282.7	1294.8
	S	0.6247	1.4825	1.4936	1.5096	1.5244	1.5383	1.5514	1.5637	1.5755
420 (449.40)	V	0.0194	1.1057	1.130	1.173	1.215	1.254	1.293	1.330	1.366
	U	428.05	1118.8	1125.3	1137.0	1148.0	1158.5	1168.6	1178.3	1187.8
	H	429.56	1204.7	1213.1	1228.2	1242.4	1256.0	1269.1	1281.7	1293.9
	S	0.6276	1.4802	1.4894	1.5056	1.5206	1.5347	1.5479	1.5603	1.5722
430 (451.74)	V	0.0195	1.0800	1.099	1.142	1.183	1.222	1.259	1.296	1.331
	U	430.64	1118.8	1123.9	1135.8	1147.0	1157.6	1167.8	1177.6	1187.1
	H	432.19	1204.8	1211.4	1226.6	1241.1	1254.8	1268.0	1280.7	1293.0
	S	0.6304	1.4781	1.4853	1.5017	1.5169	1.5311	1.5444	1.5570	1.5689
440 (454.03)	V	0.0195	1.0554	1.069	1.111	1.152	1.190	1.227	1.263	1.298
	U	433.19	1118.8	1122.6	1134.6	1145.9	1156.7	1167.0	1176.9	1186.4
	H	434.77	1204.8	1209.6	1225.1	1239.7	1253.6	1266.9	1279.7	1292.1
	S	0.6332	1.4759	1.4812	1.4979	1.5132	1.5276	1.5410	1.5537	1.5657

	V	0.0195	1.0318	1.040	1.082	1.122	1.160	1.197	1.232	1.266
450	U	435.69	1118.9	1121.2	1133.4	1144.9	1155.8	1166.1	1176.1	1185.7
(456.28)	H	437.32	1204.8	1207.8	1223.5	1238.3	1252.4	1265.8	1278.7	1291.2
	S	0.6360	1.4738	1.4771	1.4940	1.5096	1.5241	1.5377	1.5505	1.5626
	V	0.0196	1.0092	1.012	1.054	1.094	1.132	1.168	1.203	1.236
460	U	438.17	1118.9	1119.8	1132.2	1143.8	1154.8	1165.3	1175.4	1185.1
(458.50)	H	439.83	1204.8	1206.0	1222.0	1236.9	1251.1	1264.7	1277.7	1290.3
	S	0.6387	1.4718	1.4731	1.4903	1.5060	1.5207	1.5344	1.5473	1.5595
	V	0.0196	0.9876	....	1.028	1.067	1.104	1.140	1.174	1.207
470	U	440.60	1118.9	....	1131.0	1142.8	1153.9	1164.5	1174.6	1184.4
(460.68)	H	442.31	1204.8	....	1220.4	1235.5	1249.9	1263.6	1276.7	1289.4
	S	0.6413	1.4697	....	1.4865	1.5025	1.5173	1.5311	1.5441	1.5564
	V	0.0197	0.9668	....	1.002	1.041	1.078	1.113	1.147	1.180
480	U	443.00	1118.9	....	1129.8	1141.7	1152.9	1163.6	1173.8	1183.7
(462.82)	H	444.75	1204.8	....	1218.8	1234.1	1248.6	1262.4	1275.7	1288.5
	S	0.6439	1.4677	....	1.4828	1.4990	1.5139	1.5279	1.5410	1.5534
	V	0.0197	0.9468	....	0.9774	1.016	1.052	1.087	1.121	1.153
490	U	445.36	1118.9	....	1128.5	1140.6	1151.9	1162.7	1173.1	1183.0
(464.93)	H	447.15	1204.7	....	1217.1	1232.7	1247.4	1261.3	1274.7	1287.5
	S	0.6465	1.4658	....	1.4791	1.4955	1.5106	1.5247	1.5380	1.5504
	V	0.0197	0.9276	....	0.9537	0.9919	1.028	1.062	1.095	1.127
500	U	447.70	1118.8	....	1127.2	1139.5	1151.0	1161.9	1172.3	1182.3
(467.01)	H	449.52	1204.7	....	1215.5	1231.2	1246.1	1260.2	1273.6	1286.6
	S	0.6490	1.4639	....	1.4755	1.4921	1.5074	1.5216	1.5349	1.5475
	V	0.0198	0.9091	....	0.9310	0.9688	1.005	1.039	1.071	1.103
510	U	450.00	1118.8	....	1126.0	1138.4	1150.0	1161.0	1171.5	1181.6
(469.05)	H	451.87	1204.6	....	1213.8	1229.8	1244.8	1259.0	1272.6	1285.7
	S	0.6515	1.4620	....	1.4718	1.4886	1.5041	1.5185	1.5319	1.5446
	V	0.0198	0.8914	....	0.9090	0.9466	0.9820	1.016	1.048	1.079
520	U	452.27	1118.8	....	1124.7	1137.2	1149.0	1160.1	1170.7	1180.9
(471.07)	H	454.18	1204.5	....	1212.1	1228.3	1243.5	1257.8	1271.5	1284.7
	S	0.6539	1.4601	....	1.4682	1.4853	1.5009	1.5154	1.5290	1.5418
	V	0.0199	0.8742	....	0.8878	0.9252	0.9603	0.9937	1.026	1.056
530	U	454.51	1118.7	....	1123.4	1136.1	1148.0	1159.2	1169.9	1180.1
(473.05)	H	456.46	1204.5	....	1210.4	1226.8	1242.2	1256.7	1270.5	1283.8
	S	0.6564	1.4583	....	1.4646	1.4819	1.4977	1.5124	1.5261	1.5390

TABLE C.4. SUPERHEATED STEAM ENGLISH UNITS (Continued)

ABS PRESS PSIA (SAT TEMP)		SAT WATER	SAT STEAM	TEMPERATURE, DEG F 600	700	800	900	1000	1100	1200
370 (437.04)	V	0.0192	1.2546	1.605	1.790	1.967	2.140	2.310	2.478	2.645
	U	414.41	1118.4	1199.9	1242.5	1283.6	1324.6	1366.0	1408.0	1450.7
	H	415.73	1204.3	1309.8	1365.1	1418.3	1471.1	1524.1	1577.7	1631.8
	S	0.6125	1.4918	1.6004	1.6503	1.6943	1.7346	1.7723	1.8077	1.8414
380 (439.61)	V	0.0193	1.2218	1.560	1.741	1.914	2.082	2.248	2.412	2.575
	U	417.24	1118.5	1199.3	1242.1	1283.3	1324.3	1365.7	1407.8	1450.6
	H	418.59	1204.4	1309.0	1364.5	1417.9	1470.8	1523.8	1577.4	1631.6
	S	0.6156	1.4894	1.5969	1.6470	1.6911	1.7315	1.7692	1.8047	1.8384
390 (442.13)	V	0.0193	1.1906	1.517	1.694	1.863	2.028	2.190	2.350	2.508
	U	420.01	1118.6	1198.8	1241.7	1283.0	1324.1	1365.5	1407.6	1450.4
	H	421.40	1204.5	1308.2	1364.0	1417.5	1470.4	1523.5	1577.2	1631.4
	S	0.6187	1.4870	1.5935	1.6437	1.6880	1.7285	1.7662	1.8017	1.8354
400 (444.60)	V	0.0193	1.1610	1.476	1.650	1.815	1.976	2.134	2.290	2.445
	U	422.74	1118.7	1198.2	1241.3	1282.7	1323.8	1365.3	1407.4	1450.2
	H	424.17	1204.6	1307.4	1363.4	1417.0	1470.1	1523.3	1576.9	1631.2
	S	0.6217	1.4847	1.5901	1.6406	1.6850	1.7255	1.7632	1.7988	1.8325
410 (447.02)	V	0.0194	1.1327	1.438	1.608	1.769	1.926	2.081	2.233	2.385
	U	425.41	1118.7	1197.6	1240.8	1282.4	1323.6	1365.1	1407.2	1450.1
	H	426.89	1204.7	1306.6	1362.8	1416.6	1469.7	1523.0	1576.7	1631.0
	S	0.6247	1.4825	1.5868	1.6375	1.6820	1.7226	1.7603	1.7959	1.8297
420 (449.40)	V	0.0194	1.1057	1.401	1.568	1.726	1.879	2.030	2.180	2.327
	U	428.05	1118.8	1196.9	1240.4	1282.0	1323.3	1364.9	1407.0	1449.9
	H	429.56	1204.7	1305.8	1362.3	1416.2	1469.4	1522.7	1576.4	1630.8
	S	0.6276	1.4802	1.5835	1.6345	1.6791	1.7197	1.7575	1.7932	1.8269
430 (451.74)	V	0.0195	1.0800	1.366	1.529	1.684	1.835	1.982	2.128	2.273
	U	430.64	1118.8	1196.3	1240.0	1281.7	1323.0	1364.6	1406.8	1449.7
	H	432.19	1204.8	1305.0	1361.7	1415.7	1469.0	1522.4	1576.2	1630.6
	S	0.6304	1.4781	1.5804	1.6315	1.6762	1.7169	1.7548	1.7904	1.8242
440 (454.03)	V	0.0195	1.0554	1.332	1.493	1.644	1.792	1.936	2.079	2.220
	U	433.19	1118.8	1195.7	1239.6	1281.4	1322.8	1364.4	1406.6	1449.6
	H	434.77	1204.8	1304.2	1361.1	1415.3	1468.7	1522.1	1575.9	1630.4
	S	0.6332	1.4759	1.5772	1.6286	1.6734	1.7142	1.7521	1.7878	1.8216

450 (456.28)	V	0.0195	1.0318	1.300	1.458	1.607	1.751	1.892	2.032	2.170
	U	435.69	1118.9	1195.1	1239.2	1281.1	1322.5	1364.2	1406.5	1449.4
	H	437.32	1204.8	1303.3	1360.6	1414.9	1468.3	1521.8	1575.7	1630.1
	S	0.6360	1.4738	1.5742	1.6258	1.6707	1.7115	1.7495	1.7852	1.8190
460 (458.50)	V	0.0196	1.0092	1.269	1.424	1.570	1.712	1.850	1.987	2.123
	U	438.17	1118.9	1194.5	1238.8	1280.8	1322.3	1364.0	1406.3	1449.3
	H	439.83	1204.8	1302.5	1360.0	1414.4	1468.0	1521.5	1575.4	1629.9
	S	0.6387	1.4718	1.5711	1.6230	1.6680	1.7089	1.7469	1.7826	1.8165
470 (460.68)	V	0.0196	0.9875	1.240	1.392	1.536	1.674	1.810	1.944	2.077
	U	440.60	1118.9	1193.9	1238.3	1280.4	1322.0	1363.8	1406.1	1449.1
	H	442.31	1204.8	1301.7	1359.4	1414.0	1467.6	1521.2	1575.2	1629.7
	S	0.6413	1.4697	1.5681	1.6202	1.6654	1.7064	1.7444	1.7802	1.8141
480 (462.82)	V	0.0197	0.9668	1.211	1.361	1.502	1.638	1.772	1.903	2.033
	U	443.00	1118.9	1193.2	1237.9	1280.1	1321.7	1363.5	1405.9	1448.9
	H	444.75	1204.8	1300.8	1358.8	1413.6	1467.3	1520.9	1574.9	1629.5
	S	0.6439	1.4677	1.5652	1.6176	1.6628	1.7038	1.7419	1.7777	1.8116
490 (464.93)	V	0.0197	0.9468	1.184	1.332	1.470	1.604	1.735	1.864	1.991
	U	445.36	1118.9	1192.6	1237.5	1279.8	1321.5	1363.3	1405.7	1448.8
	H	447.15	1204.7	1300.0	1358.3	1413.1	1466.9	1520.6	1574.7	1629.3
	S	0.6465	1.4658	1.5623	1.6149	1.6603	1.7014	1.7395	1.7753	1.8093
500 (467.01)	V	0.0197	0.9276	1.158	1.304	1.440	1.571	1.699	1.826	1.951
	U	447.70	1118.8	1192.0	1237.1	1279.5	1321.2	1363.1	1405.5	1448.6
	H	449.52	1204.7	1299.1	1357.7	1412.7	1466.6	1520.3	1574.4	1629.1
	S	0.6490	1.4639	1.5595	1.6123	1.6578	1.6990	1.7371	1.7730	1.8069
510 (469.05)	V	0.0198	0.9091	1.133	1.277	1.410	1.539	1.665	1.789	1.912
	U	450.00	1118.8	1191.3	1236.6	1279.2	1321.0	1362.9	1405.3	1448.4
	H	451.87	1204.6	1298.3	1357.1	1412.2	1466.2	1520.0	1574.2	1628.9
	S	0.6515	1.4620	1.5567	1.6097	1.6554	1.6966	1.7348	1.7707	1.8047
520 (471.07)	V	0.0198	0.8914	1.109	1.250	1.382	1.509	1.632	1.754	1.875
	U	452.27	1118.8	1190.7	1236.2	1278.8	1320.7	1362.7	1405.1	1448.3
	H	454.18	1204.5	1297.4	1356.5	1411.8	1465.9	1519.7	1573.9	1628.7
	S	0.6539	1.4601	1.5539	1.6072	1.6530	1.6943	1.7325	1.7684	1.8024
530 (473.05)	V	0.0199	0.8742	1.086	1.225	1.355	1.479	1.601	1.720	1.839
	U	454.51	1118.7	1190.0	1235.8	1278.5	1320.4	1362.4	1404.9	1448.1
	H	456.46	1204.5	1296.5	1355.9	1411.4	1465.5	1519.4	1573.7	1628.4
	S	0.6564	1.4583	1.5512	1.6047	1.6506	1.6920	1.7302	1.7662	1.8002

TABLE C.4. SUPERHEATED STEAM ENGLISH UNITS (Continued)

ABS PRESS PSIA (SAT TEMP)		SAT WATER	SAT STEAM	TEMPERATURE, DEG F 500	520	540	560	580	600	650
	V	0.0199	0.8577	0.9045	0.9394	0.9725	1.004	1.035	1.064	1.134
540	U	456.72	1118.7	1134.9	1147.0	1158.3	1169.1	1179.4	1189.4	1213.0
(475.01)	H	458.71	1204.4	1225.3	1240.8	1255.5	1269.4	1282.8	1295.7	1326.3
	S	0.6587	1.4565	1.4786	1.4946	1.5094	1.5232	1.5362	1.5484	1.5767
	V	0.0199	0.8418	0.8846	0.9192	0.9520	0.9833	1.013	1.042	1.112
550	U	458.91	1118.6	1133.8	1145.9	1157.4	1168.3	1178.7	1188.7	1212.4
(476.94)	H	460.94	1204.3	1223.8	1239.5	1254.3	1268.4	1281.8	1294.8	1325.6
	S	0.6611	1.4547	1.4753	1.4915	1.5064	1.5203	1.5334	1.5458	1.5742
	V	0.0200	0.8264	0.8653	0.8997	0.9322	0.9632	0.9930	1.022	1.090
560	U	461.07	1118.5	1132.6	1144.9	1156.5	1167.5	1178.0	1188.0	1211.9
(478.84)	H	463.14	1204.2	1222.2	1238.1	1253.1	1267.3	1280.9	1293.9	1324.9
	S	0.6634	1.4529	1.4720	1.4884	1.5035	1.5175	1.5307	1.5431	1.5717
	V	0.0200	0.8115	0.8467	0.8808	0.9131	0.9438	0.9733	1.002	1.069
570	U	463.20	1118.5	1131.4	1143.9	1155.6	1166.6	1177.2	1187.4	1211.4
(480.72)	H	465.32	1204.1	1220.7	1236.8	1251.9	1266.2	1279.9	1293.0	1324.2
	S	0.6657	1.4512	1.4687	1.4853	1.5005	1.5147	1.5280	1.5405	1.5693
	V	0.0201	0.7971	0.8287	0.8626	0.8946	0.9251	0.9542	0.9824	1.049
580	U	465.31	1118.4	1130.2	1142.8	1154.6	1165.8	1176.5	1186.7	1210.8
(482.57)	H	467.47	1203.9	1219.1	1235.4	1250.7	1265.1	1278.9	1292.1	1323.4
	S	0.6679	1.4495	1.4654	1.4822	1.4976	1.5120	1.5254	1.5380	1.5668
	V	0.0201	0.7832	0.8112	0.8450	0.8768	0.9069	0.9358	0.9637	1.030
590	U	467.40	1118.3	1129.0	1141.7	1153.7	1165.0	1175.7	1186.0	1210.3
(484.40)	H	469.59	1203.8	1217.5	1234.0	1249.4	1264.0	1277.9	1291.2	1322.7
	S	0.6701	1.4478	1.4622	1.4792	1.4948	1.5092	1.5227	1.5354	1.5645
	V	0.0201	0.7697	0.7944	0.8279	0.8595	0.8894	0.9180	0.9456	1.011
600	U	469.46	1118.2	1127.7	1140.7	1152.8	1164.1	1175.0	1185.3	1209.8
(486.20)	H	471.70	1203.7	1215.9	1232.6	1248.2	1262.9	1276.9	1290.3	1322.0
	S	0.6723	1.4461	1.4590	1.4762	1.4919	1.5065	1.5201	1.5329	1.5621
	V	0.0202	0.7567	0.7780	0.8114	0.8427	0.8724	0.9008	0.9281	0.9927
610	U	471.50	1118.1	1126.5	1139.6	1151.8	1163.3	1174.2	1184.7	1209.2
(487.98)	H	473.78	1203.5	1214.3	1231.2	1246.9	1261.8	1275.9	1289.4	1321.3
	S	0.6745	1.4445	1.4558	1.4732	1.4891	1.5038	1.5175	1.5304	1.5598

	V	0.0202	0.7441	0.7621	0.7954	0.8265	0.8560	0.8841	0.9112	0.9751
620	U	473.52	1118.0	1125.2	1138.5	1150.8	1162.4	1173.5	1184.0	1208.7
(489.74)	H	475.84	1203.4	1212.7	1229.7	1245.7	1260.7	1274.9	1288.5	1320.5
	S	0.6766	1.4428	1.4526	1.4702	1.4863	1.5011	1.5150	1.5279	1.5575
	V	0.0202	0.7318	0.7467	0.7798	0.8108	0.8401	0.8680	0.8948	0.9580
630	U	475.52	1117.9	1123.9	1137.4	1149.9	1161.6	1172.7	1183.3	1208.1
(491.48)	H	477.88	1203.2	1211.0	1228.3	1244.4	1259.5	1273.9	1287.6	1319.8
	S	0.6787	1.4412	1.4494	1.4672	1.4835	1.4985	1.5124	1.5255	1.5552
	V	0.0203	0.7200	0.7318	0.7648	0.7956	0.8246	0.8523	0.8788	0.9415
640	U	477.49	1117.8	1122.7	1136.3	1148.9	1160.7	1171.9	1182.6	1207.6
(493.19)	H	479.89	1203.0	1209.3	1226.8	1243.1	1258.4	1272.8	1286.7	1319.1
	S	0.6808	1.4396	1.4462	1.4643	1.4807	1.4959	1.5099	1.5231	1.5530
	V	0.0203	0.7084	0.7173	0.7501	0.7808	0.8096	0.8371	0.8634	0.9254
650	U	479.45	1117.6	1121.3	1135.1	1147.9	1159.8	1171.1	1181.9	1207.0
(494.89)	H	481.89	1202.8	1207.6	1225.4	1241.8	1257.2	1271.8	1285.7	1318.3
	S	0.6828	1.4381	1.4430	1.4614	1.4780	1.4932	1.5074	1.5207	1.5507
	V	0.0204	0.6972	0.7031	0.7359	0.7664	0.7951	0.8224	0.8485	0.9098
660	U	481.38	1117.5	1120.0	1134.0	1146.9	1159.0	1170.3	1181.2	1206.5
(496.57)	H	483.87	1202.7	1205.9	1223.9	1240.5	1256.1	1270.8	1284.8	1317.6
	S	0.6849	1.4365	1.4399	1.4584	1.4752	1.4907	1.5049	1.5183	1.5485
	V	0.0204	0.6864	0.6894	0.7221	0.7525	0.7810	0.8080	0.8339	0.8947
670	U	483.30	1117.4	1118.7	1132.8	1145.9	1158.1	1169.6	1180.5	1205.9
(498.22)	H	485.83	1202.5	1204.2	1222.4	1239.2	1254.9	1269.7	1283.9	1316.8
	S	0.6869	1.4350	1.4367	1.4555	1.4725	1.4881	1.5025	1.5159	1.5463
	V	0.0204	0.6758	0.6760	0.7087	0.7389	0.7673	0.7941	0.8198	0.8801
680	U	485.20	1117.2	1117.3	1131.7	1144.9	1157.2	1168.8	1179.8	1205.3
(499.86)	H	487.77	1202.3	1202.4	1220.8	1237.9	1253.7	1268.7	1282.9	1316.1
	S	0.6889	1.4334	1.4336	1.4526	1.4698	1.4855	1.5000	1.5136	1.5442
	V	0.0205	0.6655	....	0.6956	0.7257	0.7539	0.7806	0.8061	0.8658
690	U	487.08	1117.1	....	1130.5	1143.9	1156.3	1168.0	1179.0	1204.8
(501.48)	H	489.70	1202.1	....	1219.3	1236.5	1252.5	1267.6	1282.0	1315.3
	S	0.6908	1.4319	....	1.4497	1.4671	1.4830	1.4976	1.5113	1.5421
	V	0.0205	0.6556	....	0.6829	0.7129	0.7409	0.7675	0.7928	0.8520
700	U	488.95	1116.9	....	1129.3	1142.8	1155.4	1167.1	1178.3	1204.2
(503.08)	H	491.60	1201.8	....	1217.8	1235.2	1251.3	1266.6	1281.0	1314.6
	S	0.6928	1.4304	....	1.4468	1.4644	1.4805	1.4952	1.5090	1.5399

TABLE C.4. SUPERHEATED STEAM ENGLISH UNITS (Continued)

ABS PRESS PSIA (SAT TEMP)		SAT WATER	SAT STEAM	700	750	800	900	1000	1100	1200
				TEMPERATURE, DEG F						
540 (475.01)	V	0.0199	0.8577	1.201	1.266	1.328	1.451	1.570	1.688	1.804
	U	456.72	1118.7	1235.3	1257.0	1278.2	1320.2	1362.2	1404.8	1447.9
	H	458.71	1204.4	1355.3	1383.4	1410.9	1465.1	1519.1	1573.4	1628.2
	S	0.6587	1.4565	1.6023	1.6260	1.6483	1.6897	1.7280	1.7640	1.7981
550 (476.94)	V	0.0199	0.8418	1.178	1.241	1.303	1.424	1.541	1.657	1.771
	U	458.91	1118.6	1234.9	1256.6	1277.9	1319.9	1362.0	1404.6	1447.8
	H	460.94	1204.3	1354.7	1382.9	1410.5	1464.8	1518.9	1573.2	1628.0
	S	0.6611	1.4547	1.5999	1.6237	1.6460	1.6875	1.7259	1.7619	1.7959
560 (478.84)	V	0.0200	0.8264	1.155	1.218	1.279	1.397	1.513	1.627	1.739
	U	461.07	1118.5	1234.4	1256.2	1277.5	1319.6	1361.8	1404.4	1447.6
	H	463.14	1204.2	1354.2	1382.4	1410.0	1464.4	1518.6	1572.9	1627.8
	S	0.6634	1.4529	1.5975	1.6214	1.6438	1.6853	1.7237	1.7598	1.7939
570 (480.72)	V	0.0200	0.8115	1.133	1.195	1.255	1.372	1.486	1.597	1.708
	U	463.20	1118.5	1234.0	1255.8	1277.2	1319.4	1361.6	1404.2	1447.5
	H	465.32	1204.1	1353.6	1381.9	1409.6	1464.1	1518.3	1572.7	1627.6
	S	0.6657	1.4512	1.5952	1.6191	1.6415	1.6832	1.7216	1.7577	1.7918
580 (482.57)	V	0.0201	0.7971	1.112	1.173	1.232	1.347	1.459	1.569	1.678
	U	465.31	1118.4	1233.6	1255.5	1276.9	1319.1	1361.3	1404.0	1447.3
	H	467.47	1203.9	1353.0	1381.4	1409.2	1463.7	1518.0	1572.4	1627.4
	S	0.6679	1.4495	1.5929	1.6169	1.6394	1.6811	1.7196	1.7556	1.7898
590 (484.40)	V	0.0201	0.7832	1.092	1.152	1.210	1.324	1.434	1.542	1.649
	U	467.40	1118.3	1233.1	1255.1	1276.5	1318.9	1361.1	1403.8	1447.1
	H	469.59	1203.8	1352.4	1380.9	1408.7	1463.4	1517.7	1572.2	1627.2
	S	0.6701	1.4478	1.5906	1.6147	1.6372	1.6790	1.7175	1.7536	1.7878
600 (486.20)	V	0.0201	0.7697	1.073	1.132	1.189	1.301	1.409	1.516	1.621
	U	469.46	1118.2	1232.7	1254.7	1276.2	1318.6	1360.9	1403.6	1447.0
	H	471.70	1203.7	1351.8	1380.4	1408.3	1463.0	1517.4	1571.9	1627.0
	S	0.6723	1.4461	1.5884	1.6125	1.6351	1.6769	1.7155	1.7517	1.7859
610 (487.98)	V	0.0202	0.7567	1.054	1.112	1.169	1.279	1.386	1.491	1.594
	U	471.50	1118.1	1232.2	1254.3	1275.9	1318.3	1360.7	1403.4	1446.8
	H	473.78	1203.5	1351.2	1379.9	1407.8	1462.7	1517.1	1571.7	1626.7
	S	0.6745	1.4445	1.5861	1.6104	1.6330	1.6749	1.7135	1.7497	1.7839

620 (489.74)	V	0.0202	0.7441	1.035	1.093	1.149	1.257	1.363	1.466	1.568
	U	473.52	1118.0	1231.8	1253.9	1275.6	1318.1	1360.5	1403.2	1446.6
	H	475.84	1203.4	1350.6	1379.3	1407.4	1462.3	1516.8	1571.4	1626.5
	S	0.6766	1.4428	1.5839	1.6082	1.6310	1.6729	1.7116	1.7478	1.7820
630 (491.48)	V	0.0202	0.7318	1.017	1.074	1.130	1.236	1.340	1.442	1.543
	U	475.52	1117.9	1231.3	1253.6	1275.2	1317.8	1360.2	1403.1	1446.5
	H	477.88	1203.2	1350.0	1378.8	1406.9	1461.9	1516.5	1571.2	1626.3
	S	0.6787	1.4412	1.5818	1.6062	1.6289	1.6710	1.7097	1.7459	1.7802
640 (493.19)	V	0.0203	0.7200	1.000	1.056	1.111	1.216	1.319	1.419	1.518
	U	477.49	1117.8	1230.9	1253.2	1274.9	1317.5	1360.0	1402.9	1446.3
	H	479.89	1203.0	1349.3	1378.3	1406.5	1461.6	1516.2	1570.9	1626.1
	S	0.6808	1.4396	1.5797	1.6041	1.6269	1.6690	1.7078	1.7441	1.7783
650 (494.89)	V	0.0203	0.7084	0.9835	1.039	1.093	1.197	1.298	1.397	1.494
	U	479.45	1117.6	1230.4	1252.8	1274.6	1317.3	1359.8	1402.7	1446.1
	H	481.89	1202.8	1348.7	1377.8	1406.0	1461.2	1515.9	1570.7	1625.9
	S	0.6828	1.4381	1.5775	1.6021	1.6249	1.6671	1.7059	1.7422	1.7765
660 (496.57)	V	0.0204	0.6972	0.9673	1.022	1.075	1.178	1.278	1.375	1.471
	U	481.38	1117.5	1230.0	1252.4	1274.2	1317.0	1359.6	1402.5	1446.0
	H	483.87	1202.7	1348.1	1377.3	1405.6	1460.9	1515.6	1570.4	1625.7
	S	0.6849	1.4365	1.5755	1.6001	1.6230	1.6652	1.7041	1.7404	1.7748
670 (498.22)	V	0.0204	0.6864	0.9516	1.006	1.058	1.160	1.258	1.354	1.449
	U	483.30	1117.4	1229.5	1252.0	1273.9	1316.7	1359.3	1402.3	1445.8
	H	485.83	1202.5	1347.5	1376.7	1405.1	1460.5	1515.3	1570.2	1625.5
	S	0.6869	1.4350	1.5734	1.5981	1.6211	1.6634	1.7023	1.7387	1.7730
680 (499.86)	V	0.0204	0.6758	0.9364	0.9900	1.042	1.142	1.239	1.334	1.427
	U	485.20	1117.2	1229.1	1251.6	1273.6	1316.5	1359.1	1402.1	1445.7
	H	487.77	1202.3	1346.9	1376.2	1404.7	1460.2	1515.0	1569.9	1625.3
	S	0.6889	1.4334	1.5714	1.5961	1.6192	1.6616	1.7005	1.7369	1.7713
690 (501.48)	V	0.0205	0.6655	0.9216	0.9746	1.026	1.125	1.220	1.314	1.406
	U	487.08	1117.1	1228.6	1251.3	1273.2	1316.2	1358.9	1401.9	1445.5
	H	489.70	1202.1	1346.3	1375.7	1404.2	1459.8	1514.7	1569.7	1625.0
	S	0.6908	1.4319	1.5693	1.5942	1.6173	1.6598	1.6987	1.7352	1.7696
700 (503.08)	V	0.0205	0.6556	0.9072	0.9596	1.010	1.108	1.202	1.295	1.386
	U	488.95	1116.9	1228.1	1250.9	1272.9	1315.9	1358.7	1401.7	1445.3
	H	491.60	1201.8	1345.6	1375.2	1403.7	1459.4	1514.4	1569.4	1624.8
	S	0.6928	1.4304	1.5673	1.5923	1.6154	1.6580	1.6970	1.7335	1.7679

TABLE C.4. SUPERHEATED STEAM ENGLISH UNITS (Continued)

ABS PRESS PSIA (SAT TEMP)		SAT WATER	SAT STEAM	TEMPERATURE, DEG F 520	540	560	580	600	620	650
725 (507.01)	V	0.0206	0.6318	0.6525	0.6823	0.7100	0.7362	0.7610	0.7848	0.8190
	U	493.5	1116.5	1126.3	1140.2	1153.1	1165.1	1176.5	1187.3	1202.8
	H	496.3	1201.3	1213.8	1231.7	1248.3	1263.9	1278.6	1292.6	1312.6
	S	0.6975	1.4268	1.4396	1.4578	1.4742	1.4893	1.5033	1.5164	1.5347
750 (510.84)	V	0.0207	0.6095	0.6240	0.6536	0.6811	0.7069	0.7313	0.7547	0.7882
	U	498.0	1116.1	1123.1	1137.5	1150.7	1163.0	1174.6	1185.6	1201.3
	H	500.9	1200.7	1209.7	1228.2	1245.2	1261.1	1276.1	1290.4	1310.7
	S	0.7022	1.4232	1.4325	1.4511	1.4680	1.4835	1.4977	1.5111	1.5296
775 (514.57)	V	0.0208	0.5886	0.5971	0.6267	0.6539	0.6794	0.7035	0.7265	0.7594
	U	502.4	1115.6	1119.9	1134.7	1148.3	1160.9	1172.7	1183.9	1199.9
	H	505.4	1200.1	1205.6	1224.6	1242.1	1258.3	1273.6	1288.1	1308.8
	S	0.7067	1.4197	1.4253	1.4446	1.4619	1.4777	1.4923	1.5058	1.5247
800 (518.21)	V	0.0209	0.5690	0.5717	0.6013	0.6283	0.6536	0.6774	0.7000	0.7323
	U	506.7	1115.2	1116.6	1131.9	1145.9	1158.8	1170.8	1182.2	1198.4
	H	509.8	1199.4	1201.2	1220.9	1238.9	1255.5	1271.1	1285.9	1306.8
	S	0.7111	1.4163	1.4182	1.4381	1.4558	1.4720	1.4868	1.5007	1.5198
825 (521.76)	V	0.0210	0.5505	....	0.5773	0.6042	0.6293	0.6528	0.6751	0.7069
	U	510.9	1114.6	....	1129.0	1143.4	1156.6	1168.9	1180.5	1196.9
	H	514.1	1198.7	....	1217.1	1235.6	1252.6	1268.5	1283.6	1304.8
	S	0.7155	1.4129	....	1.4315	1.4498	1.4664	1.4815	1.4956	1.5150
850 (525.24)	V	0.0211	0.5330	....	0.5546	0.5815	0.6063	0.6296	0.6516	0.6829
	U	515.1	1114.1	....	1126.0	1140.8	1154.3	1166.9	1178.7	1195.3
	H	518.4	1198.0	....	1213.3	1232.2	1249.7	1265.9	1281.2	1302.8
	S	0.7197	1.4096	....	1.4250	1.4439	1.4608	1.4763	1.4906	1.5102
875 (528.63)	V	0.0211	0.5165	....	0.5330	0.5599	0.5846	0.6077	0.6294	0.6602
	U	519.2	1113.6	....	1123.0	1138.2	1152.0	1164.9	1176.9	1193.8
	H	522.6	1197.2	....	1209.3	1228.8	1246.7	1263.3	1278.8	1300.7
	S	0.7238	1.4064	....	1.4185	1.4379	1.4553	1.4711	1.4856	1.5056
900 (531.95)	V	0.0212	0.5009	....	0.5126	0.5394	0.5640	0.5869	0.6084	0.6388
	U	523.2	1113.0	....	1119.8	1135.5	1149.7	1162.8	1175.1	1192.2
	H	526.7	1196.4	....	1205.2	1225.3	1243.6	1260.6	1276.4	1298.6
	S	0.7279	1.4032	....	1.4120	1.4320	1.4498	1.4659	1.4807	1.5010

925 (535.21)	V	0.0213	0.4861	....	0.4930	0.5200	0.5445	0.5672	0.5885	0.6186
	U	527.1	1112.4	....	1116.5	1132.7	1147.3	1160.8	1173.2	1190.7
	H	530.8	1195.6	....	1200.9	1221.7	1240.5	1257.8	1274.0	1296.6
	S	0.7319	1.4001	....	1.4054	1.4260	1.4443	1.4608	1.4759	1.4965
950 (538.39)	V	0.0214	0.4721	....	0.4744	0.5014	0.5259	0.5485	0.5696	0.5993
	U	531.0	1111.7	....	1113.2	1129.9	1144.9	1158.6	1171.4	1189.1
	H	534.7	1194.7	....	1196.6	1218.0	1237.4	1255.1	1271.5	1294.4
	S	0.7358	1.3970	....	1.3988	1.4201	1.4389	1.4557	1.4711	1.4921
975 (541.52)	V	0.0215	0.4587	....	....	0.4837	0.5082	0.5307	0.5517	0.5810
	U	534.8	1111.1	....	....	1127.0	1142.4	1156.5	1169.5	1187.5
	H	538.7	1193.8	....	....	1214.3	1234.1	1252.2	1269.0	1292.3
	S	0.7396	1.3940	....	....	1.4142	1.4335	1.4507	1.4664	1.4877
1000 (544.58)	V	0.0216	0.4460	....	....	0.4668	0.4913	0.5137	0.5346	0.5636
	U	538.6	1110.4	....	....	1124.0	1139.9	1154.3	1167.5	1185.8
	H	542.6	1192.9	....	....	1210.4	1230.8	1249.3	1266.5	1290.1
	S	0.7434	1.3910	....	....	1.4082	1.4281	1.4457	1.4617	1.4833
1025 (547.58)	V	0.0217	0.4338	....	....	0.4506	0.4752	0.4975	0.5183	0.5471
	U	542.3	1109.7	....	....	1120.9	1137.3	1152.0	1165.6	1184.2
	H	546.4	1192.0	....	....	1206.4	1227.4	1246.4	1263.9	1287.9
	S	0.7471	1.3880	....	....	1.4022	1.4227	1.4407	1.4571	1.4791
1050 (550.53)	V	0.0218	0.4222	....	....	0.4350	0.4597	0.4821	0.5027	0.5312
	U	545.9	1109.0	....	....	1117.8	1134.7	1149.8	1163.6	1182.5
	H	550.1	1191.0	....	....	1202.3	1224.0	1243.4	1261.2	1285.7
	S	0.7507	1.3851	....	....	1.3962	1.4173	1.4358	1.4524	1.4748
1075 (553.43)	V	0.0219	0.4112	....	....	0.4200	0.4449	0.4673	0.4878	0.5161
	U	549.5	1108.3	....	....	1114.5	1131.9	1147.4	1161.5	1180.8
	H	553.9	1190.1	....	....	1198.1	1220.4	1240.4	1258.6	1283.5
	S	0.7543	1.3822	....	....	1.3901	1.4118	1.4308	1.4479	1.4706
1100 (556.28)	V	0.0220	0.4006	....	....	0.4056	0.4307	0.4531	0.4735	0.5017
	U	553.1	1107.5	....	....	1111.2	1129.1	1145.1	1159.5	1179.1
	H	557.5	1189.1	....	....	1193.7	1216.8	1237.3	1255.9	1281.2
	S	0.7578	1.3794	....	....	1.3840	1.4064	1.4259	1.4433	1.4664
1125 (559.07)	V	0.0220	0.3904	....	....	0.3917	0.4170	0.4394	0.4599	0.4879
	U	556.6	1106.8	....	....	1107.7	1126.3	1142.6	1157.4	1177.3
	H	561.2	1188.0	....	....	1189.2	1213.1	1234.1	1253.1	1278.9
	S	0.7613	1.3766	....	....	1.3778	1.4009	1.4210	1.4387	1.4623

TABLE C.4. SUPERHEATED STEAM ENGLISH UNITS (Continued)

ABS PRESS PSIA (SAT TEMP)		SAT WATER	SAT STEAM	TEMPERATURE, DEG F 700	750	800	900	1000	1100	1200
725 (507.01)	V	0.0206	0.6318	0.8729	0.9240	0.9732	1.068	1.159	1.249	1.337
	U	493.5	1116.5	1227.0	1249.9	1272.0	1315.3	1358.1	1401.3	1444.9
	H	496.3	1201.3	1344.1	1373.8	1402.6	1458.5	1513.7	1568.8	1624.3
	S	0.6975	1.4268	1.5624	1.5876	1.6109	1.6536	1.6927	1.7293	1.7638
750 (510.84)	V	0.0207	0.6095	0.8409	0.8907	0.9386	1.031	1.119	1.206	1.292
	U	498.0	1116.1	1225.8	1248.9	1271.2	1314.6	1357.6	1400.8	1444.5
	H	500.9	1200.7	1342.5	1372.5	1401.5	1457.6	1512.9	1568.2	1623.8
	S	0.7022	1.4232	1.5577	1.5830	1.6065	1.6494	1.6886	1.7252	1.7598
775 (514.57)	V	0.0208	0.5886	0.8109	0.8595	0.9062	0.9957	1.082	1.166	1.249
	U	502.4	1115.6	1224.6	1247.9	1270.3	1313.9	1357.0	1400.3	1444.1
	H	505.4	1200.1	1340.9	1371.2	1400.3	1456.7	1512.2	1567.6	1623.2
	S	0.7067	1.4197	1.5530	1.5786	1.6022	1.6453	1.6846	1.7213	1.7559
800 (518.21)	V	0.0209	0.5690	0.7828	0.8303	0.8759	0.9631	1.047	1.129	1.209
	U	506.7	1115.2	1223.4	1246.9	1269.5	1313.2	1356.4	1399.8	1443.7
	H	509.8	1199.4	1339.3	1369.8	1399.1	1455.8	1511.4	1566.9	1622.7
	S	0.7111	1.4163	1.5484	1.5742	1.5980	1.6413	1.6807	1.7175	1.7522
825 (521.76)	V	0.0210	0.5505	0.7564	0.8029	0.8473	0.9323	1.014	1.094	1.172
	U	510.9	1114.6	1222.2	1245.9	1268.6	1312.6	1355.9	1399.3	1443.3
	H	514.1	1198.7	1337.7	1368.5	1398.0	1454.9	1510.7	1566.3	1622.2
	S	0.7155	1.4129	1.5440	1.5700	1.5939	1.6374	1.6770	1.7138	1.7485
850 (525.24)	V	0.0211	0.5330	0.7315	0.7770	0.8205	0.9034	0.9830	1.061	1.137
	U	515.1	1114.1	1221.0	1244.9	1267.7	1311.9	1355.3	1398.9	1442.9
	H	518.4	1198.0	1336.0	1367.1	1396.8	1454.0	1510.0	1565.7	1621.6
	S	0.7197	1.4096	1.5396	1.5658	1.5899	1.6336	1.6733	1.7102	1.7450
875 (528.63)	V	0.0211	0.5165	0.7080	0.7526	0.7952	0.8762	0.9538	1.029	1.103
	U	519.2	1113.6	1219.7	1243.9	1266.9	1311.2	1354.8	1398.4	1442.5
	H	522.6	1197.2	1334.4	1365.7	1395.6	1453.1	1509.2	1565.1	1621.1
	S	0.7238	1.4064	1.5353	1.5618	1.5860	1.6299	1.6697	1.7067	1.7416
900 (531.95)	V	0.0212	0.5009	0.6858	0.7296	0.7713	0.8504	0.9262	0.9998	1.072
	U	523.2	1113.0	1218.5	1242.8	1266.0	1310.5	1354.2	1397.9	1442.0
	H	526.7	1196.4	1332.7	1364.3	1394.4	1452.2	1508.5	1564.4	1620.6
	S	0.7279	1.4032	1.5311	1.5578	1.5822	1.6263	1.6662	1.7033	1.7382

925 (535.21)	V	0.0213	0.4861	0.6648	0.7078	0.7486	0.8261	0.9001	0.9719	1.042
	U	527.1	1112.4	1217.2	1241.8	1265.1	1309.8	1353.6	1397.4	1441.6
	H	530.8	1195.6	1331.0	1362.9	1393.2	1451.2	1507.7	1563.8	1620.0
	S	0.7319	1.4001	1.5269	1.5539	1.5784	1.6227	1.6628	1.7000	1.7349
950 (538.39)	V	0.0214	0.4721	0.6449	0.6871	0.7272	0.8030	0.8753	0.9455	1.014
	U	531.0	1111.7	1216.0	1240.7	1264.2	1309.1	1353.1	1397.0	1441.2
	H	534.7	1194.7	1329.3	1361.5	1392.0	1450.3	1507.0	1563.2	1619.5
	S	0.7358	1.3970	1.5228	1.5500	1.5748	1.6193	1.6595	1.6967	1.7317
975 (541.52)	V	0.0215	0.4587	0.6259	0.6675	0.7068	0.7811	0.8518	0.9204	0.9875
	U	534.8	1111.1	1214.7	1239.7	1263.3	1308.5	1352.5	1396.5	1440.8
	H	538.7	1193.8	1327.6	1360.1	1390.8	1449.4	1506.2	1562.5	1619.0
	S	0.7396	1.3940	1.5188	1.5463	1.5712	1.6159	1.6562	1.6936	1.7286
1000 (544.58)	V	0.0216	0.4460	0.6080	0.6489	0.6875	0.7603	0.8295	0.8966	0.9621
	U	538.6	1110.4	1213.4	1238.6	1262.4	1307.8	1351.9	1396.0	1440.4
	H	542.6	1192.9	1325.9	1358.7	1389.6	1448.5	1505.4	1561.9	1618.4
	S	0.7434	1.3910	1.5149	1.5426	1.5677	1.6126	1.6530	1.6905	1.7256
1025 (547.58)	V	0.0217	0.4338	0.5908	0.6311	0.6690	0.7405	0.8083	0.8739	0.9380
	U	542.3	1109.7	1212.1	1237.5	1261.5	1307.1	1351.4	1395.5	1440.0
	H	546.4	1192.0	1324.2	1357.3	1388.4	1447.5	1504.7	1561.3	1617.9
	S	0.7471	1.3880	1.5110	1.5389	1.5642	1.6094	1.6499	1.6874	1.7226
1050 (550.53)	V	0.0218	0.4222	0.5745	0.6142	0.6515	0.7216	0.7881	0.8524	0.9151
	U	545.9	1109.0	1210.8	1236.5	1260.6	1306.4	1350.8	1395.0	1439.6
	H	550.1	1191.0	1322.4	1355.8	1387.2	1446.6	1503.9	1560.7	1617.4
	S	0.7507	1.3851	1.5072	1.5354	1.5608	1.6062	1.6469	1.6845	1.7197
1075 (553.43)	V	0.0219	0.4112	0.5589	0.5981	0.6348	0.7037	0.7688	0.8318	0.8932
	U	549.5	1108.3	1209.4	1235.4	1259.7	1305.7	1350.2	1394.6	1439.2
	H	553.9	1190.1	1320.6	1354.4	1386.0	1445.7	1503.2	1560.0	1616.8
	S	0.7543	1.3822	1.5034	1.5319	1.5575	1.6031	1.6439	1.6816	1.7169
1100 (556.28)	V	0.0220	0.4006	0.5440	0.5826	0.6188	0.6865	0.7505	0.8121	0.8723
	U	553.1	1107.5	1208.1	1234.3	1258.8	1305.0	1349.7	1394.1	1438.7
	H	557.5	1189.1	1318.8	1352.9	1384.7	1444.7	1502.4	1559.4	1616.3
	S	0.7578	1.3794	1.4996	1.5284	1.5542	1.6000	1.6410	1.6787	1.7141
1125 (559.07)	V	0.0220	0.3904	0.5298	0.5679	0.6035	0.6701	0.7329	0.7934	0.8523
	U	556.6	1106.8	1206.7	1233.2	1257.8	1304.3	1349.1	1393.6	1438.3
	H	561.2	1188.0	1317.0	1351.4	1383.5	1443.8	1501.7	1558.8	1615.8
	S	0.7613	1.3766	1.4959	1.5250	1.5509	1.5970	1.6381	1.6759	1.7114

TABLE C.4. SUPERHEATED STEAM ENGLISH UNITS (Continued)

ABS PRESS PSIA (SAT TEMP)		SAT WATER	SAT STEAM	580	600	620	640	660	680	700
				TEMPERATURE, DEG F						
1150 (561.82)	V	0.0221	0.3807	0.4038	0.4263	0.4468	0.4656	0.4833	0.5001	0.5162
	U	560.1	1106.0	1123.3	1140.2	1155.2	1169.0	1181.9	1193.9	1205.4
	H	564.8	1187.0	1209.3	1230.9	1250.3	1268.1	1284.7	1300.3	1315.2
	S	0.7647	1.3738	1.3954	1.4160	1.4342	1.4506	1.4655	1.4793	1.4923
1175 (564.53)	V	0.0222	0.3714	0.3911	0.4137	0.4342	0.4530	0.4706	0.4872	0.5031
	U	563.5	1105.2	1120.3	1137.6	1153.1	1167.1	1180.2	1192.4	1204.0
	H	568.3	1185.9	1205.3	1227.6	1247.5	1265.6	1282.5	1298.3	1313.4
	S	0.7681	1.3711	1.3899	1.4111	1.4297	1.4463	1.4615	1.4756	1.4887
1200 (567.19)	V	0.0223	0.3624	0.3788	0.4016	0.4220	0.4408	0.4583	0.4748	0.4905
	U	566.9	1104.3	1117.2	1135.0	1150.9	1165.2	1178.5	1190.9	1202.6
	H	571.9	1184.8	1201.3	1224.2	1244.6	1263.1	1280.2	1296.3	1311.5
	S	0.7714	1.3683	1.3843	1.4061	1.4252	1.4422	1.4576	1.4718	1.4851
1225 (569.80)	V	0.0224	0.3538	0.3669	0.3899	0.4104	0.4291	0.4465	0.4629	0.4785
	U	570.3	1103.5	1114.0	1132.4	1148.6	1163.2	1176.7	1189.3	1201.2
	H	575.3	1183.7	1197.2	1220.8	1241.6	1260.5	1277.9	1294.2	1309.7
	S	0.7747	1.3656	1.3787	1.4011	1.4206	1.4380	1.4537	1.4681	1.4815
1250 (572.38)	V	0.0225	0.3456	0.3553	0.3785	0.3991	0.4178	0.4352	0.4514	0.4669
	U	573.6	1102.6	1110.7	1129.7	1146.3	1161.3	1175.0	1187.7	1199.8
	H	578.8	1182.6	1192.9	1217.2	1238.6	1257.9	1275.6	1292.2	1307.8
	S	0.7780	1.3630	1.3729	1.3961	1.4161	1.4338	1.4498	1.4644	1.4780
1275 (574.92)	V	0.0226	0.3376	0.3442	0.3676	0.3882	0.4069	0.4242	0.4404	0.4558
	U	576.9	1101.7	1107.3	1126.9	1144.0	1159.2	1173.2	1186.1	1198.3
	H	582.2	1181.4	1188.5	1213.6	1235.6	1255.2	1273.3	1290.1	1305.9
	S	0.7812	1.3603	1.3671	1.3911	1.4116	1.4297	1.4459	1.4608	1.4745
1300 (577.42)	V	0.0227	0.3299	0.3333	0.3570	0.3778	0.3965	0.4137	0.4298	0.4451
	U	580.1	1100.9	1103.7	1124.0	1141.6	1157.2	1171.4	1184.5	1196.9
	H	585.6	1180.2	1183.9	1209.9	1232.4	1252.5	1270.9	1287.9	1303.9
	S	0.7843	1.3577	1.3612	1.3860	1.4071	1.4255	1.4421	1.4572	1.4711
1325 (579.89)	V	0.0228	0.3225	0.3227	0.3467	0.3676	0.3863	0.4035	0.4196	0.4347
	U	583.3	1099.9	1100.1	1121.1	1139.1	1155.1	1169.5	1182.9	1195.4
	H	588.9	1179.0	1179.2	1206.1	1229.3	1249.8	1268.5	1285.8	1302.0
	S	0.7875	1.3551	1.3552	1.3809	1.4025	1.4214	1.4382	1.4535	1.4677

1350 (582.31)	V	0.0229	0.3154	....	0.3367	0.3578	0.3766	0.3937	0.4097	0.4248
	U	586.5	1099.0	....	1118.1	1136.6	1152.9	1167.7	1181.3	1193.9
	H	592.3	1177.8	....	1202.2	1226.0	1247.0	1266.1	1283.6	1300.1
	S	0.7906	1.3525	....	1.3757	1.3980	1.4173	1.4344	1.4500	1.4643
1375 (584.71)	V	0.0230	0.3085	....	0.3270	0.3483	0.3671	0.3843	0.4002	0.4151
	U	589.7	1098.1	....	1115.0	1134.1	1150.8	1165.8	1179.6	1192.4
	H	595.6	1176.6	....	1198.2	1222.7	1244.2	1263.6	1281.4	1298.1
	S	0.7936	1.3499	....	1.3705	1.3934	1.4131	1.4306	1.4464	1.4609
1400 (587.07)	V	0.0231	0.3018	....	0.3176	0.3390	0.3579	0.3751	0.3910	0.4059
	U	592.9	1097.1	....	1111.8	1131.5	1148.6	1163.9	1177.9	1190.9
	H	598.8	1175.3	....	1194.1	1219.3	1241.3	1261.1	1279.2	1296.1
	S	0.7966	1.3474	....	1.3652	1.3888	1.4090	1.4268	1.4428	1.4575
1425 (589.40)	V	0.0232	0.2953	....	0.3084	0.3300	0.3491	0.3662	0.3821	0.3969
	U	596.0	1096.2	....	1108.5	1128.8	1146.3	1162.0	1176.2	1189.4
	H	602.1	1174.0	....	1189.8	1215.8	1238.4	1258.5	1277.0	1294.1
	S	0.7996	1.3448	....	1.3598	1.3841	1.4048	1.4230	1.4393	1.4542
1450 (591.69)	V	0.0233	0.2891	....	0.2994	0.3213	0.3404	0.3576	0.3734	0.3882
	U	599.1	1095.2	....	1105.1	1126.1	1144.0	1160.0	1174.5	1187.9
	H	605.3	1172.7	....	1185.5	1212.3	1235.4	1255.9	1274.7	1292.0
	S	0.8026	1.3423	....	1.3544	1.3794	1.4006	1.4192	1.4358	1.4509
1475 (593.96)	V	0.0234	0.2830	....	0.2906	0.3128	0.3321	0.3493	0.3651	0.3798
	U	602.1	1094.2	....	1101.6	1123.2	1141.7	1158.0	1172.7	1186.3
	H	608.5	1171.4	....	1180.9	1208.6	1232.3	1253.3	1272.4	1290.0
	S	0.8055	1.3398	....	1.3488	1.3747	1.3965	1.4154	1.4322	1.4476
1500 (596.20)	V	0.0235	0.2772	....	0.2820	0.3046	0.3240	0.3412	0.3570	0.3717
	U	605.2	1093.1	....	1098.0	1120.4	1139.3	1155.9	1170.9	1184.7
	H	611.7	1170.1	....	1176.3	1204.9	1229.2	1250.7	1270.0	1287.9
	S	0.8085	1.3373	....	1.3431	1.3699	1.3923	1.4116	1.4287	1.4443
1525 (598.41)	V	0.0236	0.2715	....	0.2735	0.2965	0.3161	0.3334	0.3492	0.3638
	U	608.2	1092.1	....	1094.2	1117.4	1136.9	1153.9	1169.2	1183.1
	H	614.8	1168.7	....	1171.4	1201.1	1226.1	1248.0	1267.7	1285.8
	S	0.8113	1.3348	....	1.3373	1.3651	1.3880	1.4078	1.4252	1.4410
1550 (600.59)	V	0.0237	0.2660	....	....	0.2886	0.3084	0.3258	0.3415	0.3561
	U	611.2	1091.1	....	....	1114.4	1134.4	1151.8	1167.3	1181.5
	H	618.0	1167.3	....	....	1197.2	1222.8	1245.2	1265.3	1283.7
	S	0.8142	1.3323	....	....	1.3602	1.3838	1.4040	1.4217	1.4377

TABLE C.4. SUPERHEATED STEAM ENGLISH UNITS (Continued)

ABS PRESS PSIA (SAT TEMP)		SAT WATER	SAT STEAM	TEMPERATURE, DEG F 720	750	800	900	1000	1100	1200
1150 (561.82)	V	0.0221	0.3807	0.5316	0.5538	0.5889	0.6544	0.7161	0.7754	0.8332
	U	560.1	1106.0	1216.3	1232.1	1256.9	1303.6	1348.5	1393.1	1437.9
	H	564.8	1187.0	1329.5	1349.9	1382.2	1442.8	1500.9	1558.1	1615.2
	S	0.7647	1.3738	1.5045	1.5216	1.5478	1.5941	1.6353	1.6732	1.7087
1175 (564.53)	V	0.0222	0.3714	0.5183	0.5403	0.5749	0.6394	0.7000	0.7582	0.8149
	U	563.5	1105.2	1215.1	1231.0	1256.0	1302.8	1347.9	1392.6	1437.5
	H	568.3	1185.9	1327.8	1348.4	1381.0	1441.9	1500.1	1557.5	1614.7
	S	0.7681	1.3711	1.5010	1.5183	1.5446	1.5912	1.6325	1.6705	1.7061
1200 (567.19)	V	0.0223	0.3624	0.5056	0.5273	0.5615	0.6250	0.6845	0.7418	0.7974
	U	566.9	1104.3	1213.8	1229.8	1255.0	1302.1	1347.4	1392.1	1437.1
	H	571.9	1184.8	1326.1	1346.9	1379.7	1440.9	1499.4	1556.9	1614.2
	S	0.7714	1.3683	1.4975	1.5150	1.5415	1.5883	1.6298	1.6679	1.7035
1225 (569.80)	V	0.0224	0.3538	0.4934	0.5149	0.5486	0.6111	0.6697	0.7260	0.7806
	U	570.3	1103.5	1212.5	1228.7	1254.1	1301.4	1346.8	1391.7	1436.7
	H	575.3	1183.7	1324.4	1345.4	1378.4	1440.0	1498.6	1556.2	1613.6
	S	0.7747	1.3656	1.4941	1.5117	1.5385	1.5855	1.6271	1.6653	1.7010
1250 (572.38)	V	0.0225	0.3456	0.4817	0.5029	0.5362	0.5979	0.6555	0.7108	0.7645
	U	573.6	1102.6	1211.2	1227.5	1253.1	1300.7	1346.2	1391.2	1436.3
	H	578.8	1182.6	1322.6	1343.9	1377.2	1439.0	1497.8	1555.6	1613.1
	S	0.7780	1.3630	1.4907	1.5085	1.5355	1.5827	1.6245	1.6628	1.6985
1275 (574.92)	V	0.0226	0.3376	0.4705	0.4914	0.5244	0.5851	0.6419	0.6962	0.7490
	U	576.9	1101.7	1209.9	1226.4	1252.2	1300.0	1345.6	1390.7	1435.8
	H	582.2	1181.4	1320.9	1342.3	1375.9	1438.0	1497.1	1554.9	1612.6
	S	0.7812	1.3603	1.4874	1.5053	1.5325	1.5800	1.6219	1.6603	1.6961
1300 (577.42)	V	0.0227	0.3299	0.4596	0.4804	0.5129	0.5729	0.6287	0.6822	0.7341
	U	580.1	1100.9	1208.6	1225.2	1251.2	1299.3	1345.0	1390.2	1435.4
	H	585.6	1180.2	1319.2	1340.8	1374.6	1437.1	1496.3	1554.3	1612.0
	S	0.7843	1.3577	1.4841	1.5022	1.5296	1.5773	1.6194	1.6578	1.6937
1325 (579.89)	V	0.0228	0.3225	0.4492	0.4697	0.5019	0.5611	0.6161	0.6687	0.7197
	U	583.3	1099.9	1207.3	1224.1	1250.2	1298.5	1344.5	1389.7	1435.0
	H	588.9	1179.0	1317.4	1339.2	1373.3	1436.1	1495.5	1553.7	1611.5
	S	0.7875	1.3551	1.4808	1.4991	1.5267	1.5747	1.6169	1.6554	1.6913

1350 (582.31)	V	0.0229	0.3154	0.4391	0.4595	0.4913	0.5497	0.6039	0.6557	0.7059
	U	586.5	1099.0	1205.9	1222.9	1249.2	1297.8	1343.9	1389.2	1434.6
	H	592.3	1177.8	1315.6	1337.7	1372.0	1435.1	1494.7	1553.0	1610.9
	S	0.7906	1.3525	1.4776	1.4960	1.5238	1.5721	1.6144	1.6530	1.6890
1375 (584.71)	V	0.0230	0.3085	0.4294	0.4496	0.4811	0.5387	0.5922	0.6432	0.6926
	U	589.7	1098.1	1204.6	1221.7	1248.2	1297.1	1343.3	1388.7	1434.2
	H	595.6	1176.6	1313.8	1336.1	1370.7	1434.1	1494.0	1552.4	1610.4
	S	0.7936	1.3499	1.4743	1.4930	1.5210	1.5695	1.6120	1.6507	1.6868
1400 (587.07)	V	0.0231	0.3018	0.4200	0.4400	0.4712	0.5282	0.5809	0.6311	0.6798
	U	592.9	1097.1	1203.2	1220.5	1247.3	1296.3	1342.7	1388.2	1433.8
	H	598.8	1175.3	1312.0	1334.5	1369.3	1433.2	1493.2	1551.8	1609.9
	S	0.7966	1.3474	1.4711	1.4900	1.5182	1.5670	1.6096	1.6484	1.6845
1425 (589.40)	V	0.0232	0.2953	0.4109	0.4308	0.4617	0.5180	0.5700	0.6195	0.6674
	U	596.0	1096.2	1201.8	1219.3	1246.3	1295.6	1342.1	1387.8	1433.3
	H	602.1	1174.0	1310.2	1332.9	1368.0	1432.2	1492.4	1551.1	1609.3
	S	0.7996	1.3448	1.4680	1.4870	1.5154	1.5645	1.6072	1.6461	1.6823
1450 (591.69)	V	0.0233	0.2891	0.4021	0.4219	0.4525	0.5081	0.5595	0.6083	0.6554
	U	599.1	1095.2	1200.4	1218.1	1245.3	1294.9	1341.5	1387.3	1432.9
	H	605.3	1172.7	1308.3	1331.3	1366.7	1431.2	1491.6	1550.5	1608.8
	S	0.8026	1.3423	1.4648	1.4840	1.5127	1.5620	1.6049	1.6439	1.6802
1475 (593.96)	V	0.0234	0.2830	0.3937	0.4133	0.4436	0.4986	0.5493	0.5974	0.6439
	U	602.1	1094.2	1199.0	1216.9	1244.3	1294.1	1340.9	1386.8	1432.5
	H	608.5	1171.4	1306.5	1329.6	1365.3	1430.2	1490.9	1549.8	1608.3
	S	0.8055	1.3398	1.4617	1.4811	1.5100	1.5596	1.6026	1.6417	1.6780
1500 (596.20)	V	0.0235	0.2772	0.3855	0.4049	0.4350	0.4894	0.5394	0.5869	0.6327
	U	605.2	1093.1	1197.6	1215.6	1243.2	1293.4	1340.3	1386.3	1432.1
	H	611.7	1170.1	1304.6	1328.0	1364.0	1429.2	1490.1	1549.2	1607.7
	S	0.8085	1.3373	1.4586	1.4782	1.5073	1.5572	1.6004	1.6395	1.6759
1525 (598.41)	V	0.0236	0.2715	0.3775	0.3968	0.4266	0.4805	0.5299	0.5767	0.6219
	U	608.2	1092.1	1196.2	1214.4	1242.2	1292.6	1339.8	1385.8	1431.7
	H	614.8	1168.7	1302.7	1326.4	1362.6	1428.2	1489.3	1548.6	1607.2
	S	0.8113	1.3348	1.4555	1.4753	1.5046	1.5548	1.5981	1.6374	1.6738
1550 (600.59)	V	0.0237	0.2660	0.3698	0.3890	0.4186	0.4719	0.5207	0.5669	0.6114
	U	611.2	1091.1	1194.7	1213.1	1241.2	1291.9	1339.2	1385.3	1431.3
	H	618.0	1167.3	1300.8	1324.7	1361.3	1427.2	1488.5	1547.9	1606.6
	S	0.8142	1.3323	1.4524	1.4724	1.5020	1.5524	1.5959	1.6353	1.6718

TABLE C.4. SUPERHEATED STEAM ENGLISH UNITS (Continued)

ABS PRESS PSIA (SAT TEMP)		SAT WATER	SAT STEAM	TEMPERATURE, DEG F 620	640	660	680	700	720	740
1575 (602.74)	V	0.0238	0.2606	0.2809	0.3009	0.3184	0.3342	0.3487	0.3623	0.3752
	U	614.2	1090.0	1111.3	1131.8	1149.6	1165.5	1179.9	1193.3	1205.8
	H	621.1	1165.9	1193.2	1219.5	1242.4	1262.9	1281.5	1298.9	1315.2
	S	0.8170	1.3299	1.3553	1.3795	1.4001	1.4182	1.4345	1.4493	1.4630
1600 (604.87)	V	0.0239	0.2555	0.2734	0.2936	0.3112	0.3270	0.3415	0.3551	0.3679
	U	617.1	1088.9	1108.1	1129.3	1147.5	1163.6	1178.3	1191.8	1204.5
	H	624.2	1164.5	1189.0	1216.2	1239.6	1260.4	1279.4	1296.9	1313.4
	S	0.8199	1.3274	1.3503	1.3752	1.3963	1.4147	1.4312	1.4463	1.4601
1625 (606.97)	V	0.0240	0.2504	0.2660	0.2864	0.3041	0.3200	0.3345	0.3480	0.3608
	U	620.1	1087.8	1104.8	1126.6	1145.2	1161.7	1176.6	1190.3	1203.2
	H	627.3	1163.1	1184.8	1212.7	1236.7	1257.9	1277.2	1295.0	1311.7
	S	0.8226	1.3250	1.3452	1.3708	1.3925	1.4113	1.4280	1.4432	1.4572
1650 (609.05)	V	0.0241	0.2455	0.2587	0.2795	0.2973	0.3132	0.3277	0.3412	0.3539
	U	623.0	1086.7	1101.4	1123.9	1143.0	1159.8	1174.9	1188.8	1201.8
	H	630.4	1161.6	1180.4	1209.2	1233.8	1255.4	1275.0	1293.0	1309.9
	S	0.8254	1.3225	1.3400	1.3665	1.3886	1.4078	1.4248	1.4402	1.4544
1675 (611.10)	V	0.0242	0.2407	0.2516	0.2726	0.2906	0.3066	0.3211	0.3346	0.3472
	U	625.9	1085.5	1097.8	1121.1	1140.7	1157.8	1173.2	1187.3	1200.4
	H	633.4	1160.1	1175.8	1205.6	1230.8	1252.9	1272.7	1291.0	1308.1
	S	0.8282	1.3201	1.3347	1.3620	1.3847	1.4042	1.4215	1.4372	1.4515
1700 (613.13)	V	0.0243	0.2361	0.2446	0.2660	0.2841	0.3001	0.3147	0.3281	0.3407
	U	628.8	1084.4	1094.2	1118.2	1138.3	1155.8	1171.5	1185.8	1199.1
	H	636.5	1158.6	1171.1	1201.9	1227.7	1250.3	1270.5	1289.0	1306.3
	S	0.8309	1.3176	1.3293	1.3575	1.3808	1.4007	1.4183	1.4342	1.4487
1725 (615.13)	V	0.0244	0.2315	0.2377	0.2594	0.2778	0.2939	0.3084	0.3218	0.3344
	U	631.7	1083.2	1090.4	1115.3	1135.9	1153.8	1169.7	1184.2	1197.7
	H	639.5	1157.1	1166.3	1198.1	1224.6	1247.6	1268.2	1287.0	1304.4
	S	0.8336	1.3152	1.3237	1.3530	1.3768	1.3972	1.4151	1.4312	1.4458
1750 (617.12)	V	0.0245	0.2271	0.2308	0.2530	0.2716	0.2877	0.3023	0.3157	0.3282
	U	634.6	1082.0	1086.4	1112.3	1133.5	1151.8	1168.0	1182.7	1196.3
	H	642.5	1155.5	1161.2	1194.3	1221.4	1244.9	1265.9	1284.9	1302.6
	S	0.8363	1.3128	1.3180	1.3484	1.3729	1.3937	1.4119	1.4282	1.4430

1775 (619.08)	V	0.0246	0.2228	0.2240	0.2468	0.2655	0.2818	0.2964	0.3097	0.3222
	U	637.4	1080.8	1082.2	1109.2	1131.0	1149.7	1166.2	1181.1	1194.9
	H	645.5	1153.9	1155.8	1190.3	1218.2	1242.2	1263.5	1282.8	1300.7
	S	0.8390	1.3103	1.3121	1.3437	1.3689	1.3901	1.4087	1.4252	1.4402
1800 (621.02)	V	0.0247	0.2186	....	0.2406	0.2595	0.2759	0.2906	0.3039	0.3164
	U	640.3	1079.5	....	1106.1	1128.5	1147.5	1164.3	1179.5	1193.4
	H	648.5	1152.3	....	1186.2	1214.9	1239.5	1261.1	1280.7	1298.8
	S	0.8417	1.3079	....	1.3390	1.3649	1.3866	1.4054	1.4222	1.4374
1825 (622.93)	V	0.0248	0.2145	....	0.2345	0.2537	0.2702	0.2849	0.2983	0.3107
	U	643.1	1078.2	....	1102.8	1125.9	1145.4	1162.5	1177.9	1192.0
	H	651.5	1150.7	....	1182.0	1211.5	1236.7	1258.7	1278.6	1296.9
	S	0.8443	1.3055	....	1.3342	1.3608	1.3830	1.4022	1.4192	1.4346
1850 (624.83)	V	0.0249	0.2105	....	0.2285	0.2480	0.2647	0.2794	0.2928	0.3052
	U	645.9	1077.0	....	1099.5	1123.2	1143.2	1160.6	1176.3	1190.6
	H	654.4	1149.0	....	1177.7	1208.1	1233.8	1256.3	1276.5	1295.0
	S	0.8469	1.3030	....	1.3293	1.3567	1.3794	1.3990	1.4163	1.4319
1875 (626.71)	V	0.0251	0.2066	....	0.2226	0.2424	0.2592	0.2740	0.2874	0.2998
	U	648.7	1075.6	....	1096.0	1120.5	1141.0	1158.7	1174.6	1189.1
	H	657.4	1147.3	....	1173.2	1204.6	1230.9	1253.8	1274.3	1293.1
	S	0.8496	1.3005	....	1.3243	1.3525	1.3758	1.3958	1.4133	1.4291
1900 (628.56)	V	0.0252	0.2028	....	0.2168	0.2369	0.2539	0.2687	0.2821	0.2945
	U	651.5	1074.3	....	1092.4	1117.7	1138.7	1156.8	1172.9	1187.6
	H	660.4	1145.6	....	1168.6	1201.0	1227.9	1251.3	1272.1	1291.2
	S	0.8522	1.2981	....	1.3192	1.3483	1.3722	1.3925	1.4103	1.4263
1925 (630.40)	V	0.0253	0.1990	....	0.2110	0.2315	0.2486	0.2636	0.2770	0.2894
	U	654.3	1072.9	....	1088.7	1114.8	1136.3	1154.9	1171.2	1186.1
	H	663.3	1143.8	....	1163.9	1197.3	1224.9	1248.8	1269.9	1289.2
	S	0.8548	1.2956	....	1.3139	1.3441	1.3685	1.3892	1.4074	1.4236
1950 (632.22)	V	0.0254	0.1954	....	0.2053	0.2262	0.2435	0.2585	0.2720	0.2844
	U	657.1	1071.5	....	1084.8	1111.9	1134.0	1152.9	1169.5	1184.6
	H	666.2	1142.0	....	1158.9	1193.5	1221.9	1246.2	1267.7	1287.2
	S	0.8574	1.2931	....	1.3085	1.3398	1.3648	1.3860	1.4044	1.4208
1975 (634.02)	V	0.0255	0.1918	....	0.1996	0.2210	0.2385	0.2536	0.2671	0.2795
	U	659.9	1070.1	....	1080.8	1108.9	1131.6	1150.9	1167.8	1183.1
	H	669.2	1140.2	....	1153.7	1189.7	1218.7	1243.6	1265.4	1285.2
	S	0.8599	1.2906	....	1.3030	1.3354	1.3611	1.3827	1.4014	1.4181

TABLE C.4. SUPERHEATED STEAM ENGLISH UNITS (Continued)

ABS PRESS PSIA (SAT TEMP)		SAT WATER	SAT STEAM	760	780	800	900	1000	1100	1200
				TEMPERATURE, DEG F						
1575 (602.74)	V	0.0238	0.2606	0.3875	0.3993	0.4108	0.4635	0.5118	0.5574	0.6013
	U	614.2	1090.0	1217.8	1229.2	1240.2	1291.1	1338.6	1384.8	1430.8
	H	621.1	1165.9	1330.7	1345.6	1359.9	1426.2	1487.7	1547.3	1606.1
	S	0.8170	1.3299	1.4758	1.4879	1.4994	1.5501	1.5938	1.6332	1.6698
1600 (604.87)	V	0.0239	0.2555	0.3801	0.3919	0.4032	0.4555	0.5031	0.5482	0.5915
	U	617.1	1088.9	1216.5	1228.0	1239.1	1290.4	1338.0	1384.3	1430.4
	H	624.2	1164.5	1329.1	1344.1	1358.5	1425.2	1486.9	1546.6	1605.6
	S	0.8199	1.3274	1.4731	1.4852	1.4968	1.5478	1.5916	1.6312	1.6678
1625 (606.97)	V	0.0240	0.2504	0.3729	0.3846	0.3958	0.4476	0.4948	0.5392	0.5820
	U	620.1	1087.8	1215.3	1226.9	1238.1	1289.6	1337.4	1383.8	1430.0
	H	627.3	1163.1	1327.5	1342.6	1357.1	1424.2	1486.2	1546.0	1605.0
	S	0.8226	1.3250	1.4703	1.4826	1.4942	1.5455	1.5895	1.6291	1.6658
1650 (609.05)	V	0.0241	0.2455	0.3660	0.3776	0.3887	0.4400	0.4866	0.5306	0.5728
	U	623.0	1086.7	1214.1	1225.8	1237.0	1288.8	1336.8	1383.3	1429.6
	H	630.4	1161.6	1325.8	1341.1	1355.7	1423.2	1485.4	1545.3	1604.5
	S	0.8254	1.3225	1.4676	1.4800	1.4917	1.5433	1.5874	1.6271	1.6639
1675 (611.10)	V	0.0242	0.2407	0.3592	0.3707	0.3818	0.4326	0.4788	0.5222	0.5639
	U	625.9	1085.5	1212.8	1224.6	1236.0	1288.1	1336.2	1382.8	1429.2
	H	633.4	1160.1	1324.2	1339.6	1354.3	1422.2	1484.6	1544.7	1603.9
	S	0.8282	1.3201	1.4648	1.4773	1.4892	1.5410	1.5853	1.6252	1.6620
1700 (613.13)	V	0.0243	0.2361	0.3527	0.3641	0.3751	0.4255	0.4711	0.5140	0.5552
	U	628.8	1084.4	1211.6	1223.5	1234.9	1287.3	1335.6	1382.3	1428.7
	H	636.5	1158.6	1322.5	1338.0	1352.9	1421.2	1483.8	1544.0	1603.4
	S	0.8309	1.3176	1.4621	1.4747	1.4867	1.5388	1.5833	1.6232	1.6601
1725 (615.13)	V	0.0244	0.2315	0.3463	0.3576	0.3686	0.4185	0.4637	0.5061	0.5468
	U	631.7	1083.2	1210.3	1222.3	1233.9	1286.5	1335.0	1381.8	1428.3
	H	639.5	1157.1	1320.9	1336.5	1351.5	1420.1	1483.0	1543.4	1602.9
	S	0.8336	1.3152	1.4594	1.4722	1.4842	1.5366	1.5812	1.6213	1.6582
1750 (617.12)	V	0.0245	0.2271	0.3401	0.3514	0.3622	0.4118	0.4565	0.4984	0.5386
	U	634.6	1082.0	1209.1	1221.2	1232.8	1285.8	1334.4	1381.3	1427.9
	H	642.5	1155.5	1319.2	1335.0	1350.1	1419.1	1482.2	1542.7	1602.3
	S	0.8363	1.3128	1.4568	1.4696	1.4817	1.5345	1.5792	1.6194	1.6564

1775 (619.08)	V	0.0246	0.2228	0.3340	0.3453	0.3560	0.4052	0.4494	0.4909	0.5306
	U	637.4	1080.8	1207.8	1220.0	1231.7	1285.0	1333.8	1380.8	1427.5
	H	645.5	1153.9	1317.5	1333.4	1348.6	1418.1	1481.4	1542.1	1601.8
	S	0.8390	1.3103	1.4541	1.4670	1.4792	1.5323	1.5773	1.6175	1.6546
1800 (621.02)	V	0.0247	0.2186	0.3281	0.3393	0.3500	0.3988	0.4426	0.4836	0.5229
	U	640.3	1079.5	1206.5	1218.8	1230.6	1284.2	1333.2	1380.3	1427.1
	H	648.5	1152.3	1315.8	1331.9	1347.2	1417.1	1480.6	1541.4	1601.2
	S	0.8417	1.3079	1.4514	1.4645	1.4768	1.5302	1.5753	1.6156	1.6528
1825 (622.93)	V	0.0248	0.2145	0.3224	0.3335	0.3442	0.3926	0.4360	0.4766	0.5154
	U	643.1	1078.2	1205.2	1217.6	1229.5	1283.4	1332.5	1379.8	1426.6
	H	651.5	1150.7	1314.1	1330.3	1345.8	1416.0	1479.8	1540.8	1600.7
	S	0.8443	1.3055	1.4488	1.4620	1.4744	1.5281	1.5734	1.6138	1.6510
1850 (624.83)	V	0.0249	0.2105	0.3168	0.3279	0.3385	0.3865	0.4295	0.4697	0.5081
	U	645.9	1077.0	1203.9	1216.4	1228.4	1282.7	1331.9	1379.3	1426.2
	H	654.4	1149.0	1312.3	1328.7	1344.3	1415.0	1479.0	1540.1	1600.1
	S	0.8469	1.3030	1.4462	1.4595	1.4720	1.5260	1.5714	1.6120	1.6493
1875 (626.71)	V	0.0251	0.2066	0.3114	0.3224	0.3329	0.3806	0.4232	0.4630	0.5009
	U	648.7	1075.6	1202.6	1215.2	1227.3	1281.9	1331.3	1378.8	1425.8
	H	657.4	1147.3	1310.6	1327.1	1342.8	1413.9	1478.2	1539.5	1599.6
	S	0.8496	1.3005	1.4436	1.4570	1.4696	1.5239	1.5695	1.6102	1.6475
1900 (628.56)	V	0.0252	0.2028	0.3061	0.3171	0.3275	0.3749	0.4171	0.4565	0.4940
	U	651.5	1074.3	1201.2	1214.0	1226.2	1281.1	1330.7	1378.3	1425.4
	H	660.4	1145.6	1308.9	1325.5	1341.4	1412.9	1477.4	1538.8	1599.1
	S	0.8522	1.2981	1.4409	1.4545	1.4672	1.5219	1.5677	1.6084	1.6458
1925 (630.40)	V	0.0253	0.1990	0.3009	0.3119	0.3223	0.3693	0.4112	0.4501	0.4872
	U	654.3	1072.9	1199.9	1212.8	1225.1	1280.3	1330.1	1377.8	1424.9
	H	663.3	1143.8	1307.1	1323.9	1339.9	1411.8	1476.6	1538.2	1598.5
	S	0.8548	1.2956	1.4383	1.4520	1.4648	1.5198	1.5658	1.6066	1.6441
1950 (632.22)	V	0.0254	0.1954	0.2959	0.3068	0.3171	0.3639	0.4053	0.4439	0.4807
	U	657.1	1071.5	1198.5	1211.6	1224.0	1279.5	1329.5	1377.3	1424.5
	H	666.2	1142.0	1305.3	1322.3	1338.4	1410.8	1475.8	1537.5	1598.0
	S	0.8574	1.2931	1.4358	1.4496	1.4625	1.5178	1.5639	1.6049	1.6424
1975 (634.02)	V	0.0255	0.1918	0.2910	0.3018	0.3121	0.3585	0.3997	0.4379	0.4742
	U	659.9	1070.1	1197.2	1210.4	1222.8	1278.7	1328.9	1376.8	1424.1
	H	669.2	1140.2	1303.5	1320.7	1336.9	1409.7	1474.9	1536.9	1597.4
	S	0.8599	1.2906	1.4332	1.4471	1.4601	1.5158	1.5621	1.6031	1.6408

TABLE C.4. SUPERHEATED STEAM ENGLISH UNITS (Continued)

ABS PRESS PSIA (SAT TEMP)		SAT WATER	SAT STEAM	TEMPERATURE, DEG F 660	680	700	720	740	760	780
2000 (635.80)	V	0.0256	0.1883	0.2159	0.2336	0.2488	0.2623	0.2747	0.2862	0.2970
	U	662.6	1068.6	1105.8	1129.1	1148.8	1166.1	1181.6	1195.8	1209.1
	H	672.1	1138.3	1185.7	1215.5	1240.9	1263.1	1283.2	1301.7	1319.0
	S	0.8625	1.2881	1.3309	1.3574	1.3794	1.3984	1.4153	1.4306	1.4447
2025 (637.57)	V	0.0258	0.1849	0.2108	0.2287	0.2441	0.2576	0.2700	0.2815	0.2922
	U	665.4	1067.2	1102.7	1126.6	1146.7	1164.3	1180.0	1194.4	1207.9
	H	675.0	1136.4	1181.6	1212.3	1238.2	1260.8	1281.2	1299.9	1317.4
	S	0.8651	1.2856	1.3264	1.3536	1.3761	1.3955	1.4126	1.4280	1.4422
2050 (639.31)	V	0.0259	0.1815	0.2058	0.2240	0.2394	0.2531	0.2654	0.2769	0.2876
	U	668.1	1065.6	1099.4	1124.0	1144.6	1162.5	1178.4	1193.0	1206.6
	H	678.0	1134.5	1177.5	1209.0	1235.5	1258.5	1279.1	1298.1	1315.7
	S	0.8676	1.2831	1.3218	1.3497	1.3728	1.3925	1.4098	1.4255	1.4398
2075 (641.04)	V	0.0260	0.1782	0.2008	0.2193	0.2349	0.2486	0.2610	0.2724	0.2831
	U	670.9	1064.1	1096.0	1121.4	1142.5	1160.7	1176.9	1191.6	1205.3
	H	680.9	1132.5	1173.1	1205.6	1232.7	1256.1	1277.1	1296.2	1314.0
	S	0.8702	1.2805	1.3171	1.3459	1.3694	1.3895	1.4071	1.4229	1.4374
2100 (642.76)	V	0.0262	0.1750	0.1959	0.2147	0.2304	0.2442	0.2566	0.2680	0.2787
	U	673.6	1062.5	1092.6	1118.7	1140.3	1158.8	1175.3	1190.2	1204.0
	H	683.8	1130.5	1168.7	1202.1	1229.8	1253.7	1275.0	1294.3	1312.3
	S	0.8727	1.2780	1.3124	1.3420	1.3661	1.3865	1.4043	1.4204	1.4350
2125 (644.45)	V	0.0263	0.1718	0.1911	0.2102	0.2261	0.2399	0.2523	0.2637	0.2744
	U	676.4	1060.9	1089.0	1116.0	1138.1	1157.0	1173.6	1188.8	1202.8
	H	686.7	1128.5	1164.1	1198.6	1227.0	1251.3	1272.9	1292.5	1310.7
	S	0.8752	1.2754	1.3075	1.3380	1.3627	1.3835	1.4016	1.4178	1.4326
2150 (646.13)	V	0.0264	0.1687	0.1863	0.2057	0.2218	0.2357	0.2481	0.2595	0.2702
	U	679.1	1059.3	1085.3	1113.2	1135.8	1155.1	1172.0	1187.3	1201.5
	H	689.6	1126.4	1159.4	1195.0	1224.0	1248.8	1270.7	1290.6	1308.9
	S	0.8778	1.2728	1.3025	1.3340	1.3592	1.3804	1.3989	1.4153	1.4302
2175 (647.80)	V	0.0266	0.1657	0.1815	0.2013	0.2176	0.2315	0.2440	0.2554	0.2660
	U	681.9	1057.6	1081.5	1110.3	1133.5	1153.1	1170.4	1185.9	1200.2
	H	692.5	1124.3	1154.5	1191.3	1221.1	1246.3	1268.6	1288.7	1307.2
	S	0.8803	1.2702	1.2973	1.3299	1.3558	1.3774	1.3961	1.4127	1.4278

2200 (649.45)	V	0.0267	0.1627	0.1767	0.1970	0.2134	0.2275	0.2400	0.2514	0.2620
	U	684.6	1055.9	1077.5	1107.3	1131.1	1151.2	1168.7	1184.4	1198.8
	H	695.5	1122.2	1149.4	1187.5	1218.0	1243.8	1266.4	1286.7	1305.5
	S	0.8828	1.2676	1.2920	1.3258	1.3523	1.3744	1.3934	1.4102	1.4254
2225 (651.08)	V	0.0268	0.1598	0.1720	0.1927	0.2093	0.2235	0.2360	0.2475	0.2581
	U	687.3	1054.2	1073.3	1104.3	1128.8	1149.2	1167.0	1182.9	1197.5
	H	698.4	1120.0	1144.1	1183.6	1214.9	1241.2	1264.2	1284.8	1303.8
	S	0.8853	1.2649	1.2866	1.3216	1.3488	1.3713	1.3906	1.4077	1.4231
2250 (652.70)	V	0.0270	0.1569	0.1672	0.1885	0.2053	0.2196	0.2322	0.2436	0.2542
	U	690.1	1052.4	1068.9	1101.2	1126.3	1147.2	1165.3	1181.4	1196.2
	H	701.3	1117.8	1138.5	1179.7	1211.8	1238.6	1262.0	1282.8	1302.0
	S	0.8879	1.2623	1.2809	1.3173	1.3453	1.3682	1.3878	1.4051	1.4207
2275 (654.30)	V	0.0271	0.1541	0.1624	0.1843	0.2014	0.2157	0.2284	0.2398	0.2504
	U	692.8	1050.6	1064.2	1098.0	1123.8	1145.2	1163.6	1179.9	1194.8
	H	704.2	1115.5	1132.6	1175.6	1208.6	1236.0	1259.7	1280.9	1300.2
	S	0.8904	1.2596	1.2749	1.3130	1.3417	1.3652	1.3851	1.4026	1.4183
2300 (655.89)	V	0.0273	0.1513	0.1576	0.1802	0.1975	0.2120	0.2247	0.2361	0.2467
	U	695.6	1048.8	1059.3	1094.7	1121.3	1143.1	1161.8	1178.4	1193.5
	H	707.2	1113.2	1126.3	1171.4	1205.3	1233.3	1257.4	1278.9	1298.5
	S	0.8929	1.2569	1.2687	1.3086	1.3381	1.3620	1.3823	1.4000	1.4160
2325 (657.47)	V	0.0274	0.1486	0.1527	0.1761	0.1936	0.2083	0.2210	0.2325	0.2430
	U	698.3	1046.9	1053.9	1091.4	1118.7	1141.0	1160.0	1176.8	1192.1
	H	710.1	1110.9	1119.6	1167.1	1202.0	1230.6	1255.1	1276.9	1296.7
	S	0.8954	1.2542	1.2620	1.3041	1.3345	1.3589	1.3795	1.3975	1.4136
2350 (659.03)	V	0.0276	0.1460	0.1476	0.1720	0.1899	0.2046	0.2174	0.2289	0.2395
	U	701.1	1045.1	1048.0	1087.9	1116.0	1138.9	1158.2	1175.3	1190.7
	H	713.1	1108.5	1112.1	1162.7	1198.6	1227.8	1252.8	1274.8	1294.9
	S	0.8980	1.2515	1.2547	1.2995	1.3308	1.3558	1.3767	1.3950	1.4113
2375 (660.57)	V	0.0277	0.1433	....	0.1680	0.1861	0.2010	0.2139	0.2254	0.2360
	U	703.9	1043.1	....	1084.3	1113.3	1136.7	1156.4	1173.7	1189.3
	H	716.0	1106.1	....	1158.1	1195.1	1225.0	1250.4	1272.8	1293.0
	S	0.9005	1.2488	....	1.2948	1.3270	1.3526	1.3739	1.3924	1.4089
2400 (662.11)	V	0.0279	0.1408	....	0.1640	0.1824	0.1975	0.2105	0.2220	0.2326
	U	706.6	1041.2	....	1080.6	1110.6	1134.5	1154.6	1172.1	1187.9
	H	719.0	1103.7	....	1153.4	1191.6	1222.2	1248.0	1270.7	1291.2
	S	0.9031	1.2460	....	1.2900	1.3232	1.3494	1.3711	1.3899	1.4066

TABLE C.4. SUPERHEATED STEAM ENGLISH UNITS (Continued)

ABS PRESS PSIA (SAT TEMP)		SAT WATER	SAT STEAM	TEMPERATURE, DEG F 800	820	850	900	1000	1100	1200
2000	V	0.0256	0.1883	0.3072	0.3171	0.3312	0.3534	0.3942	0.4320	0.4680
(635.80)	U	662.6	1068.6	1221.7	1233.7	1250.9	1277.9	1328.2	1376.3	1423.7
	H	672.1	1138.3	1335.4	1351.1	1373.5	1408.7	1474.1	1536.2	1596.9
	S	0.8625	1.2881	1.4578	1.4701	1.4874	1.5138	1.5603	1.6014	1.6391
2025	V	0.0258	0.1849	0.3025	0.3123	0.3263	0.3483	0.3888	0.4263	0.4619
(637.57)	U	665.4	1067.2	1220.5	1232.6	1249.9	1277.1	1327.6	1375.8	1423.3
	H	675.0	1136.4	1333.9	1349.7	1372.2	1407.6	1473.3	1535.6	1596.3
	S	0.8651	1.2856	1.4555	1.4679	1.4853	1.5118	1.5585	1.5997	1.6375
2050	V	0.0259	0.1815	0.2978	0.3076	0.3215	0.3434	0.3835	0.4207	0.4559
(639.31)	U	668.1	1065.6	1219.4	1231.6	1249.0	1276.3	1327.0	1375.3	1422.8
	H	678.0	1134.5	1332.4	1348.3	1371.0	1406.5	1472.5	1534.9	1595.8
	S	0.8676	1.2831	1.4532	1.4657	1.4832	1.5099	1.5567	1.5981	1.6359
2075	V	0.0260	0.1782	0.2933	0.3030	0.3168	0.3386	0.3784	0.4152	0.4501
(641.04)	U	670.9	1064.1	1218.2	1230.5	1248.0	1275.5	1326.4	1374.8	1422.4
	H	680.9	1132.5	1330.8	1346.9	1369.7	1405.5	1471.7	1534.2	1595.2
	S	0.8702	1.2805	1.4509	1.4635	1.4811	1.5079	1.5549	1.5964	1.6343
2100	V	0.0262	0.1750	0.2888	0.2985	0.3123	0.3339	0.3734	0.4099	0.4445
(642.76)	U	673.6	1062.5	1217.1	1229.4	1247.1	1274.6	1325.8	1374.3	1422.0
	H	683.8	1130.5	1329.3	1345.4	1368.4	1404.4	1470.9	1533.6	1594.7
	S	0.8727	1.2780	1.4486	1.4613	1.4790	1.5060	1.5532	1.5948	1.6327
2125	V	0.0263	0.1718	0.2845	0.2941	0.3078	0.3293	0.3685	0.4046	0.4389
(644.45)	U	676.4	1060.9	1215.9	1228.4	1246.1	1273.8	1325.1	1373.8	1421.6
	H	686.7	1128.5	1327.8	1344.0	1367.2	1403.3	1470.0	1532.9	1594.2
	S	0.8752	1.2754	1.4463	1.4591	1.4770	1.5041	1.5514	1.5931	1.6312
2150	V	0.0264	0.1687	0.2802	0.2898	0.3035	0.3248	0.3637	0.3996	0.4335
(646.13)	U	679.1	1059.3	1214.7	1227.3	1245.1	1273.0	1324.5	1373.3	1421.1
	H	689.6	1126.4	1326.2	1342.6	1365.9	1402.2	1469.2	1532.3	1593.6
	S	0.8778	1.2728	1.4440	1.4569	1.4749	1.5022	1.5497	1.5915	1.6296
2175	V	0.0266	0.1657	0.2761	0.2856	0.2992	0.3204	0.3590	0.3946	0.4282
(647.80)	U	681.9	1057.6	1213.5	1226.2	1244.2	1272.2	1323.9	1372.8	1420.7
	H	692.5	1124.3	1324.6	1341.1	1364.6	1401.1	1468.4	1531.6	1593.1
	S	0.8803	1.2702	1.4418	1.4548	1.4729	1.5003	1.5480	1.5899	1.6281

2200 (649.45)	V	0.0267	0.1627	0.2720	0.2815	0.2950	0.3161	0.3545	0.3897	0.4231
	U	684.6	1055.9	1212.3	1225.1	1243.2	1271.4	1323.3	1372.3	1420.3
	H	695.5	1122.2	1323.1	1339.7	1363.3	1400.0	1467.6	1530.9	1592.5
	S	0.8828	1.2676	1.4395	1.4526	1.4708	1.4984	1.5463	1.5883	1.6266
2225 (651.08)	V	0.0268	0.1598	0.2680	0.2775	0.2910	0.3119	0.3500	0.3850	0.4180
	U	687.3	1054.2	1211.1	1224.0	1242.2	1270.5	1322.6	1371.8	1419.8
	H	698.4	1120.0	1321.5	1338.2	1362.0	1399.0	1466.7	1530.3	1592.0
	S	0.8853	1.2649	1.4372	1.4505	1.4688	1.4965	1.5446	1.5867	1.6251
2250 (652.70)	V	0.0270	0.1569	0.2641	0.2736	0.2870	0.3078	0.3456	0.3803	0.4131
	U	690.1	1052.4	1209.9	1222.9	1241.2	1269.7	1322.0	1371.3	1419.4
	H	701.3	1117.8	1319.9	1336.8	1360.7	1397.9	1465.9	1529.6	1591.4
	S	0.8879	1.2623	1.4350	1.4483	1.4668	1.4946	1.5430	1.5852	1.6236
2275 (654.30)	V	0.0271	0.1541	0.2603	0.2697	0.2831	0.3038	0.3414	0.3758	0.4083
	U	692.8	1050.6	1208.7	1221.8	1240.2	1268.9	1321.4	1370.7	1419.0
	H	704.2	1115.5	1318.3	1335.3	1359.4	1396.8	1465.1	1528.9	1590.9
	S	0.8904	1.2596	1.4328	1.4462	1.4648	1.4928	1.5413	1.5836	1.6221
2300 (655.89)	V	0.0273	0.1513	0.2566	0.2660	0.2793	0.2999	0.3372	0.3714	0.4035
	U	695.6	1048.8	1207.5	1220.6	1239.2	1268.0	1320.7	1370.2	1418.6
	H	707.2	1113.2	1316.7	1333.8	1358.1	1395.7	1464.2	1528.3	1590.3
	S	0.8929	1.2569	1.4305	1.4441	1.4628	1.4910	1.5397	1.5821	1.6207
2325 (657.47)	V	0.0274	0.1486	0.2529	0.2623	0.2755	0.2960	0.3331	0.3670	0.3989
	U	698.3	1046.9	1206.2	1219.5	1238.2	1267.2	1320.1	1369.7	1418.1
	H	710.1	1110.9	1315.1	1332.4	1356.8	1394.5	1463.4	1527.6	1589.8
	S	0.8954	1.2542	1.4283	1.4420	1.4608	1.4891	1.5380	1.5806	1.6192
2350 (659.03)	V	0.0276	0.1460	0.2493	0.2587	0.2719	0.2923	0.3291	0.3627	0.3944
	U	701.1	1045.1	1205.0	1218.4	1237.2	1266.3	1319.4	1369.2	1417.7
	H	713.1	1108.5	1313.4	1330.9	1355.5	1393.4	1462.6	1527.0	1589.2
	S	0.8980	1.2515	1.4261	1.4398	1.4588	1.4873	1.5364	1.5791	1.6178
2375 (660.57)	V	0.0277	0.1433	0.2458	0.2551	0.2683	0.2886	0.3252	0.3586	0.3900
	U	703.9	1043.1	1203.7	1217.2	1236.2	1265.5	1318.8	1368.7	1417.3
	H	716.0	1106.1	1311.8	1329.4	1354.1	1392.3	1461.7	1526.3	1588.7
	S	0.9005	1.2488	1.4239	1.4377	1.4569	1.4855	1.5348	1.5776	1.6164
2400 (662.11)	V	0.0279	0.1408	0.2424	0.2517	0.2648	0.2850	0.3214	0.3545	0.3856
	U	706.6	1041.2	1202.5	1216.1	1235.2	1264.6	1318.2	1368.2	1416.8
	H	719.0	1103.7	1310.1	1327.9	1352.8	1391.2	1460.9	1525.6	1588.1
	S	0.9031	1.2460	1.4217	1.4357	1.4549	1.4837	1.5332	1.5761	1.6149

TABLE C.4. SUPERHEATED STEAM ENGLISH UNITS (Continued)

ABS PRESS PSIA (SAT TEMP)		SAT WATER	SAT STEAM	TEMPERATURE, DEG F 680	700	720	740	760	780	800
2450 (665.14)	V	0.0282	0.1357	0.1560	0.1752	0.1906	0.2037	0.2153	0.2259	0.2357
	U	712.5	1037.1	1072.8	1104.9	1130.0	1150.8	1168.9	1185.1	1199.9
	H	725.3	1098.6	1143.6	1184.3	1216.4	1243.2	1266.5	1287.5	1306.8
	S	0.9085	1.2403	1.2801	1.3155	1.3429	1.3655	1.3848	1.4019	1.4173
2500 (668.11)	V	0.0286	0.1307	0.1481	0.1681	0.1839	0.1972	0.2089	0.2195	0.2293
	U	718.5	1032.9	1064.4	1098.9	1125.3	1147.0	1165.6	1182.2	1197.4
	H	731.7	1093.3	1132.9	1176.7	1210.4	1238.2	1262.3	1283.7	1303.4
	S	0.9139	1.2345	1.2695	1.3076	1.3364	1.3598	1.3797	1.3972	1.4129
2550 (671.04)	V	0.0290	0.1258	0.1401	0.1612	0.1774	0.1909	0.2027	0.2133	0.2231
	U	724.4	1028.4	1055.2	1092.6	1120.4	1143.0	1162.3	1179.3	1194.7
	H	738.1	1087.8	1121.3	1168.7	1204.1	1233.1	1257.9	1279.9	1300.0
	S	0.9194	1.2286	1.2581	1.2993	1.3297	1.3540	1.3745	1.3925	1.4085
2600 (673.91)	V	0.0294	0.1211	0.0313	0.1544	0.1711	0.1848	0.1967	0.2073	0.2171
	U	730.3	1023.8	747.9	1085.9	1115.4	1138.9	1158.8	1176.3	1192.1
	H	744.5	1082.0	762.9	1160.2	1197.7	1227.8	1253.5	1276.0	1296.5
	S	0.9247	1.2225	0.9410	1.2908	1.3228	1.3482	1.3694	1.3878	1.4042
2650 (676.74)	V	0.0298	0.1165	0.0308	0.1477	0.1649	0.1789	0.1909	0.2016	0.2113
	U	736.2	1018.9	743.9	1078.9	1110.1	1134.7	1155.3	1173.3	1189.4
	H	750.9	1076.0	759.0	1151.4	1191.0	1222.4	1248.9	1272.1	1293.0
	S	0.9301	1.2162	0.9373	1.2820	1.3158	1.3423	1.3642	1.3830	1.3998
2700 (679.53)	V	0.0303	0.1119	0.0304	0.1411	0.1588	0.1731	0.1852	0.1960	0.2058
	U	742.2	1013.7	740.5	1071.5	1104.7	1130.4	1151.7	1170.2	1186.7
	H	757.3	1069.7	755.7	1142.0	1184.0	1216.9	1244.3	1268.1	1289.5
	S	0.9356	1.2097	0.9342	1.2727	1.3087	1.3363	1.3590	1.3783	1.3954
2750 (682.26)	V	0.0308	0.1075	....	0.1345	0.1529	0.1675	0.1798	0.1906	0.2004
	U	748.2	1008.3	....	1063.6	1099.0	1126.0	1148.0	1167.0	1183.9
	H	763.9	1063.0	....	1132.0	1176.8	1211.2	1239.5	1264.0	1285.9
	S	0.9411	1.2029	....	1.2631	1.3013	1.3302	1.3537	1.3736	1.3911
2800 (684.96)	V	0.0313	0.1030	....	0.1278	0.1471	0.1620	0.1745	0.1853	0.1952
	U	754.4	1002.4	....	1055.0	1093.0	1121.4	1144.3	1163.8	1181.1
	H	770.7	1055.8	....	1121.2	1169.3	1205.3	1234.7	1259.8	1282.2
	S	0.9468	1.1958	....	1.2527	1.2938	1.3241	1.3484	1.3688	1.3867

2850 (687.61)	V	0.0319	0.0986	…	0.1209	0.1414	0.1567	0.1693	0.1803	0.1901
	U	760.9	996.1	…	1045.2	1086.8	1116.6	1140.4	1160.5	1178.2
	H	777.7	1048.2	…	1109.0	1161.4	1199.2	1229.7	1255.6	1278.5
	S	0.9526	1.1883	…	1.2411	1.2860	1.3178	1.3430	1.3640	1.3824
2900 (690.22)	V	0.0326	0.0942	…	0.1138	0.1358	0.1515	0.1643	0.1754	0.1853
	U	767.6	989.3	…	1034.2	1080.3	1111.7	1136.4	1157.2	1175.3
	H	785.1	1039.8	…	1095.2	1153.2	1193.0	1224.6	1251.3	1274.7
	S	0.9588	1.1803	…	1.2283	1.2779	1.3114	1.3375	1.3592	1.3780
2950 (692.79)	V	0.0334	0.0897	…	0.1063	0.1303	0.1464	0.1594	0.1706	0.1805
	U	774.9	981.7	…	1021.4	1073.5	1106.6	1132.4	1153.8	1172.4
	H	793.1	1030.6	…	1079.5	1144.6	1186.5	1219.4	1246.9	1270.9
	S	0.9655	1.1716	…	1.2139	1.2696	1.3048	1.3320	1.3544	1.3736
3000 (695.33)	V	0.0343	0.0850	…	0.0982	0.1248	0.1414	0.1546	0.1659	0.1759
	U	782.8	973.1	…	1006.0	1066.3	1101.3	1128.2	1150.3	1169.4
	H	801.8	1020.3	…	1060.5	1135.6	1179.8	1214.0	1242.4	1267.0
	S	0.9728	1.1619	…	1.1966	1.2610	1.2982	1.3265	1.3495	1.3692
3050 (697.82)	V	0.0354	0.0800	…	0.0884	0.1193	0.1365	0.1500	0.1614	0.1715
	U	791.9	963.0	…	985.2	1058.6	1095.9	1123.9	1146.7	1166.3
	H	811.8	1008.2	…	1035.1	1126.0	1172.9	1208.6	1237.9	1263.1
	S	0.9812	1.1508	…	1.1740	1.2518	1.2913	1.3208	1.3446	1.3648
3100 (700.28)	V	0.0368	0.0745	…	…	0.1137	0.1317	0.1455	0.1570	0.1671
	U	802.9	950.6	…	…	1050.1	1090.2	1119.5	1143.1	1163.2
	H	824.0	993.3	…	…	1115.4	1165.7	1202.9	1233.2	1259.1
	S	0.9914	1.1373	…	…	1.2419	1.2843	1.3151	1.3397	1.3604
3150 (702.70)	V	0.0390	0.0678	…	…	0.1081	0.1270	0.1411	0.1528	0.1629
	U	817.8	933.4	…	…	1040.9	1084.3	1114.9	1139.4	1160.1
	H	840.5	972.9	…	…	1103.9	1158.3	1197.2	1228.4	1255.0
	S	1.0053	1.1192	…	…	1.2314	1.2771	1.3093	1.3347	1.3560
3200 (705.08)	V	0.0447	0.0566	…	…	0.1024	0.1224	0.1367	0.1486	0.1588
	U	849.0	898.1	…	…	1030.7	1078.2	1110.3	1135.6	1156.9
	H	875.5	931.7	…	…	1091.3	1150.6	1191.2	1223.6	1250.9
	S	1.0351	1.0832	…	…	1.2199	1.2698	1.3033	1.3297	1.3515
3208.2 (705.47)	V	0.0508	0.0508	…	…	0.1014	0.1216	0.1360	0.1479	0.1582
	U	875.9	875.9	…	…	1028.9	1077.1	1109.5	1135.0	1156.3
	H	906.0	906.0	…	…	1089.1	1149.3	1190.2	1222.8	1250.2
	S	1.0612	1.0612	…	…	1.2178	1.2685	1.3024	1.3288	1.3508

TABLE C.4. SUPERHEATED STEAM ENGLISH UNITS (Continued)

ABS PRESS PSIA (SAT TEMP)		SAT WATER	SAT STEAM	TEMPERATURE, DEG F 820	850	900	950	1000	1100	1200
2450 (665.14)	V	0.0282	0.1357	0.2449	0.2579	0.2780	0.2965	0.3139	0.3466	0.3772
	U	712.5	1037.1	1213.8	1233.2	1262.9	1290.6	1316.9	1367.1	1416.0
	H	725.3	1098.6	1324.8	1350.1	1389.0	1425.0	1459.2	1524.3	1587.0
	S	0.9085	1.2403	1.4315	1.4510	1.4801	1.5062	1.5300	1.5732	1.6122
2500 (668.11)	V	0.0286	0.1307	0.2385	0.2514	0.2712	0.2896	0.3068	0.3390	0.3692
	U	718.5	1032.9	1211.4	1231.1	1261.2	1289.1	1315.6	1366.1	1415.1
	H	731.7	1093.3	1321.8	1347.4	1386.7	1423.1	1457.5	1522.9	1585.9
	S	0.9139	1.2345	1.4273	1.4472	1.4766	1.5029	1.5269	1.5703	1.6094
2550 (671.04)	V	0.0290	0.1258	0.2322	0.2451	0.2648	0.2829	0.2999	0.3317	0.3614
	U	724.4	1028.4	1209.1	1229.0	1259.5	1287.6	1314.3	1365.1	1414.3
	H	738.1	1087.8	1318.7	1344.7	1384.4	1421.1	1455.8	1521.6	1584.8
	S	0.9194	1.2286	1.4232	1.4433	1.4731	1.4996	1.5238	1.5674	1.6067
2600 (673.91)	V	0.0294	0.1211	0.2262	0.2390	0.2585	0.2765	0.2933	0.3247	0.3540
	U	730.3	1023.8	1206.7	1226.9	1257.7	1286.1	1313.0	1364.0	1413.4
	H	744.5	1082.0	1315.5	1341.9	1382.1	1419.2	1454.1	1520.2	1583.7
	S	0.9247	1.2225	1.4191	1.4395	1.4696	1.4964	1.5208	1.5646	1.6040
2650 (676.74)	V	0.0298	0.1165	0.2204	0.2332	0.2526	0.2703	0.2870	0.3179	0.3468
	U	736.2	1018.9	1204.2	1224.8	1256.0	1284.6	1311.7	1363.0	1412.5
	H	750.9	1076.0	1312.4	1339.1	1379.8	1417.2	1452.4	1518.9	1582.6
	S	0.9301	1.2162	1.4150	1.4357	1.4662	1.4932	1.5177	1.5618	1.6014
2700 (679.53)	V	0.0303	0.1119	0.2149	0.2275	0.2468	0.2644	0.2809	0.3114	0.3399
	U	742.2	1013.7	1201.8	1222.7	1254.2	1283.1	1310.3	1361.9	1411.7
	H	757.3	1069.7	1309.1	1336.3	1377.5	1415.2	1450.7	1517.5	1581.5
	S	0.9356	1.2097	1.4109	1.4319	1.4628	1.4900	1.5148	1.5591	1.5988
2750 (682.26)	V	0.0308	0.1075	0.2095	0.2221	0.2412	0.2587	0.2750	0.3052	0.3333
	U	748.2	1008.3	1199.3	1220.5	1252.4	1281.6	1309.0	1360.9	1410.8
	H	763.9	1063.0	1305.9	1333.5	1375.1	1413.2	1449.0	1516.2	1580.4
	S	0.9411	1.2029	1.4069	1.4282	1.4594	1.4869	1.5118	1.5564	1.5963
2800 (684.96)	V	0.0313	0.1030	0.2043	0.2168	0.2358	0.2531	0.2693	0.2991	0.3268
	U	754.4	1002.4	1196.8	1218.3	1250.6	1280.1	1307.7	1359.8	1409.9
	H	770.7	1055.8	1302.6	1330.7	1372.8	1411.2	1447.2	1514.8	1579.3
	S	0.9468	1.1958	1.4028	1.4245	1.4561	1.4838	1.5089	1.5537	1.5938

2850 (687.61)	V	0.0319	0.0986	0.1992	0.2118	0.2306	0.2478	0.2638	0.2933	0.3207
	U	760.9	996.1	1194.2	1216.1	1248.8	1278.5	1306.4	1358.8	1409.1
	H	777.7	1048.2	1299.3	1327.8	1370.4	1409.2	1445.5	1513.5	1578.2
	S	0.9526	1.1883	1.3988	1.4208	1.4527	1.4807	1.5060	1.5511	1.5913
2900 (690.22)	V	0.0326	0.0942	0.1943	0.2068	0.2256	0.2427	0.2585	0.2877	0.3147
	U	767.6	989.3	1191.7	1213.9	1247.0	1277.0	1305.0	1357.7	1408.2
	H	785.1	1039.8	1296.0	1324.9	1368.0	1407.2	1443.7	1512.1	1577.0
	S	0.9588	1.1803	1.3947	1.4171	1.4494	1.4777	1.5032	1.5485	1.5889
2950 (692.79)	V	0.0334	0.0897	0.1896	0.2021	0.2208	0.2377	0.2534	0.2822	0.3089
	U	774.9	981.7	1189.1	1211.6	1245.1	1275.4	1303.7	1356.7	1407.3
	H	793.1	1030.6	1292.6	1322.0	1365.6	1405.2	1442.0	1510.7	1575.9
	S	0.9655	1.1716	1.3907	1.4134	1.4461	1.4747	1.5004	1.5459	1.5865
3000 (695.33)	V	0.0343	0.0850	0.1850	0.1975	0.2161	0.2329	0.2484	0.2770	0.3033
	U	782.8	973.1	1186.4	1209.4	1243.3	1273.8	1302.3	1355.6	1406.4
	H	801.8	1020.3	1289.1	1319.0	1363.2	1403.1	1440.2	1509.4	1574.8
	S	0.9728	1.1619	1.3866	1.4097	1.4429	1.4717	1.4976	1.5434	1.5841
3050 (697.82)	V	0.0354	0.0800	0.1806	0.1930	0.2115	0.2282	0.2436	0.2719	0.2979
	U	791.9	963.0	1183.7	1207.1	1241.4	1272.3	1301.0	1354.5	1405.5
	H	811.8	1008.2	1285.6	1316.0	1360.8	1401.1	1438.5	1508.0	1573.7
	S	0.9812	1.1508	1.3826	1.4061	1.4396	1.4687	1.4948	1.5409	1.5817
3100 (700.28)	V	0.0368	0.0745	0.1763	0.1887	0.2071	0.2237	0.2390	0.2670	0.2927
	U	802.9	950.6	1181.0	1204.7	1239.5	1270.7	1299.6	1353.5	1404.7
	H	824.0	993.3	1282.1	1313.0	1358.4	1399.0	1436.7	1506.6	1572.6
	S	0.9914	1.1373	1.3786	1.4024	1.4364	1.4658	1.4920	1.5384	1.5794
3150 (702.70)	V	0.0390	0.0678	0.1721	0.1845	0.2029	0.2193	0.2345	0.2622	0.2876
	U	817.8	933.4	1178.2	1202.4	1237.6	1269.1	1298.2	1352.4	1403.8
	H	840.5	972.9	1278.6	1310.0	1355.9	1396.9	1434.9	1505.2	1571.5
	S	1.0053	1.1192	1.3745	1.3988	1.4332	1.4629	1.4893	1.5359	1.5771
3200 (705.08)	V	0.0447	0.0566	0.1680	0.1804	0.1987	0.2151	0.2301	0.2576	0.2827
	U	849.0	898.1	1175.4	1200.0	1235.7	1267.5	1296.8	1351.3	1402.9
	H	875.5	931.7	1274.9	1306.9	1353.4	1394.9	1433.1	1503.8	1570.3
	S	1.0351	1.0832	1.3705	1.3951	1.4300	1.4600	1.4866	1.5335	1.5749
3208.2 (705.47)	V	0.0508	0.0508	0.1674	0.1798	0.1981	0.2144	0.2294	0.2568	0.2820
	U	875.9	875.9	1175.0	1199.6	1235.4	1267.2	1296.6	1351.1	1402.8
	H	906.0	906.0	1274.3	1306.4	1353.0	1394.5	1432.8	1503.6	1570.2
	S	1.0612	1.0612	1.3698	1.3945	1.4295	1.4595	1.4862	1.5331	1.5745

APPENDIX

D

THE UNIFAC METHOD

The UNIQUAC equation† treats $g \equiv G^E/RT$ as comprised of two additive parts, a "combinatorial" part g^C to account for molecular size and shape differences, and a "residual" part g^R to account for molecular interactions:

$$g = g^C + g^R \tag{D.1}$$

Function g^C contains pure-species parameters only, whereas function g^R includes two interaction parameters for each pair of constituent molecules. For a multicomponent system,

$$g^C = \sum_i x_i \ln \frac{\Phi_i}{x_i} + 5 \sum_i q_i x_i \ln \frac{\theta_i}{\Phi_i} \tag{D.2}$$

and

$$g^R = -\sum_i q_i x_i \ln \left(\sum_j \theta_j \tau_{ji} \right) \tag{D.3}$$

where

$$\Phi_i \equiv \frac{x_i r_i}{\sum_j x_j r_j} \qquad \text{and} \qquad \theta_i \equiv \frac{x_i q_i}{\sum_j x_j q_j}$$

Subscript i identifies species, and j is a dummy index. All summations run over

† D. S. Abrams and J. M. Prausnitz, *AIChE J.*, **21**: 116, 1975.

all species, and $\tau_{ji} = 1$ for $i = j$. In these equations r_i (a relative molecular volume) and q_i (a relative molecular surface area) are the pure-species parameters. The temperature dependence of g enters in Eq. (D.3) through the τ_{ji}, which are given by

$$\tau_{ji} = \exp -\left(\frac{u_{ji} - u_{ii}}{RT}\right) \tag{D.4}$$

The interaction parameters for the UNIQUAC equation are the differences $(u_{ji} - u_{ii})$.

An expression for $\ln \gamma_i$ is found by application of Eq. (11.62) to the UNIQUAC equation for g [Eqs. (D.1) through (D.3)]. The result is given by the following equations:

$$\ln \gamma_i = \ln \gamma_i^C + \ln \gamma_i^R \tag{D.5}$$

$$\ln \gamma_i^C = 1 - J_i + \ln J_i - 5q_i\left(1 - \frac{J_i}{L_i} + \ln \frac{J_i}{L_i}\right) \tag{D.6}$$

and

$$\ln \gamma_i^R = q_i(1 - \ln L_i) - \sum_j \left(\theta \frac{s_{ji}}{\eta_j} - q_i \ln \frac{s_{ji}}{\eta_j}\right) \tag{D.7}$$

where

$$J_i = \frac{r_i}{\sum_j r_j x_j} \tag{D.8}$$

$$L_i = \frac{q_i}{\sum_j q_j x_j} \tag{D.9}$$

$$\theta = \sum_i q_i x_i \tag{D.10}$$

$$s_{ji} = \sum_m q_i \tau_{mj} \tag{D.11}$$

$$\eta_j = \sum_i s_{ji} x_i \tag{D.12}$$

and

$$\tau_{mj} = \exp \frac{-(u_{mj} - u_{ii})}{RT} \tag{D.13}$$

Again subscript i identifies species, and j and m are dummy indices. All summations are over all species, and $\tau_{mj} = 1$ for $m = j$. Values for the parameters r_i, q_i, and $(u_{mj} - u_{jj})$ are given by Gmehling et al.†

† J. Gmehling, U. Onken, and W. Arlt, *Vapor-Liquid Equilibrium Data Collection*, Chemistry Data Series, vol. I, parts 1-8, DECHEMA, Frankfurt/Main, 1977–1984.

Table D.1 UNIFAC-VLE subgroup volume and surface-area parameters†

Main group	Subgroup	k	R_k	Q_k	Examples of molecules and their constituent groups	
1 "CH_2"	CH_3	1	0.9011	0.848		
	CH_2	2	0.6744	0.540	*n*-Butane:	$2CH_3, 2CH_2$
	CH	3	0.4469	0.228	Isobutane:	$3CH_3, 1CH$
	C	4	0.2195	0.000	2,2-Dimethyl propane:	$4CH_3, 1C$
3 "ACH" (AC = aromatic carbon)	ACH	10	0.5313	0.400	Benzene:	6ACH
4 "$ACCH_2$"	$ACCH_3$	12	1.2663	0.968	Toluene:	$5ACH, 1ACCH_3$
	$ACCH_2$	13	1.0396	0.660	Ethylbenzene:	$1CH_3, 5ACH, 1ACCH_2$
5 "OH"	OH	15	1.0000	1.200	Ethanol:	$1CH_3, 1CH_2, 1OH$
6 "H_2O"	H_2O	17	0.9200	1.400	Water:	$1H_2O$
9 "CH_2CO"	CH_3CO	19	1.6724	1.488	Used when CO is attached to CH_3	
					Acetone:	$1CH_3CO, 1CH_3$
	CH_2CO	20	1.4457	1.180	3-Pentanone:	$2CH_3, 1CH_2CO, 1CH_2$
13 "CH_2O"	CH_3O	25	1.1450	1.088	Dimethyl ether:	$1CH_3, 1CH_3O$
	CH_2O	26	0.9183	0.780	Diethyl ether:	$2CH_3, 1CH_2, 1CH_2O$
	CH—O	27	0.6908	0.468	Diisopropyl ether:	$4CH_3$, 1CH, 1CH—O
15 "CNH"	CH_3NH	32	1.4337	1.244	Dimethylamine:	$1CH_3, 1CH_3NH$
	CH_2NH	33	1.2070	0.936	Diethylamine:	$2CH_3, 1CH_2, 1CH_2NH$
	CHNH	34	0.9795	0.624	Diisopropylamine:	$4CH_3$, 1CH, 1CHNH
19 "CCN"	CH_3CN	41	1.8701	1.724	Acetonitrile:	$1CH_3CN$
	CH_2CN	42	1.6434	1.416	Propionitrile:	$1CH_3, 1CH_2CN$

† J. Gmehling, P. Rasmussen, and Aa. Fredenslund, *Ind. Eng. Chem. Process Des. Dev.*, **21**: 118, 1982.

The UNIFAC method for evaluation of activity coefficients† depends on the concept that a liquid mixture may be considered a solution of the structural units from which the molecules are formed rather than a solution of the molecules themselves. These structural units are called *subgroups*, and a few of them are listed in the second column of Table D.1. An identifying number, represented by k, is associated with each subgroup. The relative volume R_k and relative surface area Q_k are properties of the subgroups, and values are listed in columns

compositions of molecular species. When it is possible to construct a molecule
4 and 5 of Table D.1. Also shown (column 6) are examples of the subgroup

† Aa. Fredenslund, R. L. Jones, and J. M. Prausnitz, *AIChE J.*, **21**: 1086, 1975.

from more than one set of subgroups, the set containing the least number of *different* subgroups is the correct set. The great advantage of the UNIFAC method is that a relatively small number of subgroups combine to form a very large number of molecules.

Activity coefficients depend not only on the subgroup properties R_k and Q_k, but also on interactions between subgroups. Here, similar subgroups are assigned to a main group, as shown in the first column of Table D.1. The designations of these groups, such as "CH_2," "ACH," etc., are descriptive only. All subgroups belonging to the same main group are considered identical with respect to group interactions. Therefore parameters characterizing group interactions are identified with pairs of *main* groups. Parameter values a_{mk} for a few such pairs are given in Table D.2.

The UNIFAC method is based on the UNIQUAC equation, for which the activity coefficients are given by Eq. (D.5). When applied to a solution of groups, Eqs. (D.6) and (D.7) are written:

$$\ln \gamma_i^C = 1 - J_i + \ln J_i - 5q_i\left(1 - \frac{J_i}{L_i} + \ln\frac{J_i}{L_i}\right) \tag{D.14}$$

and

$$\ln \gamma_i^R = q_i(1 - \ln L_i) - \sum_k\left(\theta_k\frac{s_{ki}}{\eta_k} - G_{ki}\ln\frac{s_{ki}}{\eta_k}\right) \tag{D.15}$$

The quantities J_i and L_i are still given by Eqs. (D.8) and (D.9). In addition, the following definitions apply:

$$r_i = \sum_k \nu_k^{(i)} R_k \tag{D.16}$$

$$q_i = \sum_k \nu_k^{(i)} Q_k \tag{D.17}$$

$$G_{ki} = \nu_k^{(i)} Q_k \tag{D.18}$$

$$\theta_k = \sum_i G_{ki} x_i \tag{D.19}$$

$$s_{ki} = \sum_m G_{mi}\tau_{mk} \tag{D.20}$$

$$\eta_k = \sum_i s_{ki} x_i \tag{D.21}$$

and

$$\tau_{mk} = \exp\frac{-a_{mk}}{T} \tag{D.22}$$

Subscript i identifies species, and j is a dummy index running over all species. Subscript k identifies subgroups, and m is a dummy index running over all subgroups. The quantity $\nu_k^{(i)}$ is the number of subgroups of type k in a molecule of species i. Values of the subgroup parameters R_k and Q_k and of the group

Table D.2 UNIFAC-VLE interaction parameters, a_{mk}, in kelvins†

(n.a. = not available)

	1	3	4	5	7	9	13	15	19
1 CH_2	0.0	61.13	76.50	986.50	1,318.00	476.40	251.50	255.70	597.00
3 ACH	−11.12	0.0	167.00	636.10	903.80	25.77	32.14	122.80	212.50
4 $ACCH_2$	−69.70	−146.80	0.0	803.20	5,695.00	−52.10	213.10	−49.29	6,096.00
5 OH	156.40	89.60	25.82	0.0	353.50	84.00	28.06	42.70	6.712
7 H_2O	300.00	362.30	377.60	−229.10	0.0	−195.40	540.50	168.00	112.60
9 CH_2CO	26.76	140.10	365.80	164.50	472.50	0.0	5.202	n.a.	481.70
13 CH_2O	83.36	52.13	65.69	237.70	−314.70	52.38	0.0	141.70	n.a.
15 CNH	65.33	−22.31	223.00	−150.00	−448.20	n.a.	−49.30	0.0	n.a.
19 CCN	24.82	−22.97	−138.40	185.40	242.80	−287.50	n.a.	n.a.	0.0

† J. Gmehling, P. Rasmussen, and Aa. Fredenslund, *Ind. Eng. Chem. Process Des. Dev.*, **21**: 118, 1982.

interaction parameters a_{mk} come from tabulations in the literature.† Tables D.1 and D.2 show a few parameter values; the number designations of the complete tables are retained.

The equations for the UNIFAC method are presented here in a form convenient for computer programming. In the following example we run through a set of hand calculations to demonstrate their application.

Example D.1 For the binary system diethylamine(1)/*n*-heptane(2) at 308.15 K, find γ_1 and γ_2 when $x_1 = 0.4$ and $x_2 = 0.6$.

SOLUTION The subgroups involved are indicated by the chemical formulas:

$$CH_3{-}CH_2NH{-}CH_2{-}CH_3(1)/CH_3{-}(CH_2)_5{-}CH_3$$

The following table shows the subgroups, their identification numbers, values of parameters R_k and Q_k (from Table D.1), and the numbers of each subgroup in each molecule:

	k	R_k	Q_k	$\nu_k^{(1)}$	$\nu_k^{(2)}$
CH_3	1	0.9011	0.848	2	2
CH_2	2	0.6744	0.540	1	5
CH_2NH	33	1.2070	0.936	1	0

By Eq. (D.16),

$$r_1 = (2)(0.9011) + (1)(0.6744) + (1)(1.2070)$$

Similarly,

$$r_2 = (2)(0.9011) + (5)(0.6744)$$

Thus,

$$r_1 = 3.6836 \qquad r_2 = 5.1742$$

In like manner by Eq. (D.17),

$$q_1 = 3.1720 \qquad q_2 = 4.3960$$

Application of Eq. (D.8) for $i = 1$ now gives

$$J_1 = \frac{3.6836}{(3.6836)(0.4) + (5.1742)(0.6)}$$

Quantity J_2 is expressed similarly; thus,

$$J_1 = 0.8046 \qquad J_2 = 1.1302$$

Equation (D.9), applied in exactly the same way, yields

$$L_1 = 0.8120 \qquad L_2 = 1.1253$$

† J. Gmehling, P. Rasmussen, and Aa. Fredenslund, *Ind. Eng. Chem. Process Des. Dev.*, **21**: 118, 1982.

The results of substitution into Eq. (D.18) are as follows:

	G_{ki}	
k	$i = 1$	$i = 2$
1	1.696	1.696
2	0.540	2.700
33	0.936	0.000

By Eq. (D.19), we have

$$\theta_1 = (1.696)(0.4) + (1.696)(0.6)$$

This and similar expressions for θ_2 and θ_{33} give

$$\theta_1 = 1.696 \qquad \theta_2 = 1.836 \qquad \theta_{33} = 0.3744$$

The following interaction parameters are found from Table D.2:

$$a_{1,1} = a_{1,2} = a_{2,1} = a_{2,2} = a_{33,33} = 0 \text{ K}$$

$$a_{1,33} = a_{2,33} = 255.7 \text{ K}$$

$$a_{33,1} = a_{33,2} = 65.33 \text{ K}$$

Substitution of these values into Eq. (D.22) with $T = 308.15$ K gives

$$\tau_{1,1} = \tau_{1,2} = \tau_{2,1} = \tau_{2,2} = \tau_{33,33} = 1$$

$$\tau_{1,33} = \tau_{2,33} = 0.4361$$

$$\tau_{33,1} = \tau_{33,2} = 0.8090$$

The results of application of Eq. (D.20) are as follows:

	s_{ki}	
k	$i = 1$	$i = 2$
1	2.993	4.396
2	2.993	4.396
33	1.911	1.917

Application of Eq. (D.21) for $k = 1$ gives

$$\eta_1 = (2.993)(0.4) + (4.396)(0.6)$$

This and similar expressions for η_2 and η_{33} give

$$\eta_1 = 3.835$$

$$\eta_2 = 3.835$$

$$\eta_{33} = 1.915$$

The activity coefficients may now be calculated. By Eq. (D.14),

$$\ln \gamma_1^C = -0.0213 \qquad \text{and} \qquad \ln \gamma_2^C = -0.0076$$

By Eq. (D.15),

$$\ln \gamma_1^R = 0.1463 \qquad \text{and} \qquad \ln \gamma_2^R = 0.0537$$

Finally, by Eq. (D.5)

$$\gamma_1 = 1.133 \qquad \text{and} \qquad \gamma_2 = 1.047$$

APPENDIX

E

NEWTON'S METHOD

Numerical techniques are sometimes required in the solution of thermodynamics problems. Particularly useful is an iteration procedure that generates a sequence of approximations which rapidly converges on the exact solution of an equation. One such procedure is *Newton's method*, a technique for finding a root $X = X_r$ of the equation

$$Y(X) = 0 \tag{E.1}$$

The basis for the method is shown graphically by Fig. E.1, a sketch of Y vs. X for the region that includes the point where $Y(X) = 0$. We let $X = X_0$ be an initial estimate of a solution to Eq. (E.1), and by construction we identify the corresponding value of $Y_0 = Y(X_0)$. A tangent drawn to the curve at (X_0, Y_0) determines by its intersection with the X axis a new estimate X_1 of the solution $X = X_r$. The value of X_1 is directly related to the slope of the tangent line, which is equal to the derivative of Y with respect to X. Thus, at (X_0, Y_0), we may write

$$\left(\frac{dY}{dX}\right)_0 = \frac{0 - Y_0}{X_1 - X_0}$$

from which

$$X_1 = X_0 - \frac{Y_0}{(dY/dX)_0} \tag{A}$$

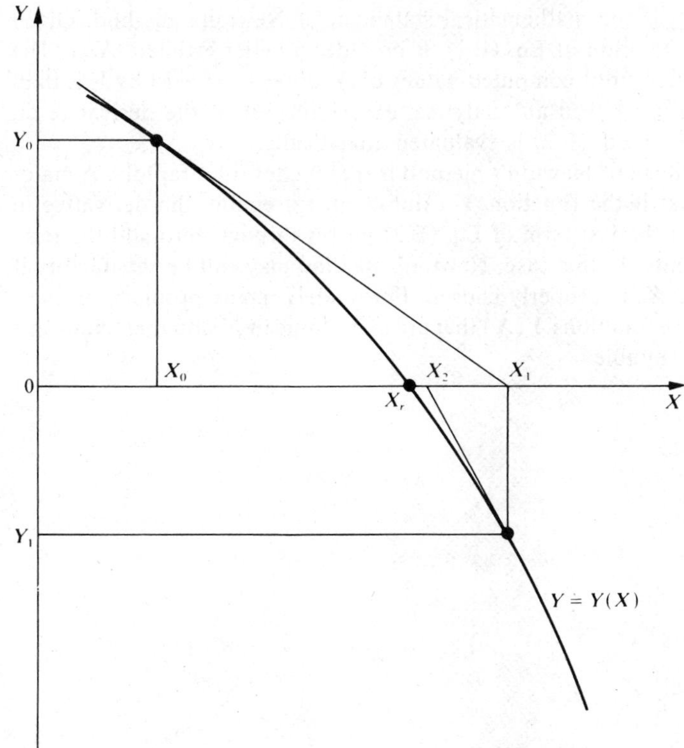

Figure E.1 Graphical construction showing Newton's method for finding a root X_r of the equation $Y(X) = 0$.

Another application of this procedure yields a better estimate X_2 of the solution $X = X_r$. Thus, at (X_1, Y_1), we write

$$\left(\frac{dY}{dX}\right)_1 = \frac{0 - Y_1}{X_2 - X_1}$$

from which

$$X_2 = X_1 - \frac{Y_1}{(dY/dX)_1} \tag{B}$$

By induction, we obtain as a generalization of Eqs. (A) and (B) the following recursion formula:

$$\boxed{X_{j+1} = X_j - \frac{Y_j}{(dY/dX)_j}} \tag{E.2}$$

Equation (E.2) is the mathematical statement of Newton's method. Given estimate X_j to the solution of Eq. (E.1), it provides a better estimate X_{j+1}. The procedure is repeated until computed values of Y differ from zero by less than some preset tolerance. Given an analytical expression for Y, the derivative on the right-hand side of Eq. (E.2) is evaluated analytically.

A major advantage of Newton's method is that it converges rapidly. A major disadvantage is that if the function Y exhibits an extremum, the derivative in the denominator of the last term of Eq. (E.2) passes through zero and the term itself becomes infinite. In this case, Newton's method may still be satisfactory if the initial estimate X_0 is properly chosen. Fortunately, many problems in thermodynamics involve functions $Y(X)$ that are monotonic in X, and then Newton's method is usually suitable.

INDEX

NAME INDEX

SUBJECT INDEX

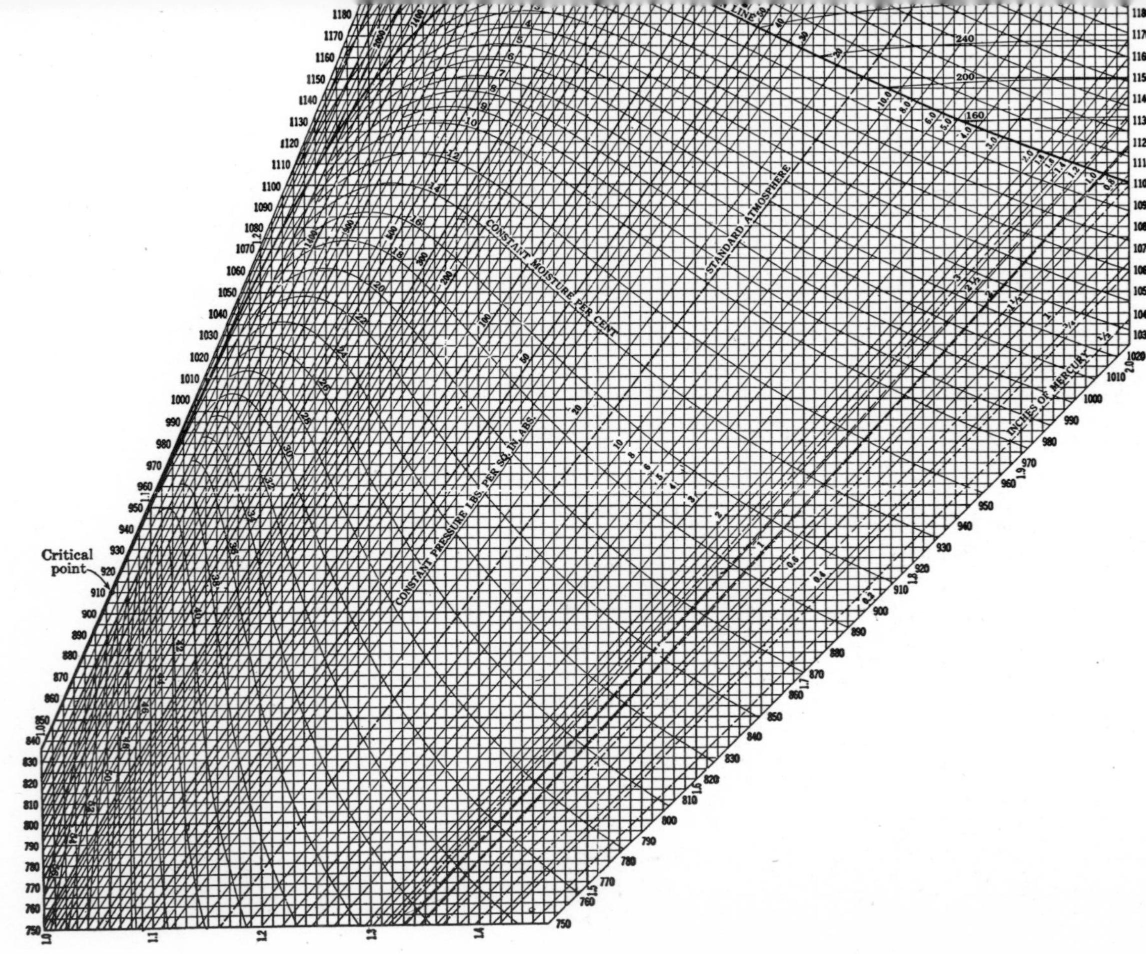
Critical point
CONSTANT PRESSURE LBS. PER SQ. IN. ABS.
CONSTANT MOISTURE PER CENT
STANDARD ATMOSPHERE
INCHES OF MERCURY

Mollier diagram for steam. ***(Reproduced by permission from "Steam Tables: Properties of Saturated and Superheated Steam," copyright 1940, Combustion Engineering, Inc.)***

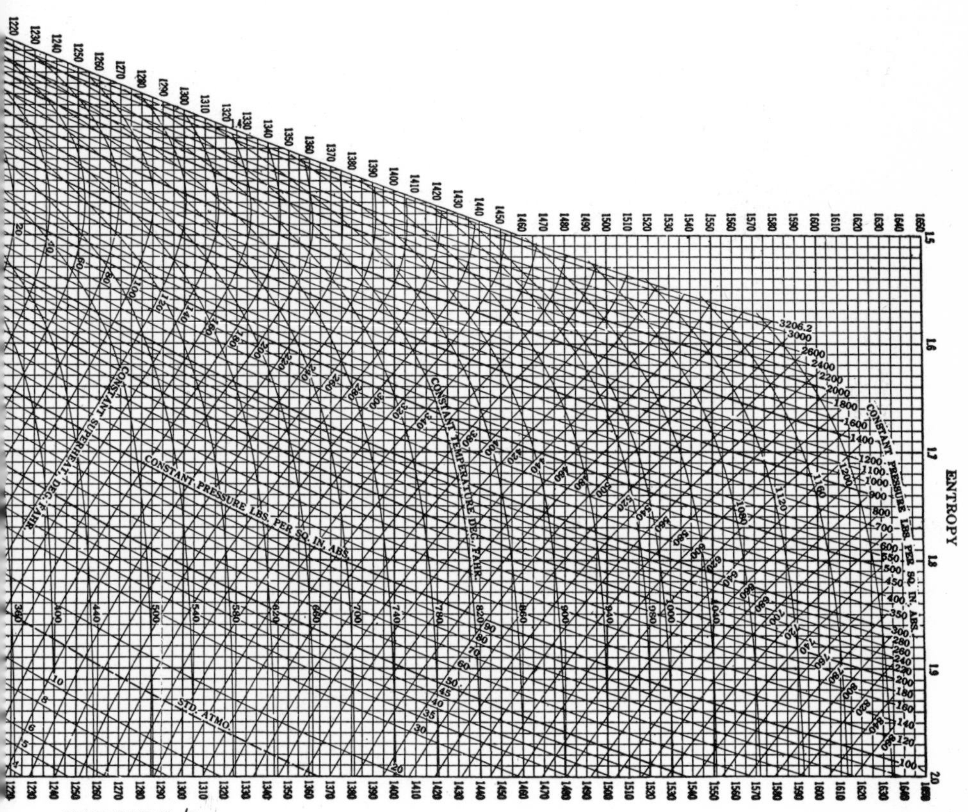